The undamped free response: $m\ddot{x} + kx = 0$ for initial conditions $x(0) = x_0$, $\dot{x}(0) = v_0$

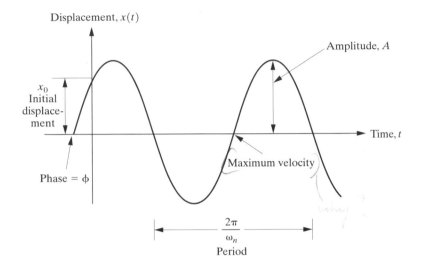

$$x(t) = \left(x_0^2 + \frac{v_0^2}{\omega_n^2}\right)^{1/2} \sin(\omega_n t + \phi), \quad \phi = \tan^{-1}\left(\frac{\omega_n x_0}{v_0}\right), \quad \omega_n = \sqrt{k/m}$$

The underdamped free response: $m\ddot{x} + c\dot{x} + kx = 0$ for initial conditions $x(0) = x_0$, $\dot{x}(0) = v_0$

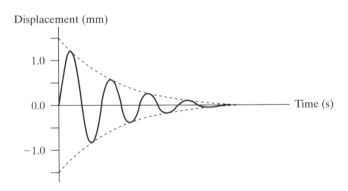

$$x(t) = Ae^{-\zeta\omega_n t}\sin(\omega_d t + \phi), \quad \omega_d = \omega_n\sqrt{1 - \zeta^2}, \quad \omega_n = \sqrt{k/m}, \quad \zeta = \frac{c}{2\sqrt{km}}$$

$$A = \sqrt{\frac{(v_0 + \omega_n\zeta x_0)^2 + (x_0\omega_d)^2}{\omega_d^2}}, \quad \phi = \tan^{-1}\left[\frac{x_0\omega_d}{v_0 + \omega_n\zeta x_0}\right]$$

The forced response of $m\ddot{x} + c\dot{x} + kx = F_0 \sin \omega t$ is $x(t) = X \sin(\omega t + \phi)$ where the normalized amplitude Xk/F_0 is given by $Xk/F_0 = \dfrac{1}{\sqrt{(1 - r^2) + (2\zeta r)^2}}$ plotted below

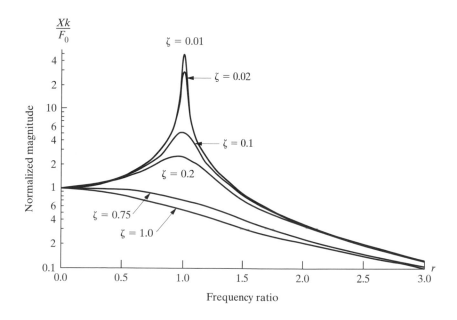

$$r = \omega/\omega_n, \ \omega_n = \sqrt{k/m}, \ \zeta = \frac{c}{2\sqrt{km}}$$

and where the phase ϕ is given by $\phi = \tan^{-1}\left[\dfrac{2\zeta r}{1 - r^2}\right]$ plotted below

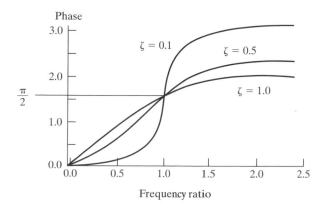

UNITS

QUANTITY	ENGLISH SYSTEM	S.I. SYSTEM
force	1 lb	4.448 Newtons (N)
mass	$1 \text{ lb} \cdot \text{sec}^2/\text{ft}$ (slug)	14.59 kg (kilogram)
length	1 ft	0.3048 meters (m)
mass density	slug/ft^3	$515.38 \text{ kg}/\text{m}^3$
torque or moment	$1 \text{ lb} \cdot \text{in}$	$0.113 \text{ N} \cdot \text{m}$
acceleration	$1 \text{ ft}/\text{sec}^2$	$0.3048 \text{ m}/\text{s}^2$
accel. of gravity	$32.2 \text{ ft}/\text{s}^2 = 386 \text{ in.}/\text{sec}^2$	$9.81 \text{ m}/\text{s}^2$
spring constant k	$1 \text{ lb}/\text{in.}$	$175.1 \text{ N}/\text{m}$
rot. spring constant k	$1 \text{ lb} \cdot \text{in.}/\text{rad}$	$0.113 \text{ N} \cdot \text{m}/\text{rad}$
damping constant c	$1 \text{ lb} \cdot \text{sec}/\text{in.}$	$175.1 \text{ n} \cdot \text{s}/\text{m}$
mass moment of inertia	1 lb.in.sec^2	$0.1129 \text{ kg}/\text{m}^2$
modulus of elasticity	10^6 lb.in.^2	$6.895 \times 10^9 \text{ N}/\text{m}^2$
modulus of elasticity of steel	$29 \times 10^6 \text{ lb}/\text{in.}^2$	$200 \times 10^9 \text{ N}/\text{m}^2$
angle	1 degree	157.3 radian

SI UNITS PREFIXES

MULTIPLICATION FACTOR	PREFIX	SYMBOL
$1\,000\,000\,000\,000 = 10^{12}$	terra	T
$1\,000\,000\,000 = 10^{9}$	giga	G
$1\,000\,000 = 10^{6}$	mega	M
$1\,000 = 10^{3}$	kilo	k
$100 = 10^{2}$	hecto	h
$10 = 10$	deka	da
$0.1 = 10^{-1}$	deci	d
$0.01 = 10^{-2}$	centi	c
$0.001 = 10^{-3}$	milli	m
$0.000\,001 = 10^{-6}$	micro	μ
$0.000\,000\,001 = 10^{-9}$	nano	n
$0.000\,000\,000\,001 = 10^{-12}$	pico	p

	RECTILINEAR SYSTEM			ROTATIONAL SYSTEM		
		UNIT			UNIT	
Quantity	Symbol	U.S. Customary	SI Unit	Symbol	English	SI Unit
Time	t	sec	s	t	sec	s
Displacement	x	in.	m	θ	rad	rad
Velocity	$\dot{x}$	in./sec	m/s	$\dot{\theta}$	rad/sec	rad/s
Acceleration	$\ddot{x}$	in./sec^2	m/s^2	$\ddot{\theta}$	rad/sec^2	rad/s^2
Mass, moment of inertia	m	lb$_f$-sec^2/in.	kg	J	in.-lb$_f$-sec^2	m$^2 \cdot$ kg
Damping factor	c	lb$_f$-sec/in.	s $\cdot$ N/m	c	in.lb$_f$-sec/rad	m $\cdot$ s $\cdot$ N/rad
Spring constant	k	lb$_f$/in.	N/m	k	in.-lb$_f$/rad	m $\cdot$ N/rad
Force, torque	$F = m\ddot{x}$	lb$_f$	N = m $\cdot$ kg/s^2	$T = J\ddot{\theta}$	in.-lb$_f$	m $\cdot$ N = m$^2 \cdot$ kg/s^2
Momentum	$m\dot{x}$	lb$_f$-sec	s $\cdot$ N = m $\cdot$ kg/s	$J\dot{\theta}$	in.-lb$_f$-sec	m$^2 \cdot$ kg $\cdot$ rad/s
Impulse	Ft	lb$_f$-sec	s $\cdot$ N	Tt	in.lb$_f$-sec	m$^2 \cdot$ kg $\cdot$ rad/s
Kinetic energy	$T = \frac{1}{2} m\dot{x}^2$	in.lb$_f$	J	$T = \frac{1}{2} J\dot{\theta}^2$	in.-lb$_f$	J
Potential energy	$U = \frac{1}{2} kx^2$	in.lb$_f$	J	$U = \frac{1}{2} k\theta^2$	in.-lb$_f$	J
Work	$\int F dx$	in.-lb$_f$	J = m $\cdot$ N = m$^2 \cdot$ kg/s^2	$\int T d\theta$	in.-lb$_f$	J = m $\cdot$ N = m$^2 \cdot$ kg/s^2
Natural frequency	$\omega_n = \sqrt{k/m}$	rad/sec	rad/s	$\omega_n = \sqrt{k/J}$	rad/sec	rad/s
	$f_n = \dfrac{\omega_n}{2\pi}$	Hz	Hz	$f_n = \dfrac{\omega_n}{2\pi}$	Hz	Hz

Engineering Vibration

Fourth Edition

DANIEL J. INMAN
University of Michigan

International Editions contributions by

RAMESH CHANDRA SINGH
Delhi Technological University

Boston   Columbus   Indianapolis   New York   San Francisco   Upper Saddle River
Amsterdam   Cape Town   Dubai   London   Madrid   Milan   Munich   Paris   Montréal   Toronto
Delhi   Mexico City   São Paulo   Sydney   Hong Kong   Seoul   Singapore   Taipei   Tokyo

Editorial Director, Computer Science, Engineering, and Advanced Mathematics: *Marcia J. Horton*
Acquisitions Editor: *Norrin Dias*
Editorial Assistant: *Jennifer Lonschein*
Senior Managing Editor: *Scott Disanno*
Publisher, International Edition: *Angshuman Chakraborty*
Publishing Administrator and Business Analyst, International Edition: *Shokhi Shah Khandelwal*
Associate Print and Media Editor, International Edition: *Anuprova Dey Chowdhuri*

Acquisitions Editor, International Edition: *Shivangi Ramachandran*
Publishing Administrator, International Edition: *Hema Mehta*
Project Editor, International Edition: *Karthik Subramanian*
Senior Manufacturing Controller, Production, International Edition: *Trudy Kimber*
Art Director: *Jayne Conte*
Cover Designer: *Bruce Kenselaar*
Art Editor: *Greg Dulles*
Manufacturing Buyer: *Lisa McDowell*
Marketing Manager: *Tim Galligan*

Pearson Education Limited
Edinburgh Gate
Harlow
Essex CM20 2JE
England

and Associated Companies throughout the world

Visit us on the World Wide Web at:
www.pearsoninternationaleditions.com

© Pearson Education Limited 2014

ISBN 10: 0-273-76844-1
ISBN 13: 978-0-273-76844-9

British Library Cataloguing-in-Publication Data
A catalogue record for this book is available from the British Library

10 9 8 7 6 5 4 3 2
14 13

Typeset in Times New Roman 10/12 by Integra Software Services.
Printed and bound by Courier Westford in The United States of America

Contents

PREFACE   8

1 INTRODUCTION TO VIBRATION AND THE FREE RESPONSE   13

1.1 Introduction to Free Vibration   14

1.2 Harmonic Motion   25

1.3 Viscous Damping   33

1.4 Modeling and Energy Methods   43

1.5 Stiffness   58

1.6 Measurement   70

1.7 Design Considerations   75

1.8 Stability   80

1.9 Numerical Simulation of the Time Response   84

1.10 Coulomb Friction and the Pendulum   93

Problems   107

MATLAB *Engineering Vibration Toolbox*   127

Toolbox Problems   128

2 RESPONSE TO HARMONIC EXCITATION   129

2.1 Harmonic Excitation of Undamped Systems   130

2.2 Harmonic Excitation of Damped Systems   142

2.3 Alternative Representations   156

2.4 Base Excitation   163

2.5 Rotating Unbalance   172

2.6 Measurement Devices   178

2.7   Other Forms of Damping   182

2.8   Numerical Simulation and Design   192

2.9   Nonlinear Response Properties   200

       Problems   209

       MATLAB *Engineering Vibration Toolbox*   226

       Toolbox Problems   226

3   GENERAL FORCED RESPONSE   **228**

3.1   Impulse Response Function   229

3.2   Response to an Arbitrary Input   238

3.3   Response to an Arbitrary Periodic Input   247

3.4   Transform Methods   254

3.5   Response to Random Inputs   259

3.6   Shock Spectrum   267

3.7   Measurement via Transfer Functions   272

3.8   Stability   274

3.9   Numerical Simulation of the Response   279

3.10   Nonlinear Response Properties   291

       Problems   299

       MATLAB *Engineering Vibration Toolbox*   313

       Toolbox Problems   313

4   MULTIPLE-DEGREE-OF-FREEDOM SYSTEMS   **315**

4.1   Two-Degree-of-Freedom Model (Undamped)   316

4.2   Eigenvalues and Natural Frequencies   330

4.3   Modal Analysis   344

4.4   More Than Two Degrees of Freedom   352

4.5   Systems with Viscous Damping   368

4.6   Modal Analysis of the Forced Response   374

4.7   Lagrange's Equations   381

4.8   Examples   389

4.9   Computational Eigenvalue Problems for Vibration   401

4.10 Numerical Simulation of the Time Response   419

      Problems   427

      MATLAB *Engineering Vibration Toolbox*   445

      Toolbox Problems   445

5   DESIGN FOR VIBRATION SUPPRESSION   **447**

5.1   Acceptable Levels of Vibration   448

5.2   Vibration Isolation   454

5.3   Vibration Absorbers   467

5.4   Damping in Vibration Absorption   475

5.5   Optimization   483

5.6   Viscoelastic Damping Treatments   491

5.7   Critical Speeds of Rotating Disks   497

      Problems   503

      MATLAB *Engineering Vibration Toolbox*   513

      Toolbox Problems   513

6   DISTRIBUTED-PARAMETER SYSTEMS   **514**

6.1   Vibration of a String or Cable   516

6.2   Modes and Natural Frequencies   520

6.3   Vibration of Rods and Bars   531

6.4   Torsional Vibration   537

6.5   Bending Vibration of a Beam   544

6.6   Vibration of Membranes and Plates   556

6.7   Models of Damping   562

6.8   Modal Analysis of the Forced Response   568

Discussed
not

Problems   578

MATLAB *Engineering Vibration Toolbox*   584

Toolbox Problems   584

7   VIBRATION TESTING AND EXPERIMENTAL MODAL ANALYSIS   585

7.1   Measurement Hardware   587

7.2   Digital Signal Processing   591

7.3   Random Signal Analysis in Testing   596

7.4   Modal Data Extraction   600

7.5   Modal Parameters by Circle Fitting   603

7.6   Mode Shape Measurement   608

7.7   Vibration Testing for Endurance and Diagnostics   618

7.8   Operational Deflection Shape Measurement   621

Problems   623

MATLAB *Engineering Vibration Toolbox*   627

Toolbox Problems   628

8   FINITE ELEMENT METHOD   629

8.1   Example: The Bar   631

8.2   Three-Element Bar   637

8.3   Beam Elements   642

8.4   Lumped-Mass Matrices   650

8.5   Trusses   653

8.6   Model Reduction   658

Problems   661

MATLAB *Engineering Vibration Toolbox*   668

Toolbox Problems   668

APPENDIX A COMPLEX NUMBERS AND FUNCTIONS   669

APPENDIX B LAPLACE TRANSFORMS   675

APPENDIX C MATRIX BASICS — **680**

APPENDIX D THE VIBRATION LITERATURE — **692**

APPENDIX E LIST OF SYMBOLS — **694**

APPENDIX F CODES AND WEB SITES — **699**

**APPENDIX G ENGINEERING VIBRATION TOOLBOX
AND WEB SUPPORT** — **700**

REFERENCES — **702**

ANSWERS TO SELECTED PROBLEMS — **704**

INDEX — **711**

Preface

This book is intended for use in a first course in vibrations or structural dynamics for undergraduates in mechanical, civil, and aerospace engineering or engineering mechanics. The text contains the topics normally found in such courses in accredited engineering departments as set out initially by Den Hartog and refined by Thompson. In addition, topics on design, measurement, and computation are addressed.

Pedagogy

Originally, a major difference between the pedagogy of this text and competing texts is the use of high level computing codes. Since then, the other authors of vibrations texts have started to embrace use of these codes. While the book is written so that the codes do not have to be used, I strongly encourage their use. These codes (Mathcad®, MATLAB®, and Mathematica®) are very easy to use, at the level of a programmable calculator, and hence do not require any prerequisite courses or training. Of course, it is easier if the students have used one or the other of the codes before, but it is not necessary. In fact, the MATLAB® codes can be copied directly and will run as listed. The use of these codes greatly enhances the student's understanding of the fundamentals of vibration. Just as a picture is worth a thousand words, a numerical simulation or plot can enable a completely dynamic understanding of vibration phenomena. Computer calculations and simulations are presented at the end of each of the first four chapters. After that, many of the problems assume that codes are second nature in solving vibration problems.

Another unique feature of this text is the use of "windows," which are distributed throughout the book and provide reminders of essential information pertinent to the text material at hand. The windows are placed in the text at points where such prior information is required. The windows are also used to summarize essential information. The book attempts to make strong connections to previous course work in a typical engineering curriculum. In particular, reference is made to calculus, differential equations, statics, dynamics, and strength of materials course work.

WHAT'S NEW IN THIS EDITION

Most of the changes made in this edition are the result of comments sent to me by students and faculty who have used the 3rd edition. These changes consist of improved clarity in explanations, the addition of some new examples that clarify concepts, and enhanced problem statements. In addition, some text material deemed outdated and not useful has been removed. The computer codes have also been updated. However, software companies update their codes much faster than the publishers can update their texts, so users should consult the web for updates in syntax, commands, etc. One consistent request from students has been not to reference data appearing previously in other examples or problems. This has been addressed by providing all of the relevant data in the problem statements. Three undergraduate engineering students (one in Engineering Mechanics, one in Biological Systems Engineering, and one in Mechanical Engineering) who had the prerequisite courses, but had not yet had courses in vibrations, read the manuscript for clarity. Their suggestions prompted us to make the following changes in order to improve readability from the student's perspective:

- Improved clarity in explanations added in 47 different passages in the text. In addition, two new windows have been added.
- Twelve new examples that clarify concepts and enhanced problem statements have been added, and ten examples have been modified to improve clarity.
- Text material deemed outdated and not useful has been removed. Two sections have been dropped and two sections have been completely rewritten.
- All computer codes have been updated to agree with the latest syntax changes made in MATLAB, Mathematica, and Mathcad.
- Fifty-four new problems have been added and 94 problems have been modified for clarity and numerical changes.
- Eight new figures have been added and three previous figures have been modified.
- Four new equations have been added.

Chapter 1: Changes include new examples, equations, and problems. New textual explanations have been added and/or modified to improve clarity based on student suggestions. Modifications have been made to problems to make the problem statement clear by not referring to data from previous problems or examples. All of the codes have been updated to current syntax, and older, obsolete commands have been replaced.

Chapter 2: New examples and figures have been added, while previous examples and figures have been modified for clarity. New textual explanations have also been added and/or modified. New problems have been added and older problems modified to make the problem statement clear by not referring to data from previous problems or examples. All of the codes have been updated to current syntax, and older, obsolete commands have been replaced.

Chapter 3: New examples and equations have been added, as well as new problems. In particular, the explanation of impulse has been expanded. In addition, previous problems have been rewritten for clarity and precision. All examples and problems that referred to prior information in the text have been modified to present a more self-contained statement. All of the codes have been updated to current syntax, and older, obsolete commands have been replaced.

Chapter 4: Along with the addition of an entirely new example, many of the examples have been changed and modified for clarity and to include improved information. A new window has been added to clarify matrix information. A figure has been removed and a new figure added. New problems have been added and older problems have been modified with the goal of making all problems and examples more self-contained. All of the codes have been updated to current syntax, and older, obsolete commands have been replaced. Several new plots intermixed in the codes have been redone to reflect issues with Mathematica and MATLAB's automated time step which proves to be inaccurate when using singularity functions. Several explanations have been modified according to students' suggestions.

Chapter 5: Section 5.1 has been changed, the figure replaced, and the example changed for clarity. The problems are largely the same but many have been changed or modified with different details and to make the problems more self-contained. Section 5.8 (Active Vibration Suppression) and Section 5.9 (Practical Isolation Design) have been removed, along with the associated problems, to make room for added material in the earlier chapters without lengthening the book. According to user surveys, these sections are not usually covered.

Chapter 6: Section 6.8 has been rewritten for clarity and a window has been added to summarize modal analysis of the forced response. New problems have been added and many older problems restated for clarity. Further details have been added to several examples. A number of small additions have been made to the to the text for clarity.

Chapters 7 and 8: These chapters were not changed, except to make minor corrections and additions as suggested by users.

Units

This book uses SI units. The 1st edition used a mixture of US Customary and SI, but at the insistence of the editor all units were changed to SI. I have stayed with SI in this edition because of the increasing international arena that our engineering graduates compete in. The engineering community is now completely global. For instance, GE Corporate Research has more engineers in its research center in India than it does in the US. Engineering in the US is in danger of becoming the 'garment' workers of the next decade if we do not recognize the global work place. Our engineers need to work in SI to be competitive in this increasingly international work place.

Instructor Support

This text comes with a bit of support. In particular, MS PowerPoint presentations are available for each chapter along with some instructive movies. The solutions manual is available in both MS Word and PDF format (sorry, instructors only). Sample tests are available. The MS Word solutions manual can be cut and pasted into presentation slides, tests, or other class enhancements. These resources can be found at www.pearsoninternationaleditions.com/inman and will be updated often. Please also email me at daninman@umich.edu with corrections, typos, questions, and suggestions. The book is reprinted often, and at each reprint I have the option to fix typos, so please report any you find to me, as others as well as I will appreciate it.

Student Support

The best place to get help in studying this material is from your instructor, as there is nothing more educational than a verbal exchange. However, the book was written as much as possible from a student's perspective. Many students critiqued the original manuscript, and many of the changes in text have been the result of suggestions from students trying to learn from the material, so please feel free to email me (daninman@umich.edu) should you have questions about explanations. Also I would appreciate knowing about any corrections or typos and, in particular, if you find an explanation hard to follow. My goal in writing this was to provide a useful resource for students learning vibration for the first time.

ACKNOWLEDGEMENTS

The cover photo of the unmanned air vehicle is provided courtesy of General Atomics Aeronautical Systems, Inc., all rights reserved. Each chapter starts with two photos of different systems that vibrate to remind the reader that the material in this text has broad application across numerous sectors of human activity. These photographs were taken by friends, students, colleagues, relatives, and some by me. I am greatly appreciative of Robert Hargreaves (guitar), P. Timothy Wade (wind mill, Presidential helicopter), General Atomics (Predator), Roy Trifilio (bridge), Catherine Little (damper), Alex Pankonien (FEM graphic), and Jochen Faber of Liebherr Aerospace (landing gear). Alan Giles of General Atomics gave me an informative tour of their facilities which resulted in the photos of their products.

Many colleagues and students have contributed to the revision of this text through suggestions and questions. In particular, Daniel J. Inman, II; Kaitlyn DeLisi; Kevin Crowely; and Emily Armentrout provided many useful comments from the perspective of students reading the material for the first time. Kaitlyn and Kevin checked all the computer codes by copying them out of the book to make sure they ran. My former PhD students Ya Wang, Mana Afshari, and Amin Karami checked many of the new problems and examples. Dr. Scott Larwood and the students in his vibrations class at the University of the Pacific sent many suggestions and corrections that helped give the book the perspective of a nonresearch insitution. I have implemented many of their suggestions, and I believe the book's explanations are much clearer due to their input. Other professors using the book, Cetin Cetinkaya of Clarkson University, Mike Anderson of the University of Idaho, Joe Slater of Wright State University, Ronnie Pendersen of Aalborg University Esbjerg, Sondi Adhikari of the Universty of Wales, David Che of Geneva College, Tim Crippen of the University of Texas at Tyler, and Nejat Olgac of the University of Conneticut, have provided discussions via email that have led to improvements in the text, all of which are greatly appreciated. I would like to thank the reviewers: Cetin Cetinkaya, Clarkson University; Dr. Nesrin Sarigul-Klijn, University of California–Davis; and David Che, Geneva College.

Many of my former PhD students who are now academics cotaught this course with me and also offered many suggestions. Alper Erturk (Georgia Tech), Henry Sodano (University of Florida), Pablo Tarazaga (Virginia Tech), Onur Bilgen (Old Dominian University), Mike Seigler (University of Kentucky), and Armaghan Salehian (University of Waterloo) all contributed to clarity in this text for which I am grateful. I have been lucky to have wonderful PhD students to work with. I learned much from them.

I would also like to thank Prof. Joseph Slater of Wright State for reviewing some of the new materials, for writing and managing the associated toolbox, and constantly sending suggestions. Several colleagues from government labs and companies have also written with suggestions which have been very helpful from that perspective of practice.

I have also had the good fortune of being sponsored by numerous companies and federal agencies over the last 32 years to study, design, test, and analyze a large variety of vibrating structures and machines. Without these projects, I would not have been able to write this book nor revise it with the appreciation for the practice of vibration, which I hope permeates the text.

Last, I wish to thank my family for moral support, a sense of purpose, and for putting up with my absence while writing.

DANIEL J. INMAN
Ann Arbor, Michigan

The publishers wish to thank Nilamber Kumar Singh of Birla Institute of Technology, Mesra for reviewing the content of the International Edition.

1 Introduction to Vibration and the Free Response

Vibration is the subdiscipline of dynamics that deals with repetitive motion. Most of the examples in this text are mechanical or structural elements. However, vibration is prevalent in biological systems and is in fact at the source of communication (the ear vibrates to hear and the tongue and vocal cords vibrate to speak). In the case of music, vibrations, say of a stringed instrument such as a guitar, are desired. On the other hand, in most mechanical systems and structures, vibration is unwanted and even destructive. For example, vibration in an aircraft frame causes fatigue and can eventually lead to failure. An example of fatigue crack is illustrated in the circle in the photo on the bottom left. Everyday experiences are full of vibration and usually ways of mitigating vibration. Automobiles, trains, and even some bicycles have devices to reduce the vibration induced by motion and transmitted to the driver.

The task of this text is to teach the reader how to analyze vibration using principles of dynamics. This requires the use of mathematics. In fact, the sine function provides the fundamental means of analyzing vibration phenomena.

The basic concepts of understanding vibration, analyzing vibration, and predicting the behavior of vibrating systems form the topics of this text. The concepts and formulations presented in the following chapters are intended to provide the skills needed for designing vibrating systems with desired properties that enhance vibration when it is wanted and reduce vibration when it is not.

This first chapter examines vibration in its simplest form in which no external force is present (free vibration). This chapter introduces both the important concept of natural frequency and how to model vibration mathematically.

The Internet is a great source for examples of vibration, and the reader is encouraged to search for movies of vibrating systems and other examples that can be found there.

1.1 INTRODUCTION TO FREE VIBRATION

Vibration is the study of the repetitive motion of objects relative to a stationary frame of reference or nominal position (usually equilibrium). Vibration is evident everywhere and in many cases greatly affects the nature of engineering designs. The vibrational properties of engineering devices are often limiting factors in their performance. When harmful, vibration should be avoided, but it can also be extremely useful. In either case, knowledge about vibration—how to analyze, measure, and control it—is beneficial and forms the topic of this book.

Typical examples of vibration familiar to most include the motion of a guitar string, the ride quality of an automobile or motorcycle, the motion of an airplane's wings, and the swaying of a large building due to wind or an earthquake. In the chapters that follow, vibration is modeled mathematically based on fundamental principles, such as Newton's laws, and analyzed using results from calculus and differential equations. Techniques used to measure the vibration of a system are then developed. In addition, information and methods are given that are useful for designing particular systems to have specific vibrational responses.

The physical explanation of the phenomena of vibration concerns the interplay between potential energy and kinetic energy. A vibrating system must have a component that stores potential energy and releases it as kinetic energy in the form of motion (vibration) of a mass. The motion of the mass then gives up kinetic energy to the potential-energy storing device.

Engineering is built on a foundation of previous knowledge and the subject of vibration is no exception. In particular, the topic of vibration builds on previous courses in dynamics, system dynamics, strength of materials, differential equations, and some matrix analysis. In most accredited engineering programs, these courses are prerequisites for a course in vibration. Thus, the material that follows draws information and methods from these courses. Vibration analysis is based on a coalescence of mathematics and physical observation. For example, consider a simple pendulum. You may have seen one in a science museum, in a grandfather clock, or you might make a simple one with a string and a marble. As the pendulum swings back and forth, observe that its motion as a function of time can be described very nicely by the sine function from trigonometry. Even more interesting, if you make a free-body diagram of the pendulum and apply Newtonian mechanics to get the equation of motion (summing moments in this case), the resulting equation of motion has the sine function as its solution. Further, the equation of motion predicts the time it takes for the pendulum to repeat its motion. In this example, dynamics, observation, and mathematics all come into agreement to produce a predictive model of the motion of a pendulum, which is easily verified by experiment (physical observation).

This pendulum example tells the story of this text. We propose a series of steps to build on the modeling skills developed in your first courses in statics, dynamics, and strength of materials combined with system dynamics to find equations of motion of successively more complicated systems. Then we will use the techniques of differential equations and numerical integration to solve these equations of motion to predict how various mechanical systems and structures vibrate. The following example illustrates the importance of recalling the methods learned in the first course in dynamics.

Example 1.1.1

Derive the equation of motion of the pendulum in Figure 1.1.

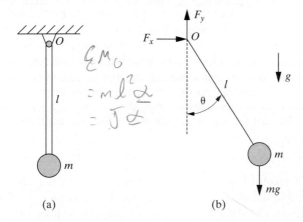

(a)   (b)

Figure 1.1   (a) A schematic of a pendulum. (b) The free-body diagram of (a).

Solution   Consider the schematic of a pendulum in Figure 1.1(a). In this case, the mass of the rod will be ignored as well as any friction in the hinge. Typically, one starts with a photograph or sketch of the part or structure of interest and is immediately faced with having to make assumptions. This is the "art" or experience side of vibration analysis and modeling. The general philosophy is to start with the simplest model possible (hence, here we ignore friction and the mass of the rod and assume the motion remains in a plane) and try to answer the relevant engineering questions. If the simple model doesn't agree with the experiment, then make it more complex by relaxing the assumptions until the model successfully predicts physical observation. With the assumptions in mind, the next step is to create a free-body diagram of the system, as indicated in Figure 1.1(b), in order to identify all of the relevant forces. With all the modeled forces identified, Newton's second law and Euler's second law are used to derive the equations of motion.

In this example Euler's second law takes the form of summing moments about point O. This yields

$$\Sigma \mathbf{M}_O = J\alpha$$

where $\mathbf{M}_O$ denotes moments about the point O, $J = ml^2$ is the mass moment of inertia of the mass m about the point O, l is the length of the massless rod, and $\boldsymbol{\alpha}$ is the angular acceleration vector. Since the problem is really in one dimension, the vector sum of moments equation becomes the single scalar equation

$$J\alpha(t) = -mgl\sin\theta(t) \quad \text{or} \quad ml^2\ddot{\theta}(t) + mgl\sin\theta(t) = 0$$

Here the moment arm for the force mg is the horizontal distance $l\sin\theta$, and the two overdots indicate two differentiations with respect to the time, t. This is a second-order ordinary differential equation, which governs the time response of the pendulum. This is exactly the procedure used in the first course in dynamics to obtain equations of motion.

The equation of motion is nonlinear because of the appearance of the $\sin(\theta)$ and hence difficult to solve. The nonlinear term can be made linear by approximating the sine for small values of $\theta(t)$ as $\sin\theta \approx \theta$. Then the equation of motion becomes

$$\ddot{\theta}(t) + \frac{g}{l}\theta(t) = 0$$

This is a linear, second-order ordinary differential equation with constant coefficients and is commonly solved in the first course of differential equations (usually the third course in the calculus sequence). As we will see later in this chapter, this linear equation of motion and its solution predict the period of oscillation for a simple pendulum quite accurately. The last section of this chapter revisits the nonlinear version of the pendulum equation.

□

Since Newton's second law for a constant mass system is stated in terms of force, which is equated to the mass multiplied by acceleration, an equation of motion with two time derivatives will always result. Such equations require two constants of integration to solve. Euler's second law for constant mass systems also yields two time derivatives. Hence the initial position for $\theta(0)$ and velocity of $\dot{\theta}(0)$ must be specified in order to solve for $\theta(t)$ in Example 1.1.1. The term $mgl\sin\theta$ is called the *restoring force*. In Example 1.1.1, the restoring force is gravity, which provides a potential-energy storing mechanism. However, in most structures and machine parts the restoring force is elastic. This establishes the need for background in strength of materials when studying vibrations of structures and machines.

As mentioned in the example, when modeling a structure or machine it is best to start with the simplest possible model. In this chapter, we model only systems that can be described by a single degree of freedom, that is, systems for which Newtonian mechanics result in a single scalar equation with one displacement coordinate. The degree of freedom of a system is the minimum number of displacement coordinates needed to represent the position of the system's mass at any instant of time. For instance, if the mass of the pendulum in Example 1.1.1 were a rigid body, free to rotate about the end of the pendulum as the pendulum swings, the angle of rotation of the mass would define an additional degree of freedom. The problem would then require two coordinates to determine the position of the mass in space, hence two degrees of freedom. On the other hand, if the rod in Figure 1.1 is flexible, its distributed mass must be considered, effectively resulting in an infinite number of degrees of freedom. Systems with more than one degree of freedom are discussed in Chapter 4, and systems with distributed mass and flexibility are discussed in Chapter 6.

The next important classification of vibration problems after degree of freedom is the nature of the input or stimulus to the system. In this chapter, only the free response of the system is considered. Free response refers to analyzing the vibration of a system resulting from a nonzero initial displacement and/or velocity of the system with no external force or moment applied. In Chapter 2, the response of a single-degree-of-freedom system to a harmonic input (i.e., a sinusoidal applied force) is discussed. Chapter 3 examines the response of a system to a general forcing function (impulse or shock loads, step functions, random inputs, etc.), building on information learned in a course in system dynamics. In the remaining chapters, the models of vibration and methods of analysis become more complex.

The following sections analyze equations similar to the linear version of the pendulum equation given in Example 1.1.1. In addition, energy dissipation is introduced, and details of elastic restoring forces are presented. Introductions to design, measurement, and simulation are also presented. The chapter ends with the introduction of high-level computer codes (MATLAB®, Mathematica, and Mathcad) as a means to visualize the response of a vibrating system and for making the calculations required to solve vibration problems more efficiently. In addition, numerical simulation is introduced in order to solve nonlinear vibration problems.

1.1.1 The Spring–Mass Model

From introductory physics and dynamics, the fundamental kinematical quantities used to describe the motion of a particle are displacement, velocity, and acceleration vectors. In addition, the laws of physics state that the motion of a mass with changing velocity is determined by the net force acting on the mass. An easy device to use in thinking about vibration is a spring (such as the one used to pull a storm door shut, or an automobile spring) with one end attached to a fixed object and a mass attached to the other end. A schematic of this arrangement is given in Figure 1.2.

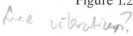

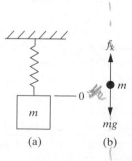

Figure 1.2   A schematic of (a) a single-degree-of-freedom spring–mass oscillator and (b) its free-body diagram.

Ignoring the mass of the spring itself, the forces acting on the mass consist of the force of gravity pulling down (mg) and the elastic-restoring force of the spring pulling back up (f_k). Note that in this case the force vectors are collinear, reducing the static equilibrium equation to one dimension easily treated as a scalar. The nature of the spring force can be deduced by performing a simple static experiment. With no mass attached, the spring stretches to the position labeled $x_0 = 0$ in Figure 1.3. As successively more mass is attached to the spring, the force of gravity causes the spring to stretch further. If the value of the mass is recorded, along with the value of the displacement of the end of the spring each time more mass is added, the plot of the force (mass, denoted by m, times the acceleration due to gravity, denoted by g) versus this displacement, denoted by x, yields a curve similar to that illustrated in Figure 1.4. Note that in the region of values for x between 0 and about 20 mm (millimeters), the curve is a straight line. This indicates that for deflections less than 20 mm and forces less than 1000 N (newtons), the force that is applied by the spring to the mass is proportional to the stretch of the spring. The constant of proportionality is the slope of the straight line between 0 and 20 mm. For the particular spring of Figure 1.4, the constant is 50 N/mm, or 5×10^4 N/m. Thus, the equation that describes the force applied by the spring, denoted by f_k, to the mass is the linear relationship

$$f_k = kx \tag{1.1}$$

The value of the slope, denoted by k, is called the *stiffness* of the spring and is a property that characterizes the spring for all situations for which the displacement is less than 20 mm. From strength-of-materials considerations, a linear spring of stiffness k stores potential energy of the amount $\frac{1}{2}kx^2$.

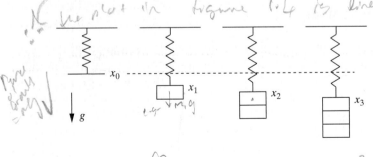

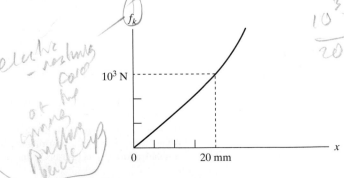

Figure 1.3   A schematic of a massless spring with no mass attached showing its static equilibrium position, followed by increments of increasing added mass illustrating the corresponding deflections.

Figure 1.4   The static deflection curve for the spring of Figure 1.3.

Note that the relationship between f_k and x of equation (1.1) is *linear* (i.e., the curve is linear and f_k depends linearly on x). If the displacement of the spring is larger than 20 mm, the relationship between f_k and x becomes *nonlinear*, as indicated in Figure 1.4. Nonlinear systems are much more difficult to analyze and form the topic of Section 1.10. In this and all other chapters, it is assumed that displacements (and forces) are limited to be in the linear range unless specified otherwise.

Next, consider a free-body diagram of the mass in Figure 1.5, with the massless spring elongated from its rest (equilibrium or unstretched) position. As in the earlier figures, the mass of the object is taken to be m and the stiffness of the spring is taken to be k. Assuming that the mass moves on a frictionless surface along the x direction, the only force acting on the mass in the x direction is the spring force. As long as the motion of the spring does not exceed its linear range, the sum of the forces in the x direction must equal the product of mass and acceleration.

Summing the forces on the free-body diagram in Figure 1.5 along the x direction yields

$$m\ddot{x}(t) = -kx(t) \qquad \text{or} \qquad m\ddot{x}(t) + kx(t) = 0 \tag{1.2}$$

where $\ddot{x}(t)$ denotes the second time derivative of the displacement (i.e., the acceleration). Note that the direction of the spring force is opposite that of the deflection ($+$ is marked to the right in the figure). As in Example 1.1.1, the displacement vector and acceleration vector are reduced to scalars, since the net force in the y direction is zero ($N = mg$) and the force in the x direction is collinear with the inertial force. Both the displacement and acceleration are functions of the elapsed time t, as denoted in equation (1.2). Window 1.1 illustrates three types of mechanical systems, which for small oscillations can be described by equation (1.2): a spring–mass system, a rotating shaft, and a swinging pendulum (Example 1.1.1). Other examples are given in Section 1.4 and throughout the book.

One of the goals of vibration analysis is to be able to predict the response, or motion, of a vibrating system. Thus it is desirable to calculate the solution to equation (1.2). Fortunately, the differential equation of (1.2) is well known and is covered extensively in introductory calculus and physics texts, as well as in texts on differential equations. In fact, there are a variety of ways to calculate this solution. These are all discussed in some detail in the next section. For now, it is sufficient to present a solution based on physical observation. From experience

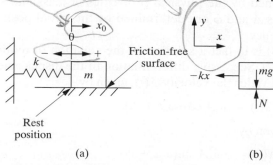

Rest position

(a)

(b)

Figure 1.5 (a) A single spring–mass system given an initial displacement of x_0 from its rest, or equilibrium, position and zero initial velocity. (b) The system's free-body diagram.

Window 1.1
Examples of Single-Degree-of-Freedom Systems (for small displacements)

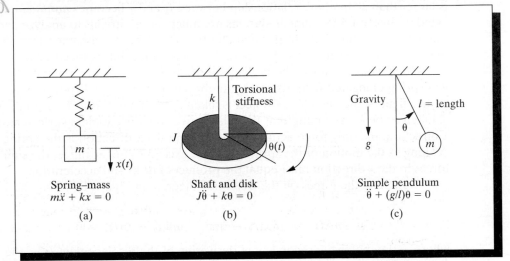

Spring–mass
$m\ddot{x} + kx = 0$

(a)

Shaft and disk
$J\ddot{\theta} + k\theta = 0$

(b)

Simple pendulum
$\ddot{\theta} + (g/l)\theta = 0$

(c)

watching a spring, such as the one in Figure 1.5 (or a pendulum), it is guessed that the motion is periodic, of the form

$$x(t) = A\sin(\omega_n t + \phi) \tag{1.3}$$

This choice is made because the sine function describes oscillation. Equation (1.3) is the sine function in its most general form, where the constant A is the *amplitude*, or maximum value, of the displacement; ω_n, the *angular natural frequency*, determines the interval in time during which the function repeats itself; and ϕ, called the *phase*, determines the initial value of the sine function. As will be discussed in the following sections, the phase and amplitude are determined by the initial state of the system (see Figure 1.7). It is standard to measure the time t in seconds (s). The phase is measured in radians (rad), and the frequency is measured in radians per second (rad/s). As derived in the following equation, the frequency ω_n is determined by the physical properties of mass and stiffness (m and k), and the constants A and ϕ are determined by the initial position and velocity as well as the frequency.

To see if equation (1.3) is in fact a solution of the equation of motion, it is substituted into equation (1.2). Successive differentiation of the displacement, $x(t)$ in the form of equation (1.3), yields the velocity, $\dot{x}(t)$, given by

$$\dot{x}(t) = \omega_n A\cos(\omega_n t + \phi) \tag{1.4}$$

and the acceleration, $\ddot{x}(t)$, given by

$$\ddot{x}(t) = -\omega_n^2 A\sin(\omega_n t + \phi) \tag{1.5}$$

Substitution of equations (1.5) and (1.3) into (1.2) yields

$$-m\omega_n^2 A \sin(\omega_n t + \phi) = -kA \sin(\omega_n t + \phi)$$

Dividing by A and m yields the fact that this last equation is satisfied if

$$\omega_n^2 = \frac{k}{m}, \quad \text{or} \quad \omega_n = \sqrt{\frac{k}{m}} \qquad (1.6)$$

Hence, equation (1.3) is a solution of the equation of motion. The constant ω_n characterizes the spring–mass system, as well as the frequency at which the motion repeats itself, and hence is called the system's *natural frequency*. A plot of the solution $x(t)$ versus time t is given in Figure 1.6. It remains to interpret the constants A and ϕ.

The units associated with the notation ω_n are rad/s and in older texts natural frequency in these units is often referred to as the *circular natural frequency* or *circular frequency* to emphasize that the units are consistent with trigonometric functions and to distinguish this from frequency stated in units of hertz (Hz) or cycles per second, denoted by f_n, and commonly used in discussing frequency. The two are related by $f_n = \omega_n/2\pi$ as discussed in Section 1.2. In practice, the phrase *natural frequency* is used to refer to either f_n or ω_n, and the units are stated explicitly to avoid confusion. For example, a common statement is: the natural frequency is 10 Hz, or the natural frequency is 20π rad/s.

Recall from differential equations that because the equation of motion is of second order, solving equation (1.2) involves integrating twice. Thus there are two constants of integration to evaluate. These are the constants A and ϕ. The physical significance, or interpretation, of these constants is that they are determined by the initial state of motion of the spring–mass system. Again, recall Newton's laws, if no force is imparted to the mass, it will stay at rest. If, however, the mass is displaced to a position of x_0 at time $t = 0$, the force kx_0 in the spring will result in motion. Also, if the mass is given an initial velocity of v_0 at time $t = 0$, motion will result because

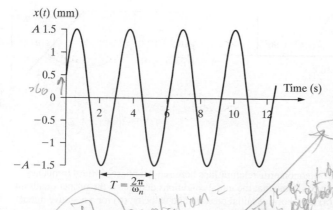

Figure 1.6   The response of a simple spring–mass system to an initial displacement of $x_0 = 0.5$ mm and an initial velocity of $v_0 = 2\sqrt{2}$ mm/s. The natural frequency is 2 rad/s and the amplitude is 1.5 mm. The period is $T = 2\pi/\omega_n = 2\pi/2 = \pi$s.

of the induced change in momentum. These are called *initial conditions* and when substituted into the solution (1.3) yield

$$x_0 = x(0) = A\sin(\omega_n 0 + \phi) = A\sin\phi \tag{1.7}$$

and

$$v_0 = \dot{x}(0) = \omega_n A\cos(\omega_n 0 + \phi) = \omega_n A\cos\phi \tag{1.8}$$

Solving these two simultaneous equations for the two unknowns A and ϕ yields

$$A = \frac{\sqrt{\omega_n^2 x_0^2 + v_0^2}}{\omega_n} \quad \text{and} \quad \phi = \tan^{-1}\frac{\omega_n x_0}{v_0} \tag{1.9}$$

as illustrated in Figure 1.7. Here the phase ϕ must lie in the proper quadrant, so care must be taken in evaluating the arc tangent. Thus, the solution of the equation of motion for the spring–mass system is given by

$$x(t) = \frac{\sqrt{\omega_n^2 x_0^2 + v_0^2}}{\omega_n}\sin\left(\omega_n t + \tan^{-1}\frac{\omega_n x_0}{v_0}\right) \tag{1.10}$$

and is plotted in Figure 1.6. This solution is called the *free response* of the system, because no force external to the system is applied after $t = 0$. The motion of the spring–mass system is called *simple harmonic motion* or *oscillatory motion* and is discussed in detail in the following section. The spring–mass system is also referred to as a *simple harmonic oscillator*, as well as an *undamped single-degree-of-freedom system*.

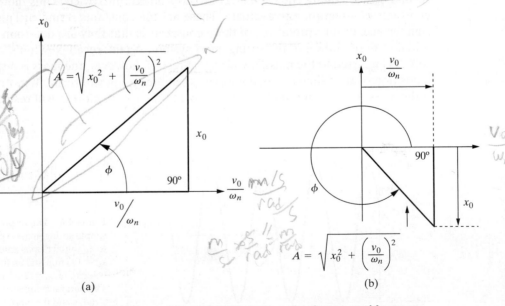

Figure 1.7 The trigonometric relationships between the phase, natural frequency, and initial conditions. Note that the initial conditions determine the proper quadrant for the phase: (a) for a positive initial position and velocity, (b) for a negative initial position and a positive initial velocity.

Cont. reading.

Example 1.1.2

The phase angle ϕ describes the relative shift in the sinusoidal vibration of the spring–mass system resulting from the initial displacement, x_0. Verify that equation (1.10) satisfies the initial condition $x(0) = x_0$. $= A \sin \phi$

Solution   Substitution of $t = 0$ in equation (1.10) yields

$$x(0) = A \sin \phi = \frac{\sqrt{\omega_n^2 x_0^2 + v_0^2}}{\omega_n} \sin\left(\tan^{-1} \frac{\omega_n x_0}{v_0}\right)$$

Figure 1.7 illustrates the phase angle ϕ defined by equation (1.9). This right triangle is used to define the sine and tangent of the angle ϕ. From the geometry of a right triangle, and the definitions of the sine and tangent functions, the value of $x(0)$ is computed to be

$$x(0) = \frac{\sqrt{\omega_n^2 x_0^2 + v_0^2}}{\omega_n} \frac{\omega_n x_0}{\sqrt{\omega_n^2 x_0^2 + v_0^2}} = x_0$$

which verifies that the solution given by equation (1.10) is consistent with the initial displacement condition.

$\square$

Example 1.1.3

A vehicle wheel, tire, and suspension assembly can be modeled crudely as a single-degree-of-freedom spring–mass system. The (unsprung) mass of the assembly is measured to be about 30 kilograms (kg). Its frequency of oscillation is observed to be 10 Hz. What is the approximate stiffness of the suspension assembly?

Solution   The relationship between frequency, mass, and stiffness is $\omega_n = \sqrt{k/m}$, so that

$$k = m\omega_n^2 = (30 \text{ kg})\left(10 \frac{\text{cycle}}{\text{s}} \cdot \frac{2\pi \text{ rad}}{\text{cycle}}\right)^2 = 1.184 \times 10^5 \text{ N/m}$$

This provides one simple way to estimate the stiffness of a complicated device. This stiffness could also be estimated by using a static deflection experiment similar to that suggested by Figures 1.3 and 1.4.

$\omega_n = \sqrt{\dfrac{1200}{7}} = 10$ $\square$

Example 1.1.4

Compute the amplitude and phase of the response of a system with a mass of 2 kg and a stiffness of 200 N/m, to the following initial conditions:

a) $x_0 = 2$ mm and $v_0 = 1$ mm/s
b) $x_0 = -2$ mm and $v_0 = 1$ mm/s
c) $x_0 = 2$ mm and $v_0 = -1$ mm/s

$x = a \sin(\omega t + \phi)$

Compare the results of these calculations.

Solution   First, compute the natural frequency, as this does not depend on the initial conditions and will be the same in each case. From equation (1.6):

$$\omega_n = \sqrt{\frac{k}{m}} = \sqrt{\frac{200 \text{ N/m}}{2 \text{ kg}}} = 10 \text{ rad/s}$$

Next, compute the amplitude, as it depends on the squares of the initial conditions and will be the same in each case. From equation (1.9):

$$A = \frac{\sqrt{\omega_n^2 x_0^2 + v_0^2}}{\omega_n} = \frac{\sqrt{10^2 \cdot 2^2 + 1^2}}{10} = 2.0025 \text{ mm}$$

Thus the difference between the three responses in this example is determined only by the phase. Using equation (1.9) and referring to Figure 1.7 to determine the proper quadrant, the following yields the phase information for each case:

a) $\phi = \tan^{-1}\left(\dfrac{\omega_n x_0}{v_0}\right) = \tan^{-1}\left(\dfrac{(10 \text{ rad/s}) (2 \text{ mm})}{1 \text{ mm/s}}\right) = 1.521 \text{ rad (or } 87.147°)$

which is in the first quadrant.

b) $\phi = \tan^{-1}\left(\dfrac{\omega_n x_0}{v_0}\right) = \tan^{-1}\left(\dfrac{(10 \text{ rad/s}) (-2 \text{ mm})}{1 \text{ mm/s}}\right) = -1.521 \text{ rad (or } -87.147°)$

which is in the fourth quadrant.

c) $\phi = \tan^{-1}\left(\dfrac{\omega_n x_0}{v_0}\right) = \tan^{-1}\left(\dfrac{(10 \text{ rad/s})(2 \text{ mm})}{-1 \text{ mm/s}}\right) = (-1.521 + \pi) \text{ rad (or } 92.85°)$

which is in the second quadrant (position positive, velocity negative places the angle in the second quadrant in Figure 1.7 requiring that the raw calculation be shifted 180°).

Note that if equation (1.9) is used without regard to Figure 1.7, parts b and c would result in the same answer (which makes no sense physically as the responses each have different starting points). Thus in computing the phase it is important to consider which quadrant the angle should lie in. Fortunately, some calculators and some codes use an arc tangent function, which corrects for the quadrant (for instance, MATLAB uses the atan2(w0*x0, v0) command).

The tan(ϕ) can be positive or negative. If the tangent is positive, the phase angle is in the first or third quadrant. If the sign of the initial displacement is positive, the phase angle is in the first quadrant. If the sign is negative or the initial displacement is negative, the phase angle is in the third quadrant. If on the other hand the tangent is negative, the phase angle is in the second or fourth quadrant. As in the previous case, by examining the sign of the initial displacement, the proper quadrant can be determined. That is, if the sign is positive, the phase angle is in the second quadrant, and if the sign is negative, the phase angle is in the fourth quadrant. The remaining possibility is that the tangent is equal to zero. In this case, the phase angle is either zero or 180°. The initial velocity determines which quadrant is correct. If the initial displacement is zero and if the initial velocity is zero, then the phase angle is zero. If on the other hand the initial velocity is negative, the phase angle is 180°.

□

The main point of this section is summarized in Window 1.2. This illustrates harmonic motion and how the initial conditions determine the response of such a system.

Window 1.2
Summary of the Description of Simple Harmonic Motion

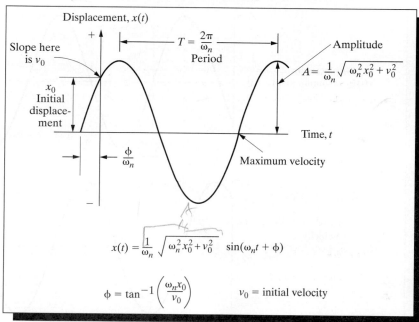

Displacement, $x(t)$

Slope here is v_0

x_0 Initial displacement

$T = \dfrac{2\pi}{\omega_n}$
Period

$\dfrac{\phi}{\omega_n}$

Amplitude

$A = \dfrac{1}{\omega_n}\sqrt{\omega_n^2 x_0^2 + v_0^2}$

Time, t

Maximum velocity

$$x(t) = \frac{1}{\omega_n}\sqrt{\omega_n^2 x_0^2 + v_0^2}\ \sin(\omega_n t + \phi)$$

$$\phi = \tan^{-1}\left(\frac{\omega_n x_0}{v_0}\right)\qquad v_0 = \text{initial velocity}$$

1.2 HARMONIC MOTION

The fundamental kinematic properties of a particle moving in one dimension are displacement, velocity, and acceleration. For the harmonic motion of a simple spring–mass system, these are given by equations (1.3), (1.4), and (1.5), respectively. These equations reveal the different relative amplitudes of each quantity. For systems with natural frequency larger than 1 rad/s, the relative amplitude of the velocity response is larger than that of the displacement response by a multiple of ω_n, and the acceleration response is larger by a multiple of ω_n^2. For systems with frequency less than 1, the velocity and acceleration have smaller relative amplitudes than the displacement. Also note that the velocity is 90° (or $\pi/2$ radians) out of phase with the position [i.e., $\sin(\omega_n t + \pi/2 + \phi) = \cos(\omega_n t + \phi)$], while the acceleration is 180° out of phase with the position and 90° out of phase with the velocity. This is summarized and illustrated in Window 1.3.

Window 1.3
The Relationship between Displacement, Velocity, and Acceleration
for Simple Harmonic Motion

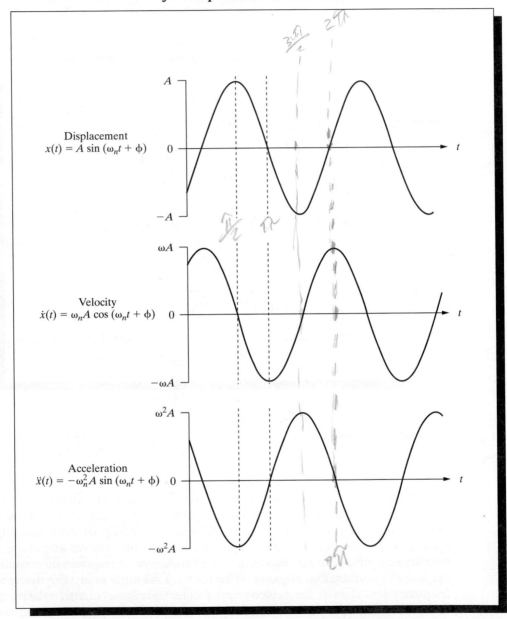

Displacement
$x(t) = A \sin(\omega_n t + \phi)$

Velocity
$\dot{x}(t) = \omega_n A \cos(\omega_n t + \phi)$

Acceleration
$\ddot{x}(t) = -\omega_n^2 A \sin(\omega_n t + \phi)$

The angular natural frequency, ω_n, used in equations (1.3) and (1.10), is measured in radians per second and describes the repetitiveness of the oscillation. As indicated in Window 1.2, the time the cycle takes to repeat itself is the *period, T,* which is related to the natural frequency by

$$T = \frac{2\pi \text{ rad}}{\omega_n \text{ rad/s}} = \frac{2\pi}{\omega_n} \text{ s} \qquad (1.11)$$

This results from the elementary definition of the period of a sine function. The frequency in hertz (Hz), denoted by f_n, is related to the frequency in radians per second, denoted by ω_n:

$$f_n = \frac{\omega_n}{2\pi} = \frac{\omega_n \text{ rad/s}}{2\pi \text{ rad/cycle}} = \frac{\omega_n \text{ cycles}}{2\pi \text{ s}} = \frac{\omega_n}{2\pi} \text{ (Hz)} \qquad (1.12)$$

Equation (1.2) is exactly the same form of differential equation as the linear pendulum equation of Example 1.1.1 and of the shaft and disk of Window 1.1(b). As such, the pendulum will have exactly the same form of solution as equation (1.3), with frequency

$$\omega_n = \sqrt{\frac{g}{l}} \text{ rad/s}$$

The solution of the pendulum equation thus predicts that the period of oscillation of the pendulum is

$$T = \frac{2\pi}{\omega_n} = 2\pi \sqrt{\frac{l}{g}} \text{ s}$$

where the non-italic s denotes seconds. This analytical value of the period can be checked by measuring the period of oscillation of a pendulum with a simple stopwatch. The period of the disk and shaft system of Window 1.1 will have a frequency and period of

$$\omega_n = \sqrt{\frac{k}{J}} \text{ rad/s} \qquad \text{and} \qquad T = 2\pi \sqrt{\frac{J}{k}} \text{ s}$$

respectively. The concept of frequency of vibration of a mechanical system is the single most important physical concept (and number) in vibration analysis. Measurement of either the period or the frequency allows validation of the analytical model. (If you made a 1-meter pendulum, the period would be about 2 s. This is something you could try at home.)

As long as the only disturbance to these systems is a set of nonzero initial conditions, the system will respond by oscillating with frequency ω_n and period T. For the case of the pendulum, the longer the pendulum, the smaller the frequency and the longer the period. That's why in museum demonstrations of a pendulum, the length is usually very large so that T is large and one can easily see the period (also a pendulum is usually used to illustrate the earth's precession; Google the phrase Foucault Pendulum).

Example 1.2.1

Consider a small spring about 30 mm (or 1.18 in) long, welded to a stationary table (ground) so that it is fixed at the point of contact, with a 12-mm (or 0.47-in) bolt welded to the other end, which is free to move. The mass of this system is about 49.2×10^{-3} kg (equivalent to about 1.73 ounces). The spring stiffness can be measured using the method suggested in Figure 1.4 and yields a spring constant of $k = 857.8$ N/m. Calculate the natural frequency and period. Also determine the maximum amplitude of the response if the spring is initially deflected 10 mm. Assume that the spring is oriented along the direction of gravity as in Window 1.1. (Ignore the effect of gravity; see below.)

Solution   From equation (1.6) the natural frequency is

$$\omega_n = \sqrt{\frac{k}{m}} = \sqrt{\frac{857.8 \text{ N/m}}{49.2 \times 10^{-3} \text{ kg}}} = 132 \text{ rad/s}$$

In hertz, this becomes

$$f_n = \frac{\omega_n}{2\pi} = 21 \text{ Hz}$$

The period is

$$T = \frac{2\pi}{\omega_n} = \frac{1}{f_n} = 0.0476 \text{ s}$$

To determine the maximum value of the displacement response, note from Figure 1.6 that this corresponds to the value of the constant A. Assuming that no initial velocity is given to the spring ($v_0 = 0$), equation (1.9) yields

$$x(t)_{max} = A = \frac{\sqrt{\omega_n^2 x_0^2 + v_0^2}}{\omega_n} = x_0 = 10 \text{ mm}$$

Note that the maximum value of the velocity response is $\omega_n A$ or $\omega_n x_0 = 1320$ mm/s and the acceleration response has maximum value

$$\omega_n^2 A = \omega_n^2 x_0 = 174.24 \times 10^3 \text{ mm/s}^2$$

Since $v_0 = 0$, the phase is $\phi = \tan^{-1}(\omega_n x_0 / 0) = \pi/2$, or $90°$. Hence, in this case, the response is $x(t) = 10 \sin(132t + \pi/2) = 10 \cos(132t)$ mm.   □

Does gravity matter in spring problems? The answer is no, if the system oscillates in the linear region. Consider the spring of Figure 1.3 and let a mass of value m extend

5 eq. of motion?

the spring. Let Δ denote the distance deflected in this static experiment (Δ is called the static deflection); then the force acting upon the mass is $k\Delta$. From static equilibrium the forces acting on the mass must be zero so that (taking positive down in the figure)

$$mg - k\Delta = 0$$

Next, sum the forces along the vertical for the mass at some point x and apply Newton's law to get

$$m\ddot{x}(t) = -k(x + \Delta) + mg = -kx + mg - \Delta k$$

Note the sign on the spring term is negative because the spring force opposes the motion, which is taken here as positive down. The last two terms add to zero ($mg - k\Delta = 0$) because of the static equilibrium condition, and the equation of motion becomes

$$m\ddot{x}(t) + kx(t) = 0 \quad = mg - k\Delta$$

Thus gravity does not affect the dynamic response. Note $x(t)$ is measured from the elongated (or compressed if upside down) position of the spring–mass system, that is, from its rest position. This is discussed again using energy methods in Figure 1.14.

Example 1.2.2

(a) A pendulum in Brussels swings with a period of 3 seconds. Compute the length of the pendulum. (b) At another location, assume the length of the pendulum is known to be 2 meters and suppose the period is measured to be 2.839 seconds. What is the acceleration due to gravity at that location?

Solution   The relationship between period and natural frequency is given in equation (1.11). (a) Substitution of the value of natural frequency for a pendulum and solving for the length of the pendulum yields

$$T = \frac{2\pi}{\omega_n} \Rightarrow \omega_n^2 = \frac{g}{l} = \frac{4\pi^2}{T^2} \Rightarrow l = \frac{gT^2}{4\pi^2} = \frac{(9.811 \text{ m/s}^2)(3)^2 s^2}{4\pi^2} = 2.237 \text{ m}$$

Here the value of $g = 9.811$ m/s^2 is used, as that is the value it has in Brussels (at 51° latitude and an altitude of 102 m). (b) Next, manipulate the pendulum period equation to solve for g. This yields

$$\frac{g}{l} = \frac{4\pi^2}{T^2} \Rightarrow g = \frac{4\pi^2}{T^2}l = \frac{4\pi^2}{(2.839)^2 s^2}(2)m = 9.796 \text{ m/s}^2$$

This is the value of the acceleration due to gravity in Denver, Colorado, United States (at an altitude 1638 m and latitude 40°).

These sorts of calculations are usually done in high school science classes but are repeated here to underscore the usefulness of the concept of natural frequency and period in terms of providing information about the vibration system's physical properties. In addition, this example serves to remind the reader of a familiar vibration phenomenon.

$\square$

The solution given by equation (1.10) was developed assuming that the response should be harmonic based on physical observation. The form of the response can also be derived by a more analytical approach following the theory of elementary differential equations (see, e.g., Boyce and DiPrima, 2009). This approach is reviewed here and will be generalized in later sections and chapters to solve for the response of more complicated systems.

Assume that the solution $x(t)$ is of the form

$$x(t) = ae^{\lambda t} \qquad (1.13)$$

where a and λ are nonzero constants to be determined. Upon successive differentiation, equation (1.13) becomes $\dot{x}(t) = \lambda ae^{\lambda t}$ and $\ddot{x}(t) = \lambda^2 ae^{\lambda t}$. Substitution of the assumed exponential form into equation (1.2) yields

$$m\lambda^2 ae^{\lambda t} + kae^{\lambda t} = 0 \qquad (1.14)$$

Since the term $ae^{\lambda t}$ is never zero, expression (1.14) can be divided by $ae^{\lambda t}$ to yield

$$m\lambda^2 + k = 0 \qquad (1.15)$$

Solving this algebraically results in

$$\lambda = \pm\sqrt{-\frac{k}{m}} = \pm\sqrt{\frac{k}{m}}j = \pm\omega_n j \qquad (1.16)$$

where $j = \sqrt{-1}$ is the imaginary number and $\omega_n = \sqrt{k/m}$ is the natural frequency as before. Note that there are two values for λ, $\lambda = +\omega_n j$ and $\lambda = -\omega_n j$, because the equation for λ is of second order. This implies that there must be two solutions of equation (1.2) as well. Substitution of equation (1.16) into equation (1.13) yields that the two solutions for $x(t)$ are

$$x(t) = a_1 e^{+j\omega_n t} \quad \text{and} \quad x(t) = a_2 e^{-j\omega_n t} \qquad (1.17)$$

Since equation (1.2) is linear, the sum of two solutions is also a solution; hence, the response $x(t)$ is of the form

$$x(t) = a_1 e^{+j\omega_n t} + a_2 e^{-j\omega_n t} \qquad (1.18)$$

where a_1 and a_2 are complex-valued constants of integration. The Euler relations for trigonometric functions state that $2j \sin\theta = (e^{\theta j} - e^{-\theta j})$ and $2\cos\theta = (e^{\theta j} + e^{-\theta j})$, where $j = \sqrt{-1}$. [See Appendix A, equations (A.18), (A.19), and (A.20), as well as Window 1.5.] Using the Euler relations, equation (1.18) can be written as

$$x(t) = A\sin(\omega_n t + \phi) \qquad (1.19)$$

where A and ϕ are real-valued constants of integration. Note that equation (1.19) is in agreement with the physically intuitive solution given by equation (1.3). The relationships among the various constants in equations (1.18) and (1.19) are given in Window 1.4. Window 1.5 illustrates the use of Euler relations for deriving harmonic functions from exponentials for the underdamped case.

Window 1.4
Three Equivalent Representations of Harmonic Motion

The solution of $m\ddot{x} + kx = 0$ subject to nonzero initial conditions can be written in three equivalent ways. First, the solution can be written as

$$x(t) = a_1 e^{j\omega_n t} + a_2 e^{-j\omega_n t}, \quad \omega_n = \sqrt{\frac{k}{m}}, \quad j = \sqrt{-1}$$

where a_1 and a_2 are complex-valued constants. Second, the solution can be written as

$$x(t) = A \sin(\omega_n t + \phi)$$

where A and ϕ are real-valued constants. Last, the solution can be written as

$$x(t) = A_1 \sin \omega_n t + A_2 \cos \omega_n t$$

where A_1 and A_2 are real-valued constants. Each set of two constants is determined by the initial conditions, x_0 and v_0. The various constants are related by the following:

$$A = \sqrt{A_1^2 + A_2^2} \quad \phi = \tan^{-1}\left(\frac{A_2}{A_1}\right)$$

$$A_1 = (a_1 - a_2)j \quad A_2 = a_1 + a_2$$

$$a_1 = \frac{A_2 - A_1 j}{2} \quad a_2 = \frac{A_2 + A_1 j}{2}$$

all of which follow from trigonometric identities and Euler's formulas. Note that a_1 and a_2 are a complex conjugate pair, so that A_1 and A_2 are both real numbers provided that the initial conditions are real valued, as is normally the case.

Often when computing frequencies from equation (1.16) such as $\lambda^2 = -4$, there is a temptation to write that the frequency is $\omega_n = \pm 2$. This is incorrect because the $\pm$ sign is used up when the Euler relation is used to obtain the function $\sin \omega_n t$ from the exponential form. The concept of frequency is not defined until it appears in the argument of the sine function and, as such, is always positive.

Precise terminology is useful in discussing an engineering problem, and the subject of vibration is no exception. Since the position, velocity, and acceleration change continually with time, several other quantities are used to discuss vibration. The *peak value*, defined as the maximum displacement, or magnitude A of equation (1.9), is often used to indicate the region in space in which the object vibrates. Another quantity useful in describing vibration is the *average value*, denoted by $\bar{x}$, and defined by

$$\bar{x} = \lim_{T \to \infty} \frac{1}{T} \int_0^T x(t)\, dt \qquad (1.20)$$

Note that the average value of $x(t) = A \sin \omega_n t$ over one period of oscillation is zero.

Since the square of displacement is associated with a system's potential energy, the average of the displacement squared is sometimes a useful vibration property to discuss. The mean-square value (or variance) of the displacement $x(t)$, denoted by $\overline{x^2}$, is defined by

$$\overline{x^2} = \lim_{T \to \infty} \frac{1}{T} \int_0^T x^2(t)\, dt \qquad (1.21)$$

The square root of this value, called the *root mean-square* (rms) value, is commonly used in specifying vibration. Because the peak value of the velocity and acceleration are multiples of the natural frequency times the displacement amplitude [i.e., equations (1.3)–(1.5)], these three peak values often differ in value by an order of magnitude or more. Hence, logarithmic scales are often used. A common unit of measurement for vibration amplitudes and rms values is the *decibel* (dB). The decibel was originally defined in terms of the base 10 logarithm of the power ratio of two electrical signals, or as the ratio of the square of the amplitudes of two signals. Following this idea, the decibel is defined as

$$dB \equiv 10 \log_{10}\left(\frac{x_1}{x_0}\right)^2 = 20 \log_{10} \frac{x_1}{x_0} \qquad (1.22)$$

Here the signal x_0 is a reference signal. The decibel is used to quantify how far the measured signal x_1 is above the reference signal x_0. Note that if the measured signal is equal to the reference signal, then this corresponds to 0 dB. The decibel is used extensively in acoustics to compare sound levels. Using a dB scale expands or compresses vibration response information for convenience in graphical representation.

Example 1.2.3

Consider a 2-meter long pendulum placed on the moon and given an initial angular displacement of 0.2 rad and zero initial velocity. Calculate the maximum angular velocity and the maximum angular acceleration of the swinging pendulum (note that gravity on the earth's moon is $g_m = g/6$, where g is the acceleration due to gravity on earth).

Solution   From Example 1.1.1 the equation of motion of a pendulum is

$$\ddot{\theta}(t) + \frac{g_m}{l}\theta(t) = 0$$

This equation is of the same form as equation (1.2) and hence has a solution of the form

$$\theta(t) = A\sin(\omega_n t + \phi), \quad \omega_n = \sqrt{\frac{g_m}{l}}$$

From equation (1.9) the amplitude is given by

$$A = \sqrt{\frac{\omega_n^2 x_0^2 + v_0^2}{\omega_n^2}} = x_0 = 0.2 \text{ rad}$$

From Window 1.3 the maximum velocity is just $\omega_n A$ or

$$v_{max} = \omega_n A = \sqrt{\frac{g_m}{l}}(0.2) = (0.2)\sqrt{\frac{9.8/6}{2}} = 0.18 \text{ rad/s}$$

The maximum acceleration is

$$a_{max} = \omega_n^2 A = \frac{g_m}{l}A = \frac{9.8/6}{2}(0.2) = 0.163 \text{ rad/s}^2$$

$\square$

Frequencies of concern in mechanical vibration range from fractions of a hertz to several thousand hertz. Amplitudes range from micrometers up to meters (for systems such as tall buildings). According to Mansfield (2005), human beings are more sensitive to acceleration than displacement and easily perceive vibration around 5 Hz at about 0.01 m/s^2 (about 0.01 mm). Horizontal vibration is easy to experience near 2 Hz. Work attempting to characterize comfort levels for human vibrations is still ongoing.

1.3 VISCOUS DAMPING

The response of the spring–mass model (Section 1.1) predicts that the system will oscillate indefinitely. However, everyday observation indicates that freely oscillating systems eventually die out and reduce to zero motion. This observation suggests that the model sketched in Figure 1.5 and the corresponding mathematical model given by equation (1.2) need to be modified to account for this decaying motion. The choice of a representative model for the observed decay in an oscillating system is based partially on physical observation and partially on mathematical convenience. The theory of differential equations suggests that adding a term to equation (1.2) of the form $c\dot{x}(t)$, where c is a constant, will result in a solution $x(t)$ that dies out. Physical observation agrees fairly well with this model and is used successfully to model the damping, or decay, in a variety of mechanical systems. This type of damping, called *viscous damping*, is described in detail in this section.

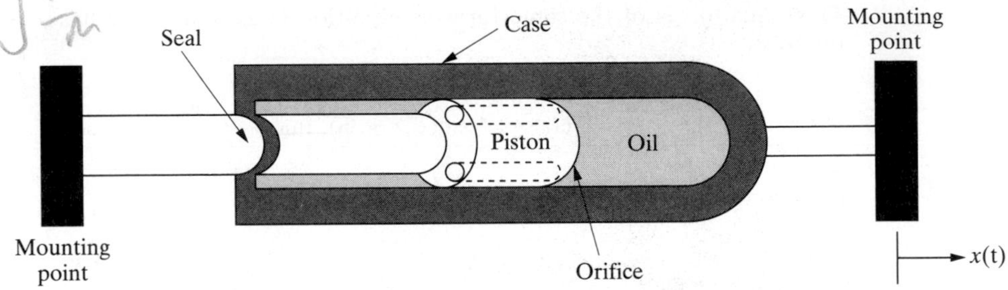

Figure 1.8   A schematic of a dashpot that produces a damping force $f_c(t) = c\dot{x}(t)$, where $x(t)$ is the motion of the case relative to the piston.

While the spring forms a physical model for storing potential energy and hence causing vibration, the *dashpot*, or *damper*, forms the physical model for dissipating energy and thus damping the response of a mechanical system. An example dashpot consists of a piston fit into a cylinder filled with oil as indicated in Figure 1.8. This piston is perforated with holes so that motion of the piston in the oil is possible. The laminar flow of the oil through the perforations as the piston moves causes a damping force on this piston. The force is proportional to the velocity of the piston in a direction opposite that of the piston motion. This damping force, denoted by f_c, has the form

$$f_c = c\dot{x}(t) \tag{1.23}$$

where c is a constant of proportionality related to the oil viscosity. The constant c, called the *damping coefficient*, has units of force per velocity, or N s/m, as it is customarily written. However, following the strict rules of SI units, the units on damping can be reduced to kg/s, which states the units on damping in terms of the fundamental (also called basic) SI units (mass, time, and length).

In the case of the oil-filled dashpot, the constant c can be determined by fluid principles. However, in most cases, f_c is provided by equivalent effects occurring in the material forming the device. A good example is a block of rubber (which also provides stiffness f_k) such as an automobile motor mount, or the effects of air flowing around an oscillating mass. In all cases in which the damping force f_c is proportional to velocity, the schematic of a dashpot is used to indicate the presence of this force. The schematic is illustrated in Figure 1.9. Unfortunately, the damping coefficient of a system cannot be measured as simply as the mass or stiffness of a system can be. This is pointed out in Section 1.6.

Using a simple force balance on the mass of Figure 1.9 in the x direction, the equation of motion for $x(t)$ becomes

$$m\ddot{x} = -f_c - f_k \tag{1.24}$$

or

$$m\ddot{x}(t) + c\dot{x}(t) + kx(t) = 0 \tag{1.25}$$

if $\ddot{x}_0$ is > 0
or deceleration if $\ddot{x} < 0$
$\ddot{x}_0 = $ acceleration

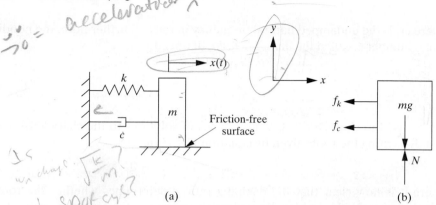

ζ

$1 <$
ω_n chap? $\sqrt{\frac{k}{m}}$
$1, $ spot cys?

(a)   (b)

Figure 1.9   (a) The schematic of a single-degree-of-freedom system with viscous damping indicated by a dashpot and (b) the corresponding free-body diagram.

W.S.
No Volumen

subject to the initial conditions $x(0) = x_0$ and $\dot{x}(0) = v_0$. The forces f_c and f_k are negative in equation (1.24) because they oppose the motion (positive to the right). Equation (1.25) and Figure 1.9, referred to as a *damped single-degree-of-freedom system*, form the topic of Chapters 1 through 3.

To solve the damped system of equation (1.25), the same method used for solving equation (1.2) is used. In fact, this provides an additional reason to choose f_c to be of the form $c\dot{x}$. Let $x(t)$ have the form given in equation (1.13), $x(t) = ae^{\lambda t}$. Substitution of this form into equation (1.25) yields

$$(m\lambda^2 + c\lambda + k)ae^{\lambda t} = 0 \tag{1.26}$$

Again, $ae^{\lambda t} \neq 0$, so that this reduces to a quadratic equation in λ of the form

$$m\lambda^2 + c\lambda + k = 0 \tag{1.27}$$

called the *characteristic equation*. This is solved using the quadratic formula to yield the two solutions

$$\lambda_{1,2} = -\frac{c}{2m} \pm \frac{1}{2m}\sqrt{c^2 - 4km} \tag{1.28}$$

Examination of this expression indicates that the roots λ will be real or complex, depending on the value of the discriminant, $c^2 - 4km$. As long as m, c, and k are positive real numbers, λ_1 and λ_2 will be distinct negative real numbers if $c^2 - 4km > 0$. On the other hand, if this discriminant is negative, the roots will be a complex conjugate pair with a negative real part. If the discriminant is zero, the two roots λ_1 and λ_2 are equal negative real numbers. Note that equation (1.15) represents the characteristic equation for the special undamped case (i.e., $c = 0$).

In examining these three cases, it is both convenient and useful to define the *critical damping coefficient*, c_{cr}, by

$$c_{cr} = 2m\omega_n = 2\sqrt{km} \tag{1.29}$$

where ω_n is the undamped natural frequency in rad/s. Furthermore, the nondimensional number ζ, called the *damping ratio*, defined by

$$\zeta = \frac{c}{c_{cr}} = \frac{c}{2m\omega_n} = \frac{c}{2\sqrt{km}} \qquad (1.30)$$

can be used to characterize the three types of solutions to the characteristic equation. Rewriting the roots given by equation (1.28) yields

$$\lambda_{1,2} = -\zeta\omega_n \pm \omega_n\sqrt{\zeta^2 - 1} \qquad (1.31)$$

where it is now clear that the damping ratio ζ determines whether the roots are complex or real. This in turn determines the nature of the response of the damped single-degree-of-freedom system. For positive mass, damping, and stiffness coefficients, there are three cases, which are delineated next.

1.3.1 Underdamped Motion

In this case, the damping ratio ζ is less than 1 $(0 < \zeta < 1)$ and the discriminant of equation (1.31) is negative, resulting in a complex conjugate pair of roots. Factoring out (-1) from the discriminant in order to clearly distinguish that the second term is imaginary yields

$$\sqrt{\zeta^2 - 1} = \sqrt{(1 - \zeta^2)(-1)} = \sqrt{1 - \zeta^2}\,j \qquad (1.32)$$

where $j = \sqrt{-1}$. Thus the two roots become

$$\lambda_1 = -\zeta\omega_n - \omega_n\sqrt{1 - \zeta^2}\,j \qquad (1.33)$$

and

$$\lambda_2 = -\zeta\omega_n + \omega_n\sqrt{1 - \zeta^2}\,j \qquad (1.34)$$

Following the same argument as that made for the undamped response of equation (1.18), the solution of (1.25) is then of the form

$$x(t) = e^{-\zeta\omega_n t}\left(a_1 e^{j\sqrt{1-\zeta^2}\omega_n t} + a_2 e^{-j\sqrt{1-\zeta^2}\omega_n t}\right) \qquad (1.35)$$

where a_1 and a_2 are arbitrary complex-valued constants of integration to be determined by the initial conditions. Using the Euler relations (see Window 1.5), this can be written as

$$x(t) = Ae^{-\zeta\omega_n t}\sin(\omega_d t + \phi) \qquad (1.36)$$

where A and ϕ are constants of integration and ω_d, called the *damped natural frequency*, is given by

$$\omega_d = \omega_n\sqrt{1 - \zeta^2} \qquad (1.37)$$

in units of rad/s.

Window 1.5
Euler Relations and the Underdamped Solution

An underdamped solution of $m\ddot{x} + c\dot{x} + kx = 0$ to nonzero initial conditions is of the form

$$x(t) = a_1 e^{\lambda_1 t} + a_2 e^{\lambda_2 t}$$

where λ_1 and λ_2 are complex numbers of the form

$$\lambda_1 = -\zeta\omega_n + \omega_d j \qquad \text{and} \qquad \lambda_2 = -\zeta\omega_n - \omega_d j$$

where $\omega_n = \sqrt{k/m}$, $\zeta = c/(2m\omega_n)$, $\omega_d = \omega_n\sqrt{1 - \zeta^2}$, and $j = \sqrt{-1}$. The two constants a_1 and a_2 are complex numbers and hence represent four unknown constants rather than the two constants of integration required to solve a second-order differential equation. This demands that the two complex numbers a_1 and a_2 be conjugate pairs so that $x(t)$ depends only on two undetermined constants. Substitution of the foregoing values of λ_i into the solution $x(t)$ yields

$$x(t) = e^{-\zeta\omega_n t}\left(a_1 e^{\omega_d jt} + a_2 e^{-\omega_d jt}\right)$$

Using the Euler relations $e^{\phi j} = \cos\phi + j\sin\phi$ and $e^{-\phi j} = \cos\phi - j\sin\phi$, $x(t)$ becomes

$$x(t) = e^{-\zeta\omega_n t}\left[(a_1 + a_2)\cos\omega_d t + j(a_1 - a_2)\sin\omega_d t\right]$$

Choosing the real numbers $A_2 = a_1 + a_2$ and $A_1 = (a_1 - a_2)j$, this becomes

$$x(t) = e^{-\zeta\omega_n t}(A_1\sin\omega_d t + A_2\cos\omega_d t)$$

which is real valued. Defining the constant $A = \sqrt{A_1^2 + A_2^2}$ and the angle $\phi = \tan^{-1}(A_2/A_1)$ so that $A_1 = A\cos\phi$ and $A_2 = A\sin\phi$, the form of $x(t)$ becomes [recall that $\sin a\cos b + \cos a\sin b = \sin(a + b)$]

$$x(t) = Ae^{-\zeta\omega_n t}\sin(\omega_d t + \phi)$$

where A and ϕ are the constants of integration to be determined from the initial conditions. Complex numbers are reviewed in Appendix A.

   The constants A and ϕ are evaluated using the initial conditions in exactly the same fashion as they were for the undamped system as indicated in equations (1.7) and (1.8). Set $t = 0$ in equation (1.36) to get $x_0 = A\sin\phi$. Differentiating (1.36) yields

$$\dot{x}(t) = -\zeta\omega_n Ae^{-\zeta\omega_n t}\sin(\omega_d t + \phi) + \omega_d Ae^{-\zeta\omega_n t}\cos(\omega_d t + \phi)$$

Let $t = 0$ and $A = x_0/\sin\phi$ in this last expression to get

$$\dot{x}(0) = v_0 = -\zeta\omega_n x_0 + x_0\omega_d\cot\phi$$

Solving this last expression for ϕ yields

$$\tan\phi = \frac{x_0\omega_d}{v_0 + \zeta\omega_n x_0}$$

With this value of ϕ, the sine becomes

$$\sin\phi = \frac{x_0\omega_d}{\sqrt{(v_0 + \zeta\omega_n x_0)^2 + (x_0\omega_d)^2}}$$

Thus the value of A and ϕ are determined to be

$$A = \sqrt{\frac{(v_0 + \zeta\omega_n x_0)^2 + (x_0\omega_d)^2}{\omega_d^2}}, \quad \phi = \tan^{-1}\frac{x_0\omega_d}{v_0 + \zeta\omega_n x_0} \quad (1.38)$$

where x_0 and v_0 are the initial displacement and velocity. A plot of $x(t)$ versus t for this underdamped case is given in Figure 1.10. Note that the motion is oscillatory with exponentially decaying amplitude. The damping ratio ζ determines the rate of decay. The response illustrated in Figure 1.10 is exhibited in many mechanical systems and constitutes the most common case. As a check to see that equation (1.38) is reasonable, note that if $\zeta = 0$ in the expressions for A and ϕ, the undamped relations of equation (1.9) result.

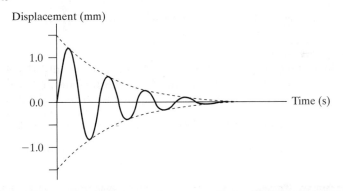

Displacement (mm)

Figure 1.10   The response of an underdamped system: $0 < \zeta < 1$.

1.3.2 Overdamped Motion

In this case, the damping ratio is greater than 1 ($\zeta > 1$). The discriminant of equation (1.31) is positive, resulting in a pair of distinct real roots. These are

$$\lambda_1 = -\zeta\omega_n - \omega_n\sqrt{\zeta^2 - 1} \quad (1.39)$$

and

$$\lambda_2 = -\zeta\omega_n + \omega_n\sqrt{\zeta^2 - 1} \quad (1.40)$$

The solution of equation (1.25) then becomes

$$x(t) = e^{-\zeta\omega_n t}\left(a_1 e^{-\omega_n\sqrt{\zeta^2-1}t} + a_2 e^{+\omega_n\sqrt{\zeta^2-1}t}\right) \tag{1.41}$$

which represents a nonoscillatory response. Again, the constants of integration a_1 and a_2 are determined by the initial conditions indicated in equations (1.7) and (1.8). In this nonoscillatory case, the constants of integration are real valued and are given by

$$a_1 = \frac{-v_0 + (-\zeta + \sqrt{\zeta^2 - 1})\omega_n x_0}{2\omega_n\sqrt{\zeta^2-1}} \tag{1.42}$$

and

$$a_2 = \frac{v_0 + (\zeta + \sqrt{\zeta^2 - 1})\omega_n x_0}{2\omega_n\sqrt{\zeta^2-1}} \tag{1.43}$$

Typical responses are plotted in Figure 1.11, where it is clear that motion does not involve oscillation. An overdamped system does not oscillate but rather returns to its rest position exponentially.

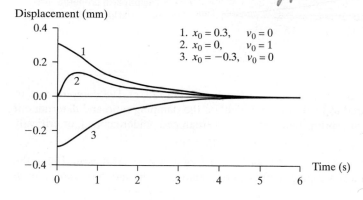

Displacement (mm)

1. $x_0 = 0.3$, $v_0 = 0$
2. $x_0 = 0$, $v_0 = 1$
3. $x_0 = -0.3$, $v_0 = 0$

Time (s)

Figure 1.11   The response of an overdamped system, $\zeta > 1$, for two different values of initial displacement (in mm) both with the initial velocity set to zero and one case with $x_0 = 0$ and $v_0 = 1$ mm/s.

1.3.3 Critically Damped Motion

In this last case, the damping ratio is exactly one ($\zeta = 1$) and the discriminant of equation (1.31) is equal to zero. This corresponds to the value of ζ that separates oscillatory motion from nonoscillatory motion. Since the roots are repeated, they have the value

$$\lambda_1 = \lambda_2 = -\omega_n \tag{1.44}$$

The solution takes the form

$$x(t) = (a_1 + a_2 t)e^{-\omega_n t} \tag{1.45}$$

where, again, the constants a_1 and a_2 are determined by the initial conditions. Substituting the initial displacement into equation (1.45) and the initial velocity into the derivative of equation (1.45) yields

$$a_1 = x_0, \quad a_2 = v_0 + \omega_n x_0 \tag{1.46}$$

Critically damped motion is plotted in Figure 1.12 for two different values of initial conditions. It should be noted that critically damped systems can be thought of in several ways. They represent systems with the smallest value of damping rate that yields nonoscillatory motion. Critical damping can also be thought of as the case that separates nonoscillation from oscillation, or the value of damping that provides the fastest return to zero without oscillation.

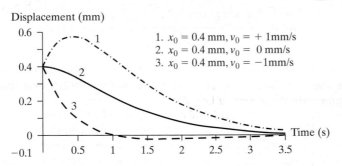

Displacement (mm)

1. $x_0 = 0.4$ mm, $v_0 = +1$ mm/s
2. $x_0 = 0.4$ mm, $v_0 = 0$ mm/s
3. $x_0 = 0.4$ mm, $v_0 = -1$ mm/s

Figure 1.12 The response of a critically damped system for three different initial velocities. The physical properties are $m = 100$ kg, $k = 225$ N/m, and $\zeta = 1$.

Example 1.3.1

Recall the small spring of Example 1.2.1 (i.e., $\omega_n = 132$ rad/s). The damping rate of the spring is measured to be 0.11 kg/s. Calculate the damping ratio and determine if the free motion of the spring–bolt system is overdamped, underdamped, or critically damped.

Solution   From Example 1.2.1, $m = 49.2 \times 10^{-3}$ kg and $k = 857.8$ N/m. Using the definition of the critical damping coefficient of equation (1.29) and these values for m and k yields

$$c_{cr} = 2\sqrt{km} = 2\sqrt{(857.8 \ N/m)(49.2 \times 10^{-3} \ kg)}$$

$$= 12.993 \ kg/s$$

If c is measured to be 0.11 kg/s, the critical damping ratio becomes

$$\zeta = \frac{c}{c_{cr}} = \frac{0.11(kg/s)}{12.993(kg/s)} = 0.0085$$

or 0.85% damping. Since ζ is less than 1, the system is underdamped. The motion resulting from giving the spring–bolt system a small displacement will be oscillatory.

The single-degree-of-freedom damped system of equation (1.25) is often written in a standard form. This is obtained by dividing equation (1.25) by the mass, m.
This yields

$$\ddot{x} + \frac{c}{m}\dot{x} + \frac{k}{m}x = 0 \qquad (1.47)$$

The coefficient of $x(t)$ is ω_n^2, the undamped natural frequency squared. A little
manipulation illustrates that the coefficient of the velocity $\dot{x}$ is $2\zeta\omega_n$. Thus equation
(1.47) can be written as

$$\ddot{x}(t) + 2\zeta\omega_n\dot{x}(t) + \omega_n^2 x(t) = 0 \qquad (1.48)$$

In this standard form, the values of the natural frequency and the damping ratio are
clear. In differential equations, equation (1.48) is said to be in monic form, meaning
that the leading coefficient (coefficient of the highest derivative) is one.

Example 1.3.2

The human leg has a measured natural frequency of around 20 Hz when in its rigid
(knee-locked) position in the longitudinal direction (i.e., along the length of the bone)
with a damping ratio of $\zeta = 0.224$. Calculate the response of the tip of the leg bone
to an initial velocity of $v_0 = 0.6$ m/s and zero initial displacement (this would correspond to the vibration induced while landing on your feet, with your knees locked
from a height of 18 mm) and plot the response. Last, calculate the maximum acceleration experienced by the leg assuming no damping.

Solution   The damping ratio is $\zeta = 0.224 < 1$, so the system is clearly underdamped.
The natural frequency is $\omega_n = \dfrac{20 \text{ cycles}}{1} \dfrac{2\pi \text{ rad}}{\text{s}} \dfrac{}{\text{cycles}} = 125.66$ rad/s. The damped natural
frequency is $\omega_d = 125.66\sqrt{1 - (0.224)^2} = 122.467$ rad/s. Using equation (1.38) with
$v_0 = 0.6$ m/s and $x_0 = 0$ yields

$$A = \frac{\sqrt{[\,0.6 + (0.224)(125.66)(0)\,]^2 + [\,(0)(122.467)\,]^2}}{122.467} = 0.005 \text{ m}$$

$$\phi = \tan^{-1}\left(\frac{(0)(\omega_d)}{v_0 + \zeta\omega_n(0)}\right) = 0$$

The response as given by equation (1.36) is

$$x(t) = 0.005e^{-28.148t}\sin(122.467t)$$

This is plotted in Figure 1.13. To find the maximum acceleration rate that the leg experiences for zero damping, use the undamped case of equation (1.9):

$$A = \sqrt{x_0^2 + \left(\frac{v_0}{\omega_n}\right)^2}, \quad \omega_n = 125.66, \ v_0 = 0.6, \ x_0 = 0$$

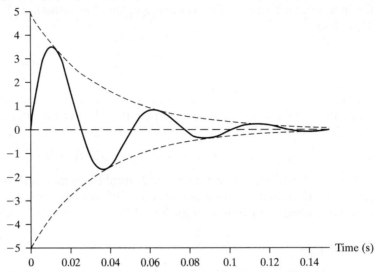

Displacement (mm)

Figure 1.13 A plot of displacement versus time for the leg bone of Example 1.3.2.

$$A = \frac{v_0}{\omega_n}\,\text{m} = \frac{0.6}{\omega_n}\,\text{m}$$

$$\max\,(\ddot{x}) = \left|-\omega_n^2 A\right| = \left|-\omega_n^2\left(\frac{0.6}{\omega_n}\right)\right| = (0.6)(125.66\ \text{m/s}^2) = 75.396\ \text{m/s}^2$$

In terms of $g = 9.81\ \text{m/s}^2$, this becomes

$$\text{maximum acceleration} = \frac{75.396\ \text{m/s}^2}{9.81\ \text{m/s}^2}\,g = 7.69\ g\text{'s}$$

Example 1.3.3

Compute the form of the response of an underdamped system using the Cartesian form of the solution given in Window 1.5.

Solution   From basic trigonometry $\sin(x + y) = \sin x \cos y + \cos x \sin y$. Applying this to equation (1.36) with $x = \omega_d t$ and $y = \phi$ yields

$$x(t) = Ae^{-\zeta\omega_n t}\sin\,(\omega_d t + \phi) = e^{-\zeta\omega_n t}(A_1 \sin \omega_d t + A_2 \cos\omega_d t)$$

where $A_1 = A \cos \phi$ and $A_2 = A \sin \phi$, as indicated in Window 1.5. Evaluating the initial conditions yields

$$x(0) = x_0 = e^0(A_1 \sin 0 + A_2 \cos 0)$$

Solving yields $A_2 = x_0$. Next, differentiate $x(t)$ to get

$$\dot{x} = -\zeta\omega_n e^{-\zeta\omega_n t}(A_1 \sin\omega_d t + A_2 \cos\omega_d t) + \omega_d e^{-\zeta\omega_n t}(A_1 \cos\omega_d t - A_2 \sin\omega_d t)$$

Applying the initial velocity condition yields

$$v_0 = \dot{x}(0) = -\zeta\omega_n(A_1 \sin 0 + x_0 \cos 0) + \omega_d(A_1 \cos 0 - x_0 \sin 0)$$

Solving this last expression yields

$$A_1 = \frac{v_0 + \zeta\omega_n x_0}{\omega_d}$$

Thus the free response in Cartesian form becomes

$$x(t) = e^{-\zeta\omega_n t}\left(\frac{v_0 + \zeta\omega_n x_0}{\omega_d} \sin \omega_d t + x_0 \cos \omega_d t\right)$$

1.4 MODELING AND ENERGY METHODS

Modeling is the art or process of writing down an equation, or system of equations, to describe the motion of a physical device. For example, equation (1.2) was obtained by modeling the spring–mass system of Figure 1.5. By summing the forces acting on the mass along the x direction and employing the experimental evidence of the mathematical model of the force in a spring given by Figure 1.4, equation (1.2) can be obtained. The success of this model is determined by how well the solution of equation (1.2) predicts the observed and measured behavior of the actual system. This comparison between the vibration response of a device and the response predicted by the analytical model is discussed in Section 1.6. The majority of this book is devoted to the analysis of vibration models. However, two methods of modeling—force balance and energy methods—are presented in this section. Newton's three laws form the basis of dynamics. Fifty years after Newton, Euler published his laws of motion. Newton's second law states: the sum of forces acting on a body is equal to the body's mass times its acceleration, and Euler's second law states: the rate of change of angular momentum is equal to the sum of external moments acting on the mass. Euler's second law can be manipulated to reveal that the sum of moments acting on a mass is equal to its rotational inertia times its angular acceleration. These two laws require the use of free-body diagrams and the proper identification of forces and moments acting on a body, forming most of the activity in the study of dynamics.

An alternative approach, studied in dynamics, is to examine the energy in the system, giving rise to what is referred to as energy methods for determining the equations of motion. The energy methods do not require free-body diagrams but rather require an understanding of the energy in a system, providing a useful alternative when forces are not easy to determine. More comprehensive treatments of modeling can be found in Doebelin (1980), Shames (1980, 1989), and Cannon (1967), for example. The best reference for modeling is the text you used to study dynamics. There are also many excellent descriptions on the Internet which can be found using a search engine such as Google.

The force summation method is used in the previous sections and should be familiar to the reader from introductory dynamics. For systems with constant mass (such as those considered here) moving in only one direction, the rate of change of momentum becomes the scalar relation

$$\frac{d}{dt}(m\dot{x}) = m\ddot{x}$$

which is often called the inertia force. The physical device of interest is examined by noting the forces acting on the device. The forces are then summed (as vectors) to produce a dynamic equation following Newton's second law. For motion along the x direction only, this becomes the scalar equation

$$\sum_i f_{xi} = m\ddot{x} \tag{1.49}$$

where f_{xi} denotes the ith force acting on the mass m along the x direction and the summation is over the number of such forces. In the first three chapters, only single-degree-of-freedom systems moving in one direction are considered; thus, Newton's law takes on a scalar nature. In more practical problems with many degrees of freedom, energy considerations can be combined with the concepts of virtual work to produce Lagrange's equations, as discussed in Section 4.7. Lagrange's equations also provide an energy-based alternative to summing forces to derive equations of motion.

For rigid bodies in plane motion (i.e., rigid bodies for which all the forces acting on them are coplanar in a plane perpendicular to a principal axis) and free to rotate, Euler's second law states that the sum of the applied torques is equal to the rate of change of angular momentum of the mass. This is expressed as

$$\sum_i M_{0i} = J\ddot{\theta} \tag{1.50}$$

where M_{0i} are the torques acting on the object about the point 0, J is the moment of inertia (also denoted by I_0) about the rotation axis, and θ is the angle of rotation. The sum of moments method was used in Example 1.1.1 to find the equation of motion of a pendulum and is discussed in more detail in Example 1.5.1.

If the forces or torques acting on an object or mechanical part are difficult to determine, an energy approach may be more efficient. In this method, the differential equation of motion is established by using the principle of energy conservation. This principle is equivalent to Newton's law for conservative systems and states that the sum of the potential energy and kinetic energy of a particle remains constant at each instant of time throughout the particle's motion:

$$T + U = \text{constant} \tag{1.51}$$

where T and U denote the total kinetic and potential energy, respectively. Conservation of energy also implies that the change in kinetic energy must equal the change in potential energy:

$$U_1 - U_2 = T_2 - T_1 \tag{1.52}$$

where U_1 and U_2 represent the particle's potential energy at the times t_1 and t_2, respectively, and T_1 and T_2 represent the particle's kinetic energy at times t_1 and t_2, respectively. For periodic motion, energy conservation also implies that

$$T_{\max} = U_{\max} \tag{1.53}$$

Since energy is a scalar quantity, using the conservation of energy principle yields a possibility of obtaining the equation of motion of a system without using force or moment summations.

   Equations (1.51), (1.52), and (1.53) are three statements of the conservation of energy. Each of these can be used to determine the equation of motion of a spring–mass system. As an illustration, consider the energy of the spring–mass system of Figure 1.14 hanging in a gravitational field of strength g. The effect of adding the mass m to the massless spring of stiffness k is to stretch the spring from its rest position at 0 to the static equilibrium position Δ. The total potential energy of the spring–mass system is the sum of the potential energy of the spring (or strain energy; see, e.g., Shames, 1989) and the gravitational potential energy. The potential energy of the spring is given by

$$U_{\text{spring}} = \tfrac{1}{2} k (\Delta + x)^2 \tag{1.54}$$

The gravitational potential energy is

$$U_{\text{gray}} = -mgx \tag{1.55}$$

where the negative sign indicates that $x = 0$ is the reference for zero potential energy. The kinetic energy of the system is

$$T = \tfrac{1}{2} m \dot{x}^2 \tag{1.56}$$

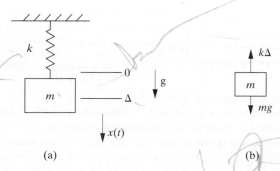

(a)

(b)

Figure 1.14   (a) A spring–mass system hanging in a gravitational field. Here Δ is the static equilibrium position and x is the displacement from equilibrium. (b) The free-body diagram for static equilibrium.

Substituting these energy expressions into equation (1.51) yields

$$\tfrac{1}{2} m\dot{x}^2 - mgx + \tfrac{1}{2} k(\Delta + x)^2 = \text{constant} \qquad (1.57)$$

Differentiating this expression with respect to time yields

$$\dot{x}(m\ddot{x} + kx) + \dot{x}(k\Delta - mg) = 0 \qquad (1.58)$$

Since the static force balance on the mass from Figure 1.14(b) yields the fact that $k\Delta = mg$, equation (1.58) becomes

$$\dot{x}(m\ddot{x} + kx) = 0 \qquad (1.59)$$

The velocity $\dot{x}$ cannot be zero for all time; otherwise, $x(t) = $ constant and no vibration would be possible. Hence equation (1.59) yields the standard equation of motion

$$m\ddot{x} + kx = 0 \qquad (1.60)$$

This procedure is called the *energy method* of obtaining the equation of motion.

The energy method can also be used to obtain the frequency of vibration directly for conservative systems that are oscillatory. The maximum value of sine (and cosine) is one. Hence, from equations (1.3) and (1.4), the maximum displacement is A and the maximum velocity is $\omega_n A$ (recall Window 1.3). Substitution of these maximum values into the expression for $U_{\max}$ and $T_{\max}$ and using the energy equation (1.53) yields

$$\tfrac{1}{2} m(\omega_n A)^2 = \tfrac{1}{2} kA^2 \qquad (1.61)$$

Solving equation (1.61) for ω_n yields the standard natural frequency relation $\omega_n = \sqrt{k/m}$.

Example 1.4.1

Figure 1.15 is a simple single-degree-of-freedom model of a wheel mounted on a spring. The friction in the system is such that the wheel rolls without slipping. Calculate the natural frequency of oscillation using the energy method. Assume that no energy is lost during the contact.

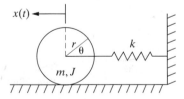

$x(t) \leftarrow$

r

θ

k

m, J

Figure 1.15 The rotational displacement of the wheel of radius r is given by $\theta(t)$ and the linear displacement is denoted by $x(t)$. The wheel has a mass m and a moment of inertia J. The spring has a stiffness k.

Solution From introductory dynamics, the rotational kinetic energy of the wheel is $T_{\text{rot}} = \tfrac{1}{2} J\dot{\theta}^2$, where J is the mass moment of inertia of the wheel and $\theta = \theta(t)$ is the angle of rotation of the wheel. This assumes that the wheel moves relative to the surface without slipping (so that no energy is lost at contact). The translational kinetic energy of the wheel is $T_T = \tfrac{1}{2} m\dot{x}^2$.

The rotation θ and the translation x are related by $x = r\theta$. Thus $\dot{x} = r\dot{\theta}$ and $T_{rot} = \frac{1}{2}J\dot{x}^2/r^2$. At maximum energy $x = A$ and $\dot{x} = \omega_n A$, so that

$$T_{max} = \frac{1}{2}m\dot{x}_{max}^2 + \frac{1}{2}\frac{J}{r^2}\dot{x}_{max}^2 = \frac{1}{2}(m + J/r^2)\omega_n^2 A^2$$

and

$$U_{max} = \frac{1}{2}kx_{max}^2 = \frac{1}{2}kA^2$$

Using conservation of energy in the form of equation (1.53) yields $T_{max} = U_{max}$, or

$$\frac{1}{2}\left(m + \frac{J}{r^2}\right)\omega_n^2 = \frac{1}{2}k$$

Solving this last expression for ω_n yields

$$\omega_n = \sqrt{\frac{k}{m + J/r^2}}$$

the desired frequency of oscillation of the suspension system.

The denominator in the frequency expression derived in this example is called the *effective mass* because the term $(m + J/r^2)$ has the same effect on the natural frequency as does a mass of value $(m + J/r^2)$.   □

Example 1.4.2

Use the energy method to determine the equation of motion of the simple pendulum (the rod l is assumed massless) shown in Example 1.1.1 and repeated in Figure 1.16.

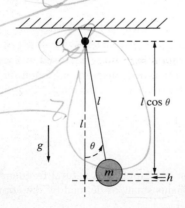

O

l

$l\cos\theta$

g

θ

m

h

Figure 1.16   The geometry of the pendulum for Example 1.4.2.

Solution   Several assumptions must first be made to ensure simple behavior (a more complicated version is considered in Example 1.4.6). Using the same assumptions given in Example 1.1.1 (massless rod, no friction in the hinge), the mass moment of inertia about point 0 is

$$J = ml^2$$

The angular displacement $\theta(t)$ is measured from the static equilibrium or rest position of the pendulum. The kinetic energy of the system is

$$T = \frac{1}{2}J\dot{\theta}^2 = \frac{1}{2}ml^2\dot{\theta}^2$$

The potential energy of the system is determined by the distance h in the figure so that

$$U = mgl(1 - \cos\theta)$$

since $h = l(1 - \cos\theta)$ is the geometric change in elevation of the pendulum mass. Substitution of these expressions for the kinetic and potential energy into equation (1.51) and differentiating yields

$$\frac{d}{dt}\left[\tfrac{1}{2}ml^2\dot{\theta}^2 + mgl(1 - \cos\theta)\right] = 0$$

or

$$ml^2\dot{\theta}\ddot{\theta} + mgl(\sin\theta)\dot{\theta} = 0$$

Factoring out $\dot{\theta}$ yields

$$\dot{\theta}(ml^2\ddot{\theta} + mgl\sin\theta) = 0$$

Since $\dot{\theta}(t)$ cannot be zero for all time, this becomes

$$ml^2\ddot{\theta} + mgl\sin\theta = 0$$

or

$$\ddot{\theta} + \frac{g}{l}\sin\theta = 0$$

This is a nonlinear equation in θ and is discussed in Section 1.10 and is derived from summing moments on a free-body diagram in Example 1.1.1. However, since $\sin\theta$ can be approximated by θ for small angles, the linear equation of motion for the pendulum becomes

$$\ddot{\theta} + \frac{g}{l}\theta = 0$$

This corresponds to an oscillation with natural frequency $\omega_n = \sqrt{g/l}$ for initial conditions such that θ remains small, as defined by the approximation $\sin\theta \approx \theta$, as discussed in Example 1.1.1.

In Example 1.4.2, it is important to not invoke the small-angle approximation before the final equation of motion is derived. For instance, if the small-angle approximation is used in the potential energy term, then $U = mgl(1 - \cos\theta) = 0$, since the small-angle approximation for $\cos\theta$ is 1. This would yield an incorrect equation of motion.

□

Example 1.4.3

Determine the equation of motion of the shaft and disk illustrated in Window 1.1 using the energy method.

Solution   The shaft and disk of Window 1.1 are modeled as a rod stiffness in twisting, resulting in torsional motion. The shaft, or rod, exhibits a torque in twisting proportional to the angle of twist $\theta(t)$. The potential energy associated with the torsional spring stiffness is $U = \frac{1}{2} k\theta^2$, where the stiffness coefficient k is determined much like the method used to determine the spring stiffness in translation, as discussed in Section 1.1. The angle $\theta(t)$ is measured from the static equilibrium, or rest, position. The kinetic energy associated with the disk of mass moment of inertia J is $T = \frac{1}{2} J\dot{\theta}^2$. This assumes that the inertia of the rod is much smaller than that of the disk and can be neglected.

Substitution of these expressions for the kinetic and potential energy into equation (1.51) and differentiating yields

$$\frac{d}{dt}\left(\tfrac{1}{2} J\dot{\theta}^2 + \tfrac{1}{2} k\theta^2\right) = \left(J\ddot{\theta} + k\theta\right)\dot{\theta} = 0$$

so that the equation of motion becomes (because $\dot{\theta} \neq 0$)

$$J\ddot{\theta} + k\theta = 0$$

This is the equation of motion for torsional vibration of a disk on a shaft. The natural frequency of vibration is $\omega_n = \sqrt{k/J}$.

□

Example 1.4.4

Model the mass of the spring in the system shown in Figure 1.17 and determine the effect of including the mass of the spring on the value of the natural frequency.

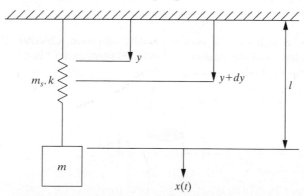

Figure 1.17   A spring–mass system with a spring of mass m_s that is too large to neglect.

Solution   One approach to considering the mass of the spring in analyzing the system vibration response is to calculate the kinetic energy of the spring. Consider the kinetic energy of the element dy of the spring. If m_s is the total mass of the spring, then $\frac{m_s}{l} dy$, is the mass of the element dy. The velocity of this element, denoted by v_{dy}, may be approximated by assuming that the velocity at any point varies linearly over the length of the spring:

$$v_{dy} = \frac{y}{l} \dot{x}(t)$$

The total kinetic energy of the spring is the kinetic energy of the element dy integrated over the length of the spring:

$$T_{spring} = \frac{1}{2} \int_0^l \frac{m_s}{l} \left[\frac{y}{l} \dot{x} \right]^2 dy$$

$$= \frac{1}{2} \left(\frac{m_s}{3} \right) \dot{x}^2$$

From the form of this expression, the effective mass of the spring is $\frac{m_s}{3}$, or one-third of that of the spring. Following the energy method, the maximum kinetic energy of the system is thus

$$T_{max} = \frac{1}{2} \left(m + \frac{m_s}{3} \right) \omega_n^2 A^2$$

Equating this to the maximum potential energy, $\frac{1}{2}kA^2$ yields the fact that the natural frequency of the system is

$$\omega_n = \sqrt{\frac{k}{m + m_s/3}}$$

Thus, including the effects of the mass of the spring in the system decreases the natural frequency. Note that if the mass of the spring is much smaller than the system mass m, the effect of the spring's mass on the natural frequency is negligible.

$\square$

Example 1.4.5

Fluid systems, as well as solid systems, exhibit vibration. Calculate the natural frequency of oscillation of the fluid in the U-shaped manometer illustrated in Figure 1.18 using the energy method.

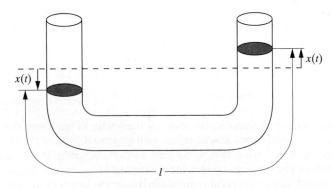

γ = weight density (volume)
A = cross-sectional area
l = length of fluid

Figure 1.18   A U-shaped manometer consisting of a fluid moving in a tube.

Solution   The fluid has weight density γ (i.e., the specific weight). The restoring force is provided by gravity. The potential energy of the fluid [(weight)(displacement of c.g.)] is $0.5(\gamma Ax)x$ in each column, so that the total change in potential energy is

$$U = U_2 - U_1 = \tfrac{1}{2}\gamma Ax^2 - \left(-\tfrac{1}{2}\gamma Ax^2\right) = \gamma Ax^2$$

The change in kinetic energy is

$$T = \frac{1}{2}\frac{Al\gamma}{g}(\dot{x}^2 - 0) = \frac{1}{2}\frac{Al\gamma}{g}\dot{x}^2$$

Equating the change in potential energy to the change in kinetic energy yields

$$\frac{1}{2}\frac{Al\gamma}{g}\dot{x}^2 = \gamma Ax^2$$

Assuming an oscillating motion of the form $x(t) = X\sin(\omega_n t + \phi)$ and evaluating this expression for maximum velocity and position yields

$$\frac{1}{2}\frac{l}{g}\omega_n^2 X^2 = X^2$$

where X is used to denote the amplitude of vibration. Solving for ω_n yields

$$\omega_n = \sqrt{\frac{2g}{l}}$$

which is the natural frequency of oscillation of the fluid in the tube. Note that it depends only on the acceleration due to gravity and the length of the fluid. Vibration of fluids inside mechanical containers (called *sloshing*) occurs in gas tanks in both automobiles and airplanes and forms an important application of vibration analysis.

$\square$

Example 1.4.6

Consider the compound pendulum of Figure 1.19 pinned to rotate around point O. Derive the equation of motion using Euler's second law (sum of moments as in Example 1.1.1). A compound pendulum is any rigid body pinned at a point other than its center of mass. If the only force acting on the system is gravity, then it will oscillate around that point and behave like a pendulum. The purpose of this example is to determine the equation of motion and to introduce the interesting dynamic property of the center of percussion.

Solution   A compound pendulum results from a simple pendulum configuration (Examples 1.1.1 and 1.4.2) if there is a significant mass distribution along its length. In the figure, G is the center of mass, O is the pivot point, and $\theta(t)$ is the angular displacement of the centerline of the pendulum of mass m and moment of inertia J measured about the z-axis at point O. Point C is the *center of percussion*, which is defined as the distance q_0 along the centerline such that a simple pendulum (a massless rod pivoted at zero with mass m at its tip, as in Example 1.4.2) of radius q_0 has the same period. Hence

$$q_0 = \frac{J}{mr}$$

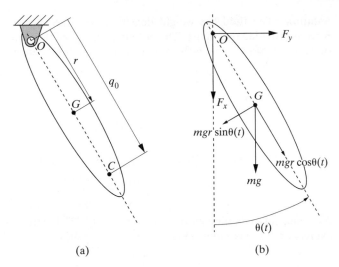

(a)   (b)

Figure 1.19   (a) A compound pendulum pivoted to swing about point O under the influence of gravity (pointing down). (b) A free-body diagram of the pendulum.

where r is the distance from the pivot point to the center of mass. Note that the pivot point O and the center of percussion C can be interchanged to produce a pendulum with the same frequency. The *radius of gyration, k_0,* is the radius of a ring that has the same moment of inertia as the rigid body does. The radius of gyration and center of percussion are related by

$$q_0 r = k_0^2$$

Consider the equation of motion of the compound pendulum. Taking moments about its pivot point O yields

$$\Sigma M_0 = J\ddot{\theta}(t) = -mgr \sin \theta(t)$$

For small $\theta(t)$ this nonlinear equation becomes ($\sin \sim \theta$)

$$J\ddot{\theta}(t) + mgr\,\theta(t) = 0$$

The natural frequency of oscillation becomes

$$\omega_n = \sqrt{\frac{mgr}{J}}$$

This frequency can be expressed in terms of the center of percussion as

$$\omega_n = \sqrt{\frac{g}{q_0}}$$

which is just the frequency of a simple pendulum of length q_0. This can be seen by examining the forces acting on the simple (massless rod) pendulum of Examples 1.1.1, 1.4.2, and Figure 1.20(a) or recalling the result obtained in these examples.

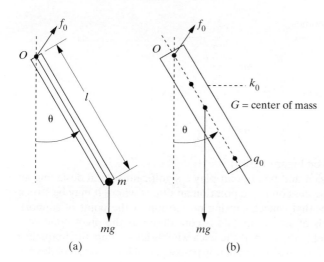

Figure 1.20 (a) A simple pendulum consisting of a massless rod pivoted at point O with a mass attached to its tip. (b) A compound pendulum consisting of a shaft with center of mass at point G. Here f_0 is the pin reaction force.

Summing moments about O yields

$$ml^2\ddot{\theta} = -mgl\sin\theta$$

or after approximating $\sin\theta$ with θ,

$$\ddot{\theta} + \frac{g}{l}\theta = 0$$

This yields the simple pendulum frequency of $\omega_n = \sqrt{g/l}$, which is equivalent to that obtained previously for the compound pendulum using $l = q_0$.

Next, consider the uniformly shaped compound pendulum of Figure 1.20(b) of length l. Here it is desired to calculate the center of percussion and radius of gyration.

The mass moment of inertia about point O is J, so that summing moments about O yields

$$J\ddot{\theta} = -mg\frac{l}{2}\sin\theta$$

since the mass is assumed to be evenly distributed and the center of mass is at $r = l/2$. The moment of inertia for a slender rod about O is $J = \frac{1}{3}ml^2$; hence, the equation of motion is

$$\frac{ml^2}{3}\ddot{\theta} + mg\frac{l}{2}\theta = 0$$

where $\sin\theta$ has again been approximated by θ, assuming small motion. This becomes

$$\ddot{\theta} + \frac{3}{2}\frac{g}{l}\theta = 0$$

so that the natural frequency is

$$\omega_n = \sqrt{\frac{3}{2}\frac{g}{l}}$$

The center of percussion becomes

$$q_0 = \frac{J}{mr} = \frac{2}{3}l$$

and the radius of gyration becomes

$$k_0 = \sqrt{q_0 r} = \frac{l}{\sqrt{3}}$$

These positions are marked on Figure 1.20(b).

   The center of percussion and pivot point play a significant role in designing an automobile. The center of percussion is the point on an object where it may be struck (impacted) producing forces that cancel, causing no motion at the point of support. The axle of the front wheels of an automobile is considered as the pivot point of a compound pendulum parallel to the road. If the back wheels hit a bump, the frequency of oscillation of the center of percussion will annoy passengers. Hence automobiles are designed such that the center of percussion falls over the axle and suspension system, away from passengers.

   The concept of center of percussion is used in many swinging, or pendulum-like, situations. This notion is sometimes used to define the "sweet spot" in a tennis racket or baseball bat and defines where the ball should be hit. If the hammer is shaped so that the impact point is at the center of percussion (i.e., the hammer's head), then ideally no force is felt if it is held at the "end" of the pendulum.

$\square$

   The energy method can be used in two ways. The first is to equate the maximum kinetic energy to the maximum potential energy [equation (1.53)] while assuming harmonic motion. This yields the natural frequency without writing out the equation of motion, as illustrated in equation (1.61). Beyond the simple calculation of frequency, this approach has limited use. However, the second use of the energy method involves deriving the equation of motion from the conservation of energy by differentiating equation (1.51) with respect to time. This concept is more useful and is illustrated in Examples 1.4.2 and 1.4.3. The concept of using energy quantities to derive the equations of motion can be extended to more complicated systems with many degrees of freedom, such as those discussed in Chapters 4 (multiple-degree-of-freedom systems) and 6 (distributed-parameter systems). The method is called Lagrange's method and is simply stated here to introduce the concept. Lagrange's method is introduced more formally in Chapter 4, where multiple-degree-of-freedom systems make the power of Lagrange's method obvious.

   Lagrange's method for conservative systems consists of defining the *Lagrangian, L,* of the system defined by $L = T - U$. Here T is the total kinetic energy of the system and U is the total potential energy in the system, both stated in terms of "generalized" coordinates. Generalized coordinates are denoted by "$q_i(t)$" and will be formally defined later. Here it is sufficient to state that q_i would be x in Example 1.4.4 and θ in Example 1.4.3. Then Lagrange's method for conservative systems states that the equations of motion for the free response of an undamped system result from

$$\frac{d}{dt}\left(\frac{\partial L}{\partial \dot{q}_i}\right) - \frac{\partial L}{\partial q_i} = 0 \tag{1.62}$$

Substitution of the expression for L into equation (1.62) yields

$$\frac{d}{dt}\left(\frac{\partial T}{\partial \dot{q}_i}\right) - \frac{\partial T}{\partial q_i} + \frac{\partial U}{\partial q_i} = 0 \tag{1.63}$$

Here one equation results for each subscript, i. In the case of the single-degree-of-freedom systems considered in this chapter, there is only one coordinate ($i = 1$) and only one equation of motion will result. The following example illustrates the use of the Lagrange approach to derive the equation of motion of a simple spring–mass system.

Example 1.4.7

Use Lagrange's method to derive the equation of motion of the simple spring–mass system of Figure 1.5. Compare this derivation to using the energy method described in Examples 1.4.2 and 1.4.3.

Solution   In the case of the simple spring–mass system, the kinetic and potential energy are, respectively,

$$T = \frac{1}{2}m\dot{x}^2 \quad \text{and} \quad U = \frac{1}{2}kx^2$$

Here the generalized coordinate $q_i(t)$ is just the displacement $x(t)$. Following the Lagrange approach, the Lagrange equation (1.63) becomes

$$\frac{d}{dt}\left(\frac{\partial T}{\partial \dot{x}}\right) - \frac{\partial T}{\partial x} + \frac{\partial U}{\partial x} = 0$$

$$\Rightarrow \frac{d}{dt}(m\dot{x}) + \frac{\partial}{\partial x}\left(\frac{1}{2}kx^2\right) = m\ddot{x} + kx = 0$$

This, of course, agrees with the approach of Newton's sum of forces. Next, consider the energy method, which starts with $T + U = $ constant. Taking the total derivative of this expression with respect to time yields

$$\frac{d}{dt}\left(\frac{1}{2}m\dot{x}^2 + \frac{1}{2}kx^2\right) = m\dot{x}\ddot{x} + kx\dot{x} = \dot{x}(m\ddot{x} + kx) = 0 \Rightarrow m\ddot{x} + kx = 0$$

since the velocity cannot be zero for all time. Thus the two energy-based approaches yield the same result and that result is equivalent to that obtained by Newton's sum of forces. Note that in order to follow the above calculations, it is important to remember the difference between total derivatives and partial derivatives and their respective rules of calculation from calculus.

□

Example 1.4.8

Use Lagrange's method to derive the equation of motion of the simple spring–mass pendulum system of Figure 1.21 and compute the system's natural frequency.

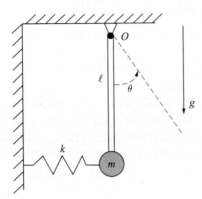

Figure 1.21   A pendulum attached to a spring.

Assume that the pendulum swings through only small angles so that the spring has negligible deflection in the vertical direction and assume that the mass of the pendulum rod is negligible.

Solution   In approaching a problem where there are several choices of variables, as in this case, it is a good idea to first write down the energy expressions in easy choices of velocities and displacements and then use a diagram to identify kinematic relationships and geometry as indicated in Figure 1.22. Referring to the figure, the kinetic energy of the mass is

$$T = \frac{1}{2} J\dot{\theta}^2 = \frac{1}{2} ml^2 \dot{\theta}^2$$

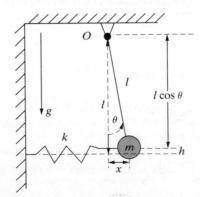

Figure 1.22   The geometry of the pendulum attached to a spring for small angles showing the kinematic relationships needed to formulate the energies in terms of a single generalized coordinate, θ.

The potential energy in the system consists of two parts, one due to the spring and one due to gravity. Thus the total potential energy is

$$U = \frac{1}{2} kx^2 + mgh$$

Next, use the figure to write each energy expression in terms of the single variable θ. From the figure, the mass moves up a distance $h = l - l \cos \theta$, and the distance the spring compresses is $x = l \sin \theta$. Thus the potential energy becomes

$$U = \frac{1}{2} l^2 \sin^2 \theta + mgl(1 - \cos \theta)$$

With the energies all stated in terms of the single generalized coordinate θ, the derivatives required in the Lagrange formulation become

$$\frac{d}{dt}\left(\frac{\partial T}{\partial \dot{\theta}}\right) = \frac{d}{dt}(ml^2\dot{\theta}) = ml^2\ddot{\theta}$$

$$\frac{\partial U}{\partial \theta} = \frac{\partial}{\partial \theta}\left(\frac{1}{2} l^2 \sin^2 \theta + mgl(1 - \cos \theta)\right) = kl^2 \sin \theta \cos \theta + mgl \sin \theta$$

Combining these expressions, the Lagrange equation (1.63) yields

$$\frac{d}{dt}\left(\frac{\partial T}{\partial \dot{\theta}}\right) - \underbrace{\frac{\partial T}{\partial \theta}}_{0} + \frac{\partial U}{\partial \theta} = ml^2\ddot{\theta} + kl^2 \sin \theta \cos \theta + mgl \sin \theta = 0$$

For small θ the equation of motion becomes

$$ml^2\ddot{\theta} + (kl^2 + mgl)\theta = 0$$

Thus the natural frequency is

$$\omega_n = \sqrt{\frac{kl + mg}{ml}}$$

Note that the equation of motion reduces to that of the pendulum given in Example 1.4.2 without the spring ($k = 0$).

☐

The Lagrange approach presented here is for the free response of undamped systems (conservative systems) and has only been applied to a single-degree-of-freedom system. However, the method is general and can be expanded to include the forced response and damping.

So far, three basic systems have been modeled: rectilinear or translational motion of a spring–mass system, torsional motion of a disk–shaft system, and the pendulum motion of a suspended mass system. Each of these motions commonly experiences energy dissipation of some form. The viscous-damping model of Section 1.3 developed for translational motion can be applied directly to both torsional and pendulum motion. In the case of torsional motion of the shaft, the energy dissipation is assumed to come from heating of the material and/or air resistance. Sometimes, as in the case of using the rod and disk to model an automobile crankshaft or camshaft, the damping is assumed to come from the oil that surrounds the disk and shaft, or bearings that support the shaft.

TABLE 1.1   A COMPARISON OF RECTILINEAR AND ROTATIONAL
SYSTEMS AND A SUMMARY OF UNITS

	Rectilinear, x (m)	Torsional/pendulum, θ (rad)
Spring force	kx	$k\theta$
Damping force	$c\dot{x}$	$c\dot{\theta}$
Inertia force	$m\ddot{x}$	$J\ddot{\theta}$
Equation of motion	$m\ddot{x} + c\dot{x} + kx = 0$	$J\ddot{\theta} + c\dot{\theta} + k\theta = 0$
Stiffness units	N/m	N·m/rad
Damping units	N·s/m, kg/s	M·N·s/rad
Inertia units	Kg	kg·m²/rad
Force/torque	$N = kg \cdot m/s^2$	$N \cdot m = kg \cdot m^2/s^2$

In all three cases, the damping is modeled as proportional to velocity (i.e., $f_c = c\dot{x}$ or $f_c = c\dot{\theta}$). The equations of motion are then of the form indicated in Table 1.1. Each of these equations can be expressed as a damped linear oscillator given in the form of equation (1.48). Hence, each of these three systems is characterized by a natural frequency and a damping ratio. Each of these three systems has a solution based on the nature of the damping ratio ζ, as discussed in Section 1.3.

1.5 STIFFNESS

The stiffness in a spring, introduced in Section 1.1, can be related more directly to material and geometric properties of the spring. This section introduces the relationships between stiffness, elastic modulus, and geometry of various types of springs and illustrates various situations that can lead to simple harmonic motion. A spring-like behavior results from a variety of configurations, including longitudinal motion (vibration in the direction of the length), transverse motion (vibration perpendicular to the length), and torsional motion (vibration rotating around the length). Consider again the stiffness of the spring introduced in Section 1.1. A spring is generally made of an elastic material. For a slender elastic material of length l, cross-sectional area A, and elastic modulus E (or Young's modulus), the stiffness of the bar for vibration along its length is given by

$$k = \frac{EA}{l} \tag{1.64}$$

This describes the spring constant for the vibration problem illustrated in Figure 1.23, where the mass of the rod is ignored (or very small relative to the mass m in the figure). The modulus E has the units of pascal (denoted by Pa), which are N/m². The modulus for several common materials is given in Table 1.2.

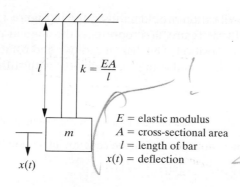

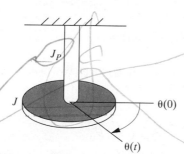

Figure 1.23   The stiffness associated with the longitudinal (along the long axis) vibration of a slender prismatic bar.

$$k = \frac{GJ_P}{l} = \text{stiffness of rod}$$

J = mass moment of inertia of the disk
G = shear modulus of rigidity of the rod
J_P = polar moment of inertia of the rod
l = length of rod
θ = angular displacement

Figure 1.24   The stiffness associated with the torsional rotation (twisting) of a shaft.

Next, consider a twisting motion with a similar rod of circular cross section, as illustrated in Figure 1.24. In this case, the rod possesses a polar moment of inertia, J_P, and (shear) modulus of rigidity, G (see Table 1.2). For the case of a wire or shaft of diameter d, $J_P = \pi d^4/32$. The modulus of rigidity has units N/m^2. The torsional stiffness is

$$k = \frac{GJ_P}{l} \tag{1.65}$$

TABLE 1.2   PHYSICAL CONSTANTS FOR SOME COMMON MATERIALS

Material	Young's modulus, $E(N/m^2)$	Density, (kg/m^3)	Shear modulus, $G(N/m^2)$
Steel	2.0×10^{11}	7.8×10^3	8.0×10^{10}
Aluminum	7.1×10^{10}	2.7×10^3	2.67×10^{10}
Brass	10.0×10^{10}	8.5×10^3	3.68×10^{10}
Copper	6.0×10^{10}	2.4×10^3	2.22×10^{10}
Concrete	3.8×10^9	1.3×10^3	—
Rubber	2.3×10^9	1.1×10^3	8.21×10^8
Plywood	5.4×10^9	6.0×10^2	—

which is used to describe the vibration problem illustrated in Figure 1.24, where the mass of the shaft is ignored. In the figure, $\theta(t)$ represents the angular position of the shaft relative to its equilibrium position. The disk of radius r and rotational moment of inertia J will vibrate around the equilibrium position $\theta(0)$ with stiffness GJ_P/l.

Example 1.5.1

Calculate the natural frequency of oscillation of the torsional system given in Figure 1.24.

Solution   Using the moment equation (1.50), the equation of motion for this system is

$$J\ddot{\theta}(t) = -k\theta(t)$$

This may be written as

$$\ddot{\theta}(t) + \frac{k}{J}\theta(t) = 0$$

This agrees with the result obtained using the energy method as indicated in Example 1.4.3. This indicates an oscillatory motion with frequency

$$\omega_n = \sqrt{\frac{k}{J}} = \sqrt{\frac{GJ_P}{lJ}}$$

Suppose that the shaft is made of steel and is 1 m long with a diameter of 5 cm. The polar moment of inertia of a rod of circular cross section is $J_P = (\pi d^4)/32$. If the disk has mass moment of inertia $J = 0.5$ kg $\cdot$ m^2 and considering that the shear modulus of steel is $G = 8 \times 10^{10}$ N/m^2, the frequency can be calculated by

$$\omega_n^2 = \frac{k}{J} = \frac{GJ_P}{lJ} = \frac{(8 \times 10^{10}\ \text{N/m}^2)\left[\frac{\pi}{32}(1 \times 10^{-2}\ \text{m})^4\right]}{(1\ \text{m})(0.5\ \text{kg} \cdot \text{m}^2)}$$

$$= 9.817 \times 10^4\,(\text{rad}^2/s^2)$$

Thus the natural frequency is $\omega_n = 313.3$ rad/s, or about 49.9 Hz. □

Consider the helical spring of Figure 1.25. In this figure the deflection of the spring is along the axis of the coil. The stiffness is actually dependent on the "twist" of the metal rod forming this spring. The stiffness is a function of the shear modulus G, the diameter of the rod, the diameter of the coils, and the number of coils. The stiffness has the value

$$k = \frac{Gd^4}{64nR^3} \tag{1.66}$$

The helical-shaped spring is very common. Some examples are the spring inside a retractable ballpoint pen and the spring contained in the front suspension of an automobile.

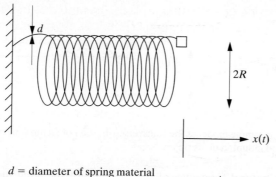

d = diameter of spring material
$2R$ = diameter of turns
n = number of turns
$x(t)$ = deflection

$$k = \frac{Gd^4}{64nR^3}$$

Figure 1.25   The stiffness associated with a helical spring.

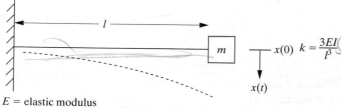

$x(0)$ $k = \dfrac{3EI}{l^3}$

E = elastic modulus
l = length of beam
I = moment of inertia of cross-sectional area about the neutral axis

Figure 1.26   The stiffness associated with the transverse (perpendicular to the long axis) vibration of the tip of a beam, also called the bending stiffness (Blevins, 1987). Assumes the mass of the beam is negligible.

Next, consider the transverse vibration of the end of a "leaf" spring illustrated in Figure 1.26. This type of spring behavior is similar to the rear suspension of an automobile as well as the wings of some aircraft. In the figure, l is the length of the beam, E is the elastic (Young's) modulus of the beam, and I is the (area) moment of inertia of the cross-sectional area. The mass m at the tip of the beam will oscillate with frequency

$$\omega_n = \sqrt{\frac{k}{m}} = \sqrt{\frac{3EI}{ml^3}} \tag{1.67}$$

in the direction perpendicular to the length of the beam $x(t)$.

Example 1.5.2

Consider an airplane wing with a fuel pod mounted at its tip as illustrated in Figure 1.27. The pod has a mass of 10 kg when it is empty and 1000 kg when it is full. Calculate the change in the natural frequency of vibration of the wing, modeled as in Figure 1.27, as the airplane uses up the fuel in the wing pod. The estimated physical parameters of the beam are $I = 5.2 \times 10^{-5} \, \text{m}^4$, $E = 6.9 \times 10^9 \, \text{N/m}^2$, and $l = 2$ m.

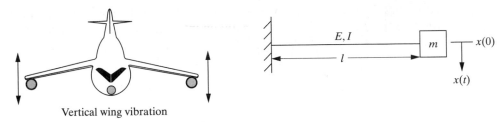

Vertical wing vibration

Figure 1.27 A simple model of the transverse vibration of an airplane wing with a fuel pod mounted on its tip.

Solution The natural frequency of the vibration of the wing modeled as a simple massless beam with a tip mass is given by equation (1.67). The natural frequency when the fuel pod is full is

$$\omega_{full} = \sqrt{\frac{3EI}{ml^3}} = \sqrt{\frac{(3)(6.9 \times 10^9)(5.2 \times 10^{-5})}{1000(2)^3}} = 11.6 \text{ rad/s}$$

which is about 1.8 Hz (1.8 cycles per second). The natural frequency for the wing when the fuel pod is empty becomes

$$\omega_{empt} = \sqrt{\frac{3EI}{ml^3}} = \sqrt{\frac{(3)(6.9 \times 10^9)(5.2 \times 10^{-5})}{10(2)^3}} = 116 \text{ rad/s}$$

or 18.5 Hz. Hence the natural frequency of the airplane wing changes by a factor of 10 (i.e., becomes 10 times larger) when the fuel pod is empty. Such a drastic change may cause changes in handling and performance characteristics of the aircraft. □

The above calculation ignores the mass of the beam. Clearly if the mass of the beam is significant compared to the tip mass, then the frequency calculation of equation (1.67) must be altered to account for the beam inertia. Similar to including the mass of the spring in Example 1.4.4, the kinetic energy of the beam itself must be considered. To estimate the kinetic energy, consider the static deflection of the beam due to a load at the tip (see any strength of materials text) as illustrated in Figure 1.28.

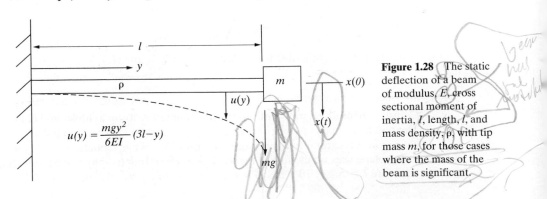

$$u(y) = \frac{mgy^2}{6EI}(3l-y)$$

Figure 1.28 The static deflection of a beam of modulus, E, cross sectional moment of inertia, I, length, l, and mass density, ρ, with tip mass m, for those cases where the mass of the beam is significant.

Figure 1.28 is roughly the same as Figure 1.25 with the static deflection indicated and the mass density of the beam, ρ, taken into consideration. The static deflection is given by

$$u(y) = \frac{mgy^2}{6EI}(3l - y) \tag{1.68}$$

Here u is the deflection of the beam perpendicular to its length and y is the distance along the length from the fixed end (left end in Figure 1.28). At the tip, the value of u is

$$u(l) = \frac{mgl^2}{6EI}(3l - l) = \frac{mgl^3}{3EI} \tag{1.69}$$

Using equation (1.69), equation (1.68) can be written in terms of the maximum deflection as

$$u(y) = \frac{u(l)y^2}{2l^3}(3l - y) \tag{1.70}$$

The maximum value of u is the displacement of the tip so it must be equal to $x(t)$. Thus the velocity of a differential element of the beam will be of the form

$$\frac{d}{dt}(u) = \frac{\dot{x}(t)y^2}{2l^3}(3l - y) \tag{1.71}$$

Substitution of this velocity into the expression for kinetic energy of the beam yields

$$T_{\text{beam}} = \frac{1}{2}\int_0^l \rho v^2\, dy = \frac{1}{2}\int_0^l \rho \dot{u}^2\, dy = \frac{1}{2}\int_0^l \rho\left[\frac{\dot{x}(t)y^2}{2l^3}(3l - y)\right]^2 dy \tag{1.72}$$

Here v is velocity and ρ is the mass per unit length of the beam. Substitution of equation (1.71) into (1.72) yields

$$T_{\text{beam}} = \frac{\rho}{2}\int_0^l \frac{\dot{x}^2}{4l^6}y^4(3l - y)^2\, dy = \frac{\rho l}{2}\frac{\dot{x}^2}{4l^6}\frac{33}{35}l^7 = \frac{1}{2}\left(\frac{33}{140}\rho l\right)\dot{x}^2(t) \tag{1.73}$$

Thus the mass of the beam, $M = \rho l$, adds this amount to the kinetic energy of the beam and tip-mass arrangement and the total kinetic energy of the system is

$$T = T_{\text{beam}} + T_m = \frac{1}{2}\left(\frac{33}{140}\rho l\right)\dot{x}^2(t) + \frac{1}{2}m\dot{x}^2(t) = \frac{1}{2}\left(\frac{33}{140}M + m\right)\dot{x}^2(t) \tag{1.74}$$

From equation (1.45), the equivalent mass of the single-degree-of-freedom system of a beam with a tip mass is

$$m_{eq} = \frac{33}{140}M + m \tag{1.75}$$

and the corresponding natural frequency is

$$\omega_n = \sqrt{\frac{k}{\frac{33}{140}M + m}} \tag{1.76}$$

This of course reduces to equation (1.67) if the mass, m, of the tip is much larger then the mass, M, of the beam.

Example 1.5.3

Referring to the airplane/wing tank of Example 1.5.2, if the wing has a mass of 500 kg, how much does this change the frequency from that calculated for a wing with a full tank (1000 kg)? Does the frequency still change by a factor of 10 when comparing full versus empty?

Solution Since the wing weight is equal to half of the fuel tank when full and 50 times the fuel tank when empty, the mass of the wing is clearly a significant factor in the frequency calculation. The two frequencies using equation (1.76) are

$$\omega_{full} = \sqrt{\frac{3EI}{\left(\frac{33}{140}M + m\right)l^3}} = \sqrt{\frac{(3)(6.9 \times 10^9)(5.2 \times 10^{-5})}{\left(\frac{33}{140}(500) + 1000\right)(2)^3}} = 10.97 \text{ rad/s}$$

$$\omega_{empty} = \sqrt{\frac{3EI}{\left(\frac{33}{140}M + m\right)l^3}} = \sqrt{\frac{(3)(6.9 \times 10^9)(5.2 \times 10^{-5})}{\left(\frac{33}{140}(500) + 10\right)(2)^3}} = 32.44 \text{ rad/s}$$

Note that modeling the wing without considering the mass of the wing is very inaccurate. Also note that the frequency shift with and without fuel is still significant (factor of 3 rather than 10).

$\square$

If the spring of Figure 1.26 is coiled in a plane as illustrated in Figure 1.29, the stiffness of the spring is greatly affected and becomes

$$k = \frac{EI}{l} \tag{1.77}$$

Several other spring arrangements and their associated stiffness values are listed in Table 1.3. Texts on solid mechanics and strength of materials should be consulted for further details.

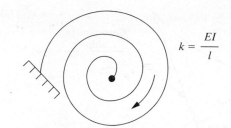

$$k = \frac{EI}{l}$$

l = total length of spring
E = elastic modulus of spring
I = moment of inertia of cross-sectional area

Figure 1.29 The stiffness associated with a spring coiled in a plane.

TABLE 1.3   SAMPLE SPRING CONSTANTS

Axial stiffness of a rod of length l, cross-sectional area A, and modulus E	$k = \dfrac{EA}{l}$
Torsional stiffness of a rod of length l, shear modulus G, and torsion constant J_P depending on the cross section ($\frac{\pi}{2}r^4$ for circle of radius r and $0.1406a^4$ for a square of side a)	$k = \dfrac{GJ_P}{l}$
Bending stiffness of a cantilevered beam of length l, modulus E, cross-sectional moment of inertia I	$k = \dfrac{3EI}{l^3}$
Axial stiffness of a tapered bar of length l, modulus E, and end diameters d_1 and d_2	$k = \dfrac{\pi E d_1 d_2}{4l}$
Torsional stiffness on a hollow uniform shaft of shear modulus G, length l, inside diameter d_1, and outside diameter d_2	$k = \dfrac{\pi G(d_2^4 - d_1^4)}{32l}$
Transverse stiffness of a pinned–pinned beam of modulus E, area moment of inertia I, and length l for a load applied at point a from its end	$k = \dfrac{3EIl}{a^2(l - a)^2}$
Transverse stiffness of a clamped–clamped beam of modulus E, area moment of inertia I, and length l for a load applied at its center	$k = \dfrac{192EI}{l^3}$

Example 1.5.4

As another example of vibration involving fluids, consider the rolling vibration of a ship in water. Figure 1.30 illustrates a schematic of a ship rolling in water. Compute the natural frequency of the ship as it rolls back and forth about the axis through M.

In the figure, G denotes the center of gravity, B denotes the center of buoyancy, M is the point of intersection of the buoyant force before and after the roll (called the metacenter), and h is the length of GM. The perpendicular line from the center of gravity to the line of action of the buoyant force is marked by the point Z. Here W denotes the weight of the ship, J denotes the mass moment of the ship about the roll axis, and $\theta(t)$ denotes the angle of roll.

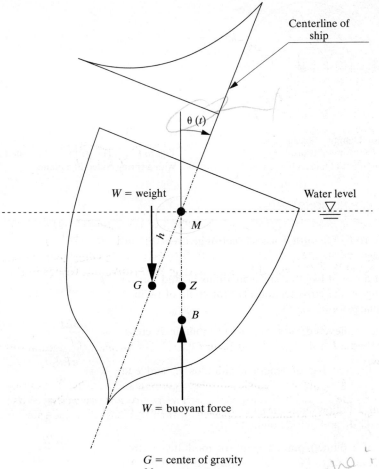

G = center of gravity
M = metacenter
B = center of buoyance

Figure 1.30   The dynamics of a ship rolling in water.

Solution   Summing moments about M yields

$$J\ddot{\theta}(t) = -W\overline{GZ} = -Wh \sin \theta(t)$$

Again, for small enough values of θ, this nonlinear equation can be approximated by

$$J\ddot{\theta}(t) + Wh\theta(t) = 0$$

Thus the natural frequency of the system is

$$\omega_n = \sqrt{\frac{hW}{J}}$$

Springs in series

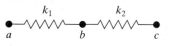

$$k_{ac} = \frac{1}{1/k_1 + 1/k_2} = \frac{k_1 k_2}{k_1 + k_2}$$

Springs in parallel

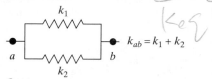

$$k_{ab} = k_1 + k_2$$

Figure 1.31   The rules for calculating the equivalent stiffness of parallel and series connections of springs.

All of the spring types mentioned are represented schematically as indicated in Figure 1.2. If more than one spring is present in a given device, the resulting stiffness of the combined spring can be calculated by two simple rules, as given in Figure 1.31. These rules can be derived by considering the equivalent forces in the system.

Example 1.5.5

Consider the spring–mass arrangement of Figure 1.32(a) and calculate the natural frequency of the system.

Solution   To find the equivalent single stiffness representation of the five-spring system given in Figure 1.32(a), the two simple rules of Figure 1.31 are applied. First, the parallel arrangement of k_1 and k_2 is replaced by the single spring, as indicated at

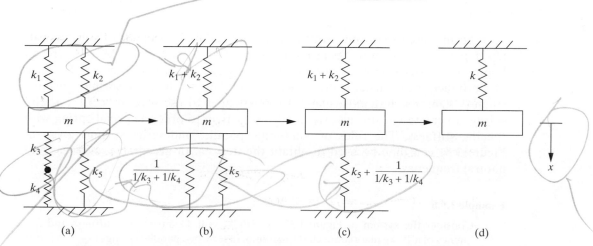

(a)   (b)   (c)   (d)

Figure 1.32   The reduction of a five-spring, one-mass system to an equivalent single-spring–mass system having the same vibration properties.

the top of Figure 1.32(b). Next, the series arrangement of k_3 and k_4 is replaced with a single spring of stiffness

$$\frac{1}{1/k_3 + 1/k_4}$$

as indicated in the bottom left side of Figure 1.32(b). These two parallel springs on the bottom of Figure 1.32(b) are next combined using the parallel spring formula to yield a single spring of stiffness

$$k_5 + \frac{1}{1/k_3 + 1/k_4} = k_5 + \frac{k_3 k_4}{k_3 + k_4}$$

as indicated in Figure 1.32(c). The final step is to realize that both the spring acting at the top of Figure 1.32(c) and the spring at the bottom attach the mass to ground and hence act in parallel. These two springs then combine to yield the single stiffness

$$k = k_1 + k_2 + k_5 + \frac{k_3 k_4}{k_3 + k_4}$$

$$= k_1 + k_2 + k_5 + \frac{k_3 k_4}{k_3 + k_4} = \frac{(k_1 + k_2 + k_5)(k_3 + k_4) + k_3 k_4}{(k_3 + k_4)}$$

as indicated symbolically in Figure 1.32(d). Hence the natural frequency of this system is

$$\omega_n = \sqrt{\frac{k_1 k_3 + k_2 k_3 + k_5 k_3 + k_1 k_4 + k_2 k_4 + k_5 k_4 + k_3 k_4}{m(k_3 + k_4)}}$$

Note that even though the system of Figure 1.32 contains five springs, it consists of only one mass moving in only one (rectilinear) direction and hence is a single-degree-of- freedom system.

□

Springs are usually manufactured in only certain increments of stiffness values depending on such things as the number of turns, material, and so on (recall Figure 1.25). Because mass production (and large sales) brings down the price of a product, the designer is often faced with a limited choice of spring constants when designing a system. It may thus be cheaper to use several "off-the-shelf" springs to create the stiffness value necessary than to order a special spring with specific stiffness. The rules of combining parallel and series springs given in Figure 1.31 can then be used to obtain the desired, or acceptable, stiffness and natural frequency.

Example 1.5.6

Consider the system of Figure 1.32(a) with $k_5 = 0$. Compare the stiffness and frequency of a 10-kg mass connected to ground, first by two parallel springs ($k_3 = k_4 = 0$, $k_1 = 1000$ N/m, and $k_2 = 3000$ N/m), then by two series springs ($k_1 = k_2 = 0$, $k_3 = 1000$ N/m, and $k_4 = 3000$ N/m).

Solution   First, consider the case of two parallel springs so that $k_3 = k_4 = 0$, $k_1 = 1000$ N/m, and $k_2 = 3000$ N/m. Then the equivalent stiffness is given by Figure 1.31 to be the simple sum given by

$$k_{eq} = 1000\,\text{N/m} + 3000\,\text{N/m} = 4000\,\text{N/m}$$

and the corresponding frequency is

$$\omega_{parallel} = \sqrt{\frac{4000\ \text{N/m}}{10\ \text{kg}}} = 20\ \text{rad/s}$$

In the case of a series connection ($k_1 = k_2 = 0$), the two springs ($k_3 = 1000$ N/m, $k_4 = 3000$ N/m) combine according to Figure 1.31 to yield

$$k_{eq} = \frac{1}{1/1000 + 1/3000} = \frac{3000}{3+1} = \frac{3000}{4} = 750\ \text{N/m}$$

The corresponding natural frequency becomes

$$\omega_{series} = \sqrt{\frac{750\ \text{N/m}}{10\ \text{kg}}} = 8.66\ \text{rad/s}$$

Note that using two identical sets of springs connected to the same mass in the two different ways produces drastically different equivalent stiffness and resulting frequency. A series connection decreases the equivalent stiffness, while a parallel connection increases the equivalent stiffness. This is useful in designing systems.   □

Example 1.5.6 illustrates that fixed values of spring constants can be used in various combinations to produce a desired value of stiffness and corresponding frequency. It is interesting to note that an identical set of physical devices can be used to create a system with drastically different frequencies simply by changing the physical arrangement of the components. This is similar to the choice of resistors in an electric circuit. The formulas of this section are intended to be aids in designing vibration systems.

In addition to understanding the effect of stiffness on the dynamics—that is, on the natural frequency—it is important not to forget static analysis when using springs. In particular, the static deflection of each spring system needs to be checked to make sure that the dynamic analysis is correctly interpreted. Recall from the discussion of Figure 1.14 that the static deflection has the value

$$\Delta = \frac{mg}{k}$$

where m is the mass supported by a spring of stiffness k in a gravitational field providing acceleration of gravity g. Static deflection is often ignored in introductory treatments but is used extensively in spring design and is essential in nonlinear analysis. Static deflection is denoted by a variety of symbols. The symbols δ, Δ, δ_s, and x_0 are all used in vibration publications to denote the deflection of a spring caused by the weight of the mass attached to it.

1.6 MEASUREMENT

Measurements associated with vibration are used for several purposes. First, the quantities required to analyze the vibrating motion of a system all require measurement. The mathematical models proposed in previous sections all require knowledge of the mass, damping, and stiffness coefficients of the device under study. These coefficients can be measured in a variety of ways, as discussed in this section. Vibration measurements are also used to verify and improve analytical models. Other uses for vibration testing techniques include reliability and durability studies, searching for damage, and testing for acceptability of the response in terms of vibration parameters. This chapter introduces some basic ideas on measurement. Further discussion of measurement can be found throughout the book, culminating with all the various concepts on measurement summarized in Chapter 7.

In many cases, the mass of an object or device is simply determined by using a scale. Mass is a relatively easy quantity to measure. However, the mass moment of inertia may require a dynamic measurement. A method of measuring the mass moment of inertia of an irregularly shaped object is to place the object on the platform of the apparatus of Figure 1.33 and measure the period of oscillation of the system, T. By using the methods of Section 1.4, it can be shown that the moment of inertia of an object, J (about a vertical axis), placed on the disk of Figure 1.33 with its mass center aligned vertically with that of the disk, is given by

$$J = \frac{gT^2 r_0^2 (m_0 + m)}{4\pi^2 l} - J_0 \tag{1.78}$$

Here m is the mass of the part being measured, m_0 is the mass of the disk, r_0 the radius of the disk, l the length of the wires, J_0 the moment of inertia of the disk, and g the acceleration due to gravity.

The stiffness of a simple spring system can be measured as suggested in Section 1.1. The elastic modulus, E, of an object can be measured in a similar

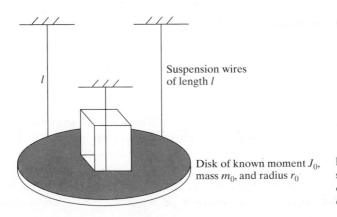

Suspension wires of length l

Disk of known moment J_0, mass m_0, and radius r_0

Figure 1.33   A Trifilar suspension system for measuring the moment of inertia of irregularly shaped objects.

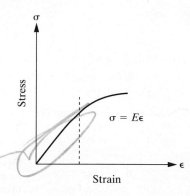

Figure 1.34   An example of a stress–strain curve of a test specimen used to determine the elastic modulus of a material.

fashion by performing a tensile test (see, e.g., Shames, 1989). In this method, a tensile test machine is employed that uses strain gauges to measure the strain, ϵ, in the test specimen as well as the stress, σ, both in the axial direction of the specimen. This produces a curve such as the one shown in Figure 1.34. The slope of the curve in the linear region defines the Young's modulus, or elastic modulus, for the test material. The relationship that the extension is proportional to the force is known as Hooke's law.

The elastic modulus can also be measured by using some of the formulas given in Section 1.5 and measurement of the vibratory response of a structure or part. For instance, consider the cantilevered arrangement of Figure 1.26. If the mass at the tip is given a small deflection, it will oscillate with frequency $\omega_n = \sqrt{k/m}$. If ω_n is measured, the modulus can be determined from equation (1.67), as illustrated in the following example.

Example 1.6.1

Consider a steel beam configuration as shown in Figure 1.26. The beam has a length $l = 1$ m and moment of inertia $I = 10^{-9}$ m^4, with a mass $m = 6$ kg attached to the tip. If the mass is given a small initial deflection in the transverse direction and oscillates with a period of $T = 0.62$ s, calculate the elastic modulus of steel.

Solution   Since $T = 2\pi/\omega_n$, equation (1.67) yields

$$T = 2\pi\sqrt{\frac{ml^3}{3EI}}$$

Solving for E yields

$$E = \frac{4\pi^2 ml^3}{3T^2 I} = \frac{4\pi^2(6\text{ kg})(1\text{ m})^3}{3(0.62\text{ s})^2(10^{-9}\text{ m}^4)} = 205 \times 10^9 \text{ N/m}^2$$

$\square$

The period T, and hence the frequency ω_n, can be measured with a stopwatch for vibrations that are large enough and last long enough to see. However, many vibrations

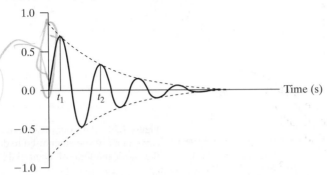

Displacement (mm)

Figure 1.35   The response of an underdamped system used to measure damping.

of interest have very small amplitudes and happen very quickly. Hence several very sophisticated devices for measuring time and frequency have been developed, requiring more sophisticated concepts presented in the chapter on measurement.

The damping coefficient or, alternatively, the damping ratio is the most difficult quantity to determine. Both mass and stiffness can be determined by static tests; however, damping requires a dynamic test to measure. A record of the displacement response of an underdamped system can be used to determine the damping ratio. One approach is to note that the decay envelope, denoted by the dashed line in Figure 1.35, for an underdamped system is $Ae^{-\zeta\omega_n t}$. The measured points $x(0)$, $x(t_1)$, $x(t_2)$, $x(t_3)$, and so on can then be curve fit to A, $Ae^{-\zeta\omega_n t_1}$, $Ae^{-\zeta\omega_n t_2}$, $Ae^{-\zeta\omega_n t_3}$, and so on. This will yield a value for the coefficient $\zeta\omega_n$. If m and k are known, ζ and c can be determined from $\zeta\omega_n$.

This approach also leads to the concept of *logarithmic decrement*, denoted by δ and defined by

$$\delta = \ln \frac{x(t)}{x(t + T)} \tag{1.79}$$

where T is the period of oscillation. Substitution of the analytical form of the underdamped response given by equation (1.36) yields

$$\delta = \ln \frac{Ae^{-\zeta\omega_n t}\sin(\omega_d t + \phi)}{Ae^{-\zeta\omega_n(t+T)}\sin(\omega_d t + \omega_d T + \phi)} \tag{1.80}$$

Since $\omega_d T = 2\pi$, the denominator becomes $e^{-\zeta\omega_n(t+T)}\sin(\omega_d t + \phi)$, and the expression for the decrement reduces to

$$\delta = \ln e^{\zeta\omega_n T} = \zeta\omega_n T \tag{1.81}$$

The period T in this case is the damped period $(2\pi/\omega_d)$ so that

$$\delta = \zeta\omega_n \frac{2\pi}{\omega_n\sqrt{1 - \zeta^2}} = \frac{2\pi\zeta}{\sqrt{1 - \zeta^2}} \tag{1.82}$$

Solving this expression for ζ yields

$$\zeta = \frac{\delta}{\sqrt{4\pi^2 + \delta^2}} \tag{1.83}$$

which determines the damping ratio given the value of the logarithmic decrement.

Thus if the value of $x(t)$ is measured from the plot of Figure 1.35 at any two successive peaks, say $x(t_1)$ and $x(t_2)$, equation (1.79) can be used to produce a measured value of δ, and equation (1.83) can be used to determine the damping ratio. The formula for the damping ratio [equations (1.29) and (1.30), also listed in the inside front cover] and knowledge of m and k subsequently yield the value of the damping coefficient c. Note that peak measurements can be used over any integer multiple of the period (see Problem 1.95) to increase the accuracy over measurements taken at adjacent peaks.

The computation in Problem 1.95 yields

$$\delta = \frac{1}{n}\ln\left(\frac{x(t)}{x(t + nT)}\right)$$

where n is any integer number of successive (positive) peaks. While this does tend to increase the accuracy of computing δ, the majority of damping measurements performed today are based on modal analysis methods (Chapters 4 and 6) presented later in Chapter 7.

Example 1.6.2

The free response of the damped single-degree-of-freedom system in Figure 1.9 with a mass of 2 kg is recorded to be of the form given in Figure 1.35. A static deflection test is performed and the stiffness is determined to be 1.5×10^3 N/m. The displacements at t_1 and t_2 are measured to be 9 and 1 mm, respectively. Calculate the damping coefficient.

Solution   From the definition of the logarithmic decrement

$$\delta = \ln\left[\frac{x(t_1)}{x(t_2)}\right] = \ln\left[\frac{9 \text{ mm}}{1 \text{ mm}}\right] = 2.1972$$

From equation (1.83),

$$\zeta = \frac{2.1972}{\sqrt{4\pi^2 + 2.1972^2}} = 0.33 \quad \text{or} \quad 33\%$$

Also,

$$c_{cr} = 2\sqrt{km} = 2\sqrt{(1.5 \times 10^3 \text{ N/m})(2 \text{ kg})} = 1.095 \times 10^2 \text{ kg/s}$$

And from equation (1.30) the damping coefficient becomes

$$c = c_{cr}\zeta = (1.095 \times 10^2)(0.33) = 36.15 \text{ kg/s}$$

□

Example 1.6.3

Mass and stiffness are usually measured in a straightforward manner as shown in Section 1.3. However, there are certain circumstances that preclude using these simple methods. In these cases, a measurement of the frequency of oscillation both before and after a known amount of mass is added can be used to determine the mass and stiffness of the original system. Suppose then that the frequency of the system in Figure 1.36(a) is measured to be 2 rad/s and the frequency of Figure 1.36(b) with an added mass of 1 kg is known to be 1 rad/s. Calculate m and k.

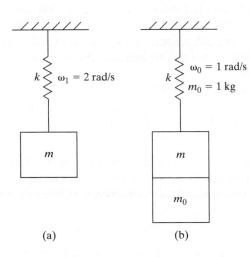

(a)   (b)

Figure 1.36   A schematic of using added mass (b) and frequency measurements to determine an unknown mass, m, and stiffness, k, of the original system (a).

Solution   From the definition of natural frequency

$$\omega_1 = 2 = \sqrt{\frac{k}{m}} \quad \text{and} \quad \omega_0 = 1 = \sqrt{\frac{k}{m+1}}$$

Solving for m and k yields

$$4m = k \quad \text{and} \quad m + 1 = k$$

or

$$m = \frac{1}{3}\,\text{kg} \quad \text{and} \quad k = \frac{4}{3}\,\text{N/m}$$

This formulation can also be used to determine changes in mass of a system. As an example, the frequency of oscillation of low amplitude vibration of a hospital patient in bed can be used to monitor the change in the patient's weight (mass) without having to move the patient from the bed. In this case the mass m_0 is considered to be the change in mass of the original system. If the original mass and frequency are known, measurement of the frequency ω_0 can be used to determine the change in mass m_0. Given that the original weight is 120 lb (54.4 kg), the original frequency is 100.4 Hz, and the frequency of the patient-bed system changes to 100 Hz, determine the change in the patient's weight.

From the two frequency relations

$$\omega_1^2 m = k$$

and

$$\omega_0^2(m + m_0) = k$$

Thus, $\omega_1^2 m = \omega_0^2(m + m_0)$. Solving for the change in mass m_0 yields

$$m_0 = m\left(\frac{\omega_1^2}{\omega_0^2} - 1\right)$$

Multiplying by g and converting the frequency to hertz yields

$$W_0 = W\left(\frac{f_1^2}{f_0^2} - 1\right)$$

or

$$W_0 = 120 \text{ lb}\left[\left(\frac{100.4 \text{ Hz}}{100 \text{ Hz}}\right)^2 - 1\right]$$

$$= 0.96 \text{ lb } (0.4\text{kg})$$

Since the frequency decreased, the patient gained almost a pound. An increase in frequency would indicate a loss of weight.

□

Measurement of m, c, k, ω_n, and ζ is used to verify the mathematical model of a system and for a variety of other reasons. Measurement of vibrating systems forms an important aspect of the activity in industry related to vibration technology. Chapter 7 is specifically devoted to measurement, however comments on vibration measurements are mentioned throughout the remaining chapters.

1.7 DESIGN CONSIDERATIONS

This section introduces the idea of designing vibration systems, which forms the topic of Chapter 5. Design in vibration refers to adjusting the physical parameters of a device to cause its vibration response to meet a specified shape or performance criteria. For instance, consider the response of the single-degree-of-freedom system of Figure 1.9. The shape of the response is somewhat determined by the value of the damping ratio in the sense that the response is either overdamped, underdamped, or critically damped ($\zeta > 1, \zeta < 1, \zeta = 1$, respectively). The damping ratio, in turn, depends on the values of m, c, and k. A designer may choose these values to produce the desired response.

Section 1.5 on stiffness considerations is actually an introduction to design as well. The formulas given there for stiffness, in terms of modulus and geometric dimensions, can be used to design a system that has a given natural frequency. Example 1.5.2 points out one of the important problems in design, that often the properties that we are interested in designing for (frequency in this case) are very sensitive to operational changes. In Example 1.5.2, the frequency changes a great deal as the airplane consumes fuel.

Another important issue in design often focuses on using devices that are already available. For example, the rules given in Figure 1.31 are design rules for producing a desired value of spring constant from a set of "available" springs by placing them in certain combinations, as illustrated in Example 1.5.6. Design work in engineering often involves using available products to produce configurations (or designs) that suit a particular application. In the case of spring stiffness, springs are usually mass produced, and hence inexpensive, in only certain discrete values of stiffness. The formulas given for parallel and series connections of springs are then used to produce the desired stiffness. If cost is not a restriction, then formulas such as those given in Table 1.3 may be used to design a single spring that meets the stated stiffness requirements. Of course, designing a spring–mass system to have a desired natural frequency may not produce a system with an acceptable static deflection. Thus, the design process becomes complicated. Design is one of the most active and exciting disciplines in engineering because it often involves compromise and choice with many acceptable solutions.

Unfortunately, the values of m, c, and k have other constraints. The size and material of which the device is made determine these parameters. Hence, the design procedure becomes a compromise. For example, geometric limitations might cause the mass of a device to be between 2 and 3 kg, and for static displacement conditions, the stiffness may be required to be greater than 200 N/m. In this case, the natural frequency must be in the interval

$$8.16 \text{ rad/s} \leq \omega_n \leq 10 \text{ rad/s} \qquad (1.84)$$

This severely limits the design of the vibration response, as illustrated in the following example.

Example 1.7.1

Consider the system of Figure 1.9 with mass and stiffness properties as summarized by inequality (1.84). Suppose that the system is subject to an initial velocity that is always less than 300 mm/s, and to an initial displacement of zero (i.e., $x_0 = 0$, $v_0 \leq 300$ mm/s). For this range of mass and stiffness, choose a value of the damping coefficient such that the amplitude of vibration is always less than 25 mm.

Solution   This is a design-oriented example, and hence, as is typical of design calculations, there is not a nice, clean formula to follow. Rather, the solution must be obtained using theory and parameter studies. First, note that for zero initial displacement, the response may be written from equation (1.38) as

$$x(t) = \frac{v_0}{\omega_d} e^{-\zeta \omega_n t} \sin(\omega_d t)$$

Also note that the amplitude of this periodic function is

$$\frac{v_0}{\omega_d}e^{-\zeta\omega_n t}$$

Thus, for small ω_d, the amplitude is larger than for larger ω_d. Hence for the range of frequencies of interest, it appears that the worst case (largest amplitude) will occur for the smallest value of the frequency ($\omega_n = 8.16$ rad/s). Also, the amplitude increases with v_0 so that using $v_0 = 300$ mm/s will ensure that amplitude is a large as possible. Now, v_0 and ω_n are fixed, so it remains to be investigated how the maximum value of $x(t)$ varies as the damping ratio is varied. One approach is to compute the amplitude of the response at the first peak. From Figure 1.10 the largest amplitude occurs at the first time the derivative of $x(t)$ is zero. Taking the derivative of $x(t)$ and setting it equal to zero yields the expression for the time to the first peak:

$$\omega_d e^{-\zeta\omega_n t}\cos(\omega_d t) - \zeta\omega_n e^{-\zeta\omega_n t}\sin(\omega_d t) = 0$$

Solving this for t and denoting this value of time by T_m yields

$$T_m = \frac{1}{\omega_d}\tan^{-1}\left(\frac{\omega_d}{\zeta\omega_n}\right) = \frac{1}{\omega_d}\tan^{-1}\left(\frac{\sqrt{1-\zeta^2}}{\zeta}\right)$$

The value of the amplitude of the first (and largest) peak is calculated by substituting the value of T_m into $x(t)$, resulting in

$$A_m(\zeta) = x(T_m) = \frac{v_0}{\omega_n\sqrt{1-\zeta^2}}e^{-\frac{\zeta}{\sqrt{1-\zeta^2}}\tan^{-1}\left(\frac{\sqrt{1-\zeta^2}}{\zeta}\right)}\sin\left(\tan^{-1}\left(\frac{\sqrt{1-\zeta^2}}{\zeta}\right)\right)$$

Simplifying yields

$$A_m(\zeta) = \frac{v_0}{\omega_n}e^{-\frac{\zeta}{\sqrt{1-\zeta^2}}\tan^{-1}\left(\frac{\sqrt{1-\zeta^2}}{\zeta}\right)}$$

For fixed initial velocity (the largest possible) and frequency (the lowest possible), this value of $A_m(\zeta)$ determines the largest value that the highest peak will have as ζ varies. The exact value of ζ that will keep this peak, and hence the response, at or below 25 mm, can be determined by numerically solving $A_m(\zeta) = 0.025$ (m) for a value of ζ. This yields $\zeta = 0.281$. Using the upper limit of the mass values ($m = 3$ kg) then yields the value of the required damping coefficient:

$$c = 2m\omega_n\zeta = 2(3)(8.16)(0.281) = 13.76 \text{ kg/s}$$

For this value of damping, the response is never larger than 25 mm. Note that if there is no damping, the same initial conditions produce a response of amplitude $A = v_0/\omega_n = 37$ mm.

$\square$

As another example of design, consider the problem of choosing a spring that will result in a spring–mass system having a desired or specified frequency. The formulas of Section 1.5 provide a means of designing a spring to have a specified stiffness in terms of the properties of the spring material (modulus) and its geometry. The following example illustrates this concept.

Example 1.7.2

Consider designing a helical spring such that when attached to a 10-kg mass, the resulting spring–mass system has a natural frequency of 10 rad/s (about 1.6 Hz).

Solution   From the definition of the natural frequency, the spring is required to have a stiffness of

$$k = \omega_n^2 m = (10)^2(10) = 10^3 \, \text{N/m}$$

The stiffness of a helical spring is given by equation (1.66) to be

$$k = 10^3 \, \text{N/m} = \frac{Gd^4}{64nR^3} \quad \text{or} \quad 6.4 \times 10^4 = \frac{Gd^4}{nR^3}$$

This expression provides the starting point for a design. The choices of variables that affect the design are: the type of material to be used (hence various values of G); the diameter of the material, d; the radius of the coils, R; and the number of turns, n. The choices of G and d are, of course, restricted by available materials, n is restricted to be an integer, and R may have restrictions dictated by the size requirements of the device. Here it is assumed that steel of 1-cm diameter is available. The shear modulus of steel is about

$$G = 8.0 \times 10^{10} \, \text{N/m}^2$$

so that the stiffness formula becomes

$$6.4 \times 10^4 \, \text{N/m} = \frac{(8.0 \times 10^{10} \, \text{N/m}^2)(10^{-2} \, \text{m})^4}{nR^3}$$

or

$$nR^3 = 1.25 \times 10^{-2}$$

If the coil radius is chosen to be 10 cm, this yields that the number of turns should be

$$n = \frac{1.25 \times 10^{-2} \, \text{m}^3}{10^{-3} \, \text{m}^3} = 12.5 \text{ or } 13$$

Thus, if 13 turns of 1-cm-diameter steel are coiled at a radius of 10 cm, the resulting spring will have the desired stiffness and the 10-kg mass will oscillate at approximately 10 rad/s. To get an exact answer, the modulus of steel must be modified. This can be done through the use of different alloys of steel, but would become expensive. So depending on the precision needed for a given application, modifying the type of steel used may or may not be practical.

$\square$

In Example 1.7.2, several variables were chosen to produce a desired design. In each case the design variables (such as d, R, etc.) are subject to constraints. Other aspects of vibration design are presented throughout the text as appropriate. There are no set rules to follow in design work. However, some organized approaches to design are presented later in Chapter 5. The following example illustrates another difficulty in design by examining what happens when operating conditions are changed after the design is over.

Example 1.7.3

As a final example, consider modeling the vertical suspension system of a small sports car, as a single-degree-of-freedom system of the form

$$m\ddot{x} + c\dot{x} + kx = 0$$

where m is the mass of the automobile and c and k are the equivalent damping and stiffness of the four-shock-absorber–spring systems. The car deflects the suspension system 0.05 m under its own weight. The suspension is chosen (designed) to have a damping ratio of 0.3. a) If the car has a mass of 1361 kg (mass of a Porsche Boxster), calculate the equivalent damping and stiffness coefficients of the suspension system. b) If two passengers, a full gas tank, and luggage totaling 290 kg are in the car, how does this affect the effective damping ratio?

Solution   The mass is 1361 kg and the natural frequency is

$$\omega_n = \sqrt{\frac{k}{1361}}$$

so that

$$k = 1361\,\omega_n^2$$

At rest, the car's springs are compressed an amount Δ, called the *static deflection*, by the weight of the car. Hence, from a force balance at static equilibrium, $mg = k\Delta$, so that

$$k = \frac{mg}{\Delta}$$

and

$$\omega_n = \sqrt{\frac{k}{m}} = \sqrt{\frac{g}{\Delta}} = \sqrt{\frac{9.8}{0.05}} = 14 \text{ rad/s}$$

The stiffness of the suspension system is thus

$$k = 1361(14)^2 = 2.668 \times 10^5 \text{ N/m}$$

For $\zeta = 0.3$, equation (1.30) becomes

$$c = 2\zeta m\omega_n = 2(0.3)(1361)(14) = 1.143 \times 10^4 \text{ kg/s}$$

Now if the passengers and luggage are added to the car, the mass increases to $1361 + 290 = 1651$ kg. Since the stiffness and damping coefficient remain the same, the new static deflection becomes

$$\Delta = \frac{mg}{k} = \frac{1651(9.8)}{2.668 \times 10^5} \approx 0.06 \text{ m}$$

The new frequency becomes

$$\omega_n = \sqrt{\frac{g}{\Delta}} = \sqrt{\frac{9.8}{0.06}} = 12.78 \text{ rad/s}$$

From equations (1.29) and (1.30), the damping ratio becomes

$$\zeta = \frac{c}{c_{cr}} = \frac{1.143 \times 10^4}{2m\omega_n} = \frac{1.143 \times 10^4}{2(1651)(12.78)} = 0.27$$

Thus the car with passengers, fuel, and luggage will exhibit less damping and hence larger amplitude vibrations in the vertical direction. The vibrations will take a little longer to die out.

□

Note that this illustrates a difficulty in design problems, in the sense that the car cannot be damped at exactly the same value for all passenger situations. In this case, even if $\zeta = 0.3$ is desirable, it really cannot be achieved. Designs that do not change dramatically when one parameter changes a small amount are said to be *robust*. This and other design concepts are discussed in greater detail in Chapter 5, as the analytical skills developed in the next few chapters are required first.

1.8 STABILITY

In the preceding sections, the physical parameters m, c, and k are all considered to be positive in equation (1.27). This allows the treatment of the solutions of equation (1.27) to be classified into three groups: overdamped, underdamped, or critically damped. The case with $c = 0$ provides a fourth class called undamped. These four solutions are all well behaved in the sense that they do not grow with time and their amplitudes are finite. There are many situations, however, in which the coefficients are not positive, and in these cases the motion is not well behaved. This situation refers to the *stability* of solutions of a system.

Recalling that the solution of the undamped case ($c = 0$) is of the form $A \sin(\omega_n t + \phi)$, it is easy to see that the undamped response is bounded. That is, if $|x(t)|$ denotes the absolute value of x, then

$$|x(t)| \leq A|\sin(\omega_n t + \phi)| = A = \frac{1}{\omega_n}\sqrt{\omega_n^2 x_0^2 + v_0^2} \tag{1.85}$$

for every value of t. Thus $|x(t)|$ is always less than some finite number for all time and for all finite choices of initial conditions. In this case, the response is well behaved and said to be *stable* (sometimes called *marginally stable*). If, on the other hand, the value of k in equation (1.2) is negative and m is positive, the solutions are of the form

$$x(t) = A \sinh \omega_n t + B \cosh \omega_n t \tag{1.86}$$

which increases without bound as t does. In this case $|x(t)|$ no longer remains finite and such solutions are called *divergent* or *unstable*. Figure 1.37 illustrates a stable response and Figure 1.38 illustrates an unstable, or divergent, response.

Consider the response of the damped system of equation (1.27) with positive coefficients. As illustrated in Figures 1.10, 1.11, 1.12, and 1.13, it is clear that $x(t)$ approaches zero as t becomes large because of the exponential-decay terms. Such solutions are called *asymptotically stable*. Again, if c or k is negative (and m is positive), the motion grows without bound and becomes unstable, as in the undamped case. In the damped case, however, the motion may be unstable in one of two ways. Similar to overdamped solutions and underdamped solutions, the motion may grow without bound and may or may not oscillate. The nonoscillatory case is called *divergent instability* and the oscillatory case is called *flutter instability*, or sometimes just *flutter*. Flutter instability is sketched in Figure 1.39. The trend of growing without bound for large t continues in Figures 1.38 and 1.39, even though the figure stops. These types of instability occur in a variety of situations, often called *self-excited vibrations*, and require some source of energy. The following example illustrates such instabilities.

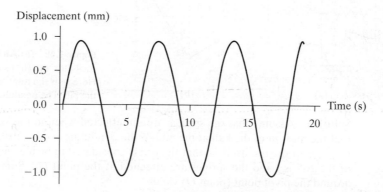

Figure 1.37   An example response of a stable single-degree-of-freedom system.

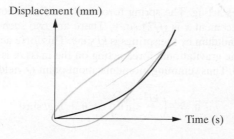

Figure 1.38   An example response of a unstable single-degree-of-freedom system (divergence).

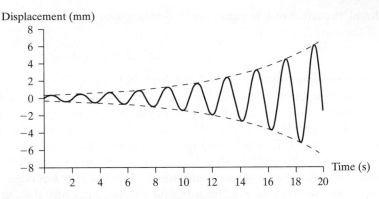

Figure 1.39 An example response of a unstable single-degree-of-freedom system which also oscillates, called flutter instability.

Example 1.8.1

Consider the inverted pendulum connected to two equal springs, shown in Figure 1.40.

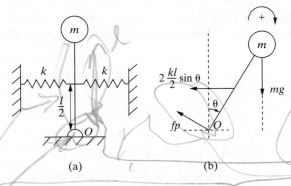

(a)   (b)

Figure 1.40 (a) An inverted pendulum oscillator and (b) its free-body diagram. Here fp is the total reaction force at the pin. The pendulum has length l.

Solution Assume that the springs are undeflected when in the vertical position and that the mass m of the ball at the end of the pendulum rod is substantially larger than the mass of the rod itself, so that the rod is considered to be massless. The total length of the rod is l and the springs are attached at the point $l/2$. Summing the moments around the pivot point (point O) yields

$$ml^2\ddot{\theta} = \sum M_0$$

There are three forces acting. The spring force is the stiffness times the displacement (kx) where the displacement x is $(l/2)\sin\theta$. There are two such springs, so the total force acting on the pendulum by the springs is $kl\sin\theta$. This force acts through a moment arm of $(l/2)\cos\theta$. The gravitational force acting on the mass m is mg acting through a moment arm of $l\sin\theta$. Thus summing moments about point O yields

$$ml^2\ddot{\theta} = -\left(\frac{kl^2}{2}\sin\theta\right)\cos\theta + mgl\sin\theta$$

and the equation of motion becomes

$$ml^2\ddot{\theta} + \left(\frac{kl^2}{2}\sin\theta\right)\cos\theta - mgl\sin\theta = 0 \tag{1.87}$$

For values of θ less than about $\pi/20$, $\sin\theta$ and $\cos\theta$ can be approximated by $\sin\theta \cong \theta$ and $\cos\theta \cong 1$. Applying this approximation to equation (1.87) yields

$$ml^2\ddot{\theta} + \frac{kl^2}{2}\theta - mgl\theta = 0$$

which upon rearranging becomes

$$2ml\ddot{\theta}(t) + (kl - 2mg)\theta(t) = 0$$

where θ is now restricted to be small (smaller than $\pi/20$). If k, l, and m are all such that the coefficient of θ, called the effective stiffness, is negative, that is, if

$$kl - 2mg < 0$$

the pendulum motion will be unstable by divergence, as illustrated in Figure 1.38.

□

Example 1.8.2

The vibration of an aircraft wing can be crudely modeled as

$$m\ddot{x} + c\dot{x} + kx = \gamma\dot{x}$$

where m, c, and k are the mass, damping, and stiffness values of the wing, respectively, modeled as a single-degree-of-freedom system, and where the term $\gamma\dot{x}$ is an approximate model of the aerodynamic forces on the wing ($\gamma > 0$ for high speed).

Solution   Rearranging the equation of motion yields

$$m\ddot{x} + (c - \gamma)\dot{x} + kx = 0$$

If γ and c are such that $c - \gamma > 0$, the system is asymptotically stable. However, if γ is such that $c - \gamma < 0$, then $\zeta = (c - \gamma)/2m\omega_n < 0$ and the solutions are of the form

$$x(t) = Ae^{-\zeta\omega_n t}\sin(\omega_d t + \phi)$$

where the exponent $(-\zeta\omega_n t) > 0$ for all $t > 0$ because of the negative damping term. Such solutions increase exponentially with time, as indicated in Figure 1.39. This is an example of flutter instability and self-excited oscillation.

□

This brief introduction to stability applies to systems that can be treated as linear and homogenous. More complex definitions of stability are required for forced systems and for nonlinear systems. The notions of stability can be thought of in terms of changing energy: stable systems having constant energy, unstable systems having increasing energy, and asymptotically stable systems having decreasing energy. Stability can also be thought of in terms of initial conditions and this is discussed in Section 1.10 where a brief introduction to nonlinear vibrations is given. An essential difference between linear and nonlinear systems lies in their respective stability properties.

1.9 NUMERICAL SIMULATION OF THE TIME RESPONSE

So far, most of the vibration problems examined have all been cast as linear differential equations that have solutions that can be determined analytically. These solutions are often plotted versus time in order to visualize the physical vibration and obtain an idea of the nature of the response. However, there are many more complex and nonlinear systems that are either difficult or impossible to solve analytically (i.e., that do not have closed-form solutions for the displacement as a function of time). The nonlinear pendulum equation given in Example 1.1.1 is "linearized" by making the approximation $\sin(\theta) = \theta$ to provide a system which is simple to solve (having the same analytical form as a linear spring–mass system). The approximation made to linearize the pendulum equation is only valid for certain, relatively small initial conditions. The approximation of $\sin(\theta) = \theta$ requires that the initial displacement and velocity are such that $\theta(t)$ remains less than about $10°$. For cases with larger initial conditions, a numerical integration routine may be used to compute and plot a solution of the nonlinear equation of motion. Numerical integration can be used to compute the solutions of a variety of difficult problems and is introduced here on simple problems that have known analytical solutions so that the nature of the approximation can be discussed. Later, numerical integration will be used for problems not having closed-form solutions.

The free response of any single-degree-of-freedom system may easily be computed by simple numerical means such as Euler's method or Runge–Kutta methods. This section examines the use of these common numerical methods for solving vibration problems that are difficult to solve in closed form. Runge–Kutta schemes can be found on calculators and in most common mathematical software packages such as Mathematica, Mathcad, Maple, and MATLAB. Alternately the numerical schemes may be programmed in more traditional languages, such as FORTRAN, or into spreadsheets. This section reviews the use of numerical methods for solving differential equations and then applies these methods to the solution of several vibration problems considered in the previous sections. These techniques are then used in the following section to analyze the response of nonlinear systems. Appendix F introduces the use of Mathematica, Mathcad, and MATLAB for numerical integration and plotting. Many modern curriculums introduce these methods and codes early in the engineering curriculum, in which case this section can be skipped or used as a quick review.

There are many schemes for numerically solving ordinary differential equations, such as those of vibration analysis. Two numerical solution schemes are presented here. The basis of numerical solutions of ordinary differential equations is to essentially undo calculus by representing each derivative by a small but finite difference (recall the definition of a derivative from calculus given in Window 1.6). A numerical solution of an ordinary differential equation is a procedure for constructing approximate discrete values: $x_1, x_2, \ldots, x_n$, of the solution $x(t)$ at the discrete values of time: $t_0 < t_1 < t_2 \ldots < t_n$. Thus a numerical procedure produces a list of discrete values $x_i = x(t_i)$ that approximate the exact solution, which is the continuous function of time $x(t)$. The initial conditions of the vibration problem of interest form the starting point of computing a numerical solution. For a single-degree-of-freedom system of the form

$$m\ddot{x} + c\dot{x} + kx = 0, \qquad x(0) = x_0 \qquad \dot{x}(0) = v_0 \tag{1.88}$$

the initial values x_0 and v_0 form the first two points of the numerical solution. Let T_f be the total length of time over which the solution is of interest (i.e., the equation is to be solved for values of t between $t = 0$ and $t = T_f$). The time interval $T_f - 0$ is then divided up into n intervals (so that $\Delta t = T_f/n$). Then equation (1.88) is calculated at the values of $t_0 = 0, t_1 = \Delta t, t_2 = 2\Delta t, \ldots, t_n = n\Delta t = T_f$ to produce an approximate representation, or simulation, of the solution.

The concept of a numerical solution is easiest to grasp by first examining the numerical solution of a first-order scalar differential equation. To this end, consider the first-order differential equation

$$\dot{x}(t) = ax(t) \qquad x(0) = x_0 \tag{1.89}$$

The Euler method proceeds from the definition of the slope form of the derivative given in Window 1.6, before the limit is taken:

$$\frac{x_{i+1} - x_i}{\Delta t} = ax_i \tag{1.90}$$

where x_i denotes $x(t_i)$, x_{i+1} denotes $x(t_{i+1})$, and Δt indicates the time interval between t_i and t_{i+1} (i.e., $\Delta t = t_{i+1} - t_i$). This expression can be manipulated to yield

$$x_{i+1} = x_i + \Delta t(ax_i) \tag{1.91}$$

This formula computes the discrete value of the response x_{i+1} from the previous value x_i, the time step Δt, and the system's parameter a. This numerical solution is called an *Euler* or *tangent line method*. The following example illustrates the use of the Euler formula for computing a solution.

Window 1.6
Definition of the Derivative

The definition of a derivative of $x(t)$ at $t = t_i$ is

$$\frac{dx(t_i)}{dt} = \lim_{\Delta t \to 0} \frac{x(t_{i+1}) - x(t_i)}{\Delta t}$$

where $t_{i+1} = t_i + \Delta t$ and $x(t)$ is continuous.

Example 1.9.1

Use the Euler formula to compute the numerical solution of $\dot{x} = -3x$, $x(0) = 1$ for various time increments in the time interval 0 to 4, and compare the results to the exact solution.

TABLE 1.4 COMPARISON OF THE EXACT SOLUTION OF $\dot{x} = -3x$, $x(0) = 1$
TO THE SOLUTION OBTAINED BY THE EULER METHOD WITH LARGE TIME
STEP ($\Delta t = 0.5$) FOR THE INTERVAL $t = 0$ TO 4

Index	Elapsed time	Exact	Euler	Absolute error
0	0	1.0000	1.0000	0
1	0.5000	0.2231	−0.5000	0.7231
2	1.0000	0.0498	0.2500	0.2002
3	1.5000	0.0111	−0.1250	0.1361
4	2.0000	0.0025	0.0625	0.0600
5	2.5000	0.0006	−0.0312	0.0318
6	3.0000	0.0001	0.0156	0.0155
7	3.5000	0.0000	−0.0078	0.0078
8	4.0000	0.0000	0.0039	0.0039

Solution   First, the exact solution can be obtained by direct integration or by assuming a solution of the form $x(t) = Ae^{\lambda t}$. Substitution of this assumed form into the equation $\dot{x} = -3x$ yields $A\lambda e^{\lambda t} = -3Ae^{\lambda t}$, or $\lambda = -3$, so that the solution is of the form $x(t) = Ae^{-3t}$. Applying the initial conditions $x(0) = 1$ yields $A = 1$. Hence the analytical solution is simply $x(t) = e^{-3t}$.

Next, consider a numerical solution using the Euler method suggested by equation (1.91). In this case the constant $a = -3$, so that $x_{i+1} = x_i + \Delta t(-3x_i)$. Suppose that a very crude time step is taken (i.e., $\Delta t = 0.5$) and the solution is formed over the interval from $t = 0$ to $t = 4$. Then Table 1.4 illustrates the values obtained from equation (1.91):

$$x_0 = 1$$

$$x_1 = x_0 + (0.5)(-3)(x_0) = -0.5$$

$$x_2 = -0.5 - (1.5)(-0.5) = 0.25$$

$$\vdots$$

forms the column marked "Euler." The column marked "Exact" is the value of e^{-3t} at the indicated elapsed time for a given index. Note that while the Euler approximation gets close to the correct final value, this value oscillates around zero while the exact value does not. This points out a possible source of error in a numerical solution. On the other hand, if Δt is taken to be very small, the difference between the solution obtained by the Euler equation and the exact solution becomes hard to see, as Figure 1.41 illustrates. Figure 1.41 is a plot of $x(t)$ obtained via the Euler formula for $\Delta t = 0.1$. Note that it looks very much like the exact solution $x(t) = e^{-3t}$.

□

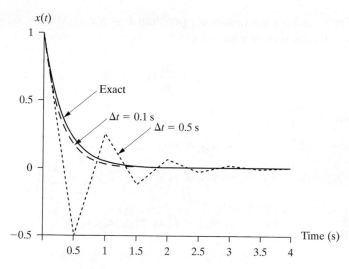

Figure 1.41   A plot of $x(t_i)$ versus t_i for $\dot{x}(t) = -3x$ using various time steps in equation (1.91) with $x(0) = 1$.

It is important to note from the example that two sources of error are present in computing the solution of a differential equation using a numerical scheme such as the Euler method. The first is called the *truncation error*, which is the difference between the exact solution and the solution obtained by the Euler approximation. This is the error indicated in the last column of Table 1.4. Note that this error accumulates as the index increases because the value at each discrete time is determined by the previous value, which is already in error. This can be somewhat controlled by the time step and the nature of the formula. The other source of error is the *round-off error* due to machine arithmetic. This is, of course, controlled by the computer and its architecture. Both sources of error can be significant. The successful use of a numerical method requires an awareness of both sources of errors in interpreting the results of a computer simulation of the solution of any vibration problem.

The Euler method can be improved upon by decreasing the step size, as Example 1.9.1 illustrates. Alternatively, a more accurate procedure can be used to improve the accuracy (smaller formula error) without decreasing the step size Δt. Several methods exist (such as the improved Euler method and various Taylor series methods) and are discussed in Boyce and DiPrima (2009), for instance. Only the Runge–Kutta method is discussed and used here.

The Runge–Kutta method was developed by two different researchers from about 1895 to 1901 (C. Runge and M. W. Kutta). The Runge–Kutta formulas (there are several) involve a weighted average of values of the right-hand side of the differential equation taken at different points between the time intervals t_i and $t_i + \Delta t$. The derivations of various Runge–Kutta formulas are tedious but straightforward and are not presented here (see Boyce and DiPrima 2009). One useful formulation can be stated for the first-order problem $\dot{x} = f(x, t), x(0) = x_0$, where f is any scalar function (linear or nonlinear) as

$$x_{n+1} = x_n + \frac{\Delta t}{6}(k_{n1} + 2k_{n2} + 2k_{n3} + k_{n4}) \tag{1.92}$$

where

$$k_{n1} = f(x_n, t_n)$$

$$k_{n2} = f\left(x_n + \frac{\Delta t}{2}k_{n1}, t_n + \frac{\Delta t}{2}\right)$$

$$k_{n3} = f\left(x_n + \frac{\Delta t}{2}k_{n2}, t_n + \frac{\Delta t}{2}\right)$$

$$k_{n4} = f(x_n + \Delta t k_{n3}, t_n + \Delta t)$$

The sum in parentheses in equation (1.92) represents the average of six numbers, each of which looks like a slope at a different time; for instance, the term k_{n1} is the slope of the function at the "left" end of the time interval.

Such formulas can be enhanced by treating Δt as a variable, Δt_i. At each time step t_i, the value of Δt_i is adjusted based on how rapidly the solution $x(t)$ is changing. If the solution is not changing very rapidly, a large value of Δt_i is allowed without increasing the formula error. On the other hand, if $x(t)$ is changing rapidly, a small Δt_i must be chosen to keep the formula error small. Such step sizes can be chosen automatically as part of the computer code for implementing the numerical solution. The Runge–Kutta and Euler formulas just listed can be applied to vibration problems by noting that the most general (damped) vibration problem can be put into a first-order form.

Returning to a damped system of the form

$$m\ddot{x}(t) + c\dot{x}(t) + kx(t) = 0 \qquad x(0) = x_0, \qquad \dot{x}(0) = \dot{x}_0 \tag{1.93}$$

the Euler method of equation (1.91) can be applied by writing this expression as two first-order equations. To this end, divide equation (1.93) by the mass m, and define two new variables by $x_1 = x(t)$ and $x_2 = \dot{x}(t)$. Then differentiate the definition of $x_1(t)$, rearrange equation (1.93), and replace x and its derivative with x_1 and x_2 to get

$$\dot{x}_1(t) = x_2(t)$$

$$\dot{x}_2(t) = -\frac{c}{m}x_2(t) - \frac{k}{m}x_1(t) \tag{1.94}$$

subject to the initial conditions $x_1(0) = x_0$ and $x_2(0) = \dot{x}_0$. The two coupled first-order differential equations given in (1.94) may be written as a single expression by using a vector and matrix form determined by first defining the vector 2×1 $\mathbf{x}(t)$ and the 2×2 matrix A by

$$
A = \begin{bmatrix} 0 & 1 \\ -\dfrac{k}{m} & -\dfrac{c}{m} \end{bmatrix} \quad \mathbf{x}(t) = \begin{bmatrix} x_1(t) \\ x_2(t) \end{bmatrix} \quad \mathbf{x}(0) = \begin{bmatrix} x_1(0) \\ x_2(0) \end{bmatrix} \tag{1.95}
$$

The matrix A defined in this way is called the *state matrix* and the vector $\mathbf{x}$ is called the *state vector*. The position x_1 and the velocity x_2 are called the *state variables*. Using these definitions (see Appendix C), the rules of vector differentiation (element by element) and multiplication of a matrix times a vector, equations (1.95) may be written as

$$
\dot{\mathbf{x}}(t) = A\mathbf{x}(t) \tag{1.96}
$$

subject to the initial condition $\mathbf{x}(0)$. Now the Euler method of numerical solution given in equation (1.91) can be applied directly to this vector-matrix formulation of Equation (1.96), by simply calling the scalar x_i, the vector $\mathbf{x}_i$, and replacing the scalar a with the matrix A to produce

$$
\mathbf{x}(t_{i+1}) = \mathbf{x}(t_i) + \Delta t A \mathbf{x}(t_i) \tag{1.97}
$$

This, along with the initial condition $\mathbf{x}(0)$, defines the Euler formula for integrating the general single-degree-of-freedom vibration problem described in equation (1.92). Equation (1.97) allows the time response to be computed and plotted.

As suggested, the Euler-formula method can be greatly improved by using a Runge–Kutta program. For instance, MATLAB has two different Runge–Kutta-based simulations: ode23 and ode45. These are automatic step-size integration methods (i.e., Δt is chosen automatically). The Engineering Vibration Toolbox has one fixed-step Runge–Kutta-based method, VTB1_3, for comparison. The M-file ode23 uses a simple second- and third-order pair of formulas for medium accuracy and ode45 uses a fourth- and fifth-order pair for greater accuracy. Each of these corresponds to a formulation similar to that expressed in equations (1.92) with more terms and a variable step size Δt. In general, the Runge–Kutta simulations are of a higher quality than those obtained by the Euler method.

Example 1.9.2

Use the ode45 function to simulate the response to $3\ddot{x} + \dot{x} + 2x = 0$ subject to the initial conditions $x(0) = 0, \dot{x}(0) = 0.25$ over the time interval $0 \le t \le 20$.

Solution   The first step is to write the equation of motion in first-order form. This yields

$$
\dot{x}_1 = x_2
$$
$$
\dot{x}_2 = -\tfrac{2}{3} x_1 - \tfrac{1}{3} x_2
$$

Next, an M-file is created to store the equations of motion. An M-file is created by choosing a name, say, sdof.m, and entering

```
function xdot = sdof(t,x);
xdot = zeros(2,1);
xdot(1) = x(2);
xdot(2) = -(2/3)*x(1) - (1/3)*x(2);
```

Next, go to the command mode and enter

```
t0 = 0;tf = 20;
x0 = [0 0.25];
[t,x] = ode45('sdof',[t0 tf],x0);
plot(t,x)
```

The first line establishes the initial, (t0), and final, (tf), times. The second line creates the vector containing the initial conditions x0. The third line creates the two vectors t, containing the time history, and x, containing the response at each time increment in *t*, by calling ode45 applied to the equations set up in sdof. The fourth line plots the vector x versus the vector *t*. This is illustrated in Figure 1.42.

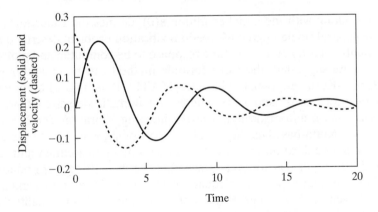

Figure 1.42 A plot of the displacement $x(t)$ of the single-degree-of freedom system of Example 1.9.2 (solid line) and the corresponding velocity $\dot{x}(t)$ (dashed line).

☐

The preceding example may also be solved using Mathematica, Mathcad, and Maple, by writing a FORTRAN routine, or by using any number of other computer codes or programmable calculators. The following example illustrates the commands required to produce the result of Example 1.9.2 using Mathematica and again using Mathcad. These approaches are then used in the next section to examine the response of certain nonlinear vibration problems.

Example 1.9.3

Solve Example 1.9.2 using the Mathematica program.

Solution   The Mathematica program uses an iterative method to compute the solution and accepts the second-order form of the equation of motion. The text after the prompt In[1]:= is typed by the user and returns the solution stored in the variable x[t]. Mathematica has several equal signs for different purposes. In the argument of the NDSolve function, the user types in the differential equation to be solved, followed by the initial conditions, the name of the variable (response), and the name of the independent variable followed by the interval over which the solution is sought. NDSolve computes the solution and stores it as an interpolating function; hence the code returns the plot following the output prompt Out[1]=. The plot command requires the name of the interpolating function returned by NDSolve, x[t] in this case, the independent variable, t, and the range of values for the independent variable.

```
In[1]:=
NDSolve[{x''[t]+(1/3)*x'[t]+(2/3)*x[t]==0,x'[0]==0.25,x[0]==0},
x,{t,0,20}];
Plot[Evaluate[x[t]/.%],{t,0,20}]
Out[1]={{x->InterpolatingFunction[{{0.,20.}},<>]}}
Out[2]=
```

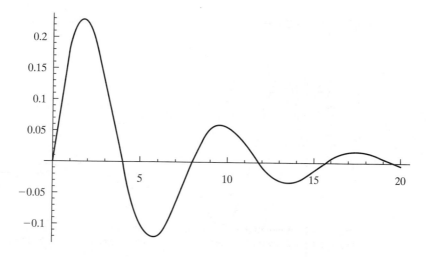

Example 1.9.4

Solve Example 1.9.2 using the Mathcad program.

Solution   The Mathcad program uses a fixed time step Runge–Kutta solution and returns the solution as a matrix with the first column consisting of the time step, the second column containing the response, and the third column containing the velocity response.

First type in the initial condition vector:

$$y: = y := \begin{bmatrix} 0 \\ 0.25 \end{bmatrix}$$

Then type in the system in first-order form:

$$D(t,y): = D(t,y) := \begin{bmatrix} y_1 \\ -\left(\frac{1}{3}y_1\right) - \frac{2}{3}y_0 \end{bmatrix}$$

Solve using Runge-Kutta:

$$Z := rkfixed(y,0,20,1000,D)$$

Name the time vector from the Runge-Kutta matrix solution:

$$t := Z^{<0>}$$

Name the displacement vector from the Runge-Kutta matrix solution:

$$x := Z^{<1>}$$

Name the velocity vector from the Runge-Kutta matrix solution:

$$dxdt := Z^{<2>}$$

Plot the solutions.

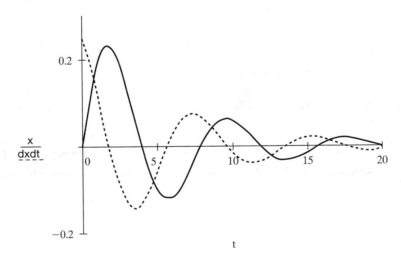

The use of these computational programs to simulate the response of a vibrating system is fairly straightforward. Further information on using each of these programs can be found in Appendix F or by consulting manuals or any one of numerous books written on using these codes to solve various math and engineering problems. You are encouraged to reproduce Example 1.9.4 and then repeat the problem for various different values of the initial conditions and coefficients. In this way, you can build some intuition and understanding of vibration phenomena and how to design a system to produce a desired response.

A note about the use of the codes presented in this text is in order. At the time of printing this edition, all the codes ran as typed. However, each year, or sometimes more frequently, companies who provide these codes update them and in so doing they often change syntax. These changes can be found on the companies' websites and should be checked if difficulty is encountered in using the codes presented here.

1.10 COULOMB FRICTION AND THE PENDULUM

In the previous sections, all of the systems considered are linear (or linearized) and have solutions that can be obtained by analytical means. In this section, two common systems are analyzed that are nonlinear and do not have simple analytical solutions. The first is a spring–mass system with sliding friction (Coulomb damping), and the second is the full nonlinear pendulum equation. In each case a solution is obtained by using the numerical integration techniques introduced in Section 1.9. The ability to compute the solution to general nonlinear systems using these numerical techniques allows us to consider vibration in more complicated configurations.

Nonlinear vibration problems are much more complex than linear systems. Their numerical solutions, however, are often fairly straightforward. Several new phenomena result when nonlinear terms are considered. Most notably, the idea of a single equilibrium point of a linear system is lost. In the case of Coulomb damping, a continuous region of equilibrium positions exists. In the case of the nonlinear pendulum, an infinite number of equilibrium points result. This single fact greatly complicates the analysis, measurement, and design of vibrating systems.

Coulomb damping is a common damping effect, often occurring in machines, that is caused by sliding friction or dry friction. It is characterized by the relation

$$f_c = F_c(\dot{x}) = \begin{cases} -\mu N & \dot{x} > 0 \\ 0 & \dot{x} = 0 \\ \mu N & \dot{x} < 0 \end{cases}$$

where f_c is the dissipation force, N is the normal force (see any introductory physics text), and μ is the coefficient of sliding friction (or kinetic friction). Figure 1.43 is a schematic of a mass m sliding on a surface and connected to a spring of stiffness k. The frictional force f_c always opposes the direction of motion causing a system with Coulomb friction to be nonlinear. Table 1.5 lists some measured values of the coefficient of kinetic friction for several different sliding objects. Summing forces

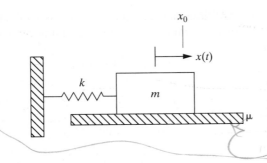

Figure 1.43 A spring–mass system sliding on a surface of kinetic coefficient of friction μ.

TABLE 1.5 APPROXIMATE COEFFICIENTS OF FRICTION FOR VARIOUS OBJECTS SLIDING TOGETHER

Material	Kinetic	Static
Metal on metal (lubricated)	0.07	0.09
Wood on wood	0.2	0.25
Steel on steel (unlubricated)	0.3	0.75
Rubber on steel	1.0	1.20

in part (a) of Figure 1.44 in the x direction yields that (note that the mass changes direction when the velocity passes through zero)

$$m\ddot{x} + kx = \mu mg \quad \text{for} \quad \dot{x} < 0 \qquad (1.98)$$

Here the sum of the forces in the vertical direction yields the fact that the normal force N is just the weight, mg, where g is the acceleration due to gravity (not the case if m is on an inclined plane as N is no longer along the same direction as W). In a similar fashion, summing forces in part (b) of Figure 1.44 yields

$$m\ddot{x} + kx = -\mu mg \quad \text{for} \quad \dot{x} > 0 \qquad (1.99)$$

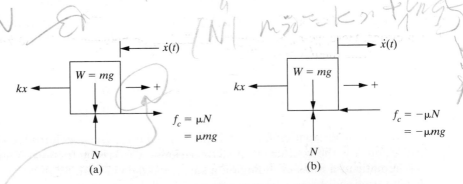

Figure 1.44 A free-body diagram of the forces acting on the sliding block system of Figure 1.43: (a) mass moving to the right ($\dot{x} < 0$), (b) mass moving to the right ($\dot{x} > 0$). From the y direction: $N = mg$.

Since the sign of $\dot{x}$ determines the direction in which the opposing frictional force acts, equations (1.98) and (1.99) can be written as the single equation

$$m\ddot{x} + \mu mg\,\text{sgn}(\dot{x}) + kx = 0 \qquad (1.100)$$

where sgn(τ) denotes the *signum function*, defined to have the value 1 for $\tau > 0$, -1 for $\tau < 0$, and 0 for $\tau = 0$. This equation cannot be solved directly using methods such as the variation of parameters or the method of undetermined coefficients. This is because equation (1.100) is a nonlinear differential equation. Rather, equation (1.100) can be solved by breaking the time intervals up into segments corresponding to the changes in direction of motion (i.e., at those time intervals separated by $\dot{x} = 0$). Alternatively, equation (1.100) can be solved numerically, as is done in the following. Because the system's equation of motion is linear in two ranges, that is, equations (1.98) and (1.99) are linear, such systems are also called *bilinear*.

The sliding block in Figure 1.44 requires nonzero initial conditions to set it in motion. Suppose first that the initial velocity is zero. The motion will result only if the initial position x_0 is such that the spring force kx_0 is large enough to overcome the static friction force $\mu_s mg$ ($kx_0 > \mu_s mg$). Here μ_s is the coefficient of static friction, which is generally larger than the kinetic or dynamic coefficient of friction for sliding surfaces. If x_0 is not large enough, no motion results. The range of values of x_0 for which no motion results defines the equilibrium position. If, on the other hand, the initial velocity is nonzero, the object will move. One of the distinguishing features of nonlinear systems is their multiple equilibrium positions. The solution of the equation of motion for the case when motion results can be obtained by considering the following cases.

With x_0 to the right of any equilibrium, the mass is moving to the left, the friction force is to the right, and equation (1.98) holds. Equation (1.98) has a solution of the form

$$x(t) = A_1 \cos \omega_n t + B_1 \sin \omega_n t + \frac{\mu mg}{k} \qquad (1.101)$$

where $\omega_n = \sqrt{k/m}$ and A_1 and B_1 are constants to be determined by the initial conditions. Here we have dropped the distinction between static and kinetic friction. Applying the initial conditions yields

$$x(0) = A_1 + \frac{\mu mg}{k} = x_0 \qquad (1.102)$$

$$\dot{x}(0) = \omega_n B_1 = 0 \qquad (1.103)$$

Hence $B_1 = 0$ and $A_1 = x_0 - \mu mg/k$ specifies the constants in equation (1.101). Thus when the mass starts from rest (at x_0) and moves to the left, it moves as

$$x(t) = \left(x_0 - \frac{\mu mg}{k} \right) \cos \omega_n t + \frac{\mu mg}{k} \qquad (1.104)$$

This motion continues until the first time $\dot{x}(t) = 0$. This happens when the derivative of equation (1.104) is zero, or when

$$\dot{x}(t) = -\omega_n \left(x_0 - \frac{\mu mg}{k} \right) \sin \omega_n t_1 = 0 \tag{1.105}$$

Thus when $t_1 = \pi/\omega_n$, $\dot{x}(t) = 0$ and the mass starts to move to the right provided that the spring force, kx, is large enough to overcome the maximum frictional force μmg. Hence equation (1.99) now describes the motion. Solving equation (1.99) yields

$$x(t) = A_2 \cos \omega_n t + B_2 \sin \omega_n t - \frac{\mu mg}{k} \tag{1.106}$$

for $\pi/\omega_n < t < t_2$, where t_2 is the second time that $\dot{x}$ becomes zero. The initial conditions for equation (1.106) are calculated from the previous solution given by equation (1.104) at t_1

$$x\left(\frac{\pi}{\omega_n}\right) = \left(x_0 - \frac{\mu mg}{k} \right) \cos \pi + \frac{\mu mg}{k} = \frac{2\mu mg}{k} - x_0 \tag{1.107}$$

$$\dot{x}\left(\frac{\pi}{\omega_n}\right) = -\omega_n \left(x_0 - \frac{\mu mg}{k} \right) \sin \pi = 0 \tag{1.108}$$

From equation (1.106) and its derivatives it follows that

$$A_2 = x_0 - \frac{3\mu mg}{k} \qquad B_2 = 0 \tag{1.109}$$

The solution for the second interval of time then becomes

$$x(t) = \left(x_0 - \frac{3\mu mg}{k} \right) \cos \omega_n t - \frac{\mu mg}{k} \qquad \frac{\pi}{\omega_n} < t < \frac{2\pi}{\omega_n} \tag{1.110}$$

This procedure is repeated until the motion stops. The motion will stop when the velocity is zero ($\dot{x} = 0$) and the spring force (kx) is insufficient to overcome the maximum frictional force (μmg). The response is plotted in Figure 1.45.

Several things can be noted about the free response with Coulomb friction versus the free response with viscous damping. First, with Coulomb damping the amplitude decays linearly with slope

$$-\frac{2\mu mg\omega_n}{\pi k} \tag{1.111}$$

rather than exponentially as does a viscously damped system. Second, the motion under Coulomb friction comes to a complete stop, at a potentially different equilibrium position than when initially at rest, whereas a viscously damped system

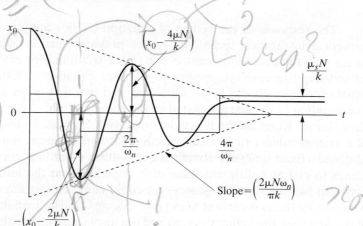

Figure 1.45 A plot of the free response, $x(t)$, of a spring–mass system with Coulomb friction.

oscillates around a single equilibrium, $x = 0$, with infinitesimally small amplitude. Finally, the frequency of oscillation of a system with Coulomb damping is the same as the undamped frequency, whereas viscous damping alters the frequency of oscillation.

Example 1.10.1

The response of a mass oscillating on a surface is measured to be of the form indicated in Figure 1.45. The initial position is measured to be 30 mm from its zero rest position, and the final position is measured to be 3.5 mm from its zero rest position after four cycles of oscillation in 1 s. Determine the coefficient of friction.

Solution   First, the frequency of motion is 4 Hz, or 25.13 rad/s, since four cycles were completed in 1 s. The slope of the line of decreasing peaks is

$$\frac{-30 + 3.5}{1} = -26.5 \text{ mm/s}$$

Therefore, from expression (1.111),

$$-26.5 \text{ mm/s} = \frac{-2\mu\, mg\omega_n}{\pi k} = \frac{-2\mu g}{\pi}\frac{\omega_n}{\omega_n^2} = \frac{-2\mu g}{\pi\omega_n}$$

Solving for μ yields

$$\mu = \frac{\pi\,(25.13 \text{ rad/s})\,(-26.5 \text{ mm/s})}{(-2)\,(9.81 \times 10^3 \text{ mm/s}^2)} = 0.107$$

This small value for μ indicates that the surface is probably very smooth or lubricated.   □

The response of the system of equation (1.100) can also be obtained by the numerical integration techniques of the previous section, which is substantially easier than the preceding construction of the solution. For example, VTB1_5 uses a fixed-step Runge–Kutta method to integrate equation (1.100). The second-order equation of motion can be reformulated into two first-order equations somewhat like equation (1.96) and integrated by the Euler method of equation (1.97), or standard Runge–Kutta methods may be employed. Figure 1.46 illustrates the response of a system subject to Coulomb friction for two different initial conditions using Mathcad's fixed-time-step Runge–Kutta routine. Note in particular that the system comes to rest at a different value of x_f depending on the initial conditions. Such a system has the same frequency, yet could come to rest anywhere in the region bounded by the two vertical lines ($x = \pm \mu m g / k$). The response will come to rest at the first time the velocity is zero and the displacement is within this region.

Comparing the response of a linear spring–mass system with viscous damping (say the underdamped response of Figure 1.10) to the response of a spring–mass system with Coulomb damping given previously, an obvious and significant difference is the rest position. These multiple rest positions constitute a major feature of nonlinear systems: the existence of more than one equilibrium position.

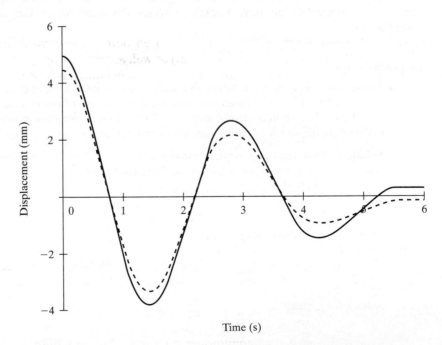

Figure 1.46   A plot of the free response (displacement versus time) of a system subject to Coulomb friction with two different initial positions (the solid line is $x_0 = 5$ mm and the dashed line is $x_0 = 4.5$ mm, both with $v_0 = 0$) for the same physical parameters ($m = 1000$ kg, $\mu = 0.3$ and $k = 5000$ N/m).

The *equilibrium point* of a system, or of a set of governing equations, may be defined best by first placing the equation of motion in state space, as was done in the previous section for the purpose of numerical integration. A general single-degree-of-freedom system may be written as

$$\ddot{x}(t) + f(x(t), \dot{x}(t)) = 0 \tag{1.112}$$

where the function f can take on any form, linear or nonlinear. For example, for a linear spring–mass system the function f is just $f(x, \dot{x}) = c\dot{x}(t) + kx(t)$, which is a linear function of the state variables of position and velocity. On the other hand, in a nonlinear system f will be some nonlinear function of the state variables. For instance, the pendulum equation derived and discussed in Example 1.1.1, $\ddot{\theta} + (g/l)\sin\theta = 0$, can be written in the form of equation (1.112) by defining f to be $f(\theta, \dot{\theta}) = (g/l)\sin(\theta)$, where θ is the displacement variable.

Using the approach following equations (1.94) and (1.95). the general state-space model of equation (1.112) is written by defining the two state variables $x_1 = x(t)$ and $x_2 = \dot{x}(t)$. Then equation (1.112) can be written as the first-order pair

$$\dot{x}_1(t) = x_2(t)$$
$$\dot{x}_2(t) = -f(x_1, x_2) \tag{1.113}$$

This state-space form of the equation is used both for numerical integration (as before for the Coulomb friction problem) as well as for formally defining an equilibrium position by defining the state vector, $\mathbf{x}$, used in equation (1.96) and a nonlinear vector function, $\mathbf{F}$, as

$$\mathbf{F} = \begin{bmatrix} x_2(t) \\ -f(x_1, x_2) \end{bmatrix} \tag{1.114}$$

Equations (1.113) may now be written in the simple form

$$\dot{\mathbf{x}} = \mathbf{F}(\mathbf{x}) \tag{1.115}$$

An equilibrium point of this system, denoted by $\mathbf{x}_e$, is defined to be any value of $\mathbf{x}$ for which $\mathbf{F}(\mathbf{x})$ is identically zero (called zero phase velocity). Thus the equilibrium point is any vector of constants that satisfies the relations

$$\mathbf{F}(\mathbf{x}_e) = 0 \tag{1.116}$$

A mechanical system is in equilibrium if its state does not change with time (i.e., $\dot{x}$ and $\ddot{x}$ are both zero).

For Coulomb friction, the equilibrium position cannot be directly determined by using the signum function (see below equation 1.100) because of the discontinuity at zero velocity. To compute the equilibrium position, consider equation (1.116) for the system of Figure 1.44. This yields

$$\begin{bmatrix} x_2 \\ \dfrac{f_c}{m} - \dfrac{k}{m} x_1 \end{bmatrix} = \begin{bmatrix} 0 \\ 0 \end{bmatrix}$$

Solving yields the two conditions:

$$x_2 = 0$$

and

$$f_c - kx_1 = 0$$

Realize that this last expression is static, so that the expression is satisfied as long as

$$-\frac{\mu_s mg}{k} < x_1 < \frac{\mu_s mg}{k}$$

As discussed earlier, the friction force is static, or in equilibrium, until the spring force kx_1 is large enough to overcome the friction force as expressed by this inequality.

This describes the condition that the velocity (x_2) is zero and the position lies within the region defined by the force of friction. Depending on the initial conditions, the response will end up at a value of $\mathbf{x}_e$ somewhere in this region. Usually, the equilibrium values are a discrete set of numbers, as the following example illustrates.

Example 1.10.2

Calculate the equilibrium position for the nonlinear system defined by $\ddot{x} + x - \beta^2 x^3 = 0$, or in state equation form, letting $x_1 = x$ as before,

$$\dot{x}_1 = x_2$$
$$\dot{x}_2 = x_1 \left(\beta^2 x_1^2 - 1 \right)$$

Solution   These equations represent the vibration of a "soft spring" and correspond to an approximation of the pendulum problem of Example 1.4.2, where $\sin x \approx x - x^3/6$. The equations for the equilibrium position are

$$x_2 = 0$$
$$x_1 \left(\beta^2 x_1^2 - 1 \right) = 0$$

There are three solutions to this set of algebraic equations corresponding to the three equilibrium positions of the soft spring. They are

$$\mathbf{x}_e = \begin{bmatrix} 0 \\ 0 \end{bmatrix}, \begin{bmatrix} \frac{1}{\beta} \\ 0 \end{bmatrix}, \begin{bmatrix} -\frac{1}{\beta} \\ 0 \end{bmatrix}$$

□

The next example considers the full nonlinear pendulum equation illustrated in Figures 1.1 and 1.47. Physically the pendulum may swing all the way around its pivot point and has equilibrium positions in both the straight-up and straight-down positions, as illustrated in Figure 1.47(b) and 1.47(c).

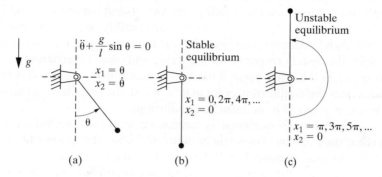

Figure 1.47   (a) A pendulum consisting of a massless rod of length l and a tip mass m. (b) The straight-down equilibrium position. (c) The straight-up equilibrium position.

Example 1.10.3

Calculate the equilibrium positions of the pendulum of Figure 1.47 with the equation of motion given in Example 1.1.1.

Solution   The pendulum equation in state-space form is given by

$$\dot{x}_1 = x_2$$

$$\dot{x}_2 = -\frac{g}{l}\sin(x_1)$$

so that the vector equation $\mathbf{F(x)} = \mathbf{0}$ yields the following equilibrium solutions:

$$x_2 = 0 \text{ and } x_1 = 0, \pi, 2\pi, 3\pi, 4\pi, 5\pi \ldots$$

since $\sin(x_1)$ is zero for any multiple of π. Note that there are an infinite number of equilibrium positions, or vectors $\mathbf{x}_e$. These are all either the up position corresponding to the odd values of π [Figure 1.47(c)], or the down position corresponding to even multiples of π [Figure 1.47(b)]. These positions form two distinct types of behavior. The response for initial conditions near the even values of π is a stable oscillation around the down position, just as in the linearized case, while the response to initial conditions near odd values of π moves away from the equilibrium position (called unstable) and the value of the response increases without bound.

□

The stability of equilibrium of a nonlinear vibration problem is very important and is based on the definitions given in Section 1.8. However, in the linear case, there is only one equilibrium value and every solution is either stable or unstable. In this case, the stability condition is said to be a *global* condition. In the nonlinear case, there is more than one equilibrium point, and the concept of stability is tied to each particular equilibrium point and is therefore referred to as *local stability*. As in the example of the nonlinear pendulum equation, some equilibrium points are stable and some are not. Furthermore, the stability of the response of a nonlinear system depends on the initial conditions. In the linear case, the initial conditions have no influence on the stability, and the system parameters and form of the equation of motion completely determine the stability of the response. To see this, look again at the pendulum of Figure 1.47. If the initial position and velocity are near the origin, the system response will be stable and oscillate around the equilibrium point at zero. On the other hand, if the same pendulum (i.e., same *l*) is given initial conditions near the equilibrium point at $\theta = \pi$ rad, the response will grow without bound. Hence, $\theta = \pi$ rad is an unstable equilibrium.

Even though nonlinear systems have multiple equilibria and more exotic behavior, their response may still be simulated using the numerical-integration methods of the previous section. This is illustrated for the pendulum in the following example, which compares the response to various initial conditions of both the nonlinear pendulum equation and its corresponding linearization treated in Examples 1.1.1 and 1.4.2.

Example 1.10.4

Compare the responses of the nonlinear and linear pendulum equations using numerical integration and the value $(g/\ell) = (0.1)^2$, for (a) the initial conditions $x_0 = 0.1$ rad and $v_0 = 0.1$ rad/s, and (b) the initial conditions $x_0 = 1$ rad and $v_0 = 1$ rad/s, by plotting the responses. Here x and v are used to denote θ and its derivative, respectively, in order to accommodate notation available in computer codes.

Solution   Depending on which program is used to integrate the solution numerically, the equations must first be put into first-order form, and then either Euler integration or Runge–Kutta routine may be implemented and the solutions plotted. Integrations in MATLAB, Mathematica, and Mathcad are presented. More details can be found in Appendix F. Note that the response to the linear system is fairly close to that of the full nonlinear system in case (a) with slightly different frequency, while case (b) with larger initial conditions is drastically different. The Mathcad solution follows.

```
First, enter the initial conditions for each response:
```

$$v_0 := 1 \qquad x_0 := 1 \qquad v1_0 := 0.1 \qquad x1_0 := 0.1$$

```
Next, define the frequency and the number of and size of the
time steps:
```

$$\omega := 0.1 \qquad N := 2000 \qquad i := 0..N \qquad \Delta := \frac{6 \cdot \pi}{\omega \cdot N}$$

```
The nonlinear Euler integration is
```

$$\begin{bmatrix} x_{i+1} \\ v_{i+1} \end{bmatrix} := \begin{bmatrix} v_i \cdot \Delta + x_i \\ -\omega^2 \cdot \sin(x_i) \cdot \Delta + v_i \end{bmatrix}$$

The linear Euler integration is

$$\begin{bmatrix} x1_{i+1} \\ v1_{i+1} \end{bmatrix} := \begin{bmatrix} v1_i \cdot \Delta + x1_i \\ -\omega^2 \cdot (\Delta) \cdot x_i 1 + v1_i \end{bmatrix}$$

The plot of these two solutions yields

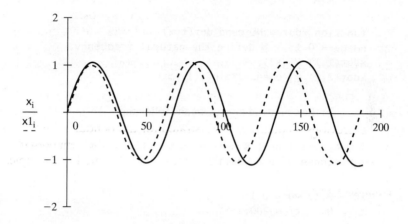

Here the dashed line is the linear solution. Next, compute these solutions again using initial conditions close to unstable equilibrium values:

$$x_0 := \pi \qquad v_0 := 0.1 \qquad x1_0 := \pi \qquad v1_0 := 0.1$$

$$\begin{bmatrix} x_{i+1} \\ v_{i+1} \end{bmatrix} := \begin{bmatrix} v_i \cdot \Delta + x_i \\ -\omega^2 \cdot \sin(x_i) \cdot \Delta + v_i \end{bmatrix} \begin{bmatrix} x1_{i+1} \\ v1_{i+1} \end{bmatrix} := \begin{bmatrix} v1_i \cdot \Delta + x1_i \\ -\omega^2 \cdot (\Delta) \cdot x_i 1 + v1_i \end{bmatrix}$$

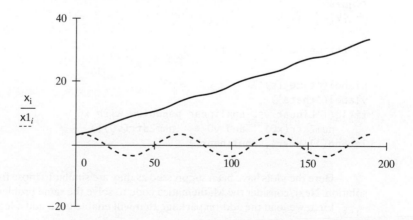

The MATLAB code for running the solutions (using Runge–Kutta this time) and plotting is obtained by first creating the appropriate M-files (named lin_pend_dot.m and NL_pend_dot.m, defining the linear and nonlinear pendulum equations, respectively).

```
function xdot = lin_pend_dot(t,x)
omega = 0.1;    % define the natural frequency
xdot(1,1) = x(2);
xdot(2,1) = -omega^2*x(1);
```

```
function xdot = NL_pend_dot(t,x)
omega = 0.1;    % define the natural frequency
xdot(1,1) = x(2);
xdot(2,1) = -omega^2*sin(x(1));
```

In the command mode type the following:

```
% Overplot linear & nonlinear simulations of the free
% response of a pendulum.

x0 = 0.1; v0 = 0.1;
ti = 0;    tf = 200;

% linear
[time_lin,sol_lin]=ode45('lin_pend_dot',[ti tf],[x0 v0]);

% nonlinear
[time_NL,sol_NL]=ode45('NL_pend_dot',[ti tf],[x0 v0]);

% overplot displacements
figure
plot(time_lin,sol_lin(:,1),'-')
hold
plot(time_NL,sol_NL(:,1),'-')

xlabel('time (s)')
ylabel('theta')
title(['Linear vs. nonlinear pendulum with x0 = ' ...
    num2str(x0) ' and v0 = ' num2str(v0)])
legend('linear','nonlinear')
```

Here the plots have been suppressed as they are similar to those from the Mathcad solution. Next, consider the Mathematica code to solve the same problem.

First, we load the add-on package that will enable us to add a legend to our plot.

```
In[1]:= <<PlotLegends'
```

Define the natural circular frequency, ω

```
In[2]:= ω=0.1;
```

The following cell solves the linear differential equation, then the nonlinear differential equation, and then produces a plot containing both responses.

```
In[3]:= xlin=NDSolve[{xl''[t]+ω²*xl[t]==0, xl[0]==.1, xl'[0]==.1},
        xl[t],{t,0,200}]
        xnonlin=NDSolve[{xnl''[t]+ω²*Sin[xnl[t]]==0, xnl[0]==.1, xnl'[0]==.1},
        xnl[t], {t,0,200}]
        Plot[{Evaluate[xl[t]/.xlin], Evaluate[xnl[t]/.xnonlin]},{t,0,200},
            PlotStyle→{Dashing[{}], Dashing[{.01,.01}]}, PlotLabel→"Linear and
            Nonlinear Response, Stable Equilibrium",
            AxesLabel→{"time,s",""}, PlotLegend→{"Linear","Non-Linear"},
            LegendPosition→{1,0}, LegendSize→{.7, .3}]
Out[3]= {{xl[t]→InterpolatingFunction[{{0., 200.}},<>][t]}}
Out[4]= {{xnl[t]→InterpolatingFunction[{{0., 200.}},<>][t]}}
Out[5]=
```

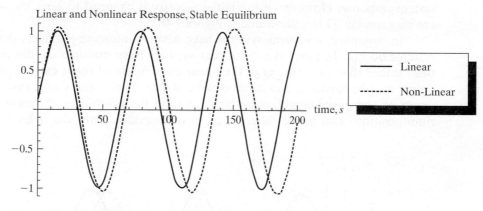

```
In[6]:= Clear[xlin, xnonlin, xl, xnl]
        xlin=NDSolve[{xl''[t]+ω²*xl[t] == 0, xl[0] ==π, xl'[0] ==.1}, xl[t],
        {t, 0, 200}]
        xnonlin=NDSolve[{xnl''[t]+ω²*Sin[xnl[t]] ==0, xnl[0] ==π, xnl'[0] ==.1},
        xnl[t], {t, 0, 200}]
        Plot[{Evaluate[xl[t]/.xlin], Evaluate[xnl[t]/.xnonlin]}, {t, 0, 200},
            PlotStyle→{Dashing[{}], Dashing[{.01,.01}]}, PlotRange→{-20, 40},
            PlotLabel→"Linear and Nonlinear Response, Unstable Equilibrium",
                AxesLabel→{"time,s", ""},
            PlotLegend→{"Linear", "Nonlinear"}, LegendPosition→{1, 0},
                LegendSize→{.7, .3}]
Out[7]= {{xl[t]→InterpolatingFunction[{{0., 200.}},<>][t]}}
Out[8]= {{xnl[t]→InterpolatingFunction[{{0., 200.}},<>][t]}}
Out[9]=
```

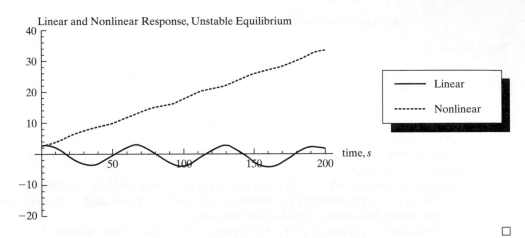

Note from the plots in Example 1.10.4 that even in the case where the initial conditions are small, the linear response is not exactly the same as the full nonlinear system response. However, if the initial velocity is changed to zero, the solutions are very similar. This is illustrated in Figure 1.48.

In summary, nonlinear systems have several interesting aspects that linear systems do not. In particular, nonlinear systems have multiple equilibrium positions rather than just one, as in the linear case. Some of these extra equilibrium points may be unstable and some may be stable. The stability of a response depends on the initial conditions, which can send the solution to different equilibrium positions and hence different types of response. Thus the behavior of the

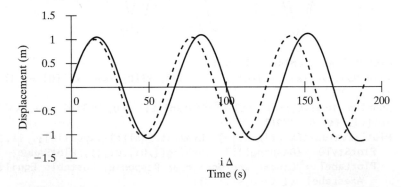

Figure 1.48 Plots of the response of both the linear (dashed line) and nonlinear (solid line) systems of Example 1.10.4 with the initial velocity set to zero in each case.

response depends on the initial conditions, not just the parameters and form of the equation, as is the case for the linear system. This is illustrated in Example 1.10.4.

Even though the response of a nonlinear system is much more complicated and closed-form analytical solutions are not always available, the response can be simulated using numerical integration. In modeling real systems, some degree of nonlinearity is always present. Whether or not it is important to include the nonlinear part of the model in computing, the response depends on the initial conditions. If the initial conditions are such that the system's nonlinearity comes into play, then these terms should be included. Otherwise a linear response is perfectly acceptable. The same can be said for whether or not to include damping in a system model. Which effects to include and which not to include when modeling and analyzing a vibrating system form one of the important aspects of engineering practice.

This section has introduced a little bit about the vibrations of systems with non-linearities. The important points are that nonlinear systems potentially have multiple equilibrium positions, each with potentially different stability behavior. Nonlinear systems typically do not have closed-form solutions so that the time history is often computed by numerical integration. Not addressed here, but nontheless very important, is that the principle of superposition, used extensively in Chapters 3 and 4, does *not* apply to nonlinear systems. The majority of this text focuses on linear vibration problems. With this brief introduction to nonlinear systems, it is important to emphasize that when solving linear problems, initial conditions must be limited such that only the linear range is excited. Students are encouraged to take a course in nonlinear systems and/or nonlinear vibrations to learn more about the analysis and behavior of nonlinear vibrations.

PROBLEMS

Those problems marked with an asterisk are intended to be solved using computational software.

Section 1.1 (Problems 1.1 through 1.26)

1.1. Consider a simple pendulum (see Example 1.1.1) and compute the magnitude of the restoring force if the mass of the pendulum is 3 kg and the length of the pendulum is 0.8 m. Assume the pendulum is at the surface of the earth at sea level.

1.2. Compute the period of oscillation of a pendulum of length 1.2 m at the North Pole where the acceleration due to gravity is measured to be 9.832 m/s^2.

1.3. The spring of Figure 1.2, repeated here as Figure P1.3, is loaded with mass of 10 kg and the corresponding (static) displacement is 0.012 m. Calculate the spring's stiffness.

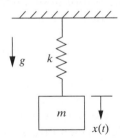

Figure P1.3

1.4. The spring of Figure P1.3 is successively loaded with mass and the corresponding (static) displacement is recorded below. Plot the data and calculate the spring's stiffness. Note that the data contain some errors. Also calculate the standard deviation.

m(kg)	10	11	12	13	14	15	16
x(m)	1.14	1.25	1.37	1.48	1.59	1.71	1.82

1.5. Consider the pendulum of Example 1.1.1 reproduced in Figure P1.5 and compute the amplitude of the restoring force if the mass of the pendulum is 2 kg and the length of the pendulum is 0.5 m if the pendulum is at the surface of the moon.

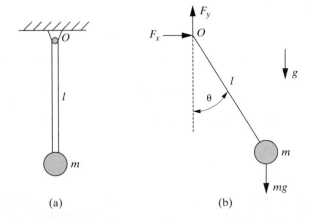

(a)   (b)

Figure P1.5 (a) The pendulum of Example 1.1.1 and (b) its free-body diagram.

1.6. Consider the pendulum of Example 1.1.1 and compute the angular natural frequency (radians per second) of vibration for the linearized system if the mass of the pendulum is 3 kg and the length of the pendulum is 0.8 m if the pendulum is at the surface of the earth. What is the period of oscillation in seconds?

1.7. Derive the solution of $m\ddot{x} + kx = 0$ and plot the result for at least two periods for the case with $\omega_n = 2$ rad/s, $x_0 = 1$ mm, and $v_0 = \sqrt{5}$ mm/s.

1.8. Solve $m\ddot{x} + kx = 0$ for $k = 4\,\text{N/m}, m = 1\,\text{kg}, x_0 = 1\,\text{mm}$, and $v_0 = 0$. Plot the solution.

1.9. The amplitude of vibration of an undamped system is measured to be 1.5 mm. The phase shift from $t = 0$ is measured to be 2 rad and the frequency is found to be 10 rad/s. Calculate the initial conditions that caused this vibration to occur. Assume the response is of the form $x(t) = A \sin(\omega_n t + \phi)$.

1.10. Determine the stiffness of a single-degree-freedom, spring-mass system with a mass of 80 kg such that the natural frequency is 12 Hz.

1.11. Find the equation of motion for the system of Figure P1.11, and find the natural frequency. In particular, using static equilibrium along with Newton's law, determine what effect gravity has on the equation of motion and the system's natural frequency. Assume the block slides without friction.

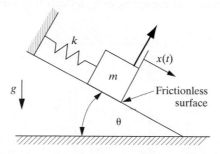

Figure P1.11

1.12. An undamped system vibrates with a frequency of 8 Hz and amplitude 1.5 mm. Calculate the maximum amplitude of the system's velocity and acceleration.

1.13. Show by calculation that $A \sin(\omega_n t + \phi)$ can be represented as $A_1 \sin \omega_n t + A_2 \cos \omega_n t$ and calculate A_1 and A_2 in terms of A and ϕ.

1.14. Using the solution of equation (1.2) in the form $x(t) = A_1 \sin \omega_n t + A_2 \cos \omega_n t$, calculate the values of A_1 and A_2 in terms of the initial conditions x_0 and v_0.

1.15. Using the drawing in Figure 1.7, verify that equation (1.10) satisfies the initial velocity condition.

1.16. A 5 kg mass is attached to a linear spring of stiffness 0.1 N/m. a) Determine the natural frequency of the system in hertz. b) Repeat this calculation for a mass of 50 kg and a stiffness of 10 N/m. Compare your result to that of part a.

1.17. Derive the solution of the single-degree-of-freedom system of Figure 1.4 by writing Newton's law, $ma = -kx$, in differential form using $a\,dx = v\,dv$ and integrating twice.

1.18. Determine the natural frequency of the two systems illustrated in Figure P1.18.

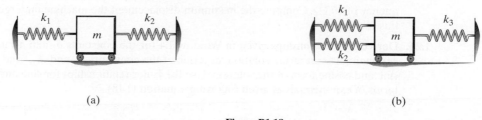

(a)   (b)

Figure P1.18

*1.19. Plot the solution given by equation (1.10) for the case $k = 1000\,\text{N/m}$ and $m = 10\,\text{kg}$ for two complete periods for each of the following sets of initial conditions: a) $x_0 = 0\,\text{m}$, $v_0 = 1\,\text{m/s}$, b) $x_0 = 0.01\,\text{m}$, $v_0 = 0\,\text{m/s}$, and c) $x_0 = 0.01\,\text{m}$, $v_0 = 1\,\text{m/s}$.

*1.20. Make a three dimensional surface plot of the amplitude A of an undamped oscillator given by equation (1.9) versus x_0 and v_0 for the range of initial conditions given by $-0.1 \leq x_0 \leq 0.1\,\text{m}$ and $-1 \leq v_0 \leq 1\,\text{m/s}$ for a system with natural frequency of $10\,\text{rad/s}$.

1.21. A machine part is modeled as a pendulum connected to a spring as illustrated in Figure P1.21. Ignore the mass of the pendulum's rod and derive the equation of motion. Then following the procedure used in Example 1.1.1, linearize the equation of motion and compute the formula for the natural frequency. Assume that the rotation is small enough so that the spring only deflects horizontally.

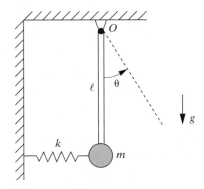

Figure P1.21

1.22. A pendulum has length of 300 mm. What is the system's natural frequency in Hertz?

1.23. The pendulum in Example 1.1.1 (see Figure P1.5) is required to oscillate once every second. What length should it be?

1.24. The approximation of $\sin\theta = \theta$, is reasonable for θ less than $10°$. If a pendulum of length 0.5 m, has an initial position of $\theta(0) = 0$, what is the maximum value of the initial angular velocity that can be given to the pendulum without violating this small-angle approximation? (Be sure to work in radians.)

1.25. A machine, modeled as a simple spring-mass system, oscillates in simple harmonic motion. Its acceleration is measured to have an amplitude of $5{,}000\,\text{mm/s}^2$ with a frequency of 10 Hz. Compute the maximum displacement the machine undergoes during this oscillation.

1.26. Derive the relationships given in Window 1.4 for the constants a_1 and a_2, used in the exponential form of the solution, in terms of the constants A_1 and A_2, used in sum of sine and cosine form of the solution. Use the Euler relationships for sine and cosine in terms of exponentials as given following equation (1.18).

Section 1.2 (Problems 1.27 through 1.40)

1.27. The acceleration of a machine part modeled as a spring–mass system is measured and recorded in Figure P1.27. Compute the amplitude of the displacement of the mass.

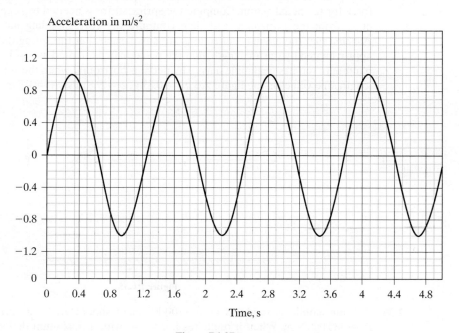

Figure P1.27

1.28. A vibrating spring and mass system has a measured acceleration amplitude of 12 mm/s^2 and measured displacement amplitude of 1.5 mm. Calculate the systems natural frequency.

1.29. A spring-mass system has measured period of 8 seconds and a known mass of 15 kg. Calculate the spring stiffness.

*1.30. Plot the solution of a linear, spring and mass system with frequency $\omega_n = 3$ rad/s, $x_0 = 1.2$ mm and $v_0 = 2.34$ mm/s, for at least two periods.

*1.31. Compute the natural frequency and plot the solution of a spring-mass system with mass of 2 kg and stiffness of 4 N/m, and initial conditions of $x_0 = 1$ mm and $v_0 = 0$ mm/s, for at least two periods.

1.32. When designing a linear spring-mass system it is often a matter of choosing a spring constant such that the resulting natural frequency has a specified value. Suppose that the mass of a system is 5 kg and the stiffness is 100 N/m. How much must the spring stiffness be changed in order to increase the natural frequency by 20%?

1.33. The pendulum in the Chicago Museum of Science and Industry has a length of 20 m, and the acceleration due to gravity at that location is known to be 9.803 m/s^2. Calculate the period of this pendulum.

1.34. Calculate the RMS values of displacement, velocity, and acceleration for the undamped single-degree-of-freedom system of equation (1.19) with zero phase.

1.35. A foot pedal mechanism for a machine is crudely modeled as a pendulum connected to a spring as illustrated in Figure P1.35. The purpose of the spring is provide a return force for the pedal action. Compute the spring stiffness needed to keep the pendulum at $1°$ from the horizontal and then compute the corresponding natural frequency. Assume that the angular deflections are small, such that the spring deflection can be approximated by the arc length, that the pedal may be treated as a point mass and that pendulum rod has negligible mass. The pedal is horizontal when the spring is at its free length. The values in the figure are $m = 0.8\,\text{kg}, g = 9.8\,\text{m/s}^2, l_1 = 0.2\,\text{m}$ and $l_2 = 0.5\,\text{m}$.

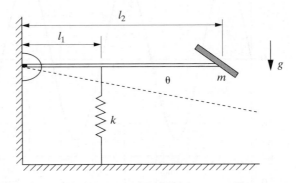

Figure P1.35

1.36. An automobile is modeled as a 1200-kg mass supported by a spring of stiffness $k = 480,000\,\text{N/m}$. When it oscillates it does so with a maximum deflection of 10 cm. When loaded with passengers, the mass increases to as much as 1000 kg. Calculate the change in frequency, velocity amplitude, and acceleration amplitude if the maximum deflection remains 10 cm.

1.37. The front suspension of some cars contains a torsion rod as illustrated in Figure P1.37 to improve the car's handling. (a) Compute the frequency of vibration of the wheel assembly

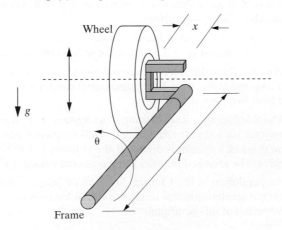

Figure P1.37

given that the torsional stiffness is 2500 N m/rad and the wheel assembly has a mass of 40 kg. Take the distance $x = 0.26$ m. (b) Sometimes owners put different wheels and tires on a car to enhance the appearance or performance. Suppose a thinner tire is put on with a larger wheel raising the mass to 45 kg. What effect does this have on the frequency?

1.38. A machine oscillates in simple harmonic motion and appears to be well modeled by an undamped single-degree-of-freedom oscillation. Its acceleration is measured to have an amplitude of 12,000 mm/s^2 at 8 Hz. What is the machine's maximum displacement?

1.39. A simple undamped spring-mass system is set into motion from rest by giving it an initial velocity of 80 mm/s. It oscillates with a maximum amplitude of 10 mm. What is its natural frequency?

1.40. An automobile exhibits a vertical oscillating displacement of maximum amplitude 5 cm and a measured maximum acceleration of 2000 cm/s^2. Assuming that the automobile can be modeled as a single-degree-of-freedom system in the vertical direction, calculate the natural frequency of the automobile.

Section 1.3 (Problems 1.41 through 1.64)

1.41. Consider a spring mass damper system, like the one in Figure 1.9, with the following values: $m = 12$ kg, $c = 4$ N/s and $k = 1200$ N/m. a) Is the system overdamped, underdamped or critically damped? b) Compute the solution if the system is given initial conditions $x_0 = 0.01$ m and $v_0 = 0$.

1.42. Consider a spring–mass–damper system with equation of motion given by $\ddot{x} + 2x + 2\dot{x} = 0$. Compute the damping ratio and determine if the system is overdamped, underdamped, or critically damped.

1.43. Consider the system for $\ddot{x} + 4\dot{x} + x = 0$ for $x_0 = 1.2$ mm, $v_0 = 0$ mm/s. Is this system overdamped, underdamped or critically damped? Compute the solution and determine which root dominates as time goes on (that is, one root will die out quickly and the other will persist).

1.44. Compute the solution to $\ddot{x} + 2\dot{x} + 2x = 0$ for $x_0 = 0$ mm, $v_0 = 1$ mm/s and write down the closed-form expression for the response.

1.45. Derive the form of λ_1 and λ_2 given by equation (1.31) from equation (1.28) and the definition of the damping ratio.

1.46. Use the Euler formulas to derive equation (1.36) from equation (1.35) and to determine the relationships listed in Window 1.4.

1.47. Using equation (1.35) as the form of the solution of the underdamped system, calculate the values for the constants a_1 and a_2 in terms of the initial conditions x_0 and v_0.

1.48. Calculate the constants A and ϕ in terms of the initial conditions and thus verify equation (1.38) for the underdamped case.

1.49. Calculate the constants a_1 and a_2 in terms of the initial conditions and thus verify equations (1.42) and (1.43) for the overdamped case.

1.50. Calculate the constants a_1 and a_2 in terms of the initial conditions and thus verify equation (1.46) for the critically damped case.

1.51. Using the definition of the damping ratio and the undamped natural frequency, derive equation (1.48) from (1.47).

1.52. For a damped system, m, c, and k are known to be $m = 1.5\,\text{kg}$, $c = 2.2\,\text{kg/s}$, $k = 12\,\text{N/m}$. Calculate the value of ζ and ω_n. Is the system overdamped, underdamped, or critically damped?

1.53. Plot $x(t)$ for a damped system of natural frequency $\omega_n = 2\,\text{rad/s}$ and initial conditions $x_0 = 1\,\text{mm}$, $v_0 = 1\,\text{mm}$, for the following values of the damping ratio:

$$\zeta = 0.01, \zeta = 0.2, \zeta = 0.1, \zeta = 0.4, \text{ and } \zeta = 0.8.$$

1.54. Plot the response $x(t)$ of an underdamped system with $\omega_n = 2\,\text{rad/s}$, $\zeta = 0.1$, and $v_0 = 0$ for the following initial displacements: $x_0 = 10\,\text{mm}$ and $x_0 = 100\,\text{mm}$.

1.55. Calculate the solution to $\ddot{x} + \dot{x} + x = 0$ with $x_0 = 1$ and $v_0 = 0$ for $x(t)$ and sketch the response.

1.56. A spring-mass-damper system has mass of 120 kg, stiffness of 3600 N/m and damping coefficient of 330 kg/s. Calculate the undamped natural frequency, the damping ratio and the damped natural frequency. Does the solution oscillate?

1.57. A rough sketch of a valve-and-rocker-arm system for an internal combustion engine is give in Figure P1.57. Model the system as a pendulum attached to a spring and a mass and assume the oil provides viscous damping in the range of $\zeta = 0.01$. Determine the equations of motion and calculate an expression for the natural frequency and the damped natural frequency. Here J is the rotational inertia of the rocker arm about its pivot point, k is the stiffness of the valve spring, and m is the mass of the valve and stem. Ignore the mass of the spring.

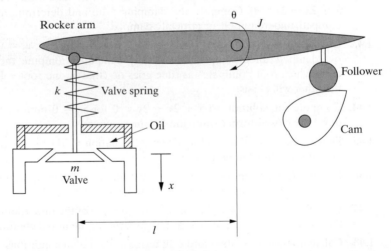

Figure P1.57

1.58. A spring-mass-damper system has mass of 160 kg, stiffness of 2000 N/m and damping coefficient of 250 kg/s. Calculate the undamped natural frequency, the damping ratio and the damped natural frequency. Is the system overdamped, underdamped or critically damped? Does the solution oscillate?

*1.59. The spring mass system of 120 kg mass, stiffness of 3200 N/m and damping coefficient of 350 Ns/m is given a zero initial velocity and an initial displacement of 0.12 m. Calculate the form of the response and plot it for as long as it takes to die out.

*1.60. The spring mass system of 180 kg mass, stiffness of 1800 N/m and damping coefficient of 250 Ns/m is given an initial velocity of 12 mm/s and an initial displacement of − 6 mm. Calculate the form of the response and plot it for as long as it takes to die out. How long does it take to die out?

*1.61. Choose the damping coefficient of a spring-mass-damper system with mass of 160 kg and stiffness of 2500 N/m such that it's response will die out after about 2 s, given a zero initial position and an initial velocity of 12 mm/s.

1.62. Derive the equation of motion of the system in Figure P1.62 and discuss the effect of gravity on the natural frequency and the damping ratio.

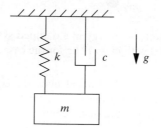

Figure P1.62

1.63. Derive the equation of motion of the system in Figure P1.63 and discuss the effect of gravity on the natural frequency and the damping ratio. You may have to make some approximations of the cosine. Assume the bearings provide a viscous-damping force only in the vertical direction. (From A. Diaz-Jimenez, *South African Mechanical Engineer*, Vol. 26, pp. 65–69, 1976) (1976)

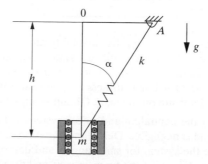

Figure P1.63

1.64. Consider the response of an underdamped system given by

$$x(t) = e^{-\zeta\omega_n t}A \sin(\omega_d t + \phi)$$

where A and ϕ are given in terms of the initial conditions $x_0 = 0$, and $v_0 \neq 0$. Determine the maximum value that the acceleration will experience in terms of v_0.

Section 1.4 (Problems 1.65 through 1.81)

1.65. Calculate the frequency of the compound pendulum of Figure P1.65 if a mass m_T is added to the tip, by using the energy method. Assume the mass of the pendulum is evenly distributed so that its center of gravity is in the middle of the pendulum of length l.

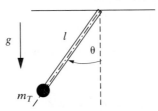

Figure P1.65 A compound pendulum with a tip mass.

1.66. Calculate the total energy in a damped system with frequency 2.5 rad/s and damping ratio $\zeta = 0.02$ with mass 12 kg for the case $x_0 = 0.1$ m and $v_0 = 0$. Plot the total energy versus time.

1.67. Use the energy method to calculate the equation of motion and natural frequency of an airplane's steering mechanism for the nose wheel of its landing gear. The mechanism is modeled as the single-degree-of-freedom system illustrated in Figure P1.67.

(Steering wheel)

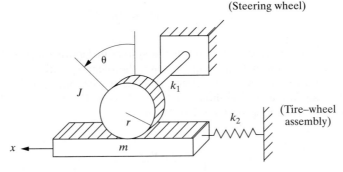

Figure P1.67

The steering wheel and tire assembly are modeled as being fixed at ground for this calculation. The steering rod gear system is modeled as a linear spring–mass system (m, k_2) oscillating in the x direction. The shaft-gear mechanism is modeled as the disk of inertia J and torsional stiffness k_2. The gear J turns through the angle θ such that the disk does not slip on the mass. Obtain an equation in the linear motion x.

1.68. Consider the pendulum-and-spring system of Figure P1.68. Here the mass of the pendulum rod is negligible. Derive the equation of motion using the energy method. Then linearize the system for small angles and determine the natural frequency. The length of the pendulum is l, the tip mass is m, and the spring stiffness is k.

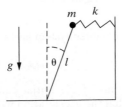

Figure P1.68 A simple pendulum connected to a spring.

1.69. A control pedal of an aircraft can be modeled as the single-degree-of-freedom system of Figure P1.69. Consider the lever as a massless shaft and the pedal as a lumped mass at the end of the shaft. Use the energy method to determine the equation of motion in θ and calculate the natural frequency of the system. Assume the spring to be unstretched at θ = 0 and gravity points down.

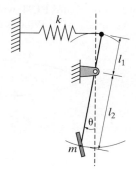

Figure P1.69

1.70. To save space, two large pipes are shipped, one stacked inside the other, as indicated in Figure P1.70. Calculate the natural frequency of vibration of the smaller pipe (of radius R_1) rolling back and forth inside the larger pipe (of radius R). Use the energy method and assume that the inside pipe rolls without slipping and has a mass of m.

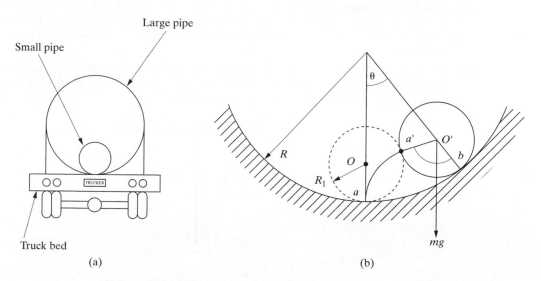

Figure P1.70 (a) Pipes stacked in a truck bed. (b) Vibration model of the inside pipe.

1.71. Consider the example of a simple pendulum given in Example 1.4.2. The pendulum motion is observed to decay with a damping ratio of ζ = 0.005. Determine a damping coefficient and add a viscous damping term to the pendulum equation.

1.72. Determine a damping coefficient for the disk-rod system of Example 1.4.3. Assuming that the damping is due to the material properties of the rod, determine c for the rod if it is observed to have a damping ratio of $\zeta = 0.008$.

1.73. The rod and disk of Window 1.1 are in torsional vibration. Calculate the damped natural frequency if $J = 1200\,\text{m}^2 \cdot \text{kg}$, $c = 25\,\text{N} \cdot \text{m} \cdot \text{s/rad}$, and $k = 500\,\text{N} \cdot \text{m/rad}$.

1.74. Consider the system of Figure P1.74, which represents a simple model of an aircraft landing system. Assume, $x = r\theta$. What is the damped natural frequency?

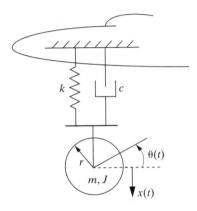

Figure P1.74

1.75. Consider Problem 1.74 with $k = 500{,}000\,\text{N/m}$, $m = 2000\,\text{kg}$, $J = 200\,\text{m}^2 \cdot \text{kg/rad}$, $r = 30\,\text{cm}$, and $c = 8000\,\text{kg/s}$. Calculate the damping ratio and the damped natural frequency. How much effect does the rotational inertia have on the undamped natural frequency?

1.76. Use Lagrange's formulation to calculate the equation of motion and the natural frequency of the system of Figure P1.76. Model each of the brackets as a spring of stiffness k, and assume the inertia of the pulleys is negligible.

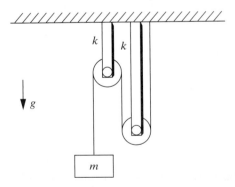

Figure P1.76

1.77. Use Lagrange's formulation to calculate the equation of motion and the natural frequency of the system of Figure P1.77. This figure represents a simplified model of a jet engine mounted to a wing through a mechanism which acts as a spring of stiffness

k and mass m_s. Assume the engine has inertia J and mass m and that the rotation of the engine is related to the vertical displacement of the engine, $x(t)$ by the "radius" r_0 (i.e., $x = r_0\theta$).

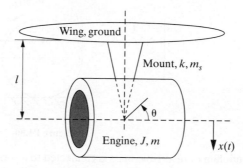

Figure P1.77

1.78. Consider the inverted simple pendulum connected to a spring of Figure P1.68. Use Lagrange's formulation to derive the equation of motion.

1.79. Lagrange's formulation can also be used for non-conservative systems by adding the applied non-conservative term to the right side of equation (1.63) to get

$$\frac{d}{dt}\left(\frac{\partial T}{\partial \dot{q}_i}\right) - \frac{\partial T}{\partial q_i} + \frac{\partial U}{\partial q_i} + \frac{\partial R_i}{\partial \dot{q}_i} = 0$$

Here R_i is the *Rayleigh dissipation function* defined in the case of a viscous damper attached to ground by

$$R_i = \frac{1}{2}c\dot{q}_i^2$$

Use this extended Lagrange formulation to derive the equation of motion of the damped automobile suspension driven by a dynamometer illustrated in Figure P1.79. Assume here that the dynamometer drives the system such that $x = r\theta$.

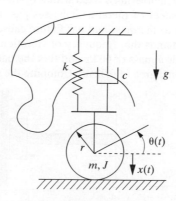

Figure P1.79

1.80. Consider the disk of Figure P1.80 connected to two springs. Use the energy method to calculate the system's natural frequency of oscillation for small angles $\theta(t)$.

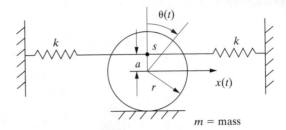

Figure P1.80

1.81. A pendulum of negligible mass is connected to a spring of stiffness k at halfway along its length, l, as illustrated in Figure P1.81. The pendulum has two masses fixed to it: one at the connection point with the spring and one at the top. Derive the equation of motion using the Lagrange formulation, linearize the equation, and compute the system's natural frequency. Assume that the angle remains small enough so that the spring only stretches significantly in the horizontal direction.

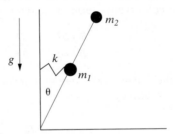

Figure P1.81

Section 1.5 (Problems 1.82 through 1.93)

1.82. A bar of negligible mass fixed with a tip mass forms part of a machine used to punch holes in a sheet of metal as it passes past the fixture as illustrated in Figure P1.82. The impact to the mass and bar fixture causes the bar to vibrate and the speed of the process demands that frequency of vibration not interfere with the process. The static design yields a mass of 60 kg and that the bar be made of steel of length 0.30 m with a cross sectional area of 0.02 m^2. Compute the system's natural frequency.

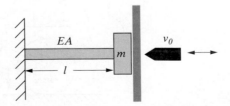

Figure P1.82   A bar model of a punch fixture.

1.83. Consider the punch fixture of Figure P1.82. If the system is giving an initial velocity of 8 m/s, what is the maximum displacement of the mass at the tip if the mass is 1200 kg and the bar is made of steel of length 0.40 m with a cross sectional area of 0.012 m^2?

1.84. Consider the punch fixture of Figure P1.82. If the punch strikes the mass off center it is possible that the steel bar may vibrate in torsion. The mass is 1200 kg and the bar 0.20 m-long, with a square cross section of 0.12 m on a side. The mass polar moment of inertia of the tip mass is 12 kg/m^2. The polar moment of inertia for a square bar is $b^4/6$, where b is the length of the side of the square. Compute both the torsion and longitudinal frequencies. Which is larger?

1.85. A helicopter landing gear consists of a metal framework rather than the coil spring based suspension system used in a fixed-wing aircraft. The vibration of the frame in the vertical direction can be modeled by a spring made of a slender bar as illustrated in Figure 1.23, where the helicopter is modeled as ground. Here $l = 0.42$ m, $E = 20 \times 10^{10}$ N/m^2, and $m = 120$ kg. Calculate the cross-sectional area that should be used if the natural frequency is to be $f_n = 520$ Hz.

1.86. The frequency of oscillation of a person on a diving board can be modeled as the transverse vibration of a beam as indicated in Figure 1.26. Let m be the mass of the diver ($m = 80$ kg) and $l = 1.5$ m. If the diver wishes to oscillate at 4 Hz, what value of EI should the diving board material have?

1.87. Consider the spring system of Figure 1.32. Let $k_1 = k_5 = k_2 = 80$ N/m, $k_3 = 40$ N/m, and $k_4 = 10$ N/m. What is the equivalent stiffness?

1.88. Springs are available in stiffness values of 10, 100, and 1000 N/m. Design a spring system using these values only, so that a 120-kg mass is connected to ground with frequency of about 1.5 rad/s.

1.89. Calculate the natural frequency of the system in Figure 1.32(a) if $k_1 = k_2 = 0$. Choose m and nonzero values of $k_3, k_4,$ and k_5 so that the natural frequency is 80 Hz.

*1.90. Example 1.4.4 examines the effect of the mass of a spring on the natural frequency of a simple spring–mass system. Use the relationship derived there and plot the natural frequency (normalized by the natural frequency, ω_n, for a massless spring) versus the percent that the spring mass is of the oscillating mass. Determine from the plot (or by algebra) the percentage where the natural frequency changes by 1% and therefore the spring's mass should not be neglected.

1.91. Calculate the natural frequency and damping ratio for the system in Figure P1.91 given the values $m = 20$ kg, $c = 120$ kg/s, $k_1 = 4800$ N/m, $k_2 = 400$ N/m and $k_3 = 800$ N/m. Assume that no friction acts on the rollers. Is the system overdamped, critically damped or underdamped?

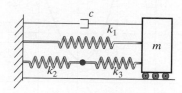

Figure P1.91

1.92. Calculate the natural frequency and damping ratio for the system in Figure P1.92. Assume that no friction acts on the rollers. Is the system overdamped, critically damped, or underdamped?

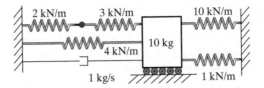

Figure P1.92

1.93. A manufacturer makes a cantilevered leaf spring from steel ($E = 2.1 \times 10^{11}\,\text{N/m}^2$) and sizes the spring so that the device has a specific frequency. Later, to save weight, the spring is made of aluminum ($E = 7.1 \times 10^{10}\,\text{N/m}^2$). Assuming that the mass of the spring is much smaller than that of the device the spring is attached to, determine if the frequency increases or decreases and by how much.

Section 1.6 (Problems 1.94 through 1.101)

1.94. The displacement of a vibrating spring–mass–damper system is recorded on an $x - y$ plotter and reproduced in Figure P1.94. The y coordinate is the displacement in cm and the x coordinate is time in seconds. From the plot determine the natural frequency, the damping ratio, and the damped natural frequency.

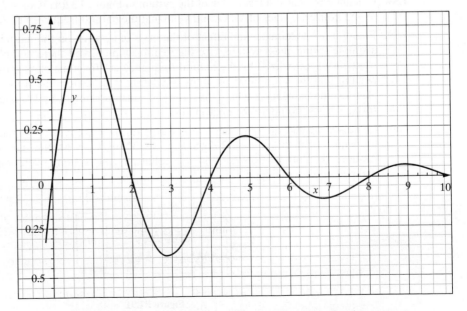

Figure P1.94   A plot of displacement versus time for a vibrating system.

1.95. Show that the logarithmic decrement is equal to

$$\delta = \frac{1}{n} \ln \frac{x_0}{x_n}$$

where x_n is the amplitude of vibration after n cycles have elapsed.

1.96. Derive the equation (1.78) for the trifalar suspension system.

1.97. A prototype composite material is formed and hence has an unknown modulus. An experiment is performed consisting of forming it into a cantilevered beam of length 1.2 m and $I = 10^{-9} \text{m}^4$ with a 6.2-kg mass attached at its end. The system is given an initial displacement and found to oscillate with a period of 0.5 s. Calculate the modulus E.

1.98. The free response of a 1200-kg car with stiffness of $k = 500,000 \text{N/m}$ is observed to be of the form given in Figure 1.35. Modeling the car as a single-degree-of-freedom oscillation in the vertical direction, determine the damping coefficient if the displacement at t_1 is measured to be 2 cm and 0.22 cm at t_2.

1.99. A pendulum decays from 8 cm to 2 cm over one period. Determine its damping ratio.

1.100. The relationship between the log decrement δ and the damping ratio ζ is often approximated as $\delta = 2\pi\zeta$. For what values of ζ would you consider this a good approximation to equation (1.82)?

1.101. A damped system is modeled as illustrated in Figure 1.9. The mass of the system is measured to be 6 kg and its spring constant is measured to be 6000 N/m. It is observed that during free vibration the amplitude decays to 0.25 of its initial value after five cycles. Calculate the viscous damping coefficient, c.

Section 1.7 (Problems 1.102 through 1.110, also see Problem Section 1.5)

1.102. Consider the system of Example 1.7.2 consisting of a helical spring of stiffness 10^3 N/m attached to a 12-kg mass Place a dashpot parallel to the spring and choose its viscous damping value so that the resulting damped natural frequency is reduced to 9 rad/s.

1.103. For an underdamped system, $x_0 = 0 \text{mm}$ and $v_0 = 8 \text{mm/s}$. Determine m, c, and k such that the amplitude is less than 1 mm.

1.104. Repeat Problem 1.103 if the mass is restricted to be between $10 \text{kg} < m < 15 \text{kg}$.

1.105. Use the formula for the torsional stiffness of a shaft from Table 1.1 to design a 1-m shaft with torsional stiffness of 10^5 N·m/rad.

1.106. Consider designing a helical spring made of aluminum, such that when it is attached to a 12-kg mass the resulting spring-mass system has a natural frequency of 10 rad/s. Thus repeat Example 1.7.2 which uses steel for the spring and note any difference.

1.107. Try to design a bar that has the same stiffness as the helical spring of Example 1.7.2 (i.e., $k = 10^3 \text{N/m}$). This amounts to computing the length of the bar with its' cross sectional area taking up about the same space at the helical spring ($R = 8 \text{cm}$). Note that the bar must remain at least 12 times as long as it is wide in order to be modeled by the stiffness formula given for the bar in Figure 1.23.

1.108. Repeat Problem 1.107 using plastic ($E = 1.40 \times 10^9$ N/m^2) and rubber ($E = 7 \times 10^6$ N/m^2). Are any of these feasible?

1.109. Consider the diving board of Figure P1.109. For divers, a certain level of static deflection is desirable, denoted by Δ. Compute a design formula for the dimensions of the board (b, h, and l) in terms of the static deflection, the average diver's mass, m, and the modulus of the board.

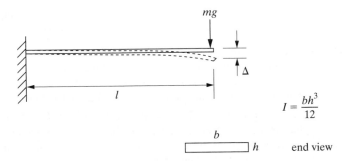

$$I = \frac{bh^3}{12}$$

end view

Figure P1.109

1.110. In designing a vehicle suspension system using a "quarter car model" consisting of a spring, mass and damper system, studies show that a desirable damping ratio is $\zeta = 0.2$. If the model has a mass of 800 kg and a frequency of 16 Hz, what should the damping coefficient be?

Section 1.8 (Problems 1.111 through 1.115)

1.111. Consider the system of Figure P1.111. (a) Write the equations of motion in terms of the angle, θ, the bar makes with the vertical. Assume linear deflections of the springs and linearize the equations of motion. (b) Discuss the stability of the linear system's solutions in terms of the physical constants, m, k, and l. Assume the mass of the rod acts at the center as indicated in the figure.

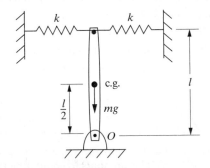

Figure P1.111

1.112. Consider the inverted pendulum of Figure 1.40 as discussed in Example 1.8.1 and repeated in Figure P1.112. Assume that a dashpot (of damping rate c) also acts on the pendulum parallel to the two springs. How does this affect the stability properties of the pendulum?

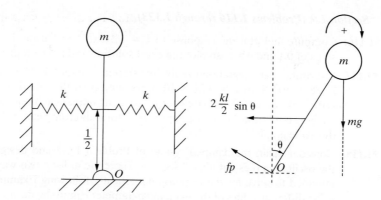

Figure P1.112   The inverted pendulum of Example 1.8.1.

1.113. Replace the massless rod of the inverted pendulum of Figure P1.112 with a solid object compound pendulum of Figure 1.20(b). Calculate the equations of vibration and discuss values of the parameter relations for which the system is stable.

1.114. A simple model of a control tab for an airplane is sketched in Figure P1.114. The equation of motion for the tab about the hinge point is written in terms of the angle θ from the centerline to be

$$J\ddot{\theta} + (c - f_d)\dot{\theta} + k\theta = 0$$

Here J is the moment of inertia of the tab, k is the rotational stiffness of the hinge, c is the rotational damping in the hinge, and $f_d\dot{\theta}$ is the negative damping provided by the aerodynamic forces (indicated by arrows in the figure). Discuss the stability of the solution in terms of the parameters c and f_d.

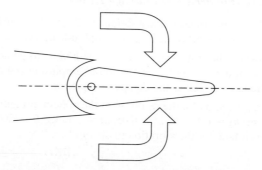

Figure P1.114   A simple model of an airplane control tab.

*1.115. In order to understand the effect of damping in design, develop some sense of how the response changes with the damping ratio by plotting the response of a single-degree-of-freedom system for a fixed amplitude, frequency, and phase as ζ changes through the following set of values $\zeta = 0.01, 0.05, 0.1, 0.2, 0.3,$ and 0.4. That is, plot the response $x(t) = e^{-10\zeta t} \sin(10\sqrt{1 - \zeta^2} t)$ for each value of ζ.

Section 1.9 (Problems 1.116 through 1.123)

*1.116. Compute and plot the response to $\dot{x} = -3x, x(0) = 1$ using Euler's method for time steps of 0.1 and 0.5. Also plot the exact solution and hence reproduce Figure 1.41.

*1.117. Use numerical integration to solve the system of Example 1.7.3 with $m = 1361$ kg, $k = 2.688 \times 10^5$ N/m, $c = 3.81 \times 10^3$ kg/s subject to the initial conditions $x(0) = 0$ and $v(0) = 0.01$ mm/s. Compare your result using numerical integration to just plotting the analytical solution (using the appropriate formula from Section 1.3) by plotting both on the same graph.

*1.118. Consider again the damped system of Problem 1.117 and design a damper such that the oscillation dies out after 2 seconds. There are at least two ways to do this. Here it is intended to solve for the response numerically, following Examples 1.9.2, 1.9.3, or 1.9.4, using different values of the damping parameter c until the desired response is achieved.

*1.119. Consider again the damped system of Example 1.9.2 and design a damper such that the oscillation dies out after 25 seconds. There are at least two ways to do this. Here it is intended to solve for the response numerically, following Examples 1.9.2, 1.9.3, or 1.9.4, using different values of the damping parameter c until the desired response is achieved. Is your result overdamped, underdamped, or critically damped?

*1.120. Repeat Problem 1.119 for the initial conditions $x(0) = 0.1$ m and $v(0) = 0.01$ mm/s.

*1.121. A spring and damper are attached to a mass of 80 kg in the arrangement given in Figure 1.9. The system is given the initial conditions $x(0) = 0.1$ m and $v(0) = 1$ mm/s. Design the spring and damper (i.e. choose k and c) such that the system will come to rest in 2 s and not oscillate more than two complete cycles. Try to keep c as small as possible. Also compute ζ.

*1.122. Repeat Example 1.7.1 by using the numerical approach of the previous 5 problems.

*1.123. Repeat Example 1.7.1 for the initial conditions $x(0) = 0.01$ m and $v(0) = 1$ mm/s.

Section 1.10 (Problems 1.124 through 1.136)

1.124. A 4-kg mass connected to a spring of stiffness 10^3 N/m has a dry sliding friction force (F_c) of 3 N. As the mass oscillates, its amplitude decreases 10 cm. How long does this take?

1.125. Consider the system of Figure 1.41 with $m = 6$ kg and $k = 8 \times 10^3$ N/m with a friction force of magnitude 6 N. If the initial amplitude is 4 cm, determine the amplitude one cycle later as well as the damped frequency.

*1.126. Compute and plot the response of the system of Figure P1.126 for the case where $x_0 = 0.1$ m, $v_0 = 0.1$ m/s, $\mu = 0.05$, $m = 250$ kg, $\theta = 20°$, and $k = 3000$ N/m. How long does it take for the vibration to die out?

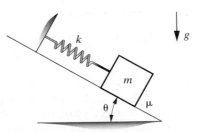

Figure P1.126

*1.127. Compute and plot the response of a system with Coulomb damping of equation (1.90) for the case where $x_0 = 0.5$ m, $v_0 = 0$, $\mu = 0.1$, $m = 80$ kg and $k = 2000$ N/m. How long does it take for the vibration to die out?

*1.128. A mass moves in a fluid against sliding friction as illustrated in Figure P1.106. Model the damping force as a slow fluid (i.e., linear viscous damping) plus Coulomb friction because of the sliding, with the following parameters: $m = 240$ kg, $\mu = 0.01$, $c = 25$ kg/s and $k = 3200$ N/m. a) Compute and plot the response to the initial conditions: $x_0 = 0.1$ m, $v_0 = 0.1$ m/s. b) Compute and plot the response to the initial conditions: $x_0 = 0.1$ m, $v_0 = 1$ m/s. How long does it take for the vibration to die out in each case?

Figure P1.128

*1.129. Consider the system of Problem 1.128 part (a), and compute a new damping coefficient, c, that will cause the vibration to die out after one oscillation.

1.130. Compute the equilibrium positions of $\ddot{x} + \omega_n^2 x + \beta x^2 = 0$. How many are there?

1.131. Compute the equilibrium positions of $\ddot{x} + \omega_n^2 x - \beta^2 x^3 + \gamma x^5 = 0$. How many are there?

*1.132. Consider the pendulum of Example 1.10.3 with length of 1 m and initial conditions of $\theta_0 = \pi/10$ rad and $\dot{\theta}_0 = 0$. Compare the difference between the response of the linear version of the pendulum equation (i.e., with $\sin(\theta) = \theta$) and the response of the nonlinear version of the pendulum equation by plotting the response of both for four periods.

*1.133. Repeat Problem 1.132 if the initial displacement is $\theta_0 = \pi/2$ rad.

1.134. If the pendulum of Example 1.10.3 is given an initial condition near the equilibrium position of $\theta_0 = \pi$ rad and $\dot{\theta}_0 = 0$, does it oscillate around this equilibrium?

*1.135. Calculate the response of the system of Problem 1.121 for the initial conditions of $x_0 = 0.02$ m, $v_0 = 0$, and a natural frequency of 4 rad/s and for $\beta = 100$, $\gamma = 0$.

*1.136. Repeat Problem 1.135 and plot the response of the linear version of the system ($\beta = 0$) on the same plot to compare the difference between the linear and nonlinear versions of this equation of motion.

MATLAB® ENGINEERING VIBRATION TOOLBOX

Dr. Joseph C. Slater of Wright State University has authored a MATLAB Toolbox keyed to this text. The *Engineering Vibration Toolbox* (EVT) is organized by chapter and may be used to solve the Toolbox problems found at the end of each chapter. In addition, the EVT may be used to solve those homework problems suggested for computer usage in Sections 1.9 and 1.10, rather than using MATLAB directly. MATLAB and the EVT are interactive and are intended to assist in analysis, parametric studies, and design, as well as in solving homework problems. The Engineering Vibration Toolbox is licensed free of charge for educational use. For professional use, users should contact the Engineering Vibration Toolbox author directly.

The EVT is updated and improved regularly and can be downloaded for free. To download, update, or obtain information on usage or current revision, go to the Engineering Vibration Toolbox home page at

http://www.cs.wright.edu/~vtoolbox

This site includes links to editions that run on earlier versions of MATLAB, as well as the most recent version. An email list of instructors who use the EVT is maintained so users can receive email notification of the latest updates. The EVT is designed to run on any platform supported by MATLAB (including Macintosh and VMS) and is regularly updated to maintain compatibility with the current version of MATLAB. A brief introduction to MATLAB and UNIX is available on the home page as well. Please read the file Readme.txt to get started and type help vtoolbox after installation to obtain an overview. Once it is installed, typing vtbud will display the current revision status of your installation and attempt to download the current revision status from the anonymous FTP site. Updates can then be downloaded incrementally as desired. Please see Appendix G for further information.

TOOLBOX PROBLEMS

TB1.1. Fix [your choice or use the values from Example 1.3.1 with $x(0) = 1$ mm] the values of m, c, k, and $x(0)$ and plot the responses $x(t)$ for a range of values of the initial velocity $\dot{x}(0)$ to see how the response depends on the initial velocity. Remember to use numbers with consistent units.

TB1.2. Using the values from Problem TB1.1 and $\dot{x}(0) = 0$, plot the response $x(t)$ for a range of values of $x(0)$ to see how the response depends on the initial displacement.

TB1.3. Reproduce Figures 1.10, 1.11, and 1.12.

TB1.4. Consider solving Problem 1.53 and compare the time for each response to reach and stay below 0.01 mm.

TB1.5. Solve Problems 1.121, 1.122, and 1.123 using the Engineering Vibration Toolbox.

TB1.6. Solve Problems 1.126, 1.127, and 1.128 using the Engineering Vibration Toolbox.

2 Response to Harmonic Excitation

This chapter focuses on the most fundamental concept in vibration analysis: the concept of resonance. Resonance occurs when a periodic external force is applied to a system having a natural frequency equal to the driving frequency. This often happens when the excitation force is derived from some rotating part, such as the helicopter shown in the top left. The rotating blade causes a harmonic force to be applied to the body of the helicopter. If the frequency of the blade rotation corresponds to the natural frequency of the body, resonance will occur as described in Section 2.1. Resonance causes large deflections, which may exceed the elastic limits and cause the structure to fail. An example familiar to most is the resonance caused by an out-of-balance tire on a car (bottom photo). The speed of tire rotation corresponds to the driving frequency. At a certain speed, the out-of-balance tire causes resonance, which is felt as shaking of the steering-wheel column. If the car is driven slower, or faster, the frequency moves away from the resonance condition and the shaking stops.

This chapter considers the response to harmonic excitation of the single-degree-of-freedom spring–mass–damper system presented in Chapter 1. Harmonic excitation refers to a sinusoidal external force of a single frequency applied to the system. Recall from introductory physics that resonance is the tendency of a system to absorb more energy when the driving frequency matches the system's natural frequency of vibration. This phenomenon commonly occurs in mechanical, acoustic, biological, and electrical systems. Examples include acoustic resonance in musical instruments, the tidal resonance in bays, basilar membranes in biological transduction of auditory input, and shaking of the front suspension of a car caused by an out-of-balance wheel. As a child, one discovers resonance when learning to swing on a playground swing (modeled as the pendulum considered in Chapter 1).

Harmonic excitations are a common source of external force applied to machines and structures. Rotating machines such as fans, electric motors, and reciprocating engines transmit a sinusoidally varying force to adjacent components. In addition, the Fourier theorem indicates that many other forcing functions can be expressed as an infinite series of harmonic terms. Since the equations of motion considered here are linear, knowing the response to individual terms in the series allows the total response to be represented as the sum of the response to the individual terms. This is the principle of superposition. In this way, knowing the response to a single harmonic input allows the calculation of the response to a variety of other input disturbances of periodic nature. General periodic disturbances are discussed in Chapter 3.

A harmonic input is also chosen for study because it can be solved mathematically with straightforward techniques. In addition, the response of a single-degree-of-freedom system to a harmonic input forms the foundation of vibration measurement, the design of devices intended to protect machines from unwanted oscillation, and the design of transducers used in measuring vibration. Harmonic excitations are simple to produce in laboratories, hence they are very useful in studying damping and stiffness properties.

2.1 HARMONIC EXCITATION OF UNDAMPED SYSTEMS

Consider the system of Figure 2.1 for the case of negligible damping ($c = 0$). There are several ways to model the harmonic nature of the applied force, $F(t)$. A harmonic function can be represented as a sine, a cosine, or a complex exponential. In the following, the driving force $F(t)$ is chosen to be of the form

$$F(t) = F_0 \cos \omega t \qquad (2.1)$$

where F_0 represents the magnitude, or maximum amplitude, of the applied force and ω denotes the frequency of the applied force. The frequency ω is also called the *input frequency*, or *driving frequency*, or *forcing frequency* and has units of rad/s. Note that some texts use Ω rather than ω to denote the driving frequency.

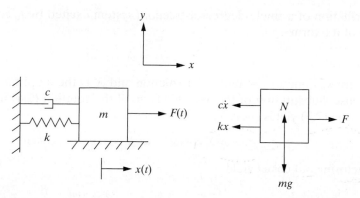

Figure 2.1 A schematic of a single-degree-of-freedom system acted on by an external force, $F(t)$, and sliding on a friction-free surface. The figure on the right is a free-body diagram of the friction-free spring–mass–damper system.

Alternately, the harmonic forcing function can be represented as the sinusoid

$$F(t) = F_0 \sin \omega t$$

or as the complex exponential

$$F(t) = F_0 e^{j\omega t}$$

where j is the imaginary unit. Each of these three forms of $F(t)$ yields the same phenomenon, but in some cases one form may be easier to manipulate than others. Each is used in the following.

From the right side of Figure 2.1, the sum of the forces in the y direction yields $N = mg$, with the result of no motion in that direction. Summing forces on the mass of Figure 2.1 in the x direction for the undamped case yields the result that the displacement $x(t)$ must satisfy

$$m\ddot{x}(t) + kx(t) = F_0 \cos \omega t \tag{2.2}$$

where the harmonic driving force is represented by the cosine function. Note that this expression is a linear equation in the variable $x(t)$. As in the homogeneous (unforced) case of Chapter 1, it is convenient to divide this expression by the mass, m, to yield

$$\ddot{x}(t) + \omega_n^2 x(t) = f_0 \cos \omega t \tag{2.3}$$

where $f_0 = F_0/m$. The magnitude f_0 is called the mass-normalized force with units N/kg. A variety of techniques can be used to solve this equation, which are commonly studied in a first course in differential equations.

First, recall from the study of differential equations that equation (2.3) is a linear nonhomogeneous equation and that its solution is therefore the sum of the homogeneous solution (i.e., the solution for the case $f_0 = 0$) and a particular solution. The particular solution can often be found by assuming that it has the same form as the forcing function. This is also consistent with observation. That is, the oscillation of a single-degree-of-freedom system excited by $f_0 \cos \omega t$ is observed to be of the form

$$x_p(t) = X \cos \omega t \tag{2.4}$$

where x_p denotes the particular solution and X is the amplitude of the forced response. Substitution of the assumed form of the solution (2.4) into the equation of motion (2.3) yields (note $\ddot{x}_p = -\omega^2 X \cos \omega t$)

$$-\omega^2 X \cos \omega t + \omega_n^2 X \cos \omega t = f_0 \cos \omega t \tag{2.5}$$

Factoring out $\cos \omega t$ yields

$$(-\omega^2 X + \omega_n^2 X - f_0) \cos \omega t = 0$$

Since $\cos \omega t$ cannot be zero for all $t > 0$, the coefficient of $\cos \omega t$ must vanish. Setting the coefficient to zero and solving for X yields

$$X = \frac{f_0}{\omega_n^2 - \omega^2} \tag{2.6}$$

provided that $\omega_n \neq \omega$. Thus, as long as the driving frequency and natural frequency are different (i.e., as long as $\omega_n \neq \omega$), the particular solution will be of the form

$$x_p(t) = \frac{f_0}{\omega_n^2 - \omega^2} \cos \omega t \tag{2.7}$$

This approach, of assuming that $x_p = X \cos \omega t$, to determine the particular solution is called the *method of undetermined coefficients* in differential equations courses.

Since the system is linear, the total solution $x(t)$ is the sum of the particular solution of equation (2.7) plus the homogeneous solution given by equation (1.19). Recalling that $A \sin(\omega_n t + \phi)$ can be represented as $A_1 \sin \omega_n t + A_2 \cos \omega_n t$ (see Window 2.1), the total solution can be expressed in the form

$$x(t) = A_1 \sin \omega_n t + A_2 \cos \omega_n t + \frac{f_0}{\omega_n^2 - \omega^2} \cos \omega t \tag{2.8}$$

where it remains to determine the values of the coefficients A_1 and A_2. These are determined by enforcing the initial conditions. Let the initial position and velocity be given by the constants x_0 and v_0 as before. Then equation (2.8) yields

$$x(0) = A_2 + \frac{f_0}{\omega_n^2 - \omega^2} = x_0 \tag{2.9}$$

and

$$\dot{x}(0) = \omega_n A_1 = v_0 \tag{2.10}$$

Window 2.1
Review of the Solution of the an Undamped Homogeneous Vibration Problem from Chapter 1

$$m\ddot{x} + kx = 0 \quad \text{subject to} \quad x(0) = x_0, \quad \dot{x}(0) = v_0$$

has solution $x(t) = A \sin(\omega_n t + \phi)$, which becomes, after evaluating the constants A and ϕ in terms of the initial conditions,

$$x(t) = \frac{\sqrt{x_0^2 \omega_n^2 + v_0^2}}{\omega_n} \sin\left(\omega_n t + \tan^{-1} \frac{x_0 \omega_n}{v_0}\right)$$

where $\omega_n = \sqrt{k/m}$ is the natural frequency. Via some simple trigonometry, this solution can also be written as

$$x(t) = A_1 \sin \omega_n t + A_2 \cos \omega_n t = \frac{v_0}{\omega_n} \sin \omega_n t + x_0 \cos \omega_n t$$

where the constants A_1, A_2, A, and ϕ are related by

$$A = \sqrt{A_1^2 + A_2^2}, \quad \tan \phi = \frac{A_2}{A_1}$$

Solving equations (2.9) and (2.10) for A_1 and A_2 and substituting these values into equation (2.8) yields the total response

$$x(t) = \frac{v_0}{\omega_n} \sin \omega_n t + \left(x_0 - \frac{f_0}{\omega_n^2 - \omega^2}\right) \cos \omega_n t + \frac{f_0}{\omega_n^2 - \omega^2} \cos \omega t \qquad (2.11)$$

Note that the coefficients A_1 and A_2 for the *total* response given in equation (2.11) are different than those given for the *free* response as reviewed in Window 2.1. Also note that if the driving force is zero, $f_0 = 0$ in equation (2.11), then A_1 and A_2 for the total response reduce to the values for the free response. Figure 2.2 illustrates a plot of the total response of an undamped system to a harmonic excitation and specified initial conditions.

Note that both the second and third terms in equation (2.11) are not valid if the driving frequency happens to be equal to the natural frequency (i.e., if $\omega = \omega_n$). Also note that as the driving frequency gets close to the natural frequency the amplitude of the resulting vibration gets very large. This large increase in amplitude defines the phenomenon of *resonance*, perhaps the most important concept in vibration analysis. Resonance is defined and discussed in detail in the paragraphs following the examples.

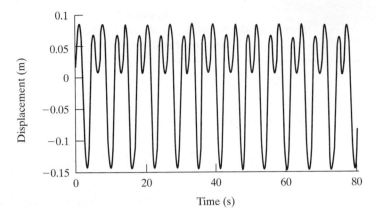

Figure 2.2   The response of an undamped system with $\omega_n = 1$ rad/s to harmonic excitation at $\omega = 2$ rad/s and nonzero initial conditions of $x_0 = 0.01$ m and $v_0 = 0.01$ m/s and magnitude $f_0 = 0.1$ N/kg. The motion is the sum of two sine curves of different frequencies.

Example 2.1.1

Compute and plot the response of a spring–mass system modeled by equation (2.2) to a force of magnitude 23 N, driving frequency of twice the natural frequency, and initial conditions given by $x_0 = 0$ m and $v_0 = 0.2$ m/s. The mass of the system is 10 kg and the spring stiffness is 1000 N/m.

Solution   First, compute the various coefficients for the response as given in equation (2.11). The natural frequency, driving frequency, and mass-normalized force magnitude are

$$\omega_n = \sqrt{\frac{1000\,\text{N/m}}{10\,\text{kg}}} = 10\,\text{rad/s}, \omega = 2\omega_n = 20\,\text{rad/s}, f_0 = \frac{23\,\text{N}}{10\,\text{kg}} = 2.3\,\text{N/kg}$$

The coefficients of the three terms in the response (note $x_0 = 0$) become

$$\frac{v_0}{\omega_n} = \frac{0.2\,\text{m/s}}{10\,\text{rad/s}} = 0.02\,\text{m}, \frac{f_0}{\omega_n^2 - \omega^2} = \frac{2.3\,\text{N/kg}}{(10^2 - 20^2)\,\text{rad}^2/\text{s}^2} = -7.6667 \times 10^{-3}\,\text{m}$$

With these values, equation (2.11) becomes

$$x(t) = 0.02\sin 10t + 7.667 \times 10^{-3}(\cos 10t - \cos 20t)\,\text{m}$$

The plot of the time response is given in Figure 2.3. Any of the software packages described in Section 1.9 or Appendix G can be used to generate this time history.

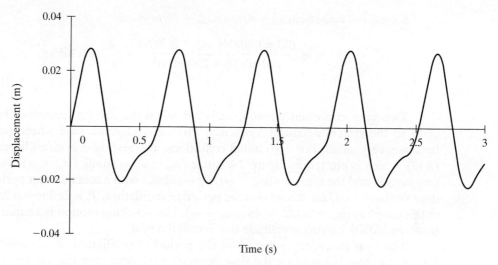

Time (s)

Figure 2.3   The time response of the undamped system of Example 2.1.1 illustrating the effect of both initial conditions and the forcing function on the response.

☐

Example 2.1.2

Consider the forced vibration of a mass m connected to a spring of stiffness 2000 N/m being driven by a 20-N harmonic force at 10 Hz (20π rad/s). The maximum amplitude of vibration is measured to be 0.1 m and the motion is assumed to have started from rest ($x_0 = v_0 = 0$). Calculate the mass of the system.

Solution   From equation (2.11) the response with $x_0 = v_0 = 0$ becomes

$$x(t) = \frac{f_0}{\omega_n^2 - \omega^2}(\cos\omega t - \cos\omega_n t) \tag{2.12}$$

Using the trigonometric identity

$$\cos u - \cos v = 2\sin\left(\frac{v-u}{2}\right)\sin\left(\frac{v+u}{2}\right)$$

equation (2.12) becomes

$$x(t) = \frac{2f_0}{\omega_n^2 - \omega^2}\sin\left(\frac{\omega_n - \omega}{2}t\right)\sin\left(\frac{\omega_n + \omega}{2}t\right) \tag{2.13}$$

The maximum value of the total response is evident from (2.13) so that

$$\frac{2f_0}{\omega_n^2 - \omega^2} = 0.1\,\text{m}$$

Solving this for m from $\omega_n^2 = k/m$ and $f_0 = F_0/m$ yields

$$m = \frac{(0.1\text{ m})(2000\text{ N/m}) - 2(20\text{ N})}{(0.1\text{ m})(10 \times 2\pi\text{ rad/s})^2} = \frac{4}{\pi^2} = 0.405\text{ kg}$$

$\square$

Two very important phenomena occur when the driving frequency becomes close to the system's natural frequency. First, consider the case where $(\omega_n - \omega)$ becomes very small. For zero initial conditions, the response is given by equation (2.13), which is plotted in Figure 2.4. Since $(\omega_n - \omega)$ is small, $(\omega_n + \omega)$ is large by comparison and the term $\sin[(\omega_n - \omega)/2]t$ oscillates with a much longer period than does $\sin[(\omega_n + \omega)/2]t$. Recall that the period of oscillation, T, is defined as $2\pi/\omega$, or in this case $2\pi/(\omega_n + \omega)/2 = 4\pi/(\omega_n + \omega)$. The resulting motion is a rapid oscillation with slowly varying amplitude that is called a *beat*.

The beat frequency is based on the period of oscillation of the solid line in Figure 2.4. This is based on the time between two successive maximums, which is half the time for one complete oscillation of the dashed line in Figure 2.4, or

$$\omega_{\text{beat}} = |\omega_n - \omega|$$

To see this more clearly, note that the mathematical definition of a period, T, is the smallest time such that $f(t + T) = f(t)$. From the solid line in Figure 2.4, this

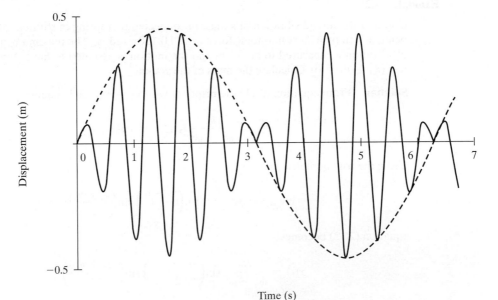

Figure 2.4 The response of an undamped system of equation (2.13) for small $\omega_n - \omega$ illustrating the phenomenon of beats. Here $f_0 = 10$ N, $\omega_n = 10$ rad/s, and $\omega = 1.1\,\omega_n$ rad/s. The dashed line is a plot of $\dfrac{2f_0}{\omega_n^2 - \omega^2}\sin\left(\dfrac{\omega_n - \omega}{2}t\right)$.

occurs at half the period of the dashed line. This period corresponds to a frequency of $|\omega_n - \omega|$.

As ω becomes exactly equal to the system's natural frequency, the solution given in equation (2.11) is no longer valid. In this case, the choice of the function $X \cos \omega t$ for a particular solution fails because it is also a solution of the homogeneous equation. Therefore, the particular solution is of the form

$$x_p(t) = tX \sin \omega t \qquad (2.14)$$

as explained in Boyce and DiPrima (2009). Substitution of (2.14) into equation (2.3) and solving for X yields

$$x_p(t) = \frac{f_0}{2\omega} t \sin \omega t \qquad (2.15)$$

Thus the total solution is now of the form ($\omega = \omega_n$)

$$x(t) = A_1 \sin \omega t + A_2 \cos \omega t + \frac{f_0}{2\omega} t \sin \omega t \qquad (2.16)$$

Evaluating the initial displacement x_0 and velocity v_0 as before yields

$$x(t) = \frac{v_0}{\omega} \sin \omega t + x_0 \cos \omega t + \frac{f_0}{2\omega} t \sin \omega t \qquad (2.17)$$

A plot of $x(t)$ is given in Figure 2.5, where it can be seen that $x(t)$ grows without bound. This defines the phenomenon of *resonance* (i.e., that the amplitude of vibration becomes unbounded at $\omega = \omega_n = \sqrt{k/m}$). This would cause the spring to break and fail.

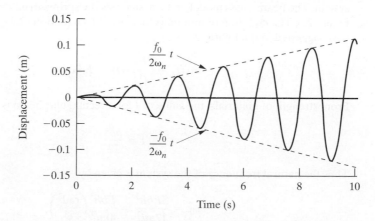

Figure 2.5   The forced response of a spring–mass system driven harmonically at its natural frequency ($\omega = \omega_n$), called resonance.

Example 2.1.3

A security camera is to be mounted on a building in an alley and will be subject to wind loads producing an applied force of $F_0 \cos \omega t$, where the largest value of F_0 is measured to be 15 N. This is illustrated on the left side of Figure 2.6. It is desired to design a mount such that the camera will experience a maximum deflection of 0.01 m when it vibrates under this load. The wind frequency is known to be 10 Hz and the camera mass is 3 kg. The mounting bracket is made of a solid piece of aluminum, 0.02×0.02 m in cross section. Compute the length of the mounting bracket that will keep the vibration amplitude less than the desired 0.01 m (ignore torsional vibration and assume the initial conditions are both zero). Note that the length must be at least 0.2 m in order to have a clear view.

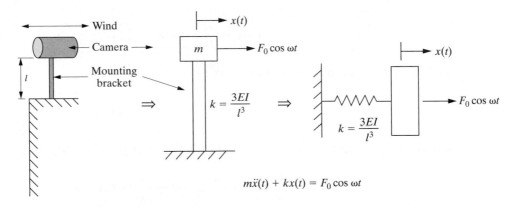

$$m\ddot{x}(t) + kx(t) = F_0 \cos \omega t$$

Figure 2.6 Simple models of a camera and mounting bracket subject to a harmonic wind load. The sketch is on the left, the strength-of-materials model is in the middle, and the spring–mass model is on the right.

Solution The sketch on the left in Figure 2.6 suggests the beam–mass, transverse-vibration model of Figure 1.26, repeated in the middle of Figure 2.6, for modeling this system. The beam–mass model, in turn, suggests the spring–mass system on the right in Figure 2.6. The equation of motion is then given by equation (2.2) with the beam stiffness suggested in the figure or

$$m\ddot{x}(t) + \frac{3EI}{l^3} x(t) = F_0 \cos \omega t$$

From strength of materials, the value of I for a rectangular beam is

$$I = \frac{bh^3}{12}$$

Thus the natural frequency of the system is given by

$$\omega_n^2 = \frac{3Ebh^3}{12ml^3} = \frac{Ebh^3}{4ml^3} \left(\frac{\text{rad}^2}{\text{s}^2} \right)$$

Note that the length l is the quantity we need to solve for in the design.

Next, consider the expression for the maximum deflection of the response computed in Example 2.1.2. Requiring this amplitude to be less than 0.01 m yields the following two cases:

$$\left| \frac{2f_0}{\omega_n^2 - \omega^2} \right| < 0.01 \Rightarrow (a) \ -0.01 < \frac{2f_0}{\omega_n^2 - \omega^2} \quad \text{and} \quad (b) \ \frac{2f_0}{\omega_n^2 - \omega^2} < 0.01$$

First, consider case (a), which holds for $\omega_n^2 - \omega^2 < 0$:

$$-0.01 < \frac{2f_0}{\omega_n^2 - \omega^2} \Rightarrow 2f_0 < 0.01\omega^2 - 0.01\omega_n^2 \Rightarrow 0.01\omega^2 - 2f_0 > 0.01\frac{Ebh^3}{4ml^3}$$

$$\Rightarrow l^3 > 0.01\frac{Ebh^3}{4m(0.01\omega^2 - 2f_0)} = 0.02 \Rightarrow l > 0.272 \text{ m}$$

Next, consider case (b), which holds for $\omega_n^2 - \omega^2 > 0$:

$$\frac{2f_0}{\omega_n^2 - \omega^2} < 0.01 \Rightarrow 2f_0 < 0.01\omega_n^2 - 0.01\omega^2 \Rightarrow 2f_0 + 0.01\omega^2 < 0.01\frac{Ebh^3}{4ml^3}$$

$$\Rightarrow l^3 < 0.01\frac{Ebh3}{4m(2f_0 + 0.01\omega^2)} = 0.012 \Rightarrow l < 0.229 \text{ m}$$

Here the value of E for aluminum is taken from Table 1.2 $(7.1 \times 10^{10} \text{ N/m}^2)$ and ω is changed to rad/s. To conserve material, case (b) is chosen. Given the constraint that the length of the bracket must be at least 0.2 m, $0.2 < l < 0.229$ so the value of $l = 0.22$ is chosen for the solution.

Next, a couple of simple checks are performed to make sure the assumptions made in solving the problem are reasonable. First, compute the value of ω_n for this value of l to see that the solution is consistent with the inequality:

$$\omega_n^2 - \omega^2 = \frac{3Ebh^3}{12ml^3} - (20\pi)^2 = (74.543)^2 - (20\pi)^2 = 1609 > 0$$

Thus case (b) is satisfied.

Note that the mass of the mounting bracket was neglected. It is always important to check assumptions. Using the density for aluminum given in Table 1.2, the mass of the bracket is

$$m = \rho lbh = (2.7 \times 10^3)(0.22)(0.01)(0.01) = 0.149 \text{ kg}$$

This is less than the mass of the camera (about 5%), so according to Example 1.4.4 it is reasonable to ignore the spring's mass in these calculations.

□

Example 2.1.4

In solving the security camera design problem of Example 2.1.3, torsional vibration, illustrated in Figure 2.7, was not considered. The purpose of this example is to examine if the assumption of ignoring torsion is correct or not. To decide this, determine if a wind load of 15 N vibrating at 10 Hz causes the end to move more than 0.01 m.

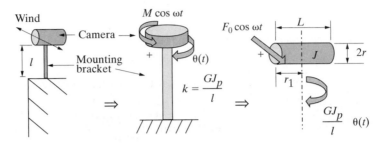

Figure 2.7 A torsional model of a camera and mounting bracket subject to a harmonic wind load. The sketch on the left is the system of Figure 2.6 showing blowing across the side of the camera causing an applied moment resulting in torsional motion. The schematic is shown in the middle and the free-body diagram is on the right.

Solution   Here we model the wind load as acting at a point on the tip of the camera, a distance $r_1 = 0.09$ m from its center, creating an applied moment of $M(t) = r_1F_0 \cos 20\pi t$ Nm. Summing the moments diagram of in Figure 2.6, the equation of motion is

$$J\ddot{\theta}(t) + k\theta(t) = r_1F_0 \cos(20\pi t)$$

Here θ is the rotational displacement of the camera about the center where the bracket connects to the camera. Modeling the camera as a solid cylinder, the mass moment of inertia is (see a dynamics text or perform a Google search)

$$J = \frac{m}{12}\left(3r^2 + L^2\right)$$

where m is the mass, $r = 0.05$ m is the radius, and $L = 2r_1 = 0.18$ m is the length of the cylinder. From Figure 1.24, the torsional stiffness of the mounting bracket is

$$k = \frac{GJ_p}{l}$$

Here G is the shear modulus of aluminum ($G = 2.67 \times 10^{10}$ N/m², from Table 1.2) and the polar moment of inertia of a square "rod" is $J_p = 0.1406\,a^4$, where $a = 0.01$ m is the length of the side of the square bracket (from Table 1.3). The value of l is taken from the solution of Example 2.1.3 to be $l = 0.22$ m. Substitution of the appropriate values yields

$$k = \frac{GJ_p}{l} = \frac{\left(2.67 \times 10^{10}\right)\left(\dfrac{N}{m^2}\right)\left((0.1406)\,(0.01^4)\right)(m^4)}{0.22 \cdot m} = 2.73 \times 10^3 \text{ Nm}$$

$$\omega_n = \sqrt{\frac{k}{J}} = \sqrt{\frac{2.73 \times 10^3 \text{ Nm}}{9.975 \times 10^{-3} \text{ kg} \cdot m^2}} = 523.67 \text{ rad/s}$$

Following the expression for the maximum value of the response for zero initial conditions given in Example 2.1.2 applied to the torsional vibration problem yields

$$\theta_{max} = \left| \frac{2\dfrac{r_1 F_0}{J}}{\omega_n^2 - \omega^2} \right| = \left| \frac{2\dfrac{(0.09)(15)}{9.975 \times 10^{-3}}}{(523.167)^2 - (62.832)^2} \right| = 1.003 \times 10^{-3}\,\text{rad}$$

The maximum linear displacement of the tip is then

$$X_{max} = r_1 \theta_{max} = (0.09)(1.003 \times 10^{-3}) = 9.031 \times 10^{-5}\,\text{m}$$

This is much less than the required 0.01 m so that the assumption of ignoring torsion was a reasonable one. This is largely because the torsional frequency (523 rad/s) is much higher then the bending frequency (75 rad/s), making the denominator in the maximum deflection calculation larger and hence the deflection smaller.

□

The previous development assumed that the harmonic forcing function was described by the cosine function. As pointed out earlier, the harmonic forcing function may also be represented in terms of the sine function. In this case, equation (2.2) becomes

$$m\ddot{x}(t) + kx(t) = F_0 \sin \omega t \quad \text{or} \quad \ddot{x}(t) + \omega_n^2 x(t) = f_0 \sin \omega t \tag{2.18}$$

Proceeding as before, using the method of undetermined coefficients, the particular solution for the case of sinusoidal excitation becomes

$$x_p(t) = X \sin \omega t \tag{2.19}$$

Substitution of this assumed solution form into equation (2.18) yields

$$-\omega^2 X \sin \omega t + \omega_n^2 X \sin \omega t = f_0 \sin \omega t \tag{2.20}$$

Factoring out $\sin \omega t \, (\neq 0)$ and solving for X yields

$$X = \frac{f_0}{\omega_n^2 - \omega^2} \tag{2.21}$$

so that the particular solution is

$$x_p(t) = \frac{f_0}{\omega_n^2 - \omega^2} \sin \omega t \tag{2.22}$$

The total solution is the sum of the homogenous solution and the particular solution, or

$$x(t) = A_1 \sin \omega_n t + A_2 \cos \omega_n t + \frac{f_0}{\omega_n^2 - \omega^2} \sin \omega t \tag{2.23}$$

It remains to evaluate the constants A_1 and A_2 in terms of the given initial conditions x_0 and v_0. To this end set $t = 0$ in equation (2.23) and its first derivative to get

$$x(0) = x_0 = A_2 \quad \text{and} \quad \dot{x}(0) = \omega_n A_1 + \frac{\omega f_0}{\omega_n^2 - \omega^2} = v_0 \qquad (2.24)$$

$$\Rightarrow A_1 = \frac{v_0}{\omega_n} - \frac{\omega}{\omega_n} \frac{f_0}{\omega_n^2 - \omega^2} \quad \text{and} \quad A_2 = x_0$$

The total solution for a sinusoidal harmonic input is thus

$$x(t) = x_0 \cos \omega_n t + \left(\frac{v_0}{\omega_n} - \frac{\omega}{\omega_n} \frac{f_0}{\omega_n^2 - \omega^2} \right) \sin \omega_n t + \frac{f_0}{\omega_n^2 - \omega^2} \sin \omega t \qquad (2.25)$$

It is important to note the differences between equations (2.11) and (2.25). Equation (2.11) is the response due to a cosine input, and equation (2.25) is the response due to a sine input. In particular, note that the input force term modifies the initial condition response differently in each case.

Note that equation (2.17) can also be obtained from equation (2.11) by taking the limit as $\omega \rightarrow \omega_n$ using the limit theorems from calculus.

2.2 HARMONIC EXCITATION OF DAMPED SYSTEMS

As noted in Chapter 1, some sort of damping or energy dissipation is always present (see Window 2.2). In this section, the response of a viscously damped single-degree-of-freedom system subjected to harmonic excitation is considered. Summing forces on the mass of Figure 2.1 in the x direction yields

$$m\ddot{x} + c\dot{x} + kx = F_0 \cos \omega t \qquad (2.26)$$

Dividing by the mass m yields

$$\ddot{x} + 2\zeta \omega_n \dot{x} + \omega_n^2 x = f_0 \cos \omega t \qquad (2.27)$$

where $\omega_n = \sqrt{k/m}$, $\zeta = c/(2m\omega_n)$, and $f_0 = F_0/m$. The calculation of the particular solution for the damped case is similar to that of the undamped case and follows the method of undetermined coefficients.

From differential equations it is known that the forced response of a damped system is of the form of a harmonic function of the same frequency as the driving force with a different amplitude and phase. The phase shift is expected because

Window 2.2
Review of the Solution of the Damped Homogeneous Vibration Problem ($0 < \zeta < 1$) from Chapter 1

$m\ddot{x} + c\dot{x} + kx = 0$ subject to $x(0) = x_0$, $\dot{x}(0) = v_0$ has the solution

$$x(t) = Ae^{-\zeta\omega_n t} \sin(\omega_d t + \phi)$$

where

$$\omega_n = \sqrt{\frac{k}{m}} \text{ is the undamped natural frequency}$$

$$\zeta = \frac{2}{2m\omega_n} \text{ is the damping ratio}$$

$$\omega_d = \omega_n \sqrt{1 - \zeta^2} \text{ is the damped natural frequency}$$

and the constants A and ϕ are determined by the initial conditions to be

$$A = \sqrt{x_0^2 + \left(\frac{v_0 + \zeta\omega_n x_0}{\omega_d}\right)^2}$$

$$\phi = \tan^{-1}\frac{x_0\omega_d}{v_0 + \zeta\omega_n x_0}$$

Alternately, the solution can be written as

$$x(t) = e^{-\zeta\omega_n t}\left[\frac{v_0 + \zeta\omega_n x_0}{\omega_d}\sin\omega_d t + x_0\cos\omega_d t\right]$$

of the effect of the damping force. Following the method of undetermined coefficients, the particular solution is assumed to be of the form

$$x_p(t) = X\cos(\omega t - \theta) \tag{2.28}$$

To make the computations easy to follow, this is written in the equivalent form

$$x_p(t) = A_s\cos\omega t + B_s\sin\omega t \tag{2.29}$$

where the constants $A_s = X\cos\theta$ and $B_s = X\sin\theta$ satisfying

$$X = \sqrt{A_s^2 + B_s^2} \quad \text{and} \quad \theta = \tan^{-1}\frac{B_s}{A_s} \tag{2.30}$$

are the undetermined constant coefficients.

Taking derivatives of the assumed form of the solution given by (2.29) yields

$$\dot{x}_p(t) = -\omega A_s \sin \omega t + \omega B_s \cos \omega t \qquad (2.31)$$

and

$$\ddot{x}_p(t) = -\omega^2(A_s \cos \omega t + B_s \sin \omega t) \qquad (2.32)$$

Substitution of x_p, $\dot{x}_p$, and $\ddot{x}_p$ into the equation of motion given by equation (2.27) and grouping terms as coefficients of $\sin \omega t$ and $\cos \omega t$ yields

$$\left(-\omega^2 A_s + 2\zeta\omega_n\omega B_s + \omega_n^2 A_s - f_0\right) \cos \omega t + \left(-\omega^2 B_s - 2\zeta\omega_n \omega A_s + \omega_n^2 B_s\right) \sin \omega t = 0 \qquad (2.33)$$

This equation must hold for all time, in particular for $t = \pi/2\omega$, so that the coefficient of $\sin \omega t$ must vanish. Similarly, for $t = 0$ the coefficient of $\cos \omega t$ must vanish. This yields the two equations

$$\left(\omega_n^2 - \omega^2\right) A_s + \left(2\zeta\omega_n\omega\right) B_s = f_0 \qquad (2.34)$$
$$\left(-2\zeta\omega_n\omega\right) A_s + \left(\omega_n^2 - \omega^2\right) B_s = 0$$

in the two undetermined coefficients A_s and B_s. These two linear equations may be written as the single matrix equation

$$\begin{bmatrix} \omega_n^2 - \omega^2 & 2\zeta\omega_n\omega \\ -2\zeta\omega_n\omega & \omega_n^2 - \omega^2 \end{bmatrix} \begin{bmatrix} A_s \\ B_s \end{bmatrix} = \begin{bmatrix} f_0 \\ 0 \end{bmatrix}$$

which has solution (compute the matrix inverse and multiply; see Appendix C)

$$A_s = \frac{\left(\omega_n^2 - \omega^2\right) f_0}{\left(\omega_n^2 - \omega^2\right)^2 + \left(2\zeta\omega_n\omega\right)^2} \qquad (2.35)$$

$$B_s = \frac{2\zeta\omega_n\omega f_0}{\left(\omega_n^2 - \omega^2\right)^2 + \left(2\zeta\omega_n\omega\right)^2}$$

Substitution of these values into equations (2.30) and (2.28) yields that the particular solution is

$$x_p(t) = \frac{f_0}{\sqrt{\left(\omega_n^2 - \omega^2\right)^2 + \left(2\zeta\omega_n\omega\right)^2}} \cos\left(\omega t - \tan^{-1}\frac{2\zeta\omega_n\omega}{\omega_n^2 - \omega^2}\right) \qquad (2.36)$$

The total solution is again the sum of the particular solution and the homogeneous solution obtained in Section 1.3. For the underdamped case $(0 < \zeta < 1)$ this becomes

$$x(t) = Ae^{-\zeta\omega_n t}\sin(\omega_d t + \phi) + X\cos(\omega t - \theta) \qquad (2.37)$$

where X and θ are the coefficients of the particular solution as defined by equation (2.36), and A and ϕ (different from those of Window 2.2) are determined by the initial conditions. Note that for large values of t, the first term, or homogeneous solution, approaches zero and the total solution approaches the particular solution. Thus $x_p(t)$ is called the *steady-state response* and the first term in equation (2.37) is called the *transient response*.

The values of the constants of integration in equation (2.37) can be found from the initial conditions by the same procedure used in the development for computing the response of an undamped system given in equation (2.11). Following the development from equation (2.29), write the transient term as $A\sin\omega_d t + C\cos\omega_d t$ rather than writing the constants of integration as a magnitude and phase and the particular solution in the form given in equation (2.37). The resulting response for the underdamped case is

$$x(t) = e^{-\zeta\omega_n t}\left\{\left(x_0 - \frac{f_0(\omega_n^2 - \omega^2)}{(\omega_n^2 - \omega^2)^2 + (2\zeta\omega_n\omega)^2}\right)\cos\omega_d t\right.$$

$$+ \left(\frac{\zeta\omega_n}{\omega_d}\left(x_0 - \frac{f_0(\omega_n^2 - \omega^2)}{(\omega_n^2 - \omega^2)^2 + (2\zeta\omega_n\omega)^2}\right)\right.$$

$$\left.\left. - \frac{2\zeta\omega_n\omega^2 f_0}{\omega_d[(\omega_n^2 - \omega^2)^2 + (2\zeta\omega_n\omega)^2]} + \frac{v_0}{\omega_d}\right)\sin\omega_d t\right\} \qquad (2.38)$$

$$+ \frac{f_0}{(\omega_n^2 - \omega^2)^2 + (2\zeta\omega_n\omega)^2}\left[(\omega_n^2 - \omega^2)\cos\omega t + 2\zeta\omega_n\omega\sin\omega t\right]$$

This is an alternate form of equation (2.37) showing the direct influence of the forcing function on the transient part of the response (i.e., the coefficient of the exponential term). Note that the expression for the forced response of an underdamped system given here collapses to the forced response of an undamped system given by equation (2.11) when the damping is set to zero in equation (2.38). Problem 2.20 requires the calculation of the constants A and ϕ for equation (2.37). A summary of phase and amplitude formulas is given in Window 2.3 for both the free and forced response.

Note that A and ϕ, the constants describing the transient response in equation (2.37), will be different from those calculated for the free-response case given in equation (1.38) or Window 2.2. This is because part of the transient term in

Window 2.3
Summary of Phase and Amplitude Relationships for Undamped and Underdamped Single-Degree-of-Freedom Systems for Both the Free Response, $F(t) = 0$, and for the Forced Response, $F(t) = F_0 \cos \omega t$, Cases

The general response for the undamped case has the form

$$x(t) = A \sin (\omega_n t + \phi) + X \cos \omega t$$

where for the free response

$$\phi = \tan^{-1} \frac{\omega_n x_0}{v_0}, \quad A = \sqrt{x_0^2 + \frac{v_0^2}{\omega_n^2}}, \quad X = 0$$

and for the forced response:

$$\phi = \tan^{-1} \frac{\omega_n (x_0 - X)}{v_0}, \quad A = \sqrt{\left(\frac{v_0}{\omega_n}\right)^2 + (x_0 - X)^2}, \quad X = \frac{f_0}{\omega_n^2 - \omega^2}$$

The general response in the underdamped case has the form

$$x(t) = A e^{-\zeta \omega_n t} \sin (\omega_d t + \phi) + X \cos (\omega t - \theta)$$

where the free response:

$$\phi = \tan^{-1} \frac{x_0 \omega_d}{v_0 + \zeta \omega_n x_0}, \quad A = \frac{1}{\omega_d} \sqrt{(v_0 + \zeta \omega_n x_0)^2 + (x_0 \omega_d)^2}, \quad X = 0$$

and for the forced response

$$\theta = \tan^{-1} \frac{2\zeta \omega_n \omega}{\omega_n^2 - \omega^2}, \quad X = \frac{f_0}{\sqrt{(\omega_n^2 - \omega^2)^2 + (2\zeta \omega_n \omega)^2}},$$

$$\phi = \tan^{-1} \frac{\omega_d (x_0 - X \cos \theta)}{v_0 + (x_0 - X \cos \theta) \zeta \omega_n - \omega X \sin \theta} \quad \text{and} \quad A = \frac{x_0 - X \cos \theta}{\sin \phi}$$

equation (2.37) is due to the magnitude of the excitation force and part is due to the initial conditions as indicated in equation (2.38).

Example 2.2.1

Examine the units for computing the forced response of a damped system. Often the equation-of-motion quantities (forces) are given in Newtons, whereas the initial displacement and velocity are given in mm. It is important to write the initial conditions in the correct units.

Solution   First, examine units in the mass-normalized equation of motion as given in equation (2.27) repeated here:

$$\ddot{x} + 2\zeta\omega_n\dot{x} + \omega_n^2 x = f_0 \cos \omega t$$

The units for f_0 are $N/kg = m/s^2$, the units of acceleration, agreeing with the first term in the equation. The damping ratio ζ has no units, so the units of the damping term are those of ω_n or rad/s, which when multiplied by the velocity yields m/s^2. Likewise, the units of the natural frequency squared are rad^2/s^2 so the stiffness term also has units of m/s^2. Thus equation (2.27) is consistent in terms of units.

Next, consider the solution. Since the amplitude of the particular solution, X, has the units of m (the units of f_0 are N/Kg or m/s^2)

$$X = \frac{f_0}{\sqrt{(\omega_n^2 - \omega^2)^2 + (2\zeta\omega_n\omega)^2}} \left(\frac{m/s^2}{rad/s^2}\right) = \frac{f_0}{\sqrt{(\omega_n^2 - \omega^2)^2 + (2\zeta\omega_n\omega)^2}}\, m$$

the initial condition x_0 must also be given in m because the value of amplitude A contains the numerator term $x_0 - X\cos(\theta)$. The same is true for the phase angle ϕ, which also contains the initial velocity added to $X\zeta\omega_n$ which will have units of m/s. Thus in solving for the force response of a damped system it is important that the initial conditions are stated in terms of the same units that the equation of motion is expressed in.

□

Example 2.2.2

A damped spring–mass system with values of $c = 100$ kg/s, $m = 100$ kg, and $k = 910$ N/m, is subject to a force of $10 \cos (3t)$ N. The system is also subject to initial conditions: $x_0 = 1$ mm and $v_0 = 20$ mm/s. Compute the total response, $x(t)$, of the system.

Solution   Following equation (2.26) with the values given here, the system to be solved is

$$100\ddot{x}(t) + 100\dot{x}(t) + 910x(t) = 10 \cos 3t, \ x_0 = 0.001 \text{ m}, \ v_0 = 0.02 \text{ m/s}$$

where the units (and thus numerical values) of the initial conditions have been changed to agree with the equation of motion per the previous example. Dividing by the mass yields the vibration properties

$$f_0 = \frac{F_0}{m} = \frac{10}{100} = 0.1 \frac{m}{s^2}, \ \omega_n = \sqrt{\frac{k}{m}} = \sqrt{\frac{910}{100}} = 3.017 \frac{rad}{s},$$

$$\zeta = \frac{c}{2\sqrt{mk}} = \frac{100}{2\sqrt{100\cdot910}} = 0.166,$$

$$\omega_d = \omega_n\sqrt{1 - \zeta^2} = 3.017\sqrt{1 - 0.166^2} = 2.975 \text{ rad/s}$$

Since $\omega = 3$ rad/s, the system is near resonance. Computing the amplitude and phase for the particular solution from the values given in Window 2.3 yields

$$X = \frac{f_0}{\sqrt{(\omega_n^2 - \omega^2)^2 + (2\zeta\omega_n\omega)^2}} = \frac{0.1}{\sqrt{(3.017^2 - 3^2)^2 + (2\cdot0.166\cdot3.017\cdot3)^2}} = 0.033 \text{ m}$$

$$\theta = \tan^{-1} \frac{2\zeta\omega_n\omega}{\omega_n^2 - \omega^2} = \tan^{-1} \frac{2 \cdot 0.166 \cdot 3.017 \cdot 3}{3.017^2 - 3^2} = 1.537 \text{ rad}$$

Next, compute phase for the transient response

$$\phi = \tan^{-1} \frac{\omega_d (x_0 - X \cos\theta)}{v_0 + (x_0 - X \cos\theta)\zeta\omega_n - \omega X \sin\theta}$$

$$= \frac{2.975(0.001 - 0.033 \cdot 0.033)}{0.02 + (0.001 - 0.033 \cdot 0.033)0.166 \cdot 3.017 - 3 \cdot 0.033 \cdot 0.999} = 4.089 \times 10^{-3} \text{ rad}$$

The amplitude of the transient is

$$A = \frac{x_0 - X \cos\theta}{\sin\phi} = \frac{0.001 - 0.033 \cdot 0.0333}{4.089 \times 10^{-3}} = -0.027 \text{ m}$$

The total response is then written from equation (2.37) as

$$x(t) = Ae^{-\zeta\omega_n t} \sin(\omega_d t + \phi) + X \cos(\omega t - \theta)$$
$$= -0.027e^{-0.5t} \sin(2.975t + 4.089 \times 10^{-3}) + 0.033 \cos(3t - 1.537) \text{ m}$$

Example 2.2.3

Compute the constants of integration A and ϕ of equation (2.37) and compare these values to the values of A and ϕ for the unforced case given in Window 2.2 for the parameters $\omega_n = 10$ rad/s, $\omega = 5$ rad/s, $\zeta = 0.01$, $F_0 = 1000$ N, $m = 100$ kg, and the initial conditions $x_0 = 0.05$ m and $v_0 = 0$.

Solution First, compute the values of X and θ from equation (2.36) for the particular solution (forced response). These are

$$X = \frac{f_0}{\sqrt{(\omega_n^2 - \omega^2)^2 + (2\zeta\omega_n\omega)^2}} = 0.133 \text{ m} \quad \text{and} \quad \theta = \tan^{-1}\left(\frac{2\zeta\omega_n\omega}{\omega_n^2 - \omega^2}\right) = 0.013 \text{ rad}$$

so that the phase for the particular solution is nearly zero (0.76°). Thus the solution given by equation (2.37) is of the form

$$x(t) = Ae^{-0.1t} \sin(9.999t + \phi) + 0.133 \cos(5t - 0.013)$$

Differentiating this solution yields the velocity expression

$$v(t) = -0.1Ae^{-0.1t} \sin(9.999t + \phi) + 9.999Ae^{-0.1t} \cos(9.999t + \phi) - 0.665 \sin(5t - 0.013)$$

Setting $t = 0$ and $x(0) = 0.05$ in the expression for $x(t)$ and solving for A yields

$$A = \frac{0.005 - 0.133 \cos(-0.013)}{\sin(\phi)} = \frac{-0.083}{\sin(\phi)}$$

Substitution of this value of A, setting $t = 0$ and $v(0) = 0$ in the expression for the velocity yields

$$0 = -0.1(-0.083) + 9.999(-0.083)\cot(\phi) + 0.665\sin(0.013)$$

Solving for the value of ϕ yields $\phi = 1.55$ rad (88.8°) and thus $A = -0.083$ m, the values of the amplitude and phase of the transient part of the solution (including the effects of the initial conditions and the applied force).

Next, consider the coefficients A and ϕ evaluated for the homogenous case $F_0 = 0$, but keeping the same initial conditions. Using these values (see Window 2.2), the incorrect magnitude and phase become

$$A = \sqrt{(0.05)^2 + \left[\frac{(0.01)(10)(0.05)}{10\sqrt{1-(0.01)^2}}\right]^2} = 0.05\,\text{m}$$

and

$$\phi = \tan^{-1}\left(\frac{9.999}{0.05(10)}\right) = 1.521\,\text{rad} \ (87.137°)$$

Comparing these two sets of values for the magnitude A and the two sets of values for the phase ϕ, we see that they are very different. Thus the values of the constants of integration are greatly affected by the forcing term. In particular, the amplitude of the transient is greatly increased and its phase is reduced by the influence of the driving force. ☐

It is common to ignore the transient part of the total solution given by equation (2.37) and to focus only on the steady-state response: $X\cos(\omega t - \theta)$. The rationale for considering only the steady-state response is based on the value of the damping ratio ζ. If the system has relatively large damping, the term $e^{-\zeta\omega_n t}$ causes the transient response to die out very quickly—perhaps in a fraction of a second. If, on the other hand, the system is lightly damped (ζ is very small), the transient part of the solution may last long enough to be significant and should not be ignored. The decision whether to ignore the transient part of the solution should also be based on the application. In fact, in some applications (such as earthquake analysis or satellite analysis) the transient response may become even more important than the steady-state response. An example is the Hubble space telescope, which originally experienced a transient vibration that lasted over 10 minutes, causing the telescope to be unusable every time it passed out of the earth's shadow, until the system was corrected.

The transient response can also be very important if it has a relatively large amplitude. Usually, devices are designed and analyzed based on the steady-state response, but the transient should always be checked to determine whether it is reasonable to ignore it or if it should be considered seriously.

With this caveat in mind, it is of interest to consider the magnitude, X, and the phase, θ, of the steady-state response as a function of the driving frequency. Examining the form of equation (2.36) and comparing it to the assumed form $X \cos(\omega_d t - \theta)$ yields the fact that the amplitude, X, and the phase, θ, are

$$X = \frac{f_0}{\sqrt{(\omega_n^2 - \omega^2) + (2\zeta\omega_n\omega)^2}}, \quad \theta = \tan^{-1}\frac{2\zeta\omega_n\omega}{\omega_n^2 - \omega^2} \qquad (2.39)$$

After some manipulation (i.e., factoring out ω_n^2 and dividing the magnitude by F_0/m), these expressions for the magnitude and phase can be written as

$$\frac{Xk}{F_0} = \frac{X\omega_n^2}{f_0} = \frac{1}{\sqrt{(1 - r^2)^2 + (2\zeta r)^2}}, \quad \theta = \tan^{-1}\frac{2\zeta r}{1 - r^2} \qquad (2.40)$$

Here r is the frequency ratio $r = \omega/\omega_n$, a dimensionless quantity. Equations (2.40) for the magnitude and phase are plotted versus the frequency ratio r in Figure 2.8 for several values of the damping ratio ζ. Note that as the driving frequency approaches the undamped natural frequency $(r \to 1)$, the magnitude approaches a maximum value for those curves corresponding to light damping $(\zeta \leq 0.1)$. Also note that as the driving frequency approaches the undamped natural frequency, the phase shift crosses through 90°. The phase lies between zero and π, as discussed in Window 2.4. This defines *resonance* for the damped case. These two observations have important uses in both vibration design and measurement. As ω approaches zero, the amplitude approaches f_0/ω_n^2 and as ω becomes very large, the amplitude approaches zero asymptotically.

It is also important from the design point of view to note how the amplitude of steady-state vibration is affected by changing the damping ratio. This is illustrated in Figure 2.9 and Example 2.2.4. Figure 2.9 is a repeat of the magnitude curve presented in Figure 2.8, except that here the magnitude is plotted on a log scale so that the curves for both small and large values of damping may be detailed on the same graph. The use of a log scale for vibration amplitude is common in vibration analysis and measurement. Note that as the damping ratio is

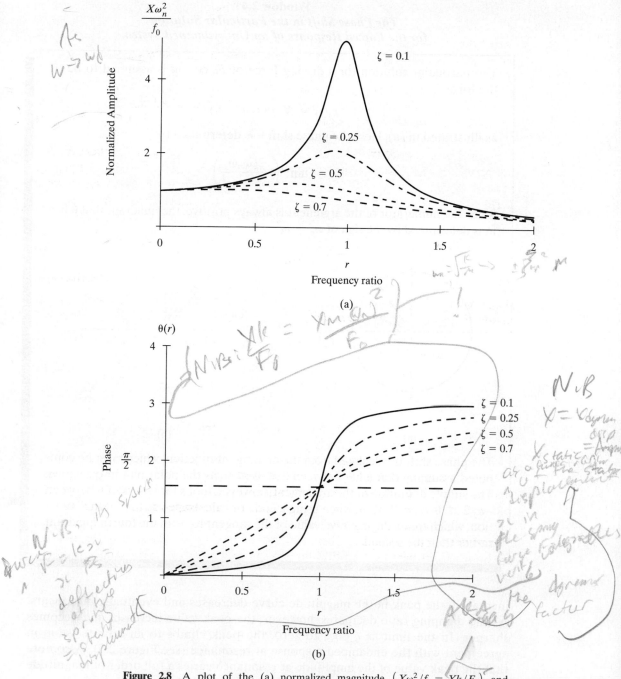

Figure 2.8 A plot of the (a) normalized magnitude $\left(X\omega_n^2/f_0 = Xk/F_0\right)$ and (b) phase of the steady-state response of a damped system versus the frequency ratio for several different values of the damping ratio ζ as determined by equation (2.40).

Window 2.4
The Phase Shift in the Particular Solution
for the Forced Response of an Underdamped System

The particular solution for a driving force of $F_0 \cos \omega t$ is assumed to be of the form

$$x_p(t) = X \cos(\omega t - \theta)$$

as illustrated in (a). Here the phase shift θ is determined by

$$\theta = \tan^{-1}\left(\frac{2\zeta\omega_n\omega}{\omega_n^2 - \omega^2}\right)$$

Since the numerator of the argument is always positive, the quadrant that θ lies in is determined by the sign of $\omega_n^2 - \omega^2$.

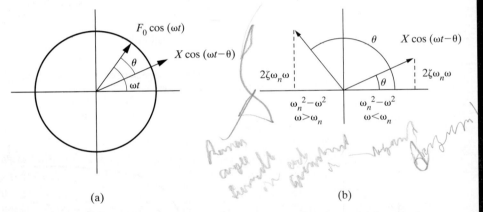

(a)   (b)

The phase shift is calculated from the arctangent function, which must be computed assuming that θ lies between $0 \le \theta \le \pi$, as the polar plot in (b) shows. The simple definition of the arctangent, however, looks for values of θ between $-\pi/2 \le \theta \le \pi/2$. Thus, when using a code or calculator, use the "atan2" function, which treats the negative value of the tangent to be in the fourth quadrant rather than the second.

increased, the peak in the magnitude curve decreases and eventually disappears. As the damping ratio decreases, however, the peak value increases and becomes sharper. In the limit as ζ goes to zero, the peak climbs to an infinite value in agreement with the undamped response at resonance (see Figure 2.5). Also note that the peak value of the amplitude at resonance varies a full order of magnitude as the damping changes.

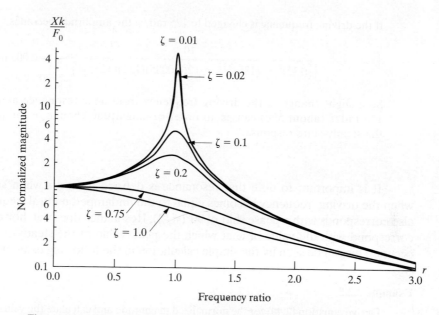

Figure 2.9 The magnitude (log scale) of the steady-state response versus the frequency ratio for several values of the damping ratio ζ.

Example 2.2.4

Consider a simple spring–mass–damper system with $m = 49.2 \times 10^{-3}$ kg, $c = 0.11$ kg/s, and $k = 857.8$ N/m. Calculate the value of the steady-state response if $\omega = 132$ rad/s for $f_0 = 10$ N/kg. Calculate the change in amplitude if the driving frequency changes to $\omega = 125$ rad/s.

Solution The frequency and damping ratio are determined from the given values as

$$\omega_n = \sqrt{\frac{k}{m}} = 132 \text{ rad/s}, \quad \zeta = \frac{c}{2\sqrt{mk}} = 0.0085$$

respectively. From equation (2.39) the magnitude of $x_p(t)$ is

$$|x_p(t)| = X = \frac{f_0}{\sqrt{(\omega_n^2 - \omega^2) + (2\zeta\omega_n\omega)^2}}$$

$$= \frac{10}{\{[(132)^2 - (132)^2]^2 + [2(0.0085)(132)(132)]^2\}^{1/2}}$$

$$= \frac{10}{2(0.0085)(132)^2}$$

$$= 0.034 \text{ m}$$

(handwritten in left margin: The mag of the particular solution)

If the driving frequency is changed to 125 rad/s, the amplitude becomes

$$\frac{10}{\{[(132)^2 - (125)^2]^2 + [2(0.0085)(132)(125)]^2\}^{1/2}} = 0.005 \text{ m}$$

So a slight change in the driving frequency from near resonance at 132 rad/s to 125 rad/s (about 5%) causes an order-of-magnitude change in the amplitude of the steady-state response.

□

It is important to note that resonance is defined to occur when $\omega = \omega_n$ (i.e., when the driving frequency becomes equal to the undamped natural frequency). This also corresponds with a phase shift of $90°(\pi/2)$. Resonance does not, however, exactly correspond with the value of ω at which the peak value of the steady-state response occurs. This can be seen by the simple calculation in the following example.

Example 2.2.5

Derive equation (2.40) for the normalized magnitude and calculate the value of $r = \omega/\omega_n$ for which the amplitude of the steady-state response takes on its maximum value.

Solution   From equation (2.36) the magnitude of the steady-state response is

$$X = \frac{f_0}{\sqrt{(\omega_n^2 - \omega^2)^2 + (2\zeta\omega_n\omega)^2}} = \frac{F_0/m}{\sqrt{(\omega_n^2 - \omega^2) + (2\zeta\omega_n\omega)^2}}$$

Factoring ω_n^2 out of the denominator and recalling that $\omega_n^2 = k/m$ yields

$$X = \frac{F_0/m}{\omega_n^2\sqrt{\left[1 - \left(\frac{\omega}{\omega_n}\right)^2\right]^2 + \left[2\zeta\left(\frac{\omega}{\omega_n}\right)\right]^2}} = \frac{F_0/k}{\sqrt{(1 - r^2)^2 + (2\zeta r)^2}}$$

where $r = \omega/\omega_n$. Dividing both sides by F_0/k yields equation (2.40). The maximum value of X will occur where the first derivative of X/F_0 vanishes, that is

$$\frac{d}{dr}\left(\frac{Xk}{F_0}\right) = \frac{d}{dr}\{[(1 - r^2)^2 + (2\zeta r)^2]^{-1/2}\} = 0$$

Thus

$$r_{peak} = \sqrt{1 - 2\zeta^2} = \frac{\omega_p}{\omega_n} \tag{2.41}$$

defines the value of the driving frequency, ω_p at which the peak value of the magnitude occurs. This holds only for underdamped systems for which $\zeta < 1/\sqrt{2}$.

Otherwise, the magnitude does not have a maximum value or peak for any value of $\omega > 0$ because $\sqrt{1 - 2\zeta^2}$ becomes an imaginary number for values of ζ larger than $1/\sqrt{2}$. Note also that this peak occurs a little to the left of, or before, resonance $(r = 1)$ since

$$r_{peak} = \sqrt{1 - 2\zeta^2} < 1$$

This can be seen in both Figures 2.8 and 2.9. The value of the magnitude at r_{peak} is

$$\frac{Xk}{F_0} = \frac{1}{2\zeta\sqrt{1 - \zeta^2}} \tag{2.42}$$

which is obtained simply by substituting $r_{peak} = \sqrt{1 - 2\zeta^2}$ into the expression for the normalized magnitude Xk/F_0. □

Note that for damped systems resonance is usually defined, as in the undamped case, by $r = 1$ or $\omega_n = \omega$. However, this condition does not define precisely the peak value of the magnitude of the steady-state response as defined by equation (2.40) and as plotted in Figure 2.9. This is the point of Example 2.2.5, which illustrates that the maximum value of Xk/F_0 occurs at $r = \sqrt{1 - 2\zeta^2}$ if $0 \le \zeta < 1/\sqrt{2}$ and at $r = 0$ if $\zeta > 1/\sqrt{2}$. For the small damping case $(\zeta < 1/\sqrt{2})$ the value of the driving frequency corresponding to the maximum value of Xk/F_0 is called the *peak frequency*, denoted by ω_p, which has the value derived previously:

$$\omega_p = \omega_n\sqrt{1 - 2\zeta^2} \quad \text{for} \quad 0 \le \zeta \le \frac{1}{\sqrt{2}} \tag{2.43}$$

Note that as the damping decreases, ω_p approaches ω_n, resulting in the usual undamped resonance condition. As ζ increased from zero, the curves in Figure 2.9 have peaks that occur farther and farther away from the vertical line $r = 1$. Eventually, the damping ratio increases past the value $1/\sqrt{2}$ and the largest value of Xk/F_0 occurs at $r = 0$ In many applications ζ is small, so that the value $\sqrt{1 - 2\zeta^2}$ is very close to 1. Hence the undamped resonance condition $\omega = \omega_n$ (i.e., $r = 1$) is often used for resonance in the (lightly) damped case as well. As an example, for $\zeta = 0.1$ a system with an undamped natural frequency of 200 Hz would have a peak value of 198 Hz, which is less than a 1% error (i.e., $r = 0.9899$ instead of $r = 1$). Hence, in practice, the value of the frequency corresponding to the peak is often taken to be simply the undamped natural frequency.

It is interesting to examine the peak value of the magnitude response as a function of the damping ratio for underdamped systems. The bottom plot in Figure 2.10 shows how the damping affects the peak amplitude by plotting Xk/F_0 as a function of ζ as given by equation (2.41). The plot in the upper right shows how the peak frequency ratio changes with the damping ratio. Note from the figure that the magnitude varies across three orders of magnitude as the damping is increased $(0 \le \zeta \le 0.707)$. Also note that the peak value moves to the left of $r = 1$ for high values of the damping ratio.

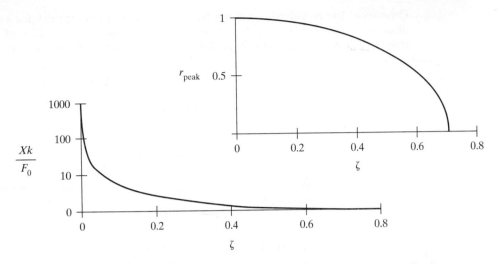

Figure 2.10 The lower graph is a plot of equation (2.41) on a log scale indicating how the increased damping reduces the peak response. The plot on the upper right shows how much the peak amplitude shifts from $r = 1$ as the damping increases. (Note that these plots are only defined for $\zeta \leq 0.707$.)

To explain physically the phenomena of resonance, consider the steady-state forced response of the system where the applied force is $F_0 \cos(\omega t)$, the displacement is $x_p(t) = X \cos(\omega t - \theta)$, and the velocity is $\dot{x}_p(t) = -\omega X \sin(\omega t - \theta)$. At resonance, $\theta = \pi/2$. Thus $\dot{x}_p(t)_{(resonance)} = \omega X \cos(\omega t)$. This shows that at resonance the velocity and the force are exactly in phase but have different magnitudes. Physically, this means that the force is always pushing in the direction of the velocity and that the force changes magnitude and direction just as the velocity does. This condition will cause the vibration amplitude of the system to reach its maximum value because at resonance the external force never opposes the velocity.

2.3 ALTERNATIVE REPRESENTATIONS

As you may recall from the theory of differential equations, there are a variety of methods useful for calculating solutions of a spring–mass–damper system excited by a harmonic force as described by equation (2.26). In Section 2.2, the method of undetermined coefficients was used. In this section, three other approaches to solve this problem are discussed: a geometric approach, a frequency response approach, and a transform approach.

2.3.1 Geometric Method

The geometric approach consists of solving equation (2.26) by treating each force as a vector. Recall that x_p, $\dot{x}_p$, and $\ddot{x}_p$ will each be 90° out of phase with each other. Each of these are plotted in Figure 2.11 for the assumed solution $x_p = X\cos(\omega t - \theta)$, $\dot{x}_p = \omega X\cos(\omega t - \theta + 90°)$, and $\ddot{x}_p = -\omega^2 X\cos(\omega t - \theta)$. Adding these three quantities as vectors indicates that X can be solved in terms of F_0 by combining the sides of the right triangle ABC to yield

$$F_0^2 = (k - m\omega^2)^2 X^2 + (c\omega)^2 X^2 \tag{2.44}$$

or

$$X = \frac{F_0}{\sqrt{(k - m\omega^2)^2 + (c\omega)^2}} \tag{2.45}$$

and

$$\theta = \tan^{-1}\frac{c\omega}{k - m\omega^2} \tag{2.46}$$

Substituting in the values for $F_0 = mf_0$ and $c = 2m\omega_n\zeta$ illustrates that this solution is identical with equation (2.36) derived by the method of undetermined coefficients. Note from the figure that at resonance $\omega^2 = k/m$ causing line AB to be zero, lines CD and AE become the same length and the angle θ becomes a right angle. Thus at resonance the phase shift is 90°.

The graphical method of solving equation (2.26) is more illustrative than useful, as it is difficult to extend to other forms of the forcing function or to more complicated problems. It is presented here because the method potentially helps clarify and illustrate the forced vibration of a simple single-degree-of-freedom system. In particular, Figure 2.11 makes it easy to see that at resonance ($\theta = 90°$) the applied force and damping force are acting in the same direction and the stiffness force is equal and opposite to the inertial force.

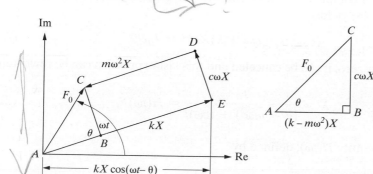

Figure 2.11 A graphical representation of the solution of equation (2.26).

An alternative method that is similar to the graphical approach is to treat the solution of equation (2.26) as a complex function. This leads to a frequency response description of forced harmonic motion and is more useful for complicated problems involving many degrees of freedom. Complex functions are reviewed in Appendix A.

2.3.2 Complex Response Method

Euler's formula for trigonometric functions relates the exponential function to harmonic motion by the complex relation

$$Ae^{j\omega t} = A\cos\omega t + (A\sin\omega t)j \tag{2.47}$$

where $j = \sqrt{-1}$. Thus $Ae^{j\omega t}$ is a complex function with a real part ($A\cos\omega t$) and an imaginary part ($A\sin\omega t$). Appendix A reviews complex numbers and functions. With this notation in mind, $Ae^{j\omega t}$ represents a harmonic function and can be used to discuss forced harmonic motion by rewriting the equation of motion (2.26)

$$m\ddot{x} + c\dot{x} + kx = F_0\cos\omega t$$

as the complex equation

$$m\ddot{x}(t) + c\dot{x}(t) + kx(t) = F_0 e^{j\omega t} \tag{2.48}$$

Here, the real part of the complex solution corresponds to the physical solution $x(t)$. This representation is extremely useful in solving multiple-degree-of-freedom systems (Chapter 4), as well as in understanding vibration measurement systems (Chapter 7).

This method proceeds by assuming that the complex particular solution of equation (2.48) is of the exponential form

$$x_p(t) = Xe^{j\omega t} \tag{2.49}$$

where X is now a complex-valued constant to be determined. Substitution of this into equation (2.48) yields

$$(-\omega^2 m + cj\omega + k)Xe^{j\omega t} = F_0 e^{j\omega t} \tag{2.50}$$

Since $e^{j\omega t}$ is never zero, it can be canceled and this last expression can be rewritten as

$$X = \frac{F_0}{(k - m\omega^2) + (c\omega)j} = H(j\omega)F_0 \tag{2.51}$$

The complex quantity $H(j\omega)$, defined by

$$H(j\omega) = \frac{1}{(k - m\omega^2) + (c\omega j)} \tag{2.52}$$

is called the (complex) *frequency response function.* Following the rules for manipulating complex numbers (i.e., multiplying by the complex conjugate over itself
and taking the modulus of the result as outlined in Appendix A) yields

$$X = \frac{F_0}{[(k - m\omega^2)^2 + (c\omega)^2]^{1/2}} e^{-j\theta} \tag{2.53}$$

where

$$\theta = \tan^{-1} \frac{c\omega}{(k - m\omega^2)} \tag{2.54}$$

Substituting the value for X into equation (2.49) yields the solution

$$x_p(t) = \frac{F_0}{[(k - m\omega^2)^2 + (c\omega)^2]^{1/2}} e^{-j(\omega t - \theta)} \tag{2.55}$$

The real part of this expression corresponds to the solution given in equation (2.36) obtained by the method of undetermined coefficients. The complex exponential approach
for obtaining the forced harmonic response corresponds to the graphical approach
described in Figure 2.11 by labeling the x axis as the real part of $e^{j\omega t}$ and the y axis as the
complex part.

Example 2.3.1

Use the frequency response approach to compute the amplitude of the particular solution for the undamped system of equation (2.2) defined by

$$m\ddot{x}(t) + kx(t) = F_0 \cos \omega t$$

Solution   First, write equation (2.2) with the forcing function modeled as a complex
exponential:

$$m\ddot{x}(t) + kx(t) = F_0 e^{j\omega t}$$

Dividing by the mass, m, yields the monic form

$$\ddot{x}(t) + \omega_n^2 x(t) = f_0 e^{j\omega t}$$

Assume a particular solution of the exponential form given in equation (2.49) and substitute into the last expression to get

$$(-\omega^2 + \omega_n^2) X e^{j\omega t} = f_0 X e^{j\omega t}$$

Solving for X yields

$$X = \frac{f_0}{\omega_n^2 - \omega^2}$$

This is in perfect agreement with equation (2.7) derived using the cosine representation of
the forcing function and (2.21) derived using the sine representation. This also agrees with
solution given in equation (2.36) for the damped case, by setting $\zeta = 0$ in that expression.

□

The method used here to derive the solution for the forced response and the resulting frequency response function is very similar to the eigenvalue approach to solving vibration problems. This approach, introduced in Section 1.3, is used extensively in Chapter 4 and consists of assuming solutions with exponential time dependence, as illustrated previously.

2.3.3 Transfer Function Method

Next consider using the Laplace transform (see Appendix B and Section 3.4 for a review) approach to solve for the particular solution of equation (2.26). The Laplace transform method is a powerful approach that can be used for a variety of forcing functions (see Section 3.4) and can be readily applied to multiple-degree-of-freedom systems. Taking the Laplace transform of the equation of motion (2.26)

$$m\ddot{x}(t) + c\dot{x}(t) + kx(t) = F_0 \cos \omega t$$

assuming that the initial conditions are zero, yields

$$(ms^2 + cs + k) X(s) = \frac{F_0 s}{s^2 + \omega^2} \tag{2.56}$$

where s is the complex transform variable and $X(s)$ denotes the Laplace transform of the unknown function $x(t)$. Solving algebraically for the unknown function $X(s)$ yields

$$X(s) = \frac{F_0 s}{(ms^2 + cs + k)(s^2 + \omega^2)} \tag{2.57}$$

which represents the transformed solution. To calculate the inverse Laplace transform of $X(s)$, the right side of equation (2.57) can be found in a table of Laplace transform pairs, or the method of partial fractions can be used to reduce the right-hand side of equation (2.57) to simpler quantities for which the inverse Laplace transform is known. The solution obtained by the inversion procedure is, of course, equivalent to the solution given in (2.36) and again in (2.55). This solution technique is discussed in more detail in Section 3.4.

Of particular use is the frequency response function defined by equation (2.52). This function is related to the Laplace transform for a vibrating system. Consider equation (2.26) and its Laplace transform

$$(ms^2 + cs + k) X(s) = F(s) \tag{2.58}$$

where $F(s)$ symbolically denotes the Laplace transform of the driving function [i.e., the right-hand side of (2.26)]. Manipulating equation (2.58) yields

$$\frac{X(s)}{F(s)} = \frac{1}{(ms^2 + cs + k)} = H(s) \tag{2.59}$$

which expresses the ratio of the Laplace transform of the output (response) to the Laplace transform of the input (driving force) for the case of zero initial conditions. This ratio, denoted by $H(s)$, is called the *transfer function* of the system and provides an important tool for vibration analysis, design, and measurement as discussed in the remaining chapters.

Recall that the Laplace transform variable s is a complex number. If the value of s is restricted to lie along the imaginary axis in the complex plane (i.e., if $s = j\omega$), the transfer function becomes

$$H(j\omega) = \frac{1}{k - m\omega^2 + c\omega j} \qquad (2.60)$$

which, upon comparison with equation (2.52), is the frequency response function of the system. Hence the frequency response function of the system is the transfer function of the system evaluated along $s = j\omega$. Both the transfer function and the frequency response function are used in Chapters 4 and 7.

Example 2.3.2

Consider the system of Figure 2.12. Let J denote the inertia of the wheel and hub about the shaft, and let k denote the torsional stiffness of the system. The suspension system is subjected to a harmonic excitation as indicated. Compute an expression for the forced response using the Laplace transform method (assuming zero initial conditions and that the tire is not touching the ground).

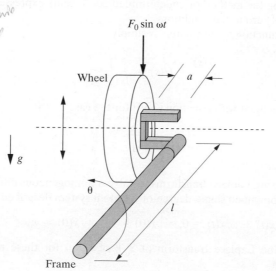

Figure 2.12   A schematic of a torsional suspension.

Solution   Modeling the system as a torsional vibration problem and summing moments about the shaft, the equation of motion becomes

$$J\ddot{\theta} + k\theta = aF_0 \sin \omega t$$

Taking the Laplace transform (see Appendix B) of the equation of motion yields

$$Js^2 X(s) + kX(s) = aF_0 \frac{\omega}{s^2 + \omega^2}$$

where $X(s)$ is the Laplace transform of $\theta(t)$. Algebraically solving for $X(s)$ yields

$$X(s) = a\omega F_0 \frac{1}{(s^2 + \omega^2)(Js^2 + k)}$$

Next use a table (see Appendix B) to compute the inverse Laplace transform to get

$$\theta(t) = L^{-1}(X(s)) = a\omega F_0 L^{-1}\left(\frac{1}{(s^2 + \omega^2)(Js^2 + k)}\right)$$

$$= \frac{a\omega F_0}{J} L^{-1}\left(\frac{1}{(s^2 + \omega^2)(s^2 + \omega_n^2)}\right) = \frac{a\omega F_0}{J} \frac{1}{\omega^2 - \omega_n^2}\left(\frac{1}{\omega}\sin\omega t - \frac{1}{\omega_n}\sin\omega_n t\right)$$

Here L^{-1} denotes the inverse Laplace transform and the natural frequency is

$$\omega_n = \sqrt{\frac{k}{J}}$$

Note that the solution computed here using the Laplace transform agrees with the solution obtained using the method of undetermined coefficients expressed in equation (2.25) for the case of zero initial conditions.

The transfer function for the system is simply

$$\frac{X(s)}{F(s)} = H(s) = \frac{1}{Js^2 + k}$$

This result is in agreement with equation (2.59) for the case $c = 0$.

□

Example 2.3.3

As an example of using Laplace transforms to solve a homogeneous differential equation, consider the undamped single-degree-of-freedom system described by

$$\ddot{x}(t) + \omega_n^2 x(t) = 0, \qquad x(0) = x_0, \qquad \dot{x}(0) = v_0$$

Solution Taking the Laplace transform of $\ddot{x} + \omega_n^2 x = 0$ for these nonzero initial conditions results in

$$s^2 X(s) - sx_0 - v_0 + \omega_n^2 X(s) = 0$$

by direct application of the definition given in Appendix B and the linear nature of the Laplace transform. Algebraically solving this last expression for $X(s)$ yields

$$X(s) = \frac{x_0 + sv_0}{s^2 + \omega_n^2}$$

Using $L^{-1}[X(s)] = x(t)$ and entries (6) and (5) of Table B.1 yields that the solution is

$$x(t) = x_0 \cos \omega_n t + \frac{v_0}{\omega_n} \sin \omega_n t$$

This is, of course, in total agreement with the solution obtained in Chapter 1.

$\square$

This section presented three alternative methods for calculating the particular solution for a harmonically excited system. Each was shown to yield the same result. The concepts presented in these solution techniques are generalized and used for more complicated problems in later chapters.

2.4 BASE EXCITATION

Often, machines, or parts of machines, are harmonically excited through elastic mountings, which may be modeled by springs and dashpots. For example, an automobile suspension system is excited harmonically by a road surface through a shock absorber, which may be modeled by a linear spring in parallel with a viscous damper. Other examples are the rubber motor mounts that separate an automobile engine from its frame or an airplane's engine from its wing or tail section. Such systems can be modeled by considering the system to be excited by the motion of its support. This forms the *base-excitation* or *support-motion* problem modeled in Figure 2.13.

Summing the relevant forces on the mass, m, Figure 2.13 yields (i.e., the inertial force $m\ddot{x}$ is equal to the sum of the two forces acting on m, and the gravitational force is balanced against the static deflection of the spring as before)

$$m\ddot{x} + c(\dot{x} - \dot{y}) + k(x - y) = 0 \qquad (2.61)$$

Here note that the spring deflects a distance $(x - y)$ and the damper experiences a velocity of $(\dot{x} - \dot{y})$. For the base-excitation problem it is assumed that the base moves harmonically, that is, that

$$y(t) = Y \sin \omega_b t \qquad (2.62)$$

where Y denotes the amplitude of the base motion and ω_b represents the frequency of the base oscillation. Substitution of $y(t)$ from equation (2.62) into the equation of motion given in (2.61) yields, after some rearrangement,

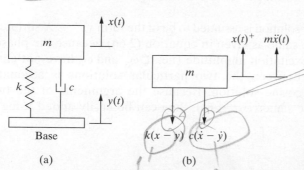

$x(t)$

m

k c

$y(t)$

Base

(a)

$x(t)^+$ $m\ddot{x}(t)$

m

$k(x - y)$ $c(\dot{x} - \dot{y})$

(b)

Figure 2.13   (a) Base-excitation problem models the motion of an object of mass m as being excited by a prescribed harmonic displacement acting through the spring and damper. (b) A free-body diagram of the base motion problem in (a).

$$m\ddot{x} + c\dot{x} + kx = cY\omega_b \cos\omega_b t + kY \sin\omega_b t \qquad (2.63)$$

This can be thought of as a spring–mass–damper system with *two* harmonic inputs. The expression is very similar to the problem stated by equation (2.26) for the forced harmonic response of a damped system with $F_0 = cY\omega_b$ and $\omega = \omega_b$, except for the "extra" forcing term $kY \sin\omega_b t$. The solution approach is to use the linearity of the equation of motion and realize that the particular solution of equation (2.63) will be the sum of the particular solution obtained by assuming an input force of $cY\omega_b \cos\omega_b t$, denoted by $x_p^{(1)}$, and the particular solution obtained by assuming an input force of $kY \sin\omega_b t$, denoted by $x_p^{(2)}$.

Calculating these particular solutions follows directly from the calculation made in Section 2.2. Dividing equation (2.63) by m and using the definitions of damping ratio and natural frequency yields

$$\ddot{x} + 2\zeta\omega_n\dot{x} + \omega_n^2 x = 2\zeta\omega_n\omega_b Y \cos\omega_b t + \omega_n^2 Y \sin\omega_b t \qquad (2.64)$$

Thus substituting $f_0 = 2\zeta\omega_n\omega_b Y$ into equation (2.36) yields that the particular solution $x_p^{(1)}$ due to the cosine excitation is

$$x_p^{(1)} = \frac{2\zeta\omega_n\omega_b Y}{\sqrt{(\omega_n^2 - \omega_b^2)^2 + (2\zeta\omega_n\omega_b)^2}} \cos(\omega_b t - \theta_1) \qquad (2.65)$$

where

$$\theta_1 = \tan^{-1}\frac{2\zeta\omega_n\omega_b}{\omega_n^2 - \omega_b^2} \qquad (2.66)$$

To calculate $x_p^{(2)}$, the method of undetermined coefficients is applied again with the harmonic input $\omega_n^2 Y \sin\omega_b t$. Following the procedures used to calculate equation (2.36) results in

$$x_p^{(2)} = \frac{\omega_n^2 Y}{\sqrt{(\omega_n^2 - \omega_b^2)^2 + (2\zeta\omega_n\omega_b)^2}} \sin(\omega_b t - \theta_1) \qquad (2.67)$$

Note that equation (2.67) with $\zeta = 0$ agrees with equation (2.22) for the undamped case, as it should.

Here the particular solution is assumed to be of the form $x_p^{(2)} = X \sin(\omega_b t - \theta_1)$. Here the angle θ_1 is the same as given in equation (2.66) because the phase angle is independent of the excitation amplitude (i.e., ζ, ω_n, and ω_b have not changed). The phase difference between the two particular solutions is accounted for by using the sine and cosine solution. Because the arguments of the two particular solutions are the same ($\omega_b t - \theta_1$), they can be easily added using simple trigonometry.

From the principle of linear superposition, the total particular solution is the sum of equations (2.65) and (2.67) (i.e., $x_p = x_p^{(1)} + x_p^{(2)}$). Adding solutions (2.65) and (2.67) yields

$$x_p t = \omega_n Y \left[\frac{\omega_n^2 + (2\zeta\omega_b)^2}{(\omega_n^2 - \omega_b^2)^2 + (2\zeta\omega_n\omega_b)^2} \right]^{1/2} \cos(\omega_b t - \theta_1 - \theta_2) \qquad (2.68)$$

where

$$\theta_2 = \tan^{-1} \frac{\omega_n}{2\zeta\omega_b} \qquad (2.69)$$

It is convenient to denote the magnitude of the particular solution, $x_p(t)$, by X so that

$$X = Y \left[\frac{1 + (2\zeta r)^2}{(1 - r^2)^2 + (2\zeta r)^2} \right]^{1/2} \qquad (2.70)$$

where the frequency ratio $r = \omega_b/\omega_n$. Dividing this last expression by the magnitude of base motion, Y, yields

$$\frac{X}{Y} = \left[\frac{1 + (2\zeta r)^2}{(1 - r^2)^2 + (2\zeta r)^2} \right]^{1/2} \qquad (2.71)$$

which expresses the ratio of the maximum response magnitude to the input displacement magnitude. This ratio is called the *displacement transmissibility* and is used to describe how motion is transmitted from the base to the mass as a function of the frequency ratio ω_b/ω_n. This ratio is plotted in Figure 2.14. Note that near $r = \omega_b/\omega_n = 1$, or resonance, the maximum amount of base motion is transferred to displacement of the mass.

Note from Figure 2.14 that for $r < \sqrt{2}$ the transmissibility ratio is greater than 1, indicating that for these values of the system's parameters (ω_n) and base frequency (ω_b), the motion of the mass is an amplification of the motion of the base. Notice also that for a given value of r, the value of the damping ratio ζ determines the level of amplification. Specifically, larger ζ yields smaller transmissibility ratios.

For values $r > \sqrt{2}$ the transmissibility ratio is always less than 1 and the motion of the mass will be of smaller amplitude than the amplitude of the exciting base motion. In this higher-frequency range, the effect of increasing damping is just the opposite of that in the low-frequency case. Increasing the damping actually increases the amplitude ratio in the higher-frequency range. However, the amplitude is always less than 1 for underdamped systems. The frequency range defined by $r > \sqrt{2}$ forms the important concept of vibration isolation discussed in detail in Section 5.2.

For a fixed amount of damping, say $\zeta = 0.01$, the important aspect of base motion is that the mass experiences larger amplitude oscillations than the base excitation provides for $r < \sqrt{2}$ and experiences smaller amplitude oscillations

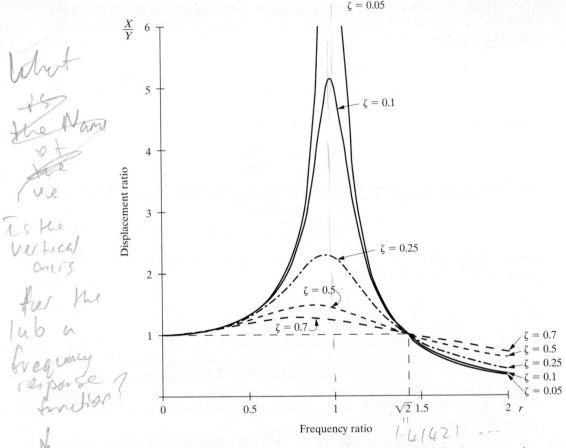

Figure 2.14 Displacement transmissibility as a function of the frequency ratio, illustrating how the dimensionless deflection X/Y varies as the frequency of the base motion increases for several different damping ratios.

than the base excitation provides for $r > \sqrt{2}$. Near resonance, most of the motion of the base is amplified into motion of the mass, causing it to have large amplitude oscillations.

It is interesting to compare the transmissibility plot of Figure 2.14 and equation (2.71) for base excitation with the steady-state magnitude plot for harmonic excitation of the mass as given in Figure 2.9 and equation (2.40). First, note that the frequency ratio, r, is independent of the damping ratio by definition in both cases. However, the dependence of the peak value, r_{peak}, on the damping ratio is different in each case (recall the computation in Example 2.2.5 for the peak value). In particular, Figure 2.10 will be different for base excitation than for harmonic excitation of the mass directly. This difference is caused by the numerator having an additional term in the base-excitation problem. This term, $2\zeta r$, comes from the load carried through the damper which is not present when the mass is excited directly, as in equation (2.40).

Another quantity of interest in the base-excitation problem is the force transmitted to the mass as the result of a harmonic displacement of the base. The force transmitted to the mass is done so through the spring and damper. Hence, the force transmitted to the mass is the sum of the force in the spring and the force in the damper, or from the free-body diagram, Figure 2.13,

$$F(t) = k(x - y) + c(\dot{x} - \dot{y}) \tag{2.72}$$

This force must balance the inertial force of the mass m; thus

$$F(t) = -m\ddot{x}(t) \tag{2.73}$$

In the steady state, the solution for x is given by equation (2.68). Differentiating equation (2.68) twice and substituting into equation (2.73) yields

$$F(t) = m\omega_b^2\omega_n Y\left[\frac{\omega_n^2 + (2\zeta\omega_b)^2}{(\omega_n^2 - \omega_b^2)^2 + (2\zeta\omega_n\omega_b)^2}\right]^{1/2} \cos(\omega_b t - \theta_1 - \theta_2) \tag{2.74}$$

Again using the frequency ratio r, this becomes

$$F(t) = F_T \cos(\omega_b t - \theta_1 - \theta_2) \tag{2.75}$$

where the magnitude of the transmitted force, F_T, is given by

$$F_T = kYr^2\left[\frac{1 + (2\zeta r)^2}{(1 - r^2)^2 + (2\zeta r)^2}\right]^{1/2} \tag{2.76}$$

Equation (2.76) is used to define *force transmissibility* by forming the ratio

$$\frac{F_T}{kY} = r^2\left[\frac{1 + (2\zeta r)^2}{(1 - r^2)^2 + (2\zeta r)^2}\right]^{1/2} \tag{2.77}$$

This force transmissibility ratio, F_T/kY, expresses a dimensionless measure of how displacement in the base of amplitude Y results in a force magnitude applied to the mass. Note from equations (2.75) and (2.68) that the force transmitted to the mass is in phase with the displacement of the mass. Figure 2.15 illustrates the force transmissibility as a function of the frequency ratio for four values of the damping ratio. Note that unlike the displacement transmissibility, the force transmitted does not necessarily fall off for $r > \sqrt{2}$. In fact, as the damping increases, the force transmitted increases dramatically for $r > \sqrt{2}$.

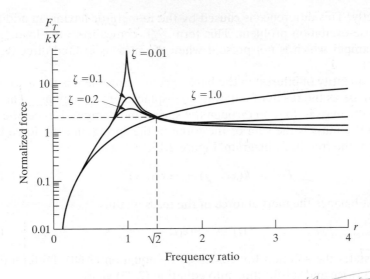

Figure 2.15 The force transmitted to the mass as a function of the frequency ratio illustrating how the dimensionless force ratio varies as the frequency of the base motion increases for four different damping ratios ($\zeta = 0.01, 0.1, 0.2,$ and 1.0).

Example 2.4.1

Consider the base-excitation problem with the following data: $m = 100$ kg, $c = 30$ kg/s, $k = 2000$ N/m, $Y = 0.03$ m, and $\omega_b = 6$ rad/s. Compute the magnitude of the transmissibility ratio and then the force transmissibility ratio.

Solution   First, define the usual vibration properties by dividing by the mass to get

$$\omega_n = \sqrt{\frac{2000}{100}} = 4.472 \text{ rad/s}, \zeta = \frac{c}{2\sqrt{mk}} = \frac{30}{2\sqrt{2} \times 10^5} = 0.034$$

$$r = \frac{\omega_b}{\omega_n} = \frac{6}{4.472} = 1.342$$

Then use equation (2.71) to compute the magnitude of the particular solution:

$$\frac{X}{Y} = \left[\frac{1 + (2\zeta r)^2}{(1 - r^2)^2 + (2\zeta r)^2}\right]^{1/2} = \left[\frac{1 + (2 \cdot 0.034 \cdot 1.342)^2}{(1 - (1.342)^2)^2 + (2 \cdot 0.034 \cdot 1.342)^2}\right] = 0.557$$

The force transmissibility ratio becomes

$$\frac{F_T}{kY} = r^2\frac{X}{Y} = (1.342)^2(0.557) = 1.003$$

Note that if the damping value is changed to $c = 300$ kg/s, the force transmissibility ratio is 1.203 and the transmissibility is 0.669. So increasing the damping increases both the force and the displacement transmissibility.

$\square$

A comparison between force transmissibility (force applied to the mass normalized by the magnitude of the displacement of the base) and displacement transmissibility (displacement of the mass normalized by the magnitude of the displacement of the base) is given in Figure 2.16.

The formulas for transmissibility of force and displacement are very useful in the design of systems to provide protection from unwanted vibration. This is discussed in detail in Section 5.2 on vibration isolation, where the transmissibility ratio is derived for a *fixed* base and compared to the development here in Window 5.1. The following example illustrates some practical values of transmissibility for the base-excitation problem.

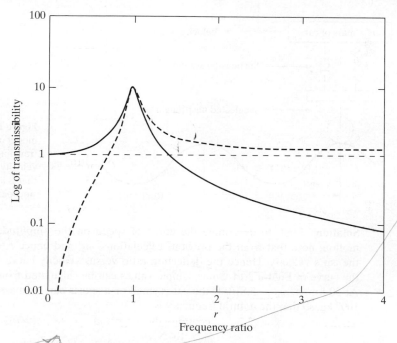

Figure 2.16 A comparison between force transmissibility (dashed line) and displacement transmissibility (solid line) for a damping ratio of $\zeta = 0.05$ on a semilog plot using equations (2.71) and (2.77).

Example 2.4.2

A common example of base motion is the single-degree-of-freedom model of an automobile driving over a road or an airplane taxiing over a runway, indicated in Figure 2.17. The road (or runway) surface is approximated as sinusoidal in cross section providing a base motion displacement of

$$y(t) = (0.01 \text{ m}) \sin \omega_b t$$

where

$$\omega_b = v(\text{km/h})\left(\frac{1}{0.006 \text{ km}}\right)\left(\frac{\text{hour}}{3600 \text{ s}}\right)\left(\frac{2\pi \text{ rad}}{\text{cycle}}\right) = 0.2909v \text{ rad/s}$$

where v denotes the vehicle's velocity in km/h. Thus the vehicle's speed determines the frequency of the base motion. Determine the effect of speed on the amplitude of displacement of the automobile as well as the effect of the value of the car's mass. Assume that the suspension system provides an equivalent stiffness of 4×10^4 N/m and damping of 20×10^2 N·s/m.

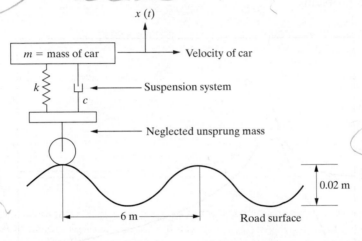

Figure 2.17 A simple model of a vehicle traveling with constant velocity on a wavy surface that is approximated as a sinusoid.

Solution First, to determine the effect of speed on the amplitude of the vehicle's motion, note that from the previous calculations, ω_b, and hence r, vary linearly with the car's velocity. Hence the deflection ratio versus velocity curve will be much like the curve of Figure 2.14. Some sample values can be calculated from equation (2.70). At 20 km/h, $\omega_b = 5.818$. If the car is small or a sports car, its mass might be around 1007 kg, so that the natural frequency is

$$\omega_n = \sqrt{\frac{4 \times 10^4 \text{ N/m}}{1007 \text{ kg}}} = 6.303 \text{ rad/s } (\approx 1 \text{ Hz})$$

so that $r = 5.818/6.303 = 0.923$ and

$$\zeta = \frac{c}{2\sqrt{km}} = \frac{2000 \text{ N·s/m}}{2\sqrt{(4 \times 10^4 \text{ N/m})(1007 \text{ kg})}} = 0.158$$

Equation (2.70) then yields that the deflection experienced by the car will be

$$X = (0.01 \text{ m})\sqrt{\frac{1 + \left[2(0.158)(0.923)\right]^2}{\left[1 - (0.923)^2\right]^2 + \left[2(0.158)(0.923)\right]^2}} = 0.0319$$

This means that a 1-cm bump in the road is transmitted into a 3.2-cm "bump" experienced by the chassis and subsequently transmitted to the occupants. Hence the suspension system amplifies the rough road bumps in this circumstance and is not desirable.

Table 2.1 lists several different values of the vehicle displacement for two different vehicles traveling at four different speeds over the same 1-cm bump. Car 1 with frequency ratio r_1 is a 1007-kg sports car, while car 2 is a 1585-kg sedan with frequency ratio r_2. The same suspension system was used on both cars to illustrate the need to design suspension systems based on a given vehicle's specifications (see Chapter 5). Note that with higher speed, negligible vibration is experienced by the occupants of the car. Also, notice that the suspension system parameters chosen (k and c) work better in general for the larger car except at very low speeds.

TABLE 2.1   COMPARISON OF CAR VELOCITY, FREQUENCY, AND DISPLACEMENT FOR TWO DIFFERENT CARS

Speed (km/h)	ω_b	r_1	r_2	x_1 (cm)	x_2 (cm)
20	5.817	0.923	1.158	3.19	2.32
80	23.271	3.692	4.632	0.12	0.07
100	29.088	4.615	5.79	0.09	0.05
150	43.633	6.923	8.686	0.05	0.03

Example 2.4.3

A large rotating machine causes the floor of a factory to oscillate sinusoidally. A punch press is to be mounted on the same floor (Figure 2.18). The displacement of the floor at the point where the punch press is to be mounted is measured to be $y(t) = 0.1 \sin \omega_b t$ (cm). Using the base support model of this section, calculate the maximum force transmitted to the punch press at resonance if the press is mounted on a rubber fitting of stiffness $k = 40,000$ N/m; damping, $c = 900$ N·s/m; and mass, $m = 3000$ kg.

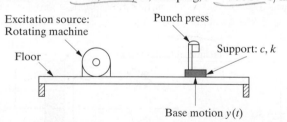

Excitation source: Rotating machine

Floor

Punch press

Support: c, k

Base motion $y(t)$

Figure 2.18   A model of a machine causing support motion.

Solution   The force transmitted to the punch press is given by equation (2.77). At resonance, $r = 1$, so that equation (2.77) becomes

$$\frac{F_T}{kY} = \left[\frac{1 + (2\zeta)^2}{(2\zeta)^2} \right]^{1/2}$$

or

$$F_T = \frac{kY}{2\zeta} \left(1 + 4\zeta^2 \right)^{1/2}$$

From the definition of ζ and the values given previously for m, c, and k,

$$\zeta = \frac{c}{2\sqrt{km}} = \frac{900}{2\left[(40,000)(3000)\right]^{1/2}} \cong 0.04$$

From the measured excitation $Y = 0.001$ m, so that

$$F_T = \frac{kY}{2\zeta}\left(1 + 4\zeta^2\right)^{1/2} = \frac{(40,000\ \text{N/m})(0.001\ \text{m})}{2(0.04)}\left[1 + 4(0.04)^2\right]^{1/2}$$

$$= 501.6\ \text{N}$$

□

The analysis presented here for base motion is very useful in design. This forms the topic of Chapter 5, which includes as Section 5.2 a more detailed analysis of base motion in the context of the vibration isolation problem for both fixed-base and moving-base models. Section 5.2 also includes a discussion of shock isolation. Some may prefer to jump to Section 5.2 at this point to examine how the concepts of transmissibility are applied to the base isolation problem.

2.5 ROTATING UNBALANCE

A common source of troublesome vibration is rotating machinery. Many machines and devices have rotating components, usually driven by electric motors. Small irregularities in the distribution of the mass in the rotating component can cause substantial vibration. This is called a *rotating unbalance*. A schematic of such a rotating unbalance of mass, m_0, a distance e from the center of rotation is given in Figure 2.19.

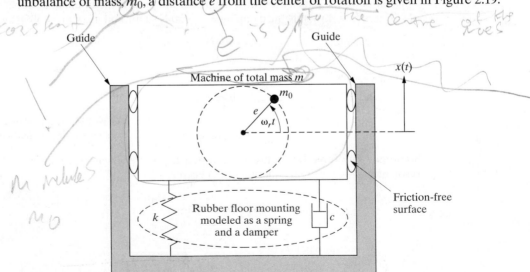

Figure 2.19   A model of a machine causing support motion.

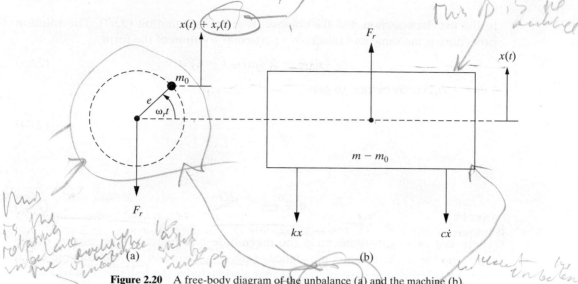

Figure 2.20   A free-body diagram of the unbalance (a) and the machine (b).

The frequency of rotation of the machine is denoted by ω_r. Summing forces in the vertical direction (x) from the free-body diagram of the out-of-balance mass given in Figure 2.20(a) yields

$$m_0(\ddot{x} + \ddot{x}_r) = -F_r \tag{2.78}$$

Summing forces from the free-body diagram of the machine given in Figure 2.20(b) yields

$$(m - m_0)\ddot{x} = F_r - c\dot{x} - kx \tag{2.79}$$

Combining equations (2.78) and (2.79) yields

$$m\ddot{x} + m_0\ddot{x}_r + c\dot{x} + kx = 0 \tag{2.80}$$

The forces in the horizontal direction are canceled by the guides and are not considered here.

Assuming that the machine rotates with a constant frequency, ω_r, the x component of the motion of the mass, m_0, is $x_r = e \sin \omega_r t$, so that

$$\ddot{x}_r = -e\omega_r^2 \sin \omega_r t \tag{2.81}$$

Substitution of equation (2.81) into (2.80) yields

$$m\ddot{x} + c\dot{x} + kx = m_0 e\omega_r^2 \sin \omega_r t \tag{2.82}$$

after rearranging the terms. Equation (2.82) is similar to equation (2.26) with $F_0 = m_0 e\omega_r^2$, with the exception of the phase shift of the forcing function (i.e., $\sin \omega_r t$ instead of $\cos \omega t$). The sine excitation is discussed as the second particular solution in the previous section and the solution is given in equation (2.67). The solution procedure is the same and results in a particular solution of the form

$$x_p(t) = X \sin(\omega_r t - \theta) \qquad (2.83)$$

Let $r = \omega_r/\omega_n$, as before, to get

$$X = \frac{m_0 e}{m} \frac{r^2}{\sqrt{(1 - r^2)^2 + (2\zeta r)^2}} \qquad (2.84)$$

and

$$\theta = \tan^{-1} \frac{2\zeta r}{1 - r^2} \qquad (2.85)$$

These last two expressions yield the magnitude and phase of the motion of the mass, m, due to the rotating unbalance of mass m_0. Note that the mass m in equation (2.84) is the total mass of the machine and includes the unbalance mass m_0.

The magnitude of the steady-state displacement, X, as a function of the rotating speed (frequency) is examined by plotting the dimensionless displacement magnitude $mX/m_0 e$ versus r, as indicated in Figure 2.21 for various values of the damping ratio ζ. Note that equation (2.84) is similar to the magnitude analyzed in Example 2.2.4. From the form of the denominator, which is identical to that of Example 2.2.4, it is observed that the maximum deflection is less than or equal to 1 for any system with $\zeta > 1$. This indicates that the increase in amplification of the amplitude caused by the unbalance can be eliminated by increasing the damping in

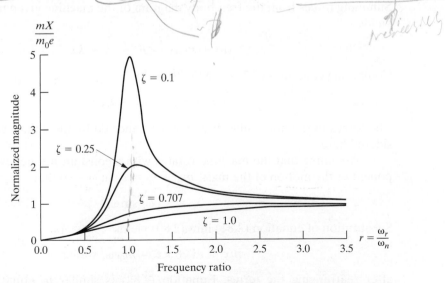

Figure 2.21 Magnitude of the dimensionless displacement versus frequency ratio caused by a rotating unbalance of mass m_0 and radius e.

the system. However, large damping is not always practical. Note from Figure 2.21 that the magnitude of the dimensionless displacement approaches unity if r is large. Hence, if the running frequency ω_r is such that $r \gg 1$ the effect of the unbalance is limited. For large values of r, all the magnitude curves for each value of ζ approach unity, so that the choice of damping coefficient for large r is not important. These results can be obtained from examining the plots of Figure 2.21 or from investigating the limit of mX/m_0e as r goes to infinity. These observations have important implications in the design of rotating machines.

The rotating unbalanced model can also be used to explain the behavior of an automobile with an out-of-balance wheel and tire. Here ω_r is determined by the speed of the car and e by the diameter of the wheel. The deflection x_p can be felt through the steering mechanism as shaking of the steering wheel. This usually only happens at a certain speed (near $r = 1$). As the driver increases or decreases speed, the shaking effect in the steering wheel reduces. This change in speed is equivalent to operating conditions on either side of the peak in Figure 2.21.

Example 2.5.1

Consider a machine with rotating unbalance as described in Figure 2.19. At resonance, the maximum deflection is measured to be 0.1 m. From a free decay of the system, the damping ratio is estimated to be $\zeta = 0.05$. From manufacturing data, the out-of-balance mass, m_0, is estimated to be 10%. Estimate the radius e and hence the approximate location of the unbalanced mass. Also determine how much mass should be added (uniformly) to the system to reduce the deflection at resonance to 0.01 m.

Solution   At resonance, $r = 1$, so that

$$\frac{mX}{m_0e} = \frac{1}{2\zeta} = \frac{1}{2(0.05)}$$

Hence

$$(10)\frac{(0.1\,\text{m})}{e} = \frac{1}{2\zeta} = \frac{1}{0.1} = 10$$

so that $e = 0.1$ m. Again at resonance

$$\frac{m}{m_0}\left(\frac{X}{0.1\,\text{m}}\right) = 10$$

If it is desired to change m, say by Δm, so that $X = 0.01$ m, the foregoing resonance expression becomes

$$\frac{m + \Delta m}{m_0}\left(\frac{0.01}{0.1}\right) = 10 \quad \text{or} \quad \frac{m + \Delta m}{(0.1)m} = 100$$

which implies that $\Delta m = 9m$. Thus the total mass must be increased by a factor of 9 in order to reduce the deflection to a centimeter.

$\square$

Example 2.5.2

Rotating unbalance is also important in rotorcraft such as helicopters and prop planes. The tail rotor of a helicopter (the small rotor rotating in a vertical plane at the back of a helicopter used to provide yaw control and torque balance) as sketched in Figure 2.22 can be modeled as a rotating-unbalance problem discussed in this section with stiffness $k = 1 \times 10^5$ N/m (provided by the tail section in the vertical direction) and mass of 20 kg. The tail section providing the vertical stiffness has a mass of 60 kg. Suppose that a 500-g mass is stuck on one of the blades at a distance of 15 cm from the axis of rotation. Calculate the magnitude of the deflection of the tail section of the helicopter as the tail rotor rotates at 1500 rpm. Assume a damping ratio of 0.01. At what rotor speed is the deflection at maximum? Calculate the maximum deflection.

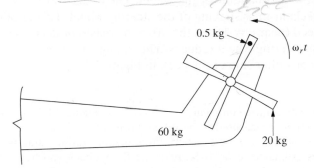

Figure 2.22 A schematic of a helicopter tail section illustrating a tail rotor. The tail rotor provides a "counterclockwise" thrust (when looking at the top of the helicopter) to counteract the "clockwise" thrust created by the main rotor, which provides lift and horizontal motion. An out-of-balance rotor can cause damaging vibrations and limit the helicopter's performance.

Solution   The rotor system is modeled as a machine of mass 20.5 kg attached to a spring, as indicated in Figure 2.23. Here only the vibration of the tail section in the vertical direction is modeled and the helicopter body is modeled as ground. The spring

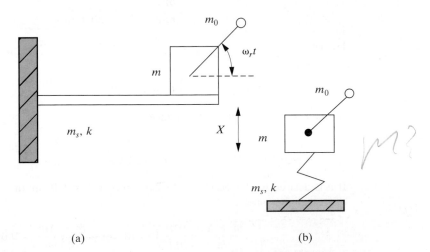

(a)   (b)

Figure 2.23   (a) The vertical vibration model of a tail section modeled as a spring consisting of a long, slender bar with machine mounted on it with a rotational unbalance. (b) This sketch is the equivalent spring–mass model used for unbalance problems (note that to be consistent with Figure 2.19, m includes m_0).

used to represent the tail section has significant mass, so equation (1.76) of Section 1.5 for a heavy beam is used to find the equivalent mass of the system. Using the equivalent mass concept yields that the natural frequency is

$$\omega_n = \sqrt{\frac{k}{m + \frac{33}{140}m_s}} = \sqrt{\frac{10^5 \, \text{N/m}}{20.5 + \frac{33}{140}60 \, \text{kg}}} = 53.727 \, \text{rad/s}$$

The frequency of rotation in rad/s is

$$\omega_r = 1500 \, \text{rpm} = 1500 \, \frac{\text{rev}}{\text{min}} \frac{\text{min}}{60 \, \text{s}} \frac{2\pi \, \text{rad}}{\text{rev}} = 157 \, \text{rad/s}$$

Hence, the frequency ratio, r, becomes

$$r = \frac{\omega_r}{\omega_n} = \frac{157 \, \text{rad/s}}{53.727 \, \text{rad/s}} = 2.92$$

With $r = 2.92$ rad/s and $\zeta = 0.01$, equation (2.84) yields that the magnitude of oscillation of the tail rotor is

$$X = \frac{m_0 e}{m} \frac{r^2}{\sqrt{(1 - r^2)^2 + (2\zeta r)^2}}$$

$$= \frac{(0.5 \, \text{kg})(0.15 \, \text{m})}{34.64 \, \text{kg}} \frac{(2.92)^2}{\sqrt{[1 - (2.92)^2]^2 - [2(0.01)(2.92)]^2}} = 0.002 \, \text{m}$$

Here the equivalent mass is $m_{eq} = m + m_s = 34.64$ kg.
The maximum deflection occurs at about $r = 1$ or

$$\omega_r = \omega_n = 53.72 \, \text{rad/s} = 53.72 \, \frac{\text{rad}}{\text{s}} \frac{\text{revs}}{2\pi \, \text{rad}} \frac{60 \, \text{s}}{\text{min}} = 513.1 \, \text{rpm}$$

In this case, the (maximum) deflection becomes

$$X = \frac{(0.5 \, \text{kg})(0.15)}{34.34 \, \text{kg}} \frac{1}{2(0.01)} = 0.108 \, \text{m} = 10.8 \, \text{cm}$$

which represents a large unacceptable deflection of the rotor. Thus the tail rotor should not be allowed to rotate at 513.1 rpm.

More on the special nature of rotating systems is discussed in Section 5.5. Additional discussion of vibration problems associated with rotating machinery is given in Section 5.7 on critical speeds. Some treatments include the discussion of critical speeds in rotating shafts immediately following the discussion of unbalance, and it is possible to skip to Section 5.7 before continuing. Vibration problems associated with rotor dynamics are both important and vast enough to study as a separate course.

2.6 MEASUREMENT DEVICES

An important application of the forced-harmonic-vibration analysis and base-excitation problem presented in Sections 2.2 and 2.4, respectively, is in the design of devices used to measure vibration. A device that changes mechanical motion into a voltage (or vice versa) is called a *transducer*. Several transducers are sketched in Figures 2.24 to 2.26. Each of these devices changes mechanical vibration into a voltage proportional to acceleration.

Referring to the accelerometer of Figure 2.24, a balance of forces on the seismic mass m yields

$$m\ddot{x} = -c(\dot{x} - \dot{y}) - k(x - y) \tag{2.86}$$

Here it is assumed that the base that is mounted to the structure being measured undergoes a motion of $y = Y \cos \omega_b t$ (i.e., that the structure being measured is undergoing simple harmonic motion). The motion of the accelerometer mass relative to the base, denoted by $z(t)$, is defined by

$$z(t) = x(t) - y(t) \tag{2.87}$$

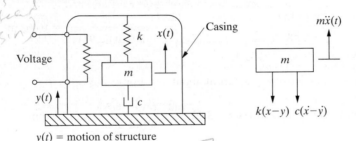

Figure 2.24 A schematic of an accelerometer mounted on a structure. The insert indicates the relevant forces acting on the mass m. The force $k(x - y)$ is actually parallel to the damping force because they are both connected to ground.

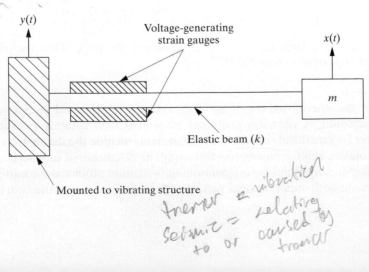

Figure 2.25 A schematic of a seismic accelerometer made of a small beam.

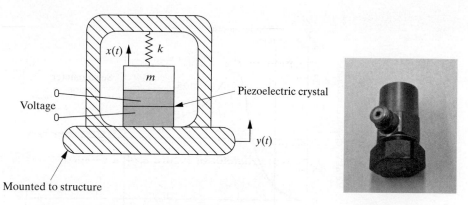

Figure 2.26   A schematic of a piezoelectric accelerometer and a photograph of a commercially available version.

Equation (2.86) can then be written in terms of the relative displacement $z(t)$. Hence equation (2.86) becomes the familiar expression

$$m\ddot{z} + c\dot{z} + kz = m\omega_b^2 Y \cos\omega_b t \tag{2.88}$$

This expression has exactly the same form as equation (2.26). Thus the solution in steady state will be of the same form as given by equation (2.36) or

$$z(t) = \frac{\omega_b^2 Y}{\sqrt{(\omega_n^2 - \omega_b^2)^2 + (2\zeta\omega_n\omega_b)^2}} \cos\left[\omega_b t + \left(-\tan^{-1}\frac{2\zeta\omega_n\omega_b}{\omega_n^2 - \omega_b^2}\right)\right] \tag{2.89}$$

The difference between equations (2.36) and (2.89) is that the latter is for the relative displacement (z) and the former is for the absolute displacement (x). Further manipulation of the magnitude of equation (2.89) yields

$$\frac{Z}{Y} = \frac{r^2}{\sqrt{(1 - r^2)^2 + (2\zeta r)^2}} \tag{2.90}$$

for the amplitude ratio as a function of the frequency ratio $r = \omega_b/\omega_n$, and

$$\theta = \tan^{-1}\frac{2\zeta r}{1 - r^2} \tag{2.91}$$

for the phase shift.

Consider the plot of the magnitude of Z/Y versus the frequency ratio as given in Figure 2.27. Note that for larger values of r (i.e., for $r \geq 3$) the magnitude ratio approaches unity, so that $Z/Y = 1$ or $Z = Y$ and the relative displacement and the displacement of the base have the same magnitude. Hence the accelerometer of Figure 2.24 can be used to measure harmonic base displacement if the frequency of the base displacement is at least three times the accelerometer's natural frequency.

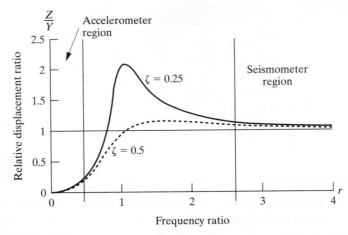

Figure 2.27 The magnitude versus frequency of the relative displacement for a transducer used to measure acceleration and for seismic measurements.

Next, consider the equation of the system in Figure 2.24 for the case when r is small. Factoring ω_n^2 out of the denominator, equation (2.89) can be written as

$$\omega_n^2 z(t) = \frac{1}{\sqrt{(1 - r^2)^2 + (2\zeta r)^2}} \, \omega_b^2 Y \cos(\omega_b t - \theta) \qquad (2.92)$$

Since $y = Y \cos(\omega_b t - \theta)$, the last term is recognized to be $-\ddot{y}(t)$ so that

$$\omega_n^2 z(t) = \frac{-1}{\sqrt{(1 - r^2)^2 + (2\zeta r)^2}} \, \ddot{y}(t) \qquad (2.93)$$

This expression illustrates that, for small values of r, the quantity $\omega_n^2 z(t)$ is proportional to the base acceleration, $\ddot{y}(t)$, since

$$\lim_{r \to 0} \frac{1}{\sqrt{(1 - r^2)^2 + (2\zeta r)^2}} = 1 \qquad (2.94)$$

In practice, this coefficient is taken as close to 1 for any value of $r < 0.5$. This indicates that, for these frequencies of base motion, the relative position $z(t)$ is proportional to the base acceleration. The effect of the accelerometer internal damping, ζ, in the constant of proportionality between the relative displacement and the base acceleration is illustrated in Figure 2.28, which consists of a plot of this constant versus the frequency ratio for a variety of values of ζ, for values of $r < 1$. Note from the figure that the curve corresponding to $\zeta = 0.7$ is closest to being constant at unity over the largest range of $r < 1$. For this curve, the magnitude is relatively flat for values of r between zero and about 0.2. In fact, within this region, the curve varies

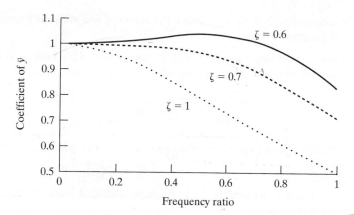

Figure 2.28 The effect of damping on the constant of proportionality between base acceleration and the relative displacement (voltage) for an accelerometer.

from one by less than one percent. This defines the useful range of operation for the accelerometer:

$$0 < \frac{\omega_b}{\omega_n} < 0.2 \qquad (2.95)$$

where ω_n is the natural frequency of the device. Multiplying this inequality by the device frequency yields

$$0 < \omega_b < 0.2\omega_n \qquad \text{or} \qquad 0 < f_m < 0.2f_n \qquad (2.96)$$

where f_m is the frequency to be measured by the accelerometer in hertz. For the mechanical accelerometer of Figure 2.24, the device frequency may be on the order of 100 Hz. Thus inequality (2.96) indicates that the highest frequency that can be effectively measured by the device would be 20 Hz (0.2×100).

Many structures and machines vibrate at frequencies larger than 20 Hz. The piezoelectric accelerometer design indicated in Figure 2.26 provides a device with a natural frequency of about 8×10^4 Hz. In this case, the inequality predicts that vibration with frequency content up to 16,000 Hz can be measured. Vibration measurement is discussed in more detail in Chapter 7, where practical problems, such as phase and amplitude distribution of signals measured with accelerometers, are discussed.

Example 2.6.1

This example illustrates how an independent measurement of acceleration can provide a measurement of a transducer's mechanical properties. An accelerometer is used to measure the oscillation of an airplane wing caused by the plane's engine operating at 6000 rpm (628 rad/s). At this engine speed the wing is known, from other measurements, to experience 1.0-g acceleration. The accelerometer measures an acceleration of 10 m/s². If the accelerometer has a 0.01-kg moving mass and a damped natural frequency of 100 Hz (628 rad/s), the difference between the measured and the known acceleration is used to calculate the damping and stiffness parameters associated with the accelerometer.

Solution   From equation (2.93), the amplitude of the measured values of acceleration $|\omega_n^2 z(t)|$ is related to the actual values of acceleration $|\ddot{y}(t)|$ by

$$\frac{|\omega_n^2 z(t)|}{|\ddot{y}(t)|} = \frac{1}{\sqrt{(1 - r^2)^2 + (2\zeta r)^2}} = \frac{10 \text{ m/s}^2}{9.8 \text{ m/s}^2} = 1.02$$

Rewriting this expression yields one equation in ζ and r:

$$(1 - r^2)^2 + (2\zeta r)^2 = 0.96$$

A second expression in ζ and r can be obtained from the definition of the damped natural frequency:

$$\frac{\omega_b}{\omega_d} = \frac{\omega_b}{\omega_n}\frac{1}{\sqrt{1 - \zeta^2}} = r\frac{1}{\sqrt{1 - \zeta^2}} = \frac{628 \text{ rad/s}}{628 \text{ rad/s}} = 1$$

Thus $r = \sqrt{1 - \zeta^2}$, providing a second equation in ζ and r. This can be manipulated to yield $\zeta^2 = (1 - r^2)$, which when substituted with the preceding expression for r and ζ yields

$$\zeta^4 + 4\zeta^2(1 - \zeta^2) = 0.96$$

This is a quadratic equation in ζ^2:

$$3\zeta^4 - 4\zeta^2 + 0.96 = 0$$

This quadratic expression yields the two roots $\zeta = 0.56, 1.01$. Using $\zeta = 0.56$, the damping constant is $(\sqrt{1 - \zeta^2} = 0.83, \omega_n = \omega_d/\sqrt{1 - \zeta^2} = 758.0 \text{ rad/s})$

$$c = 2m\omega_n\zeta = 2(0.01)(758.0)(0.56) = 8.49 \text{ N·s/m}$$

Similarly, the stiffness in the accelerometer is

$$k = m\omega_n^2 = (0.01)(758.0)^2 = 5745.6 \text{ N/m}$$

$\square$

2.7 OTHER FORMS OF DAMPING

The damping used in previous sections has been treated as linear-viscous damping, with the exception of the treatment of Coulomb damping in Section 1.10. In this section, the discussion of Coulomb damping is continued and other forms of damping are introduced. Because damping is both difficult to model mathematically and difficult to measure, choosing the correct form of damping is not an easy task. Hence damping is often approximated as a linear dependence on velocity, as done in the previous sections. Other models, though not mathematically convenient, may provide a more accurate description of the damping in a vibrating system. Coulomb, or friction, damping is one example of such a form. A number of other mathematical forms of damping exist. In this section, these damping forms are all treated in the forced-response case by examining an equivalent linear system based on the energy dissipated during vibration. In Section 2.9, these systems are numerically integrated in their nonlinear form and compared to the equivalent linear response discussed here.

First, consider the response of a system with Coulomb damping, introduced in Section 1.10, to a harmonic driving force. Recall the systems of Figures 1.43 and 1.44 described in the free-response case by equation (1.101). The equation of motion in the forced-response case becomes

$$m\ddot{x} + \mu mg\,\text{sgn}(\dot{x}) + kx = F_0 \sin \omega t \qquad (2.97)$$

Rather than solving this equation directly, one can approximate the solution of equation (2.97) with the solution of a viscously damped system that dissipates an equivalent amount of energy per cycle. This is a reasonable assumption if the magnitude of the applied force is much larger than the Coulomb force ($F_0 \gg \mu mg$). This approximation is accomplished by again assuming that the steady-state response will be of the form

$$x_{ss}(t) = X \sin \omega t \qquad (2.98)$$

The energy dissipated, ΔE, in a viscously damped system per cycle with viscous-damping coefficient c is given by

$$\Delta E = \oint F_d dx = \int_0^{2\pi/\omega} c\dot{x}\frac{dx}{dt}dt = \int_0^{2\pi/\omega} c\dot{x}^2 dt \qquad (2.99)$$

At steady state, $x = X \sin \omega t$, $\dot{x} = \omega X \cos \omega t$, and equation (2.99) becomes

$$\Delta E = c \int_0^{2\pi/\omega} (\omega^2 X^2 \cos^2 \omega t)dt = \pi c\omega X^2 \qquad (2.100)$$

This is the energy dissipated per cycle by a viscous damper. On the other hand, the energy dissipated by the Coulomb friction on a horizontal surface per cycle is

$$\Delta E = \mu mg \int_0^{2\pi/\omega} \left[\text{sgn}(\dot{x})\dot{x} \right] dt \qquad (2.101)$$

Substitution of the steady-state velocity into this expression and splitting the integrations up into segments corresponding to the sign change in $\dot{x}$ yields

$$\Delta E = \mu mgX \left(\int_0^{\pi/2} \cos u\, du - \int_{\pi/2}^{3\pi/2} \cos u\, du + \int_{3\pi/2}^{2\pi} \cos u\, du \right) \qquad (2.102)$$

where $u = \omega t$ and $du = \omega dt$. Completing the integration yields that the energy dissipated by Coulomb friction is

$$\Delta E = 4\mu mg X \tag{2.103}$$

To create a viscously damped system of equivalent energy loss, the energy-loss expression for viscous damping of equation (2.100) is equated to the energy loss associated with Coulomb friction, given by equation (2.103) to yield

$$\pi c_{eq}\omega X^2 = 4\mu mg X \tag{2.104}$$

where c_{eq} denotes the equivalent viscous-damping coefficient. Solving for c_{eq} yields

$$c_{eq} = \frac{4\mu mg}{\pi\omega X} \tag{2.105}$$

In terms of an equivalent damping ratio, ζ_{eq}, equation (2.105) must also equal $2\zeta_{eq}\omega_n m$, so that

$$\zeta_{eq} = \frac{2\mu g}{\pi\omega_n\omega X} \tag{2.106}$$

Thus the viscously damped system described by

$$\ddot{x} + 2\zeta_{eq}\omega_n\dot{x} + \omega_n^2 x = f_0 \sin\omega t \tag{2.107}$$

will dissipate as much energy as does the Coulomb system described by equation (2.97). Considering (2.107) as an approximation of equation (2.97), the approximate magnitude and phase of the steady-state response of equation (2.97) can be calculated. Substitution of the equivalent viscous-damping ratio given in equation (2.106) into the magnitude of equation (2.40) yields the result that the magnitude X of the steady-state response is

$$X = \frac{F_0/k}{\sqrt{(1-r^2)^2 + (2\zeta_{eq}r)^2}} = \frac{F_0/k}{[(1-r^2)^2 + (4\mu mg/\pi kX)^2]^{1/2}} \tag{2.108}$$

Solving this expression for the amplitude X yields

$$X = \frac{F_0}{k}\frac{\sqrt{1 - (4\mu mg/\pi F_0)^2}}{|(1-r^2)|} \tag{2.109}$$

with phase shift given by equation (2.40) as

$$\theta = \tan^{-1}\frac{2\zeta_{eq}r}{1-r^2} = \tan^{-1}\frac{4\mu mg}{\pi kX(1-r^2)} \tag{2.110}$$

The expression for the phase can be further examined by substituting for the value of X from equation (2.109). This yields

$$\theta = \tan^{-1} \frac{\pm 4\mu mg}{\pi F_0 \sqrt{1 - (4\mu mg/\pi F_0)^2}} \qquad (2.111)$$

where the $\pm$ originates from the absolute value in equation (2.109). Thus θ is positive if $r < 1$ and negative if $r > 1$. Also note from equation (2.111) that θ is constant for a given F_0 and μ and is independent of the driving frequency.

Several differences are apparent in the behavior of the approximate phase and amplitude of the response with Coulomb friction compared with that of viscous friction. First, at resonance, $r = 1$, the magnitude in equation (2.109) becomes infinite, unlike the viscously damped case. Second, the phase is discontinuous at resonance, rather than passing through 90° as in the viscous case. Note also from equation (2.111) that the approximation is good only if the argument in the radical of (2.109) is positive, that is, if

$$4\mu mg < \pi F_0 \qquad (2.112)$$

This confirms and quantifies the physical statement made at the outset (i.e., that the applied force must be larger in magnitude than the sliding friction force in order to overcome the friction to provide motion).

Example 2.7.1

Consider a spring–mass system with sliding friction described by equation (2.97) with stiffness $k = 1.5 \times 10^4$ N/m, driving harmonically a 10-kg mass by a force of 90 N at 25 Hz. Calculate the approximate amplitude of steady-state motion assuming that both the mass and the surface are steel (unlubricated).

Solution   First, look up the coefficient of friction in Table 1.5, which is $\mu = 0.3$. Then from inequality (2.112),

$$4\mu mg = 4(0.3)(10\,\text{kg})(9.8\,\text{m/s}^2) = 117.6\,\text{N}$$

$$< (90\,\text{N})(3.1415) = 282.74 = \pi F_0$$

so that the approximation developed previously for the steady-state-response amplitude is valid. Converting 25 Hz to 157 rad/s and using equation (2.109) then yields

$$X = \frac{90\,\text{N}}{1.5 \times 10^4\,\text{N/m}} \frac{\sqrt{1 - (117.6\,\text{N}/282.74\,\text{N})^2}}{|1 - (2.467 \times 10^4/1.5 \times 10^3)|} = 3.53 \times 10^{-4}\,\text{m}$$

Thus the amplitude of oscillation will be less than 1 mm.

$\square$

Several other forms of damping are available for modeling a particular mechanical device or structure in addition to viscous and Coulomb damping. It is common to study damping mechanisms by examining the energy dissipated per cycle under a harmonic loading. Often, force versus displacement curves, or stress versus strain curves, are used to measure the energy lost and hence determine a measure of the damping in the system.

The energy lost per cycle, given in equation (2.100) to be $\pi c \omega X^2$, is used to define the *specific damping capacity* as the energy lost per cycle divided by the peak potential energy, $\Delta E / U$. A more commonly used quantity is the energy lost per radian divided by the peak-potential, or strain, energy U_{max}. This is defined to be the *loss factor* or *loss coefficient*, denoted by η, and given by

$$\eta = \frac{\Delta E}{2\pi U_{max}} \tag{2.113}$$

where U_{max} is defined as the potential energy at maximum displacement of X (or strain energy).

The loss factor is related to the damping ratio of a viscously damped system at resonance. To see this, substitute the value for ΔE from (2.100) into (2.113) to get

$$\eta = 1\frac{\pi c \omega X^2}{2\pi \left(\frac{1}{2} kX^2\right)} \tag{2.114}$$

At resonance, $\omega = \omega_n = \sqrt{k/m}$ so that (2.113) becomes

$$\eta = \frac{c}{\sqrt{km}} = 2\zeta \tag{2.115}$$

Hence, at resonance, the loss factor is twice the damping ratio.

Next, consider a force-displacement curve for a system with viscous damping. The force required to displace the mass is that force required to overcome the spring and damper forces, or

$$F = kx + c\dot{x} \tag{2.116}$$

At steady state, as given by equation (2.98), this becomes

$$F = kx + cX \omega \cos \omega t \tag{2.117}$$

Using a trigonometric identity ($\sin^2\phi + \cos^2\phi = 1$) on the cos ωt term yields

$$F = kx \pm c\omega X(1 - \sin^2 \omega t)^{1/2}$$
$$= kx \pm c\omega \left[X^2 - (X \sin \omega t)^2\right]^{1/2} \tag{2.118}$$
$$= kx \pm c\omega \sqrt{X^2 - x^2}$$

Squaring this expression yields, upon rearrangement,

$$F^2 + (c^2\omega^2 + k^2)x^2 - (2k)xF - c^2\omega^2X^2 = 0 \tag{2.119}$$

which can be recognized as the general equation for an ellipse ($c^2\omega^2 > 0$) rotated about the origin in the F-x plane (see a precalculus text). This is plotted in Figure 2.29.

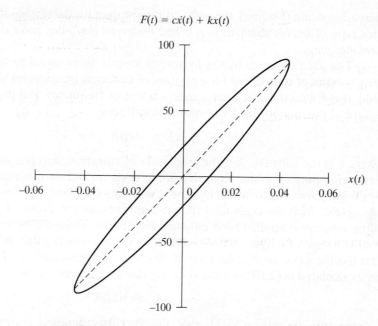

$$F(t) = c\dot{x}(t) + kx(t)$$

Figure 2.29   A plot of force versus displacement defining the hysteresis loop for a viscously damped system.

The ellipse in the Figure 2.29 is called a hysteresis loop, and the area enclosed is the energy lost per cycle as calculated in equation (2.100) and is equal to $\pi c\omega X^2$. Note that if $c = 0$, the ellipse of Figure 2.29 collapses to the straight line of slope k indicated by the dashed line in the figure.

Materials are often tested by measuring stress (force) and strain (displacement) under carefully controlled steady-state harmonic loading. Many materials exhibit internal friction between various planes of material as the material is deformed. Such tests produce hysteresis loops of the form shown in Figure 2.30. Note that, for

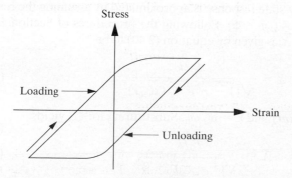

Figure 2.30   An experimental stress–strain plot for one cycle of harmonically loaded material for steady state illustrating a hysteresis loop associated with internal damping.

increasing strain (loading), the path is different than for decreasing strain (unloading). This type of energy dissipation is called *hysteretic damping, solid damping*, or *structural damping*.

The area enclosed by the hysteresis loop is again equal to the energy loss. If the experiment is repeated for a number of different frequencies at constant amplitude, it is found that the area is independent of frequency and proportional to the square of the amplitude of vibration and stiffness:

$$\Delta E = \pi k \beta X^2 \tag{2.120}$$

where k is the stiffness, X is the amplitude of vibration, and β is defined as the *hysteretic damping constant*. Note that some texts formulate this equation differently by defining $h = k\beta$ to be the hysteretic damping constant.

Next, apply the equivalent viscous-damping concept used for Coulomb damping. If this concept is applied here, equating the energy dissipated by a viscously damped system to that of a hysteretic system is equivalent to finding the ellipse of Figure 2.29 that has the same area as the loop of Figure 2.20. Thus equating (2.120) with the energy calculated in (2.100) for the viscously damped system yields

$$\pi c_{eq} \omega X^2 = \pi k \beta X^2 \tag{2.121}$$

Solving this expression yields that the viscously damped system dissipating the same amount of energy per cycle as the hysteretic system will have the equivalent damping constant given by

$$c_{eq} = \frac{k\beta}{\omega} \tag{2.122}$$

where β is determined experimentally from the hysteresis loop.

The approximate steady-state response of a system with hysteretic damping can be determined from substitution of this equivalent damping expression into the equation of motion to yield

$$m\ddot{x} + \frac{\beta k}{\omega}\dot{x} + kx = F_0 \cos \omega t \tag{2.123}$$

In this case, the steady-state response is approximated by assuming the response is of the form $x_{ss}(t) = X \cos(\omega t - \phi)$. Following the procedures of Section 2.2, the magnitude of the response, X, is given by equation (2.40) to be

$$X = \frac{F_0/k}{\sqrt{(1 - r^2)^2 + (2\zeta_{eq}r)^2}} \tag{2.124}$$

where ζ_{eq} is now $c_{eq}/(2\sqrt{km})$ and $r = \omega/\omega_n$. Substituting for ζ_{eq} yields

$$X = \frac{F_0/k}{\sqrt{(1 - r^2)^2 + \beta^2}} \tag{2.125}$$

Similarly, the phase becomes

$$\phi = \tan^{-1}\frac{\beta}{1 - r^2} \tag{2.126}$$

These are plotted in Figure 2.31 for several values of β.

Compared to a viscously damped system, the hysteretic system's magnitude obtains a maximum of $F_0/\beta k$. This is obtained by setting $r = 1$ in equation (2.125) so that the maximum value is obtained at the resonant frequency rather than below it, as is the case for viscous damping. Examination of the phase shift shows that the response of a hysteretic system is never in phase with the applied force, which is not true for viscous damping.

In Section 2.3, the complex exponential was used to represent a harmonic input. Using equation (2.48), the equivalent hysteretic system can be written as

$$m\ddot{x} + \frac{\beta k}{\omega}\dot{x} + kx = F_0 e^{j\omega t} \tag{2.127}$$

Substitution of the assumed form of the solution given by $x(t) = Xe^{j\omega t}$ for just the velocity term yields

$$m\ddot{x} + k(1 + j\beta)x = F_0 e^{j\omega t} \tag{2.128}$$

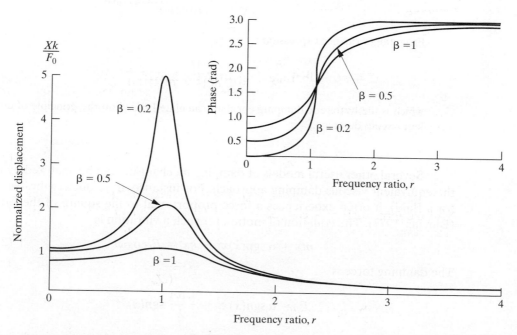

Figure 2.31 Steady-state magnitude and phase versus frequency ratio for a system with hysteretic damping coefficient β approximated by a system with viscous damping.

This gives rise to the notion of *complex stiffness* or *complex modulus*. The damped problem is represented in equation (2.128) as an undamped problem with complex stiffness coefficient $k(1 + j\beta)$. This approach is very popular in the material engineering literature on damping.

Example 2.7.2

An experiment is performed on a hysteretic system with known spring stiffness of $k = 4 \times 10^4\,\text{N/m}$. The system is driven at resonance, the area of the hysteresis loop is measured to be $\Delta E = 30\,\text{N} \cdot \text{m}$, and the amplitude, X, is measured to be $X = 0.02\,\text{m}$. Calculate the magnitude of the driving force and the hysteretic damping constant.

Solution   At resonance, equation (2.125) yields

$$X = \frac{F_0}{k\beta}$$

or $k\beta = F_0/X$. The area enclosed by the hysteresis loop is equal to $\pi k\beta X^2$, so that

$$30\,\text{N} \cdot \text{m} = \pi k\beta X^2 = \pi \frac{F_0}{X} X^2$$

and hence

$$F_0 = \frac{30\,\text{N} \cdot \text{m}}{\pi X} = \frac{30\,\text{N} \cdot \text{m}}{\pi(0.02\,\text{m})} = 477.5\,\text{N}$$

From the resonance expression,

$$\beta = \frac{F_0}{Xk} = \frac{477.5}{(0.02\,\text{m})(4 \times 10^4\,\text{N/m})} = 0.60$$

which is the hysteretic damping constant calculated based on the principle of equivalent viscous damping.

□

Several other useful models of damping mechanisms can be analyzed by using the equivalent viscous-damping approach. For instance, if an object vibrates in air (or a fluid), it often experiences a force proportional to the square of the velocity (Blevins, 1977). The equation of motion for such a vibration is

$$m\ddot{x} + \alpha\,\text{sgn}(\dot{x})\dot{x}^2 + kx = F_0 \cos \omega t \tag{2.129}$$

The damping force is

$$F_d = \alpha\,\text{sgn}(\dot{x})\dot{x}^2 = \frac{C\rho A}{2}\,\text{sgn}(\dot{x})\dot{x}^2$$

which opposes the direction of motion, similar to Coulomb friction, and depends on the square of the velocity, $\dot{x}$; the drag coefficient of the mass, C; the density of the fluid, ρ; and the cross-sectional area, A, of the mass. This type of damping is referred to as *air damping, quadratic damping,* or *velocity-squared damping.* As in the Coulomb friction case, this is a nonlinear equation that does not have a convenient closed-form solution to analyze. While it can be solved numerically (see Section 2.8), an approximation to the behavior of the solution during steady-state harmonic excitation can be made by the equivalent viscous-damping method. Assuming a solution of the form $x = X \sin \omega t$ and computing the energy integrals following the steps taken in equation (2.102) yields that the energy dissipated per cycle is

$$\Delta E = \frac{8}{3} \alpha X^3 \omega^2 \tag{2.130}$$

Again, equating this to the energy dissipated by a viscously damped system given in equation (2.100) yields

$$c_{eq} = \frac{8}{3\pi} \alpha \omega X \tag{2.131}$$

This equivalent viscous-damping value can then be used in the amplitude and phase formulas for a linear viscously damped forced harmonic motion to approximate the steady-state response.

Example 2.7.3

Calculate the approximate amplitude at resonance for velocity-squared damping.

Solution   Using the magnitude expression for viscous damping at resonance ($r = 1$ and $\omega = \omega_n$), equation (2.40) and the expression for the damping ratio given in equation (1.30) yields

$$X = \frac{f_0}{2\zeta\omega\omega_n} = \frac{mf_0}{c_{eq}\omega}$$

Substitution of (2.131) yields

$$X = \frac{mf_0}{(8/3\pi)\alpha\omega^2 X}$$

so that

$$X = \sqrt{\frac{3\pi mf_0}{8\alpha\omega^2}} = \sqrt{\frac{3\pi f_0 m^2}{8kC\rho A}}$$

As expected, for larger values of the mass density of the fluid, the drag coefficient of the cross-sectional area produces a smaller amplitude at resonance.

□

If several forms of damping are present, one approach to examining the harmonic response is to calculate the energy lost per cycle of each form of damping present, add them up, and compare them to the energy loss from a single viscous damper. Then the formulas for magnitude and phase from equation (2.40) are used

TABLE 2.2   DAMPING MODELS

Name	Damping Force	c_{eq}	Source
Linear-viscous damping	$c\dot{x}$	c	Slow fluid
Air damping	$a\,\mathrm{sgn}(\dot{x})\dot{x}^2$	$\dfrac{8a\omega X}{3\pi}$	Fast fluid
Coulomb damping	$\beta\,\mathrm{sgn}\,\dot{x}$	$\dfrac{4\beta}{\pi\omega X}$	Sliding friction
Displacement-squared damping	$d\,\mathrm{sgn}(\dot{x})x^2$	$\dfrac{4dX}{3\pi\omega}$	Material damping
Solid, or structural, damping	$b\,\mathrm{sgn}(\dot{x})|x|$	$\dfrac{2b}{\pi\omega}$	Internal damping

to approximate the response. For n damping mechanisms dissipating energy per cycle of ΔE_i, for the ith mechanism, the equivalent viscous-damping constant is

$$c_{eq} = \frac{\sum_{i=1}^{n}\Delta E_i}{\pi\omega X^2} \tag{2.132}$$

A study of various damping mechanisms is presented by Bandstra (1983) and summarized in Table 2.2.

2.8 NUMERICAL SIMULATION AND DESIGN

In the previous sections, great effort was put forth to derive analytical expressions for the response of various single-degree-of-freedom systems driven by a harmonic load. These analytical expressions are extremely useful for design and for understanding some of the physical phenomena. Plots of the time response and of the steady-state magnitude and phase were constructed to realize the nature and features of the response. Rather than plotting the analytical function describing the response, the time response may also be computed numerically using an Euler or Runge–Kutta integration and computational software packages as introduced in Section 1.8. While numerical solutions such as these are not exact, they do allow nonlinear terms to be considered. In addition, these packages may be used to generate all of the plots of phase and magnitude and the time responses given in the previous sections (in fact, all plots in this text are generated using MATLAB or, in some cases, Mathcad). The computational packages will also help you derive expressions, such as equation (2.38), by using symbolic algebra. Perhaps the most advantageous use of computational software is the ability to resolve quickly the time response for various values of parameters. The ability to plot the solution quickly allows engineers to examine what would happen if the damping changes or the input force level changes. Such parametric studies of the time response are useful for design and for building intuition about a given system.

In order to solve for the forced response to a harmonic input numerically, equation (1.97) needs to be modified slightly to incorporate the applied force. The first-order, or state-space, form of equation (2.27) becomes

$$\dot{x}_1(t) = x_2(t)$$

$$\dot{x}_2(t) = -2\zeta\omega_n x_2(t) - \omega_n^2 x_1(t) + f_0 \cos\omega t \qquad (2.133)$$

where x_2 denotes the velocity $\dot{x}(t)$ and x_1 denotes the position $x(t)$, as before. This is subject to the initial conditions $x_1(0) = x_0$ and $x_2(0) = v_0$. Given ω, ω_n, ζ, f_0, x_0, and v_0, the solution of equation (2.133) can be determined numerically. The matrix form of equation (2.133) becomes

$$\dot{\mathbf{x}}(t) = A\mathbf{x}(t) + \mathbf{f}(t) \qquad (2.134)$$

where $\mathbf{x}$ and A are the state vector and state matrix as defined previously in equation (1.96). The vector $\mathbf{f}$ is the applied force and takes the form

$$\mathbf{f}(t) = \begin{bmatrix} 0 \\ f_0 \cos\omega t \end{bmatrix} \qquad (2.135)$$

The Euler form of equation (2.134) is

$$\mathbf{x}(t_i + 1) = \mathbf{x}(t_i) + A\mathbf{x}(t_i)\Delta t + \mathbf{f}(t_i)\Delta t \qquad (2.136)$$

This expression can also be adapted to the Runge–Kutta formulation, and most of the codes mentioned in Appendix G have built-in commands for a Runge–Kutta solution.

The numerical integration to determine the response of a system is an approximation, whereas the plotting of the analytical solution is exact. So why bother to integrate numerically to find the solution? Because closed form solutions often do not exist, such as in the treatment of nonlinear terms as was illustrated in Section 1.10. This section discusses the solution of the forced response using numerical integration in an environment where the exact solution is available for comparison. The examples in this section introduce numerical integration to compute the forced response and to compare these to the exact solution.

The following example illustrates the use of various programs to compute and plot the solution.

Example 2.8.1

Numerically integrate and plot the response of an underdamped system determined by $m = 100$ kg, $k = 2000$ N/m, and $c = 200$ kg/s, subject to the initial conditions of $x_0 = 0.01$ m and $v_0 = 0.1$ m/s, and the applied force $F(t) = 150 \cos 10t$. Then plot the exact response as computed by equation (2.38). Compare the plot of the exact solution to the numerical simulation.

Solution   This is first worked out in Mathcad. The equivalent commands for MATLAB and Mathematica are given at the end of this example. Start by entering the relevant numbers.

$$x0 := 0.01 \qquad v0 := 0.1 \qquad m := 100 \qquad c := 200 \qquad k := 2000$$

$$\omega n := \sqrt{\frac{k}{m}} \qquad \zeta := \frac{c}{2\sqrt{k \cdot m}} \qquad F0 := 150 \qquad \omega := 10$$

$$\omega d := \omega n \cdot \sqrt{1 - \zeta^2} \qquad f0 := \frac{F0}{m}$$

$$B := \frac{f0}{(\omega n^2 - \omega^2)^2 + (2 \cdot \zeta \cdot \omega n \cdot \omega)^2} \qquad \boxed{\text{Defines one coefficient}}$$

$$xTc(t) := e^{-\zeta \cdot \omega n \cdot t} \cdot \left[x0 - (\omega n^2 - \omega^2) \cdot B \right] \cdot \cos(\omega d \cdot t) \qquad \boxed{\begin{array}{l}\text{Write out equation}\\ \text{(2.38) in parts}\end{array}}$$

$$xTs(t) := e^{-\zeta \cdot \omega n \cdot t} \cdot \left[\left[x0 - (\omega n^2 - \omega^2) \cdot B \right] \cdot \frac{\zeta \cdot \omega n}{\omega d} - 2 \cdot \zeta \cdot \omega n \cdot \frac{\omega^2}{\omega d} \cdot B + \frac{v0}{\omega d} \right] \cdot \sin(\omega d \cdot t)$$

$$xSS(t) := B \cdot \left[(\omega n^2 - \omega^2) \cdot \cos(\omega \cdot t) + 2 \cdot \zeta \cdot \omega n \cdot \omega \cdot \sin(\omega d \cdot t) \right]$$

$$x(t) := xTc(t) + xTs(t) + xSS(t)$$

Next compute the same response using Runge–Kutta by setting up a state-space representation, use rkfixed to solve, and save the solution in x, t vectors

$$X := \begin{bmatrix} x0 \\ v0 \end{bmatrix} \qquad D(t,X) := \begin{bmatrix} X_1 \\ -2 \cdot \zeta \cdot \omega n \cdot X_1 - \omega n^2 \cdot X_0 + f0 \cdot \cos(\omega \cdot t) \end{bmatrix}$$

$$Z := \text{rkfixed}(X, 0, 6, 2000, D)$$

$$t := Z^{<0>} \qquad xs := Z^{<1>}$$

$$x := \overrightarrow{x(t)} \qquad \boxed{\text{Change exact solution into a vector for plotting}}$$

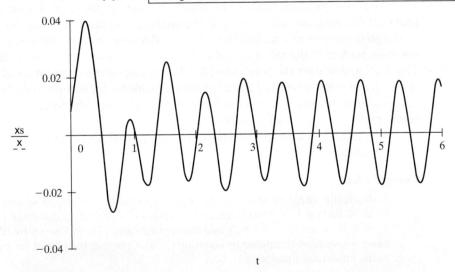

Figure 2.32 The exact solution (dashed line) and a Runge–Kutta solution (solid line) plotted on the same graph.

From Figure 2.32, note that the numerical solution and the exact solution are the same. However, it is always important to remember that the numerical integration yields only an approximate solution.

Next the MATLAB code for computing these plots is given. First an M-file is created with the equation of motion given in first-order form.

```
--------------------
function v=f(t,x)
m=100; k=2000; c=200; Fo=150; w=10;
v=[x(2); x(1).*-k/m+x(2).*-c/m+Fo/m*cos(w*t)];
---------------------------
```

Then the following is entered into the command window:

```
clear all


xo=0.01; vo=0.1; m=100; c=200; k=2000;
Fo=150; w=10;
t=0:0.01:5;


wn=sqrt(k/m);
z=c/(2*sqrt(k*m));
wd=wn*sqrt(1-z^2);
fo=Fo/m;


% Defines one coefficient
B=fo/((wn^2-w^2)^2+(2*z*wn*w)^2);


for i=1:max(length(t))


%Write out equation (2.38) in parts
xTc(i)=exp(-z*wn*t(i)) * (xo-(wn^2-w^2)*B)*cos(wd*t(i));
xTs(i)=exp(-z*wn*t(i)) * ((xo-(wn^2-w^2)*B)*z*wn/wd -
2*z*wn*w^2/wd*B + vo/wd) * sin(wd*t(i));
xSs(i)=B*((wn^2-w^2)*cos(w*t(i)) + 2*z*wn*w*sin(w*t(i)));
x(i)=xTc(i)+xTs(i)+xSs(i);
end


figure(1)
plot(t,x)


clear all


xo=[0.01; 0.1];
ts=[0 5];
[t,x]=ode45('f',ts,xo);


figure (2)
plot(t,x(:,1))
```

In Mathematica, the exact solution and numerical solution are computed and plotted by the following list of commands.

```
In[1] := m = 100;
         k = 2000;
         c = 200;
         x0 = .01;
         v0 = .1;
```
$$\omega n = \sqrt{\frac{k}{m}};$$
$$\zeta = \frac{c}{2\sqrt{k * m}};$$
```
         ω = 10
```
$$\omega d = \omega n * \sqrt{1 - \zeta^2};$$
```
         F0 = 150;
```
$$f0 = \frac{F0}{m};$$
$$X = \frac{f0}{\sqrt{(\omega n^2 - \omega^2)^2 + (2 * \zeta * \omega n * \omega)^2}};$$
```
         θ = ArcTan[ωn² - ω², 2 * ζ * ωn * ω];
         φ = ArcTan[v0 - X * ω * Sin[θ] + ζ * ωn *
             (x0 - X * Cos[θ]), ωd * (x0 - X * Cos[θ])];
```
$$A = \frac{x0 - X * Cos[\theta]}{Sin[\phi]};$$
```
         xanal[t_] = A * Exp[-ζ * ωn * t] * Sin[ωd * t + φ] + X
             * Cos[ω * t - θ];

         numerical = NDSolve[{x''[t] + 2 * ζ * ωn * x'[t] + ωn² *
             x[t] == f0 * Cos[ω * t], x[0] == x0, x'[0] == v0},
             x[t], {t, 0, 5}];

         Plot[{Evaluate[x[t] /. numerical], xanal[t]}, {t, 0, 5},
             PlotRange → {-0.04, 0.04}]
```

□

With the ability to compute numerical solutions, either by solving a differential equation and plotting it or by plotting the analytical solution, comes the ability to perform parametric studies of the response quickly. Once a response plot is written into a code, it is a trivial matter to reproduce the plot with new values of the physical parameters (mass, damping, stiffness, initial conditions, and the driving force magnitude and frequency). Such parametric studies can be used both to understand the physical nature of the response and to design. Here *design* refers to choosing the physical parameters to obtain a more desirable response. Chapter 5 focuses on design. The following example illustrates how the computer may be used to determine design parameters.

Example 2.8.2

An electronics module is mounted on a machine and is modeled as a single-degree-of-freedom spring, mass, and damper. During normal operation, the module (having a mass of 100 kg) is subject to a harmonic force of 150 N at 5 rad/s. Because of material considerations and static deflection, the stiffness is fixed at 500 N/m and the natural damping in the system is 10 kg/s. The machine starts and stops during its normal operation, providing initial conditions to the module of $x_0 = 0.01$ m and $v_0 = 0.5$ m/s. The module must not have an amplitude of vibration larger than 0.2 m even during the transient stage. First, compute the response by numerical simulation to see if the constraint is satisfied. If the constraint is not satisfied, find the smallest value of damping that will keep the deflection less than 0.2 m.

Solution This requires the numerical integration of a second-order differential equation. Codes for these are given in Example 2.8.1. Use either equation (2.136) or a Runge–Kutta equivalent to integrate numerically the equation of motion and plot the result to see if the response is larger than 0.2 m. Here Mathcad is used to generate the response plot of Figure 2.33.

$$x0 := 0.01 \qquad v0 := 0.5 \qquad m := 100 \qquad k := 500 \qquad c := 10$$

$$F0 := 150 \qquad \omega n := \sqrt{\frac{k}{m}} \qquad \zeta := \frac{c}{2\sqrt{k \cdot m}}$$

$$f0 := \frac{F0}{m}$$

$$\zeta = 0.022$$

$$\omega := 5 \qquad \omega n = 2.236$$

$$X := \begin{bmatrix} x0 \\ v0 \end{bmatrix}$$

$$D(t, X) := \begin{bmatrix} X_1 \\ -2 \cdot \zeta \cdot \omega n \cdot X_1 - \omega n^2 \cdot X_0 + f0 \cdot \cos(\omega \cdot t) \end{bmatrix}$$

$$Z := \text{rkfixed}(X, 0, 40, 4000, D)$$

$$t := Z^{<0>}$$

$$x := Z^{<1>}$$

Note from the simulated response that the transient term is larger than the steady state and has violated the constraint that $x(t) \leq 0.2$ m. Thus the damping must be increased to bring down the amplitude of the transient response. The design is performed by simply increasing the value of c in the preceding code and running it again. This is repeated until the response falls below 0.2 m. Because damping is expensive to add to a system, the increment of damping at each iteration is very small. This "design" procedure produces the plot shown in Figure 2.34 for a damping coefficient of $c = 195$ kg/s ($\zeta = 0.436$).

A value of the damping coefficient that is a few kg/s less than 195 kg/s will produce a response larger than the desired 0.2 m. This is a fairly large value of damping, a source of concern for the designer.

Next the MATLAB code for computing these plots is given. First an M-file is created with the equation of motion given in first-order form.

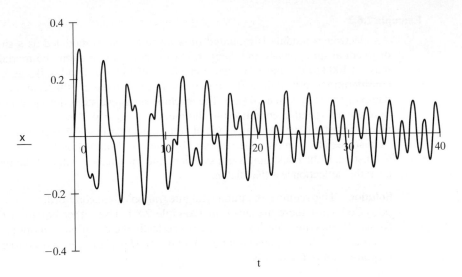

Figure 2.33 The simulated response for $c = 10$ kg/s illustrating that the transient response exceeds 0.2 m.

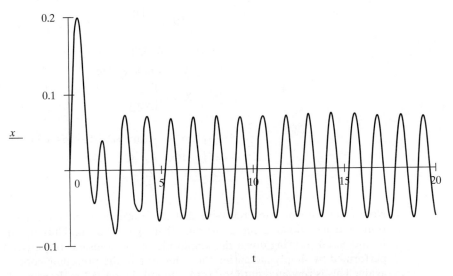

Figure 2.34 The simulated response for $c = 195$ kg/s illustrating that the transient does not exceed 0.2 m.

```
-------------------------------------------------
function v=f(t,x)
m=100; k=500; c=10; Fo=150; w=5;
v=[x(2); x(1).*-k/m+x(2).*-c/m + Fo/m*cos(w*t)];
-------------------------------------------------
```

Then the following is typed in the command window:

```
clear all

xo=[0.01; 0.5];
ts=[0 40];
[t,x]=ode45('f',ts,xo);

figure(1)
plot(t,x(:,1))
```

This is repeated with different values of damping until the desired amplitude is reached.
    In Mathematica, the exact solution and numerical solution are computed and plotted by the following list of commands:

$$\text{In[1]} := m = 100;$$
$$k = 500;$$
$$c = 10;$$
$$x0 = .01;$$
$$v0 = .5;$$
$$\omega n = \sqrt{\frac{k}{m}};$$
$$\zeta = \frac{c}{2 * \sqrt{k * m}};$$
$$\omega = 5;$$
$$\omega d = \omega n * \sqrt{1 - \zeta^2};$$
$$F0 = 150;$$
$$f0 = \frac{F0}{m};$$
$$X = \frac{f0}{\sqrt{(\omega n^2 - \omega^2)^2 + (2 * \zeta * \omega n * \omega)^2}};$$

$$\theta = \text{ArcTan}[\omega n^2 - \omega^2, \ 2 * \zeta * \omega n * \omega];$$
$$\phi = \text{ArcTan}[v0 - X * \omega * \text{Sin}[\theta] + \zeta * \omega n * (x0 - X * \text{Cos}[\theta]),$$
$$\omega d * (x0 - X * \text{Cos}[\theta])];$$
$$A = \frac{x0 - X * \text{Cos}[\theta]}{\text{Sin}[\phi]};$$

```
xanal[t_] = A * Exp[-ζ * ωn * t] * Sin[ωd * t + φ] + X
    * Cos[ω * t - θ];
numerical = NDSolve[{x"[t] + 2 * ζ * ωn * x'[t] + ωn² * x[t] == f0
    * Cos[ω * t], x[0] == x0, x'[0] == v0}; x[t], {t,0,40}];
Plot[{Evaluate[x[t]/.numerical], xanal[t]}, {t,0,40},
    PlotRange → {-4, .4},
    PlotStyle → {RGBColor[1,0,0], RGBColor[0,1,0]}]
```

□

2.9 NONLINEAR RESPONSE PROPERTIES

The use of numerical integration as introduced in the previous section allows us to consider the effects of various nonlinear terms in the equation of motion. As noted in the free-response case discussed in Section 1.10, the introduction of nonlinear terms generally results in an inability to find exact solutions, so we must rely on numerical integration and qualitative analysis to understand the response. Several important differences between linear and nonlinear systems are as follows:

1. A nonlinear system has more than one equilibrium point and each may be either stable or unstable.
2. Steady-state behavior of a nonlinear system does not always exist, and the nature of the solution is strongly dependent on the value of the initial conditions.
3. The period of oscillation of a nonlinear system depends on the initial conditions, the amplitude of excitation, and the physical parameters, unlike the linear response, which depends only on mass, damping, and stiffness values and is independent from the initial conditions.
4. Resonance in nonlinear systems may occur at excitation frequencies that are not equal to the linear system's natural frequency.
5. We cannot use the idea of superposition, used in Section 2.4, in a nonlinear system.
6. A harmonic excitation may cause a nonlinear system to respond in a nonperiodic, or chaotic, motion.

Many of these phenomena are very complex and require analysis skills beyond the scope of a first course in vibration. However, some initial understanding of nonlinear effects in vibration analysis can be observed by using the numerical solutions covered in the previous section. In this section, several simulations of the response of nonlinear systems are numerically computed and compared to their linear counterparts.

Recall from Section 1.10 that, if the equations of motion are nonlinear, the general single-degree-of-freedom system may be written as

$$\ddot{x}(t) + f\big[x(t), \dot{x}(t)\big] = 0 \tag{2.137}$$

where the function, f, can take on any form, linear or nonlinear. In the forced response case considered in this chapter, the equation of motion becomes

$$\ddot{x}(t) + f\big[x(t), \dot{x}(t)\big] = f_0 \cos \omega t \tag{2.138}$$

Formulating this last expression into the state-space, or first-order, equation (2.138) takes on the form

$$\dot{x}_1(t) = x_2(t)$$
$$\dot{x}_2(t) = -f(x_1, x_2) + f_0 \cos \omega t \tag{2.139}$$

This state-space form of the equation is used for numerical simulation in several of the codes. By defining the state vector, $\mathbf{x} = [x_1(t), x_2(t)]^T$, used in equation (1.96) and a nonlinear vector function $\mathbf{F}$ as

$$\mathbf{F}(\mathbf{x}) = \begin{bmatrix} x_2(t) \\ -f(x_1, x_2) \end{bmatrix} \tag{2.140}$$

equations (2.139) may now be written in the first-order vector form

$$\dot{\mathbf{x}} = \mathbf{F}(\mathbf{x}) + \mathbf{f}(t) \tag{2.141}$$

Equation (2.141) is the forced version of equation (1.116). Here $\mathbf{f}(t)$ is simply

$$\mathbf{f}(t) = \begin{bmatrix} 0 \\ f_0 \cos \omega t \end{bmatrix} \tag{2.142}$$

Then the Euler integration method for the equations of motion in the first-order form becomes

$$\mathbf{x}(t_{i+1}) = \mathbf{x}(t_i) + \mathbf{F}[\mathbf{x}(t_i)]\Delta t + \mathbf{f}(t_i)\Delta t \tag{2.143}$$

This expression forms a basic approach to numerically integrating to compute the forced response of a nonlinear system and is the nonlinear, forced-response version of equations (1.100) and (2.134).

     Nonlinear systems are difficult to analyze numerically as well as analytically. For this reason, the results of a numerical simulation must be examined carefully. In fact, using a more sophisticated integration method, such as Runge–Kutta, is recommended for nonlinear systems. In addition, checks on the numerical results using qualitative behavior should also be performed whenever possible.

     In the following we consider the single-degree-of-freedom system illustrated in Figure 2.35, with nonlinear spring or damping elements. A series of examples is presented using numerical simulation to examine the behavior of nonlinear systems and to compare them to the corresponding linear systems.

Example 2.9.1

Compute the response of the system in Figure 2.35 for the case that the damping is linear and viscous and the spring is a nonlinear softening spring of the form

$$f_k(x) = kx - k_1 x^3$$

and the system is subject to a harmonic excitation of 1500 N at a frequency of approximately one-third the natural frequency ($\omega = \omega_n/2.964$) and initial conditions of $x_0 = 0.01$ m and

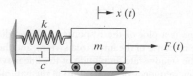

Figure 2.35 A spring–mass–damper system with potentially nonlinear elements.

$v_0 = 0.1$ m/s. The system has a mass of 100 kg, a damping coefficient of 170 kg/s, and a linear stiffness coefficient of 2000 N/m. The value of k_1 is taken to be 520 N/m³. (a) Compute the solution and compare it to the linear solution ($k_1 = 0$). (b) Examine the response for the case that the driving force is near linear resonance ($\omega = \omega_n/1.09$).

Solution   The equation of motion becomes

$$m\ddot{x}(t) + c\dot{x}(t) + kx(t) - k_1x^3(t) = 1500\cos\omega t$$

Dividing by the mass yields

$$\ddot{x}(t) + 2\zeta\omega_n\dot{x}(t) + \omega_n^2x(t) - \alpha x^3(t) = 15\cos\omega t$$

Next, write this equation in state-space form to get

$$\dot{x}_1(t) = x_2(t)$$

$$\dot{x}_2(t) = -2\zeta\omega_n x_2(t) - \omega_n^2 x_1(t) + \alpha x_1^3(t) + 150\cos\omega t$$

This last set of equations can be used in MATLAB or Mathcad to integrate numerically for the time response. Mathematica uses the second-order equation directly. Figures 2.36 and 2.37 illustrate three plots. The straight line in each is the magnitude of the linear steady-state response for the parameters given as computed by equation (2.39). The solid line in each is the response of the nonlinear system, while the dashed line is the response of the linear system. In case (a), the response of the nonlinear system exceeds that of the linear system and appears to be in resonance even though the driving frequency is almost one-third that of the natural frequency. However, in case (b) for the system near resonance (Figure 2.37), the nonlinear system response

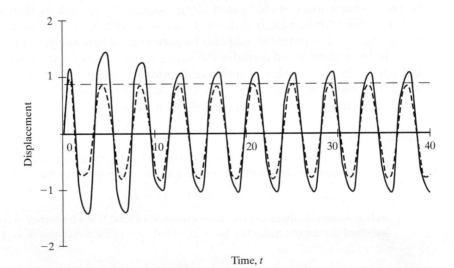

Time, t

Figure 2.36   The solid line is a plot of the response of the nonlinear system, the dashed line is a plot of the response of the linearized system, and the straight dashed line is the magnitude of the steady-state amplitude of the linear system as given by equation (2.39) for a driving frequency near one-third of the natural frequency.

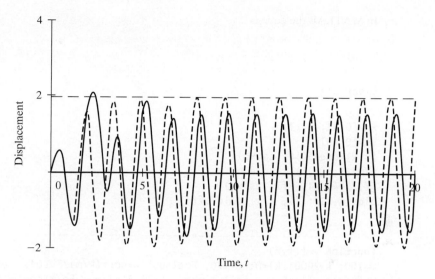

Figure 2.37 The solid line is a plot of the response of the nonlinear system, the dashed line is a plot of the response of the linearized system, and the straight dashed line is the magnitude of the steady-state amplitude of the linear system as given by equation (2.39) for a driving frequency near the natural frequency of the linear system.

is lower in amplitude than the linear system response and appears to be oscillating at two frequencies. An essential difference between linear and nonlinear systems is that a harmonically excited nonlinear system may oscillate at frequencies other than the frequency of excitation.

The codes for numerically simulating and plotting the curves given in Figures 2.36 and 2.37 are given next. In Mathcad the code is

$$x0 := 0.01 \qquad v0 := 0.1 \qquad m := 100 \qquad k := 2000 \qquad c := 170$$

$$\alpha := 5.2 \qquad F0 := 1500$$

$$\omega n := \sqrt{\frac{k}{m}} \qquad \zeta := \frac{c}{2\sqrt{k \cdot m}} \qquad f0 := \frac{F0}{m} \qquad \omega := \frac{\omega n}{2.964}$$

$$X := \begin{bmatrix} x0 \\ v0 \end{bmatrix} \qquad Y := X$$

$$D(t,X) := \begin{bmatrix} X_1 \\ -2 \cdot \zeta \cdot \omega n \cdot X_1 - \omega n^2 \cdot X_0 + \alpha \cdot (X_0)^3 + f0 \cdot \cos(\omega \cdot t) \end{bmatrix}$$

$$L(t,Y) := \begin{bmatrix} Y_1 \\ (-2 \cdot \zeta \cdot \omega n \cdot Y_1 - \omega n^2 \cdot Y_0) + f0 \cdot \cos(\omega \cdot t) \end{bmatrix}$$

$$Z := rkfixed(X,0,40,4000,D)$$

$$t := Z^{<0>} \qquad x := Z^{<1>} \qquad W := rkfixed(Y,0,40,4000,L)$$

$$xL := W^{<1>} \qquad d(t) := \frac{f0}{\sqrt{(\omega n^2 - \omega^2)^2 + (2 \cdot \zeta \cdot \omega n \cdot \omega)^2}} \qquad F := \overrightarrow{d(t)}$$

In MATLAB the code is

```
% (a)

clear all

xo=[0.01; 0.1];
ts=[0 40];
[t,x]=ode45('f',ts,xo);
plot(t,x(:,1)); hold on
[t,x1]=ode45('f1',ts,xo);
plot(t,x1(:,1),'r'); hold off


%-----------------------------------------------------
function v=f(t,x)
m=100; k=2000; k1=0; c=170; Fo=1500; w=sqrt(k/m)/2.964;
v=[x(2); x(1).*-k/m+x(2).*-c/m + x(1)^3*k1/m + Fo/m*cos(w*t)];


%-----------------------------------------------------
function v=f1(t,x)
m=100; k=2000; k1=520; c=170; Fo=1500; w=sqrt(k/m)/2.964;
v=[x(2); x(1).*-k/m+x(2).*-c/m + x(1)^3*k1/m + Fo/m*cos(w*t)];


%(b)

clear all

xo=[0.01; 0.1];
ts=[0 20];
[t,x]=ode45('f',ts,xo);
figure(2)
plot(t,x(:,1)); hold on
[t,x1]=ode45('f1',ts,xo);
plot(t,x1(:,1),'r'); hold off


%-----------------------------------------------------
function v=f(t,x)
m=100; k=2000; k1=0; c=170; Fo=1500; w=sqrt(k/m)/1.09;
v=[x(2); x(1).*-k/m + x(2).*-c/m + x(1)^3*k1/m + Fo/m*cos(w*t)];


%-----------------------------------------------------
function v=f1(t,x)
m=100; k=2000; k1=520; c=170; Fo=1500; w=sqrt(k/m)/1.09;
v=[x(2); x(1).*-k/m+x(2).*-c/m + x(1)^3*k1/m + Fo/m*cos(w*t)];
```

In Mathematica the code is

```
In[1] := <<PlotLegends'
In[2] := m = 100;
         k = 2000;
         c = 170;
         F0 = 1500;
         F0 = F0/m;
         α = 5.2;
         ωn = √(k/m);
         ζ = c/(2 * √(k * m));
         x0 = 0.01;
         v0 = 0.1;
         ω = ωn/2.964;
         ssmagnitude = f0/√((ωn² − ω²)² + (2 * ζ * ωn * ω)²);
In[14] := nonlinear = NDSolve[{x"[t] + 2 * ζ * ωn * x'[t] + ωn²
              * x[t] − α * (x[t])³ == f0 * Cos[ω * t], x[0] == x0,
              x'[0] == v0}, x[t], {t, 0, 40}, MaxSteps → 2000];
          linear = NDSolve[{x1"[t]+2 * ζ * ωn * x1'[t]+ωn²]
              * x1[t] == f0 * Cos[ω * t],
              x1[0] == x0, x1'[0] == v0}, x1[t], {t, 0, 40},
              MaxSteps → 2000];
          Plot[{Evaluate[x[t] /. nonlinear],
              Evaluate[x1[t] /. linear], ssmagnitude}, {t, 0, 40},
              PlotRange → {-2,2},
          PlotStyle → {RGBColor[1,0,0], RGBColor[0, 1, 0],
              RGBColor[0, 0, 1]},
          PlotLegend → {Nonlinear", "Linear",
              "Steady State Amp."}, LegendPosition → {1, 0},
              LegemdSize → {1, .3}]
```

A very important point is that the initial conditions are critical in determining the nature of the response (recall the pendulum of Example 1.10.4). If the initial position and/or velocity are increased, the nonlinear solution will grow without bound and the numerical integration will fail. On the other hand, the linear solution will still oscillate with amplitude less than the straight line in Figures 2.36 and 2.37.

□

Example 2.9.2

Compare the forced response of a system with velocity-squared damping as defined in equation (2.129) using numerical simulation of the nonlinear equation to that of the response of the linear system obtained using equivalent viscous damping as defined by equation (2.131).

Solution   Velocity-squared damping with a linear spring and harmonic input is described by equation (2.129), repeated here:

$$m\ddot{x} + \alpha\,\text{sgn}(\dot{x})\dot{x}^2 + kx = F_0\cos\omega t$$

The equivalent viscous-damping coefficient is calculated in equation (2.131) to be

$$c_{eq} = \frac{8}{3\pi}\alpha\omega X$$

The value of the magnitude, X, can be approximated for near resonance conditions. The value is computed in Example 2.7.3 to be

$$X = \sqrt{\frac{3\pi m f_0}{8\alpha\omega^2}}$$

Combining these last two expressions yields an equivalent viscous-damping value of

$$c_{eq} = \sqrt{\frac{8m\alpha f_0}{3\pi}}$$

Using this value as the damping coefficient results in a linear system of Figure 2.34 that approximates equation (2.131). Figures 2.38 and 2.39 are plots of the linear system with equivalent viscous damping and a numerical simulation of the full nonlinear equation (2.129) for two different values of the parameter α depending on the drag coefficient. Several conclusions can be made from these two plots.

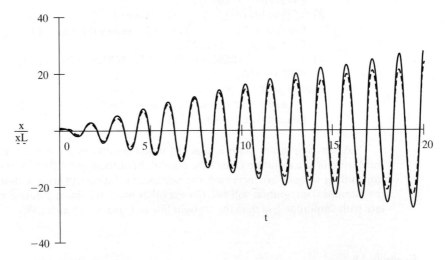

Figure 2.38 The displacement of the equivalent viscous damping (dashed line) and the displacement of the nonlinear system (solid line) versus time for the case of $\alpha = 0.005$.

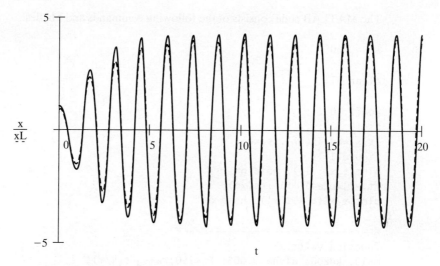

Figure 2.39   The displacement of the equivalent viscous damping (dashed line) and the displacement of the nonlinear system (solid line) versus time for the case of $\alpha = 0.5$.

First, the larger the drag coefficient, the greater the error is in using the concept of equivalent viscous damping. Second, the frequency of the response looks similar, but the amplitude of oscillation is greatly overestimated by the equivalent viscous-damping technique.

The computer codes for solving and plotting both the linear and nonlinear equations follow.

The code in Mathcad is

$$m := 10 \qquad k := 200 \qquad \alpha := .0050 \qquad F0 := 150$$

$$\omega n := \sqrt{\frac{k}{m}} \qquad f0 := \frac{F0}{m} \qquad \omega := 1 \cdot \omega n \qquad ceq := \sqrt{\frac{8 \cdot \alpha \cdot m}{3 \cdot \pi}} \cdot f0$$

$$\zeta := \frac{ceq}{2\sqrt{k \cdot m}} \qquad X := \begin{bmatrix} 1 \\ 0.1 \end{bmatrix} \qquad Y := X$$

$$D(t, X) := \begin{bmatrix} X_1 \\ -\omega n^2 \cdot X_0 - \dfrac{\alpha}{m} \cdot (X_1)^2 \cdot \dfrac{X_1}{|X_1|} + f0 \cdot \cos(\omega \cdot t) \end{bmatrix}$$

$$L(t, Y) := \begin{bmatrix} Y_1 \\ (-2 \cdot \zeta \cdot \omega n \cdot Y_1 - \omega n^2 \cdot Y_0) + f0 \cdot \cos(\omega \cdot t) \end{bmatrix}$$

$$Z := \text{rkfixed}(X, 0, 40, 2000, D)$$

$$t: = Z^{<0>} \qquad x := Z^{<1>} \qquad W := \text{rkfixed}(Y, 0, 40, 2000, L)$$

$$xL := W^{<1>}$$

The MATLAB code consists of the following commands and M-files:

```
% α=0.005

clear all

xo=[1; 0.1];
ts=[0 20];
[t,x]=ode45('f',ts,xo);
figure(1)
plot(t,x(:,1)); hold on
[t,x1]=ode45('f1',ts,xo);
plot(t,x1(:,1),'r'); hold off


%-------------------------------
function v=f(t,x)
m=10; k=200; alpha=0.005; Fo=150; w=sqrt(k/m);
ceq=sqrt(8*m*alpha*Fo/m/3/pi);
v=[x(2); x(1).*-k/m+x(2).*-ceq/m + Fo/m*cos(w*t)];


%-------------------------------
function v=f1(t,x)
m=10; k=200; alpha=0.005; Fo=150; w=sqrt(k/m);
v=[x(2); x(1).*-k/m + x(2)^2.*-alpha/m * sign(x(2)) + Fo/m*cos(w*t)];
```

The Mathematica code is

```
In[1] := <<PlotLegends'
In[2] := m = 10;
         k = 200;
         F0 = 150;
```

$$f0 = \frac{F0}{m};$$

$$\alpha = .005;$$

$$\omega n = \sqrt{\frac{k}{m}};$$

$$ceq = \sqrt{\frac{8*\alpha*m}{3*\pi}} * f0;$$

$$\zeta = \frac{ceq}{2*\sqrt{k*m}};$$

$$x0 = 1;$$

$$v0 = 0.1;$$

$$\omega = \omega n;$$

```
In[13] := velsquared = NDSolve[{x''[t] + α/m * Sign[x'[t]] * (x'[t])²
              + ωn² * x[t] == f0 * Cos[ω * t], x[0] == x0,
              x'[0] == v0}, x[t], {t, 0, 20}, MaxSteps → 2000];
```

```
equivdamping = NDSolve[{xeq''[t] + 2 * ζ * ωn * xeq'[t] + ωn²
    * xeq[t] == f0 * Cos[ω * t],
    xeq[0] == x0, xeq'[0] == v0}, xeq[t], {t, 0, 20},
    MaxSteps → 2000];
Plot[{Evaluate[x[t] /. velsquared],
    Evaluate[xeq[t] /. equivdamping]}, {t, 0, 20},
    PlotRange → {-40, 40},
    PlotStyle → {RGBColor[1, 0, 0], RGBColor[0, 1, 0],
        RGBColor[0, 0, 1]},
    PlotLegend → {"Velocity Squared", "Equivalent Damping",
        LegendPosition → {1, 0}, LegendSize → {1, .3}]
```

□

PROBLEMS

Those problems marked with an asterisk are intended to be solved using computational software.

Section 2.1 (Problems 2.1 through 2.19)

2.1. The forced response of a single-degree-of-freedom, spring-mass system is modeled by (assume the units are Newtons)

$$6\ddot{x}(t) + 24x(t) = 6 \cos \omega t$$

Compute the magnitude of the forced response for the two cases $\omega = 2.1$ rad/s and $\omega = 2.5$ rad/sec. Comment on why one value is larger then the other.

2.2. Consider the forced response of a single-degree-of-freedom, spring–mass system that is modeled by (assume the units are Newtons)

$$3\ddot{x}(t) + 12x(t) = 3 \cos \omega t$$

Compute the total response of the system if the driving frequency is 5.0 rad/s and the initial position and velocity are both zero.

2.3. Compute the response of a spring–mass system modeled by equation (2.2) to a force of magnitude 25 N, driving frequency of twice the natural frequency, and initial conditions given by $x_0 = 0$ m and $v_0 = 0.2$ m/s. The mass of the system is 10 kg, the spring stiffness is 1000 N/m, and the mass of the spring is considered and known to be 1 kg. What percent does the natural frequency change if the mass of the spring is not taken into consideration?

2.4. Show that the solution $x(t) = \dfrac{f_0}{\omega_n^2 - \omega^2} \left[\cos \omega t - \cos \omega_n t \right]$ can be written

$$x(t) = \frac{f_0}{2(\omega_n^2 - \omega^2)} \sin \frac{\omega t + \omega_n t}{2} \sin \frac{\omega_n t - \omega t}{2}.$$

2.5. A spring-mass system is driven from rest harmonically such that the displacement response exhibits a beat of period of 0.4π s. The period of oscillation is measured to be 0.04π s. Calculate the natural frequency and the driving frequency of the system.

2.6. An airplane wing modeled as a spring-mass system with natural frequency 50 Hz is driven harmonically by the rotation of its engines at 49.9 Hz. Calculate the period of the resulting beat.

2.7. Compute the total response of a spring–mass system with the following values: $k = 1500\,\text{N/m}$, $m = 15\,\text{kg}$, subject to a harmonic force of magnitude $F_0 = 150\,\text{N}$ and frequency of $8.162\,\text{rad/s}$, and initial conditions given by $x_0 = 0.01\,\text{m}$ and $v_0 = 0.01\,\text{m/s}$. Plot the response.

2.8. Consider the system in Figure P2.8, write the equation of motion and calculate the response assuming (a) that the system is initially at rest, and (b) that the system has an initial displacement of 0.04 m.

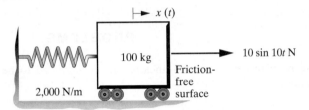

Figure P2.8

2.9. Consider the system in Figure P2.9, write the equation of motion and calculate the response assuming that the system is initially at rest for the values $k_1 = 200\,\text{N/m}$, $k_2 = 1000\,\text{N/m}$ and $m = 178\,\text{kg}$.

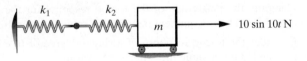

Figure P2.9

2.10. Consider the system in Figure P2.10, write the equation of motion and calculate the response assuming that the system is initially at rest for the values $\theta = 30°$, $k = 1500\,\text{N/m}$ and $m = 60\,\text{kg}$.

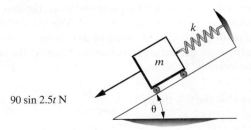

Figure P2.10

2.11. Compute the initial conditions such that the response of

$$m\ddot{x} + kx = F_0 \cos \omega t$$

oscillates at only one frequency, (ω).

2.12. The natural frequency of a 75-kg person illustrated in Figure P2.12 is measured along vertical, or longitudinal direction to be 4.5 Hz. (a) What is the effective stiffness of this person in the longitudinal direction? (b) If the person, 1.86 m in length and 0.60 m^2 in cross sectional area, is modeled as a thin bar, what is the modulus of elasticity for this system?

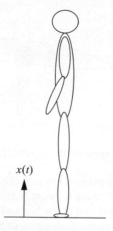

$x(t)$

Figure P2.12   The longitudinal vibration of a person.

2.13. If the person in Problem 2.12 is standing on a floor, vibrating at 3.5 Hz with an amplitude of 1 N (very small), what longitudinal displacement would the person "feel"? Assume that the initial conditions are zero.

2.14. Vibration of body parts is a significant problem in designing machines and structures. A jackhammer provides a harmonic input to the operator's arm. To model this situation, treat the forearm as a compound pendulum subject to a harmonic excitation (say of mass 5 kg and length 30 cm) as illustrated in Figure P2.14. Consider point O as a fixed pivot. Compute the maximum deflection of the hand end of the arm if the jackhammer applies a force of 10 N at 2 Hz.

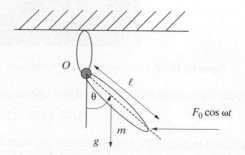

O

θ

ℓ

$F_0 \cos \omega t$

m

g

Figure P2.14   A vibration model of a forearm driven by a jackhammer.

2.15. An airfoil is mounted in a wind tunnel for the purpose of studying the aerodynamic properties of the airfoil's shape. A simple model of this is illustrated in Figure P2.15 as a rigid inertial body mounted on a rotational spring, fixed to the floor with a rigid support. Find a design relationship for the spring stiffness k in terms of the rotational inertia, J, the magnitude of the applied moment, M_0, and the driving frequency, ω, that will keep the magnitude of the angular deflection less then $6°$. Assume that the initial conditions are zero and that the driving frequency is such that $\omega_n^2 - \omega^2 > 0$.

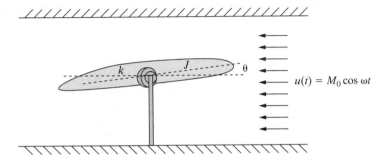

Figure P2.15   A vibration model of a wing in a wind tunnel.

2.16. The spar of an airplane wing is a relatively rigid beam extending along the length of the wing inside the wing to provide strength. It is typical to model a spar as a cantilever beam with the fixed end at the body of the aircraft. An example is given in Figure P2.16. Using the modeling methods given in Section 1.5, determine a single-degree-of-freedom model for the spar and compute its natural frequency. The spar here is modeled as a cantilever beam of dimensions length 560 mm, width 38 mm, and thickness 3.175 mm, and has a mass of 13.975 grams. The beam's Young's modulus is 10.29 GPa and its shear modulus is 1.65 GPa.

Figure P2.16   A small, unmanned air vehicle with a rigid spar, modeled as a beam.

2.17. Compute the response of a shaft and disc system to an applied moment of

$$M = 12\sin 312t$$

as indicated in Figure P2.17. Assume that the shaft is initially at rest (zero initial conditions) and $J = 0.5\,\text{kg m}^2$, the shear modulus is $G = 8.2 \times 10^{10}\,\text{N/m}^2$, the shaft is 1.2 m long, of diameter 6 cm and made of steel.

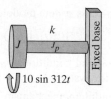

Figure P2.17

2.18. Consider a spring-mass system with zero initial conditions described by

$$\ddot{x}(t) + 4x(t) = 10 \cos 2t, \quad x(0) = 0, \quad \dot{x}(0) = 0$$

and compute the form of the response of the system.

2.19. Consider a spring mass system with zero initial conditions described by

$$\ddot{x}(t) + 4x(t) = 10 \sin 4t, \quad x(0) = 0, \quad \dot{x}(0) = 0$$

and compute the form of the response of the system.

Section 2.2 (Problems 2.20 through 2.38)

2.20. Calculate the constants A and ϕ for arbitrary initial conditions, x_0 and v_0, in the case of the forced response given by

$$x(t) = Ae^{-\zeta \omega_n t} \sin (\omega_d t + \phi) + X \cos (\omega t - \theta)$$

Compare this solution to the transient response obtained in the case of no forcing function (i.e., $F_0 = 0$).

2.21. Consider the spring-mass-damper system defined by (use basic SI units)

$$\ddot{x}(t) + 6\dot{x}(t) + 25x(t) = 4 \cos 5t$$

First determine if the system is underdamped, critically damped or overdamped. Then compute the magnitude and phase of the steady state response.

2.22. Show that the following two expressions are equivalent:

$$x_p(t) = X \cos (\omega t - \theta) \text{ and } x_p(t) = A_s \cos \omega t + B_s \sin \omega t$$

2.23. Calculate the total solution of

$$\ddot{x} + 2\zeta \omega_n \dot{x} + \omega_n^2 x = f_0 \cos \omega t$$

for the case that $m = 1.2 \, \text{kg}$, $\zeta = 0.01$, $\omega_n = 2.5 \, \text{rad/s}$. $f_0 = 5 \, \text{N/kg}$, and $\omega = 10 \, \text{rad/s}$, with initial conditions $x_0 = 1 \, \text{m}$ and $v_0 = 1 \, \text{m/s}$, and then plot the response.

2.24. A 120 kg mass is suspended by a spring of stiffness $32 \times 10^3 \, \text{N/m}$ with a viscous damping constant of 1200 Ns/m. The mass is initially at rest and in equilibrium. Calculate the steady-state displacement amplitude and phase if the mass is excited by a harmonic force of 100 N at 3 Hz.

2.25. Plot the total solution of the system of Problem 2.24 including the transient.

2.26. A damped spring–mass system modeled by (units are Newtons)

$$10\ddot{x}(t) + \dot{x}(t) + 170x(t) = 100 \cos 4t$$

is also subject to initial conditions: $x_0 = 1\,$mm and $v_0 = 20\,$mm/s. Compute the total response, $x(t)$, of the system.

2.27. Consider the pendulum mechanism of Figure P2.27 which is pivoted at point O. Calculate both the damped and undamped natural frequency of the system for small angles. Assume that the mass of the rod, spring, and damper are negligible. What driving frequency will cause resonance?

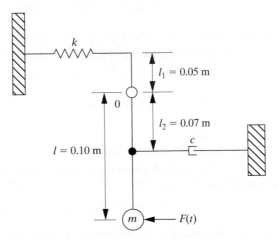

$l_1 = 0.05$ m

$l_2 = 0.07$ m

$l = 0.10$ m

c

$F(t)$

Figure P2.27

2.28. Consider the pivoted mechanism of Figure P2.27 with $k = 4 \times 10^3\,$N/m, $l_1 = 0.06\,$m, $l_2 = 0.09\,$m, and $l = 0.12\,$m. and $m = 40\,$kg. The mass of the beam is 40 kg; it is pivoted at point 0 and assumed to be rigid. Design the dashpot (i.e. calculate c) so that the damping ratio of the system is 0.2. Also determine the amplitude of vibration of the steady-state response if a 15-N force is applied to the mass, as indicated in the figure, at a frequency of 10 rad/s.

2.29. Compute the response of a shaft and disc system to an applied moment of

$$M = 10 \sin 312t$$

as indicated in Figure P2.29. Assume that the shaft is initially at rest (zero initial conditions) and $J = 0.6\,$kg m², the shear modulus is $G = 8.2 \times 10^{10}\,$N/m², the shaft is 1 m long, of diameter 5 cm and made of steel. Assume the damping ratio of steel is $\zeta = 0.01$.

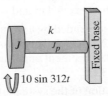

k

J J_p

$10 \sin 312t$

Fixed base

Figure P2.29

2.30. Compute the forced response of a spring-mass-damper system with the following values: $c = 250\,\text{kg/s}$, $k = 2500\,\text{N/m}$, $m = 125\,\text{kg}$, subject to a harmonic force of magnitude $F_0 = 15\,\text{N}$ and frequency of 10 rad/s and initial conditions of $x_0 = 0.01\,\text{m}$ and $v_0 = 0.1\,\text{m/s}$. Plot the response. How long does it take for the transient part to die off?

2.31. Compute a value of the damping coefficient, c, such that the steady-state response amplitude of the system in Figure P2.31 is 0.01 m.

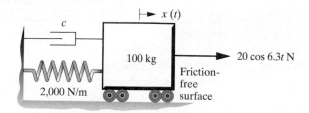

$\blacktriangleright x(t)$

100 kg $\blacktriangleright$ 20 cos 6.3t N

Friction-free surface

2,000 N/m

Figure P2.31

2.32. Consider a spring-mass-damper systems like the one in Figure P2.31 with the following values: $m = 80\,\text{kg}$, $c = 80\,\text{kg/s}$, $k = 2400\,\text{N/m}$ and $F_0 = 20\,\text{N}$, and the driving frequency $\omega = 4.3$ rad/s. Compute the magnitude of the steady-state response and compare it to the magnitude of the forced response of an undamped system.

2.33. Compute the response of the system in Figure P2.33 if the system is initially at rest for the values $k_1 = 200\,\text{N/m}$, $k_2 = 1000\,\text{N/m}$, $c = 20\,\text{kg/s}$, and $m = 175$ kg.

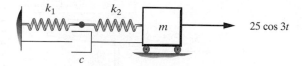

k_1 k_2 m 25 cos 3t

c

Figure P2.33

2.34. Write the equation of motion for the system given in Figure P2.34 for the case that $F(t) = F\cos\omega t$ and the surface is friction free. Does the angle θ affect the magnitude of oscillation?

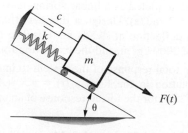

c k m $F(t)$ θ

Figure P2.34

2.35. A foot pedal for a musical instrument is modeled by the sketch in Figure P2.35. With $k = 2400\,\text{N/m}$, $c = 30\,\text{kg/s}$, $m = 32\,\text{kg}$ and $F(t) = 50\cos 2\pi t\,\text{N}$, compute the steady state response assuming the system starts from rest. Also use the small angle approximation.

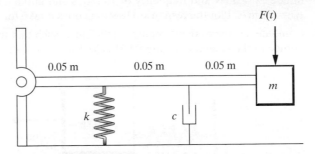

0.05 m   0.05 m   0.05 m

k c m $F(t)$

Figure P2.35

2.36. Consider the system of Problem 2.15, repeated here as Figure P2.36 with the effects of damping indicated. The physical constants are $J = 24 \, \text{kg m}^2$, $k = 2500 \, \text{Nm/rad}$, and the applied moment is 5 Nm at 1.432 Hz acting through the distance $r = 0.5 \, \text{m}$. Compute the magnitude of the steady state response if the measured damping ratio of the spring system is $\zeta = 0.01$. Compare this to the response for the case where the damping is not modeled ($\zeta = 0$).

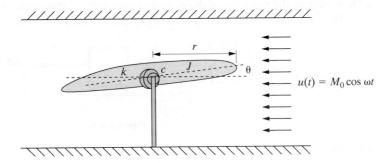

r

k c J θ

$u(t) = M_0 \cos \omega t$

Figure P2.36   Model of an airfoil in a wind tunnel including the effects of damping.

2.37. A machine, modeled as a linear spring–mass–damper system, is driven at resonance ($\omega_n = \omega = 2 \, \text{rad/s}$). Design a damper (that is, choose a value of c) such that the maximum deflection at steady state is 0.05 m. The machine is modeled as having a stiffness of 2000 kg/m, and the excitation force has a magnitude of 100 N.

2.38. Derive the total response of the system to initial conditions x_0 and v_0 using the homogenous solution in the form $x_h(t) = e^{-\zeta \omega_n t}(A_1 \sin \omega_d t + A_2 \cos \omega_d t)$ and hence verify equation (2.38) for the forced response of an underdamped system.

Section 2.3 (Problems 2.39 through 2.44)

2.39. Referring to Figure 2.11, draw the solution for the magnitude X for the case $m = 80 \, \text{kg}$, $c = 3200 \, \text{N s/m}$, and $k = 8{,}000 \, \text{N/m}$. Assume that the system is driven at resonance by a 10-N force.

2.40. Use the graphical method to compute the phase shift for the system with $m = 100$ kg, $c = 4000$ N s/m, $k = 10,000$ N/m, and $F_0 = 10$ N, if $\omega = \omega_n/2$ and again for the case $\omega = 2\omega_n$.

2.41. A body of mass 80 kg is suspended by a spring of stiffness of 25 kN/m and dashpot of damping constant 800 N s/m. Vibration is excited by a harmonic force of amplitude 60 N and a frequency of 3 Hz. Calculate the amplitude of the displacement for the vibration and the phase angle between the displacement and the excitation force using the graphical method.

2.42. Calculate the real part of equation (2.55)

$$x_p(t) = \frac{F_0}{\left[(k - m\omega^2)^2 + (c\omega)^2\right]^{1/2}} e^{j(\omega t - \theta)}$$

to verify that this is consistent with the equation (2.36)

$$X_p = \frac{f_0}{\sqrt{\left(\omega_n^2 - \omega^2\right)^2 + (2\zeta\omega_n\omega)^2}}$$

and hence establish the equivalence of the exponential approach to solving the damped vibration problem with method of undetermined coefficients.

2.43. Referring to equation (2.56)

$$\left(ms^2 + cs + k\right)X(s) = \frac{F_0 s}{s^2 + \omega^2}$$

and a table of Laplace transforms (see Appendix B), calculate the solution $x(t)$ by using a table of Laplace transform pairs, and show that the solution obtained this way is equivalent to (2.36).

2.44. Solve the following system using the Laplace transform method and the table in Appendix B:

$$m\ddot{x}(t) + kx(t) = F_0 \cos \omega t, x(0) = x_0, \dot{x}(0) = v_0$$

Check your solution against equation (2.11) obtained via the method of undetermined coefficients.

Section 2.4 (Problems 2.45 through 2.60)

2.45. For a base motion system described by

$$m\ddot{x} + c\dot{x} + kx = cY\omega_b \cos \omega_b t + kY \sin \omega_b t$$

with $m = 100$ kg, $c = 50$ kg/s, $k = 1000$ N/m, $Y = 0.03$ m, and $\omega_b = 3$ rad/s, compute the magnitude of the particular solution. Last, compute the transmissibility ratio.

2.46. For a base motion system described by

$$m\ddot{x} + c\dot{x} + kx = cY\omega_b \cos \omega_b t + kY \sin \omega_b t$$

with $m = 100$ kg, $c = 50$ N/m, $Y = 0.03$ m, and $\omega_b = 3$ rad/s, find largest value of the stiffness k and that makes the transmissibility ratio less than 0.75.

2.47. A machine weighing 1800 N rests on a support as illustrated in Figure P2.47. The support deflects about 4 cm as a result of the weight of the machine. The floor under the support is somewhat flexible and moves, because of the motion of a nearby machine, harmonically near resonance ($r = 1$) with an amplitude of 0.2 cm. Model the floor as base motion, and assume a damping ratio of $\zeta = 0.01$, and calculate the transmitted force and the amplitude of the transmitted displacement.

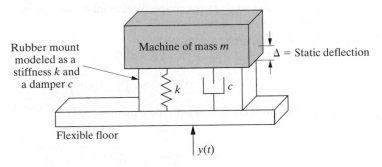

Rubber mount modeled as a stiffness k and a damper c

Machine of mass m

Δ = Static deflection

Flexible floor

$y(t)$

Figure P2.47

2.48. Derive equation (2.70)

$$X = Y\left[\frac{1 + (2\zeta r)^2}{(1 - r^2)^2 + (2\zeta r)^2}\right]^{1/2}$$

from (2.68)

$$x_p(t) = \omega_n Y\left[\frac{\omega_n^2 + (2\zeta\omega_b)^2}{(\omega_n^2 - \omega_b^2)^2 + (2\zeta\omega_n\omega_b)^2}\right]^{1/2} \cos(\omega_b t - \theta_1 - \theta_2)$$

to see if the author has done it correctly.

2.49. From the equation describing Figure 2.14, show that the point $(\sqrt{2}, 1)$ corresponds to the value TR > 1 (i.e., for all $r < \sqrt{2}$, TR > 1).

2.50. Consider the base-excitation problem for the configuration shown in Figure P2.50. In this case, the base motion is a displacement transmitted through a dashpot or pure damping element. Derive an expression for the force transmitted to the support in steady state.

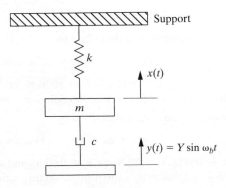

Support

k

$x(t)$

m

c

$y(t) = Y \sin \omega_b t$

Figure P2.50

2.51. A very common example of base motion is the single-degree-of-freedom model of an automobile driving over a rough road. The road is modeled as providing a base motion displacement of $y(t) = (0.01)\sin(5.818t)$ m. The suspension provides an equivalent stiffness of $k = 3.273 \times 10^4$ N/m, a damping coefficient of $c = 231$ kg/s, and a mass of 1007 kg. Determine the amplitude of the absolute displacement of the automobile mass.

2.52. A vibrating mass of 250 kg, mounted on a massless support by a spring of stiffness 32,000 N/m and a damper of unknown damping coefficient, is observed to vibrate with a 10-mm amplitude while the support vibration has a maximum amplitude of only 2.5 mm (at resonance). Calculate the damping constant and the amplitude of the force on the base.

2.53. Referring to Example 2.4.2, at what speed does Car 1 experience resonance? At what speed does Car 2 experience resonance? Calculate the maximum deflection of both cars at resonance.

2.54. For cars of Example 2.4.2, calculate the best choice of the damping coefficient so that the transmissibility is as small as possible by comparing the magnitude of $\zeta = 0.01$, $\zeta = 0.1$, and $\zeta = 0.2$ for the case $r = 2$. What happens if the road "frequency" changes?

2.55. A system modeled by Figure 2.13, has a mass of 200 kg with a spring stiffness of 3.0×10^4 N/m. Calculate the damping coefficient given that the system has a deflection (X) of 0.7 cm when driven at its natural frequency while the base amplitude (Y) is measured to be 0.3 cm.

2.56. Consider Example 2.4.2 for Car 1 illustrated in Figure P2.56 if three passengers totaling 200 kg are riding in the car. Calculate the effect of the mass of the passengers on the deflection at 20, 80, 100, and 150 km/h. What is the effect of the added passenger mass on Car 2?

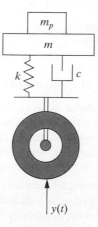

Figure P2.56 A model of a car suspension with the mass of the occupants, m_p, included.

2.57. Consider Example 2.4.2. Choose values of c and k for the suspension system for Car 2 (the sedan) such that the amplitude transmitted to the passenger compartment is as small as possible for the 1-cm bump at 50 km/h. Also calculate the deflection at 100 km/h for your values of c and k.

2.58. Consider the base motion problem of Figure 2.13. (a) Compute the damping ratio needed to keep the displacement magnitude transmissibility less then 0.50 for a frequency ratio of $r = 1.5$. (b) What is the value of the force transmissibility ratio for this system?

2.59. Consider the effect of variable mass on an aircraft landing suspension system by modeling the landing gear as a moving base problem similar to that shown in Figure P2.56 for a car suspension. The mass of a regional jet is 13,236 kg empty and its maximum takeoff mass is 21,523 kg. Compare the maximum deflection for a wheel motion of magnitude 0.50 m and frequency of 35 rad/s for these two different masses. Take the damping ratio to be $\zeta = 0.1$ and the stiffness to be 4.22×10^6 N/m.

2.60. Consider the simple model of a building subject to ground motion suggested in Figure P2.60. The building is modeled as a single-degree-of-freedom spring–mass system where the building mass is lumped atop two beams used to model the walls of the building in bending. Assume the ground motion is modeled as having amplitude of 0.1 m at a frequency of 7.5 rad/s. Approximate the building mass by 10^5 kg and the stiffness of each wall by 3.519×10^6 N/m. Compute the magnitude of the deflection of the top of the building.

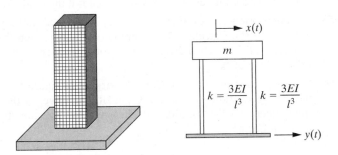

Figure P2.60   A simple model of a building subject to ground motion, such as an earthquake.

Section 2.5 (Problems 2.61 through 2.68)

2.61. A lathe can be modeled as an electric motor mounted on a steel table. The table plus the motor have a mass of 60 kg. The rotating parts of the lathe have a mass of 4 kg at a distance 0.12 m from the center. The damping ratio of the system is measured to be $\zeta = 0.06$ (viscous damping) and its natural frequency is 7.5 Hz. Calculate the amplitude of the steady-state displacement of the motor, assuming $\omega_r = 30$ Hz.

2.62. The system of Figure 2.19 produces a forced oscillation of varying frequency. As the frequency is changed, it is noted that at resonance the amplitude of the displacement is 10 mm. As the frequency is increased several decades past resonance, the amplitude of the displacement remains fixed at 1 mm. Estimate the damping ratio for the system.

2.63. An electric motor (Figure P2.63) has an eccentric mass of 12 kg (12% of the total mass of 100 kg) and is set on two identical springs ($k = 3000$ N/m). The motor runs at 1800 rpm, and the mass eccentricity is 100 mm from the center. The springs are mounted 250 mm apart with the motor shaft in the center. Neglect damping and determine the amplitude of the vertical vibration.

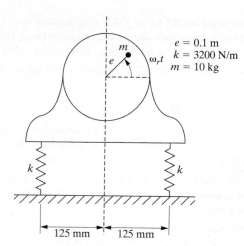

$e = 0.1$ m
$k = 3200$ N/m
$m = 10$ kg

Figure P2.63   A vibration model for an electric motor with an unbalance.

125 mm   125 mm

2.64. Consider a system with rotating unbalance as illustrated in Figure P2.63. Suppose the deflection at 1800 rpm is measured to be 0.05 m and the damping ratio is measured to be $\zeta = 0.1$. The out-of-balance mass is estimated to be 10%. Locate the unbalanced mass by computing e.

2.65. A fan of 45 kg has an unbalance that creates a harmonic force. A spring-damper system is designed to minimize the force transmitted to the base of the fan. A damper is used having a damping ratio of $\zeta = 0.2$. Calculate the required spring stiffness so that only 10% of the force is transmitted to the ground when the fan is running at 10,000 rpm.

2.66. Plot the normalized displacement magnitude versus the frequency ratio for the out-of-balance problem (i.e., repeat Figure 2.21) for the case of $\zeta = 0.05$.

2.67. Consider a typical unbalanced machine problem as given in Figure P2.67 with a machine mass of 150 kg, a mount stiffness of 1000 kN/m and a damping value of 600 kg/s.

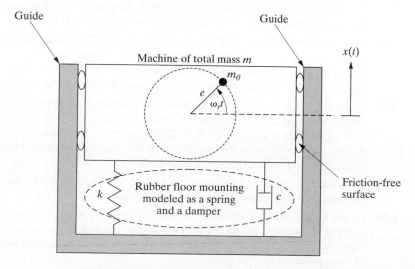

Guide   Guide

Machine of total mass m $x(t)$

m_0

e

$\omega_r t$

Rubber floor mounting
modeled as a spring
and a damper

k c

Friction-free
surface

Figure P2.67   A typical unbalance machine problem.

The out of balance force is measured to be 374 N at a running speed of 3000 rev/min. (a) Determine the amplitude of motion due to the out of balance. (b) If the out of balance mass is estimated to be 1% of the total mass, estimate the value of the e.

2.68. Plot the response of the mass in Problem 2.67 assuming zero initial conditions.

Section 2.6 (Problems 2.69 through 2.72)

2.69. Calculate damping and stiffness coefficients for the accelerometer of Figure 2.24 with moving mass of 0.04 kg such that the accelerometer is able to measure vibration between 0 and 50 Hz within 5%. (*Hint:* For an accelerometer it is desirable for $Z/\omega_b^2 Y = $ constant.)

2.70. The damping constant for a particular accelerometer of the type illustrated in Figure 2.26 is 50 N s/m. It is desired to design the accelerometer (i.e., choose m and k) for a maximum error of 3% over the frequency range 0 to 75 Hz.

2.71. The accelerometer of Figure 2.24 has a natural frequency of 120 kHz and a damping ratio of 0.2. Calculate the error in measurement of a sinusoidal vibration at 60 kHz.

2.72. Design an accelerometer (i.e., choose m, c, and k) configured as in Figure 2.24 with very small mass that will be accurate to 1% over the frequency range 0 to 50 Hz.

Section 2.7 (Problems 2.73 through 2.89)

2.73. Consider a spring-mass sliding along a surface providing Coulomb friction, with stiffness 1.25×10^4 N/m and mass 10 kg, driven harmonically by a force of 50 N at 10 Hz. Calculate the approximate amplitude of steady-state motion assuming that both the mass and the surface that it slides on, are made of lubricated steel.

2.74. A spring-mass system with Coulomb damping of 10 kg, stiffness of 2400 N/m, and coefficient of friction of 0.15 is driven harmonically at 10 Hz. The amplitude at steady state is 4 cm. Calculate the magnitude of the driving force.

2.75. A system of mass 10 kg and stiffness 1.8×10^4 N/m is subject to Coulomb damping. If the mass is driven harmonically by a 90-N force at 25 Hz, determine the equivalent viscous damping coefficient if the coefficient of friction is 0.12.

2.76. **a.** Plot the free response of the system of Problem 2.75 to initial conditions of $x(0) = 0$ and $\dot{x}(0) = |F_0/m| = 9$ m/s using the solution in Section 1.10.
b. Use the equivalent viscous-damping coefficient calculated in Problem 2.75 and plot the free response of the "equivalent" viscously damped system to the same initial conditions.

2.77. Referring to the system of Example 2.7.1; a spring–mass system with sliding friction described by equation (2.97) with stiffness $k = 1.5 \times 10^4$ N/m, driving harmonically a 10-kg mass by a force of 90 N at 25 Hz, calculate how large the magnitude of the driving force must be to sustain motion if the steel is lubricated. How large must this magnitude be if the lubrication is removed?

2.78. Calculate the phase shift between the driving force and the response for the system of Problem 2.77 using the equivalent viscous-damping approximation.

2.79. Derive the equation of vibration for the system of Figure P2.79 assuming that a viscous dashpot of damping constant c is connected in parallel to the spring. Calculate the energy loss and determine the magnitude and phase relationships for the forced response of the equivalent viscous system.

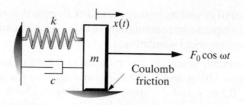

Figure P2.79

2.80. A system of unknown damping mechanism is driven harmonically at 10 Hz with an adjustable magnitude. The magnitude is changed, and the energy lost per cycle and amplitudes are measured for five different magnitudes. The measured quantities are

$\Delta E(J)$	0.25	0.45	0.8	1.16	3.0
$X(M)$	0.01	0.02	0.04	0.08	0.15

Is the damping viscous or Coulomb?

2.81. Calculate the equivalent loss factor for a system with Coulomb damping.

2.82. A spring-mass system ($m = 10\,\text{kg}$, $k = 4 \times 10^3\,\text{N/m}$) vibrates horizontally on a surface with coefficient of friction $\mu = 0.18$. When excited harmonically at 5 Hz, the steady-state displacement of the mass is 5 cm. Calculate the amplitude of the harmonic force applied.

2.83. Calculate the displacement for a system with air damping using the equivalent viscous-damping method.

2.84. Calculate the semimajor and semiminor axis of the ellipse of equation (2.119). Then calculate the area of the ellipse. Use $c = 10\,\text{kg/s}$, $\omega = 2\,\text{rad/s}$, and $X = 0.01$ m.

2.85. The area of a force deflection curve of Figure 2.29 is measured to be 2.5 N · m, and the maximum deflection is measured to be 8 mm. From the "slope" of the ellipse, the stiffness is estimated to be 5×10^4 N/m. Calculate the hysteretic damping coefficient. What is the equivalent viscous damping if the system is driven at 10 Hz?

2.86. The area of the hysteresis loop of a hysterically damped system is measured to be 5 N · m and the maximum deflection is measured to be 1 cm. Calculate the equivalent viscous damping coefficient for a 24-Hz driving force. Plot c_{eq} versus ω for $2\pi \le \omega \le 100\pi\,\text{rad/s}$.

2.87. Calculate the nonconservative energy of a system subject to both viscous and hysteretic damping.

2.88. Derive a formula for equivalent viscous damping for the damping force of the form, $F_d = c(\dot{x})^n$, where n is an integer.

2.89. Using the equivalent viscous-damping formulation, determine an expression for the steady-state amplitude under harmonic excitation for a system with both Coulomb and viscous damping present.

Section 2.8 (Problems 2.90 through 2.96)

*2.90. Numerically integrate and plot the response of an underdamped system determined by $m = 100\,\text{kg}$, $k = 20,000\,\text{N/m}$, and $c = 200\,\text{kg/s}$, subject to the initial conditions of

$x_0 = 0.01$ m and $v_0 = 0.1$ m/s, and the applied force $F(t) = 160 \cos 5t$. Then plot the exact response as computed by equation (2.33). Compare the plot of the exact solution to the numerical simulation.

*2.91. Numerically integrate and plot the response of an underdamped system determined by $m = 120$ kg, and $k = 3200$ N/m subject to the initial conditions of $x_0 = 0.01$ m and $v_0 = 0.1$ m/s, and the applied force $F(t) = 15 \cos 10t$, for various values of the damping coefficient. Use this "program" to determine a value of damping that causes the transient term to die out with in 3 seconds. Try to find the smallest such value of damping remembering that added damping is usually expensive.

*2.92. Compute the total response of a spring-mass system with the following values: $k = 1000$ N/m, $m = 12$ kg, subject to a harmonic force of magnitude $F_0 = 100$ N and frequency of 8.162 rad/s, and initial conditions given by $x_0 = 0.01$ m and $v_0 = 0.01$ m/s, by numerically integrating rather than using analytical expressions as was done in Problem 2.8. Plot the response.

*2.93. A foot pedal for a musical instrument is modeled by the sketch in Figure P2.93. With $k = 2000$ N/m, $c = 25$ kg/s, $m = 25$ kg, and $F(t) = 50 \cos 2\pi t$ N, numerically simulate the response of the system assuming the system starts from rest. Use the small-angle approximation.

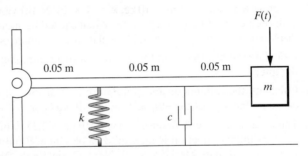

Figure P2.93

*2.94. Numerically integrate and plot the response of an underdamped system determined by $m = 100$ kg, $k = 2000$ N/m, and $c = 200$ kg/s, subject to the applied force $F(t) = 150 \cos 10t$ for the following sets of initial conditions:
   (a) $x_0 = 0.0$ m and $v_0 = 0.1$ m/s
   (b) $x_0 = 0.01$ m and $v_0 = 0.0$ m/s
   (c) $x_0 = 0.05$ m and $v_0 = 0.0$ m/s
   (d) $x_0 = 0.0$ m and $v_0 = 0.5$ m/s

Plot these responses on the same graph and note the effects of the initial conditions on the transient part of the response.

*2.95. A DVD drive is mounted on a chassis and is modeled as a single-degree-of-freedom spring, mass, and damper. During normal operation, the drive (having a mass of 0.4 kg) is subject to a harmonic force of 1 N at 10 rad/s. Because of material considerations and static deflection, the stiffness is fixed at 500 N/m and the natural damping in the system is 10 kg/s. The DVD player starts and stops during its normal operation providing initial conditions to the module of $x_0 = 0.001$ m and $v_0 = 0.5$ m/s. The DVD drive must not have an amplitude of vibration larger then 0.008 m even during the transient stage. First, compute the response by numerical simulation to see if the constraint is satisfied. If the constraint is not satisfied, find the smallest value of damping that will keep the deflection less than 0.008 m.

2.96. Use a plotting routine to examine the base motion problem (see Figure 2.13) by plotting the particular solution (for an undamped system) for the three cases $k = 1500$ N/m, $k = 2500$ N/m, and $k = 700$ N/m. Also note the values of the three frequency ratios and the corresponding amplitude of vibration of each case compared to the input. Use the following values: $\omega_b = 4.4$ rad/s, $m = 100$ kg, and $Y = 0.05$ m.

Section 2.9 (Problems 2.97 through 2.102)

*2.97. Compute the response of the system in Figure P2.93 for the case that the damping is linear viscous, the spring is a nonlinear soft spring of the form

$$k(x) = kx - k_1 x^3$$

and the system is subject to a harmonic excitation of 300 N at a frequency of approximately one third the natural frequency ($\omega = \omega_n/3$) and initial conditions of $x_0 = 0.01$ m and $v_0 = 0.1$ m/s. The system has a mass of 100 kg, a damping coefficient of 170 kg/s, and a linear stiffness coefficient of 2000 N/m. The value of k_1 is taken to be 10,000 N/m^3. Compute the solution and compare it to the linear solution ($k_1 = 0$). Which system has the largest magnitude?

*2.98. Compute the response of the system in Figure P2.97 for the case that the damping is linear viscous, the spring is a nonlinear hard spring of the form

$$k(x) = kx + k_1 x^3$$

and the system is subject to a harmonic excitation of 300 N at a frequency equal to the natural frequency ($\omega = \omega_n$) and initial conditions of $x_0 = 0.01$ m and $v_0 = 0.1$ m/s. The system has a mass of 100 kg, a damping coefficient of 170 kg/s, and a linear stiffness coefficient of 2000 N/m. The value of k_1 is taken to be 10,000 N/m^3. Compute the solution and compare it to the linear solution ($k_1 = 0$). Which system has the largest magnitude?

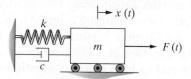

Figure P2.97

*2.99. Compute the response of the system in Figure P2.97 for the case that the damping is linear viscous, the spring is a nonlinear soft spring of the form

$$k(x) = kx - k_1 x^3$$

and the system is subject to a harmonic excitation of 300 N at a frequency equal to the natural frequency ($\omega = \omega_n$) and initial conditions of $x_0 = 0.01$ m and $v_0 = 0.1$ m/s. The system has a mass of 100 kg, a damping coefficient of 15 kg/s, and a linear stiffness coefficient of 2000 N/m. The value of k_1 is taken to be 100 N/m^3. Compute the solution and compare it to the hard spring solution ($k(x) = kx + k_1 x^3$).

*2.100. Compute the response of the system in Figure P2.97 for the case that the damping is linear viscous, the spring is a nonlinear soft spring of the form

$$k(x) = kx - k_1x^3$$

and the system is subject to a harmonic excitation of 300 N at a frequency equal to the natural frequency ($\omega = \omega_n$) and initial conditions of $x_0 = 0.01$ m and $v_0 = 0.1$ m/s. The system has a mass of 100 kg, a damping coefficient of 15 kg/s, and a linear stiffness coefficient of 2000 N/m. The value of k_1 is taken to be 1000 N/m^3. Compute the solution and compare it to the quadratic soft spring ($k(x) = kx + k_1x^2$).

*2.101. Compare the forced response of a system with velocity-squared damping with equation of motion given by

$$m\ddot{x} + \alpha \, \text{sgn}(\dot{x})\dot{x}^2 + kx = F_0 \cos \omega t$$

using numerical simulation of the nonlinear equation to that of the response of the linear system obtained using equivalent viscous damping as defined by equation (2.131)

$$c_{eq} = \frac{8}{3\pi} \alpha \omega X$$

Use as initial conditions, $x_0 = 0.01$ m and $v_0 = 0.1$ m/s with a mass of 10 kg, stiffness of 25 N/m, applied force of 150 cos ($\omega_n t$), and drag coefficient of $\alpha = 250$.

*2.102. Compare the forced response of a system with structural damping (see Table 2.2) using numerical simulation of the nonlinear equation to that of the response of the linear system obtained using equivalent viscous damping as defined in Table 2.2. Use as initial conditions, $x_0 = 0.01$ m and $v_0 = 0.1$ m/s with a mass of 10 kg, stiffness of 25 N/m, applied force of 150 cos ($\omega_n t$), and solid damping coefficient of $b = 25$.

MATLAB® ENGINEERING VIBRATION TOOLBOX

If you did not use the *Engineering Vibration Toolbox* for Chapter 1, refer to that section for information regarding using MATLAB files or refer to Appendix G.

The files for Chapter 2, entitled VTB2_1, VTB2_2, and so on, can be found in folder VTB2. The files in VTB2 can be used to help solve the preceding homework problems and to help gain information about the nature of the response of single-degree-of-freedom systems to harmonic inputs. The following problems are intended to help you gain some experience with the concepts in this chapter.

TOOLBOX PROBLEMS

TB2.1. Using file VTB2_1, reproduce Figure 2.2.

TB2.2. Carefully investigate the response of an undamped system near resonance by trying several values of ω near ω_n for the values of Figure 2.2. Do you get the beats of Figure 2.3?

TB2.3. Using file VTB2_3, reproduce Figure 2.9. Also plot Xk/f_0 versus r for the values given in Example 2.2.3 and plot the associated time response, $x_p(t)$, for a value of $r = 0.5$ using VTB2_2. Do these plots again for $\zeta = 0.01$ and $\zeta = 0.1$ and comment on how the time response changes as the damping ratio, ζ, changes by an order of magnitude.

TB2.4. Using file VTB2_5 for rotating unbalance, make a plot of x versus r for the helicopter of Example 2.4.2.

TB2.5. Using file VTB2_6 for damping mechanisms, compare the time response of a system (with physical parameters of $m = 10$, $k = 100$, $\alpha = 0.05$, $X = 1$) with air damping as given by equation (2.129) with initial conditions $x_0 = 1$ and $v_0 = 0$ to that of an equivalent viscously damped system using equation (2.131) for an input of $10 \sin 3t$.

3 General Forced Response

This chapter starts out by considering the response to systems subject to a shock loading or impulse. An example of such a load occurs during landing an airplane. The aircraft landing gear, pictured at the top, provides stiffness and damping designed (see Chapter 5) to mitigate the effect of the shock on the aircraft. The input to the structure is not completely periodic, as examined in Chapter 2, but has random shock and other components, as discussed in this chapter.

Another source of vibration that is not at a single frequency (as in Chapter 2) is the human heart. The heart vibrates at a variety of different rates depending on the level of activity and emotional state of the person. Some hearts need regulating by using a device such as the pacemaker, pictured at the bottom. These are run by batteries, which require replacement every seven to ten years, involving major surgery. Vibration researchers have recently developed energy harvesting devices that transduce the heart-induced, chest-cavity vibrations into electrical energy. This energy is then used to recharge the pacemaker battery. Fitting such a device inside the pacemaker required a basic understanding of vibration and, in particular models, of the vibration response of a structure (the harvester in this case) to inputs that have energy at many different frequencies, as discussed in this chapter.

In Chapter 2 the forced response of a single-degree-of-freedom system was considered for the special case of a harmonic driving force. *Harmonic excitation* refers to an applied force that is sinusoidal of a single frequency. In this chapter, the response of a system to a variety of different types of forces is considered, as well as a general formulation for calculating the forced response for any type of applied force. If the system considered is linear, the principle of superposition can be used to calculate the response to various combinations of forces based on the individual response to a specific force.

Superposition refers to the fact that for a linear equation of motion, say $\ddot{x} + \omega_n^2 x = 0$, if x_1 and x_2 are both solutions of the equation, then $x = a_1 x_1 + a_2 x_2$ is also a solution where a_1 and a_2 are any constants. This concept also implies that if x_1 is a particular solution to $\ddot{x} + \omega_n^2 x = f_1$ and x_2 is a solution to $\ddot{x} + \omega_n^2 x = f_2$, then $x_1 + x_2$ is a solution of $\ddot{x} + \omega_n^2 x = f_1 + f_2$. Thus this method of superposition can be used to construct the solution to a complicated forcing function by solving a series of simpler problems. Superposition was used to solve the base-excitation problem of Section 2.3. The principle of superposition in linear systems is a very powerful technique and is used extensively.

A variety of forces are applied to mechanical systems that result in vibration. Earthquake forces are sometimes modeled as sums of decaying periodic or harmonic forces. High winds can be a source of impulsive or step loadings to structures. Rough roads provide a variety of forcing conditions to automobiles. The ocean waves and wind provide forces to ships at sea. Various manufacturing processes produce applied forces that are random, periodic, nonperiodic, or transient in nature. Air and relative motion provide forces to the wing of an aircraft that can cause it to oscillate. All of these forces can cause vibration.

Periodic forces are those that repeat in time. An example is an applied force consisting of the sum of two harmonic forces at different frequencies. A nonperiodic force is one that does not repeat itself in time. A step function is an example of a force that is a nonperiodic excitation. A transient force is one that reduces to zero after a finite, usually small, time. An impulse or a shock are examples of transient excitations. All of the aforementioned classes of excitation are deterministic (i.e., they are known precisely as a function of time). On the other hand, a random excitation is one that is unpredictable in time and must be described in terms of probability and statistics. This chapter introduces a sample of these various classes of force excitations and how to calculate and analyze the resulting motion when applied to a single-degree-of-freedom spring–mass–damper system.

3.1 IMPULSE RESPONSE FUNCTION

A very common source of vibration is the sudden application of a short-duration force called an impulse. An impulse excitation is a force that is applied for a very short, or infinitesimal, length of time and represents one example of a *shock loading.*

An impulse is a nonperiodic force. The response of a system to an impulse is identical to the free response of the system to certain initial conditions, as shown in this section. In many useful situations the applied force $F(t)$ is impulsive in nature (i.e., acts with large magnitude for a very short period of time).

First, consider a mathematical model of an impulse excitation. A graphical time history of a model of the impulse is given in Figure 3.1. This is a rectangular pulse of very large magnitude and very small width (duration).

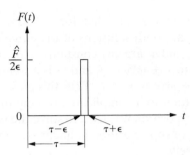

Figure 3.1 The time history of an impulse force used to model impulsive loading consisting of a large magnitude applied over a short time interval.

The rule of describing the force in Figure 3.1 is stated symbolically as

$$F(t) = \begin{cases} 0 & t \leq \tau - \varepsilon \\ \dfrac{\hat{F}}{2\varepsilon} & \tau - \varepsilon < t < \tau + \varepsilon \\ 0 & t \geq \tau + \varepsilon \end{cases} \tag{3.1}$$

where ε is a small positive number. This simple rule, $F(t)$, can be integrated to define the *impulse*. The impulse of the force $F(t)$ is defined by the integral, denoted by $I(\varepsilon)$, by

$$I(\varepsilon) = \int_{\tau-\varepsilon}^{\tau+\varepsilon} F(t)\, dt$$

which provides a measure of the strength of the forcing function, $F(t)$. Since the rule, $F(t)$, is zero outside the time interval from $\tau - \varepsilon$ to $\tau + \varepsilon$, the limits of integration on $I(\varepsilon)$ can be extended to yield

$$I(\varepsilon) = \int_{-\infty}^{\infty} F(t)\, dt \tag{3.2}$$

which has the units of $N \cdot s$.

In this case, the integral of equation (3.2) is evaluated by calculating the area under the curve using equation (3.1), which becomes

$$I(\varepsilon) = \int_{-\infty}^{\infty} F(t) dt = \frac{\hat{F}}{2\varepsilon} 2\varepsilon = \hat{F} \tag{3.3}$$

independent of the value of ε as long as $\varepsilon \neq 0$. In the limit as $\varepsilon \rightarrow 0$ (but $\varepsilon \neq 0$), the integral takes the value $I(\varepsilon) = \hat{F}$. This is used to define the *impulse function* as the function $F(t)$ with the two properties

$$F(t - \tau) = 0 \qquad t \neq \tau \tag{3.4}$$

and

$$\int_{-\infty}^{\infty} F(t - \tau) \, dt = \hat{F} \tag{3.5}$$

If the magnitude of $\hat{F}$ is unity, this becomes the definition of the *unit impulse function*, denoted by $\delta(t)$, also called the *Dirac delta function* (Boyce and DiPrima, 2009).

The solution for response of the single-degree-of-freedom system (see Window 3.1)

$$m\ddot{x}(t) + c\dot{x}(t) + kx(t) = \hat{F}(t), \quad x(0) = 0, \quad \dot{x}(0) = 0$$

to an impulsive load for the system initially at rest is calculated by recalling from physics that an impulse imparts a change in momentum to a body. For the sake

Window 3.1
Review of the Free Response of the Single-Degree-of-Freedom System of Chapter 1

$$m\ddot{x} + c\dot{x} + kx = F(t)$$

$$x(0) = x_0 \qquad \dot{x}(0) = v_o$$

$$\ddot{x} + 2\zeta\omega_n\dot{x} + \omega_n^2 x = f(t)$$

This system has free response [i.e., $f(t) = 0$] in the underdamped case (i.e., $0 < \zeta < 1$) given by

$$x(t) = \frac{\sqrt{(v_0 + \zeta\omega_n x_0)^2 + (x_0\omega_d)^2}}{\omega_n\sqrt{1-\zeta^2}} \ e^{-\zeta\omega_n t} \sin(\omega_d t + \phi)$$

where

$$\omega_d = \omega_n\sqrt{1-\zeta^2} \qquad \text{and} \qquad \phi = \tan^{-1}\frac{x_0\omega_d}{v_0 + \zeta\omega_n x_0}$$

Here $\omega_n = \sqrt{k/m}$, $\zeta = c/(2m\omega_n)$, and $0 < \zeta < 1$ must hold for the preceding solution to be valid [from equations (1.36) to (1.38)].

of simplicity, take $\tau = 0$ in the definition of an impulse. Consider the mass to be at rest just prior to the application of an impulse force. This instant of time is denoted by 0^-. Likewise the instant of time just after $t = 0$ is denoted by 0^+. The initial conditions are both zero, so that $x(0^-) = \dot{x}(0^-) = 0$, since the system is initially at rest. However, the velocity just after the impulse is $\dot{x}(0^+)$, denoted here as v_0. Thus the change in momentum at impact is $m\dot{x}(0^+) - m\dot{x}(0^-) = mv_0$, so that $\hat{F} = F\Delta t = mv_0 - 0 = mv_0$, while the initial displacement remains at zero. By this physical line of thought, an impulse applied to a single-degree-of-freedom spring–mass–damper system is the same as applying the initial conditions of zero displacement and an initial velocity of $v_0 = F\Delta t/m$.

Referring to Window 3.1, the response of an underdamped single-degree-of-freedom system ($0 < \zeta < 1$) with zero initial displacement ($x_0 = 0$) is just

$$x(t) = \frac{v_0}{\omega_d} e^{-\zeta\omega_n t} \sin \omega_d t$$

Substitution of $v_0 = \hat{F}/m$ ($\hat{F} = F\Delta t$, with units of N s) into this last expression yields

$$x(t) = \frac{\hat{F}e^{-\zeta\omega_n t}}{m\omega_d} \sin \omega_d t \tag{3.6}$$

as predicted by equations (1.36) and (1.38), repeated in Window 3.1. It is convenient to write this solution in the form

$$x(t) = \hat{F}h(t) \tag{3.7}$$

where $h(t)$ is defined by

$$h(t) = \frac{1}{m\omega_d} e^{-\zeta\omega_n t} \sin \omega_d t \tag{3.8}$$

Note that the function $h(t)$ is the response to a unit impulse applied at time $t = 0$. If applied at time $t = \tau$, $\tau \neq 0$, this can also be written as (replace t in the foregoing with $t - \tau$)

$$h(t - \tau) = \frac{1}{m\omega_d} e^{-\zeta\omega_n(t-\tau)} \sin \omega_d(t - \tau) \quad t > \tau \tag{3.9}$$

and zero for the interval $0 < t < \tau$. The functions $h(t)$ and $h(t - \tau)$ are each called the *impulse response function* of the system.

While the impulse is a mathematical abstraction of an infinite force applied over an infinitesimal time, in applications it presents an excellent model of a large force applied over a short period of time. The impulse response is physically interpreted as the response to an initial velocity with no initial displacement (hence no phase shift for $\tau = 0$). The impulse response function, combined with the principle of superposition, is also useful for calculating the response of a system to a general applied force excitation as discussed in Section 3.2.

A common occurrence that causes an impulse excitation is an impact. In vibration testing, a mechanical device under test is often given an impact, and the response is measured to determine the system's vibration properties. The impact is often created by hitting the test specimen with a hammer containing a device for measuring the force of the impact. Use of the impulse response for vibration testing is also discussed in Chapter 7. In practice, a force is considered to be an impulse if its duration (Δt) is very short compared with the period, $T = 2\pi/\omega_n$, associated with the structure's undamped natural frequency. In typical vibration tests, Δt is on the order of 10^{-3} s.

Example 3.1.1

Consider a spring–mass–damper system with $m = 100$ kg, $c = 20$ kg/s, and $k = 2000$ N/m with an impulse force applied to it of 1000 N for 0.01 s. Compute the resulting response.

Solution   A 1000 N force acting over 0.01 s provides (area under the curve) a value of $\hat{F} = F\Delta t = 1000 \cdot 0.01 = 10$ N · s. Using the values given, the equation of motion is

$$100\ddot{x}(t) + 20\dot{x}(t) + 2000x(t) = 10\delta(t)$$

Thus the natural frequency, damping ratio, and damped natural frequency are

$$\omega_n = \sqrt{\frac{2000}{100}} = 4.427 \text{ rad/s}, \zeta = \frac{20}{2\sqrt{100 \cdot 2000}} = 0.022,$$

$$\omega_d = 4.472\sqrt{1 - 0.022^2} = 4.471 \text{ rad/s}$$

Using equation (3.6), the response becomes

$$x(t) = \frac{\hat{F}e^{-\zeta\omega_n t}}{m\omega_d} \sin \omega_d t = 0.022e^{-0.1t} \sin(4.471t)$$

□

Example 3.1.2

Suppose a 1-kg bird flies into the 3-kg security camera of Example 2.1.3, repeated in Figure 3.2. If the bird is flying at 72 kmph, compute the maximum deflection the impact causes based on the design given in Example 2.1.3. Does the maximum deflection violate the design constraint? Ignore damping.

Solution   From the design solution of Example 2.1.3, the stiffness of the camera's mounting bracket is ($I = bh^3/12$):

$$k = \frac{3Ebh^3}{12l^3} = \frac{(7.1 \times 10^{10} \text{N/m}^2)(0.02\,\text{m})(0.02\,\text{m})^3}{4(0.55\,\text{m})^3} = 1.707 \times 10^4 \text{ N/m}$$

The mass of the camera is $m_c = 3$ kg, so the natural frequency is 75.43 rad/s. Combining equations (3.7) and (3.8) for $\zeta = 0$, the response is

$$x(t) = \frac{F\Delta t}{m_c\omega_n} \sin \omega_n t = \frac{m_b v}{m_c\omega_n} \sin \omega_n t$$

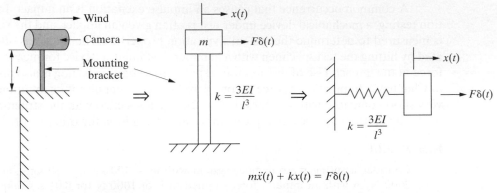

$$m\ddot{x}(t) + kx(t) = F\delta(t)$$

Figure 3.2   A vibration model of a security camera and mount.

where $m_b v$ is the linear momentum of the bird, which imparts the impact force F. The impulse is thus

$$m_b v = 1 \text{ kg} \cdot 72 \, \frac{\text{km}}{\text{hour}} \cdot \frac{1000 \text{ m}}{\text{km}} \cdot \frac{\text{hour}}{3600 \text{ s}} = 20 \text{ kg} \cdot \text{m/s}$$

This has maximum amplitude of

$$X = \left| \frac{F\Delta t}{m_c \omega_n} \right| = \left| \frac{m_b v}{m_c \omega_n} \right| = \left| \frac{20 \text{ kg} \cdot \text{m/s}}{3 \text{ kg} \cdot 75.43 \text{ rad/s}} \right| = 0.088 \text{ m}$$

Thus, the design constraint of holding the vibration of the camera within 0.01 m required in Example 2.1.3 is violated under a bird strike.

$\square$

Example 3.1.3

In vibration testing, an instrumented hammer is often used to hit a device to excite it and to measure the impact force simultaneously. If the device being tested is a single-degree-of-freedom system, plot the response given that $m = 1 \text{ kg}, c = 0.5 \text{ kg/s}, k = 4 \text{ N/m}$, and $\hat{F} = 0.2 \text{ N} \cdot \text{s}$. It is often difficult to provide a single impact with a hammer. Sometimes a "double hit" occurs, so the exciting force may have the form

$$F(t) = 0.2\delta(t) + 0.1\delta(t - \tau)$$

Plot the response of the same system with a double hit and compare it with the response to a single impact. Assume that the initial conditions are zero.

Solution   The solution to the single unit impact and time $t = 0$ is given by equations (3.7) and (3.8) with $\omega_n = \sqrt{4} = 2 \text{ rad/s}$ and $\zeta = c/(2m\omega_n) = 0.125$. Thus, with $\hat{F} = 0.2\delta(t)$

$$x_1(t) = \frac{0.2}{(1)\left(2\sqrt{1 - (0.125)^2}\right)} e^{-(0.125)(2)t} \sin 2\sqrt{1 - (0.125)^2} t$$

$$= 0.1008 e^{-0.25t} \sin(1.984t) m$$

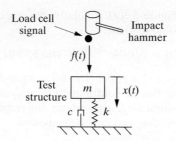

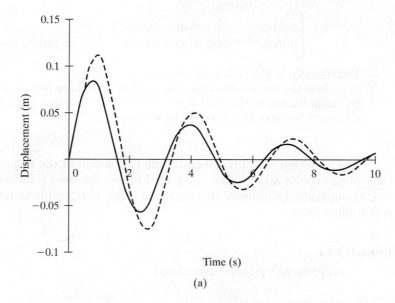

(a)

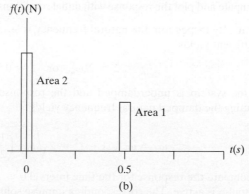

(b)

Figure 3.3   (a) The response of a single-degree-of-freedom system to a single impact (solid line) and a double impact for $\tau = 0.5$ s (dashed line). Plot (b) indicates the force-versus time curve for the applied double impact. Note the larger amplitude of the double hit.

which is plotted in Figure 3.3(a). Similarly, the response to $0.1\delta(t - \tau)$ is calculated from equations (3.7) and (3.9) as

$$x_2(t) = 0.0504e^{-0.25(t-\tau)} \quad \sin 1.984(t - \tau)m \quad t > \tau$$

and $x_2(t) = 0$ for $0 < t < \tau = 0.5$. The force input $f(t)$ is indicated in Figure 3.3(b). It is important to note that no contribution from x_2 occurs until time $t = \tau$. Using the principle of superposition for linear systems, the response to the "double impact" will be the sum of the preceding two impulse responses:

$$x(t) = x_1(t) + x_2(t)$$

$$= \begin{cases} 0.1008e^{-0.25t} \sin(1.984t) & 0 < t < \tau \\ 0.1008e^{-0.25t} \sin(1.984t) + 0.0504e^{-0.25(t-\tau)} \sin 1.984(t - \tau) & t > \tau \end{cases}$$

This is plotted in Figure 3.3(a) for the value $\tau = 0.5$ s.

Note that the obvious difference between the two responses is that the "double-hit" response has a "spike" at $\tau = t = 0.5$, causing a larger amplitude. The time, τ, represents the time delay between the two hits.

☐

The following example illustrates the calculation of the response due to both an applied impulse and initial conditions, forming the total response of the system. The example also introduces the concept of using a Heaviside step function to represent the response.

Example 3.1.4

Consider the system (mass normalized)

$$\ddot{x}(t) + 2\dot{x}(t) + 4x(t) = \delta(t) - \delta(t - 4)$$

and compute and plot the response with initial conditions $x_0 = 1$ mm and $v_0 = -1$ mm/s.

Solution   By inspection, the natural frequency is $\omega_n = 2\text{rad/s}$. Examining the velocity coefficient yields

$$2 = 2\zeta\omega_n \quad \text{or} \quad \zeta = 0.5$$

Thus, the system is underdamped and the response given in Window 3.1 applies. Computing the damped natural frequency yields

$$\omega_d = \omega_n\sqrt{1 - \zeta^2} = 2\sqrt{1 - \left(\frac{1}{2}\right)^2} = \sqrt{3}$$

First, compute the response for the time interval $0 \leq t \leq 4$ s. In this interval, only the first impulse is active. The corresponding impulse solution is, by equation (3.6),

$$x_1(t) = \frac{\hat{F}}{m\omega_d}e^{-\zeta\omega_n t} \sin \omega_d t = \frac{1}{\sqrt{3}}e^{-t} \sin \sqrt{3}t, \quad 0 \leq t < 4$$

The total solution for the first time interval is then equal to the sum of the homogeneous and impulse solutions. The homogeneous solution is

$$x_h(t) = e^{-t}(A \sin \omega_d t + B \cos \omega_d t), \qquad 0 \le t < 4$$

where A and B are the constants of integration to be determined by the initial conditions and the subscript h denotes the solution due to the initial conditions. Differentiating the displacement yields the velocity:

$$\dot{x}_h(t) = -e^{-t}(A \sin \sqrt{3}t + B \cos \sqrt{3}t)$$
$$+ e^{-t}(\sqrt{3}A \cos \sqrt{3}t - \sqrt{3}B \sin \sqrt{3}t)$$

Setting $t = 0$ in these last two expressions and using the initial conditions yields the following two equations:

$$x_h(0) = 1 = B$$
$$\dot{x}_h(0) = -1 = -B + \sqrt{3}A$$

Solving for A and B yields $A = 0$ and $B = 1$, so that $x_h(t) = e^{-t} \cos \sqrt{3}t$. Next, compute the response due to the impulse at $t = 0$, which is equivalent to solving the initial value problem for $x_I(0) = 0$ and $\dot{x}_I(0) = 1$. Following the same procedure to compute the constants of integration for the impulse yields

$$B = 1 \text{ and } A = \frac{1}{\sqrt{3}} \text{ so that } x_I(t) = \frac{e^{-t}}{\sqrt{3}} \sin \sqrt{3}t$$

Adding the homogenous response and the impulse response yields

$$x_1(t) = e^{-t}\left(\cos \sqrt{3}t + \frac{1}{\sqrt{3}} \sin \sqrt{3}t \right), \qquad 0 \le t < 4$$

Next, compute the response of the system to the second impulse, which starts at $t = 4$s. Using equation (3.9) with $\tau = 4$s, the response to the second impulse is

$$x_2(t) = \frac{\hat{F}}{m\omega_d} e^{-\zeta \omega_n(t-\tau)} \sin \omega_d(t - \tau) = -\frac{1}{\sqrt{3}} e^{-t+4} \sin \sqrt{3}(t - 4), \qquad t > 4$$

The *Heaviside step function* defined by

$$\Phi(t - \tau) = \begin{cases} 0, & t < \tau \\ 1, & t \ge \tau \end{cases}$$

is perfect for writing functions that "turn on" after some time has evolved. Heaviside functions are also denoted by $H(t - \tau)$. Using superposition, the total solution is

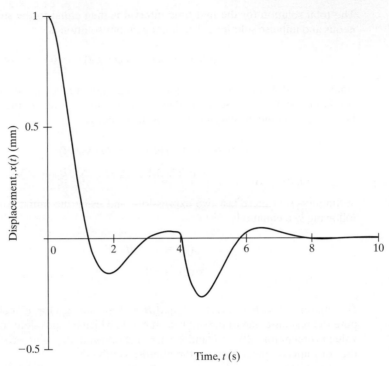

Figure 3.4　A plot of displacement versus time for a double impact, with the second impact applied at $t = 4$ s.

$x = x_1 + x_2$, and the Heaviside function is used to indicate that x_2 "starts" after $\tau = 4$. The solution can be written as

$$x(t) = e^{-t}\left(\cos\sqrt{3}t + \frac{1}{\sqrt{3}}\sin\sqrt{3}t\right) - \left[\frac{e^{-(t-4)}}{\sqrt{3}}\sin\sqrt{3}(t-4)\right]\Phi(t-4)\ \text{mm}$$

This is plotted in Figure 3.4. Note the sharp change in the response as the second impact is applied. This is in contrast to the double hit in the previous example. In the previous example, the second impulse occurs in the same "direction" as the current response. However, in this example, the second impact occurs out of phase with the response of the first impact and causes an abrupt change in direction.

□

3.2 RESPONSE TO AN ARBITRARY INPUT

The response of a single-degree-of-freedom system to an arbitrary, general excitation is examined in this section. The response of a single-degree-of-freedom system to an arbitrary force of varying magnitude can be calculated from the concept of

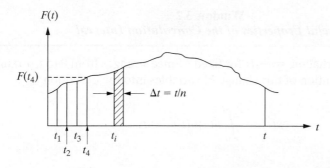

Figure 3.5   An arbitrary excitation force $F(t)$ split up into n impulse forces.

the impulse response defined in Section 3.1. The procedure is to divide the exciting force up into impulses of infinitesimal area, calculate the responses to these individual impulses, and add the individual responses to calculate the total response using the concept of superposition. This is best shown in Figure 3.5, which illustrates an arbitrary applied force $F(t)$ divided into n time intervals of length Δt so that each time increment is defined by $\Delta t = t/n$. At each time interval t_i, the solution can be calculated by considering the response to be due to an impulse Δt in duration and of force magnitude $F(t)$ [i.e., an impulse of magnitude $F(t_i)\Delta t$].

The part of the response due to the impulse acting during the time interval between t_i and t_{i+1} is then given by equation (3.7) as the increment

$$\Delta x(t_i) = F(t_i)h(t - t_i)\Delta t \tag{3.10}$$

so that the total response after n intervals is the sum

$$x(t_n) = \sum_{i=1}^{n} F(t_i)h(t - t_i)\Delta t \tag{3.11}$$

This again uses the fact that the equation of motion is linear, so that the principle of superposition applies. Forming the sequence of partial sums and finding the limit as $\Delta t \to 0$ $(n \to \infty)$ yields

$$x(t) = \int_0^t F(\tau)h(t - \tau)d\tau \tag{3.12}$$

from the first fundamental theorem of integral calculus. The integral in equation (3.12) is called the *convolution integral*. A convolution integral is simply the integral of the product of two functions, one of which is shifted by the variable of integration. Convolution is used again in Section 3.4 as a useful technique in using transforms. Additional properties of the convolution integral are given in Window 3.2.

For an underdamped single-degree-of-freedom system, the impulse response function $h(t - \tau)$ is given by equation (3.9). Substitution of the impulse response function of equation (3.9) into equation (3.12) then yields the result that the response of an underdamped system to an arbitrary input $F(t)$ of the form

$$m\ddot{x}(t) + c\dot{x}(t) + kx(t) = F(t), x_0 = 0, v_0 = 0$$

Window 3.2
Useful Properties of the Convolution Integral

Let $\alpha = t - \tau$, so that $d\alpha = -d\tau$ for fixed t. Since τ ranges from 0 to t, α ranges from t to 0. Substitution of this change of variables into the definition

$$x(t) = \int_0^t F(\tau)h(t - \tau)d\tau$$

yields

$$x(t) = -\int_t^0 F(t - \alpha)h(\alpha)d\alpha = \int_0^t F(t - \alpha)h(\alpha)d\alpha$$

Thus,

$$\int_0^t F(\tau)h(t - \tau)d\tau = \int_0^t F(t - \tau)h(\tau)d\tau$$

is given by

$$x(t) = \frac{1}{m\omega_d} e^{-\zeta\omega_n t} \int_0^t [F(\tau)e^{\zeta\omega_n \tau} \sin \omega_d(t - \tau)]d\tau$$

$$= \frac{1}{m\omega_d} \int_0^t F(t - \tau)e^{-\zeta\omega_n \tau} \sin \omega_d \tau\, d\tau \tag{3.13}$$

as long as the initial conditions are zero. The integral in equation (3.13) is a convolution integral where one of the functions is the impulse response function—hence the shift in the integral. A convolution integral used to compute a system response is called a *Duhamel integral*, after the French mathematician J. M. C. Duhamel (1797–1872). The Duhamel integral can be used to calculate the response to an arbitrary input as long as it satisfies certain mathematical conditions. The following example illustrates the procedure.

Example 3.2.1

Consider an excitation force of the form given in Figure 3.6. The force is zero until time t_0, when it jumps to a constant level, F_0. This is called the *step function*, and when used to excite a single-degree-of-freedom system, it might model some machine operation or an automobile running over a surface that changes level (such as a curb). The step function

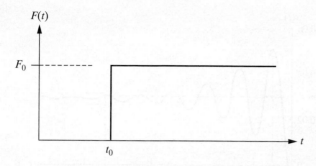

Figure 3.6   Step function of magnitude F_0 applied at time $t = t_0$.

of unit magnitude is called the Heaviside step function as defined in Example 3.1.4. Calculate the solution of

$$m\ddot{x} + c\dot{x} + kx = F(t) = \begin{cases} 0, & t_0 > t > 0 \\ F_0, & t \geq t_0 \end{cases} \tag{3.14}$$

with $x_0 = v_0 = 0$ and $F(t)$ as described in Figure 3.6. Here it is assumed that the values of m, c, and k are such that the system is underdamped ($0 < \zeta < 1$).

Solution   Applying the convolution integral given by equation (3.13) directly yields

$$x(t) = \frac{1}{m\omega_d} e^{-\zeta\omega_n t} \left[\int_0^{t_0} (0)e^{\zeta\omega_n \tau} \sin \omega_d(t - \tau)d\tau + \int_{t_0}^{t} F_0 e^{\zeta\omega_n \tau} \sin \omega_d(t - \tau)d\tau \right]$$

$$= \frac{F_0}{m\omega_d} e^{-\zeta\omega_n t} \int_{t_0}^{t} e^{\zeta\omega_n \tau} \sin \omega_d(t - \tau)d\tau$$

Using a table of integrals to evaluate this expression yields

$$x(t) = \frac{F_0}{k} - \frac{F_0}{k\sqrt{1-\zeta^2}} e^{-\zeta\omega_n(t-t_0)}\cos[\omega_d(t - t_0) - \theta] \quad t \geq t_0 \tag{3.15}$$

where

$$\theta = \tan^{-1} \frac{\zeta}{\sqrt{1 - \zeta^2}} \tag{3.16}$$

Note that if $t_0 = 0$, equation (3.15) becomes just

$$x(t) = \frac{F_0}{k} - \frac{F_0}{k\sqrt{1 - \zeta^2}} e^{-\zeta\omega_n t}\cos(\omega_d t - \theta) \tag{3.17}$$

and if there is no damping ($\zeta = 0$), this expression simplifies further to

$$x(t) = \frac{F_0}{k}(1 - \cos \omega_n t) \tag{3.18}$$

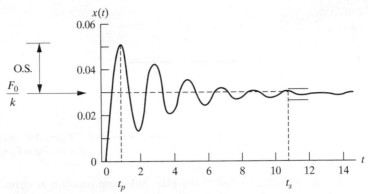

Figure 3.7 The response of an underdamped system to the step excitation of Figure 3.6 for $\zeta = 0.1$ and $\omega_n = 3.16$ rad/s (with $F_0 = 30$ N, $k = 1000$ N/m, $t_0 = 0$).

Examining the damped response given in equation (3.17), it is obvious that for large time the second term of the response dies out and the steady state is just

$$x_{ss}(t) = \frac{F_0}{k} \tag{3.19}$$

In fact, the underdamped step response given by equations (3.15) and (3.17) consists of the constant function F_0/k minus a decaying oscillation, as illustrated in Figure 3.7.

Often in the design of vibrating systems subject to a step input, the time it takes for the response to reach the largest value, called the *time to peak* and denoted t_p in Figure 3.7, is used as a measure of the quality of the response. Other quantities used to measure the character of the step response are the *overshoot*, denoted by O.S. in Figure 3.7, which is the largest value of the response "over" the steady-state value, and the *settling time*, denoted by t_s in Figure 3.7. The settling time is the time it takes for the response to get and stay within a certain percentage of the steady-state response. For the case of $t_0 = 0$, t_p and t_s are given by $t_p = \pi/\omega_d$ and $t_s = 3.5/\zeta\omega_n$. The peak time is exact (see Problem 3.25), and the settling time is an approximation of when the response stays within 3% of the steady-state value.

□

Example 3.2.2

Another common excitation in vibration is a constant force that is applied for a short period of time and then removed. A rough model of such a force is given in Figure 3.8. Calculate the response of an underdamped system to this excitation.

Solution This pulse-like loading can be written as a combination of step functions calculated in Example 3.2.1, as illustrated in Figure 3.8. The response of a single-degree-of-freedom system to $F(t) = F_1(t) + F_2(t)$ is just the sum of the response to $F_1(t)$ and the response of $F_2(t)$, because the system is linear. First, consider the response of an underdamped system to $F_1(t)$. This response is just that calculated in Example 3.2.1 for $t_0 = 0$ and given by equation (3.17). Next, consider the response of

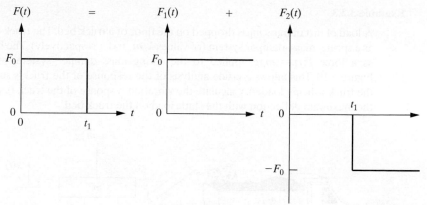

Figure 3.8   Square-pulse excitation of magnitude F_0 lasting for t_1 seconds can be written as the sum of a step function starting at zero of magnitude F_0 and a step function starting at t_1 of magnitude $-F_0$ [i.e., $F(t) = F_1(t) + F_2(t)$].

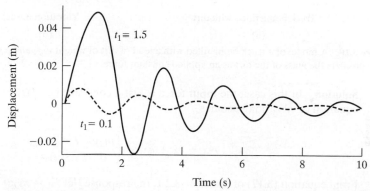

Figure 3.9   Response of an underdamped system to a pulse input of width t_1. The dashed line is for $t_1 = 0.1 < \pi/\omega_n$ and the solid line is for $t_1 = 1.5 > \pi/\omega_n$. Both plots are for the case $F_0 = 30$ N, $k = 1000$ N/m, $\zeta = 0.1$, and $\omega_n = 3.16$ rad/s.

the system to $F_2(t)$. This is just the response given by equation (3.15) with F_0 replaced by $-F_0$ and t_0 replaced by t_1. Hence, subtracting equation (3.15) from equation (3.17) yields the result that the response to the pulse of Figure 3.8 is

$$x(t) = \frac{F_0 e^{-\zeta \omega_n t}}{k\sqrt{1 - \zeta^2}} \{e^{\zeta \omega_n t_1} \cos[\omega_d(t - t_1) - \theta] - \cos(\omega_d t - \theta)\}, \qquad t > t_1$$

where θ is as defined in equation (3.16). A plot of this response is given in Figure 3.9 for different pulse widths t_1. Note that the response is much different for $t_1 > \pi/\omega_n$ and has a maximum magnitude of about five times the maximum magnitude of the time response for $t_1 < \pi/\omega_n$. Also, note that the steady-state response (i.e., the response for large time) is zero in this case.

☐

Example 3.2.3

A load of dirt of mass m_d is dropped on the floor of a truck bed. The truck bed is modeled as a spring–mass–damper system (of values k, m, and c, respectively). The load is modeled as a force $F(t) = m_d g$ applied to the spring–mass–damper system, as illustrated in Figure 3.10. This allows a crude analysis of the response of the truck's suspension when the truck is being loaded. Calculate the vibration response of the truck bed, and compare the maximum deflection with the static load on the truck bed.

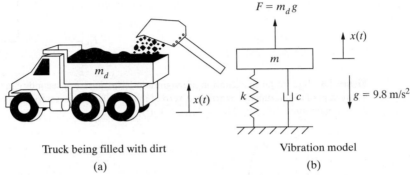

Truck being filled with dirt   Vibration model

(a)   (b)

Figure 3.10   A model of a truck being filled with a load of dirt of weight $m_d g$ and a vibration model that considers the mass of the dirt as an applied constant force.

Solution   In this case, the input force is just a constant [i.e., $F(t) = m_d g$], so that the equation of vibration becomes

$$m\ddot{x} + c\dot{x} + kx = \begin{cases} m_d g & t > 0 \\ 0 & t \leq 0 \end{cases}$$

From equation (3.17) of Example 3.2.1, the response (let $F_0 = m_d g$) is just

$$x(t) = \frac{m_d g}{k}\left[1 - \frac{1}{\sqrt{1 - \zeta^2}}e^{-\zeta\omega_n t}\cos(\omega_d t - \theta)\right]$$

To obtain a rough idea about the nature of this expression, its undamped value is

$$x(t) = \frac{m_d g}{k}(1 - \cos\omega_n t)$$

which has a maximum amplitude (when t is such that $\cos\omega_n t = -1$) of

$$x_{max} = 2\frac{m_d g}{k}$$

This is twice the static displacement (i.e., twice the distance the truck would be deflected if the dirt were placed gently and slowly onto it). Thus, if the truck were designed with springs based only on the static load, with no margins of safety, the springs in the truck would potentially break, or permanently deform, when subjected to the same mass applied dynamically (i.e., dropped) to the truck. Hence, it is important to consider the vibration (dynamic) response in designing structures that could be loaded dynamically.

☐

It should be noted that the response of a single-degree-of-freedom system to an arbitrary input can be calculated numerically, even if the integral in equation (3.12) cannot be evaluated in closed form as done in the preceding examples. Such general numerical procedures, based roughly on equation (3.10), are discussed in Section 3.8, in which the numerical solutions discussed in Section 2.8 are applied. Numerical integration is often used to solve vibration problems with arbitrary forcing functions.

The response calculations for a general external disturbance (input) force do not include the response that might exist because of nonzero initial conditions. The total response for an impulse disturbance with nonzero initial conditions is given in Example 3.1.4. The total response to an arbitrary input force as well as nonzero initial conditions for an undamped system is given using the convolution integral by

$$x(t) = x_0 \cos \omega_n t + \frac{v_0}{\omega_n} \sin \omega_n t + \int_0^t h(t - \tau) F(\tau) \, d\tau$$

Note that the effect of the applied force on the homogeneous response is zero because the value of the convolution form of the particular solution and its derivative (velocity) are both zero at time $t = 0$. On the other hand, if the particular solution $x_p(t)$ or its derivative do not vanish at $t = 0$, the form of the total response is

$$x(t) = A \cos \omega_n t + B \sin \omega_n t + x_p(t)$$

Here the constants A and B must be determined by the initial conditions and will be affected by the value of x_p and its derivative at $t = 0$. In this last case, the constants of integration A and B become

$$A = x_0 - x_p(0) \quad \text{and} \quad B = \frac{v_0 - \dot{x}_p(0)}{\omega_n}$$

A similar result holds for damped systems.

The analytical calculations made with the convolution integral are not always easy to evaluate, and in many cases must be evaluated numerically. Laplace transform methods are often useful in convolution-type evaluations but, in practice, solutions are often found through numerical integration and simulation. While numerical simulation is used in practice, the *concept* of convolution is essential to understanding signal processing and for understanding the results of numerical simulations.

Example 3.2.4

Solve $\ddot{x}(t) + 16x(t) = \cos 2t$ for the response to arbitrary initial conditions x_0 and v_0 using the convolution integral. Next, compare this to the result obtained by solving this problem using the method of undetermined coefficients explained in Section 2.1 and given in equation (2.11).

Solution   From the equation of motion, $m = 1$, $\omega_n = 4$, $\omega = 2$, and $F_0 = f_0 = 1$, where the units are assumed to be consistent. Using the convolution expression, equation (3.12), the particular solution has the form

$$x_p(t) = \int_0^t h(t - \tau)F(\tau)d\tau$$

The impulse response function, $h(t - \tau)$, for an undamped system is found from equation (3.8) with $\zeta = 0$. With the values given above for mass and frequency, the impulse response function is

$$h(t - \tau) = \frac{1}{4} \sin(4t - 4\tau)$$

Thus the convolution expression for the particular solution is

$$x_p(t) = \frac{1}{4} \int_0^t \sin(4t - 4\tau) \cos(2\tau)d\tau$$

Integrating (using a symbolic code or repeated use of trig identities) yields

$$x_p(t) = \frac{1}{4}\left(\frac{\cos(4t - 2\tau)}{4} + \frac{\cos(4t - 6\tau)}{12} \right)_0^t = \frac{1}{12}(\cos 2t - \cos 4t)$$

The total solution is of the form

$$x(t) = A \sin 4t + B \cos 4t + \frac{1}{12}(\cos 2t - \cos 4t)$$

Using the initial conditions to evaluate the constants of integration A and B yields

$$x(0) = x_0 = A \sin(0) + B \cos(0) + \frac{1}{12}(\cos(0) - \cos(0)),$$

$$\dot{x}(0) = v_0 = 4A \cos(0) + 4x_0 \sin(0) - \frac{2}{12} \sin(0) + \frac{4}{12} \sin(0)$$

Solving this set of equations for A and B yields

$$A = \frac{v_0}{4} \quad \text{and} \quad B = x_0$$

The total solution is then

$$x(t) = \frac{v_0}{4} \sin 4t + (x_0 - \frac{1}{12}) \cos 4t + \frac{1}{12} \cos 2t$$

This is in total agreement with the solution given in equation (2.11) with $m = 1$, $\omega_n = 4$, $\omega = 2$, $F_0 = f_0 = 1$, and $\omega_n^2 - \omega^2 = 12$.

The purpose of this is example is to show that two different methods yield the same answer, as they should. The method of undetermined coefficients is a much simpler calculation to make, but only works for harmonic forcing functions. The convolution approach is more complicated, but can be used for *any* forcing function and is thus a more general approach.

$\square$

3.3 RESPONSE TO AN ARBITRARY PERIODIC INPUT

The specific case of periodic inputs is considered in this section. The response to periodic inputs can be calculated by the methods of Section 3.2. However, periodic disturbances that occur quite often merit special consideration. In Chapter 2, the response to a harmonic input is considered. The term *harmonic input* refers to a sinusoidal driving function at a single frequency. Here, the response to any periodic input is considered. A periodic function is any function that repeats itself in time [i.e., any function for which there exists a fixed time, T, called the period, such that $f(t) = f(t + T)$ for all values of t]. A simple example is a forcing function that is the sum of two sinusoids of different frequency with a rational frequency ratio. An example of a general periodic forcing function, $F(t)$, of period T is given in Figure 3.11. Note from the figure that the periodic force does not look periodic at all if examined in an interval less than the period T. However, the forcing function does repeat itself every T seconds.

According to the theory developed by Fourier, any periodic function $F(t)$, with period T, may be represented by an infinite series of the form

$$F(t) = \frac{a_0}{2} + \sum_{n=1}^{\infty} (a_n \cos n\omega_T t + b_n \sin n\omega_T t) \tag{3.20}$$

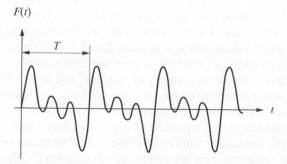

$F(t)$

T

t

Figure 3.11 An example of a general periodic function of period T.

where $\omega_T = 2\pi/T$ and where the coefficients a_0, a_n, and b_n for a given periodic function $F(t)$ are calculated by the formulas

$$a_0 = \frac{2}{T} \int_0^T F(t)dt \tag{3.21}$$

$$a_n = \frac{2}{T} \int_0^T F(t)\cos n\omega_T t \, dt \quad n = 1, 2, \ldots \tag{3.22}$$

$$b_n = \frac{2}{T} \int_0^T F(t)\sin n\omega_T t \, dt \quad n = 1, 2, \ldots \tag{3.23}$$

Note that the first coefficient a_0 is twice the average of the function $F(t)$ over one cycle. The coefficients a_0, a_n, and b_n, are called *Fourier coefficients*. The series of equation (3.20) is the *Fourier series*. A more complete discussion of Fourier series can be found in most introductory differential equation texts (e.g., Boyce and DiPrima, 2009).

The Fourier series is useful and relatively straightforward to work with because of a special property of the trigonometric functions used in the series. This special property, called *orthogonality*, can be stated as follows:

$$\int_0^T \sin n\omega_T t \sin m\omega_T t \, dt = \begin{cases} 0 & m \neq n \\ T/2 & m = n \end{cases} \tag{3.24}$$

$$\int_0^T \cos n\omega_T t \cos m\omega_T t \, dt = \begin{cases} 0 & m \neq n \\ T/2 & m = n \end{cases} \tag{3.25}$$

and

$$\int_0^T \cos n\omega_T t \sin m\omega_T t \, dt = 0 \tag{3.26}$$

The m and n here are integers. The truth of these three orthogonality conditions follows from direct integration. The orthogonality property (i.e., the integral of the product of two functions is zero) is used repeatedly in vibration analysis. In particular, orthogonality is used extensively in Chapters 4, 6, 7, and 8. Orthogonality is also used in statics and dynamics (i.e., the unit vectors are orthogonal).

In Fourier analysis, the orthogonality of the sine and cosine functions on the interval $0 < t < T$ is used to derive the formulas given in equations (3.21), (3.22), and (3.23). These coefficient values are derived as follows. The Fourier coefficients a_n are determined by multiplying equation (3.20) by $\cos m\omega_T t$ and integrating over the period T. Similarly, the coefficients b_n are determined by multiplying by $\sin m\omega_T t$ and integrating. The summation on the right side of equation (3.20) vanishes except for one term because of the orthogonality properties of the trigonometric function. Using the orthogonality conditions given previously determines that all terms of the integrated product $\int_0^T F(t) \sin m\omega_T t \, dt$ will be zero except for the term containing b_m. This yields equation (3.23). Likewise, all the terms in the series $\int_0^T F(t) \cos m\omega_T t \, dt$ are zero, except for the term containing a_m. This yields equation (3.22). Furthermore, this procedure can be repeated for each of the values of n in the summation in the Fourier series. The procedure for calculating the Fourier coefficient of a simple force is illustrated in the following example.

Example 3.3.1

A triangular wave of period T is illustrated in Figure 3.12 and is described by

$$F(t) = \begin{cases} \dfrac{4}{T}t - 1 & 0 \le t \le \dfrac{T}{2} \\[3mm] 1 - \dfrac{4}{T}\left(t - \dfrac{T}{2}\right) & \dfrac{T}{2} \le t \le T \end{cases}$$

Determine the Fourier coefficients for this function.

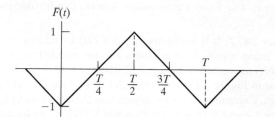

Figure 3.12 Plot of a triangular wave of period T.

Solution   Straightforward integration of equation (3.21) yields

$$a_0 = \frac{2}{T}\int_0^{T/2}\left(\frac{4}{T}t - 1\right)dt + \frac{2}{T}\int_{T/2}^{T}\left[1 - \frac{4}{T}\left(t - \frac{T}{2}\right)\right]dt = 0$$

which is also the average value of the triangular wave over one period. Similarly, integration of equation (3.23) yields the result that $b_n = 0$ for every n. Equation (3.22) yields

$$a_n = \frac{2}{T}\int_0^{T/2}\left(\frac{4}{T}t - 1\right)\cos n\omega_T t \, dt + \frac{2}{T}\int_{T/2}^{T}\left[1 - \frac{4}{T}\left(t - \frac{T}{2}\right)\right]\cos n\omega_T t \, dt$$

$$= \begin{cases} 0 & n \text{ even} \\[3mm] \dfrac{-8}{\pi^2 n^2} & n \text{ odd} \end{cases}$$

Thus the Fourier representation of this function becomes

$$F(t) = -\frac{8}{\pi^2}\left[\cos\frac{2\pi}{T}t + \frac{1}{9}\cos\frac{6\pi}{T}t + \frac{1}{25}\cos\frac{10\pi}{T}t \cdots\right]$$

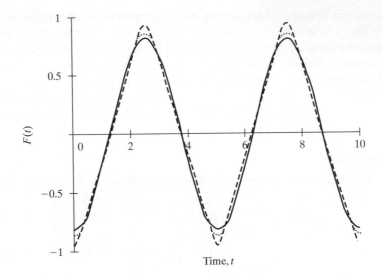

Figure 3.13 Plots of $F(t)$ for one (long dashed line), two (short dashed line), and four (solid line) terms of the Fourier series indicate how close each series gets to the plot of Figure 3.12.

which has frequency $2\pi/T$. It is instructive to plot $F(t)$ by adding one term at a time to make clear how many terms of the infinite series are needed to obtain a reasonable representation of $F(t)$ as plotted in Figure 3.12. (Run VTB3_3 to observe this convergence.) This is done in Figure 3.13, which is a plot of $F(t)$ for one, two, and four terms of the Fourier series. Computer codes for computing the series and plotting the results are given in Section 3.8 and VTB3_3. Toolbox file VTB3_3 can be used to obtain the coefficients of an arbitrary signal and for plotting the results. Substitution of the values a_n and b_n into VTB3_5 will visually verify the result.

□

Note that when a Fourier series is used to approximate a periodic function with discontinuities, an overshoot (or ringing) of the Fourier series occurs at the discontinuity. This overshoot is called Gibbs phenomenon.

Since a general periodic force can be represented as a sum of sines and cosines, and since the system under consideration is linear, the response of a single-degree-of-freedom system is calculated by computing the response to the individual terms of the Fourier series and adding the results. This is similar to the procedure used to solve the base-excitation problem of equation (2.63), where the input to a single-degree-of-freedom system consisted of the sum of a single sine term and a single cosine term. This is how superposition and Fourier series are used together to compute the solution for any periodic input. Thus, the particular solution $x(t)$ of

$$m\ddot{x}(t) + c\dot{x}(t) + kx(t) = F(t) \tag{3.27}$$

where $F(t)$ is periodic, can be written as

$$x_p(t) = x_1(t) + \sum_{n=1}^{\infty}[x_{cn}(t) + x_{sn}(t)] \tag{3.28}$$

Here the particular solution $x_1(t)$ satisfies the equation

$$m\ddot{x}_1(t) + c\dot{x}_1(t) + kx_1(t) = \frac{a_0}{2} \tag{3.29}$$

the particular solution $x_{cn}(t)$ satisfies the equation

$$m\ddot{x}_{cn}(t) + c\dot{x}_{cn}(t) + kx_{cn}(t) = a_n \cos n\omega_T t \tag{3.30}$$

for all values of n, and the particular solution $x_{sn}(t)$ satisfies the equation

$$m\ddot{x}_{sn}(t) + c\dot{x}_{sn}(t) + kx_{sn}(t) = b_n \sin n\omega_T t \tag{3.31}$$

for all values of n. The solutions to equations (3.30) and (3.31) are calculated in Section 2.2, and the solution to equation (3.29) is calculated in Section 3.2. If the system is subject to nonzero initial conditions, this must also be taken into consideration.

The particular solution to equation (3.29) is that of the step response calculated in equation (3.17) with $F_0 = a_0/2$. This yields

$$x_1(t) = \frac{a_0}{2k} \tag{3.32}$$

The particular solution of equation (3.30) is calculated in equation (2.36) to be

$$x_{cn}(t) = \frac{a_n/m}{[[\omega_n^2 - (n\omega_T)^2]^2 + (2\zeta\omega_n n\omega_T)^2]^{1/2}} \cos(n\omega_T t - \theta_n) \tag{3.33}$$

where

$$\theta_n = \tan^{-1}\frac{2\zeta\omega_n n\omega_T}{\omega_n^2 - (n\omega_T)^2}$$

Similarly, the particular solution of equation (3.31) is calculated to be

$$x_{sn}(t) = \frac{b_n/m}{[[\omega_n^2 - (n\omega_T)^2]^2 + (2\zeta\omega_n n\omega_T)^2]^{1/2}} \sin(n\omega_T t - \theta_n) \tag{3.34}$$

The total particular solution of equation (3.27) is then given by the sum of equations (3.32), (3.33), and (3.34) as indicated by equation (3.28). The total solution $x(t)$ is the sum of the particular solution $x_p(t)$ calculated previously and the homogeneous solution obtained in Section 1.3. For the underdamped case ($0 < \zeta < 1$), this becomes

$$x(t) = Ae^{-\zeta\omega_n t} \sin(\omega_d t + \phi) + \frac{a_0}{2k} + \sum_{n=1}^{\infty}[x_{cn}(t) + x_{sn}(t)] \tag{3.35}$$

where A and ϕ are determined by the initial conditions. As in the case of a simple harmonic input as described in equation (2.37), the constants A and ϕ describing the transient response will be different than those calculated for the free-response case given in equation (1.38). This is because part of the transient term of equation (3.35) is the result of initial conditions and part is due to the excitation force $F(t)$.

Example 3.3.2

Consider the base-excitation problem of Section 2.4 (see Window 3.3), and calculate the total response of the system to initial conditions $x_0 = 0.01$ m and $v_0 = 3.0$ m/s. Assume that $\omega_b = 3$ rad/s, $m = 1$ kg, $c = 10$ kg/s, $k = 1000$ N/m, and $Y = 0.05$ m.

Window 3.3
Review of the Base-Excitation Problem of Section 2.4

The base-excitation problem is to solve the expression

$$\ddot{x} + 2\zeta\omega_n\dot{x} + \omega_n^2 x = 2\zeta\omega_n\omega_b Y \cos \omega_b t + \omega_n^2 Y \sin \omega_b t$$

for the motion of a mass, $x(t)$, excited by a harmonic displacement of frequency ω_b and amplitude Y through its spring–damper connections. This has the particular solution indicated by the second term on the right-hand side of equation (3.37).

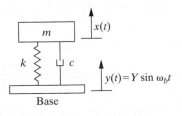

Base

Solution   The equation of motion is given by equation (2.63), which has a periodic input of

$$F(t) = cY\omega_b \cos \omega_b t + kY \sin \omega_b t \tag{3.36}$$

Comparing coefficients with the Fourier expansion of equation (3.20) yields $a_0 = 0$, $a_n = b_n = 0$, for all $n > 1$, and

$$a_1 = cY\omega_b = (10 \text{ kg/s})(0.05 \text{ m})(3 \text{ rad/s}) = 1.5 \text{ N}$$
$$b_1 = kY = (1000 \text{ N/m})(0.05 \text{ m}) = 50 \text{ N}$$

The solution for $x_{c1}(t)$ from equation (3.30) is given by equations (2.65) and (2.66), and the solution for $x_{s1}(t)$ from equation (3.31) is given by equations (2.66) and (2.67). The summation of solutions indicated in equation (3.28) is then given by equation (2.68), and the total solution becomes

$$x(t) = Ae^{-\zeta\omega_n t} \sin(\omega_d t + \phi) + \omega_n Y \left[\frac{\omega_n^2 + (2\zeta\omega_b)^2}{(\omega_n^2 - \omega_b^2)^2 + (2\zeta\omega_n\omega_b)^2} \right]^{1/2} \cos(\omega_b t - \theta_1 - \theta_2) \quad (3.37)$$

where A and ϕ are to be determined by the initial conditions, and θ_1 and θ_2 are as defined by equations (2.66) and (2.69). Since $\zeta = c/(2\sqrt{km}) = 0.158$ and $\omega_n = \sqrt{k/m} = 31.62$ rad/s, these phase angles become

$$\theta_1 = \tan^{-1}\left(\frac{2(0.158)(31.62)(3)}{(31.62)^2 - (3)^2}\right) = 0.03 \text{ rad}$$

$$\theta_2 = \tan^{-1}\left(\frac{31.62}{(2)(0.158)(3)}\right) = 1.541 \text{ rad}$$

and the magnitude becomes

$$\omega_n Y \left[\frac{\omega_n^2 + (2\zeta\omega_b)^2}{(\omega_n^2 + \omega_b^2)^2 + (2\zeta\omega_n\omega_b)^2} \right]^{1/2} = (31.62)(0.05)\left\{ \frac{(31.62)^2 + [2(0.158)(3)]^2}{[(31.62)^2 - (3)^2]^2 + [2(0.158)(3)(31.62)]^2} \right\}^{1/2}$$

$$= 0.05 \text{ m}$$

The solution given in equation (3.37) takes the form

$$x(t) = Ae^{-5t} \sin(31.22t + \phi) + 0.05\cos(3t - 1.571) \quad (3.38)$$

where $\omega_d = \omega_n\sqrt{1 - \zeta^2} = 31.225$ rad/s. At $t = 0$, this becomes

$$x(0) = A\sin(\phi) + 0.05\cos(-1.571)$$

or

$$0.01 \text{ m} = A\sin\phi + (0.05)(-0.00204) \quad (3.39)$$

Differentiating $x(t)$ yields

$$\dot{x}(t) = Ae^{-5t}\cos(31.225t + \phi)(31.225) - 5Ae^{-5t}\sin(31.225t + \phi) - 0.15\sin(3t - 1.571)$$

At $t = 0$, this becomes

$$3 = (31.225)A\cos(\phi) - 5A\sin(\phi) - 0.15\sin(-1.571) \quad (3.40)$$

Equations (3.39) and (3.40) represent two equations in the two unknown constants of integration, A and ϕ. Solving these yields $A = -0.0096$ m and $\phi = 0.1083$ rad, so that the total solution is

$$x(t) = 0.0096e^{-5t}\sin(31.225t + 0.1083) + 0.05\cos(3t - 1.571)$$

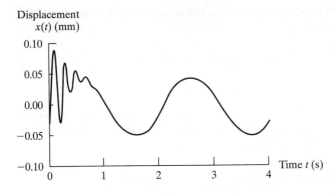

Figure 3.14   The total time response of a spring–mass–damper system under base excitation as calculated in Example 3.3.1.

This is plotted in Figure 3.14. Note that the transient term is not noticeable after 1 s. A comparison with a numerical solution of the same problem is given in Example 3.8.3 of Section 3.8.

☐

The computation of the response to complicated inputs becomes tedious when using the analytical approach. With the advent of computational software, practicing engineers are more likely to use a numerical approach to compute the solution. While numerical approaches are approximations, they do allow quick calculation of the response to systems with complicated inputs consisting of step functions and long periodic disturbances. Numerical approaches are discussed in Section 3.9.

3.4 TRANSFORM METHODS

The Laplace transform was introduced briefly in Section 2.3 as an alternative method of solving for the forced harmonic response of a single-degree-of-freedom system. The Laplace transform technique is even more useful for calculating the responses of systems to a variety of force excitations, both periodic and nonperiodic. The usefulness of the Laplace transform technique of solving differential equations and, in particular, solving for the forced response lies in the availability of tabulated Laplace transform pairs. Using tabulated Laplace transform pairs reduces the solution of forced vibration problems to algebraic manipulations and table "lookup." In addition, the Laplace transform approach provides certain theoretical advantages and leads to a formulation that is very useful for experimental vibration measurements.

The definition of a Laplace transform of the function of time $f(t)$ is

$$L[f(t)] = F(s) = \int_0^\infty f(t)e^{-st}dt \tag{3.41}$$

for an integrable function $f(t)$ such that $f(t) = 0$ for $t < 0$. The variable s is complex valued. The Laplace transform changes the domain of the function from the positive real-number line (t) to the complex-number plane (s). The integration in the Laplace transform changes differentiation into multiplication, as the following example illustrates.

Example 3.4.1

Calculate the Laplace transform of the derivative $\dot{f}(t)$.

Solution

$$L\big[\dot{f}(t)\big] = \int_0^\infty \dot{f}(t)e^{-st}dt = \int_0^\infty e^{-st}\frac{d[f(t)]}{dt}dt$$

Integration by parts yields

$$L\big[\dot{f}(t)\big] = e^{-st}f(t)\Big|_0^\infty + s\int_0^\infty e^{-st}f(t)\,dt$$

Recognizing that the integral in the last term of the preceding equation is the definition of $F(s)$ yields

$$L\big[\dot{f}(t)\big] = sF(s) - f(0)$$

where $F(s)$ denotes the Laplace transform of $f(t)$. Repeating this procedure on $\ddot{f}(t)$ yields

$$L\big[\ddot{f}(t)\big] = s^2F(s) - sf(0) - \dot{f}(0).$$

□

Example 3.4.2

Calculate the Laplace transform of the unit step function defined by the right-hand side of equation (3.41) and denoted by $\Phi(t)$ for the case $t_0 = 0$.

Solution

$$L\big[\Phi(t)\big] = \int_0^\infty e^{-st}dt = -\frac{e^{-st}}{s}\Big|_0^\infty = -\frac{e^{-\infty}}{s} + \frac{e^{-0}}{s} = \frac{1}{s}$$

□

The procedure for solving for the forced response of a mechanical system is first to take the Laplace transform of the equation of motion. Next, the transformed expression is algebraically solved for $X(s)$, the Laplace transform of the response. The inverse transform of this expression is found by using a table of Laplace transforms to yield the desired time history of the response $x(t)$. This is illustrated in the following example. A sample table of Laplace transform pairs is given in Table 3.1.

TABLE 3.1   COMMON LAPLACE TRANSFORMS FOR ZERO INITIAL CONDITIONS[a]

$F(s)$	$f(t)$
1. 1	$\delta(0)$ unit impulse
2. $1/s$	1, unit step $\Phi(t)$
3. $\dfrac{1}{s + a}$	e^{-at}
4. $\dfrac{1}{(s + a)(s + b)}$	$\dfrac{1}{b - a}(e^{-at} - e^{-bt})$
5. $\dfrac{\omega_n}{s^2 + \omega_n^2}$	$\sin \omega_n t$
6. $\dfrac{s}{s^2 + \omega_n^2}$	$\cos \omega_n t$
7. $\dfrac{1}{s(s^2 + \omega_n^2)}$	$\dfrac{1}{\omega_n^2}(1 - \cos \omega_n t)$
8. $\dfrac{1}{s^2 + 2\zeta\omega_n s + \omega_n^2}$	$\dfrac{1}{\omega_d} e^{-\zeta\omega_n t} \sin \omega_d t,\ \zeta < 1,\ \omega_d = \omega_n \sqrt{1 - \zeta^2}$
9. $\dfrac{\omega_n^2}{s(s^2 + 2\zeta\omega_n s + \omega_n^2)}$	$1 - \dfrac{\omega_n}{\omega_d} e^{-\zeta\omega_n t} \sin(\omega_d t + \phi),\ \phi = \cos^{-1}\zeta,\ \zeta < 1$
10. e^{-as}	$\delta(t - a)$
11. $F(s - a)$	$e^{at}f(t) \geq 0$
12. $e^{-as}F(s)$	$f(t - a)\Phi(t - a)$

[a]A more complete table appears in Appendix B. Here the Heaviside step function or unit step function is denoted by Φ. Other notations for this function include μ and H.

Example 3.4.3

Calculate the forced response of an undamped spring–mass system to a unit step function. Assume that both initial conditions are zero.

Solution   The equation of motion is

$$m\ddot{x}(t) + kx(t) = \Phi(t)$$

Taking the Laplace transform of this equation yields

$$(ms^2 + k)X(s) = \frac{1}{s}$$

Solving algebraically for $X(s)$ yields

$$X(s) = \frac{1}{s(ms^2 + k)} = \frac{1/m}{s(s^2 + \omega_n^2)}$$

Examining the definition of the Laplace transform, note that the coefficient $1/m$ passes through the transform. The time function corresponding to the value of $X(s)$ in the preceding equation can be found as entry 7 in Table 3.1. This implies that

$$x(t) = \frac{1/m}{\omega_n^2}(1 - \cos\omega_n t) = \frac{1}{k}(1 - \cos\omega_n t)$$

which, of course, agrees with the solution given by equation (3.18) with $F_0 = 1$.

□

Example 3.4.4

Calculate the response of an underdamped spring–mass system to a unit impulse. Assume zero initial conditions.

Solution   The equation of motion is

$$m\ddot{x} + c\dot{x} + kx = \delta(t)$$

Taking the Laplace transform of both sides of this expression using the results of Example 3.4.1 and entry 1 in Table 3.1 yields

$$\left(ms^2 + cs + k\right)X(s) = 1$$

Solving for $X(s)$ yields

$$X(s) = \frac{1/m}{s^2 + 2\zeta\omega_n s + \omega_n^2}$$

Assuming that $\zeta < 1$ and consulting entry 8 of Table 3.1 yields

$$x(t) = \frac{1/m}{\omega_n\sqrt{1 - \zeta^2}}e^{-\zeta\omega_n t}\sin\left(\omega_n\sqrt{1 - \zeta^2}t\right) = \frac{1}{m\omega_d}e^{-\zeta\omega_n t}\sin\omega_d t$$

in agreement with equation (3.6).

□

Example 3.4.5

Compute the solution of the spring–mass–damper system subject to an impulse at time $t = \pi$ s defined by the following equation of motion:

$$\ddot{x}(t) + 2\dot{x}(t) + 2x(t) = \delta(t - \pi), x_0 = v_0 = 0$$

Solution   From the coefficients

$$\omega_n = \sqrt{2}\,\text{rad/s}, \zeta = \frac{2}{2\sqrt{2}} = \frac{1}{\sqrt{2}} \text{ and } \omega_d = \sqrt{2}\sqrt{1 - (1/\sqrt{2})^2} = 1\,\text{rad/s}$$

Taking the Laplace transform of the equation of motion yields

$$(s^2 + 2s + 2)X(s) = e^{-\pi s}$$

Solving algebraically for $X(s)$ yields

$$X(s) = \frac{e^{-\pi s}}{s^2 + 2s + 2} = (e^{-\pi s})\left(\frac{1}{s^2 + 2s + 2}\right)$$

The inverse Laplace transform of the last term is (from entry 8 in Table 3.1)

$$L^{-1}\left(\frac{1}{s^2 + 2s + 1}\right) = e^{-t}\sin t$$

From entry 12 of Table 3.1, the inverse Laplace transform of $X(s)$ then becomes

$$x(t) = e^{-(t-\pi)}\sin(t-\pi)\Phi(t-\pi) = \begin{cases} 0, & t < \pi \\ e^{-(t-\pi)}\sin(t-\pi), & t \geq \pi \end{cases}$$

$\square$

It should be noted that the Laplace transform may be used for problems with untabulated pairs by inverting the integration indicated in equation (3.41). The inversion integral is

$$x(t) = \frac{1}{2\pi j}\int_{-\infty}^{\infty} X(s)e^{st}ds \tag{3.42}$$

where $j = \sqrt{-1}$. The inverse Laplace transform is discussed in greater detail in Appendix B.

An often-used tool in Laplace transform analysis is the idea of convolution, introduced in Section 3.2. In fact, the convolution integral is often defined first in terms of the Laplace transform. Consider the response written as the convolution integral as given in equation (3.12), and take the Laplace transform assuming zero initial conditions. This yields

$$X(s) = F(s)H(s) \tag{3.43}$$

called Borel's theorem. Here $F(s)$ is the Laplace transform of the driving force, $f(t)$, and $H(s)$ is the Laplace transform of the impulse response function $h(t)$. Taking the transform of $h(t)$ defined by equation (3.8) and using entry 8 of Table 3.1 yields

$$H(s) = \frac{1}{s^2 + 2\zeta\omega_n s + \omega_n^2} \tag{3.44}$$

which is the transfer function of a single-degree-of-freedom oscillator as defined in Section 2.3, equation (2.59).

A related transform is the *Fourier transform*, which arises from considering the Fourier series of a nonperiodic function. The Fourier transform of a function $x(t)$ is denoted by $X(\omega)$ and is defined by

$$X(\omega) = \frac{1}{2\pi}\int_{-\infty}^{\infty} x(t)e^{-j\omega t}dt \tag{3.45}$$

which transforms the variable $x(t)$ from a function of time into a function of frequency ω. The inversion of this transform is performed by the integral

$$x(t) = \int_{-\infty}^{\infty} X(\omega)e^{j\omega t}d\omega \tag{3.46}$$

The Fourier transform integral defined by equation (3.45) arises from the Fourier series representation of a function described by equation (3.20) by writing the series in complex form and allowing the period to go to infinity. [See Newland (1993), page 39, for details.]

Note that the definitions of the Fourier transform and the Laplace transform are similar. In fact, the form of the Fourier transform pairs given in equations (3.45) and (3.46) can be obtained by substituting $s = j\omega$ into the Laplace transform pair given by equations (3.41) and (3.42). Although this does not constitute a rigorous definition, it does provide a connection between the two types of transforms.

Fourier transforms are not used as frequently for solving vibration problems as are Laplace transforms. However, the Fourier transform is used extensively in discussing random vibration problems and in the measurement of vibration parameters. Appendix B discusses additional details of transforms. A rigorous description of the use of various transforms, their properties, and their applications can be found in Churchill (1972).

3.5 RESPONSE TO RANDOM INPUTS

So far, all the driving forces considered have been deterministic functions of time. That is, given a value of the time t, the value of $F(t)$ is precisely known. Here the response of a system subject to a random force input $F(t)$ is investigated. Disturbances are often characterized as random if the value of $F(t)$ for a given value of t is known only statistically. That is, a random signal has no obvious pattern. For random signals it is not possible to focus on the details of the signal, as it is with a pure deterministic signal. Hence random signals are classified and manipulated in terms of their statistical properties.

Randomness in vibration analysis can be thought of as the result of a series of experiments, all performed in an identical fashion under identical circumstances, each of which produces a different response. One record or time history is not enough to describe such a vibration; rather, a statistical description of all possible responses is required. In this case, a vibration response $x(t)$ should not be thought of as a single signal, but rather as a collection, or ensemble, of possible time histories resulting from the same conditions (i.e., same system, same controlled environment, same length of time). A single element of such an ensemble is called a *sample* function (or response).

Consider a random signal $x(t)$, or sample, as pictured in Figure 3.15(d). The first distinction to be made about a random time history is whether or not the signal is *stationary*. A random signal is *stationary* if its statistical properties (usually

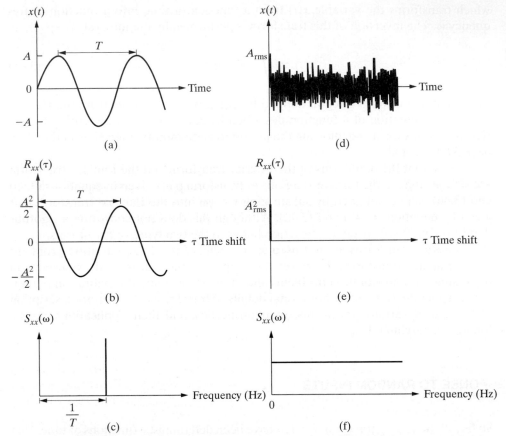

Figure 3.15 (a) A simple sine function; (b) its autocorrelation; (c) its power spectral density. (d) A random signal; (e) its autocorrelation; (f) its power spectral density.

its average or mean square) do not change with time. The *average* of the random signal $x(t)$ is defined and denoted by

$$\bar{x} = \lim_{T \to \infty} \frac{1}{T} \int_0^T x(t)dt \tag{3.47}$$

as introduced in Section 1.2, equation (1.20), for deterministic signals. Here, it is convenient to consider signals with a zero average or mean [i.e., $\bar{x}(t) = 0$]. This is not too restrictive an assumption, since if $\bar{x}(t) \neq 0$, a new variable $x' = x - \bar{x}$ can be defined. The new variable x' now has zero mean.

The *mean-square* value of the random variable $x(t)$ is denoted by $\overline{x^2}$ and is defined by

$$\overline{x^2} = \lim_{T \to \infty} \frac{1}{T} \int_0^T x^2(t)\, dt \tag{3.48}$$

as introduced in Section 1.2, equation (1.21), for deterministic signals. In the case of random signals, this is also called the *variance* and provides a measure of the magnitude of the fluctuations in the signal $x(t)$. A related quantity, called the *root-mean-square (rms)* value, is just the square root of the variance:

$$x_{rms} = \sqrt{\overline{x^2}} \tag{3.49}$$

This definition can be applied to the value of a single response over its time history or to an ensemble value at a fixed time.

Another measure of interest in random variables is how fast the value of the variable changes. This addresses the issue of how long it takes to measure enough samples of the variable before a meaningful statistical value can be calculated. Many measured vibration signals are random and, as such, an indication of how quickly a variable changes is very useful. The *autocorrelation function,* denoted by $R_{xx}(\tau)$ and defined by

$$R_{xx}(\tau) = \lim_{T \to \infty} \frac{1}{T} \int_0^T x(t)x(t + \tau)dt \tag{3.50}$$

provides a measure of how fast the signal $x(t)$ is changing. The value τ is the time difference between the values at which the signal $x(t)$ is sampled. The prefix *auto* refers to the fact that the term $x(t)x(t + \tau)$ is the product of values of the same sample at two different times. The autocorrelation is a function of the time difference τ only in the special case of stationary random signals. Figure 3.15(e) illustrates the autocorrelation of a random signal, and Figure 3.15(b) illustrates that of a sine function. The Fourier transform of the autocorrelation function defines the *power spectral density* (PSD). Denoting the PSD by $S_{xx}(\omega)$ and repeating the definition of equation (3.45) results in

$$S_{xx}(\omega) = \frac{1}{2\pi} \int_{-\infty}^{\infty} R_{xx}(\omega)e^{-j\omega\tau}d\tau \tag{3.51}$$

Note that this integral of $R_{xx}(\tau)$ changes the real number τ into a frequency-domain value ω. Figure 3.15(c) illustrates the PSD of a pure sine signal, and Figure 3.15(f) illustrates the PSD of a random signal. The autocorrelation and power spectral density, defined by equations (3.50) and (3.51), respectively, can be used to examine the response of a spring–mass system to a random excitation.

Recall from Section 3.2 that the response $x(t)$ of a spring–mass–damper system to an arbitrary forcing function $F(t)$ can be represented by using the impulse response function $h(t - \tau)$ given by equation (3.9) for underdamped systems. The Fourier transform of the function $h(t - \tau)$ can be used to relate the PSD of the random input of an underdamped system to the PSD of the system's response. First note

from equation (3.8), Example 3.4.4, and entry 8 of Table 3.1 that the Laplace transform of $h(t)$ for a single-degree-of-freedom system is

$$L[h(t)] = L\left[\frac{1}{m\omega_d} e^{-\zeta\omega_n t} \sin \omega_d t\right] = \frac{1}{m\omega_d} L\left[e^{-\zeta\omega_n t} \sin \omega_d t\right] \qquad (3.52)$$

$$= \frac{1}{ms^2 + cs + k} = H(s)$$

where $H(s)$ is the system transfer function as defined by equation (2.59). In this
case, the Fourier transform of $h(t)$ can be obtained from the Laplace transform by
setting $s = j\omega$ in equation (3.52). This yields simply

$$H(j\omega) = \frac{1}{k - \omega^2 m + c\omega j} \qquad (3.53)$$

which, upon comparison with equations (2.60) and (2.52), is also the frequency
response function for the single-degree-of-freedom oscillator. Let $X(\omega)$ denote
the Fourier transform of the impulse response function, $h(t)$; then, from equations
(3.45) and (3.53)

$$X(\omega) = \frac{1}{2\pi} \int_{-\infty}^{\infty} h(t)e^{-j\omega_n t} dt = H(\omega) \qquad (3.54)$$

where the j is dropped from the argument of H for convenience. Thus the frequency response function of Section 2.3 can be related to the Fourier transform
of the impulse response function. This becomes extremely significant in vibration
measurement as discussed in Chapter 7.

Next, recall the formulation of the solution of a vibration problem using the
impulse response function. From equation (3.12), the response $x(t)$ to a driving
force $F(t)$ is simply

$$x(t) = \int_0^t F(\tau)h(t - \tau)d\tau \qquad (3.55)$$

Note that since $h(t - \tau) = 0$ for $t < T$, the upper limit can be extended to plus
infinity. Since $F(t) = 0$ for $t < 0$, the lower limit can be extended to minus infinity.
Thus, expression (3.55) can be rewritten as

$$x(t) = \int_{-\infty}^{\infty} F(\tau)h(t - \tau)d\tau \qquad (3.56)$$

Next, the variable of integration τ can be changed to θ by using $\tau = t - \theta$, and hence
$d\tau = -d\theta$. Using this change of variables, the previous integral can be written

$$x(t) = -\int_{\infty}^{-\infty} F(t - \theta)h(\theta)d(\theta) = \int_{-\infty}^{\infty} F(t - \theta)h(\theta)d\theta \qquad (3.57)$$

which provides an alternative form of the solution of a forced vibration problem in terms of the impulse response function.

Finally, consider the PSD of the response $x(t)$ given by equation (3.51) as

$$S_{xx}(\omega) = \frac{1}{2\pi} \int_{-\infty}^{\infty} R_{xx}(\tau) e^{-j\omega\tau} d\tau \tag{3.58}$$

Upon substitution of the definition of $R_{xx}(\tau)$ from equation (3.50), this becomes

$$S_{xx}(\omega) = \frac{1}{2\pi} \int_{-\infty}^{\infty} \left[\lim_{T \to \infty} \frac{1}{T} \int_{0}^{T} x(\sigma)x(\sigma + \tau) d\sigma \right] e^{-j\omega\tau} d\tau \tag{3.59}$$

The expressions for $x(t)$ in the integral are evaluated next using equation (3.57), which results in

$$S_{xx}(\omega) =$$

$$\frac{1}{2\pi} \int_{-\infty}^{\infty} \left[\lim_{T \to \infty} \frac{1}{T} \int_{0}^{T} \left[\int_{-\infty}^{\infty} F(\sigma - \theta) h(\theta) d\theta \int_{-\infty}^{\infty} F(\sigma - \theta + \tau) h(\theta) d\theta \right] d\sigma \right] e^{-j\omega\tau} d\tau \tag{3.60}$$

$$= \frac{1}{2\pi} \int_{-\infty}^{\infty} \lim_{T \to \infty} \frac{1}{T} \int_{0}^{T} \left[F(\hat{t}) F(\hat{t} + \tau) \int_{-\infty}^{\infty} h(\theta) e^{-j\omega\theta} d\theta \int_{-\infty}^{\infty} h(\theta) e^{j\omega\theta} d\theta \right] d\sigma e^{-j\omega\tau} d\tau \tag{3.61}$$

where $e^{(\hat{t} - \hat{t})j\omega} = 1$ has been inserted inside the inner integrals and a subsequent change of variables ($\hat{t} = \sigma - \theta$) has been performed on the argument of F, which is subsequently moved outside the integral. The two integrals inside the brackets in equation (3.61) are $H(\omega)$ and its complex conjugate $H(-\omega)$, according to equation (3.54). Recognizing the frequency response functions $H(\omega)$ and $H(-\omega)$ in equation (3.61), this expression can be rewritten as

$$S_{xx}(\omega) = |H(\omega)|^2 \left[\frac{1}{2\pi} \int_{-\infty}^{\infty} R_{ff}(\tau) e^{-j\omega\tau} d\tau \right]$$

or simply

$$S_{xx}(\omega) = |H(\omega)|^2 S_{ff}(\omega) \tag{3.62}$$

Here R_{ff} denotes the autocorrelation function for $F(t)$ and S_{ff} denotes the PSD of the forcing function $F(t)$. The notation $|H(\omega)|^2$ indicates the square of the magnitude of the complex frequency response function. A more rigorous derivation of the result can be found in Newland (1993). It is more important to study the result [i.e., equation (3.62)] than the derivations at this level.

Equation (3.62) represents an important connection between the power spectral density of the driving force, the dynamics of the structure, and the power spectral density of the response. In the deterministic case, a solution was obtained relating the harmonic force applied to the system and the resulting response (Chapter 2). In the case where the input is a random excitation, the statement equivalent to a solution is equation (3.62), which indicates how fast the response $x(t)$ changes in relation to how fast the (random) driving force is changing. In the deterministic case of a sinusoidal driving force, the solution allowed the conclusion that the response was also sinusoidal with a new magnitude and phase, but of the same frequency as the driving force. In a way, equation (3.62) makes the equivalent statement for a random excitation (see Window 3.4). It states that when the excitation is a stationary random process, the response will be a stationary random process and the response changes as rapidly as the driving force, but with a modified amplitude. In both the deterministic case and the random case, the amplitude of the response is related to the frequency response function of the structure.

Example 3.5.1

Consider the single-degree-of-freedom system of Window 3.1 subject to a random (white noise) force input $F(t)$. Calculate the power spectral density of the response $x(t)$ given that the PSD of the applied force is the constant value S_0.

Solution   The equation of motion is

$$m\ddot{x} + c\dot{x} + kx = F(t)$$

From equation (2.59) or equation (3.53), the frequency response function is

$$H(\omega) = \frac{1}{k - m\omega^2 + c\omega j}$$

Thus,

$$|H(\omega)|^2 = \left| \frac{1}{k - m\omega^2 + c\omega j} \right|^2 = \frac{1}{(k - m\omega^2) + c\omega j} \cdot \frac{1}{(k - m\omega^2) - c\omega j}$$

$$= \frac{1}{(k - m\omega^2)^2 + c^2\omega^2}$$

From equation (3.62), the PSD of the response becomes

$$S_{xx} = |H(\omega)|^2 S_{ff} = \frac{S_0}{(k - m\omega^2)^2 + c^2\omega^2}$$

This states that if a single-degree-of-freedom system is excited by a stationary random force (of constant mean and rms value) that has a constant power spectral density of value S_0, the response of the system will also be random with nonconstant (i.e., frequency-dependent) PSD of $S_{xx}(\omega) = S_0/[(k - m\omega^2)^2 + c^2\omega^2]$.

□

Another useful quantity in discussing the response of a system to random vibration is the expected value. The *expected value* (or more appropriately, the *ensemble average*) of $x(t)$ is denoted by $E[x]$ and defined by

$$E[x] = \lim_{T \to \infty} \int_0^T \frac{x(t)}{T} \, dt \tag{3.63}$$

which, from equation (3.47), is also the mean value, $\bar{x}$. The expected value is also related to the probability that $x(t)$ lies in a given interval through the *probability density function* $p(x)$. An example of $p(x)$ is the familiar Gaussian distribution function (bell-shaped curve). In terms of the probability density function, the expected value is defined by

$$E[x] = \int_{-\infty}^{\infty} x p(x) \, dx \tag{3.64}$$

The average of the product of the two functions $x(t)$ and $x(t + \tau)$ describes how the function $x(t)$ changes with time and, for a stationary random process, is the autocorrelation function

$$E[x(t)x(t + \tau)] = \lim_{T \to \infty} \frac{1}{T} \int_0^T x(t)x(t + \tau) \, dt = R_{xx}(\tau) \tag{3.65}$$

upon comparison with equation (3.50). From equation (3.48), the mean-square value becomes

$$\bar{x}^2 = R_{xx}(0) = E[x^2] \tag{3.66}$$

The mean-square value can, in turn, be related to the power spectral density function by inverting equation (3.51) using the Fourier transform pair of equations (3.45) and (3.46). This yields

$$R_{xx}(\tau)_{\tau=0} = \int_{-\infty}^{\infty} S_{xx}(\omega) \, d\omega = E[x^2] \tag{3.67}$$

Equation (3.62) relates the PSD of the response $x(t)$ to the PSD of the driving force $F(t)$ and the frequency response function. Combining equations (3.62) and (3.67) yields

$$E[x^2] = \int_{-\infty}^{\infty} |H(\omega)|^2 S_{ff}(\omega) \, d\omega \tag{3.68}$$

This expression relates the mean-square value of the response to the PSD of the (random) driving force and the dynamics of the system. Equations (3.68) and (3.62) form the basis for random vibration analysis for stationary random driving forces. These expressions represent the equivalent of using the impulse response function and frequency response functions to describe deterministic vibration excitations (see Window 3.4).

Window 3.4
Comparison between Calculations for the Response of a
Spring–Mass–Damper System and Deterministic and Random Excitations

Transfer function: $G(s) = \dfrac{1}{ms^2 + cs + k}$

Frequency response function: $G(j\omega) = H(\omega) = \dfrac{1}{k - m\omega^2 + c\omega j}$

Impulse response function: $h(t) = \dfrac{1}{m\omega_d} e^{-\zeta\omega_n t} \sin \omega_d t$

The Laplace transform of the impulse response function is

$$L[h(t)] = \frac{1}{ms^2 + cs + k} = G(s)$$

and the Fourier transform of the impulse response function is just the frequency response function $H(\omega)$. These quantities relate the input and response by

For deterministic $f(t)$:   For random $f(t)$:

$$X(s) = G(s)F(s) \qquad\longleftrightarrow\qquad S_{xx}(\omega) = |H(\omega)|^2 S_{ff}(\omega)$$

$$x(t) = \int_0^t h(t - \tau)f(\tau)d\tau \qquad\longleftrightarrow\qquad E[x2] = \int_{-\infty}^{\infty} |H(\omega)|^2 S_{ff}(\omega)d\omega$$

To use equation (3.68), the integral involving $|H(\omega)|^2$ must be evaluated. In many useful cases, $S_{ff}(\omega)$ is constant. Hence, values of $\int |H(\omega)|^2$ have been tabulated (see Newland, 1993). For example

$$\int_{-\infty}^{\infty} \left| \frac{B_0}{A_0 + j\omega A_1} \right|^2 d\omega = \frac{\pi B_0^2}{A_0 A_1} \tag{3.69}$$

and

$$\int_{-\infty}^{\infty} \left| \frac{B_0 + j\omega B_1}{A_0 + j\omega A_1 - \omega^2 A_2} \right|^2 d\omega = \frac{\pi \left(A_0 B_1^2 + A_2 B_0^2 \right)}{A_0 A_1 A_2} \tag{3.70}$$

Such integrals, along with equation (3.68), allow computation of the expected value, as the following example illustrates.

Example 3.5.2

Calculate the mean-square value of the response of the system described in Example 3.5.1 with equation of motion $m\ddot{x} + c\dot{x} + kx = F(t)$, where the PSD of the applied force is the constant value S_0.

Solution   Since the PSD of the forcing function is the constant S_0, equation (3.68) becomes

$$E[x^2] = S_0 \int_{-\infty}^{\infty} \left| \frac{1}{k - m\omega_n^2 + jc\omega} \right|^2 d\omega$$

Comparison with equation (3.70) yields $B_0 = 1$, $B_1 = 0$, $A_0 = k$, $A_1 = c$, and $A_2 = m$. Thus,

$$E[x^2] = S_0 \frac{\pi m}{kcm} = \frac{\pi S_0}{kc}$$

Hence, if a spring–mass–damper system is excited by a random force described by a constant PSD, S_0, it will have a random response, $x(t)$, with mean-square value $\pi S_0/kc$.
□

Two basic relationships used in analyzing spring–mass–damper systems excited by random inputs are illustrated in this section. The output or response of a randomly excited system is also random and, unlike deterministic systems, cannot be exactly predicted. Hence, the response is related to the driving force through the statistical quantities of power spectral density and mean-square values. See Window 3.4 for a comparison of response calculations for deterministic and random inputs. In deterministic vibrations, the concern in design is usually to compute the magnitude and phase of the response to a known deterministic forcing function. This section addressed the same problem when the forcing function has a random nature. Given a statistical property of the forcing function, say the average magnitude, then the best we can do is to compute the average value of the magnitude of the response.

3.6 SHOCK SPECTRUM

Many disturbances are abrupt or sudden in nature. The impulse is an example of a force applied suddenly. Such a sudden application of a force or other form of disturbance resulting in a transient response is referred to as a *shock*. Because of the common occurrence of shock inputs, a special characterization of the response to a shock has developed as a standard design and analysis tool. This characterization is called the *response spectrum* and consists of a plot of the maximum absolute value of the system's time response versus the natural frequency of the system.

The impulse response discussed in Section 3.1 provides a mechanism for studying the response of a system to a shock input. Recall that the impulse response function, $h(t)$, was derived from considering a force input, $\delta(t)$, of large magnitude and short duration and can be used to calculate the response of a system to any input. The impulse response function forms the basis for calculating the response spectrum introduced here.

Recall from equation (3.12) that the response of a system to an arbitrary input $F(t)$ can be written as

$$x(t) = \int_0^t F(\tau)h(t - \tau)d\tau \tag{3.71}$$

where $h(t - \tau)$ is the impulse response function for the system. For an underdamped system, $h(t - \tau)$ is given by equation (3.9):

$$h(t - \tau) = \frac{1}{m\omega_d} e^{-\zeta\omega_n(t - \tau)} \sin \omega_d(t - \tau) \qquad t > \tau \tag{3.72}$$

which becomes

$$h(t - \tau) = \frac{1}{m\omega_n} \sin \omega_n(t - \tau) \tag{3.73}$$

in the undamped case. The response spectrum is defined to be a plot of the peak or maximum value of the response versus frequency. For an undamped system, equations (3.71) and (3.73) can be combined to yield the maximum value of the displacement response as

$$x(t)_{\text{max}} = \frac{1}{m\omega_n} \left| \int_0^t F(\tau) \sin [\omega_n(t - \tau)]d\tau \right|_{\text{max}} \tag{3.74}$$

Calculating a response spectrum then involves substitution of the appropriate $F(t)$ into equation (3.74) and plotting $x(t)_{\text{max}}$ versus the undamped natural frequency. This is usually done numerically on a computer; however, the following example illustrates the procedure by hand calculation.

Example 3.6.1

Calculate the response spectrum for the forcing function given in Figure 3.16 applied to the linear spring–mass system. The abruptness of the response is characterized by the time t_1.

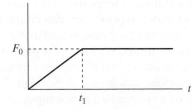

Figure 3.16   A step disturbance with rise time of t_1 seconds.

Solution   As in Example 3.2.2, the forcing function $F(t)$ sketched in Figure 3.16 can be written as the sum of two other simple functions. In this case, the input is the sum of

$$F_1(t) = \frac{t}{t_1} F_0$$

and

$$F_2(t) = \begin{cases} 0 & 0 < t < t_1 \\ -\dfrac{t - t_1}{t_1} F_0 & t \geq t_1 \end{cases}$$

Following the steps taken in Example 3.2.2, the response is calculated by evaluating the response to $F_1(t)$ and separately to $F_2(t)$. Linearity is then used to obtain the total response to $F(t) = F_1(t) + F_2(t)$. The response to $F_1(t)$, denoted by $x_1(t)$, calculated using equations (3.71) and (3.73), becomes

$$x_1(t) = \frac{\omega_n}{k} \int_0^t \frac{F_0 \tau}{t_1} \sin \omega_n(t - \tau)\, d\tau = \frac{F_0}{k} \left(\frac{t}{t_1} - \frac{\sin \omega_n t}{\omega_n t_1} \right) \tag{3.75}$$

Similarly, the response to $F_2(t)$, denoted by $x_2(t)$, becomes

$$x_2(t) = \int_0^t F_2(\tau) \frac{1}{m \omega_n} \sin \omega_n(t - \tau) d\tau = \frac{-F_0}{m \omega_n} \int_{t_1}^t \frac{\tau - t_1}{t_1} \sin \omega_n(t - \tau) d\tau$$

which becomes

$$x_2(t) = -\frac{F_0}{k} \left[\frac{t - t_1}{t_1} - \frac{\sin \omega_n(t - t_1)}{\omega_n t_1} \right] \tag{3.76}$$

so that the total response becomes the sum $x(t) = x_1(t) + x_2(t)$:

$$x(t) = \begin{cases} \dfrac{F_0}{k} \left(\dfrac{t}{t_1} - \dfrac{\sin \omega_n t}{\omega_n t_1} \right) & t < t_1 \\[3mm] \dfrac{F_0}{k \omega_n t_1} [\omega_n t_1 - \sin \omega_n t + \sin \omega_n(t - t_1)] & t \geq t_1 \end{cases} \tag{3.77}$$

Alternately, the Heaviside step function may be used to write this solution as

$$x(t) = \frac{F_0}{k} \left(\frac{t}{t_1} - \frac{\sin \omega_n t}{\omega_n t_1} \right) - \frac{F_0}{k} \left(\frac{t - t_1}{t_1} - \frac{\sin \omega_n(t - t_1)}{\omega_n t_1} \right) \Phi(t - t_1) \tag{3.78}$$

Equation (3.77) is the response of an undamped system to the excitation of Figure 3.16. To find the maximum response, the derivative of equation (3.77) is set equal to zero and solved for the time t_p at which the maximum occurs. This time t_p is then substituted into the response $x(t_p)$ given by equation (3.77) to yield the maximum response $x(t_p)$. Differentiating equation (3.77) for $t > t_1$ yields $\dot{x}(t_p) = 0$ or

$$-\cos \omega_n t_p + \cos \omega_n(t_p - t_1) = 0 \tag{3.79}$$

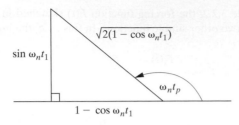

Figure 3.17   A graphical representation of equations (3.80) and (3.81).

Using simple trigonometry formulas and solving for $\omega_n t_p$ yields

$$\tan \omega_n t_p = \frac{1 - \cos \omega_n t_1}{\sin \omega_n t_1} \quad \text{or} \quad \omega_n t_p = \tan^{-1}\!\left(\frac{1 - \cos \omega_n t_1}{\sin \omega_n t_1}\right) \tag{3.80}$$

where t_p denotes the time to the first peak [i.e., the time for which the maximum value of equation (3.77) occurs]. Expression (3.80) corresponds to a right triangle of sides $(1 - \cos \omega_n t_1)$, and $\sin \omega_n t_1$, and hypotenuse

$$\sqrt{\sin^2 \omega_n t_1 + (1 - \cos \omega_n t_1)^2} = \sqrt{2(1 - \cos \omega_n t_1)} \tag{3.81}$$

This relationship is illustrated in Figure 3.17. Hence, $\sin \omega_n t_p$ can be calculated from

$$\sin \omega_n t_p = -\sqrt{\frac{1}{2}(1 - \cos \omega_n t_1)} \tag{3.82}$$

and

$$\cos \omega_n t_p = \frac{-\sin \omega_n t_1}{\sqrt{2(1 - \cos \omega_n t_1)}} \tag{3.83}$$

Substitution of this expression into solution (3.77) evaluated at t_p yields, after some manipulation [here $x(t_p) = x_{\max}$],

$$\frac{x_{\max} k}{F_0} = 1 + \frac{1}{\omega_n t_1} \sqrt{2(1 - \cos \omega_n t_1)} \tag{3.84}$$

where the left side represents the dimensionless maximum displacement. It is customary to plot the response spectrum (dimensionless) versus the dimensionless frequency

$$\frac{t_1}{T} = \frac{\omega_n t_1}{2\pi} \tag{3.85}$$

where T is the structure's natural period. This provides a scale related to the characteristic time, t_1, of the input. Figure 3.18 is a plot of the response spectrum for the ramp input force of Figure 3.16. Note that each point on the plot corresponds to a different rise time, t_1, of the excitation. The vertical scale is an indication of the relationship between the structure and the rise time of the excitation.

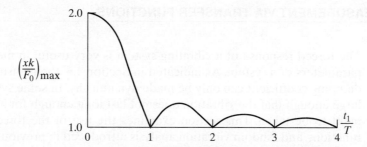

Figure 3.18 Response spectrum for the input force of Figure 3.16. The vertical axis is the dimensionless maximum response, and the horizontal axis is the dimensionless frequency (or delay time).

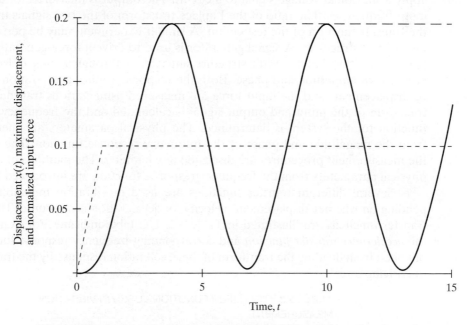

Figure 3.19 A plot of the time response (solid line) of an undamped system to the input given in Figure 3.16. Also shown are the maximum magnitude (long dashed line) and the input function (short dashed line) for the parameters $k = 10$ N/m, $m = 10$ kg, $F_0 = 1$ N, and $t_1 = 1$ s.

The response is plotted using equation (3.77) along with the maximum magnitude as given by equation (3.84) and the ramp input function in Figure 3.19. Note from these plots that the amplitude of the response is magnified, or larger than the level of the input force. If t_1 is chosen to be near a period (the minimum in Figure 3.18), then the response is lower than this value and the maximum response will be equal to the input level. The effects of the various parameters form the topic of shock isolation (Section 5.2, which can be read now) and are examined numerically in Section 3.9.

□

3.7 MEASUREMENT VIA TRANSFER FUNCTIONS

The forced response of a vibrating system is very useful in measuring the physical parameters of a system. As indicated in Section 1.6, the measurement of a system's damping coefficient can only be made dynamically. In some systems the damping is large enough that the vibration does not last long enough for a free decay measurement to be taken. This section examines the use of the forced response, transfer functions, and random vibration analysis introduced in previous sections to measure the mass, damping, and stiffness of a system.

The use of transfer functions to measure the properties of structures comes from electrical engineering. In circuit applications, a function generator is used to apply a sinusoidal voltage signal to a circuit. The output is measured for a range of input frequencies. The ratio of the Laplace transform of the two signals then yields the transfer function of the test circuit. A similar experiment may be performed on mechanical structures. A signal generator is used to drive a force-generating device (called a *shaker*) that drives the structure sinusoidally through a range of frequencies at a known amplitude and phase. Both the response (either acceleration, velocity, or displacement) and the input force are measured using various transducers. The transform of the input and output signal is calculated and the frequency response function for the system is determined. The physical parameters are then derived from the magnitude and phase of the frequency response function. The details of the measurement procedures are discussed in Chapter 7. The methods of extracting physical parameters from the frequency response function are introduced here.

Several different transfer functions are used in vibration measurement, depending on whether displacement, velocity, or acceleration is measured. The various transfer functions are illustrated in Table 3.2. The table indicates, for example, that the *accelerance transfer function* and corresponding frequency response function are obtained from dividing the transform of the acceleration response by the transform of the driving force.

TABLE 3.2 TRANSFER FUNCTIONS USED IN VIBRATION
MEASUREMENT

Response Measurement	Transfer Function	Inverse Transfer Function
Acceleration	Accelerance	Apparent mass
Velocity	Mobility	Impedance
Displacement	Receptance	Dynamic stiffness

The three transfer functions given in Table 3.2 are related to each other by simple multiplications of the transform variable s, since this corresponds to differentiation. Thus, with the *receptance transfer function* (also called the *compliance* or *admittance*) denoted by

$$\frac{X(s)}{F(s)} = H(s) = \frac{1}{ms^2 + cs + k} \tag{3.86}$$

the *mobility transfer function* becomes

$$\frac{s\,X(s)}{F(s)} = sH(s) = \frac{s}{ms^2 + cs + k} \tag{3.87}$$

because $sX(s)$ is the transform of the velocity. Similarly, $s^2X(s)$ is the transform of the acceleration, and the *accelerance transfer function (inertance)* becomes

$$\frac{s^2X(s)}{F(s)} = s^2H(s) = \frac{s^2}{ms^2 + cs + k} \tag{3.88}$$

Each of these also defines the corresponding frequency response function by substituting $s = j\omega$.

Consider calculating the magnitude of the complex compliance $H(j\omega)$ from equation (2.53) or (3.86). As expected from equation (2.70), this yields

$$|H(j\omega)| = \frac{1}{\sqrt{(k - m\omega^2)^2 + (c\omega)^2}} \tag{3.89}$$

Note that the largest value of this magnitude occurs near $k - m\omega^2 = 0$, or when the driving frequency is equal to the undamped natural frequency, $\omega = \omega_n = \sqrt{k/m}$. Recall from Section 2.2 that this also corresponds to a phase shift of $90°$. This argument is used in testing to determine the natural frequency of vibration of a test particle from a measured magnitude plot of the system transfer function. This is illustrated in Figure 3.20. The exact value of the peak frequency is derived in Example 2.2.5.

In principle, each of the physical parameters in the transfer function can be determined from the experimental plot of the frequency response function's magnitude. The natural frequency, ω_n, is determined from the position of the

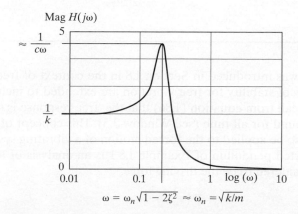

Figure 3.20 A magnitude plot for a spring–mass–damper system for the compliance transfer function indicating the determination of the natural frequency and stiffness.

peak. The damping constant c is approximated from the value of the frequency and a measurement of the magnitude $|H(j\omega)|$ at $\omega = \sqrt{k/m}$, since, from equation (3.89)

$$\left| H\left(j\sqrt{\frac{k}{m}} \right) \right| = \frac{1}{\sqrt{(c\omega_n)^2}} = \frac{1}{c\omega_n} \tag{3.90}$$

This formulation for measuring the damping coefficient provides an alternative to the logarithmic decrement technique presented in Section 1.6. Next, the stiffness can be determined from the zero frequency point. For $\omega = 0$, equation (3.87) yields

$$|H(0)| = \frac{1}{\sqrt{k^2}} = \frac{1}{k} \tag{3.91}$$

Since $\omega_n = \sqrt{k/m}$, knowledge of ω_n and k yields the value for m. In this way, m, c, k, ω_n, and ζ can all be determined from measurements of $|H(j\omega)|$. More practical methods are discussed in Chapter 7. Of course, m and k can usually be measured by static experiments as well, for comparison.

The preceding analysis all depends on the experimentally determined function $|H(j\omega)|$. Most experiments contain several sources of noise, so that a clean plot of $|H(j\omega)|$ is hard to get. The common approach is to repeat the experiment several times and essentially average the data (i.e., use ensemble averages). In practice, matched sets of input force time histories, $f(t)$, and response time histories, $x(t)$, are averaged to produce $R_{xx}(t)$ and $R_{ff}(t)$ using equation (3.48). The Fourier transform of these averages is then taken using equation (3.51) to get the PSD functions $S_{xx}(\omega)$ and $S_{ff}(\omega)$. Equation (3.62) is then used to calculate $|H(j\omega)|$ from the PSD values of the measured input and response. This procedure works for averaging noisy data as well as for the case of using a random excitation (zero mean) as the driving force. The transforms and computations required to calculate $|H(j\omega)|$ are usually made digitally in a dedicated computer used for vibration testing. This is discussed in Chapter 7 in more detail.

3.8 STABILITY

The concept of stability was introduced in Section 1.8 in the context of free vibration. Here the definitions of stability for free vibration are extended to include the forced-response case. Recall from equation (1.86) that the free response is stable if it stays within a finite bound for all time (see Window 3.5). This concept of a well-behaved response can also be applied to the forced motion of a vibrating system. In fact, in a sense, the inverted pendulum of Example 1.8.1 is an analysis of a forced response if gravity is considered to be the driving force.

Window 3.5
Review of Stability of the Free Response from Section 1.8

A solution $x(t)$ is *stable* if there exists some finite number M such that

$$|x(t)| < M$$

for all $t > 0$. If this bound cannot be satisfied, the response $x(t)$ is said to be *unstable*.

    If a response $x(t)$ is stable and $x(t)$ approaches zero as t gets large, the solution $x(t)$ is said to be *asymptotically stable*. An undamped spring–mass system is stable as long as m and k are positive and the value of M is just the amplitude A [i.e., $M = A$, where $x(t) = A \sin(\omega_n + \phi)$]. A damped system is asymptotically stable if m, c, and k are all positive [i.e., $x(t) = Ae^{-\zeta\omega_n t}\sin(\omega_d t + \phi)$ goes to zero as t increases].

    The stability of the forced response of a system can be defined by considering the nature of the applied force or input. The system

$$m\ddot{x} + c\dot{x} + kx = F(t) \tag{3.92}$$

is defined to be *bounded-input, bounded-output stable* (or simply BIBO stable) if, for any bounded input, $F(t)$, the output, or response $x(t)$, is bounded for any arbitrary set of initial conditions. Systems that are BIBO stable are manageable at resonance and do not "blow up."

    Note that the undamped version of equation (3.92) is not BIBO stable. To see this, note that if $F(t)$ is chosen to be $F(t) = \sin[(k/m)^{1/2}t]$ for the case $c = 0$, the response $x(t)$ is clearly not bounded as indicated in Figure 2.5. Also recall that the magnitude of the forced response of an undamped system is $F_0/[\omega_n^2 - \omega^2]$, which approaches infinity as ω approaches ω_n (see Window 3.6). However, the input force $F(t)$ is bounded since

$$|F(t)| = \left|\sin\left(\sqrt{\frac{k}{m}}t\right)\right| \leq 1 \tag{3.93}$$

for all time. Thus, there is some bounded force for which the response is not bounded and the definition of BIBO stable is violated. This situation corresponds to resonance. Clearly, an undamped system is poorly behaved at resonance.

    Next, consider the damped case ($c > 0$). Immediately, the preceding example of resonance is no longer unbounded. The forced-response magnitude curves given in Figure 2.8 illustrate that the response is always bounded for any bounded periodic driving force. To see that the response of an underdamped system is bounded

Window 3.6
Review of the Response
of a Single-Degree-of-Freedom System to Harmonic Excitation

The undamped system

$$m\ddot{x} + kx = F_0 \cos \omega t \qquad x(0) = x_0, \qquad \dot{x}(0) = v_0$$

has the solution

$$x(t) = \frac{v_0}{\omega_n} \sin \omega_n t + \left(x_0 - \frac{f_0}{\omega_n^2 - \omega^2} \right) \cos \omega_n t + \frac{f_0}{\omega_n^2 - \omega^2} \cos \omega t \qquad (2.11)$$

where $f_0 = F_0/m$, and $\omega_n = \sqrt{k/m}$, and x_0 and v_0 are initial conditions. The underdamped system

$$m\ddot{x} + c\dot{x} + kx = F_0 \cos \omega t$$

has the steady-state solution

$$x(t) = \frac{f_0}{\sqrt{(\omega_n^2 - \omega^2)^2 + (2\zeta\omega_n\omega)^2}} \cos \left(\omega t - \tan^{-1} \frac{2\zeta\omega_n\omega}{\omega_n^2 - \omega^2} \right) \qquad (2.36)$$

where the damping ratio ζ satisfies $0 < \zeta < 1$.

for any bounded input, recall that the solution for an arbitrary driving force is given in terms of the impulse response in equation (3.12) to be

$$x(t) = \int_0^t f(\tau)h(t - \tau)d\tau \qquad (3.94)$$

where $f(t) = F(t)/m$. Taking the absolute value of both sides of this expression yields

$$|x(t)| = \left| \int_0^t f(\tau)h(t - \tau)d\tau \right| \le \int_0^t |f(\tau)h(t - \tau)| d\tau \qquad (3.95)$$

where the inequality results from the definition of integrals as a limit of summations. Noting that $|hf| \le |h||f|$ yields

$$|x(t)| \le \int_0^t |f(\tau)||h(t - \tau)| d\tau \le M \int_0^t |h(t - \tau)| d\tau \qquad (3.96)$$

where $f(t)$ [and hence $F(t)$] is assumed to be bounded by M [i.e., $|f(t)| < M$]. Note that the choice of the constant M is arbitrary and is always chosen as a matter of convenience. Next consider evaluating the integral on the right of inequality (3.96) for the underdamped case. The impulse response for an underdamped system is given by equation (3.9). Substitution of equation (3.9) into (3.96) yields

$$|x(t)| \leq M \int_0^t \frac{1}{m\omega_d} \left| e^{-\zeta\omega_n(t-\tau)} \right| \left| \sin \omega_d(t - \tau) \right| d\tau \qquad (3.97)$$

$$\leq \frac{M}{m\omega_d} e^{-\zeta\omega_n t} \int_0^t e^{\zeta\omega_n\tau} d\tau = \frac{M}{m\zeta\omega_n\omega_d}(1 - e^{-\zeta\omega_n t}) \leq M$$

since $\left| \sin \omega_d(t - \tau) \right| \leq 1$ for all $t > 0$, $m\zeta\omega_n\omega_d > 1$ and $1 - e^{-\zeta\omega_n t} < 1$. Thus

$$|x(t)| \leq M$$

Hence as long as the input force is bounded (say, by M) the preceding calculation illustrates that the response $x(t)$ of an underdamped system is also bounded, and the system is BIBO stable.

The results of Section 2.1 clearly indicate that the response of an undamped system is well behaved, or bounded, as long as the harmonic input is *not* at or near the natural frequency (see Window 3.6). In fact, the response given by equation (2.11) illustrates that the maximum magnitude will be less than some constant as long as $\omega \neq \omega_n$. To see this, take the absolute value of equation (2.11), which yields

$$|x(t)| \leq \left| \frac{v_0}{\omega_n} \right| + \left| x_0 - \frac{f_0}{\omega_n^2 - \omega^2} \right| + \left| \frac{f_0}{\omega_n^2 - \omega^2} \right| < M \qquad (3.98)$$

where M is finite since each term is finite as long as $\omega \neq \omega_n$. Here v_0 and x_0 are the initial velocity and displacement, respectively. Thus the undamped forced response is sometimes well behaved and sometimes not. Such systems are said to be *Lagrange stable*. Specifically, a system is defined to be Lagrange stable, or bounded, with respect to a *given* input if the response is bounded for any set of initial conditions. Undamped systems are Lagrange stable with respect to many inputs. This definition is useful when $F(t)$ is known completely or known to fall into a specific class of functions. Both the damped and undamped solutions given in Window 3.6 are Lagrange stable for $\omega_n \neq \omega$.

In general, if the homogeneous solution is asymptotically stable, the forced response will be BIBO stable. If the homogeneous response is stable (marginally stable), the forced response will only be Lagrange stable. The forced response of an unstable homogeneous system can still be BIBO stable, as illustrated in the following example.

Example 3.8.1

Consider the inverted pendulum of Example 1.8.1, illustrated in Figure 3.21, and discuss its stability properties.

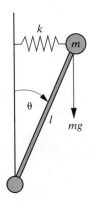

Figure 3.21   The spring-supported inverted pendulum of Example 3.8.1.

Solution   Summing the moments about the pivot point yields

$$\sum M_0 = ml^2\ddot\theta(t) = [-kl\sin\theta(t)][l\cos\theta(t)] + mg[l\sin\theta(t)]$$

Considering the small-angle approximation of the inverted pendulum equation results in the equation of motion

$$ml^2\ddot\theta(t) + kl^2\theta(t) = mgl\theta(t)$$

If $mgl\theta$ is considered to be an applied force, the homogeneous solution is stable since m, l, and k are all positive. Writing the equation of motion as a homogeneous equation yields

$$ml^2\ddot\theta(t) + [kl^2 - mgl]\theta(t) = 0$$

The forced response, however, is not bounded unless $kl > mg$ and was shown in Example 1.8.1 to be divergent (unbounded) in this case. Hence the forced response of this system is Lagrange stable for $F(t) = mgl\theta$ if $kl > mg$, and unbounded (unstable) if $kl < mg$.

□

Example 3.8.2

Consider again the inverted pendulum of Example 3.8.1. Design an applied force $F(t)$ such that the response is bounded for $kl < mg$.

Solution   The problem is to find $F(t)$ such that θ satisfying

$$ml^2\ddot\theta(t) + (kl^2 - mgl)\theta(t) = F(t)$$

is bounded. As a starting point, assume that $F(t)$ has the form

$$F(t) = -a\theta(t) - b\ddot\theta(t)$$

where a and b are to be determined by the design for stability. This form is attractive because it changes the inhomogeneous problem into a homogeneous problem. The equation of motion then becomes

$$ml^2\ddot{\theta}(t) + (kl^2 - mgl)\theta(t) = -a\theta(t) - b\dot{\theta}(t)$$

This can be written as a homogenous equation:

$$ml^2\ddot{\theta}(t) + b\dot{\theta} + (kl^2 - mgl + a)\theta = 0$$

From Section 1.8, it is known that if each of the coefficients is positive, the response is asymptotically stable, which is certainly bounded. Hence, choose $b > 0$ and a such that

$$kl^2 - mgl + a > 0$$

and the forced response will be bounded.

$\square$

An applied force can also cause a stable (or asymptotically stable) system response to become unstable. To see this, consider the system

$$\ddot{x}(t) + \dot{x}(t) + 4x(t) = f(t) \tag{3.99}$$

where $f(t) = ax(t) + b\dot{x}(t)$. If a is chosen to be 2 and $b = 2$, the equation of motion becomes

$$\ddot{x}(t) - \dot{x}(t) + 2x(t) = 0 \tag{3.100}$$

which has a solution that grows exponentially and illustrates flutter instability. At first glance, this seems to violate the earlier statement that asymptotically stable homogeneous systems are BIBO stable. This example, however, does not violate the definition because the input $f(t)$ is not bounded. The applied force is a function of the displacement and velocity, which grow without bound.

3.9 NUMERICAL SIMULATION OF THE RESPONSE

Numerical simulation, as introduced in Section 1.9 for the free response and Section 2.8 for the response to harmonic inputs, can be used to compute and plot the response to any arbitrary forcing function. Numerical simulation has become the preferred method for computing the response, as it requires a minimum amount of analysis and can be applied to any type of input force, including experimental data (such as time histories of earthquakes). Numerical solutions may also be used to check analytical work, and analytical work should be used to check numerical results as often as possible. This section presents some common codes for simulating and plotting the response of systems to a general force.

In the previous sections, great effort was put forth to derive analytical expressions for the response of various single-degree-of-freedom systems driven by a variety of forces. These analytical expressions are extremely useful for design and for understanding some of the physical phenomena. Plots of the time response were constructed to realize the nature and features of the response. Rather than plotting the analytical function describing the response, the time response may also be computed numerically using an Euler or Runge–Kutta integration and computational software packages as introduced in Sections 1.8 and 2.8. While numerical solutions such as these are not exact, they do allow nonlinear terms to be considered, as well as the solution of systems with complicated forcing terms that do not have analytical solutions.

In order to solve for the force response to an arbitrary input numerically, equation (2.133) needs to be modified slightly to incorporate an arbitrary applied force. It is possible to generate an approximate numerical solution of

$$m\ddot{x}(t) + c\dot{x}(t) + kx(t) = F(t) \tag{3.101}$$

subject to any initial conditions for *any* arbitrary force $F(t)$. For most codes, equation (3.101) must be cast into the first-order, or state-space, form by renaming x as x_1 and writing

$$\dot{x}_1(t) = x_2(t)$$
$$\dot{x}_2(t) = -2\zeta\omega_n x_2(t) - \omega_n^2 x_1(t) + f(t) \tag{3.102}$$

where x_2 denotes the velocity $\dot{x}(t)$ and x_1 denotes the position $x(t)$ as before and $f(t) = F(t)/m$. This is subject to the initial conditions $x_1(0) = x_0$ and $x_2(0) = v_0$. Given $\omega_n, \zeta, f_0, x_0, v_0$, and either the analytical or numerical form of $F(t)$, the solution of equation (3.101) can be determined numerically. The matrix form of equation (3.102) becomes

$$\dot{\mathbf{x}}(t) = A\mathbf{x}(t) + \mathbf{f}(t) \tag{3.103}$$

where $\mathbf{x}$ and A are the state vector and state matrix as defined previously in equation (1.96) and repeated here:

$$\mathbf{x} = \begin{bmatrix} x_1 \\ x_2 \end{bmatrix}, \quad \text{and} \quad A = \begin{bmatrix} 0 & 1 \\ -\omega_n^2 & -2\zeta\omega_n \end{bmatrix}$$

The vector $\mathbf{f}$ is the applied force and takes the form

$$\mathbf{f}(t) = \begin{bmatrix} 0 \\ f(t) \end{bmatrix} \tag{3.104}$$

The Euler form of equation (3.103) is

$$\mathbf{x}(t_{i+1}) = \mathbf{x}(t_i) + A\mathbf{x}(t_i)\Delta t + \mathbf{f}(t_i)\Delta t \tag{3.105}$$

In this case, where $\mathbf{f}(t)$ is an arbitrary force, either $\mathbf{f}(t_i)$ is $\mathbf{f}$ evaluated at each time instant, if the analytical form of $\mathbf{f}$ is known, or it is a discrete time history of data points, if $\mathbf{f}$ is known numerically or experimentally. This expression can also be adapted to the Runge–Kutta formulation, and most of the codes mentioned in Appendix G have built-in commands for a Runge–Kutta solution.

The numerical integration to determine the response of a system is an approximation, whereas the plotting of the analytical solution is exact. So the question arises, "Why bother to integrate numerically to find the solution?" The answer lies in the fact that many practical problems do not have exact analytical solutions, such as the treatment of nonlinear terms, as was illustrated in Section 1.10. This section discusses the solution of the forced response using numerical integration in an environment where the exact solution is available for comparison. The examples in this section introduce numerical integration to compute forced responses and to compare them with the exact solution.

The following examples illustrate the use of various programs to compute and plot the solution. Note that VTB1_3 and VTB1_4 allow the use of data points as a forcing function.

Example 3.9.1

Solve for the response of the system in Example 3.2.1, using the parameters given in Figure 3.7 numerically. Recall the equation of motion is

$$m\ddot{x} + c\dot{x} + kx = F(t) = \begin{cases} 0, & t_0 > t > 0 \\ F_0, & t \ge t_0 \end{cases}$$

[Use the values $\zeta = 0.1$ and $\omega_n = 3.16$ rad/s (with $F_0 = 30$ N, $k = 1000$ N/m, $t_0 = 0$).] Compare the numerical solution with the analytical solution given in Example 3.2.1.

Solution The analytical solution given in equation (3.15) with the parameter values given in Figure 3.7 becomes

$$x(t) = (0.03 - 0.03e^{-0.316(t-t_0)}\cos[3.144(t - t_0) - 0.101])\Phi(t - t_0) \quad (3.106)$$

where Φ denotes the Heaviside step function used to indicate that $x(t)$ is zero until t reaches t_0. The state equations for the equation of motion are written as

$$\begin{bmatrix} \dot{x}_1 \\ \dot{x}_2 \end{bmatrix} = \begin{bmatrix} 0 & 1 \\ -\dfrac{k}{m} & -\dfrac{c}{m} \end{bmatrix} \begin{bmatrix} x_1 \\ x_2 \end{bmatrix} + \begin{bmatrix} 0 \\ \dfrac{F_0}{m}\Phi(t - t_0) \end{bmatrix} \quad (3.107)$$

The result of plotting the analytical solution of equation (3.106) and numerically integrating and plotting equation (3.107) is given in Figure 3.22. Note that this is virtually identical to the plot of Figure 3.7, which is the analytical solution. The codes for producing the numerical solution follow.

In Mathcad, the code for solving equation (3.107) is

$$F0 := 30 \quad k := 1000 \quad \omega n := 3.16 \quad \zeta := 0.1 \quad t0 := 0$$

$$\theta := \text{atan}\left[\frac{\zeta}{1 - \zeta^2}\right] \quad \frac{F0}{k} = 0.03 \quad \frac{F0}{k \cdot \sqrt{1 - \zeta^2}} = 0.03015 \quad \zeta \cdot \omega n = 0.316$$

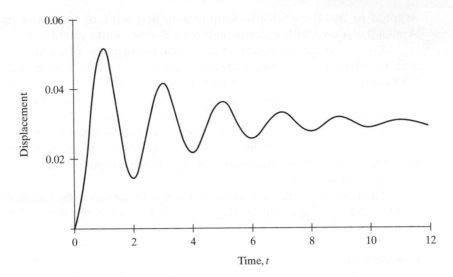

Figure 3.22   A plot of the numerical solution of the system of Examples 3.2.1 and 3.9.1.

$$\omega d := \omega n \cdot \sqrt{1 - \zeta^2} \qquad \omega d = 3.14416 \qquad \theta = 0.101 \qquad m := \left(\frac{k}{\omega n^2}\right)$$

$$xa(t) := \left[\left[\frac{F0}{k} - \frac{F0}{k \cdot \sqrt{1 - \zeta^2}} \cdot e^{-\zeta \cdot \omega n \cdot (t - t0)} \cdot \cos[\omega d \cdot (t - t0) - \theta]\right]\right] \cdot \Phi(t - t0)$$

$$X := \begin{bmatrix} 0 \\ 0 \end{bmatrix} \qquad D(t, X) := \begin{bmatrix} X_1 \\ -2 \cdot \zeta \cdot \omega n \cdot X_1 - \omega n^2 \cdot X_0 + \frac{F0}{m} \cdot \Phi(t - t0) \end{bmatrix}$$

$$Z := \text{rkfixed}(X, 0, 12, 2000, D)$$

$$t := Z^{<0>} \qquad x := \overrightarrow{xa(t)} \qquad xn := Z^{<1>}$$

The MATLAB code for computing the solution and plotting equation (3.107) is

```
clear all
%% Analytical solution

F0=30;  k=1000;  wn=3.16;  zeta=0.1;  t0=0;
theta=atan(zeta/(1-zeta^2));
wd=wn*sqrt(1-zeta^2);
t=0:0.01:12;

Heaviside=stepfun(t,t0); % define Heaviside step
function for 0 < t < 12
```

```
xt=(F0/k-F0/(k*sqrt(1-zeta^2)) * exp(-zeta*wn*(t - t0)).
*cos(wd*(t-t0)-theta)).*Heaviside;

plot(t,xt);'Hold on'

%% Numerical Solution

xo=[0; 0];
ts=[0 12];
[t, x]=ode45('f',ts,xo);
plot(t,x(:,1),'r'); hold off

%---------------------------------------------
function v=f(t,x)

Fo=30; k=1000; wn=3.16; zeta=0.1; to=0; m=k/wn^2;
v=[x(2); x(2).*-2*zeta*wn + x(1).*-wn^2 + Fo/m*stepfun(t, to)];
```

The Mathematica code for computing the solution and plotting equation (3.107) is

```
In[1]:= <<PlotLegends'
```

$$In[2]:= \mathbf{F0} = \mathbf{30};$$
$$\mathbf{k} = \mathbf{1000};$$
$$\omega\mathbf{n} = \mathbf{3.16};$$
$$\zeta = \mathbf{0.1};$$
$$\mathbf{t0} = \mathbf{0};$$
$$\theta = \mathbf{ArcTan} \ [\sqrt{1 - \zeta^2}, \ \zeta];$$
$$\mathbf{m} = \frac{\mathbf{k}}{\omega\mathbf{n}^2} \ ;$$
$$\omega\mathbf{d} = \omega\mathbf{n} \ * \ \sqrt{1 - \zeta^2};$$

$$In[10]:= \mathbf{xanal[t\_]}$$
$$= \left(\frac{\mathbf{F0}}{\mathbf{k}} - \frac{\mathbf{F0}}{\mathbf{k}*\sqrt{1 - \zeta^2}} \ *\mathbf{Exp}[-\zeta*\omega\mathbf{n}*(\mathbf{t} - \mathbf{t0})]*\mathbf{Cos}[\omega\mathbf{d}*(\mathbf{t} - \mathbf{t0})-\theta]\right)$$
$$* \ \mathbf{UnitStep[t - t0]};$$

$$In[11]:= \mathbf{xnumer=NDSolve} \ [\{\mathbf{x''[t]} + 2 * \zeta * \omega\mathbf{n} * \mathbf{x'[t]} + \omega\mathbf{n}^2 * \mathbf{x[t]}$$
$$== \frac{\mathbf{F0}}{\mathbf{m}} * \mathbf{UnitStep[t - t0]}, \ \mathbf{x[0]} == 0, \ \mathbf{x'[0]} == 0\}, \ \mathbf{x[t]},$$
$$\{\mathbf{t}, \ 0, \ 12\}];$$

```
Plot[{Evaluate[x[t] /. xnumer], xanal[t]}, {t, 0, 12},
 PlotStyle → {RGBColor[1, 0, 0], RGBColor[0, 1, 0]},
 PlotLegend → {"Numerical", "Analytical"},
 LegendSize → {1, .3}, LegendPosition→ {1, 0}]
```

□

Next, consider computing the response to systems where the force is applied for only a specified amount of time. Such problems are difficult to solve analytically because they require the use of a change of variables in the convolution integral. Alternately, a number of analytical responses must be calculated, as indicated in Example 3.2.2. In the next example, numerical integration is used to solve for the response to these types of forcing functions.

Example 3.9.2

Numerically compute the solution to Example 3.2.2 using the data given in Figure 3.9 ($F_0 = 30$ N, $k = 1000$ N/m, $\zeta = 0.1$, and $\omega_n = 3.16$ rad/s). Plot the result for the two pulse times given in the figure (i.e., for $t_1 = 0.5$ and for $t_1 = 1.5$ s).

Solution   First, write the equation of motion using a Heaviside step function to represent the driving force. The equation of motion written with Heaviside functions to describe the applied force is

$$m\ddot{x}(t) + c\dot{x}(t) + kx(t) = F_0[1 - \Phi(t - t_1)] \qquad (3.108)$$

This can be solved numerically directly in this form using Mathematica or by putting equation (3.108) into state-space form and solving via Runge–Kutta in Mathcad or MATLAB. The response for two different values of t_1 is given in Figure 3.23. The codes follow.

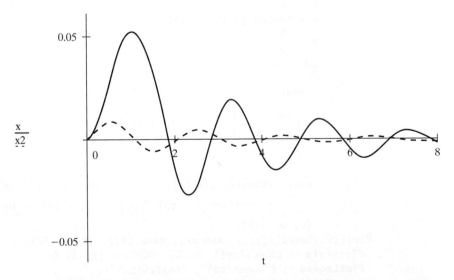

Figure 3.23   The response of a spring–mass–damper system to a square input of pulse duration $t_1 = 0.1$ s (dashed line) and $t_1 = 1.5$ s (solid line), with $k = 1000$ N/m, $m = 100.14$ kg, $\zeta = 0.1$, and $F_0 = 30$ N.

The Mathcad code for solving for the response is

$$t1 := 15 \qquad k := 1000 \qquad \omega n := 3.16 \qquad F0 := 30$$

$$\zeta := 0.1 \qquad m: = \frac{k}{\omega n^2} \qquad m = 100.144$$

$$X: = \begin{bmatrix} 0 \\ 0 \end{bmatrix} \qquad f(t) := F0 - F0 \cdot \Phi(t - t1) \qquad f2(t) := F0 - F0\ \Phi(t - 0.1)$$

$$Y := X$$

$$D(t, X) := \begin{bmatrix} X_1 \\ -\omega n^2 \cdot X_0 - 2\zeta \cdot \omega n \cdot X_1 + \dfrac{f(t)}{m} \end{bmatrix} \qquad D2(t, Y) := \begin{bmatrix} Y_1 \\ -\omega n^2 \cdot Y_0 - 2\zeta \cdot \omega n \cdot Y_1 + \dfrac{f(t)}{m} \end{bmatrix}$$

$$Z := \text{rkfixed}\ (X, 0, 8, 2000, D) \qquad Z2 := \text{rkfixed}\ (Y, 0, 8, 2000, D2)$$

$$t := Z^{<0>} \qquad x := Z^{<1>} \qquad x2 := Z2^{<1>}$$

The MATLAB code for solving for the response is

```
clear all
xo=[0; 0];
ts=[0 8];

[t,x]=ode45('f', ts, xo);
plot(t,x(:,1),'--'); hold on

[t,x]=ode45('f1', ts, xo);
plot(t,x(:,1)); hold off

%-------------------------------------------------
function v=f(t, x)
Fo=30; k=1000; wn=3.16; zeta=0.1; to=0.1; m=k/wn^2;
v=[x(2); x(2).*-2*zeta*wn + x(1).*-wn^2 + Fo/m*(1-stepfun(t, to))];

%-------------------------------------------------
function v=f1(t,x)
Fo=30; k=1000; wn=3.16; zeta=0.1; to=1.5; m=k/wn^2;
v=[x(2); x(2).*-2*zeta*wn + x(1).*-wn^2 + Fo/m*(1-stepfun(t,to))];
```

The Mathematica code for solving for the response is

```
In[1]:= <<PlotLegends'

In[2]:= FO = 30;
        k = 1000;
        ωn = 3.16;
        ζ = 0.1;
        t1 = 1.5;
```

```
                    t2 = 0.1;

                         k
                    m = ───;
                        ωn²

In[9]:= xpointone =

        NDSolve [{x1''[t] + 2 * ζ * ωn * x1'[t] + ωn² * x1[t] == FO
                                                                   ──
                                                                   m
              * (1 - UnitStep[t - t1]), x1[0] == 0, x1'[0] == 0}, x[t],
              {t, 0, 8}]
        xonepointfive =

        NDSolve [{x2''[t] + 2 * ζ * ωn * x2'[t] + ωn² * x1[t] == FO
                                                                   ──
                                                                   m
              * (1 - UnitStep[t - t2]), x2[0] == 0, x2'[0] == 0}, x2[t],
              {t, 0, 8}];
        Plot[{Evaluate[x1[t] /. xpointone], Evaluate[x2[t] /.
              xonepointfive]}, {t, 0, 8},
        PlotStyle → {RGBColor[1, 0, 0], RGBColor[0, 1, 0]},
              PlotRange → {-.06, .06},
        PlotLegend → {"t1=1.5", "t2=0.1"}, LegendSize → {1, .3},
              LegendPosition → {1, 0}];
```
□

Example 3.9.3

Consider the base-excitation problem of Example 3.3.2 and compute the response numerically. Compare the result to the analytical solution computed in Example 3.3.2 by plotting both the analytical solution and the numerical solution on the same graph.

Solution   The equation of motion to be solved in this example is

$$\ddot{x}(t) + 10\dot{x}(t) + 1000x(t) = 1.5\cos 3t + 50\sin 3t \qquad (3.109)$$

From Example 3.3.2, the analytical solution is

$$x(t) = 0.09341e^{-5t}\sin(31.225t + 0.1074) + 0.05\cos(3t - 1.571)$$

The equation of motion can be solved numerically in the form given by equation (3.109) directly in Mathematica or by putting equation (3.109) into state-space form and solving via Runge–Kutta in Mathcad or MATLAB. The plots of both the numerical solution and analytical solution are given in Figure 3.24.

The code for solving this system numerically and for plotting the analytical solution in Mathcad follows:

$$x0 := 0.01 \qquad v0 := 3 \qquad \omega b := 3 \qquad m := 1 \qquad c := 10 \qquad t0 := 0$$

$$k := 1000 \qquad \omega n := \sqrt{\frac{k}{m}} \qquad \zeta := \frac{c}{2 \cdot \sqrt{k \cdot m}} \qquad \omega n = 31.623 \qquad Y := 0.05$$

$$\omega d := \omega n \sqrt{1 - \zeta^2} \qquad \omega d = 31.22499 \qquad \theta := 1.53$$

$$A := 0.09341 \qquad \phi := -0.1074$$

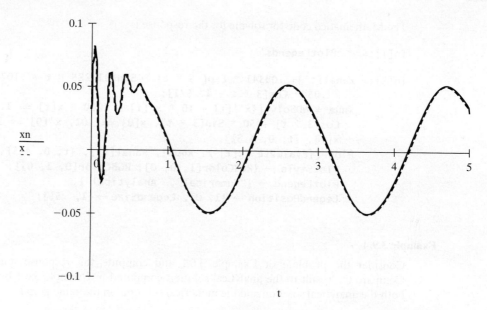

Figure 3.24 The numerical solution (solid line) and analytical solution (dashed line) for the system of Example 3.3.2 indicating almost perfect agreement.

$$xa(t) := A \cdot e^{-\zeta \cdot \omega n \cdot t} \sin(\omega d \cdot t - \phi) + 0.05 \cdot \cos(3 \cdot t - \theta) \qquad X := \begin{bmatrix} x0 \\ v0 \end{bmatrix}$$

$$D(t, X) := \begin{bmatrix} X_1 \\ -2 \cdot \zeta \cdot \omega n \cdot X_1 - \omega n^2 \cdot X_0 + \dfrac{c}{m} \cdot Y \cdot \omega b \cdot \cos(\omega b \cdot t) + \dfrac{k}{m} \cdot Y \cdot \sin(\omega b \cdot t) \end{bmatrix}$$

$$Z := rkfixed\ (X,\ 0,\ 6,\ 2000,\ D)$$

$$x := \overrightarrow{xa(t)} \qquad t := Z^{<0>} \qquad xn := Z^{<1>}$$

The MATLAB code for solving for the response is

```
clear all
%% Analytical solution
t=0:0.01:5;
xt=0.09341*exp(-5*t).*sin(31.225*t+0.1074)+0.05*cos(3*t-1.571);
plot(t, xt,'--');'hold on'

%% Numerical Solution
xo=[0.01; 3];
ts=[0 5];

[t, x]=ode45('f', ts, xo);plot(t, x(:,1)); hold off

%------------------------------------------------
function v=f(t, x)
m=1; c=10; k=1000; wb=3; wn=sqrt(k/m); zeta=c/2*sqrt(m*k);
wd=wn*sqrt(1-zeta^2); Y=0.05;
v=[x(2); x(2).*-2*zeta*wn + x(1).*-wn^2 + c/m*Y*wb*cos(wb*t) +...
k/m*Y*sin(wb*t)];
```

The Mathematica code for solving for the response is

```
In[1]:= <<PlotLegends'

In[2]:= xanal[t_]= .09341 * Exp[-5 * t] * Sin[31.225 * t + .1074]
           + .05 * Cos[3 * t - 1. 571];
         xnum = NDSolve[{x''[t] + 10 * x'[t] + 1000 * x[t] == 1.5 *
           Cos[3 * t] + 50 * Sin[3 * t], x[0] == .01, x'[0] == 3},
           x[t], {t, 0, 4.5}];
         Plot [{Evaluate [x[t] /. xnum], xanal[t]}, {t, 0, 4.5},
           PlotStyle → {RGBColor[1, 0, 0], RGBColor[0, 1, 0]},
           PlotLegend → {"Numerical", "Analytical"},
           LegendPosition → {1, 0}, LegendSize → {1, .5}];
```

□

Example 3.9.4

Consider the problem of Example 3.6.1 and compute the response numerically. Compare the result to the analytical solution computed in Example 3.6.1 by plotting both the analytical solution and the numerical solution on the same graph.

Solution The equation of motion to be solved in this example is

$$10\ddot{x}(t) + 10x(t) = \frac{t}{t_1} - \left(\frac{t - t_1}{t_1}\right)\Phi(t - t_1) \qquad (3.110)$$

where Φ is the Heaviside step function used to "turn on" the second term in the forcing function at $t > t_1$. The parameter t_1 is used to control how steep the disturbance is. See Figure 3.16 for a plot of the driving force. The analytical solution is given in equation (3.78) to be

$$x(t) = \frac{F_0}{k}\left(\frac{t}{t_1} - \frac{\sin\omega_n t}{\omega_n t_1}\right) - \frac{F_0}{k}\left(\frac{t - t_1}{t_1} - \frac{\sin\omega_n(t - t_1)}{\omega_n t_1}\right)\Phi(t - t_1) \qquad (3.111)$$

Again, the equation of motion can be solved numerically in the form given by equation (3.110) directly in Mathematica or by putting equation (3.110) into state-space form and solving Runge–Kutta in Mathcad or MATLAB. The plots of both the numerical solution and analytical solution are given in Figure 3.25.

The Mathcad code for this solution follows:

$$x0 := 0 \qquad v0 := 0 \qquad m := 10 \qquad t1 := 1$$

$$k := 10 \qquad \omega n := \sqrt{\frac{k}{m}} \qquad \omega n = 1 \qquad F0 := 1$$

$$tp := \frac{1}{\omega n} \cdot atan\left(\frac{1 - \cos(\omega n \cdot t1)}{\sin(\omega n \cdot t1)}\right) \qquad \omega n \cdot tp = 0.5 \qquad tp = 0.5$$

$$xm(t) := \frac{F0}{k} \cdot \left[1 + \frac{1}{\omega n \cdot t1} \cdot \sqrt{2 \cdot (1 - \cos(\omega n \cdot t1))}\right]$$

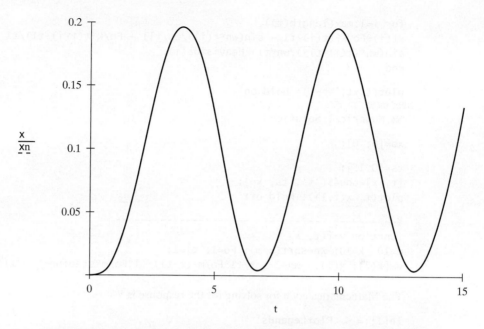

Figure 3.25   The numerical solution (solid line) and analytical solution (dashed line) for the system of Example 3.6.1 indicating almost perfect agreement.

$$xa(t) \; := \; \frac{F0}{k} \cdot \left(\frac{t}{t1} - \frac{\sin \, (\omega n \cdot t)}{\omega n \cdot t1} \right) - \frac{F0}{k} \left[\frac{t - t1}{t1} - \frac{\sin[\omega n \cdot (t - t1)]}{\omega n \cdot t1} \right] \Phi(t - t1)$$

$$f(t) \; := \; \frac{t}{t1} \cdot \frac{F0}{m} - \frac{t - t1}{t1} \cdot \frac{F0}{m} \cdot \Phi(t - t_1) \qquad X \; := \; \begin{bmatrix} 0 \\ 0 \end{bmatrix}$$

$$D(t, \, X) \; := \; \begin{bmatrix} X_1 \\ -(\omega n^2 \cdot X_0) + f(t) \end{bmatrix} \qquad xm(0) \; = \; 0.196$$

$$Z \; := \; \text{rkfixed} \; (X, \, 0, \, 15, \, 2000, \, D)$$

$$t \; := \; Z^{<0>} \qquad x \; := \; \overrightarrow{xa(t)} \qquad xn \; := \; Z^{<1>} \qquad F \; := \; \overrightarrow{f(t)} \qquad Xmax \; := \; \overrightarrow{xm(t)}$$

The MATLAB code for solving for the response is

```
clear all

%% analytical solution
t=0:0.01:15;

m=10; k=10; Fo=1; t1=1;
wn=sqrt(k/m);

Heaviside=stepfun(t, t1);% define Heaviside Step function for 0<t<15
```

```
for i=1:max(length(t)),
xt(i)=Fo/k*(t(i)/t1 - sin(wn*t(i))/wn/t1) - Fo/k*((t(i)-t1)/t1 -
sin(wn*(t(i)-t1))/wn*t1)*Heaviside(i);
end

plot(t,xt,'- -'); hold on

%% Numerical Solution

xo=[0; 0];

ts=[0 15];
[t, x]=ode45('f', ts, xo);
plot(t, x(:,1)); hold off

%---------------------------------------------
function v=f(t, x)
m=10; k=10; wn=sqrt(k/m); Fo=1; t1=1;
v=[x(2); x(1).*-wn^2 + t/t1*Fo/m-(t-t1)/t1*Fo/m*stepfun(t, t1)];
```

The Mathematica code for solving for the response is

```
In[1]:= <<PlotLegends'
```

$$In[2]:= x0 = 0;$$
$$v0 = 0;$$
$$m = 10;$$
$$k = 10;$$
$$\omega n = \sqrt{\frac{k}{m}};$$
$$t1 = 1;$$
$$F0 = 1;$$
$$tp = 0.5;$$

$$In[10]:= xanal[t\_] = \frac{F0}{k}*\left(\frac{t}{t1} - \frac{Sin[\omega n*t]}{\omega n*t1}\right) - \frac{F0}{k}*\left(\frac{t - t1}{t1} - \frac{Sin[\omega n*(t - t1)]}{\omega n*t1}\right)$$

$$* \; UnitStep[t - t1];$$

$$xnum = NDSolve[\{10 * x''[t] + 10 * x[t] == \frac{t}{t1} - \left(\frac{t - t1}{t1}\right)$$

$$* \; UnitStep[t - t1], \; x[0] == x0, \; x'[0] == v0\}, \; x[t],$$
$$\{t, \; 0, \; 15\}];$$
$$Plot[\{Evaluate[x[t] \; /. \; xnum], \; xanal[t]\}, \; \{t, \; 0, \; 15\},$$
$$PlotStyle \to \{RGBColor[1, \; 0, \; 0], \; RBColor[0, \; 1, \; 0]\},$$
$$PlotLegend \to \{"Numerical", \; "Analytical"\},$$
$$LegendPosition \to \{1, \; 0\}, \; LegendSize \to \{1, \; 0.5\}];$$

□

Using numerical integration, the response of a system to a variety of different forcing functions may be computed relatively easily. Furthermore, once the solution is programmed, it is a trivial matter to change parameters and solve the system again. By visualizing the response through simple plots, one can gain the ability to design and understand the system's dynamic behavior.

3.10 NONLINEAR RESPONSE PROPERTIES

The use of numerical integration as introduced in the previous section allows us to consider the effects of various nonlinear terms in the equation of motion. As noted in the free-response case discussed in Section 1.10 and the response to a harmonic load discussed in Section 2.9, the introduction of nonlinear terms results in an inability to find exact solutions, so we must rely on numerical integration and qualitative analysis to understand the response. Several important differences between linear and nonlinear systems are outlined in Section 2.9. In particular, when working with nonlinear systems, it is important to remember that a nonlinear system has more than one equilibrium point and each may be either stable or unstable. Furthermore, we cannot use the idea of superposition, used in all of the previous sections of this chapter, in a nonlinear system.

Many of the nonlinear phenomena are very complex and require analysis skills beyond the scope of a first course in vibration. However, some initial understanding of nonlinear effects in vibration analysis can be observed by using the numerical solutions covered in the previous section. In this section, several simulations of the response of nonlinear systems are numerically computed and compared to their linear counterparts.

Recall from Section 2.9 that if the equations of motion are nonlinear, the general single-degree-of-freedom system may be written as

$$\ddot{x}(t) + f(x(t), \dot{x}(t)) = F(t) \tag{3.112}$$

where the function f can take on any form, linear or nonlinear, and the forcing term $F(t)$ can be almost anything (periodic or not), each term of which has been divided by the mass. Formulating this last expression into the state space, or first-order, equation (3.112) takes on the form

$$\dot{x}_1(t) = x_2(t)$$
$$\dot{x}_2(t) = -f(x_1, x_2) + F(t) \tag{3.113}$$

This state-space form of the equation is used for numerical simulation in several of the codes. By defining the state vector, $\mathbf{x} = [x_1(t)\ x_2(t)]^T$, used in equation (3.113) and the nonlinear vector function $\mathbf{F}$ as

$$\mathbf{F}(\mathbf{x}) = \begin{bmatrix} x_2(t) \\ -f(x_1, x_2) \end{bmatrix} \tag{2.140}$$

equations (3.113) may now be written in the first-order vector form

$$\dot{\mathbf{x}} = \mathbf{F}(\mathbf{x}) + \mathbf{f}(t) \tag{3.114}$$

Equation (3.114) is the forced version of equation (1.115). Here $\mathbf{f}(t)$ is simply

$$\mathbf{f}(t) = \begin{bmatrix} 0 \\ F(t) \end{bmatrix} \tag{3.115}$$

Then the Euler integration method for the equations of motion in the first-order form becomes

$$\mathbf{x}(t_{i+1}) = \mathbf{x}(t_i) + \mathbf{F}(\mathbf{x}(t_i))\Delta t + \mathbf{f}(t_i)\Delta t \tag{3.116}$$

This expression forms a basic approach to integrating numerically to compute the forced response of a nonlinear system and is the nonlinear, general forced-response version of equations (1.98) and (2.134). This is basically identical to equation (2.143) except for the interpretation of the force $\mathbf{f}$.

    Nonlinear systems are difficult to analyze numerically as well as analytically. For this reason, the results of a numerical simulation must be examined carefully. In fact, use of a more sophisticated integration method, such as Runge–Kutta, is recommended for nonlinear systems. In addition, checks on the numerical results using qualitative behavior should also be performed whenever possible.

    In the following examples, consider the single-degree-of-freedom system illustrated in Figure 3.26, with a nonlinear spring element subject to a general driving force. A series of examples are presented using numerical simulation to examine the behavior of nonlinear systems and to compare them to the corresponding linear systems.

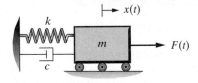

Figure 3.26 A spring–mass–damper system with potentially nonlinear elements and general applied force.

Example 3.10.1

    Compute the response of the system in Figure 3.26 for the case that the damping is linear viscous, the spring is a nonlinear "hardening" spring of the form

$$k(x) = kx + k_1 x^3 \tag{3.117}$$

and the system is subject to an applied excitation of the form

$$F(t) = 1500[\Phi(t - t_1) - \Phi(t - t_2)]\text{N} \tag{3.118}$$

and initial conditions of $x_0 = 0.01$ m and $v_0 = 1$ m/s. Here Φ denotes the Heaviside step function and the times $t_1 = 1.5$ s and $t_2 = 5$ s. This driving function is plotted in

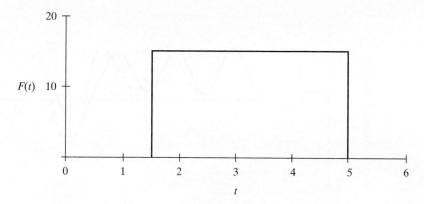

Figure 3.27   The pulse input function defined by equation (3.118).

Figure 3.27. The system has a mass of 100 kg, a damping coefficient of 20 kg/s, and a linear stiffness coefficient of 2000 N/m. The value of k_1 is taken to be 300 N/m^3. Compute the solution and compare it to the linear solution ($k_1 = 0$).

Solution   Summing forces in the horizontal direction, the equation of motion becomes

$$m\ddot{x}(t) + c\dot{x}(t) + kx(t) + k_1 x^3(t) = 1500[\Phi(t - t_1) - \Phi(t - t_2)]$$

Dividing by the mass yields

$$\ddot{x}(t) + 2\zeta\omega_n\dot{x}(t) + \omega_n^2 x(t) + \alpha x^3(t) = 15[\Phi(t - t_1) - \Phi(t - t_2)]$$

Next write this equation in state-space form to get

$$\dot{x}_1(t) = x_2(t)$$
$$\dot{x}_2(t) = -2\zeta\omega_n x_2(t) - \omega_n^2 x_1(t) - \alpha x_1^3(t) + 15[\Phi(t - t_1) - \Phi(t - t_2)]$$

This last set of equations can be used in MATLAB or Mathcad to integrate numerically for the time response. Mathematica uses the second-order equation directly. Figure 3.28 illustrates the response to both the linear and nonlinear system. The solid line is the response of the nonlinear system while the dashed line is the response of the linear system. The difference between linear and nonlinear systems is that, in this case, the nonlinear spring has smaller response amplitude than the linear system does. This is useful in design as it illustrates that the use of a hardening spring reduces the amplitude of vibration to a shock type of input.

One possibility for designing a nonlinear isolation spring is to use the numerical codes listed later in this example to vary parameters (damping, mass, and stiffness) until a desired response is obtained.

It is important to remember, however, that if designing with a nonlinear element, new equilibriums are introduced that may be unstable. Hence care must be taken to assure that additional difficulties are not introduced when using a nonlinear spring to reduce the response. The following codes can be used to generate and plot the solution just given.

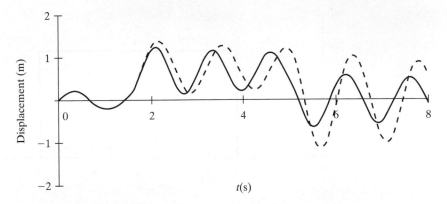

Figure 3.28 The response of the system of Figure 3.26 to the input force given in Figure 3.27. The solid line is the response of the nonlinear system while the dashed line is the response of the linear system.

In Mathcad, the code is

$$x0 := 0.01 \qquad v0 := 1 \qquad m := 100 \qquad k := 2000 \qquad c := 20$$

$$\alpha := 3 \qquad F0 := 1500 \qquad t1 := 1.5 \qquad t2 := 5$$

$$\omega n := \sqrt{\frac{k}{m}} \qquad \zeta := \frac{c}{2\sqrt{k \cdot m}} \qquad f0 := \frac{F0}{m} \qquad \zeta = 0.022$$

$$X := \begin{bmatrix} x0 \\ v0 \end{bmatrix} \qquad Y := X \qquad f(t) := f0 \cdot \Phi(t - t1) - f0 \cdot \Phi(t - t2)$$

$$D(t, X) := \begin{bmatrix} X_1 \\ -2 \cdot \zeta \cdot \omega n \cdot X_1 - \omega n^2 \cdot X_0 + [-\alpha \cdot (X_0)^3 + f(t)] \end{bmatrix}$$

$$L(t, Y) := \begin{bmatrix} Y_1 \\ -2 \cdot \zeta \cdot \omega n \cdot Y_1 - \omega n^2 \cdot Y_0) + f(t) \end{bmatrix}$$

$$Z := \text{rkfixed}(X, 0, 10, 2000, D) \qquad W := \text{rkfixed}(Y, 0, 10, 2000, L)$$

$$t := Z^{<0>} \qquad xs := Z^{<1>} \qquad xL := W^{<1>}$$

The MATLAB code is

```
clear all
xo=[0.01; 1];
ts=[0 8];

[t,x]=ode45('f',ts,xo);
plot(t, x(:,1)); hold on        % The response of nonlinear system
[t,x]=ode45('f1',ts,xo);
plot(t,x(:,1),'--'); hold off      % The response of linear system
```

```
%----------------------------------------------
function v=f(t,x)
m=100; k=2000; c=20; wn=sqrt(k/m); zeta=c/2/sqrt(m*k); Fo=1500;
alpha=3; t1=1.5; t2=5;
v=[x(2); x(2).*-2*zeta*wn + x(1).*-wn^2 - x(1)^3.*alpha+...
Fo/m*(stepfun(t,t1)-stepfun(t,t2))];


%----------------------------------------------
function v=f1(t,x)
m=100; k=2000; c=20; wn=sqrt(k/m); zeta=c/2/sqrt(m*k); Fo=1500;
alpha=0; t1=1.5; t2=5;
v=[x(2); x(2).*-2*zeta*wn + x(1).*-wn^2 - x(1)^3.*alpha+...
Fo/m*(stepfun(t,t1)-stepfun(t,t2))];
```

The Mathematica code is

```
In[1]:= <<PlotLegends`

In[2]:= x0 = .01;
        v0 = 1;
        m = 100;
        k = 2000;
        k1 = 300;
        c = 20;
```

$$\omega n = \sqrt{\frac{k}{m}};$$

$$\alpha = \frac{k1}{m};$$

```
        t1 = 1.5;
        t2 = 5;
        F0 = 1500;
```

$$f0 = \frac{F0}{m};$$

$$\zeta = \frac{c}{2*\sqrt{k*m}};$$

```
In[21]:= xlin = NDSolve [{x1''[t] + 2 * ζ * ωn * x1'[t] + ωn^2
            * x1[t] == 15 * (UnitStep[t - t1] - UnitStep[t - t2]),
            x1[0] == x0, x1'[0] == v0}, x1[t], {t, 0, 8}];
        xnon1 =
          NDSolve [{xn1''[t] + 2 * ζ * ωn * xn1'[t] + ωn^2 * xn1[t]
            + α * (xn1[t])^3 == 15 * (UnitStep[t - t1]
            - UnitStep[t - t2]), xn1[0] == x0, xn1'[0] == v0}, xn1[t],
            {t, 0, 8}, Method → "ExplicitRungeKutta"];
        Plot[{Evaluate[x1[t] /. xlin], Evaluate[xn1[t] /. xnon1]},
            {t, 0, 8}, PlotRange → {-2, 2},
```

```
PlotStyle → {RGBColor[1, 0, 0], RGBColor[0, 1, 0]},
PlotLegend → {"Linear", "NonLinear"},
LegendPosition → {1, 0}, LegendSize → {1, .5}]
```

□

Example 3.10.2

Compare the forced response of a system with velocity-squared damping, as defined in equation (2.129) using numerical simulation of the nonlinear equation, to that of the response of the linear system obtained using equivalent viscous damping, as defined by equation (2.131), where the input force is given by

$$F(t) = 150[\Phi(t - t_1) - \Phi(t - t_2)]\mathrm{m}$$

and initial conditions are $x_0 = 0$ and $v_0 = 1$ m/s. Here Φ denotes the Heaviside step function and the times $t_1 = 0.5$ s and $t_2 = 2$ s.

Solution This is essentially the same as Example 2.9.2 except that the driving force is a pulse here rather than harmonic as in Section 2.9. Velocity-squared damping with a linear spring and pulsed input is described by

$$m\ddot{x} + \alpha \, \mathrm{sgn}(\dot{x})\dot{x}^2 + kx = 150[\Phi(t - t_1) - \Phi(t - t_2)]$$

The equivalent viscous-damping coefficient is calculated in equation (2.131) to be

$$c_{eq} = \frac{8}{3\pi} \alpha \omega X$$

This value assumes that the motion is harmonic, which is somewhat violated here. The value of the magnitude, X, can be approximated for near resonance conditions. The value is computed in Example 2.9.2 to be

$$X = \sqrt{\frac{3\pi m f_0}{8\alpha \omega^2}}$$

Combining these last two expressions yields an equivalent viscous-damping value of

$$c_{eq} = \sqrt{\frac{8m\alpha f_0}{3\pi}}$$

Using this value as the damping coefficient results in a linear system that approximates equation (2.131). Figure 3.29 illustrates the linear and nonlinear response for the values $m = 10$ kg, $k = 200$ N/m, and $\alpha = 5$. Note that the linear response underestimates the actual response maximum by about 10%. For large times, the response of the linear system dies out much sooner than that of the nonlinear system, indicating a large error.

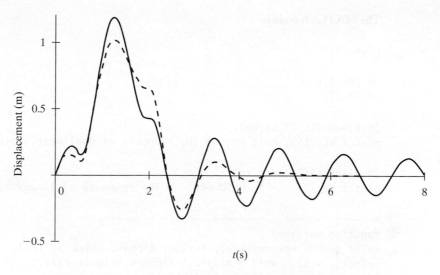

Figure 3.29   The response of a nonlinear system (solid line) and a linear system (dashed line) formed by using the concept of equivalent viscous damping.

The computer codes for generating the solutions and plotting Figure 3.29 are given next.

The Mathcad code is

$$x0 := 0 \qquad v0 := 1 \qquad m := 10 \qquad k := 200 \qquad \alpha := 5 \qquad F0 := 150$$

$$\omega n := \sqrt{\frac{k}{m}} \qquad f0 := \frac{F0}{m}$$

$$t1 := 0.5 \qquad t2 := 2 \qquad ceq := \sqrt{\frac{8 \cdot \alpha \cdot m}{3 \cdot \pi}} \cdot f0$$

$$X := \begin{bmatrix} x0 \\ v0 \end{bmatrix} \qquad Y := X \qquad \zeta := \frac{ceq}{2\sqrt{k \cdot m}} \qquad f(t) := f0 \cdot \Phi(t - t1) - f0 \cdot \Phi(t - t2)$$

$$D(t, X) := \begin{bmatrix} X_1 \\ -\omega n^2 \cdot X_0 - \frac{\alpha}{m} \cdot (X1)^2 \cdot \frac{X_1}{|X_1|} + f(t) \end{bmatrix}$$

$$L(t, Y) := \begin{bmatrix} Y_1 \\ (-2 \cdot \zeta \cdot \omega n \cdot Y_1 - \omega n^2 \cdot Y_0) + f(t) \end{bmatrix}$$

$$Z := rkfixed\ (X,\ 0,\ 20,\ 2000,\ D) \qquad W := rkfixed\ (Y,\ 0,\ 20,\ 2000,\ L)$$

$$t := Z^{<0>} \qquad x := Z^{<1>} \qquad xL := W^{<1>}$$

The MATLAB code is

```
clear all

xo=[0; 1];
ts=[0 8];

[t,t]=ode45('f',ts,xo);
plot(t,x(:,1)); hold on    % The response of nonlinear system

[t,x]=ode45('f1',ts,xo);
plot(t,x(:,1),'--'); hold off    % The response of linear system

%---------------------------------------------
function v=f(t,x)
m=10; k=200; wn=sqrt(k/m); Fo=150; alpha=5; t1=0.5; t2=2;
v=[x(2); x(1).*-wn^2 + x(2)^2.*-alpha/m.*sign(x(2))+...
Fo/m*(stepfun(t, t1)-stepfun(t,t2))];

%---------------------------------------------
function v=f1(t,x)
m=10; k=200; wn=sqrt(k/m); Fo=150; alpha=5; t1=0.5; t2=2;
ceq=sqrt(8*alpha*m/3/pi*Fo/m); zeta=ceq/2/sqrt(m*k);
v=[x(2); x(2).*-2*zeta*wn + x(1).*-wn^2+...
Fo/m*(stepfun(t,t1)-stepfun(t,t2))];
```

The Mathematica code is

In[1]:=<<**PlotLegends'**

In[2]:= **x0 = 0; v0 = 1; m = 10; k = 200; α =5; F0 = 150; f0 = $\frac{\text{F0}}{\text{m}}$;**

$\qquad$ **ωn = $\sqrt{\dfrac{\text{k}}{\text{m}}}$; t1 = 0.5; t2 = 2;**

$\qquad$ **ceq = $\sqrt{\dfrac{8*\alpha*\text{m}}{3*\pi}}$*f0;**

$\qquad$ **ζ = $\dfrac{\text{ceq}}{2*\sqrt{\text{k}*\text{m}}}$;**

In[6]:= **xnonlin = NDSolve[{x''[t] + $\dfrac{\alpha}{\text{m}}$ * Sign[x'[t]]**

$\qquad$ *** (x'[t])^2 + wn^2 * x[t] == f0 * (UnitStep[t - t1]**

$\qquad$ **- UnitStep[t - t2]), x[0] == 0, x'[0] == 1},**

$\qquad$ **x[t], {t, 0, 20}];**

```
In[7]:= xlin = NDSolve[{x''[t] + 2 * ζ * ωn * x'[t] + ωn^2
        * x[t] == f0 * (UnitStep[t - t1] - UnitStep[t - t2]),
        x[0] == 0, x'[0] == 1}, x[t], {t, 0, 20}];

In[8]:= Plot[{x[t]/.xnonlin, x[t]/.xlin},{t, 0, 8},
        PlotStyle → {GrayLevel [0], Dashing[{.03}]}]
```

□

PROBLEMS

Those problems marked with an asterisk are intended to be solved using computational software.

Section 3.1 (Problems 3.1 through 3.17)

3.1. Calculate the solution to

$$1000\ddot{x}(t) + 200\dot{x}(t) + 2000x(t) = 100\delta(t), x_0 = 0, v_0 = 0$$

3.2. Consider a spring-mass-damper system with $m = 2\,\text{kg}, c = 2\,\text{kg/s}$ and $k = 3000\,\text{N/m}$ with an impulsive force applied to it of 10,000 N for 0.01 s. Compute the resulting response.

3.3. Calculate the solution to

$$\ddot{x} + 2\dot{x} + 4x = \delta(t - \pi)$$

$$x(0) = 0 \quad \dot{x}(0) = 1$$

and plot the response.

3.4. Calculate the solution to

$$\ddot{x} + 2\dot{x} + 3x = \sin t + \delta(t - \pi)$$

$$x(0) = 0 \quad \dot{x}(0) = 1$$

and plot the response.

3.5. Calculate the response of a critically damped system to a unit impulse.

3.6. Calculate the response of an overdamped system to a unit impulse.

3.7. Derive equation (3.6) from equations (1.36) and (1.38).

3.8. Consider a simple model of an airplane wing given in Figure P3.8. The wing is approximated as vibrating back and forth in its plane, massless compared to the missile

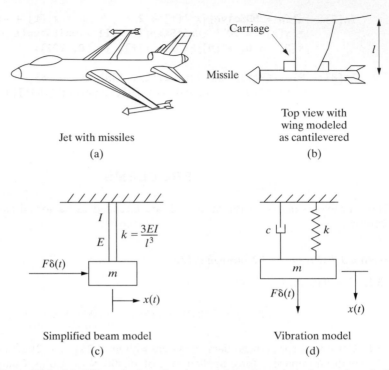

Figure P3.8 Modeling of wing vibration resulting from the release of a missile. Figure (a) is the system of interest; (b) is the simplification of the detail of interest; (c) is a crude model of the wing: a cantilevered beam section (recall Figure 1.26); and (d) is the vibration model used to calculate the response neglecting the mass of the wing.

carriage system (of mass m). The modulus and the moment of inertia of the wing are approximated by E and I, respectively, and l is the length of the wing. The wing is modeled as a simple cantilever for the purpose of estimating the vibration resulting from the release of the missile, which is approximated by the impulse function $F\delta(t)$. Calculate the response and plot your results for the case of an aluminum wing 3 m long with $m = 2000\,\text{kg}$, $\zeta = 0.02$, and $I = 1.0m^4$. Model F as 2000 N lasting for 0.1s.

3.9. A cam in a large machine can be modeled as applying a 15,000 N-force over an interval of 0.01 s. This can strike a valve that is modeled as having physical parameters: $m = 10\,\text{kg}$, $c = 20\,\text{Ns/m}$, and stiffness $k = 8000\,\text{N/m}$. The cam strikes the valve once every 1 s. Calculate the vibration response, $x(t)$, of the valve once it has been impacted by the cam. The valve is considered to be closed if the distance between its rest position and its actual position is less than 0.0001 m. Is the valve closed the very next time it is hit by the cam?

3.10. The vibration of a package dropped from a height of h meters can be approximated by considering Figure P3.10 and modeling the point of contact as an impulse applied to the system at the time of contact. Calculate the vibration of the mass m after the system falls and hits the ground. Assume that the system is underdamped.

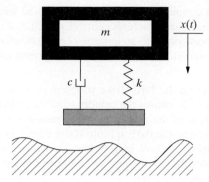

Figure P3.10 The vibration model of a package being dropped onto the ground.

3.11. Calculate the response of

$$4\ddot{x}(t) + 16\dot{x}(t) + 16x(t) = 4\delta(t)$$

for zero initial conditions. The units are in Newtons. Plot the response.

3.12. Compute the response of the system:

$$4\ddot{x}(t) + 16\dot{x}(t) + 16x(t) = 4\delta(t)$$

subject to the initial conditions $x(0) = 0.01\,\text{m}$ and $v(0) = 0$. The units are in Newtons. Plot the response.

3.13. Calculate the response of the system

$$3\ddot{x}(t) + 6\dot{x}(t) + 12x(t) = 3\delta(t) - \delta(t - 1)$$

subject to the initial conditions $x(0) = 0.02\,\text{m}$ and $v(0) = 2\,\text{m/s}$. The units are in Newtons. Plot the response.

3.14. A chassis dynamometer is used to study the unsprung mass of an automobile as illustrated in Figure P3.14, and discussed in Example 1.4.1. Compute the maximum magnitude of the center of the wheel due to an impulse of 10000 N applied over 0.01 seconds in the x direction. Assume the wheel mass is $m = 15\,\text{kg}$, the spring stiffness is $k = 500,000\,\text{N/m}$,

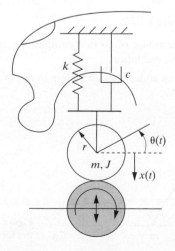

Figure P3.14 A simple model of an automobile suspension system mounted on a chassis dynamometer. The rotation of the car's wheel/tire assembly (of radius r) is given by $\theta(t)$ and is vertical deflection by $x(t)$.

the shock absorber provides a damping ratio of $\zeta = 0.3$, and the rotational inertia is $J = 2.323\,\text{kg}\,\text{m}^2$. Assume that the dynamometer is controlled such that $x = r\theta$. Compute and plot the response of the wheel system to an impulse of 5000 N over 0.01 s. Compare the undamped maximum amplitude to that of the maximum amplitude of the damped system (use $r = 0.457\,\text{m}$).

3.15. Consider the effect of damping on the bird strike problem of Example 3.1.2. Recall from the example that the bird strike causes the camera to vibrate out of limits. Adding damping will cause the magnitude of the response to decrease but may not be able to keep the camera from vibrating past the 0.01 m limit. If the damping in the aluminum is modeled as $\zeta = 0.02$, approximately how long before the camera vibration reduces to the required limit? (*Hint:* plot the time response and note the value for time after which the oscillations remain below 0.01 m).

3.16. Consider the jet engine and mount indicated in Figure P3.16 and model it as a mass on the end of a beam, as done in Figure 1.26. The mass of the engine is usually fixed. Find an expression for the value of the transverse mount stiffness, k, as a function of the relative speed of the bird, v, the bird mass, the mass of the engine, and the maximum displacement that the engine is allowed to vibrate.

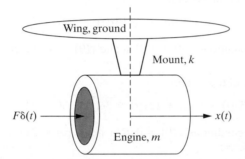

Figure P3.16 A model of a jet engine in transverse vibration due to a bird strike.

3.17. A machine part is regularly subject to a force of 350 N lasting 0.01 seconds, as part of a manufacturing process. Design a damper (i.e., choose a value of the damping constant, c, such that the part does not deflect more that 0.01 m), given that the part has a mass of 100 kg and a stiffness of 1250 N/m.

Section 3.2 (Problems 3.18 through 3.29)

3.18. Calculate the analytical response of an overdamped single-degree-of-freedom system to an arbitrary nonperiodic excitation.

3.19. Calculate the response of an underdamped system to the excitation given in Figure P3.19 where the pulse ends at π s.

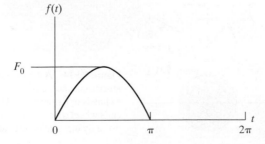

Figure P3.19 Plot of a pulse input of the form $f(t) = F_0 \sin t$.

3.20. Speed bumps are used to force drivers to slow down. Figure P3.20 is a model of a car going over a speed bump. Using the data from Example 2.4.2 and an undamped model of the suspension system (i.e., $k = 4 \times 10^5$ N/m, $m = 1007$ kg), find an expression for the maximum relative deflection of the car's mass versus the velocity of the car. Model the bump as a half sine of length 40 cm and height 20 cm. Note that this is a moving-base problem.

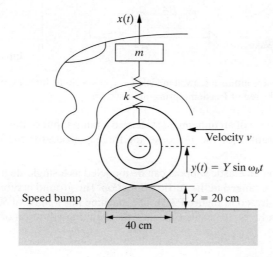

Figure P3.20   Model of a car driving over a speed bump.

3.21. Calculate and plot the response of an undamped system to a step function with a finite rise time of t_1 for the case $m = 1$ kg, $k = 1$ N/m, $t_1 = 4$ s, and $F_0 = 20$ N. This function is described by

$$F(t) = \begin{cases} \dfrac{F_0 t}{t_1} & 0 \le t \le t_1 \\ F_0 & t > t_1 \end{cases}$$

3.22. A wave consisting of the wake from a passing boat impacts a seawall. It is desired to calculate the resulting vibration. Figure P3.22 illustrates the situation and suggests a model. The force in Figure P3.22 can be expressed as

$$F(t) = \begin{cases} F_0\left(1 - \dfrac{t}{t_0}\right) & 0 \le t \le t_0 \\ 0 & t > t_0 \end{cases}$$

Calculate the response of the seawall–dike system to such a load.

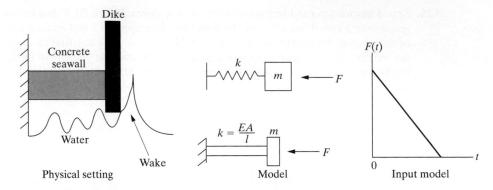

Figure P3.22   A wave hitting a seawall modeled as a nonperiodic force exciting an undamped single-degree-of-freedom, spring–mass system.

3.23. Determine the response of an undamped system to a ramp input of the form $F(t) = F_0 t$, where F_0 is a constant. Plot the response for three periods for the case $m = 2\,\text{kg}$, $k = 200\,\text{N/m}$ and $F_0 = 100\,\text{N}$.

3.24. A machine resting on an elastic support can be modeled as a single-degree-of-freedom, spring–mass system arranged in the vertical direction. The ground is subject to a motion $y(t)$ of the form illustrated in Figure P3.24. The machine has a mass of 5000 kg, and the support has stiffness $1.5 \times 10^3\,\text{N/m}$. Calculate the resulting vibration of the machine.

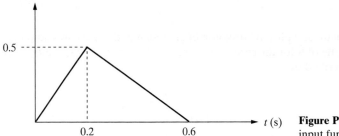

Figure P3.24   Triangular pulse input function.

3.25. Consider the step response described in Figure 3.7 and Example 3.2.1. Calculate the analytical value of t_p by noting that it occurs at the first peak, or critical point, of the curve.

3.26. Calculate the value of the overshoot (O.S.), for the system of Example 3.2.1. Note from the example that the overshoot is defined as occurring at the peak time defined by $t_p = \pi/\omega_d$ and is the difference between the value of the response at t_p and the steady-state response at t_p.

3.27. It is desired to design a system so that its step response has a settling time of 4 s and a time to peak of 1 s. Calculate the appropriate natural frequency and damping ratio to use in the design.

3.28. Plot the response of a spring–mass–damper system to a square input of magnitude $F_0 = 30\,\text{N}$, illustrated in Figure 3.8 of Example 3.2.1, for the case that the pulse width is the natural period of the system (i.e., $t_1 = \pi \omega_n$). Recall that $k = 1000 \, \text{N/m}$, $\zeta = 0.1$, and $\omega_n = 3.16 \, \text{rad/s}$.

3.29. Consider the spring-mass system described by

$$m\ddot{x}(t) + kx(t) = F_0 \sin \omega t, \quad x_0 = 0.01 \, \text{m} \quad \text{and} \quad v_0 = 0$$

Compute the response of this system for the values of $m = 100 \, \text{kg}$, $k = 2000 \, \text{N/m}$, $\omega = 10 \, \text{rad/s}$ and $F_0 = 10 \, \text{N}$, using the convolution integral approach outlined in Example 3.2.4. Check your answer using the results of equation (2.25).

Section 3.3 (Problems 3.30 through 3.38)

3.30. Derive equations (3.24), (3.25), and (3.26) and hence verify the equations for the Fourier coefficient given by equations (3.21), (3.22), and (3.23).

3.31. Calculate b_n from Example 3.3.1 for the triangular force given by

$$F(t) = \begin{cases} \dfrac{4}{T}t - 1 & 0 \le t \le \dfrac{T}{2} \\[2mm] 1 - \dfrac{4}{T}\left(t - \dfrac{T}{2}\right) & \dfrac{T}{2} \le t \le T \end{cases}$$

and show that $b_n = 0$, $n = 1,2,\ldots,\infty$. Also verify the expression a_n by completing the integration indicated. (*Hint*: Change the variable of integration from t to $x = 2\pi nt/T$.)

3.32. Determine the Fourier series for the rectangular wave illustrated in Figure P3.32.

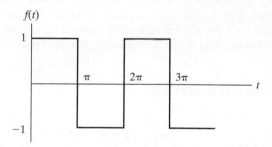

Figure P3.32 A rectangular periodic signal.

3.33. Determine the Fourier series representation of the sawtooth curve illustrated in Figure P3.33.

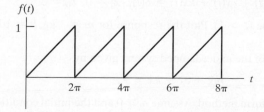

Figure P3.33 A sawtooth periodic signal.

3.34. Calculate and plot the response of the base excitation problem with base motion specified by the velocity

$$\dot{y}(t) = 4e^{-t/2}\phi(t)m/s$$

where $\Phi(t)$ is the unit step function and $m = 10\,\text{kg}$, $\zeta = 0.01$, and $k = 1000\,\text{N/m}$. Assume that the initial conditions are both zero.

3.35. Calculate and plot the total response of the spring-mass-damper system with $m = 80\,\text{kg}$, $\zeta = 0.1$ and $k = 1000\,\text{N/m}$ to the signal of defined by

$$F(t) = \begin{cases} \dfrac{4}{T}t - 1 & 0 \le t \le \dfrac{T}{2} \\[2mm] 1 - \dfrac{4}{T}\left(t - \dfrac{T}{2}\right) & \dfrac{T}{2} \le t \le T \end{cases}$$

with maximum force of 1 N. Assume that the initial conditions are zero and let $T = 2\pi\,\text{s}$.

3.36. Calculate the total response of the system of Example 3.3.2 for the case of a base motion driving frequency of $\omega_b = 1.414\,\text{rad/s}$ with amplitude $Y = 0.05\,\text{m}$ subject to initial conditions $x_0 = 0.01\,\text{m}$ and $v_0 = 3.0\,\text{m/s}$. The system is defined by $m = 1\,\text{kg}$, $c = 10\,\text{kg/s}$, and $k = 1000\,\text{N/m}$.

*3.37. Validate your solution to the square wave Problem 3.32 by calculating a_n and b_n using VTB3_3 in the Vibration Toolbox. Print the function and its Fourier series approximation for 5, 20, then 100 terms. The Toolbox makes this easy. The purpose is to illustrate the Gibbs effect in approximation by Fourier series.

*3.38. Validate your solution to the sawtooth wave of Problem 3.33 by calculating a_n and b_n using VTB3_3 in the Vibration Toolbox. Print the function and its Fourier series approximation for 5, 20, and 100 terms. The Toolbox makes this easy. The purpose is to illustrate the Gibbs effect in approximation by Fourier series.

Section 3.4 (Problems 3.39 through 3.43)

3.39. Calculate the response of

$$m\ddot{x} + c\dot{x} + kx = F_0\Phi(t)$$

where $\phi(t)$ is the unit step function for the case with $x_0 = v_0 = 0$. Use the Laplace transform method and assume that the system is underdamped.

3.40. Using the Laplace transform method, calculate the response of the system

$$m\ddot{x}(t) + c\dot{x}(t) + kx(t) = \delta(t), \quad x_0 = 0, \quad v_0 = 0$$

for the overdamped case ($\zeta > 1$). Plot the response for $m = 1\,\text{kg}$, $k = 100\,\text{N/m}$, and $\zeta = 1.5$.

3.41. Calculate the response of the underdamped system given by

$$m\ddot{x} + c\dot{x} + kx = F_0 e^{-at}$$

using the Laplace transform method. Assume $a > 0$ and the initial conditions are both zero.

3.42. Solve the following system for the response $x(t)$ using Laplace transforms:

$$100\ddot{x}(t) + 2000x(t) = 40\delta(t)$$

where the units are in Newtons and the initial conditions are both zero.

3.43. Use the Laplace transform approach to solve for the response of the spring–mass system with equation of motion and initial conditions given by

$$\ddot{x}(t) + x(t) = \sin 2t, \quad x_0 = 0, \quad v_0 = 1$$

Assume the units are consistent. (*Hint:* See the example in Appendix B.)

Section 3.5 (Problems 3.44 through 3.48)

3.44. Calculate the mean-square response of a system to an input force of constant PSD, S_0, and frequency response function $H(\omega) = 10/(5 + 3j\omega)$.

3.45. Consider the base-excitation problem of Section 2.4 as applied to an automobile model of Example 2.4.2 and illustrated in Figure 2.17. Recall that the model is a spring–mass–damper system with values $m = 1007$ kg, $c = 2000$ kg/s, $k = 40{,}000$ N/m. In this problem let the road have a random stationary cross section producing a PSD of S_0. Calculate the PSD of the response and the mean-square value of the response.

3.46. To obtain a feel for the correlation functions, compute autocorrelation $R_{xx}(\tau)$ for the deterministic signal $A \sin \omega_n t$.

3.47. The autocorrelation of a signal is given by

$$R_{xx}(\tau) = 10 + \frac{4}{3 + 2\tau + 4\tau^2}$$

Compute the mean-square value of the signal.

3.48. Verify that the average $x - \bar{x}$ is zero by using the definition given in equation (3.47) to compute the average.

Section 3.6 (Problems 3.49 through 3.50)

3.49. A power line pole with a transformer is modeled by

$$m\ddot{x} + kx = -\ddot{y}$$

where x and y are as indicated in Figure P3.49. Assuming the initial conditions are zero, calculate the response of the relative displacement $(x - y)$ if the pole is subject to an earthquake-based excitation of

$$\ddot{y}(t) = \begin{cases} A\left(1 - \dfrac{t}{t_0}\right) & 0 \le t \le 2t_0 \\ 0 & t > 2t_0 \end{cases}$$

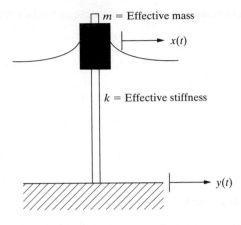

Figure P3.49 A vibration model of a power-line pole with a transformer mounted on it.

3.50. Calculate the response spectrum of an undamped system to the forcing function

$$
F(t) = \begin{cases} F_0 \sin \dfrac{\pi t}{t_1} & 0 \le t \le t_1 \\ 0 & t > t_1 \end{cases}
$$

assuming the initial conditions are zero.

Section 3.7 (Problems 3.51 through 3.58)

3.51. Using complex algebra, derive equation (3.89) from (3.86) with $s = j\omega$.

3.52. Using the plot in Figure P3.52, estimate the system's parameters m, c, and k, as well as the natural frequency.

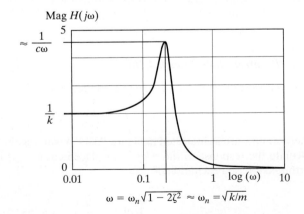

Figure P3.52 The magnitude plot of a spring–mass–damper system.

3.53. From a compliance transfer function of a spring–mass–damper system the stiffness is determined to have a value of 0.5 N/m, a natural frequency of 0.25 rad/s, and a damping coefficient of 0.087 kg/s. Plot the inertance transfer function's magnitude and phase for this system.

3.54. From a compliance transfer function of a spring–mass–damper system the stiffness is determined to have a value of 0.5 N/m, a natural frequency of 0.25 rad/s, and a damping coefficient of 0.087 kg/s. Plot the mobility transfer function's magnitude and phase for the system.

3.55. Calculate the compliance transfer function for a system described by

$$a\frac{d^4x(t)}{dt^4} + b\frac{d^3x(t)}{dt^3} + c\frac{d^2x(t)}{dt^2} + \frac{dx(t)}{dt} + ex(t) = f(t)$$

where $f(t)$ is the input force and $x(t)$ is a displacement.

3.56. Calculate the frequency response function for the compliance for the system defined by

$$a\frac{d^4x(t)}{dt^4} + b\frac{d^3x(t)}{dt^3} + c\frac{d^2x(t)}{dt^2} + \frac{dx(t)}{dt} + ex(t) = f(t).$$

*3.57. Plot the magnitude of the frequency response function for the system of Problem 3.56 for

$$a = 1, \quad b = 4, \quad c = 12, \quad d = 16, \quad \text{and} \quad e = 10.$$

3.58. An experimental (compliance) magnitude plot is illustrated in Fig. P3.58. Determine ω, ζ, c, m, and k. Assume that the units correspond to m/N along the vertical axis.

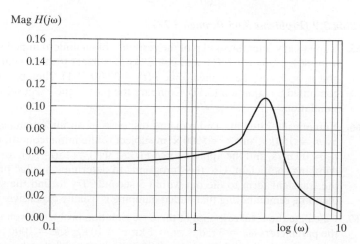

Figure P3.58   An experimentally determined compliance magnitude plot.

Section 3.8 (Problems 3.59 through 3.64)

3.59. Show that a critically damped system is bounded-input, bounded-output stable.

3.60. Show that an overdamped system is bounded-input, bounded-output stable.

3.61. Is the solution of $2\ddot{x} + 18x = 4\cos 2t + \cos t$ Lagrange stable?

3.62. Calculate the response of the system described by

$$\ddot{x}(t) + \dot{x}(t) + 4x(t) = ax(t) + b\dot{x}(t)$$

for $x_0 = 0, v_0 = 1$ for the case that $a = 4$ and $b = 0$. Is the response bounded?

3.63. A crude model of an aircraft wing can be modeled as

$$100\ddot{x}(t) + 25\dot{x}(t) + 2000x(t) = a\dot{x}(t)$$

Here the factor a is determined by the aerodynamics of the wing and is proportional to the air speed. At what value of the parameter a will the system start to flutter?

3.64. Consider the inverted pendulum of Figure P3.64 and compute the value of the stiffness k that will keep the linear system stable. Assume that the pendulum rod is massless.

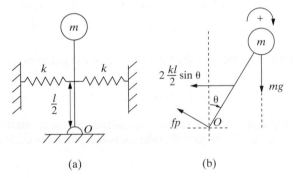

Figure P3.64 An inverted pendulum.

(a)   (b)

Section 3.9 (Problems 3.65 through 3.72)

*3.65. Numerically integrate and plot the response of an underdamped system determined by $m = 80$ kg, $k = 800$ N/m, and $c = 20$ kg/s, subject to the initial conditions of $x_0 = 0$ and $v_0 = 0$, and the applied force $F(t) = 30\Phi(t - 1)$. Then plot the exact response as computed by equation (3.17). Compare the plot of the exact solution to the numerical simulation.

*3.66. Numerically integrate and plot the response of an underdamped system determined by $m = 150$ kg, and $k = 5000$ N/m subject to the initial conditions of $x_0 = 0.01$ m and $v_0 = 0.1$ m/s, and the applied force $F(t) = \Phi(t) = 15(t - 1)$, for various values of the damping coefficient. Use this "program" to determine a value of damping that causes the transient term to die out within 3 seconds. Try to find the smallest such value of damping remembering that added damping is usually expensive.

*3.67. Calculate the total response of the base isolation problem given in Example 3.3.2, with the parameters $\omega_b = 3$ rad/s, $m = 1$ kg, $c = 10$ kg/s, $k = 1000$ N/m, and $Y = 0.05$ m, subject to initial conditions $x_0 = 0.01$ m and $v_0 = 3.0$ m/s, by numerically integrating rather than using analytical expressions. Plot the response, reproduce Figure 3.14, and compare the results to see that they are the same.

*3.68. Numerically simulate the response of the system of a single-degree-of-freedom spring–mass system subject to the motion $y(t)$ given in Figure P3.68 and plot the response. The mass is 5000 kg and the stiffness is 1.5×10^3 N/m.

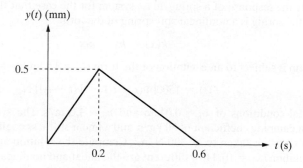

Figure P3.68   The base motion for Problem 3.68.

*3.69. Numerically simulate the response of an undamped system to a step function with a finite rise time of t_1 for the case $m = 1$ kg, $k = 1$ N/m, $t_1 = 4$ s, and $F_0 = 20$ N. This function is described by

$$F(t) = \begin{cases} \dfrac{F_0 t}{t_1} & 0 \le t \le t_1 \\ F_0 & t > t_1 \end{cases}$$

plot the response.

*3.70. Numerically simulate the response of the system of Problem 3.22 for a 2-meter concrete wall with cross section 0.03 m² and mass modeled as lumped at the end of 1000 kg. Use $F_0 = 100$ N, and plot the response for the case $t_0 = 0.25$ s.

*3.71. Numerically simulate the response of an undamped system to a ramp input of the form $F(t) = F_0\, t$, where F_0 is a constant. Plot the response for three periods for the case $m = 1$ kg, $k = 100$ N/m, and $F_0 = 50$ N.

*3.72. Compute and plot the response of the following system using numerical integration:

$$10\ddot{x}(t) + 20\dot{x}(t) + 1500x(t) = 20 \sin 25t + 10 \sin 15t + 20 \sin 2t$$

with initial conditions of $x_0 = 0.01$ m and $v_0 = 1.0$ m/s.

Section 3.10 (Problems 3.73 through 3.79)

*3.73. Compute the response of the system in Figure 3.26 for the case that the damping is linear viscous, the spring is a nonlinear soft spring of the form

$$k(x) = kx - k_1 x^3$$

the system is subject to an excitation of the form ($t_1 = 1.5$ and $t_2 = 1.6$)

$$F(t) = 1500[\Phi(t - t_1) - \Phi(t - t_2)] \text{ N}$$

and initial conditions of $x_0 = 0.01$ m and $v_0 = 1.0$ m/s. The system has a mass of 100 kg, a damping coefficient of 30 kg/s, and a linear stiffness coefficient of 2000 N/m.

The value of k_1 is taken to be 300 N/m³. Compute the solution and compare it to the linear solution ($k_1 = 0$). Which system has the largest magnitude? Compare your solution to that of Example 3.10.1.

*3.74. Compute the response of a spring–mass system for the case that the damping is linear viscous, the spring is a nonlinear soft spring of the form

$$k(x) = kx - k_1 x^3$$

the system is subject to an excitation of the form ($t_1 = 1.5$ and $t_2 = 1.6$)

$$F(t) = 1500[\Phi(t - t_1) - \Phi(t - t_2)] \, \text{N}$$

and initial conditions of $x_0 = 0.01$ m and $v_0 = 1.0$ m/s. The system has a mass of 100 kg, a damping coefficient of 30 kg/s, and a linear stiffness coefficient of 2000 N/m. The value of k_1 is taken to be 300 N/m³. Compute the solution and compare it to the linear solution ($k_1 = 0$). How different are the linear and nonlinear responses? Repeat this for $t_2 = 2$. What can you say regarding the effect of the time length of the pulse?

*3.75. Compute the response of a spring–mass–damper system for the case that the damping is linear viscous, the spring stiffness is of the form

$$k(x) = kx - k_1 x^2$$

the system is subject to an excitation of the form ($t_1 = 1.5$ and $t_2 = 2.5$)

$$F(t) = 1500[\Phi(t - t_1) - \Phi(t - t_2)] \, \text{N}$$

and initial conditions of $x_0 = 0.01$ m and $v_0 = 1$ m/s. The system has a mass of 100 kg, a damping coefficient of 30 kg/s, and a linear stiffness coefficient of 2000 N/m. The value of k_1 is taken to be 450 N/m³. Which system has the largest magnitude?

*3.76. Compute the response of a spring–mass–damper system for the case that the damping is linear viscous, the spring stiffness is of the form

$$k(x) = kx + k_1 x^2$$

the system is subject to an excitation of the form ($t_1 = 1.5$ and $t_2 = 2.5$)

$$F(t) = 1500[\Phi(t - t_1) - \Phi(t - t_2)] \, \text{N}$$

and initial conditions of $x_0 = 0.01$ m and $v_0 = 1$ m/s. The system has a mass of 100 kg, a damping coefficient of 30 kg/s, and a linear stiffness coefficient of 2000 N/m. The value of k_1 is taken to be 450 N/m³. Which system has the largest magnitude?

*3.77. Compute the response of a spring–mass–damper system for the case that the damping is linear viscous, the spring stiffness is of the form

$$k(x) = kx - k_1 x^2$$

the system is subject to an excitation of the form ($t_1 = 1.5$ and $t_2 = 2.5$)

$$F(t) = 150[\Phi(t - t_1) - \Phi(t - t_2)] \, \text{N}$$

and initial conditions of $x_0 = 0.01$ m and $v_0 = 1$ m/s. The system has a mass of 100 kg, a damping coefficient of 30 kg/s, and a linear stiffness coefficient of 2000 N/m. The value of k_1 is taken to be 5500 N/m³. Which system has the largest transient magnitude? Which has the largest magnitude in steady state?

*3.78. Compare the forced response of a system with velocity-squared damping, as defined in equation (2.129) using numerical simulation of the nonlinear equation, to that of the response of the linear system obtained using equivalent viscous damping, as defined by equation (2.131). Use the initial conditions $x_0 = 0.01$ m and $v_0 = 0.1$ m/s with a mass of 10 kg, stiffness of 25 N/m, applied force of the form ($t_1 = 1.5$ and $t_2 = 2.5$)

$$F(t) = 15[\Phi(t - t_1) - \Phi(t - t_2)]\,\text{N}$$

and drag coefficient of $\alpha = 25$.

*3.79. Compare the forced response of a system with structural damping (see Table 2.2) using numerical simulation of the nonlinear equation to that of the response of the linear system obtained using equivalent viscous damping as defined in Table 2.2. Use the initial conditions $x_0 = 0.01$ m and $v_0 = 0.1$ m/s with a mass of 10 kg, stiffness of 25 N/m, applied force of the form ($t_1 = 1.5$ and $t_2 = 2.5$)

$$F(t) = 15[\Phi(t - t_1) - \Phi(t - t_2)]\,\text{N}$$

and solid damping coefficient of $b = 8$. Does the equivalent viscous-damping linearization overestimate the response or underestimate it?

MATLAB ENGINEERING VIBRATION TOOLBOX

You may use the files contained in the *Engineering Vibration Toolbox*, first discussed at the end of Chapter 1 (immediately following the problems) and discussed in Appendix G, to help solve many of the preceding problems. The files contained in folder VTB3 may be used to help understand the nature of the general forced response of a single-degree-of-freedom system as discussed in this chapter and the dependence of this response on various parameters. VTB1_3 and VTB1_4 must be used if an arbitrary forcing function is applied (one other than a simple function call). The following problems are suggested to help build some intuition regarding the material on general forced response and to become familiar with the various formulas.

TOOLBOX PROBLEMS

TB3.1. Use file VTB3_1 to solve for the response of a system with a 10-kg mass, damping $c = 2.1$ kg/s, and stiffness $k = 2100$ N/m, subject to an impulse at time $t = 0$ of magnitude 10 N. Next, vary the value of c, first increasing it, then decreasing it, and note the effect in the responses.

TB3.2. Use file VTB3_2 to reproduce the plot of Figure 3.7. Then see what happens to the response as the damping coefficient is varied by trying an overdamped and critically damped value of ζ and examining the resulting response.

TB3.3. If you are confident with MATLAB, try using the plot command to plot (say, for $T = 6$)

$$-\frac{8}{\pi^2}\cos\frac{2\pi}{T}t, \qquad -\frac{8}{\pi^2}\left(\cos\frac{2\pi}{T}t + \frac{1}{9}\cos\frac{6\pi}{T}t\right)$$

then

$$-\frac{8}{\pi^2}\left(\cos\frac{2\pi}{T}t + \frac{1}{9}\cos\frac{6\pi}{T}t + \frac{1}{25}\cos\frac{10\pi}{T}t\right)$$

and so on, until you are satisfied that the Fourier series computed in Example 3.3.1 converges to the function plotted in Figure 3.13. If you are not familiar enough with MATLAB to try this on your own, run VTB3_3, which is a demo that does this for you.

TB3.4. Using VTB3_3, rework Problem 3.32 for first 5, then 10, and finally 50 terms.

TB3.5. Using file VTB3_4, examine the effect of varying the system's natural frequency on the response spectrum for the force given in Figure 3.16. Pick the frequencies $f = $ 10 Hz, 100 Hz, and 1000 Hz and compare the various response spectrum plots.

4 Multiple-Degree-of-Freedom Systems

This chapter introduces the analysis needed to understand the vibration of systems with more than one degree of freedom. The number of degrees of freedom of a system is determined by the number of moving parts and the number of directions in which each part can move. More than one degree of freedom means more than one natural frequency, greatly increasing the opportunity for resonance to occur. This chapter also introduces the important concept of a mode shape, and the highly used method of modal analysis for studying the response of multiple-degree-of-freedom (MDOF) systems. Most structures are modeled as MDOF systems. The all-terrain vehicle suspension shown in the photo at the top forms an example of a system that can be modeled as two or more degrees of freedom. Designers need to be able to predict the vibration response in order to improve the ride and ensure durability. The blades of a jet engine pictured at the bottom also require MDOF analysis but with a much larger number of degrees of freedom. Airplanes, satellites, automobiles, and so on, all provide examples of vibrating systems well modeled by the MDOF analysis introduced in this chapter.

In the preceding chapters, a single coordinate and single second-order differential equation sufficed to describe the vibratory motion of the mechanical device under consideration. However, many mechanical devices and structures cannot be modeled successfully by single-degree-of-freedom models. For example, the base-excitation problem of Section 2.4 requires a coordinate for the base as well as the main mass if the base motion is not prescribed, as assumed in Section 2.4. If the base motion is not prescribed and if the base has significant mass, then the coordinate, y, will also satisfy a second-order differential equation, and the system becomes a two-degree-of-freedom model. Machines with many moving parts have many degrees of freedom.

In this chapter, a two-degree-of-freedom example is first used to introduce the special phenomena associated with multiple-degree-of-freedom systems. These phenomena are then extended to systems with an arbitrary but finite number of degrees of freedom. To keep a record of each coordinate of the system, vectors are introduced and used along with matrices. This is done both for the ease of notation and to enable vibration theory to take advantage of the disciplines of matrix theory, linear algebra, and computational codes.

4.1 TWO-DEGREE-OF-FREEDOM MODEL (UNDAMPED)

This section introduces two-degree-of-freedom systems and how to solve for the response of each degree of freedom. The approach presented here is detailed because the goal is to provide background for solving systems with any number of degrees of freedom. In practice, computer methods are most commonly used to solve for the response of complex systems. This was not the case when most vibration texts were written. Hence, the approach here is a bit different than the approach found in the more traditional and older texts on vibration; the focus here is in setting up vibration problems in terms of matrices and vectors used in computer codes for solving practical problems.

In moving from single-degree-of-freedom systems to two or more degrees of freedom, two important physical phenomena result. The first important difference is that a two-degree-of-freedom system will have *two* natural frequencies. The second important phenomenon is that of a mode shape, which is not present in single-degree-of-freedom systems. A mode shape is a vector that describes the relative motion between the two masses or between two degrees of freedom. These important concepts of multiple natural frequencies and mode shapes are intimately tied to the mathematical concepts of eigenvalues and eigenvectors of computational matrix theory. This establishes the need to cast the vibration problem in terms of vectors and matrices.

Consider the two-mass system of Figure 4.1(a). This undamped system is similar to the system of Figure 2.13 except that the base motion is not prescribed in this case and the base now has mass. Figure 4.1(b) illustrates a single-mass system capable of moving in two directions and hence provides an example of a

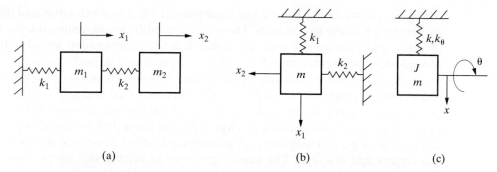

(a)   (b)   (c)

Figure 4.1 (a) A simple two-degree-of-freedom model consisting of two masses connected in series by two springs. (b) A single mass with two degrees of freedom (i.e., the mass moves along both the x_1 and x_2 directions). (c) A single mass with one translational degree of freedom and one rotational degree of freedom.

two-degree-of-freedom system as well. Figure 4.1(c) illustrates a single rigid mass that is capable of moving in translation as well as rotation about its axis. In each of these three cases, more than one coordinate is required to describe the vibration of the system. Each of the three parts of Figure 4.1 constitutes a two-degree-of-freedom system. A physical example of each system might be (a) a two-story building, (b) the vibration of a drill press, or (c) the rocking motion of an automobile or aircraft.

A free-body diagram illustrating the spring forces acting on each mass in Figure 4.1(a) is illustrated in Figure 4.2. The force of gravity is excluded following the reasoning used in Figure 1.14 (i.e., the static deflection balances the gravitational force and no friction is present). Summing forces on each mass in the horizontal direction yields

$$m_1\ddot{x}_1 = -k_1 x_1 + k_2(x_2 - x_1)$$
$$m_2\ddot{x}_2 = -k_2(x_2 - x_1)$$

(4.1)

Rearranging these two equations yields

$$m_1\ddot{x}_1 + (k_1 + k_2)x_1 - k_2 x_2 = 0$$
$$m_2\ddot{x}_2 - k_2 x_1 + k_2 x_2 = 0$$

(4.2)

Equations (4.2) consist of two coupled second-order ordinary differential equations, with constant coefficients, each of which requires two initial conditions to solve. Hence these two coupled equations are subject to the four initial conditions:

$$x_1(0) = x_{10} \quad \dot{x}_1(0) = \dot{x}_{10} \quad x_2(0) = x_{20} \quad \dot{x}_2(0) = \dot{x}_{20}$$

(4.3)

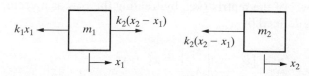

Figure 4.2 Free-body diagrams of each mass in the system of Figure 4.1(a), indicating the restoring force provided by the springs.

where the constants $\dot{x}_{10}$, $\dot{x}_{20}$ and x_{10}, x_{20} represent the initial velocities and displacements of each of the two masses. These initial conditions are assumed to be known or given and provide the four constants of integration needed to solve the two second-order differential equations for the free response of each mass.

There are several approaches available to solve equations (4.2) given (4.3) and the values of m_1, m_2, k_1, and k_2 for the responses $x_1(t)$ and $x_2(t)$. First, note that neither equation can be solved by itself because each equation contains both x_1 and x_2 (i.e., the equations are *coupled*). Physically, this states that the motion of x_1 affects the motion of x_2, and vice versa. A convenient method of solving this system is to use vectors and matrices. The vector approach to solving this simple two-degree-of-freedom problem is also readily extendable to systems with an arbitrary finite number of degrees of freedom and is compatible with computer codes. Vectors and matrices were introduced in Section 1.9 in order to enter a single-degree-of-freedom vibration problem into a numerical equation solver such as MATLAB. Vectors and matrices are reviewed here briefly and more details can be found in Appendix C. Here vectors and matrices are used to compute a solution of equation (4.1).

Define the vector $\mathbf{x}(t)$ to be the column vector consisting of the two responses of interest:

$$\mathbf{x}(t) = \begin{bmatrix} x_1(t) \\ x_2(t) \end{bmatrix} \tag{4.4}$$

This is called a displacement or response vector and is a 2×1 array of functions. Differentiation of a vector is defined here by differentiating each element so that

$$\dot{\mathbf{x}}(t) = \begin{bmatrix} \dot{x}_1(t) \\ \dot{x}_2(t) \end{bmatrix} \quad \text{and} \quad \ddot{\mathbf{x}}(t) = \begin{bmatrix} \ddot{x}_1(t) \\ \ddot{x}_2(t) \end{bmatrix} \tag{4.5}$$

are the velocity and the acceleration vectors, respectively. A square matrix is a square array of numbers, which could be made, for instance, by combining two 2×1 column vectors to produce a 2×2 matrix. An example of a 2×2 matrix is given by

$$M = \begin{bmatrix} m_1 & 0 \\ 0 & m_2 \end{bmatrix} \tag{4.6}$$

Note here that italic capital letters are used to denote matrices and bold lowercase letters are used to denote vectors.

Vectors and matrices can be multiplied together in a variety of ways. In vibration analysis the method most useful to define the product of a matrix times a vector is to define the result to be a vector with elements consisting of the dot product of the vector with each "row" of the matrix (i.e., by treating the row as a vector). The *dot product* of a vector is defined by

$$\mathbf{x}^T \mathbf{x} = \begin{bmatrix} x_1 & x_2 \end{bmatrix} \begin{bmatrix} x_1 \\ x_2 \end{bmatrix} = x_1^2 + x_2^2 \tag{4.7}$$

which is a scalar. The symbol $\mathbf{x}^T$ denotes the transpose of the vector and changes a column vector into a row vector. Equation (4.7) is also called the *inner product* or *scalar product* of the vector $\mathbf{x}$ with itself. A scalar, a, times a vector, $\mathbf{x}$, is simply defined as $a\mathbf{x} = [ax_1 \quad ax_2]^T$ (i.e., a vector of the same dimension with each element multiplied by the scalar). (Recall that a scalar is any real or complex number.) These rules for manipulating vectors should be familiar from introductory mechanics (i.e., statics and dynamics) texts.

The following example illustrates the rules for multiplying a matrix times a vector.

Example 4.1.1

Consider the product of the matrix M of equation (4.6) and the acceleration vector $\ddot{\mathbf{x}}$ of equation (4.5). This product becomes

$$M\ddot{\mathbf{x}} = \begin{bmatrix} m_1 & 0 \\ 0 & m_2 \end{bmatrix} \begin{bmatrix} \ddot{x}_1 \\ \ddot{x}_2 \end{bmatrix} = \begin{bmatrix} m_1\ddot{x}_1 + 0\ddot{x}_2 \\ 0\ddot{x}_1 + m_2\ddot{x}_2 \end{bmatrix} = \begin{bmatrix} m_1\ddot{x}_1 \\ m_2\ddot{x}_2 \end{bmatrix} \tag{4.8}$$

where the first element of the product is defined to be the dot product of the row vector $[m_1 \quad 0]$ with the column vector $\ddot{\mathbf{x}}$, and the second element is the dot product of the row vector $[0 \quad m_2]$ with $\ddot{\mathbf{x}}$. Note that the product of a matrix and a vector is a vector.

$\square$

Example 4.1.2

Consider the 2×2 matrix K defined by

$$K = \begin{bmatrix} k_1 + k_2 & -k_2 \\ -k_2 & k_2 \end{bmatrix} \tag{4.9}$$

and calculate the product $K\mathbf{x}$.

Solution   Again, the product is formed by considering the first element to be the inner product of the row vector $[k_1 + k_2 \quad -k_2]$ and the column vector $\mathbf{x}$. The second element of the product vector $K\mathbf{x}$ is formed from the inner product of the row vector $[-k_2 \quad k_2]$ and the vector $\mathbf{x}$. This yields

$$K\mathbf{x} = \begin{bmatrix} k_1 + k_2 & -k_2 \\ -k_2 & k_2 \end{bmatrix} \begin{bmatrix} x_1 \\ x_2 \end{bmatrix} = \begin{bmatrix} (k_1 + k_2)x_1 - k_2x_2 \\ -k_2x_1 + k_2x_2 \end{bmatrix} \tag{4.10}$$

$\square$

Two vectors of the same size are said to be equal if and only if each element of one vector is equal to the corresponding element in the other vector. With this in mind, consider the *vector equation*

$$M\ddot{\mathbf{x}} + K\mathbf{x} = \mathbf{0} \tag{4.11}$$

where $\mathbf{0}$ denotes the column vector of zeros:

$$\mathbf{0} = \begin{bmatrix} 0 \\ 0 \end{bmatrix}$$

Substitution of the value for M from equation (4.6) and the value for K from equation (4.9) into equation (4.11) yields

$$\begin{bmatrix} m_1 & 0 \\ 0 & m_2 \end{bmatrix}\begin{bmatrix} \ddot{x}_1 \\ \ddot{x}_2 \end{bmatrix} + \begin{bmatrix} k_1 + k_2 & -k_2 \\ -k_2 & k_2 \end{bmatrix}\begin{bmatrix} x_1 \\ x_2 \end{bmatrix} = \begin{bmatrix} 0 \\ 0 \end{bmatrix}$$

These products can be carried out as indicated in Example 4.1.1 and equation (4.10) to yield

$$\begin{bmatrix} m_1\ddot{x}_1 \\ m_2\ddot{x}_2 \end{bmatrix} + \begin{bmatrix} (k_1 + k_2)x_1 - k_2x_2 \\ -k_2x_1 + k_2x_2 \end{bmatrix} = \begin{bmatrix} 0 \\ 0 \end{bmatrix}$$

Adding the two vectors on the left side of the equation, element by element, yields

$$\begin{bmatrix} m_1\ddot{x}_1 + (k_1 + k_2)x_1 - k_2x_2 \\ m_2\ddot{x}_2 - k_2x_1 + k_2x_2 \end{bmatrix} = \begin{bmatrix} 0 \\ 0 \end{bmatrix} \tag{4.12}$$

Equating the corresponding elements of the two vectors in equation (4.12) yields

$$m_1\ddot{x}_1 + (k_1 + k_2)x_1 - k_2x_2 = 0 \tag{4.13}$$

$$m_2\ddot{x}_2 - k_2x_1 + k_2x_2 = 0$$

which are identical to equations (4.2). Hence the system of equations (4.2) can be written as the vector equation given in (4.11), where the coefficient matrices are defined by the matrices of equations (4.6) and (4.9). The matrix M defined by equation (4.6) is called the *mass matrix*, and the matrix K defined by equation (4.9) is called the *stiffness matrix*. The preceding calculation and comparison provide an extremely important connection between vibration analysis and matrix analysis. This simple connection allows computers to be used to solve large and complicated vibration problems quickly (discussed in Section 4.10). It also forms the foundation for the rest of this chapter (as well as the rest of the book).

The mass and stiffness matrices, M and K, described previously have the special property of being symmetric. A *symmetric matrix* is a matrix that is equal to its transpose. The *transpose* of a matrix, denoted by A^T, is formed from interchanging the rows and columns of a matrix. The first row of A^T is the first column of A and so on. The mass matrix M is also called the *inertia matrix*, and the force vector $M\ddot{\mathbf{x}}$ corresponds to the inertial forces in the system of Figure 4.1(a). Similarly, the force $K\mathbf{x}$ represents the elastic restoring forces of the system described in Figure 4.1(a).

Example 4.1.3

Consider the matrix A defined by

$$A = \begin{bmatrix} a & b \\ c & d \end{bmatrix}$$

where a, b, c, and d are real numbers. Calculate values of these constants such that the matrix A is symmetric.

Solution   For A to be symmetric, $A = A^T$ or

$$A = \begin{bmatrix} a & b \\ c & d \end{bmatrix} = \begin{bmatrix} a & c \\ b & d \end{bmatrix} = A^T$$

Comparing the elements of A and A^T yields that $c = b$ must hold if the matrix A is to be symmetric. Note that the elements in the c and b position of the matrix K given in equation (4.9) are equal so that $K = K^T$.

□

It is useful to note that if $\mathbf{x}$ is a column vector

$$\mathbf{x} = \begin{bmatrix} x_1 \\ x_2 \end{bmatrix}$$

then $\mathbf{x}^T$ is a row vector (i.e., $\mathbf{x}^T = [x_1 \quad x_2]$). This makes it convenient to write a column vector in one line. For example, the vector $\mathbf{x}$ can also be written as $\mathbf{x} = [x_1 \quad x_2]^T$, a column vector. The act of forming a transpose also undoes itself, so that $(A^T)^T = A$.

The initial conditions can also be written in terms of vectors as

$$\mathbf{x}_0 = \begin{bmatrix} x_1(0) \\ x_2(0) \end{bmatrix} \qquad \dot{\mathbf{x}}_0 = \begin{bmatrix} \dot{x}_1(0) \\ \dot{x}_2(0) \end{bmatrix} \qquad (4.14)$$

Here $\mathbf{x}_0$ denotes the initial displacement vector and $\dot{\mathbf{x}}_0$ denotes the initial velocity vector. Equation (4.12) can now be solved by following the procedures used for solving single-degree-of-freedom systems and incorporating a few results from matrix theory.

Recall from Section 1.2 that the single-degree-of-freedom version of equation (4.11) was solved by assuming a harmonic solution and calculating values for the constants in the assumed form. The same approach is used here. Following the argument used in equations (1.13) to (1.19), a solution is assumed of the form

$$\mathbf{x}(t) = \mathbf{u}e^{j\omega t} \qquad (4.15)$$

Here $\mathbf{u}$ is a nonzero vector of constants to be determined, ω is a constant to be determined, and $j = \sqrt{-1}$. Recall that the scalar $e^{j\omega t}$ represents harmonic motion since $e^{j\omega t} = \cos \omega t + j \sin \omega t$. The vector $\mathbf{u}$ cannot be zero; otherwise no motion results.

Substitution of this assumed form of the solution into the vector equation of motion yields

$$(-\omega^2 M + K)\mathbf{u}e^{j\omega t} = \mathbf{0} \tag{4.16}$$

where the common factor $\mathbf{u}e^{j\omega t}$ has been factored to the right side. Note that the scalar $e^{j\omega t} \neq 0$ for any value of t and hence equation (4.16) yields the fact that ω and $\mathbf{u}$ must satisfy the vector equation

$$(-M\omega^2 + K)\mathbf{u} = \mathbf{0}, \mathbf{u} \neq \mathbf{0} \tag{4.17}$$

Note that this represents two algebraic equations in the three unknown scalars: ω, u_1, and u_2 where $\mathbf{u} = [u_1 \quad u_2]^T$.

For this homogeneous set of algebraic equations to have a nonzero solution for the vector $\mathbf{u}$, the inverse of the coefficient matrix $(-M\omega^2 + K)$ must not exist. To see that this is the case, suppose that the inverse of $(-M\omega^2 + K)$ does exist. Then multiplying both sides of equation (4.17) by $(-M\omega^2 + K)^{-1}$ yields $\mathbf{u} = \mathbf{0}$, a trivial solution, as it implies no motion. Hence the solution of equation (4.11) depends in some way on the *matrix inverse*. Matrix inverses are reviewed in the following example.

Example 4.1.4

Consider the 2×2 matrix A defined by

$$A = \begin{bmatrix} a & b \\ c & d \end{bmatrix}$$

and calculate its inverse.

Solution   The inverse of a square matrix A is a matrix of the same dimension, denoted by A^{-1}, such that

$$AA^{-1} = A^{-1}A = I$$

where I is the identity matrix. In this case I has the form

$$I = \begin{bmatrix} 1 & 0 \\ 0 & 1 \end{bmatrix}$$

The inverse matrix for a general 2×2 matrix is

$$A^{-1} = \frac{1}{\det A} \begin{bmatrix} d & -b \\ -c & a \end{bmatrix} \tag{4.18}$$

provided that $\det A \neq 0$, where $\det A$ denotes the *determinant* of the matrix A. The determinant of the matrix A has the value

$$\det A = ad - bc$$

To see that equation (4.18) is in fact the inverse, note that

$$A^{-1}A = \frac{1}{ad - bc} \begin{bmatrix} d & -b \\ -c & a \end{bmatrix} \begin{bmatrix} a & b \\ c & d \end{bmatrix}$$

$$= \frac{1}{ad - bc} \begin{bmatrix} ad - bc & bd - bd \\ ac - ac & ad - bc \end{bmatrix} = \begin{bmatrix} 1 & 0 \\ 0 & 1 \end{bmatrix}$$

It is important to realize that the matrix A has an inverse if and only if det $A \neq 0$. Thus, requiring det $A = 0$ forces A not to have an inverse. Matrices that do not have an inverse are called *singular* matrices. Note that if the matrix A is symmetric, $c = b$ and A^{-1} is also symmetric.

$\square$

Applying the condition of singularity to the coefficient matrix of equation (4.17) yields the result that for a nonzero solution $\mathbf{u}$ to exist,

$$\det(-\omega^2 M + K) = 0 \tag{4.19}$$

which yields one algebraic equation in one unknown (ω^2). Substituting the values of the matrices M and K into this expression yields

$$\det \begin{bmatrix} -\omega^2 m_1 + k_1 + k_2 & -k_2 \\ -k_2 & -\omega^2 m_2 + k_2 \end{bmatrix} = 0 \tag{4.20}$$

Using the definition of the determinant yields that the unknown quantity ω^2 must satisfy

$$m_1 m_2 \omega^4 - (m_1 k_2 + m_2 k_1 + m_2 k_2)\omega^2 + k_1 k_2 = 0 \tag{4.21}$$

This expression is called the *characteristic equation* for the system and is used to determine the constants ω in the assumed form of the solution given by equation (4.15) once the values of the physical parameters m_1, m_2, k_1, and k_2 are known.

Example 4.1.5

Calculate the solutions for ω of the characteristic equation given by equation (4.21) for the case that the physical parameters have the values $m_1 = 9$ kg, $m_2 = 1$ kg, $k_1 = 24$ N/m, and $k_2 = 3$ N/m.

Solution   For these values the characteristic equation (4.21) becomes

$$\omega^4 - 6\omega^2 + 8 = (\omega^2 - 2)(\omega^2 - 4) = 0$$

so that $\omega_1^2 = 2$ and $\omega_2^2 = 4$. There are two roots and each corresponds to two values of the constant ω in the assumed form of the solution:

$$\omega_1 = \pm\sqrt{2} \text{ rad/s}, \qquad \omega_2 = \pm 2 \text{ rad/s}$$

$\square$

Note that in the statement of equation (4.17) ω^2 appears, not ω. However, in proceeding to the solution in time, the frequency of oscillation will become ω and the plus and minus signs on ω are absorbed in changing the exponential into a trigonometric function, as described in the following pages.

Once the value of ω in equation (4.15) is established, the value of the constant vector $\mathbf{u}$ can be found by solving equation (4.17) for $\mathbf{u}$ given each value of ω^2. That is, for each value of ω^2 (i.e., ω_1^2 and ω_2^2) there is a vector $\mathbf{u}$ satisfying equation (4.17). For ω_1^2, the vector $\mathbf{u}_1$ satisfies

$$(-\omega_1^2 M + K)\mathbf{u}_1 = \mathbf{0} \tag{4.22}$$

and for ω_2^2, the vector $\mathbf{u}_2$ satisfies

$$(-\omega_2^2 M + K)\mathbf{u}_2 = \mathbf{0} \tag{4.23}$$

These two expressions can be solved for the direction of the vectors $\mathbf{u}_1$ and $\mathbf{u}_2$, but not for the magnitude. To see that this is true, note that if $\mathbf{u}_1$ satisfies equation (4.22), so does the vector $a\mathbf{u}_1$, where a is any nonzero number. Hence the vectors satisfying (4.22) and (4.23) are of arbitrary magnitude. The following example illustrates one way to compute $\mathbf{u}_1$ and $\mathbf{u}_2$ for the values of Example 4.1.5.

Example 4.1.6

Calculate the vectors $\mathbf{u}_1$ and $\mathbf{u}_2$ of equations (4.22) and (4.23) for the values of ω, K, and M of Example 4.1.5.

Solution   Let $\mathbf{u}_1 = [u_{11} \quad u_{21}]^T$. Then equation (4.22) with $\omega^2 = \omega_1^2 = 2$ becomes

$$\begin{bmatrix} 27 - 9(2) & -3 \\ -3 & 3 - (2) \end{bmatrix} \begin{bmatrix} u_{11} \\ u_{21} \end{bmatrix} = \begin{bmatrix} 0 \\ 0 \end{bmatrix}$$

Performing the indicated product and enforcing the equality yields the two equations

$$9u_{11} - 3u_{21} = 0 \quad \text{and} \quad -3u_{11} + u_{21} = 0$$

Note that these two equations are dependent and yield the same solution; that is,

$$\frac{u_{11}}{u_{21}} = \frac{1}{3} \quad \text{or} \quad u_{11} = \frac{1}{3}u_{21}$$

Only the ratio of the elements is determined here [i.e., only the direction of the vector is determined by equation (4.17), not its magnitude]. As mentioned previously, this happens because if $\mathbf{u}$ satisfies equation (4.17), then so does $a\mathbf{u}$, where a is any nonzero number.

A numerical value for each element of the vector **u** may be obtained by arbitrarily assigning one of the elements. For example, let $u_{21} = 1$; then the value of $\mathbf{u}_1$ is

$$\mathbf{u}_1 = \begin{bmatrix} \dfrac{1}{3} \\ 1 \end{bmatrix}$$

This procedure is repeated using $\omega_2^2 = 4$ to yield that the elements of $\mathbf{u}_2$ must satisfy

$$-9u_{12} - 3u_{22} = 0 \quad \text{or} \quad u_{12} = -\frac{1}{3}u_{22}$$

Choosing $u_{22} = 1$ yields

$$\mathbf{u}_2 = \begin{bmatrix} -\dfrac{1}{3} \\ 1 \end{bmatrix}$$

which is the vector satisfying equation (4.23). There are several other ways of fixing the magnitude of a vector besides the one illustrated here. Some other methods are presented in Example 4.2.3 and equation (4.44). A more systematic method called normalizing will be used with larger problems and is presented in the next section.

□

The solution of equation (4.11) subject to initial conditions $\mathbf{x}_0$ and $\dot{\mathbf{x}}_0$ can be constructed in terms of the numbers $\pm\omega_1$, $\pm\omega_2$ and the vectors $\mathbf{u}_1$ and $\mathbf{u}_2$. This is similar to the construction of the solution of the single-degree-of-freedom case discussed in Section 1.2. Since the equations to be solved are linear, the sum of any two solutions is also a solution. From the preceding calculation, there are four solutions in the form of equation (4.15) made up of the four values of ω and the two vectors:

$$\mathbf{x}(t) = \mathbf{u}_1 e^{-j\omega_1 t}, \quad \mathbf{u}_1 e^{+j\omega_1 t}, \quad \mathbf{u}_2 e^{-j\omega_2 t}, \quad \text{and} \quad \mathbf{u}_2 e^{+j\omega_2 t} \tag{4.24}$$

Thus a general solution is the linear combination of these:

$$\mathbf{x}(t) = (ae^{j\omega_1 t} + be^{-j\omega_1 t})\mathbf{u}_1 + (ce^{j\omega_2 t} + de^{-j\omega_2 t})\mathbf{u}_2 \tag{4.25}$$

where a, b, c, and d are arbitrary constants of integration to be determined by the initial conditions.

Applying Euler formulas for the sine function to equation (4.25) yields an alternative form of the solution (provided neither ω_i is zero):

$$\mathbf{x}(t) = A_1 \sin(\omega_1 t + \phi_1)\mathbf{u}_1 + A_2 \sin(\omega_2 t + \phi_2)\mathbf{u}_2 \tag{4.26}$$

where the constants of integration are now in the form of two amplitudes, A_1 and A_2, and two phase shifts, ϕ_1 and ϕ_2. Recall that this is the same procedure used in equations (1.17), (1.18), and (1.19). These constants can be calculated from the initial conditions $\mathbf{x}_0$ and $\dot{\mathbf{x}}_0$. Equation (4.26) is the two-degree-of-freedom analog of equation (1.19) for a single-degree-of-freedom case.

The form of equation (4.26) gives physical meaning to the solution. It states that each mass in general oscillates at two frequencies: ω_1 and ω_2. These are called the *natural frequencies* of the system. Furthermore, suppose that the initial conditions are chosen such that $A_2 = 0$. With such initial conditions, each mass oscillates at only one frequency, ω_1, and the relative positions of the masses at any given instant of time are determined by the elements of the vector $\mathbf{u}_1$. Hence $\mathbf{u}_1$ is called the *first mode shape* of the system. Similarly, if the initial conditions are chosen such that A_1 is zero, both coordinates oscillate at frequency ω_2, with relative positions given by the vector $\mathbf{u}_2$, called the *second mode shape*. The mode shapes and natural frequencies are clarified further in the following exercises and sections. Mode shapes have become a standard in vibration engineering and are used extensively in vibration analysis. The concepts of natural frequencies and mode shapes are extremely important and form one of the major ideas used in vibration studies.

In the derivation of equation (4.26) it is assumed that neither of the values of ω is zero. One or the other may have the value zero in some applications, but then the solution takes on another form. The value of $\mathbf{u}$, however, cannot be zero. A frequency can be zero, but a mode shape cannot be zero. The zero frequency case corresponds to rigid body motion and is the topic of Problem 4.12. The concept of rigid body motion is detailed in Section 4.4.

Note that the positive and negative sign on ω resulting from the solution of equation (4.19) is used in going from equation (4.25) to equation (4.26) when invoking the Euler formula for trig functions. Thus in equation (4.26) ω is only a positive number. Equation (4.26) also provides the interpretation of ω as a frequency of vibration, which is now necessarily a positive number. This is similar to the explanation provided in the single-degree-of-freedom case following equation (1.19).

Example 4.1.7

Calculate the solution of the system of Example 4.1.5 for the initial conditions $x_1(0) = 1$ mm, $x_2(0) = 0$, and $\dot{x}_1(0) = \dot{x}_2(0) = 0$.

Solution   To solve this, equation (4.26) is written as

$$
\begin{bmatrix} x_1(t) \\ x_2(t) \end{bmatrix} = [\mathbf{u}_1 \quad \mathbf{u}_2] \begin{bmatrix} A_1 \ \sin(\omega_1 t + \phi_1) \\ A_2 \ \sin(\omega_2 t + \phi_2) \end{bmatrix}
$$

$$
= \begin{bmatrix} \dfrac{1}{3} A_1 \sin(\sqrt{2}t + \phi_1) - \dfrac{1}{3} A_2 \sin(2t + \phi_2) \\ A_1 \sin(\sqrt{2}t + \phi_1) + A_2 \sin(2t + \phi_2) \end{bmatrix} \tag{4.27}
$$

At $t = 0$ this yields

$$\begin{bmatrix} 1 \\ 0 \end{bmatrix} = \begin{bmatrix} \frac{1}{3} A_1 \sin \phi_1 - \frac{1}{3} A_2 \sin \phi_2 \\ A_1 \sin \phi_1 + A_2 \sin \phi_2 \end{bmatrix} \tag{4.28}$$

Differentiating equation (4.27) and evaluating the resulting expression at $t = 0$ yields

$$\begin{bmatrix} \dot{x}_1(0) \\ \dot{x}_2(0) \end{bmatrix} = \begin{bmatrix} 0 \\ 0 \end{bmatrix} = \begin{bmatrix} \frac{\sqrt{2}}{3} A_1 \cos \phi_1 - \frac{2}{3} A_2 \cos \phi_2 \\ \sqrt{2} A_1 \cos \phi_1 + 2 A_2 \cos \phi_2 \end{bmatrix} \tag{4.29}$$

Equations (4.28) and (4.29) represent four equations in the four unknown constants of integrations A_1, A_2, ϕ_1, and ϕ_2. Writing out these four equations yields

$$3 = A_1 \sin \phi_1 - A_2 \sin \phi_2 \tag{4.30}$$

$$0 = A_1 \sin \phi_1 + A_2 \sin \phi_2 \tag{4.31}$$

$$0 = \sqrt{2} A_1 \cos \phi_1 - 2 A_2 \cos \phi_2 \tag{4.32}$$

$$0 = \sqrt{2} A_1 \cos \phi_1 + 2 A_2 \cos \phi_2 \tag{4.33}$$

Adding equations (4.32) and (4.33) yields that

$$2\sqrt{2} A_1 \cos \phi_1 = 0$$

so that $\phi_1 = \pi/2$. Since $\phi_1 = \pi/2$, equation (4.33) reduces to

$$2 A_2 \cos \phi_2 = 0$$

so that $\phi_2 = \pi/2$. Substitution of the values of ϕ_1 and ϕ_2 into equations (4.30) and (4.31) yields

$$3 = A_1 - A_2 \quad \text{and} \quad 0 = A_1 + A_2$$

which has solutions $A_1 = 3/2$ mm, $A_2 = -3/2$ mm. Thus

$$x_1(t) = 0.5 \sin\left(\sqrt{2}t + \frac{\pi}{2}\right) + 0.5 \sin\left(2t + \frac{\pi}{2}\right) = 0.5\left(\cos \sqrt{2}t + \cos 2t\right) \text{ mm}$$
$$\tag{4.34}$$
$$x_2(t) = \frac{3}{2} \sin\left(\sqrt{2}t + \frac{\pi}{2}\right) - \frac{3}{2} \sin\left(2t + \frac{\pi}{2}\right) = 1.5\left(\cos \sqrt{2}t - \cos 2t\right) \text{ mm}$$

These are plotted in Figure 4.3. More efficient ways to calculate the solutions are presented in later sections. The numerical aspects of calculating a solution are discussed in Section 4.10.

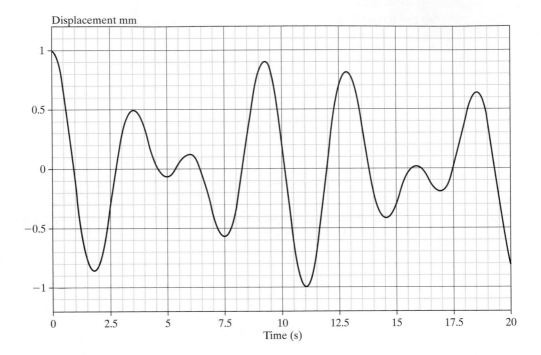

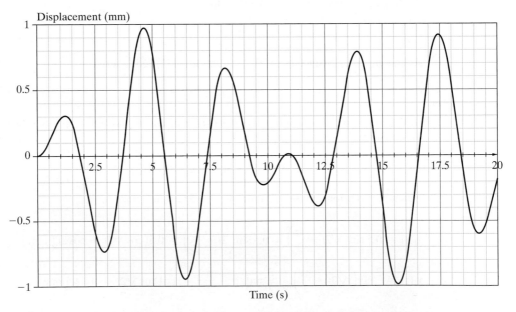

Figure 4.3   Plots of the responses of $x_1(t)$ on the top and $x_2(t)$ on the bottom for Example 4.1.7.

Note that in this case, the response of each mass contains both frequencies of the system. That is, the responses for both $x_1(t)$ and $x_2(t)$ are combinations of signals containing the two frequencies ω_1 and ω_2 (i.e., the sum of two harmonic signals). Note from the development of equation (4.34) that the mode shapes determine the relative magnitude of these two harmonic signals.

$\square$

In the previous example, the arbitrary choice of the magnitude of the mode-shape vectors $\mathbf{u}_1$ and $\mathbf{u}_2$ made in Example 4.1.6 does not affect the solution because these vectors are multiplied by the constants of integration A_1 and A_2, respectively. The initial conditions then scale the magnitude of these vectors, so that the solution given in equation (4.34) will be the same for any choice of fixing the vector magnitude made in Example 4.1.6.

Frequencies   It is interesting and important to note that the two natural frequencies ω_1 and ω_2 of the two-degree-of-freedom system are *not* equal to either of the natural frequencies of the two single-degree-of-freedom systems constructed from the same components. To see this, note that in Example 4.1.5, $\sqrt{k_1/m_1} = 1.63$, which is not equal to ω_1 or ω_2 (i.e., $\omega_1 = \sqrt{2}$, $\omega_2 = 2$). Similarly, $\sqrt{k_2/m_2} = 1.732$, which does not coincide with either frequency of the two-degree-of-freedom system composed of the same springs and masses, each attached to ground.

Beats   The beat phenomenon introduced in Example 2.1.2 for the forced response of a single-degree-of-freedom system can also exist in the free response of a two- (or more) degree-of-freedom system. If the mass and stiffness of the system of Figure 4.1(a) are such that the two frequencies are close to each other, then solutions derived in Example 4.1.7 will produce beats. In fact, a close examination of the plots of the response in Figure 4.3 shows that the response $x_2(t)$ is close to the shape of the beat illustrated in Figure 2.4. This happens because the two frequencies of Example 4.1.7 are reasonably close to each other (1.414 and 2). As the two natural frequencies become closer, the beat phenomenon will become more evident (see Problem 4.17). Thus, beats in vibrating systems can occur in two separate circumstances: first, in a forced-response case as the result of a driving frequency being close to a natural frequency (Example 2.1.2) and, second, as the result of two natural frequencies being close together in a free-response situation (Problem 4.18).

Calculations   The method used to compute the natural frequencies and mode shapes presented in this section is not the most efficient way to solve vibration problems. Nor is the approach presented here the most illuminating of the physical nature of vibration of two-degree-of-freedom systems. The calculation method presented in Examples 4.1.5 and 4.1.6 are instructive, but tedious. The approach of these examples also ignores the key issues of orthogonality of mode shapes and decoupling of the equations of motion, which are key concepts in understanding vibration analysis, design, and measurement. These issues are discussed in the following section, which connects the problem of computing natural frequencies and mode shapes to the symmetric eigenvalue and eigenvector calculations of mathematics. Once the natural frequency and mode-shape formulation is connected to the algebraic eigenvalue–eigenvector problem, then mathematical software packages can be used to compute the mode shapes and natural frequencies without going through the tedious computations of the preceding examples. This connection to the algebraic eigenvalue problem is also a key in understanding the topics of vibration testing discussed in Chapter 7.

4.2 EIGENVALUES AND NATURAL FREQUENCIES

The method of solution indicated in Section 4.1 can be extended and formalized to take advantage of the symmetric algebraic eigenvalue problem. This allows the power of mathematics to be used in solving vibration problems, allows the use of mathematical software packages, and sets the background needed for analyzing systems with an arbitrary number of degrees of freedom. In addition, the important concepts of mode shapes and natural frequencies can be generalized by connecting the undamped-vibration problem to the mathematics of the algebraic eigenvalue problem.

There are many ways to connect the solution of the vibration problem with that of the algebraic eigenvalue problem. The most productive approach is to cast the vibration problem as a symmetric eigenvalue problem because of the special properties associated with symmetry. Note that the physical nature of both the mass and stiffness matrices is that they are usually symmetric. Hence preserving this symmetry is also a natural approach to solving the vibration problem. Since M is symmetric and positive definite, it may be factored into two terms:

$$M = LL^T$$

where L is a special matrix with zeros in every position above the diagonal (called a lower triangular matrix). A matrix M is positive definite if the scalar formed from the product

$$\mathbf{x}^T M \mathbf{x} > 0$$

for every nonzero choice of the vector $\mathbf{x}$. The factorization L is called the Cholesky decomposition and is examined, along with the notion of positive definite, in more detail in Appendix C and Section 4.9. In the special case that M happens to be diagonal, as in the examples considered so far, the Cholesky decomposition just becomes the notion of a matrix square root, and the notion of positive definite just means that the diagonal elements of M are all positive, nonzero numbers.

In solving a single-degree-of-freedom system it was useful to divide the equation of motion by the mass. Hence, consider resolving the system of two equations described in matrix form by equation (4.19) by making a coordinate transformation that is equivalent to dividing the equations of motion by the mass in the system. To that end, consider the *matrix square root* defined to be the matrix $M^{1/2}$ such that $M^{1/2}M^{1/2} = M$, the mass matrix. For the simple example of the mass matrix given in equation (4.6), the mass matrix is diagonal and the matrix square root becomes simply

$$L = M^{1/2} = \begin{bmatrix} \sqrt{m_1} & 0 \\ 0 & \sqrt{m_2} \end{bmatrix} \tag{4.35}$$

This factors M into $M = M^{1/2}M^{1/2}$ (or into $M = LL^T$) in the common case of a diagonal mass matrix. If M is not diagonal, the notion of a square root is dropped in favor of using the Cholesky decomposition L, which is discussed in Section 4.9 under dynamically coupled systems. The use of the Cholesky decomposition L is preferred because it is a single command in most codes, and hence more convenient for numerical computation.

The inverse (Example 4.1.4) of the diagonal matrix $M^{1/2}$, denoted by $M^{-1/2}$, becomes simply

$$L^{-1} = M^{-1/2} = \begin{bmatrix} \dfrac{1}{\sqrt{m_1}} & 0 \\ 0 & \dfrac{1}{\sqrt{m_2}} \end{bmatrix} \tag{4.36}$$

The matrix of equation (4.36) provides a means of changing coordinate systems to one in which the vibration problem is represented by a *single symmetric* matrix. This allows the vibration problem to be cast as the symmetric eigenvalue problem described in Window 4.1. The symmetric eigenvalue problem has distinct advantages both computationally and analytically, as will be illustrated next. The analytical advantage of this change of coordinates is similar to the advantage gained in solving the inclined plane problem in statics by writing the coordinate system along the incline rather than along the horizontal.

To accomplish this transformation, or change of coordinates, let the vector $\mathbf{x}$ in equation (4.11) be replaced with

$$\mathbf{x}(t) = M^{1/2}\mathbf{q}(t) \tag{4.37}$$

and multiply the resulting equation by $M^{-1/2}$. This yields

$$M^{-1/2}MM^{-1/2}\ddot{\mathbf{q}}(t) + M^{-1/2}KM^{-1/2}\mathbf{q}(t) = \mathbf{0} \tag{4.38}$$

Since $M^{-1/2}MM^{-1/2} = I$, the identity matrix, expression (4.38) reduces to

$$I\ddot{\mathbf{q}}(t) + \tilde{K}\mathbf{q}(t) = 0 \tag{4.39}$$

The matrix $\tilde{K} = M^{-1/2}KM^{-1/2}$, like the matrix K, is a symmetric matrix. The matrix $\tilde{K}$ is called the *mass-normalized stiffness* and is analogous to the single-degree-of-freedom constant, k/m.

Window 4.1
Properties of the Symmetric Eigenvalue Problem

The algebraic eigenvalue problem is the problem of computing the scalar λ and the nonzero vector $\mathbf{v}$ satisfying

$$A\mathbf{v} = \lambda\mathbf{v}$$

Here, A is an $n \times n$ real valued, symmetric matrix, the vector $\mathbf{v}$ is $n \times 1$, and there will be n values of the scalar λ, called the *eigenvalues*, and n values of the corresponding vector $\mathbf{v}$, one for each value of λ. The vectors $\mathbf{v}$ are called the *eigenvectors* of the matrix A.

The eigenvalues of A are all real numbers.

The eigenvectors of A are real valued.

The eigenvalues of A are positive numbers if and only if A is positive definite.

The eigenvectors of A can be chosen to be orthogonal, even for repeated eigenvalues.

Symmetry also implies that the set of eigenvectors are linearly independent and can be used like a Fourier series to expand any vector into a sum of eigenvectors. This forms the basis of modal expansions used in both analysis and experiments (Chapter 7). In addition, the numerical algorithms used to compute the eigenvalues and eigenvectors of A are faster and more efficient for symmetric matrices.

Example 4.2.1

Show that $\tilde{K}$ is a symmetric matrix if K and M are symmetric. Note that if M is symmetric so is M^{-1} and $M^{-1/2}$. This is trivial in the cases used here where M is diagonal, but this is also true for fully populated symmetric matrices. Also show the matrix is symmetric using the Cholesky factors of M.

Solution   To show that a matrix is symmetric, use the rule that for any two square matrices of the same size $(AB)^T = B^T A^T$. Applying this rule twice yields

$$\tilde{K}^T = (M^{-1/2} K M^{-1/2})^T = (K M^{-1/2})^T M^{-1/2} = M^{-1/2} K^T M^{-1/2} = M^{-1/2} K M^{-1/2} = \tilde{K}$$

Thus $\tilde{K} = \tilde{K}^T$ and is a symmetric matrix.

$\square$

Equation (4.39) is solved, as before, by assuming a solution of the form $\mathbf{q}(t) = \mathbf{v}e^{j\omega t}$ where $\mathbf{v} \neq \mathbf{0}$ is a vector of constants. Substitution of this form into equation (4.39) yields

$$\tilde{K}\mathbf{v} = \omega^2\mathbf{v} \tag{4.40}$$

upon dividing by the nonzero scalar $e^{j\omega t}$. Here it is important to note that the constant vector $\mathbf{v}$ cannot be zero if motion is to result. Next let $\lambda = \omega^2$ in equation (4.40). This yields

$$\tilde{K}\mathbf{v} = \lambda\mathbf{v} \tag{4.41}$$

where $\mathbf{v} \neq \mathbf{0}$. This is precisely the statement of the algebraic eigenvalue problem. The scalar λ satisfying equation (4.41) for nonzero vectors $\mathbf{v}$ is called the *eigenvalue* and $\mathbf{v}$ is called the (corresponding) *eigenvector*. Since the matrix $\tilde{K}$ is symmetric, this is called the *symmetric eigenvalue problem*. The eigenvector $\mathbf{v}$ generalizes the concept of a mode shape $\mathbf{u}$ used in Section 4.1.

If the system being modeled has n degrees of freedom, each free to move with a single displacement labeled $x_i(t)$, the matrices M, K, and hence $\tilde{K}$ will be $n \times n$, and the vectors $\mathbf{x}(t)$, $\mathbf{q}(t)$, and $\mathbf{v}$ will be $n \times 1$ in dimension. Each subscript i denotes a single degree of freedom where i ranges from 1 to n, and the vector $\mathbf{x}(t)$ denotes the collection of the n degrees of freedom. It is also convenient to label the frequencies ω and eigenvectors $\mathbf{v}$ with subscripts i, so that ω_i and $\mathbf{v}_i$ denote the ith natural frequency and corresponding ith eigenvector, respectively. In this section, only $n = 2$ is considered, but the notation is useful and valid for any number of degrees of freedom.

Equation (4.41) connects the problem of calculating the free vibration response of a conservative system with the mathematics of symmetric eigenvalue problems. This allows the developments of mathematics to be applied directly to vibration. The theoretical advantage of this relationship is significant and is used here. These properties are summarized in Window 4.1 and reviewed in Appendix C. The computational advantage, which is substantial, is discussed in Section 4.9.

Example 4.2.2

Calculate the matrix $\tilde{K}$ for

$$M = \begin{bmatrix} 9 & 0 \\ 0 & 1 \end{bmatrix}, \quad K = \begin{bmatrix} 27 & -3 \\ -3 & 3 \end{bmatrix}$$

as given in Example 4.1.5.

Solution   Matrix products are defined here for matrices of the same size by extending the idea of a matrix times a vector outlined in Example 4.1.1. The result is a third matrix of the same size. The first column of the matrix product AB is the product of the matrix A with the first column of B considered as a vector, and so on. To illustrate this, consider the product $KM^{-1/2}$, where $M^{-1/2}$ is defined by equation (4.36).

$$KM^{-1/2} = \begin{bmatrix} 27 & -3 \\ -3 & 3 \end{bmatrix} \begin{bmatrix} \dfrac{1}{3} & 0 \\ 0 & 1 \end{bmatrix} = \begin{bmatrix} (27)\left(\dfrac{1}{3}\right) + (-3)(0) & (27)(0) + (-3)(1) \\ (-3)\left(\dfrac{1}{3}\right) + 3(0) & (-3)(1) + (3)(1) \end{bmatrix}$$

$$= \begin{bmatrix} 9 & -3 \\ -1 & 3 \end{bmatrix}$$

Multiplying this by $M^{-1/2}$ yields

$$M^{-1/2}KM^{-1/2} = \begin{bmatrix} \frac{1}{3} & 0 \\ 0 & 1 \end{bmatrix} \begin{bmatrix} 9 & -3 \\ -1 & 3 \end{bmatrix}$$

$$= \begin{bmatrix} \left(\frac{1}{3}\right)(9) + (0)(-1) & \left(\frac{1}{3}\right)(-3) + (0)(3) \\ (0)(9) + (1)(-1) & (0)(-3) + (1)(3) \end{bmatrix} = \begin{bmatrix} 3 & -1 \\ -1 & 3 \end{bmatrix}$$

Note that $(M^{-1/2}KM^{-1/2})^T = M^{-1/2}KM^{-1/2}$, so that $\tilde{K}$ is symmetric.

It is tempting to relate equation (4.11) to the algebraic eigenvalue problem by simply multiplying equation (4.17) by M^{-1} to get $\lambda \mathbf{u} = M^{-1}K\mathbf{u}$. However, as the following computation indicates, this does not yield a symmetric eigenvalue problem. Computing the product yields

$$M^{-1}K = \begin{bmatrix} \frac{1}{9} & 0 \\ 0 & 1 \end{bmatrix} \begin{bmatrix} 27 & -3 \\ -3 & 3 \end{bmatrix} = \begin{bmatrix} 3 & -\frac{1}{3} \\ -3 & 3 \end{bmatrix} \neq \begin{bmatrix} 3 & -3 \\ -\frac{1}{3} & 3 \end{bmatrix} = KM^{-1}$$

so that this matrix product is not symmetric. The use of $M^{-1}K$ also becomes computationally more expensive, as discussed in Section 4.9.

□

Alternately, the Cholesky factorization may be used as described in Window 4.2.

The symmetric eigenvalue problem has several advantages. A summary of properties of the symmetric eigenvalue problem is given in Window 4.1. For example, it can readily be shown that the solutions of equation (4.41) are real numbers. Furthermore, it can be shown that the eigenvectors satisfying equation (4.41) are orthogonal and never zero just like the unit vectors $(\hat{\mathbf{i}}, \hat{\mathbf{j}}, \hat{\mathbf{k}}, \hat{\mathbf{e}}_1, \hat{\mathbf{e}}_2, \hat{\mathbf{e}}_3)$ used in the vector analysis of forces (regardless of whether or not the eigenvalues are repeated). Two vectors $\mathbf{v}_1$ and $\mathbf{v}_2$ are defined to be *orthogonal* if their dot product is zero, that is, if

$$\mathbf{v}_1^T\mathbf{v}_2 = 0 \tag{4.42}$$

(Ortho comes from a Greek word meaning straight.) The eigenvectors satisfying (4.41) are of arbitrary length just like the vectors $\mathbf{u}_1$ and $\mathbf{u}_2$ of Section 4.1. Following the analogy of unit vectors from statics (introductory mechanics), the eigenvectors can be normalized so that their length is 1. The norm of a vector is denoted by $\|\mathbf{x}\|$ and defined by

$$\|\mathbf{x}\| = \sqrt{\mathbf{x}^T\mathbf{x}} = \left[\sum_{i=1}^{n} (x_i^2) \right]^{1/2} \tag{4.43}$$

A set of vectors that satisfies both (4.42) and $\|\mathbf{x}\| = 1$ are called *orthonormal*. The unit vectors from a Cartesian coordinate system form an *orthonormal* set of vectors (recall that $\hat{\mathbf{i}} \cdot \hat{\mathbf{i}} = 1$, $\hat{\mathbf{i}} \cdot \hat{\mathbf{j}} = 0$, etc.). A summary of vector inner products is given in Appendix C.

Window 4.2
Symmetric Eigenvalue Problem by Cholesky Factorization

The Cholesky factorization of the mass matrix is $M = LL^T$ for any symmetric M, even if it is not necessarily diagonal. Consider the substitution $\mathbf{x}(t) = (L^T)^{-1}\mathbf{z}(t)$ into equation (4.11) and multiply by L^{-1}. This yields

$$L^{-1}(LL^T)(L^T)^{-1}\ddot{\mathbf{z}}(t) + L^{-1}K(L^T)^{-1}\mathbf{z}(t) = \mathbf{0}$$

The action of taking the transpose and taking the inverse are interchangeable. This, combined with the rule that $(AB)^T = B^T A^T$, yields that the first coefficient is the identity matrix:

$$L^{-1}(LL^T)(L^T)^{-1} = (L^{-1}L)(L^T)(L^T)^{-1} = (I)(I) = I$$

The coefficient of $\mathbf{z}$ is symmetric and used as an alternate definition of the mass-normalized stiffness matrix. To see that it is symmetric, calculate its transpose:

$$\tilde{K}^T = [L^{-1}K(L^T)^{-1}]^T = [(L^T)^{-1}]^T(L^{-1}K)^T = L^{-1}K^T(L^{-1})^T = \tilde{K}$$

since $K = K^T$. Combining all three equations and substitution of $\mathbf{z} = e^{\lambda t}$ yields the symmetric eigenvalue problem $\tilde{K}\mathbf{v} = \lambda\mathbf{v}$. Using L rather the $M^{1/2}$ to form the mass-normalized stiffness matrix has numerical advantages which come into play for larger problems.

To normalize the vector $\mathbf{u}_1 = [1/3 \quad 1]^T$, or any vector for that matter, an unknown scalar α is sought such that the scaled vector $\alpha\mathbf{u}_1$ has unit norm, that is, such that

$$(\alpha\mathbf{u}_1)^T(\alpha\mathbf{u}_1) = 1$$

Writing this expression for $\mathbf{u}_1 = [1/3 \quad 1]^T$ yields

$$\alpha^2(1/9 + 1) = 1$$

or $\alpha = 3/\sqrt{10}$. Thus the new vector $\alpha\mathbf{u}_1 = [1/\sqrt{10} \quad 3/\sqrt{10}]^T$ is the normalized version of the vector $\mathbf{u}_1$. Remember that the eigenvalue problem determines only the direction of the eigenvector, leaving its magnitude arbitrary. The process of normalization is just a systematic way to scale each eigenvector or mode shape. In general, any vector $\mathbf{x}$ can be normalized simply by calculating

$$\frac{1}{\sqrt{\mathbf{x}^T\mathbf{x}}}\mathbf{x} \tag{4.44}$$

Equation (4.44) can be used to normalize any nonzero real vector of any length. Note again that since $\mathbf{x}$ here is an eigenvector, it cannot be zero so that dividing by the scalar $\mathbf{x}^T\mathbf{x}$ is always possible.

The normalizing of the eigenvectors to remove the arbitrary choice of one element in the vector is the systematic method mentioned at the end of Example 4.1.6. Normalizing is an alternative to choosing one element of the vector to have the value 1, as was done in Example 4.1.6.

The following example illustrates the eigenvalue problem, the process of normalizing vectors, and the concept of orthogonal vectors.

Example 4.2.3

Solve the eigenvalue problem for the two-degree-of-freedom system of Example 4.2.2 where

$$\tilde{K} = \begin{bmatrix} 3 & -1 \\ -1 & 3 \end{bmatrix}$$

Normalize the eigenvectors, check if they are orthogonal, and compare them to the mode shapes of Example 4.1.6.

Solution   The eigenvalue problem is to calculate the eigenvalues λ and eigenvectors $\mathbf{v}$ that satisfy equation (4.41). Rewriting equation (4.41) yields

$$(\tilde{K} - \lambda I)\mathbf{v} = \mathbf{0}$$

or

$$\begin{bmatrix} 3 - \lambda & -1 \\ -1 & 3 - \lambda \end{bmatrix}\mathbf{v} = 0 \tag{4.45}$$

where $\mathbf{v}$ must be nonzero. Hence the matrix coefficient must be singular and therefore its determinant must be zero.

$$\det\begin{bmatrix} 3 - \lambda & -1 \\ -1 & 3 - \lambda \end{bmatrix} = \lambda^2 - 6\lambda + 8 = 0$$

This last expression is the characteristic equation and has the two roots

$$\lambda_1 = 2 \quad \text{and} \quad \lambda_1 = 4$$

which are the eigenvalues of the matrix $\tilde{K}$. Note that these are also the squares of the natural frequencies, ω_i^2, as calculated in Example 4.1.5.

The eigenvector associated with λ_1 is calculated from equation (4.41) with $\lambda = \lambda_1 = 2$ and $\mathbf{v}_1 = [v_{11} \quad v_{21}]^T$:

$$(\tilde{K} - \lambda_1 I)\mathbf{v}_1 = \mathbf{0} = \begin{bmatrix} 3 - 2 & -1 \\ -1 & 3 - 2 \end{bmatrix}\begin{bmatrix} v_{11} \\ v_{21} \end{bmatrix} = \begin{bmatrix} 0 \\ 0 \end{bmatrix}$$

This results in the two dependent scalar equations

$$v_{11} - v_{21} = 0 \quad \text{and} \quad -v_{11} + v_{21} = 0$$

Hence $v_{11} = v_{21}$, which defines the direction of the vector $\mathbf{v}_1$.

To fix a value for the elements of $\mathbf{v}_1$, the normalization condition of equation (4.44) is used to force $\mathbf{v}_1$ to have a magnitude of 1. This results in (setting $v_{11} = v_{21}$)

$$1 = \|\mathbf{v}_1\| = \sqrt{v_{21}^2 + v_{21}^2} = \sqrt{2}\,v_{21}$$

Solving for v_{21} yields

$$v_{21} = \frac{1}{\sqrt{2}}$$

so that the normalized vector $\mathbf{v}_1$ becomes

$$\mathbf{v}_1 = \frac{1}{\sqrt{2}}\begin{bmatrix} 1 \\ 1 \end{bmatrix}$$

Similarly, substitution of $\lambda_2 = 4$ into (4.41), solving for the elements of $\mathbf{v}_2$, and normalizing the result yields

$$\mathbf{v}_2 = \frac{1}{\sqrt{2}}\begin{bmatrix} 1 \\ -1 \end{bmatrix}$$

Now note that the product $\mathbf{v}_1^T\mathbf{v}_2$ yields

$$\mathbf{v}_1^T\mathbf{v}_2 = \frac{1}{\sqrt{2}}\frac{1}{\sqrt{2}}\begin{bmatrix} 1 & 1 \end{bmatrix}\begin{bmatrix} 1 \\ -1 \end{bmatrix} = \frac{1}{2}(1 - 1) = 0$$

so that the set of vectors $\mathbf{v}_1$ and $\mathbf{v}_2$ *are orthogonal* as well as normal. Hence the two vectors $\mathbf{v}_1$ and $\mathbf{v}_2$ form an orthonormal set as described in Window 4.3.

Next, consider the mode-shape vectors computed in Example 4.1.6 directly from equations (4.22) and (4.23): $\mathbf{u}_1 = [1/3 \quad 1]^T$ and $\mathbf{u}_2 = [-1/3 \quad 1]^T$. The following calculation shows that these vectors *are not orthogonal*:

$$\mathbf{u}_1^T\mathbf{u}_2 = \begin{bmatrix} \frac{1}{3} & 1 \end{bmatrix}\begin{bmatrix} -\frac{1}{3} \\ 1 \end{bmatrix} = -\frac{1}{9} + 1 = \frac{8}{9} \neq 0$$

Next, normalize $\mathbf{u}_1$ and $\mathbf{u}_2$ using equation (4.44) to get

$$\hat{\mathbf{u}}_1 = \frac{1}{\sqrt{\mathbf{u}_1^T\mathbf{u}_1}}\mathbf{u}_1 = \frac{1}{\sqrt{\frac{1}{9} + 1}}\begin{bmatrix} \frac{1}{3} \\ 1 \end{bmatrix} = \begin{bmatrix} 0.31623 \\ 0.94868 \end{bmatrix}$$

where the "hat" is used to denote a unit vector, a notation that is usually dropped in favor of renaming the normalized version $\mathbf{u}_1$ as well. Following a similar calculation, the normalized version of $\mathbf{u}_2$ becomes $\mathbf{u}_2 = [-0.31623 \quad 0.948681]^T$. Note that the normalized versions of $\mathbf{u}_1$ and $\mathbf{u}_2$ are not orthogonal either. Only the eigenvectors computed from the symmetric matrix $\tilde{K}$ are orthogonal. However, the vectors $\mathbf{u}_i$ computed in

Example 4.1.6 can be normalized and made orthogonal with respect to the mass matrix M, as discussed later. The mode shapes $\mathbf{u}_i$ and eigenvectors $\mathbf{v}_i$ are related by the square root of the mass matrix as discussed next.

☐

The previous example showed how to calculate the eigenvalues and eigenvectors of the symmetric eigenvalue problem related to the vibration problem. In comparing this to the mode shape and natural frequency calculation made earlier, the eigenvalues are exactly the squares of the natural frequencies. However, there is some difference between the mode shapes of Example 4.1.6 and the eigenvectors of the previous example. This difference is captured by the fact that the mode shapes as calculated in Example 4.1.6 are not orthogonal, but the eigenvectors are orthogonal. The property of orthogonality is extremely important in developing modal analysis (Section 4.3) because it allows the equations of motion to uncouple, reducing the analysis to that of solving several single-degree-of-freedom systems defined by scalar equations.

The eigenvectors and the mode shapes are related through equation (4.37) by

$$\mathbf{u}_1 = M^{-1/2}\mathbf{v}_1 \quad \text{and} \quad \mathbf{v}_1 = M^{1/2}\mathbf{u}_1$$

To see this, note that

$$M^{1/2}\mathbf{u}_1 = \begin{bmatrix} 3 & 0 \\ 0 & 1 \end{bmatrix} \begin{bmatrix} \frac{1}{3} \\ 1 \end{bmatrix} = \begin{bmatrix} 1 \\ 1 \end{bmatrix} = \mathbf{v}_1$$

using the values from the examples. Thus the mode shapes and eigenvectors are related by a simple matrix transformation, $M^{1/2}$. The important point to remember from this series of examples is that the eigenvalues are the squares of the natural frequencies and that the mode shapes are related to the eigenvectors by a factor of the mass matrix.

As indicated in Example 4.2.3, the eigenvectors of a symmetric matrix are orthogonal and can always be calculated to be normal. Such vectors are called *orthonormal*, as summarized in Window 4.3. This fact can be used to decouple the equations of motion of any order undamped system by making a new matrix P out of the normalized eigenvectors, such that each vector forms a column. Thus the matrix P is defined by

$$p = [\mathbf{v}_1 \quad \mathbf{v}_2 \quad \mathbf{v}_3 \ldots \mathbf{v}_n] \tag{4.46}$$

where n is the number of degrees of freedom in the system ($n = 2$ for the examples of this section). Note the matrix P has the unique property that $P^T P = I$, which follows directly from considering the matrix product definition that the ijth element of $P^T P$ is the product of the ith row of P^T with the jth column of P. Matrices that satisfy the equation $P^T P = I$ are called *orthogonal matrices*.

Window 4.3
Summary of Orthonormal Vectors

Two vectors $\mathbf{x}_1$ and $\mathbf{x}_2$ are normal if

$$\mathbf{x}_1^T\mathbf{x}_1 = 1 \quad\text{and}\quad \mathbf{x}_2^T\mathbf{x}_2 = 1$$

and orthogonal if $\mathbf{x}_1^T\mathbf{x}_2 = 0$. If $\mathbf{x}_1$ and $\mathbf{x}_2$ are normal and orthogonal, they are said to be *orthonormal*. This is abbreviated

$$\mathbf{x}_i^T\mathbf{x}_j = \delta_{ij} \quad i = 1, 2, \quad j = 1, 2$$

where δ_{ij} is the *Kronecker delta*, defined by

$$\delta_{ij} = \begin{cases} 0 \text{ if } i \neq j \\ 1 \text{ if } i = j \end{cases}$$

If a set of n vectors $\{\mathbf{x}\}_{i=1}^n$ is orthonormal, it is denoted by

$$\mathbf{x}_i^T\mathbf{x}_j = \delta_{ij} \quad i, j = 1, 2, \dots n$$

Be careful not to confuse δ_{ij}, the Kronecker delta used here, with $\delta = \ln[x(t)/x(t + T)]$, the logarithmic decrement of Section 1.6, or with $\delta(t - t_j)$, the Dirac delta or impulse function of Section 3.1, or with the static deflection δ_s of Section 5.2.

Example 4.2.4

Write out the matrix P for the system of Example 4.2.3 and calculate P^TP.

Solution   Using the values for the orthonormal vectors $\mathbf{v}_1$ and $\mathbf{v}_2$ from Example 4.2.3 yields

$$P = [\mathbf{v}_1 \quad \mathbf{v}_2] = \frac{1}{\sqrt{2}}\begin{bmatrix} 1 & 1 \\ 1 & -1 \end{bmatrix}$$

so that P^TP becomes

$$P^TP = \left(\frac{1}{\sqrt{2}}\right)\left(\frac{1}{\sqrt{2}}\right)\begin{bmatrix} 1 & 1 \\ 1 & -1 \end{bmatrix}\begin{bmatrix} 1 & 1 \\ 1 & -1 \end{bmatrix}$$

$$= \frac{1}{2}\begin{bmatrix} 1 + 1 & 1 - 1 \\ 1 - 1 & 1 + 1 \end{bmatrix} = \begin{bmatrix} 1 & 0 \\ 0 & 1 \end{bmatrix} = I$$

$\square$

Another interesting and useful matrix calculation is to consider the product of the three matrices $P^T\tilde{K}P$. It can be shown (see Appendix C) that this product results in a diagonal matrix. Furthermore, the diagonal entries are the eigenvalues of the matrix $\tilde{K}$ and the squares of the system's natural frequencies. This is denoted by

$$\Lambda = \text{diag}(\lambda_i) = P^T \tilde{K} P \tag{4.47}$$

and is called the *spectral matrix* of $\tilde{K}$. The following example illustrates this calculation.

Example 4.2.5

Calculate the matrix $P^T \tilde{K} P$ for the two-degree-of-freedom system of Example 4.2.2.

$$P^T \tilde{K} P = \frac{1}{\sqrt{2}} \begin{bmatrix} 1 & 1 \\ 1 & -1 \end{bmatrix} \begin{bmatrix} 3 & -1 \\ -1 & 3 \end{bmatrix} \frac{1}{\sqrt{2}} \begin{bmatrix} 1 & 1 \\ 1 & -1 \end{bmatrix}$$

$$= \frac{1}{2} \begin{bmatrix} 1 & 1 \\ 1 & -1 \end{bmatrix} \begin{bmatrix} 2 & 4 \\ 2 & -4 \end{bmatrix}$$

$$= \frac{1}{2} \begin{bmatrix} 4 & 0 \\ 0 & 8 \end{bmatrix} = \begin{bmatrix} 2 & 0 \\ 0 & 4 \end{bmatrix} = \Lambda$$

Note that the diagonal elements of the spectral matrix Λ are the natural frequencies ω_1 and ω_2 squared. That is, from Example 4.2.3, $\omega_1^2 = 2$ and $\omega_2^2 = 4$, so that $\Lambda = \text{diag}(\omega_1^2 \quad \omega_2^2) = \text{diag}(2 \quad 4)$.

□

Examining the solution of Example 4.2.5 and comparing it to the natural frequencies of Example 4.1.3 suggests that in general

$$\Lambda = \text{diag}(\lambda_i) = \text{diag}(\omega_i^2) \tag{4.48}$$

This expression connects the eigenvalues with the natural frequencies (i.e., $\lambda_i = \omega_i^2$). The following example illustrates the matrix methods for vibration analysis presented in this section and provides a summary.

Example 4.2.6

Consider the system of Figure 4.4. Write the dynamic equations in matrix form, calculate $\tilde{K}$, its eigenvalues and eigenvectors, and hence determine the natural frequencies of the system (use $m_1 = 1$ kg, $m_2 = 4$ kg, $k_1 = k_3 = 10$ N/m, and $k_2 = 2$ N/m). Also calculate the matrices P and Λ, and show that equation (4.47) is satisfied and that $P^T P = I$.

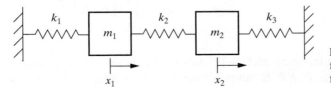

Figure 4.4 A two-degree-of-freedom model of a structure fixed at both ends.

Solution Using free-body diagrams of each of the two masses yields the following equations of motion:

$$m_1\ddot{x}_1 + (k_1 + k_2)x_1 - k_2x_2 = 0$$
$$m_2\ddot{x}_2 - k_2x_1 + (k_2 + k_3)x_2 = 0 \qquad (4.49)$$

In matrix form this becomes

$$\begin{bmatrix} m_1 & 0 \\ 0 & m_2 \end{bmatrix}\ddot{x}(t) + \begin{bmatrix} k_1 + k_2 & -k_2 \\ -k_2 & k_2 + k_3 \end{bmatrix}x(t) = 0 \qquad (4.50)$$

Using the numerical values for the physical parameters m_i and k_i yields that

$$M = \begin{bmatrix} 1 & 0 \\ 0 & 4 \end{bmatrix} \qquad K = \begin{bmatrix} 12 & -2 \\ -2 & 12 \end{bmatrix}$$

The matrix $M^{-1/2}$ becomes

$$M^{-1/2} = \begin{bmatrix} 1 & 0 \\ 0 & \frac{1}{2} \end{bmatrix}$$

so that

$$\tilde{K} = M^{-1/2}(KM^{-1/2}) = \begin{bmatrix} 1 & 0 \\ 0 & \frac{1}{2} \end{bmatrix}\begin{bmatrix} 12 & -1 \\ -2 & 6 \end{bmatrix} = \begin{bmatrix} 12 & -1 \\ -1 & 3 \end{bmatrix}$$

Note that $\tilde{K}$ is symmetric (i.e., $\tilde{K}^T = \tilde{K}$), as expected. The eigenvalues of $\tilde{K}$ are calculated from

$$\det(\tilde{K} - \lambda I) = \det\begin{bmatrix} 12 - \lambda & -1 \\ -1 & 3 - \lambda \end{bmatrix} = \lambda^2 - 15\lambda + 35 = 0$$

This quadratic equation has the solution

$$\lambda = \frac{15}{2} \pm \frac{1}{2}\sqrt{85}$$

so that

$$\lambda_1 = 2.8902 \qquad \lambda_2 = 12.1098$$

Thus $\omega_1 = \sqrt{\lambda_1} = 1.7$ and $\omega_2 = \sqrt{\lambda_2} = 3.48$ rad/s.

The eigenvectors are calculated from (for λ_1)

$$\begin{bmatrix} 12 - 2.8902 & -1 \\ -1 & 3 - 2.8902 \end{bmatrix}\begin{bmatrix} v_{11} \\ v_{21} \end{bmatrix} = \begin{bmatrix} 0 \\ 0 \end{bmatrix}$$

so that the vector $\mathbf{v}_1 = [v_{11} \quad v_{21}]^T$ satisfies

$$9.1098 v_{11} = v_{21}$$

Normalizing the vector $\mathbf{v}_1$ yields

$$1 = \|\mathbf{v}_1\| = \sqrt{v_{11}^2 + v_{21}^2} = \sqrt{v_{11}^2 + (9.1098)^2 v_{11}^2}$$

so that

$$v_{11} = \frac{1}{\sqrt{1 + (9.1098)^2}} = 0.1091$$

and

$$v_{21} = 9.1098 v_{11} = 0.9940$$

Thus the normalized eigenvector $\mathbf{v}_1 = [0.1091 \quad 0.9940]^T$.

Similarly, the vector $\mathbf{v}_2$ corresponding to the eigenvalue λ_2 in normalized form becomes $\mathbf{v}_2 = [-0.9940 \quad 0.1091]^T$. Note that $\mathbf{v}_1^T \mathbf{v}_2 = 0$ and $\sqrt{\mathbf{v}_1^T \mathbf{v}_1} = 1$. The matrix of eigenvectors P becomes

$$P = [\mathbf{v}_1 \quad \mathbf{v}_2] = \begin{bmatrix} 0.1091 & -0.9940 \\ 0.9940 & 0.1091 \end{bmatrix}$$

Thus the matrix Λ becomes

$$\Lambda = P^T \tilde{K} P = \begin{bmatrix} 0.1091 & 0.9940 \\ -0.9940 & 0.1091 \end{bmatrix} \begin{bmatrix} 12 & -1 \\ -1 & 3 \end{bmatrix} \begin{bmatrix} 0.1091 & -0.9940 \\ 0.9940 & 0.1091 \end{bmatrix} = \begin{bmatrix} 2.8402 & 0 \\ 0 & 12.1098 \end{bmatrix}$$

This shows that the matrix P transforms the mass-normalized stiffness matrix into a diagonal matrix of the squares of the natural frequencies. Furthermore,

$$P^T P = \begin{bmatrix} 0.1091 & 0.9940 \\ -0.9940 & 0.1091 \end{bmatrix} \begin{bmatrix} 0.1091 & -0.9940 \\ 0.9940 & 0.1091 \end{bmatrix} = \begin{bmatrix} 1 & 0 \\ 0 & 1 \end{bmatrix} = I$$

as it should.

$\square$

The computations made in the previous example can all be performed easily in most programmable calculators as well as in the mathematics software packages used in Sections 1.9, 2.8, and 3.9, in the Toolbox, and as illustrated in Section 4.9. It is good to work a few of the calculations for frequencies and eigenvectors by hand. However, larger problems require the accuracy of using a code.

An alternative approach to normalizing mode shapes is often used. This method is based on equation (4.17). Each vector $\mathbf{u}_i$ corresponding to each natural frequency ω_i is normalized with respect to the mass matrix M by scaling the mode-shape vector, which satisfies equation (4.17) such that the vector $\mathbf{w}_i = \alpha_i \mathbf{u}_i$ satisfies

$$(\alpha_i \mathbf{u}_i)^T M (\alpha_i \mathbf{u}_i) = 1 \tag{4.51}$$

or

$$\alpha_i^2 \mathbf{u}_i^T M \mathbf{u}_i = \mathbf{w}_i^T M \mathbf{w}_i = 1$$

This yields the special choice of $\alpha_i = 1/\sqrt{\mathbf{u}_i^T M \mathbf{u}_i}$.

The vector $\mathbf{w}_i$ is said to be *mass normalized*. Multiplying equation (4.17) by the scalar α_i yields (for $i = 1$ and 2)

$$-\omega_i^2 M \mathbf{w}_i + K \mathbf{w}_i = \mathbf{0} \tag{4.52}$$

Multiplying this by $\mathbf{w}_i^T$ yields the two scalar relations (for $i = 1$ and 2)

$$\omega_i^2 = \mathbf{w}_i^T K \mathbf{w}_i \qquad i = 1, 2 \tag{4.53}$$

where the mass normalization $\mathbf{w}_i^T M \mathbf{w}_i = 1$ of equation (4.51) was used to evaluate the left side.

Next, consider the vector $\mathbf{v}_i = M^{1/2} \mathbf{u}_i$, where $\mathbf{v}_i$ is normalized so that $\mathbf{v}_i^T \mathbf{v}_i = 1$. Then by substitution

$$\mathbf{v}_i^T \mathbf{v}_i = (M^{1/2} \mathbf{u}_i)^T M^{1/2} \mathbf{u}_i$$
$$= \mathbf{u}_i^T M^{1/2} M^{1/2} \mathbf{u}_i$$
$$= \mathbf{u}_i^T M \mathbf{u}_i = 1$$

so that the vector $\mathbf{u}_i$ is mass normalized. In this last argument, the property of the transpose is used [i.e., $(M^{1/2}\mathbf{u})^T = \mathbf{u}^T (M^{1/2})^T$] and the fact that $M^{1/2}$ is symmetric, so that $(M^{1/2})^T = M^{1/2}$.

As vibration problems become more complex, more degrees of freedom are needed to model the system's behavior. Thus the mass and stiffness matrices introduced in Section 4.1 become large and the analysis becomes more complicated, even though the basic principle of eigenvalues/frequencies and eigenvectors/mode shapes remains the same. Increased understanding of the vibration problem can be obtained by borrowing from the theory of matrices and computational linear algebra. There are three different methods of casting the undamped-vibration problem in terms of the matrix theory. They are

$$\text{(i) } \omega^2 M \mathbf{u} = K \mathbf{u} \qquad \text{(ii) } \omega^2 \mathbf{u} = M^{-1} K \mathbf{u} \qquad \text{(iii) } \omega^2 \mathbf{v} = M^{-1/2} K M^{-1/2} \mathbf{v}$$

Each of these methods results in identical natural frequencies and mode shapes to within numerical precision. Each of these three eigenvalue problems is related by a simple matrix transformation. The following summarizes their differences.

(i) *The Generalized Symmetric Eigenvalue Problem* This is the simplest method to compute by hand. However, when using a code, $\mathbf{u}$ must be further normalized with respect to the mass matrix to obtain an orthonormal set. In addition, the matrix transformation from (i) to (iii) must be used to prove that the resulting ω_i are real and to prove the existence of orthogonal mode shapes. This is the second most expensive computational algorithm, requiring $7n^3$ floating-point operations per second (flops) where n is the number of degrees of freedom.

(ii) *The Asymmetric Eigenvalue Problem* When using a code, **u** must be further normalized with respect to the mass matrix to obtain an orthonormal set. This is the most expensive computational algorithm, requiring $15n^3$ flops.

(iii) *The Symmetric Eigenvalue Problem* Because symmetry is preserved, this, or its Cholesky equivalent, is the best method to use for larger systems or when using a code. The resulting eigenvectors $\mathbf{v}_i$ form an orthonormal set. Algorithms require only n^3 flops, including transforming to the symmetric form and transforming the result to physical mode shapes (computed by a simple matrix multiplication).

The question remains: Which approach should be used? The symmetric form (iii) is used here because computational algorithms produce orthonormal eigenvectors without additional calculation and because the algorithm for (iii) is less computationally intensive. In addition, the analytical properties of the symmetric eigenvalue problem can be used directly. For two-degree-of-freedom systems, computation should be done by hand, and the most straightforward hand calculation is to proceed with the generalized eigenvalue problem $\lambda M\mathbf{u} = K\mathbf{u}$, as analyzed in Section 4.1. For problems with more than two degrees of freedom, a code should be used to avoid numerical mistakes and ensure accuracy. In these practical cases, the most efficient approach is to use the symmetric eigenvalue problem (iii) presented in this section.

The symmetric eigenvalue problem for the mass-normalized stiffness matrix $(\tilde{K})$ provides the transformation (P) that diagonalizes $\tilde{K}$ and uncouples the equations of motion. This process of transforming $\tilde{K}$ into a diagonal form is called *modal analysis* and forms the topic of the next section.

4.3 MODAL ANALYSIS

The matrix of eigenvectors P calculated in Section 4.2 can be used to decouple the equations of vibration into two separate equations. The two separate equations are then second-order-single-degree-of-freedom equations that can be solved and analyzed using the methods of Chapters 1 through 3. The matrices P and $M^{-1/2}$ can be used again to transform the solution back to the original coordinate system. The matrices P and $M^{-1/2}$ can also be called *transformations*, which is appropriate in this case because they are used to transform the vibration problem between different coordinate systems. This procedure is called *modal analysis*, because the transformation $S = M^{-1/2}P$, often called the modal matrix, is related to the mode shapes of the vibrating system.

Consider the matrix form of the equation of vibration

$$M\ddot{\mathbf{x}}(t) + K\mathbf{x}(t) = \mathbf{0} \tag{4.54}$$

subject to the initial conditions

$$\mathbf{x}(0) = \mathbf{x}_0 \qquad \dot{\mathbf{x}}(0) = \dot{\mathbf{x}}_0$$

Here $\mathbf{x}_0 = [x_1(0) \quad x_2(0)]^T$ is the vector of initial displacements, and $\dot{\mathbf{x}}_0 = [\dot{x}_1(0) \quad \dot{x}_2(0)]^T$ is the vector of initial velocities. As outlined in Section 4.2, substitution of $\mathbf{x} = M^{-1/2}\mathbf{q}(t)$ into equation (4.54) and multiplying from the left by $M^{-1/2}$ yields

$$I\ddot{\mathbf{q}}(t) + \tilde{K}\mathbf{q}(t) = \mathbf{0} \tag{4.55}$$

where $\ddot{\mathbf{x}} = M^{-1/2}\ddot{\mathbf{q}}$, since the matrix M is constant, and where $\tilde{K} = M^{-1/2}KM^{-1/2}$, as before. The transformation $M^{-1/2}$ simply transforms the problem from the co-ordinate system defined by $\mathbf{x} = [x_1(t) \quad x_2(t)]^T$ to a new coordinate system, $\mathbf{q} = [q_1(t) \quad q_2(t)]^T$, defined by

$$\mathbf{q}(t) = M^{1/2}\mathbf{x}(t) \tag{4.56}$$

Next, define a second coordinate system $\mathbf{r}(t) = [r_1(t) \quad r_2(t)]^T$ by

$$\mathbf{q}(t) = P\mathbf{r}(t) \tag{4.57}$$

where P is the matrix composed of the orthonormal eigenvectors of $\tilde{K}$ as defined in equation (4.46). Substitution of the vector $\mathbf{q} = P\mathbf{r}(t)$ into equation (4.55) and multiplying from the left by the matrix P^T yields

$$P^T P\ddot{\mathbf{r}}(t) + P^T \tilde{K}P\mathbf{r}(t) = \mathbf{0} \tag{4.58}$$

Using the result $P^T P = I$ and equation (4.47), this can be reduced to

$$I\ddot{\mathbf{r}}(t) + \Lambda\mathbf{r}(t) = \mathbf{0} \tag{4.59}$$

Equation (4.59) can be written out by performing the indicated matrix calculations as

$$\begin{bmatrix} 1 & 0 \\ 0 & 1 \end{bmatrix}\begin{bmatrix} \ddot{r}_1(t) \\ \ddot{r}_2(t) \end{bmatrix} + \begin{bmatrix} \omega_1^2 & 0 \\ 0 & \omega_2^2 \end{bmatrix}\begin{bmatrix} r_1(t) \\ r_2(t) \end{bmatrix} = \begin{bmatrix} 0 \\ 0 \end{bmatrix} \tag{4.60}$$

$$\begin{bmatrix} \ddot{r}_1(t) + \omega_1^2 r_1(t) \\ \ddot{r}_2(t) + \omega_2^2 r_2(t) \end{bmatrix} = \begin{bmatrix} 0 \\ 0 \end{bmatrix} \tag{4.61}$$

The equality of the two vectors in this last expression implies the two decoupled equations

$$\ddot{r}_1(t) + \omega_1^2 r_1(t) = 0 \tag{4.62}$$

and

$$\ddot{r}_2(t) + \omega_2^2 r_2(t) = 0 \tag{4.63}$$

These two equations are subject to initial conditions which must also be transformed into the new coordinate system $\mathbf{r}(t)$ from the original coordinate system $\mathbf{x}(t)$. Following the preceding two transformations applied to the initial displacement yields

$$\mathbf{r}_0 = \begin{bmatrix} r_{10} \\ r_{20} \end{bmatrix} = P^T \mathbf{q}(0) = P^T M^{1/2} \mathbf{x}_0 \qquad (4.64)$$

where $\mathbf{q}(t) = P\mathbf{r}(t)$ was multiplied by P^T to get $\mathbf{r}(t) = P^T \mathbf{q}(t)$, since $P^T P = I$. Likewise the initial velocity in the decoupled coordinate system, $\mathbf{r}(t)$, becomes

$$\dot{\mathbf{r}}_0 = \begin{bmatrix} \dot{r}_{10} \\ \dot{r}_{20} \end{bmatrix} = P^T \dot{\mathbf{q}}(0) = P^T M^{1/2} \dot{\mathbf{x}}_0 \qquad (4.65)$$

Equations (4.62) and (4.63) are called the *modal equations*, and the coordinate system $\mathbf{r}(t) = [r_1(t) \quad r_2(t)]^T$ is called the *modal coordinate system*. Equations (4.62) and (4.63) are said to be decoupled because each depends only on a single coordinate. Hence each equation can be solved independently by using the method of Sections 1.1 and 1.2 (see Window 4.4). Denoting the initial conditions individually by r_{10}, $\dot{r}_{10}$, r_{20}, and $\dot{r}_{20}$ and using equation (1.10), the solution of each of the modal equations (4.62) and (4.63) is simply

$$r_1(t) = \frac{\sqrt{\omega_1^2 r_{10}^2 + \dot{r}_{10}^2}}{\omega_1} \sin\left(\omega_1 t + \tan^{-1} \frac{\omega_1 r_{10}}{\dot{r}_{10}}\right) \qquad (4.66)$$

$$r_2(t) = \frac{\sqrt{\omega_2^2 r_{20}^2 + \dot{r}_{20}^2}}{\omega_2} \sin\left(\omega_2 t + \tan^{-1} \frac{\omega_2 r_{20}}{\dot{r}_{20}}\right) \qquad (4.67)$$

provided ω_1 and ω_2 are nonzero.

Window 4.4
***Review of the Solution to a Single-Degree-of-Freedom Undamped System
from Section 1.1 and Window 1.2***

The solution to $m\ddot{x} + kx = 0$ or $\ddot{x} + \omega_n^2 x = 0$ subject to $x(0) = x_0$ and $\dot{x}(0) = v_0$ is

$$x(t) = \sqrt{x_0^2 + \frac{v_0^2}{\omega_n^2}} \sin\left(\omega_n t + \tan^{-1} \frac{\omega_n x_0}{v_0}\right) \qquad (1.10)$$

where $\omega_n = \sqrt{k/m} \neq 0$.

Once the modal solutions (4.66) and (4.67) are known, the transformations $M^{1/2}$ and P can be used on the vector $\mathbf{r}(t) = [r_1(t) \quad r_2(t)]^T$ to recover the solution $\mathbf{x}(t)$ in the physical coordinates $x_1(t)$ and $x_2(t)$. To obtain the vector $\mathbf{x}$ from the vector $\mathbf{r}$, substitute equations (4.56) into $\mathbf{q}(t) = P\mathbf{r}(t)$ to get

$$\mathbf{x}(t) = M^{-1/2}\mathbf{q}(t) = M^{-1/2}P\mathbf{r}(t) \tag{4.68}$$

The matrix product $M^{-1/2}P$ is again a matrix, which is denoted by

$$S = M^{-1/2}P \tag{4.69}$$

and is the same size as M and P (2×2). The matrix S is called the *matrix of mode shapes*, each column of which is a mode-shape vector. This procedure, referred to as *modal analysis*, provides a means of calculating the solution to a two-degree-of-freedom vibration problem by performing a number of matrix calculations. The usefulness of this approach is that these matrix computations can easily be automated in a computer code (even on some calculators). In addition, the modal-analysis procedure is easily extended to systems with an arbitrary number of degrees of freedom, as developed in the next section. Figure 4.5 illustrates the coordinate transformation used in modal analysis. The following summarizes the procedure of modal analysis using the matrix transformation S. This is followed by an example that re-solves Example 4.1.6 using modal methods.

Figure 4.5 summarizes how computing the matrix of mode shapes S transforms the vibration problem from a coupled set of equations of motion into a set of single-degree-of-freedom problems. Effectively, the matrix S transforms multiple-degree-of-freedom problems that are complicated to solve into single-degree-of-freedom problems that are easy to solve (from Chapter 1). Furthermore, the single-degree-of-freedom problems

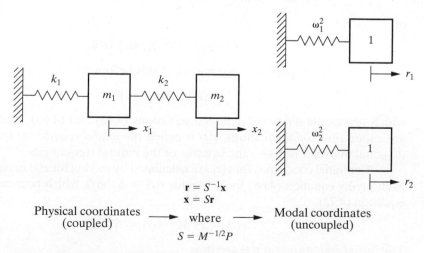

Figure 4.5   Schematic illustration of decoupling equations of motion using modal analysis and the matrix of mode shapes S.

obtained by the modal transformation S all have unit mass and each has a stiffness corresponding to one of the system's natural frequencies squared. Not only is the modal description given on the right side of Figure 4.5 easy to solve, but it forms the basis of most vibration tests, called modal testing, as discussed in Chapter 7. The idea of modal analysis is one of the foundations of vibration analysis (the others being the concepts of natural frequency and resonance), and as such the properties of the matrix S are extremely important.

Taking the transpose of equation (4.69) yields

$$S^T = (M^{1/2}P)^T = P^T M^{-1/2} \tag{4.70}$$

since $(AB)^T = B^T A^T$ (see Appendix C). In addition, the inverse of a matrix product is given by $(AB)^{-1} = B^{-1}A^{-1}$ (see Appendix C), so that

$$S^{-1} = (M^{1/2}P)^{-1} = P^{-1}M^{1/2} \tag{4.71}$$

However, the matrix P has as its inverse P^T since $P^T P = I$. Thus equation (4.71) yields that the inverse of the matrix of mode shapes is

$$S^{-1} = P^T M^{1/2} \tag{4.72}$$

These matrix results are useful for solving equation (4.54) by modal analysis.

The modal analysis of equation (4.54) starts with the substitution of $\mathbf{x}(t) = S\mathbf{r}(t)$ into equation (4.54). Multiplying the result by S^T yields

$$S^T M S \ddot{\mathbf{r}}(t) + S^T K S \mathbf{r}(t) = \mathbf{0} \tag{4.73}$$

Expanding the matrix S in equation (4.73) into its factors as given by equation (4.69) yields

$$P^T M^{-1/2} M M^{-1/2} P \ddot{\mathbf{r}}(t) + P^T M^{-1/2} K M^{-1/2} P \mathbf{r}(t) = \mathbf{0} \tag{4.74}$$

or

$$P^T P \ddot{\mathbf{r}}(t) + P^T \widetilde{K} P \mathbf{r}(t) = \mathbf{0} \tag{4.75}$$

Using the properties of the matrix P, this becomes

$$\ddot{\mathbf{r}}(t) + \Lambda \mathbf{r}(t) = \mathbf{0} \tag{4.76}$$

which represents the two decoupled equations (4.62) and (4.63). Recall that these are called the *modal equations*, $\mathbf{r}(t)$ is called the *modal coordinate system*, and the diagonal matrix Λ contains the squares of the natural frequencies.

The initial conditions for $\mathbf{r}(t)$ are calculated by solving for $\mathbf{r}(t)$ in equation (4.68). Multiplying equation (4.68) by S^{-1} yields $\mathbf{r}(t) = S^{-1}\mathbf{x}(t)$, which becomes, after using equation (4.72),

$$\mathbf{r}(t) = P^T M^{1/2} \mathbf{x}(t) = S^{-1}\mathbf{x}(t) \tag{4.77}$$

The initial conditions on $\mathbf{r}(t)$ are thus

$$\mathbf{r}(0) = P^T M^{1/2} \mathbf{x}_0 \quad \text{and} \quad \dot{\mathbf{r}}(0) = P^T M^{1/2} \dot{\mathbf{x}}_0 \tag{4.78}$$

as derived in equations (4.64) and (4.65). With the initial conditions transformed into modal coordinates by equations (4.78), the modal equations given by equation (4.76) yield the solution vector in modal coordinates $\mathbf{r}(t)$. In order to obtain the solution in the original physical coordinate system, $\mathbf{x}(t)$, the transformation S is again used. Multiplying $\mathbf{r}(t) = S^{-1}\mathbf{x}(t)$ by S to get

$$\mathbf{x}(t) = S\mathbf{r}(t) = M^{-1/2}P\mathbf{r}(t) \tag{4.79}$$

yields the solution in physical coordinates. Equation (4.79) effectively takes the solution from the right side of Figure 4.5 back to the left side. The basic idea is that the matrix S^{-1} takes the problem from the physical coordinates, where the equations of motion are coupled and hard to solve, to modal coordinates, where the equations are uncoupled and easy to solve. Then the modal matrix S takes the solution back to the physical coordinates corresponding to the original problem. These steps are summarized in Window 4.5 and illustrated in the following examples.

Window 4.5
Steps in Solving Equation (4.54) by Modal Analysis

1. Calculate $M^{-1/2}$.
2. Calculate $\tilde{K} = M^{-1/2}KM^{-1/2}$, the mass-normalized stiffness matrix.
3. Calculate the symmetric eigenvalue problem for $\tilde{K}$ to get ω_i^2 and $\mathbf{v}_i$.
4. Normalize $\mathbf{v}_i$ and form the matrix $P = [\mathbf{v}_1 \quad \mathbf{v}_2]$.
5. Calculate $S = M^{-1/2}P$ and $S^{-1} = P^T M^{1/2}$.
6. Calculate the modal initial conditions: $\mathbf{r}(0) = S^{-1}\mathbf{x}_0, \dot{\mathbf{r}}(0) = S^{-1}\dot{\mathbf{x}}_0$.
7. Substitute the components of $\mathbf{r}(0)$ and $\dot{\mathbf{r}}(0)$ into equations (4.66) and (4.67) to get the solution in modal coordinate $\mathbf{r}(t)$.
8. Multiply $\mathbf{r}(t)$ by S to get the solution $\mathbf{x}(t) = S\mathbf{r}(t)$.

Note that S is the matrix of mode shapes and P is the matrix of eigenvectors.

Example 4.3.1

Calculate the solution of the two-degree-of-freedom system given by

$$M = \begin{bmatrix} 9 & 0 \\ 0 & 1 \end{bmatrix} \quad K = \begin{bmatrix} 27 & -3 \\ -3 & 3 \end{bmatrix} \quad \mathbf{x}(0) = \begin{bmatrix} 1 \\ 0 \end{bmatrix} \quad \dot{\mathbf{x}}(0) = \mathbf{0}$$

using modal analysis. Compare the result to that obtained in Example 4.1.6 for the same system and initial conditions.

Solution From Examples 4.1.5, 4.2.2, 4.2.4, and 4.2.5 the following have been calculated

$$M^{-1/2} = \begin{bmatrix} \frac{1}{3} & 0 \\ 0 & 1 \end{bmatrix} \qquad \tilde{K} = \begin{bmatrix} 3 & -1 \\ -1 & 3 \end{bmatrix}$$

$$P = \frac{1}{\sqrt{2}}\begin{bmatrix} 1 & 1 \\ 1 & -1 \end{bmatrix} \qquad \Lambda = \mathrm{diag}\,(2, 4)$$

which provides the information required in the first three steps of Window 4.5. The next step is to calculate the matrix S and its inverse.

$$S = M^{-1/2}P = \frac{1}{\sqrt{2}}\begin{bmatrix} \frac{1}{3} & 0 \\ 0 & 1 \end{bmatrix}\begin{bmatrix} 1 & 1 \\ 1 & -1 \end{bmatrix} = \frac{1}{\sqrt{2}}\begin{bmatrix} \frac{1}{3} & \frac{1}{3} \\ 1 & -1 \end{bmatrix}$$

$$S^{-1} = P^T M^{1/2} = \frac{1}{\sqrt{2}}\begin{bmatrix} 1 & 1 \\ 1 & -1 \end{bmatrix}\begin{bmatrix} 3 & 0 \\ 0 & 1 \end{bmatrix} = \frac{1}{\sqrt{2}}\begin{bmatrix} 3 & 1 \\ 3 & -1 \end{bmatrix}$$

The reader should verify that $SS^{-1} = I$, as a check. In addition, note that $S^T MS = I$. The modal initial conditions are calculated from equations (4.78):

$$\mathbf{r}(0) = S^{-1}\mathbf{x}_0 = \frac{1}{\sqrt{2}}\begin{bmatrix} 3 & 1 \\ 3 & -1 \end{bmatrix}\begin{bmatrix} 1 \\ 0 \end{bmatrix} = \begin{bmatrix} \dfrac{3}{\sqrt{2}} \\ \dfrac{3}{\sqrt{2}} \end{bmatrix}$$

$$\dot{\mathbf{r}}(0) = S^{-1}\dot{\mathbf{x}}_0 = S^{-1}\mathbf{0} = \mathbf{0}$$

so that $r_1(0) = r_2(0) = 3/\sqrt{2}$ and $\dot{r}_1(0) = \dot{r}_2(0) = 0$. Equations (4.66) and (4.67) yield that the modal solutions are

$$r_1(t) = \frac{3}{\sqrt{2}}\sin\left(\sqrt{2}t + \frac{\pi}{2}\right) = \frac{3}{\sqrt{2}}\cos\sqrt{2}t$$

$$r_2(t) = \frac{3}{\sqrt{2}}\sin\left(2t + \frac{\pi}{2}\right) = \frac{3}{\sqrt{2}}\cos 2t$$

The solution in the physical coordinate system $\mathbf{x}(t)$ is calculated from

$$\mathbf{x}(t) = S\mathbf{r}(t) = \frac{1}{\sqrt{2}}\begin{bmatrix} \frac{1}{3} & \frac{1}{3} \\ 1 & -1 \end{bmatrix}\begin{bmatrix} \dfrac{3}{\sqrt{2}}\cos\sqrt{2}t \\ \dfrac{3}{\sqrt{2}}\cos 2t \end{bmatrix} = \begin{bmatrix} (0.5)(\cos\sqrt{2}t + \cos 2t) \\ (1.5)(\cos\sqrt{2}t - \cos 2t) \end{bmatrix}$$

This is, of course, identical to the solution obtained in Example 4.1.7 and plotted in Figure 4.3.

$\square$

Example 4.3.2

Calculate the response of the system

$$\begin{bmatrix} 1 & 0 \\ 0 & 4 \end{bmatrix} \ddot{\mathbf{x}}(t) + \begin{bmatrix} 12 & -2 \\ -2 & 12 \end{bmatrix} \mathbf{x}(t) = \mathbf{0}$$

of Example 4.2.6, illustrated in Figure 4.4, to the initial displacement $\mathbf{x}(0) = \begin{bmatrix} 1 & 1 \end{bmatrix}^T$ and $\dot{\mathbf{x}}(0) = \mathbf{0}$ using modal analysis.

Solution   Again following the steps illustrated in Window 4.5, the matrices $M^{-1/2}$ and $\tilde{K}$ become

$$M^{-1/2} = \begin{bmatrix} 1 & 0 \\ 0 & \frac{1}{2} \end{bmatrix} \quad \tilde{K} = \begin{bmatrix} 12 & -1 \\ -1 & 3 \end{bmatrix}$$

Solving the symmetric eigenvalue problem for $\tilde{K}$ (this time using a computer and commercial code as outlined in Section 4.9) yields

$$P = \begin{bmatrix} -0.1091 & -0.9940 \\ -0.9940 & 0.1091 \end{bmatrix} \quad \Lambda = \text{diag} \, (2.8902, \, 12.1098)$$

Here the arithmetic is held to eight decimal places, but only four are shown. The matrices S and S^{-1} become

$$S = \begin{bmatrix} -0.1091 & -0.4970 \\ -0.9940 & 0.0546 \end{bmatrix} \quad S^{-1} = \begin{bmatrix} -0.1091 & -0.9940 \\ -1.9881 & 0.2182 \end{bmatrix}$$

As a check, note that

$$P^T \tilde{K} P = \begin{bmatrix} 2.8902 & 0 \\ 0 & 12.1098 \end{bmatrix} \quad P^T P = I$$

The modal initial conditions become

$$\mathbf{r}(0) = S^{-1} \mathbf{x}_0 = \begin{bmatrix} -0.1091 & -0.9940 \\ -0.9881 & 0.2182 \end{bmatrix} \begin{bmatrix} 1 \\ 1 \end{bmatrix} = \begin{bmatrix} -2.0972 \\ -0.7758 \end{bmatrix}$$

$$\dot{\mathbf{r}}(0) = S^{-1} \dot{\mathbf{x}}_0 = \mathbf{0}$$

Using these values of $r_1(0)$, $r_2(0)$, $\dot{r}_1(0)$, and $\dot{r}_2(0)$ in equations (4.66) and (4.67) yields the modal solutions

$$r_1(t) = -2.0972 \cos \, (1.7001t)$$

$$r_2(t) = -0.7758 \cos \, (3.4799t)$$

Using the transformation $\mathbf{x} = S\mathbf{r}(t)$ yields that the solution in physical coordinates is

$$\mathbf{x}(t) = \begin{bmatrix} 0.2288 \cos \, (1.7001t) + 0.7712 \cos \, (3.4799t) \\ 1.0424 \cos \, (1.7001t) - 0.0424 \cos \, (3.4799t) \end{bmatrix}$$

Note that $\mathbf{x}(t)$ satisfies the initial conditions, as it should. A plot of the responses is given in Figure 4.6.

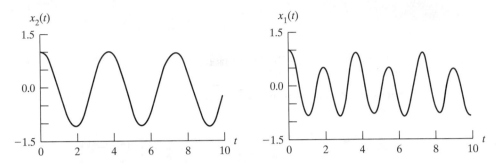

Figure 4.6 Plot of the solutions given in Example 4.3.2.

The plot of $x_2(t)$ in the figure illustrates that the mass m_2 is not much affected by the second frequency. This is because the particular initial condition does not cause the first mass to be excited very much in the second mode, i.e., at $\omega_2 = 3.4799$ rad/s [note the coefficient of $\cos(3.4799t)$ in the equation for $x_2(t)$]. However, the plot of $x_1(t)$ clearly indicates the presence of both frequencies, because the initial condition strongly excites both modes (i.e., both frequencies) in this coordinate. The effects of changing the initial conditions on the response can be examined by using the program VTB4_2 in the Engineering Vibration Toolbox to solve Problem TB4.3 at the end of the chapter. Changing initial conditions is also discussed in Section 4.9.

□

This section introduces the computations of modal analysis for a two-degree-of-freedom system. This entire approach is easily extended to any number of degrees of freedom, as discussed in the following section. Furthermore, the process of modal analysis is easily performed using any of the modern mathematical software packages, as discussed in Section 4.9, or as indicated in the Toolbox associated with this text. This section forms the foundation for the rest of this chapter and is applied to damped systems (in Section 4.5) and forced systems (in Section 4.6). The idea of modal analysis is used again in studying distributed systems (Chapter 6) and vibration testing (Chapter 7). All of this material depends on understanding the modal matrix S, how to compute it, and the physical interpretation of S given in Figure 4.5.

4.4 MORE THAN TWO DEGREES OF FREEDOM

Many structures, machines, and mechanical devices require numerous coordinates to describe their vibrational motion. For instance, an automobile suspension was modeled in earlier chapters as a single degree of freedom. However, a car has four wheels; hence a more accurate model is to use four degrees of freedom or coordinates. Since an automobile can roll, pitch, and yaw, it may be appropriate to use even more coordinates to describe the motion. Systems with any finite number of degrees of freedom can be analyzed by using the modal analysis procedure outlined in Window 4.5.

For each mass in the system and/or for each degree of freedom, there corresponds a coordinate, $x_i(t)$, describing its motion in one dimension; this gives rise to an $n \times 1$ vector $\mathbf{x}(t)$, with $n \times n$ mass matrix M and stiffness matrix K satisfying

$$M\ddot{\mathbf{x}}(t) + K\mathbf{x}(t) = \mathbf{0} \tag{4.80}$$

The form of equation (4.80) also holds if each mass is allowed to rotate or move in the y, z, or pitch and yaw directions. In this situation, the vector $\mathbf{x}$ could reflect up to six coordinates for each mass, and the mass and stiffness matrices would be modified to reflect the additional inertia and stiffness quantities. Figure 4.7 illustrates the possibilities of coordinates for a simple element. However, for the sake of simplicity of explanation, the initial discussion is confined to mass elements that are free to move in only one direction.

As a generic example, consider the n masses connected by n springs in Figure 4.8. Summing the forces on each of the n masses yields n equations of the form

$$m_i\ddot{x}_i + k_i(x_i - x_{i-1}) - k_{i+1}(x_{i+1} - x_i) = 0 \quad i = 1, 2 \ldots, n \tag{4.81}$$

where m_i denotes the ith mass and k_i the ith spring coefficient. In matrix form these equations take the form of equation (4.80) where

$$M = \text{diag}(m_1, \quad m_2, \ldots, m_n) \tag{4.82}$$

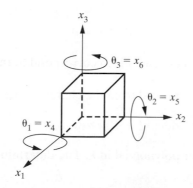

Figure 4.7 A single mass element illustrating all the possible degrees of freedom. The six degrees of freedom of the rigid body consist of three rotational and three translational motions. If the predominant forward motion of the body is in the x_2 direction, such as in an airplane, then θ_2 is called *roll*, θ_3 is called *yaw*, and θ_1 is called *pitch*.

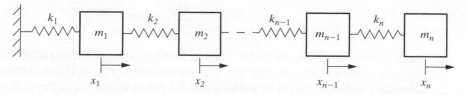

Figure 4.8 An example of an n-degree-of-freedom system.

and

$$K = \begin{bmatrix} k_1 + k_2 & -k_2 & 0 & 0 & \cdots & & 0 \\ -k_2 & k_2 + k_3 & -k_3 & & & & \vdots \\ 0 & -k_3 & k_3 + k_4 & & & & \\ \vdots & & & & & k_{n-1} + k_n & -k_n \\ 0 & & \cdots & & & -k_n & k_n \end{bmatrix} \qquad (4.83)$$

The $n \times 1$ vector $\mathbf{x}(t)$ becomes

$$\mathbf{x}(t) = \begin{bmatrix} x_1(t) \\ x_2(t) \\ \vdots \\ x_n(t) \end{bmatrix} \qquad (4.84)$$

The notation of Window 4.5 can be directly used to solve n-degree-of-freedom problems. Each of the steps is exactly the same; however, the matrix computations are all $n \times n$ and the resulting modal equation becomes the n decoupled equations

$$\begin{aligned} \ddot{r}_1(t) + \omega_1^2 r_1(t) &= 0 \\ \ddot{r}_2(t) + \omega_2^2 r_2(t) &= 0 \\ &\vdots \\ \ddot{r}_n(t) + \omega_n^2 r_n(t) &= 0 \end{aligned} \qquad (4.85)$$

There are now n natural frequencies, ω_i, which correspond to the eigenvalues of the $n \times n$ matrix $M^{-1/2}KM^{-1/2}$.

The n eigenvalues are determined from the characteristic equation given by

$$\det\left(\lambda I - \tilde{K}\right) = 0 \qquad (4.86)$$

which gives rise to an nth-order polynomial in λ. The determinant of an $n \times n$ matrix A is given by

$$\det A = \sum_{s=1}^{n} a_{ps} |A_{ps}| \qquad (4.87)$$

for any fixed value of p between 1 and n. Here a_{ps} is the element of the matrix A at the intersection of the pth row and sth column, and $|A_{ps}|$ is the determinant of the submatrix formed from A by striking out the pth row and sth column, multiplied by $(-1)^{p \pm s}$. The following example illustrates the use of equation (4.87).

Example 4.4.1

Expand equation (4.87) for $p = 1$ to calculate the following determinant:

$$\det A = \det \begin{bmatrix} 1 & 3 & -2 \\ 0 & 1 & 1 \\ 2 & 5 & 3 \end{bmatrix}$$

$$= 1[(1)(3) - (1)(5)] - 3[(0)(3) - (2)(1)] - 2[(0)(5) - (1)(2)] = 8$$

□

Once the ω_i^2 are determined from equation (4.86), the normalized eigenvectors are obtained following the methods suggested in Example 4.2.3. The key differences between using the modal approach described by Window 4.5 for multiple-degree-of-freedom systems and for a two-degree-of-freedom system is in the computation of the characteristic equation, its solution to get ω_i^2, and solving for the normalized eigenvectors. With the exception of calculating the matrix P and Λ, the rest of the procedure is simple matrix multiplication. The following example illustrates the procedure for a three-degree-of-freedom system.

Example 4.4.2

Calculate the solution of the n-degree-of-freedom system of Figure 4.8 for $n = 3$ by modal analysis. Use the values $m_1 = m_2 = m_3 = 4$ kg and $k_1 = k_2 = k_3 = 4$ N/m, and the initial condition $x_1(0) = 1$ m with all other initial displacements and velocities zero.

Solution   The mass and stiffness matrices for $n = 3$ for the values given become

$$M = 4I \quad K = \begin{bmatrix} 8 & -4 & 0 \\ -4 & 8 & -4 \\ 0 & -4 & 4 \end{bmatrix}$$

Following the steps suggested in Window 4.5 yields

1. $M^{-1/2} = \dfrac{1}{2}I$

2. $\tilde{K} = M^{-1/2}KM^{-1/2} = \dfrac{1}{4}\begin{bmatrix} 8 & -4 & 0 \\ -4 & 8 & -4 \\ 0 & -4 & 4 \end{bmatrix} = \begin{bmatrix} 2 & -1 & 0 \\ -1 & 2 & -1 \\ 0 & -1 & 1 \end{bmatrix}$

3. $\det(\lambda I - \tilde{K}) = \det\left(\begin{bmatrix} \lambda - 2 & 1 & 0 \\ 1 & \lambda - 2 & 1 \\ 0 & 1 & \lambda - 1 \end{bmatrix} \right)$

$$= (\lambda - 2)\det\left(\begin{bmatrix} \lambda - 2 & 1 \\ 1 & \lambda - 1 \end{bmatrix} \right)$$

$$- (1)\det\left(\begin{bmatrix} 1 & 1 \\ 0 & \lambda - 1 \end{bmatrix} \right) + (0)\det\left(\begin{bmatrix} 1 & \lambda - 2 \\ 0 & 1 \end{bmatrix} \right)$$

$$= (\lambda - 2)[(\lambda - 2)(\lambda - 1) - 1] - 1[(\lambda - 1) - 0]$$
$$= (\lambda - 1)(\lambda - 2)^2 - (\lambda - 2) - \lambda + 1$$
$$= \lambda^3 - 5\lambda^2 + 6\lambda - 1 = 0$$

The roots of this cubic equation are

$$\lambda_1 = 0.1981 \qquad \lambda_2 = 1.5550 \qquad \lambda_3 = 3.2470$$

Thus the system's natural frequencies are

$$\omega_1 = 0.4450 \qquad \omega_2 = 1.2470 \qquad \omega_3 = 1.8019$$

To calculate the first eigenvector, substitute $\lambda_1 = 0.1981$ into $(\tilde{K} - \lambda I)\mathbf{v}_1 = \mathbf{0}$ and solve for the vector $\mathbf{v}_1 = [v_{11} \quad v_{12} \quad v_{13}]^T$. This yields

$$\begin{bmatrix} 2 - 0.1981 & -1 & 0 \\ -1 & 2 - 0.1981 & -1 \\ 0 & -1 & 1 - 0.1981 \end{bmatrix} \begin{bmatrix} v_{11} \\ v_{21} \\ v_{31} \end{bmatrix} = \begin{bmatrix} 0 \\ 0 \\ 0 \end{bmatrix}$$

Multiplying out this last expression yields three equations, only two of which are independent:

$$(1.8019)v_{11} - v_{21} = 0$$
$$-v_{11} + (1.8019)v_{21} - v_{31} = 0$$
$$-v_{21} + (0.8019)v_{31} = 0$$

Solving the first and third equations yields

$$v_{11} = 0.4450v_{31} \quad \text{and} \quad v_{21} = 0.8019v_{31}$$

The second equation is dependent and does not yield any new information. Substituting these values into the vector $\mathbf{v}_1$ yields

$$\mathbf{v}_1 = v_{31} \begin{bmatrix} 0.4450 \\ 0.8019 \\ 1 \end{bmatrix}$$

4. Normalizing the vector yields

$$\mathbf{v}_1^T\mathbf{v}_1 = v_{31}^2 \left[(0.4450)^2 + (0.8019)^2 + 1^2 \right] = 1$$

Solving for v_{31} and substituting back into the expression for $\mathbf{v}_1$ yields the normalized version of the eigenvector $\mathbf{v}_1$ as

$$\mathbf{v}_1 = \begin{bmatrix} 0.3280 \\ 0.5910 \\ 0.7370 \end{bmatrix}$$

Similarly, $\mathbf{v}_2$ and $\mathbf{v}_3$ can be calculated and normalized to be

$$\mathbf{v}_2 = \begin{bmatrix} -0.7370 \\ -0.3280 \\ 0.5910 \end{bmatrix} \qquad \mathbf{v}_3 = \begin{bmatrix} -0.5910 \\ 0.7370 \\ -0.3280 \end{bmatrix}$$

The matrix P is then

$$P = \begin{bmatrix} 0.3280 & -0.7370 & -0.5910 \\ 0.5910 & -0.3280 & 0.7370 \\ 0.7370 & 0.5910 & -0.3280 \end{bmatrix}$$

(The reader should verify that $P^T P = I$ and $P^T \tilde{K} P = \Lambda$.)

5. The matrix $S = M^{-1/2} P = \dfrac{1}{2} IP$ or

$$S = \begin{bmatrix} 0.1640 & -0.3685 & -0.2955 \\ 0.2955 & -0.1640 & 0.3685 \\ 0.3685 & 0.2955 & -0.1640 \end{bmatrix}$$

and

$$S^{-1} = P^T M^{1/2} = 2P^T I = \begin{bmatrix} 0.6560 & 1.1820 & 1.4740 \\ -1.4740 & -0.6560 & 1.1820 \\ -1.1820 & 1.4740 & -0.6560 \end{bmatrix}$$

(Again the reader should verify that $S^{-1} S = I$.)

6. The initial conditions in modal coordinates become

$$\dot{\mathbf{r}}(0) = S^{-1} \dot{\mathbf{x}}_0 = S^{-1} \mathbf{0} = \mathbf{0}$$

and

$$\mathbf{r}(0) = S^{-1} \mathbf{x}_0 = \begin{bmatrix} 0.6560 & 1.1820 & 1.4740 \\ -1.4740 & -0.6560 & 1.1820 \\ -1.1820 & 1.4740 & -0.6560 \end{bmatrix} \begin{bmatrix} 1 \\ 0 \\ 0 \end{bmatrix} = \begin{bmatrix} 0.6560 \\ -1.4740 \\ -1.1820 \end{bmatrix}$$

7. The modal solutions of equation (4.85) are each of the form given by equation (4.67) and can now be determined as

$$r_1(t) = (0.6560) \sin\left(0.4450t + \frac{\pi}{2}\right) = 0.6560 \cos(0.4450t)$$

$$r_2(t) = (-1.4740) \sin\left(1.247t + \frac{\pi}{2}\right) = -1.4740 \cos(1.2470t)$$

$$r_3(t) = (-1.1820) \sin\left(1.8019t + \frac{\pi}{2}\right) = -1.1820 \cos(1.8019t)$$

8. The solution in physical coordinates is next calculated from

$$\mathbf{x} = S\mathbf{r}(t) = \begin{bmatrix} 0.1640 & -0.3685 & -0.2955 \\ 0.2955 & -0.1640 & 0.3685 \\ 0.3685 & 0.2955 & -0.1640 \end{bmatrix} \begin{bmatrix} 0.6560 \cos(0.4450t) \\ -1.4740 \cos(1.2470t) \\ -1.1820 \cos(1.8019t) \end{bmatrix}$$

$$
\begin{bmatrix} x_1(t) \\ x_2(t) \\ x_3(t) \end{bmatrix} = \begin{bmatrix} 0.1075\cos(0.4450t) + 0.5443\cos(1.2470t) + 0.3492\cos(1.8019t) \\ 0.1938\cos(0.4450t) + 0.2417\cos(1.2470t) - 0.4355\cos(1.8019t) \\ 0.2417\cos(0.4450t) - 0.4355\cos(1.2470t) + 0.1935\cos(1.8019t) \end{bmatrix}
$$

The calculations in this example are a bit tedious. Fortunately, they are easily made using software as done in Section 4.9 and in the Engineering Vibration Toolbox. In fact, computing the frequencies, the matrix of eigenvectors P, and subsequently the mode-shape matrix S starting with the determinant is not the recommended way to proceed. Rather, the symmetric algebraic eigenvalue problem should be solved directly, and this is best done using the software methods covered in Section 4.9. The solution in this example is plotted and compared to a numerical simulation in Section 4.10.

□

Mode Summation Method

Another approach to modal analysis is to use the *mode summation* or *expansion method*. This procedure is based on a fact from linear algebra—that the eigenvectors of a real symmetric matrix form a complete set (see Window 4.1; i.e., that any n-dimensional vector can be represented as a linear combination of the eigenvectors of an $n \times n$ symmetric matrix). Recall the symmetric statement of the vibration problem:

$$
I\ddot{\mathbf{q}}(t) + \tilde{K}\mathbf{q}(t) = \mathbf{0} \tag{4.88}
$$

Let $\mathbf{v}_i$ denote the n eigenvectors of the matrix $\tilde{K}$, and let $\lambda_i \neq 0$ denote the corresponding eigenvalues. According to the argument preceding equation (4.41), a solution of (4.88) is

$$
\mathbf{q}_i(t) = \mathbf{v}_i e^{\pm\sqrt{\lambda_i}jt} \tag{4.89}
$$

since $\lambda_i = \omega_i^2$. This represents two solutions that can be added together following the argument used for equation (1.18) to yield

$$
\mathbf{q}_i(t) = (a_i e^{-\sqrt{\lambda_i}jt} + b_i e^{\sqrt{\lambda_i}jt})\mathbf{v}_i \tag{4.90}
$$

or, using Euler's formula,

$$
\mathbf{q}_i(t) = d_i\sin(\omega_i t + \phi_i)\mathbf{v}_i \tag{4.91}
$$

where d_i and ϕ_i are constants to be determined by initial conditions. Since the set of vectors $\mathbf{v}_i$, $i = 1, 2, \ldots, n$ are eigenvectors of a symmetric matrix, a linear combination can be used to represent any $n \times 1$ vector, and in particular, the solution vector $\mathbf{q}(t)$. Hence

$$
\mathbf{q}(t) = \sum_{i=1}^{n} d_i\sin(\omega_i t + \phi_i)\mathbf{v}_i \tag{4.92}
$$

The constants d_i and ϕ_i can be evaluated from the initial conditions

$$\mathbf{q}(0) = \sum_{i=1}^{n} d_i \sin \phi_i \mathbf{v}_i \qquad (4.93)$$

and

$$\dot{\mathbf{q}}(0) = \sum_{i=1}^{n} \omega_i d_i \cos \phi_i \mathbf{v}_i \qquad (4.94)$$

Multiplying equation (4.93) by $\mathbf{v}_j^T$ and using the orthogonality (see Window 4.3) of the vector $\mathbf{v}_i$ (i.e., $\mathbf{v}_j^T \mathbf{v}_i = 0$ for all values of the summation index $i = 1, 2, \ldots, n$ except for $i = j$) yields

$$\mathbf{v}_j^T \mathbf{q}(0) = d_j \sin \phi_j \qquad (4.95)$$

for each value of $j = 1, 2, \ldots, n$. Similarly, multiplying equation (4.94) by $\mathbf{v}_j^T$ yields

$$\mathbf{v}_j^T \dot{\mathbf{q}}(0) = \omega_j d_j \cos \phi_j \qquad (4.96)$$

for each $j = 1, 2, \ldots, n$. Combining equations (4.95) and (4.96) and renaming the index yields

$$\phi_i = \tan^{-1} \frac{\omega_i \mathbf{v}_i^T \mathbf{q}(0)}{\mathbf{v}_i^T \dot{\mathbf{q}}(0)} \qquad i = 1, 2, \ldots n \qquad (4.97)$$

and (if $\phi_i \neq 0$)

$$d_i = \frac{\mathbf{v}_i^T \mathbf{q}(0)}{\sin \phi_i} \qquad i = 1, 2, \ldots n \qquad (4.98)$$

Equations (4.92), (4.97), and (4.98) represent the solution in modal summation form. Equation (4.92) is sometimes called the *expansion theorem* and is equivalent to writing a function as a Fourier series. The constants d_i are sometimes referred to as expansion coefficients.

Note as an immediate consequence of equation (4.97) that if the system has zero initial velocity, $\dot{\mathbf{q}} = \mathbf{0}$, each coordinate has a phase shift of 90°. The initial conditions in the coordinate system $\mathbf{q}(t)$ can also be chosen such that $d_i = 0$ for all $i = 2, \ldots, n$. In this case the summation in equation (4.92) reduces to the single term

$$\mathbf{q}(t) = d_1 \sin(\omega_1 t + \phi_1)\mathbf{v}_1 \qquad (4.99)$$

This states that each coordinate $q_i(t)$ oscillates with the same frequency and phase. To obtain the solution in physical coordinates, recall that $\mathbf{x} = M^{-1/2}\mathbf{q}$ so that

$$\mathbf{x}(t) = d_1 \sin(\omega_1 t + \phi_1)M^{-1/2}\mathbf{v}_1 \qquad (4.100)$$

The product of a matrix and a vector is another vector. Defining $\mathbf{u}_1 = M^{-1/2}\mathbf{v}_1$, equation (4.100) becomes

$$\mathbf{x}(t) = d_1 \sin(\omega_1 t + \phi_1)\mathbf{u}_1 \tag{4.101}$$

This states that if the system is given a set of initial conditions such that $d_2 = d_3 = \cdots = d_n = 0$, then (4.101) is that total solution and each mass oscillates at the first natural frequency (ω_1). Furthermore, the vector $\mathbf{u}_1$ specifies the relative magnitudes of oscillation of each mass with respect to the rest position. Hence $\mathbf{u}_1$ is called the *first mode shape*. Note that the first mode shape is related to the first eigenvector of $\tilde{K}$ by

$$\mathbf{u}_1 = M^{-1/2}\mathbf{v}_1 \tag{4.102}$$

This argument can be repeated for each of the indices i so that $\mathbf{u}_2 = M^{-1/2}\mathbf{v}_2$, $\mathbf{u}_3 = M^{-1/2}\mathbf{v}_3$, etc., which become the second, third, etc., mode shapes. The series solution

$$\mathbf{x}(t) = \sum_{i=1}^{n} d_i \sin(\omega_i t + \phi_i)\mathbf{u}_i \tag{4.103}$$

illustrates how each mode shape contributes to forming the total response of the system.

The constants of integration d_i represent a scaling of how each mode participates in the total response. The larger d_i is, the more the ith mode affects the response. Hence the d_i are called *modal participation factors*.

The initial condition required to excite a system into a single mode can be determined from equation (4.98). Because of the mutual orthogonality of the eigenvectors, if $\mathbf{q}(0)$ is chosen to be one of the eigenvectors, $\mathbf{v}_j$, then each d_i is zero except for the index $i = j$. Hence to excite the structure in, say, the second mode, choose $\mathbf{q}(0) = \mathbf{v}_2$ and $\dot{\mathbf{q}}(0) = \mathbf{0}$. Then the solution for equation (4.92) becomes

$$\mathbf{q}(t) = d_2 \sin\left(\omega_2 t + \frac{\pi}{2}\right)\mathbf{v}_2 \tag{4.104}$$

and each coordinate of $\mathbf{q}$ oscillates with frequency ω_2. To transform $\mathbf{q}(0) = \mathbf{v}_2$ into physical coordinates, note that $\mathbf{x} = M^{-1/2}\mathbf{q}$, so that $\mathbf{x}(0) = M^{-1/2}\mathbf{q}(0) = M^{-1/2}\mathbf{v}_2 = \mathbf{u}_2$. Hence exciting the system by imposing an initial displacement equal to the second mode shape results in each mass oscillating at the second natural frequency. The modal summation method is illustrated in the next example.

Example 4.4.3

Consider a simple model of the horizontal vibration of a four-story building as illustrated in Figure 4.9, subject to a wind that gives the building an initial displacement of $\mathbf{x}(0) = [0.001 \quad 0.010 \quad 0.020 \quad 0.025]^T$ and zero initial velocity.

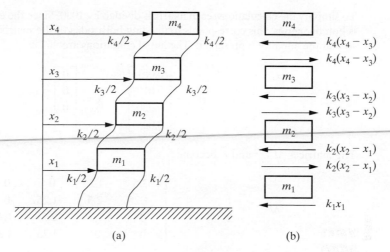

(a)   (b)

Figure 4.9   (a) A simple model of the horizontal vibration of a four-story building. Here each floor is modeled as a lumped mass, and the walls are modeled as providing horizontal stiffness. (b) The restoring forces acting on each mass (floor).

Solution   In modeling buildings, it is known that most of the mass is in the floor of each section and that the walls can be treated as massless columns providing lateral stiffness. From Figure 4.9 the equations of motion of each floor are

$$m_1\ddot{x}_1 + (k_1 + k_2)x_1 - k_2x_2 = 0$$

$$m_2\ddot{x}_2 - k_2x_1 + (k_2 + k_3)x_2 - k_3x_3 = 0$$

$$m_3\ddot{x}_3 - k_3x_2 + (k_3 + k_4)x_3 - k_4x_4 = 0$$

$$m_4\ddot{x}_4 - k_4x_3 + k_4x_4 = 0$$

In matrix form, these four equations can be written as

$$
\begin{bmatrix}
m_1 & 0 & 0 & 0 \\
0 & m_2 & 0 & 0 \\
0 & 0 & m_3 & 0 \\
0 & 0 & 0 & m_4
\end{bmatrix}
\ddot{\mathbf{x}} +
\begin{bmatrix}
k_1 + k_2 & -k_2 & 0 & 0 \\
-k_2 & k_2 + k_3 & -k_3 & 0 \\
0 & -k_3 & k_3 + k_4 & -k_4 \\
0 & 0 & -k_4 & k_4
\end{bmatrix}
\mathbf{x} = \mathbf{0}
$$

Some reasonable values for a building are $m_1 = m_2 = m_3 = m_4 = 4000$ kg and $k_1 = k_2 = k_3 = k_4 = 5000$ N/m. In this case, the numerical values of M and K become

$$
M = 4000I \qquad K =
\begin{bmatrix}
10{,}000 & -5000 & 0 & 0 \\
-5000 & 10{,}000 & -5000 & 0 \\
0 & -5000 & 10{,}000 & -5000 \\
0 & 0 & -5000 & 5000
\end{bmatrix}
$$

To simplify the calculations, each matrix is divided by 1000. Since the equation of motion is homogeneous, this corresponds to dividing both sides of the matrix equation by 1000 so that the equality is preserved. The initial conditions are

$$\mathbf{x}(0) = \begin{bmatrix} 0.001 \\ 0.010 \\ 0.020 \\ 0.025 \end{bmatrix} \qquad \dot{\mathbf{x}} = \begin{bmatrix} 0 \\ 0 \\ 0 \\ 0 \end{bmatrix}$$

The matrices $M^{-1/2}$ and $\tilde{K}$ become

$$M^{-1/2} = \frac{1}{2}I \qquad \tilde{K} = \begin{bmatrix} 2.5 & -1.25 & 0 & 0 \\ -1.25 & 2.5 & -1.25 & 0 \\ 0 & -1.25 & 2.5 & -1.25 \\ 0 & 0 & -1.25 & 1.25 \end{bmatrix}$$

The matrix $M^{1/2}$ and the initial condition on $\mathbf{q}(t)$ become

$$M^{1/2} = 2I \qquad \mathbf{q}(0) = M^{1/2}\mathbf{x}(0) = \begin{bmatrix} 0.002 \\ 0.020 \\ 0.040 \\ 0.050 \end{bmatrix} \qquad \dot{\mathbf{q}}(0) = M^{1/2}\mathbf{0} = \mathbf{0}$$

Using an eigenvalue solver (see Section 4.9 for details or use the files discussed at the end of this chapter and contained in the Toolbox), the eigenvalue problem for $\tilde{K}$ yields

$$\lambda_1 = 0.1508 \qquad \lambda_2 = 1.2500 \qquad \lambda_3 = 2.9341 \qquad \lambda_4 = 4.4151$$

$$\mathbf{v}_1 = \begin{bmatrix} 0.2280 \\ 0.4285 \\ 0.5774 \\ 0.6565 \end{bmatrix} \quad \mathbf{v}_2 = \begin{bmatrix} 0.5774 \\ 0.5774 \\ 0.0 \\ -0.5774 \end{bmatrix} \quad \mathbf{v}_3 = \begin{bmatrix} 0.6565 \\ -0.2280 \\ -0.5774 \\ 0.4285 \end{bmatrix} \quad \mathbf{v}_4 = \begin{bmatrix} -0.4285 \\ 0.6565 \\ -0.5774 \\ 0.2280 \end{bmatrix}$$

Converting this into natural frequencies and mode shapes ($\omega_i = \sqrt{\lambda_i}$ and $\mathbf{u}_i = M^{-1/2}\mathbf{v}_i$) yields $\omega_1 = 0.3883$, $\omega_2 = 1.1180$, $\omega_3 = 1.7129$, $\omega_4 = 2.1012$, and

$$\mathbf{u}_1 = \begin{bmatrix} 0.1140 \\ 0.2143 \\ 0.2887 \\ 0.3283 \end{bmatrix} \quad \mathbf{u}_2 = \begin{bmatrix} 0.2887 \\ 0.2887 \\ 0.0 \\ -0.2887 \end{bmatrix} \quad \mathbf{u}_3 = \begin{bmatrix} 0.3283 \\ -0.1140 \\ -0.2887 \\ 0.2143 \end{bmatrix} \quad \mathbf{u}_4 = \begin{bmatrix} -0.2143 \\ 0.3283 \\ -0.2887 \\ 0.1140 \end{bmatrix}$$

Since $\dot{\mathbf{q}}(0) = \mathbf{0}$, equation (4.96) yields that each of the phase shifts is $\phi_i = \pi/2$ and equation (4.98) becomes

$$d_i = \frac{\mathbf{v}_i^T\mathbf{q}(0)}{\sin(\pi/2)} = \mathbf{v}_i^T\mathbf{q}(0)$$

Substituting $\mathbf{v}_i^T$ and $\mathbf{q}(0)$ into the expansion above yields the following values for the modal participation factors:

$$d_1 = 0.065 \qquad d_2 = -0.016 \qquad d_3 = -4.9 \times 10^{-3} \qquad d_4 = 5.8 \times 10^{-4}$$

The solution given by equation (4.103) then becomes (in meters)

$$\mathbf{x}(t) = \begin{bmatrix} 0.007 \\ 0.014 \\ 0.019 \\ 0.021 \end{bmatrix} \cos{(0.3883t)} + \begin{bmatrix} -4.67 \times 10^{-3} \\ -4.67 \times 10^{-3} \\ 0 \\ 4.67 \times 10^{-3} \end{bmatrix} \cos{(1.1180t)}$$

$$+ \begin{bmatrix} -1.61 \times 10^{-3} \\ 5.60 \times 10^{-4} \\ 1.42 \times 10^{-3} \\ -1.05 \times 10^{-3} \end{bmatrix} \cos{(1.7129t)} + \begin{bmatrix} -1.24 \times 10^{-4} \\ 1.91 \times 10^{-4} \\ -1.68 \times 10^{-4} \\ 6.62 \times 10^{-5} \end{bmatrix} \cos{(2.101t)}$$

The mode shapes $\mathbf{u}_1$, $\mathbf{u}_2$, $\mathbf{u}_3$, and $\mathbf{u}_4$ are plotted in Figure 4.10. Note that the modal participation factor d_4 is much smaller than the others. Thus oscillation at 2.102 rad/s will not be too evident. The response of x_3 will be dominated almost completely by the first natural frequency.

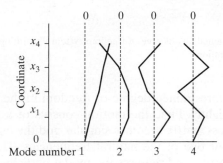

Figure 4.10 Plot of the four mode shapes associated with the solution for the system of Figure 4.9 (not to scale).

□

Nodes of a Mode

A node of a mode shape is simply the coordinate of a zero entry in the mode shape. For instance, the second mode shape in Example 4.4.3 has a zero value in the location of coordinate $x_3(t)$. Thus the third coordinate is a node of the second mode. This means that if the system is excited by an initial condition to vibrate only at the second natural frequency, the third coordinate will not move! Thus a node has a place of no motion for certain initial conditions. If a sensor were to be placed on the third mass, it would not be able to measure any vibration at the second natural frequency, because that mass does not have a response at ω_2. Nodes also make excellent mounting points for machines. Note the word *node* is also used in finite elements to mean something different (see Chapter 8).

Rigid-Body Modes

It often happens that a vibrating system is also translating, or rotating, away from its equilibrium position in one coordinate while the other coordinates are vibrating about their equilibrium point. Such systems are said to be unrestrained and technically violate the stability conditions given in Section 1.8. An example is a train (see Problem 4.13), where the coupling between each car may be modeled as a spring, the cars themselves as lumped masses. As the train rolls down a track it is moving with rigid-body, unrestrained motion while the cars vibrate relative to each other. Figure 4.11 is an example of an unrestrained, two-degree-of-freedom system.

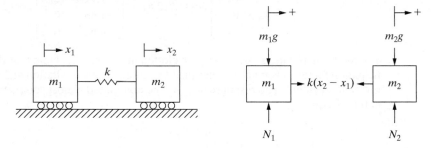

Figure 4.11   An unrestrained, two-degree-of-freedom system illustrating both rigid-body translation and vibration.

The existence of the unrestrained degree of freedom in the equations of motion changes the analysis slightly. First, the motion consists of a translation plus a vibration. Second, the stiffness matrix becomes singular and the eigenvalue problem results in a value of zero for one of the natural frequencies. The zero frequency renders equation (4.66) incorrect and requires the modal participation factors given by equations (4.95) and (4.96) to be altered. The following example illustrates how to compute the response for a system with unrestrained motion and how to correct these equations for a zero natural frequency.

Example 4.4.4

Compute the solution of the unrestrained system given in Figure 4.11 using both the eigenvector method and modal analysis. Let $m_1 = 1$, $m_2 = 4$, $k = 4$, $\mathbf{x}_0 = [1 \quad 0]^T$, and $\mathbf{v}_0 = \mathbf{0}$. Assume that the units are consistent.

Solution   Summing forces in the horizontal direction on each of the free-body diagrams given in Figure 4.11 yields

$$m_1\ddot{x}_1 = k(x_2 - x_1)$$
$$m_2\ddot{x}_2 = -k(x_2 - x_1)$$

Bringing all the forces to the left side and writing in matrix form yields

$$\begin{bmatrix} 1 & 0 \\ 0 & 4 \end{bmatrix}\begin{bmatrix} \ddot{x}_1 \\ \ddot{x}_2 \end{bmatrix} + 4\begin{bmatrix} 1 & -1 \\ -1 & 1 \end{bmatrix}\begin{bmatrix} x_1 \\ x_2 \end{bmatrix} = \begin{bmatrix} 0 \\ 0 \end{bmatrix}$$

Note that the determinant of the stiffness matrix K is zero, indicating that it is singular and hence has a zero eigenvalue (see Appendix C). Following the steps of Window 4.5 and substituting the values for M and K yields the following:

1. $M^{-1/2} = \begin{bmatrix} 1 & 0 \\ 0 & \frac{1}{2} \end{bmatrix}$

2. $\widetilde{K} = M^{-1/2} K M^{-1/2} = 4\begin{bmatrix} 1 & 0 \\ 0 & \frac{1}{2} \end{bmatrix}\begin{bmatrix} 1 & -1 \\ -1 & 1 \end{bmatrix}\begin{bmatrix} 1 & 0 \\ 0 & \frac{1}{2} \end{bmatrix} = \begin{bmatrix} 4 & -2 \\ -2 & 1 \end{bmatrix}$

3. Calculating the eigenvalue problem for 2 yields

$$\det\left(\widetilde{K} - \lambda I\right) = \det\left(\begin{bmatrix} 4 - \lambda & -2 \\ -2 & 1 - \lambda \end{bmatrix}\right) = (\lambda^2 - 5\lambda) = 0$$

This has solutions $\lambda_1 = 0$ and $\lambda_2 = 5$, so that $\omega_1 = 0$ and $\omega_2 = \sqrt{5} = 2.236$ rad/s. Note the zero eigenvalue/frequency. However, the eigenvector for λ_1 is not zero (eigenvectors are never zero) as the following calculation for the eigenvector for $\lambda_1 = 0$ yields

$$\begin{bmatrix} 4 - 0 & -2 \\ -2 & 1 - 0 \end{bmatrix}\begin{bmatrix} v_{11} \\ v_{21} \end{bmatrix} = \begin{bmatrix} 0 \\ 0 \end{bmatrix} \quad \text{or} \quad 4v_{11} - 2v_{21} = 0$$

Thus $2v_{11} = v_{21}$, or $\mathbf{v}_1 = [1\quad 2]^T$. Repeating the procedure for λ_2 yields $\mathbf{v}_2 = [2\quad -1]^T$.

4. Normalizing both eigenvectors yields

$$\mathbf{v}_1 = \begin{bmatrix} 0.4472 \\ 0.8944 \end{bmatrix} \text{ and } \mathbf{v}_2 = \begin{bmatrix} -0.8944 \\ 0.4472 \end{bmatrix}$$

Note that the eigenvector $\mathbf{v}_1$ associated with the eigenvalue $\lambda_1 = 0$ is not zero. Combining these to form the matrix of eigenvectors yields

$$P = \begin{bmatrix} 0.4472 & -0.8944 \\ 0.8944 & 0.4472 \end{bmatrix}$$

As a check note that

$$P^T P = I \text{ and } P^T \widetilde{K} P = \text{diag}[0\quad 5]$$

5. Calculating the matrix of mode shapes yields

$$S = M^{-1/2}P = \begin{bmatrix} 1 & 0 \\ 0 & \frac{1}{2} \end{bmatrix}\begin{bmatrix} 0.4472 & -0.8944 \\ 0.8944 & 0.4472 \end{bmatrix} = \begin{bmatrix} 0.4472 & -0.8944 \\ 0.4472 & 0.2236 \end{bmatrix}$$

$$S^{-1} = P^T M^{1/2} = \begin{bmatrix} 0.4472 & 1.7889 \\ -0.8944 & 0.8944 \end{bmatrix}$$

6. Calculating the modal initial conditions yields

$$\mathbf{r}(0) = S^{-1}\mathbf{x}_0 = \begin{bmatrix} 0.4472 & 1.7889 \\ -0.8944 & 0.8944 \end{bmatrix} \begin{bmatrix} 1 \\ 0 \end{bmatrix}$$

$$= \begin{bmatrix} 0.4472 \\ -0.8944 \end{bmatrix}, \quad \dot{\mathbf{r}}(0) = S^{-1}\mathbf{v}_0 = \mathbf{0}$$

7. Here is where the zero eigenvalue makes a difference as equations (4.66) and (4.67) only apply for the second natural frequency. The modal equation for the first mode becomes $\ddot{r}_1 = 0$, which has solution $r_1(t) = a + bt$. Here a and b are the constants of integration to be determined from the modal initial conditions. Applying the modal initial conditions yields the two equations

$$r_1(0) = a = 0.4472$$

$$\dot{r}_1(0) = b = 0.0$$

Thus the first modal equation has solution

$$r_1(t) = 0.4472$$

a constant. The solution for the second mode follows directly from equations (4.66) and (4.67) as before and yields

$$r_2(t) = -0.894 \cos\left(\sqrt{5}t\right)$$

Thus the modal response vector is

$$\mathbf{r}(t) = \begin{bmatrix} 0.447 \\ -0.894 \cos\left(\sqrt{5}t\right) \end{bmatrix}$$

8. Transforming back into the physical coordinates yields the solution

$$\mathbf{x}(t) = S\mathbf{r}(t) = \begin{bmatrix} 0.4472 & -0.8944 \\ 0.4472 & 0.2236 \end{bmatrix} \begin{bmatrix} 0.447 \\ -0.894 \cos\left(\sqrt{5}t\right) \end{bmatrix}$$

$$= \begin{bmatrix} 0.2 + 0.8 \cos\left(\sqrt{5}t\right) \\ 0.2 - 0.2 \cos\left(\sqrt{5}t\right) \end{bmatrix}$$

Note that each of the two physical coordinates moves a constant distance 0.2 units and then oscillates at the second natural frequency.

Next consider the effect of the zero frequency in using the mode summation method. In this case, equation (4.90) becomes

$$\mathbf{q}_1(t) = (a + bt)\mathbf{v}_1$$

and equations (4.93) and (4.94) become

$$\mathbf{q}(0) = (a + b0)\mathbf{v}_1 + \sum_{i=2}^{n} d_i \sin\phi_i\mathbf{v}_i \quad \text{and} \quad \dot{\mathbf{q}}(0) = b\mathbf{v}_i + \sum_{i=2}^{n} \omega_i d_i \cos\phi_i\mathbf{v}_i$$

Following the same steps as before and using orthogonality yields

$$a = \mathbf{v}_1^T\mathbf{q}(0) \quad \text{and} \quad b = \mathbf{v}_1^T\dot{\mathbf{q}}(0)$$

which replace the modal constants of integration d_i and ϕ_i for the zero valued mode. Computing d_2 and ϕ_2 following equations (4.97) and (4.98), and combining the modes according to equation (4.103), again yields the solution

$$\mathbf{x}(t) = a\mathbf{u}_1 + d_2 \cos(\sqrt{5}t)\mathbf{u}_2 = \begin{bmatrix} 0.2 + 0.8\cos(\sqrt{5}t) \\ 0.2 - 0.2\cos(\sqrt{5}t) \end{bmatrix}$$

The solutions are plotted in Figure 4.12.

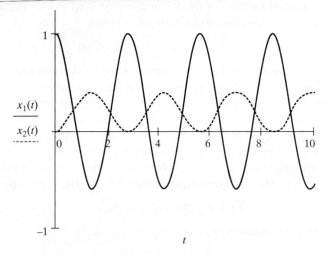

Figure 4.12   Plots of the solution versus time (s) for Example 4.4.4 showing vibration superimposed over a rigid-body mode.

Note that in all the previous examples of more than two-degrees-of freedom, the stiffness matrix K is *banded* (i.e., the matrix has nonzero elements on the diagonal and one element above and below the diagonal, the other elements being zero). This is typical of structural models but is not necessarily the case for machine parts or other mechanical devices.

The concept of mode shapes presented in this section and illustrated in Example 4.4.3 is extremely important. The language of modes, mode shapes, and natural frequencies forms the basis for discussing vibration phenomena of complex systems. The word *mode* generally refers to both the natural frequency and its corresponding mode shape. A mode shape is a mathematical description of a deflection. It forms a pattern that describes the shape of vibration if the system were to vibrate only at the corresponding natural frequency. It is neither tangible nor simple to observe; however, it provides a simple way to discuss and understand the vibration of complex objects. Its physical significance lies in the fact that every vibrational response of a system consists of some combinations of mode shapes. An entire industry has been formed around the concept of modes.

4.5 SYSTEMS WITH VISCOUS DAMPING

Viscous energy dissipation can be introduced to the modal analysis solution suggested previously in two ways. Again, as in modeling single-degree-of-freedom systems, viscous damping is introduced more as a mathematical convenience rather than a physical truth. However, viscous damping provides an excellent model in many physical situations and represents a significant improvement over the undamped model. The simplest method of modeling damping is to use *modal damping*. Modal damping places an energy dissipation term of the form

$$2\zeta_i \omega_i \dot{r}_i(t) \tag{4.105}$$

in equations (4.85). Here $\dot{r}_i(t)$ denotes the velocity of the ith modal coordinate, ω_i is the ith natural frequency, and ζ_i is the ith *modal damping ratio*. The modal damping ratios, ζ_i, are assigned by "experience" or measurement (see Chapter 7) to be some number between 0 and 1, or by making measurements of the response and estimating ζ_i. Usually, ζ_i is small unless the structure contains viscoelastic material or a hydraulic damper is present. Common values are $0 \le \zeta < 0.05$ (see Section 5.6). An automobile shock absorber, which uses fluid, may yield values as high as $\zeta = 0.5$.

Once the modal damping ratios are assigned, equations (4.85) become

$$\ddot{r}_i(t) + 2\zeta_i \omega_i \dot{r}_i(t) + \omega_i^2 r_i(t) = 0 \qquad i = 1, 2, \ldots, n \tag{4.106}$$

which have solutions of the form ($0 < \zeta_i < 1$)

$$r_i(t) = A_i e^{-\zeta_i \omega_i t} \sin(\omega_{di} t + \phi_i) \qquad i = 1, 2, \ldots, n \tag{4.107}$$

where A_i and ϕ_i are constants to be determined by the initial conditions and $\omega_{di} = \omega_i \sqrt{1 - \zeta_i^2}$ as given in Window 4.6. Once this modal solution is established,

Window 4.6
Review of a Damped Single-Degree-of-Freedom System

The solution of $m\ddot{x} + c\dot{x} + kx = 0$, $x(0) = x_0$, $\dot{x}(0) = \dot{x}_0$, or $\ddot{x} + 2\zeta\omega_n\dot{x} + \omega_n^2 x = 0$ is (for the underdamped case $0 < \zeta < 1$)

$$x(t) = Ae^{-\zeta\omega_n t} \sin(\omega_d t + \theta)$$

where $\omega_n = \sqrt{k/m}$, $\zeta = c/(2m\omega_n)$, $\omega_d = \omega_n\sqrt{1 - \zeta^2}$, and

$$A = \left[\frac{(\dot{x}_0 + \zeta\omega_n x_0)^2 + (x_0\omega_d)^2}{\omega_d^2} \right]^{1/2} \qquad \theta = \tan^{-1} \frac{x_0\omega_d}{\dot{x}_0 + \zeta\omega_n x_0}$$

from equations (1.36), (1.37), and (1.38).

the modal analysis method of solution suggested in Window 4.5 is used to transform the response into the physical coordinate system.

The only difference is that equation (4.107) replaces step 7 where $(\omega_i \neq 0)$

$$A_i = \left[\frac{(\dot{r}_{i0} + \zeta_i \omega_i r_{i0})^2 + (r_{i0} \omega_{di})^2}{\omega_{di}^2} \right]^{1/2} \tag{4.108}$$

$$\phi_i = \tan^{-1} \frac{r_{i0} \omega_{di}}{\dot{r}_{i0} + \zeta_i \omega_i r_{i0}} \tag{4.109}$$

Here r_{i0} and $\dot{r}_{i0}$ are the ith elements of $\mathbf{r}(0)$ and $\dot{\mathbf{r}}(0)$, respectively. Equations (4.108) and (4.109) are derived directly from equation (1.38) for a single-degree-of-freedom system of the same form as equation (4.107). These equations are correct only if each modal damping ratio is underdamped and no rigid-body modes are present. If a zero frequency exists, then the method of Example 4.4.4 must be used. The following example illustrates the solution technique for a system with assumed modal damping.

Example 4.5.1

Consider again the system of Example 4.3.1, which has equation of motion

$$\begin{bmatrix} 9 & 0 \\ 1 & 0 \end{bmatrix} \ddot{\mathbf{x}}(t) + \begin{bmatrix} 27 & -3 \\ -3 & 3 \end{bmatrix} \mathbf{x}(t) = \mathbf{0}, \quad \mathbf{x}(0) = \begin{bmatrix} 1 \\ 0 \end{bmatrix}, \quad \dot{\mathbf{x}}(t) = \mathbf{0}$$

and calculate the solution of the same system if modal damping of the form $\zeta_1 = 0.05$ and $\zeta_2 = 0.1$ is assumed.

Solution   From Example 4.3.1, $\omega_1 = \sqrt{2}$ and $\omega_2 = 2$. Since $\omega_{di} = \omega_i \sqrt{1 - \zeta_i^2}$ for underdamped systems, then the damped natural frequencies become

$$\omega_{d1} = \sqrt{2}[1 - (0.05)^2]^{1/2} = 1.4124 \text{ and } \omega_{d2} = 2[1 - (0.1)^2]^{1/2} = 1.9900.$$

The modal initial conditions calculated in Example 4.3.1 are $r_{10} = r_{20} = 3/\sqrt{2}$ and $\dot{r}_{10} = \dot{r}_{20} = 0$. Substitution of these values into equations (4.108) and (4.109) yields

$$A_1 = 2.1240 \qquad \phi_1 = 1.52 \text{ rad} \qquad (87.13°)$$

$$A_2 = 2.1340 \qquad \phi_2 = 1.47 \text{ rad} \qquad (84.26°)$$

Note that compared to the constant A_i and ϕ_i in the magnitude and phase of the undamped system, only a small change occurs in the amplitude. The phase, however, changes 3° and 6°, respectively, because of the damping. The solution is then of the form

$$\mathbf{x}(t) = S\mathbf{r}(t) = \frac{1}{\sqrt{2}} \begin{bmatrix} \frac{1}{3} & \frac{1}{3} \\ 1 & -1 \end{bmatrix} \begin{bmatrix} 2.1240e^{-0.0706t} \sin(1.4124t + 1.52) \\ 2.1320e^{-0.2t} \sin(1.9900t + 1.47) \end{bmatrix}$$

or

$$\mathbf{x}(t) = \begin{bmatrix} 0.5006e^{-0.0706t} \sin(1.4124t + 1.52) + 0.5025e^{-0.2t} \sin(1.9900t + 1.47) \\ 1.5019e^{-0.0706t} \sin(1.4124t + 1.52) - 1.5076e^{-0.2t} \sin(1.9900t + 1.47) \end{bmatrix}$$

A plot of $x_1(t)$ versus t and $x_2(t)$ versus t is given in Figure 4.13.

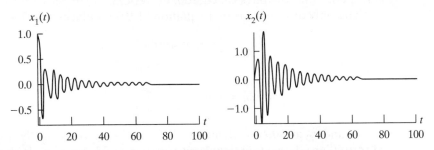

Figure 4.13   Plot of the damped response of the system of Example 4.5.1.

□

The modal damping ratio approach is also easily applicable to the mode summation method of Section 4.4. In this case, equation (4.91) is replaced by the damped version

$$\mathbf{q}_i(t) = d_i e^{-\zeta_i \omega_i t} \sin(\omega_{di} t + \phi_i) \mathbf{v}_i \tag{4.110}$$

The initial displacement condition calculation becomes

$$\mathbf{q}_i(0) = d_i \sin(\phi_i) \mathbf{v}_i \tag{4.111}$$

so that equation (4.95) still holds. However, the velocity becomes

$$\dot{\mathbf{q}}_i(t) = d_i[\omega_{di} e^{-\zeta_i \omega_i t} \cos(\omega_{di} t + \phi_i) - \zeta_i \omega_i e^{-\zeta_i \omega_i t} \sin(\omega_i t + \phi_i)] \mathbf{v}_i \tag{4.112}$$

or at $t = 0$,

$$\dot{\mathbf{q}}_i(0) = d_i(\omega_{di} \cos\phi_i - \zeta_i \omega_i \sin\phi_i) \mathbf{v}_i \tag{4.113}$$

Multiplying equations (4.111) and (4.113) by $\mathbf{v}_i^T$ from the left and solving for the constants ϕ_i and d_i yields

$$\phi_i = \tan^{-1} \frac{\omega_{di} \mathbf{v}_i^T \mathbf{q}(0)}{\mathbf{v}_i^T \dot{\mathbf{q}}(0) + \zeta_i \omega_i \mathbf{v}_i^T \mathbf{q}(0)}$$

$$d_i = \frac{\mathbf{v}_i^T \mathbf{q}(0)}{\sin\phi_i} \tag{4.114}$$

Again, the values of ζ_i are assigned based on experience or on measurement; then the calculations of Section 4.4 are used with the initial conditions given by equations (4.114). The solution given by equation (4.103) is replaced by

$$\mathbf{x}(t) = \sum_{i=1}^{n} d_i e^{-\zeta_i \omega_i t} \sin(\omega_{di} t + \phi_i) \mathbf{u}_i \tag{4.115}$$

which yields the damped response.

Example 4.5.2

Recall Example 4.4.3, and the model of building vibration defined by the equation of motion

$$4000I\ddot{\mathbf{x}}(t) + \begin{bmatrix} 10{,}000 & -5000 & 0 & 0 \\ -5000 & 10{,}000 & -5000 & 0 \\ 0 & -5000 & 10{,}000 & -5000 \\ 0 & 0 & -5000 & 5000 \end{bmatrix}\mathbf{x}(t) = \mathbf{0}$$

subject to a wind that gives the building an initial displacement of $\mathbf{x}(0) = [0.001\ \ 0.010\ \ 0.020\ \ 0.025]^T$ and zero initial velocity, and assume that the damping in the building is measured to be about $\zeta = 0.01$ in each mode.

Solution Each of the steps of the solution to Example 4.4.3 is the same until the initial conditions are calculated. From equation (4.114) with $\dot{\mathbf{q}}(0) = M^{1/2}\dot{\mathbf{x}}(0) = \mathbf{0}$, for each i,

$$\phi_i = \tan^{-1}\frac{\omega_{di}}{\zeta_i\omega_i} = \tan^{-1}\frac{\sqrt{1 - \zeta_i^2}}{\zeta_i}$$

For $\zeta_i = 0.01$, this becomes

$$\phi_i = 89.42° \qquad i = 1, 2, 3, 4 \qquad (\text{or } 1.56 \text{ rad})$$

Since $1/\sin(89.42°) = 1.00005$, the expansion coefficients d_i are taken to be $\mathbf{v}_i^T\mathbf{q}(0)$, as calculated in Example 4.4.3 because of the small damping. The natural frequencies are $\omega_1 = 0.3883$, $\omega_2 = 1.1180$, $\omega_3 = 1.7129$, and $\omega_4 = 2.1012$ rad/s. With a damping ratio of 0.01 assigned to each mode, the damped natural frequencies become ($\omega_{di} = \omega_i\sqrt{1 - \zeta_i^2}$) nearly the same to the third decimal place as the natural frequencies. The value of the exponent in the exponential decay terms become

$$-\zeta_1\omega_1 = -0.004, \quad -\zeta_2\omega_2 = -0.011, \quad -\zeta_3\omega_3 = -0.017, \quad \text{and} \quad -\zeta_4\omega_4 = -0.021$$

The mode shapes are unchanged, so the solution becomes

$$\mathbf{x}(t) = \begin{bmatrix} 0.007 \\ 0.014 \\ 0.019 \\ 0.021 \end{bmatrix}e^{-0.004t}\cos(0.388t + 1.56) + \begin{bmatrix} -4.67 \times 10^{-3} \\ -4.67 \times 10^{-3} \\ 0 \\ 4.67 \times 10^{-3} \end{bmatrix}e^{-0.011t}\cos(1.118t + 1.56)$$

$$+ \begin{bmatrix} -1.61 \times 10^{-3} \\ 5.60 \times 10^{-4} \\ 1.42 \times 10^{-3} \\ -1.05 \times 10^{-3} \end{bmatrix}e^{-0.017t}\cos(1.713t + 1.56) + \begin{bmatrix} -1.24 \times 10^{-4} \\ 1.91 \times 10^{-4} \\ -1.68 \times 10^{-4} \\ 6.62 \times 10^{-5} \end{bmatrix}e^{-0.021t}\cos(2.101t + 1.56)$$

Each coordinate of $\mathbf{x}(t)$ is plotted in Figure 4.14. Note that each plot shows the effects of multiple frequencies and is lightly damped. Window 4.7 summarizes this use of modal damping in the mode summation method.

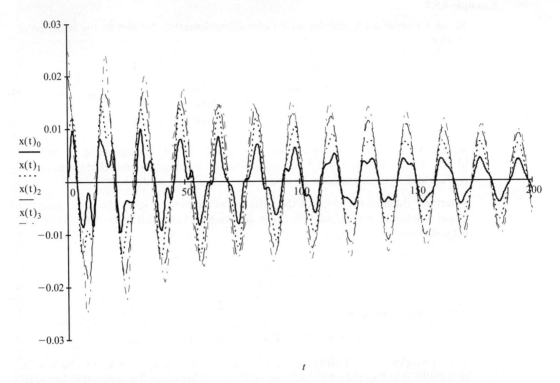

Figure 4.14 A plot of the damped response of the system of Example 4.5.2, which is the building of Example 4.4.3 with damping of $\zeta = 0.01$ in each mode. The plot was made in Mathcad so that $x_1(t)$ is denoted by x(t)$_0$, $x_2(t)$ is x(t)$_1$, etc.

☐

Damping can also be modeled directly. For instance, consider the system of Figure 4.15. The equations of motion of this system can be found from summing the forces on each mass, as before. This yields the following equations of motion in matrix form:

$$\begin{bmatrix} m_1 & 0 \\ 0 & m_2 \end{bmatrix} \ddot{\mathbf{x}} + \begin{bmatrix} c_1 + c_2 & -c_2 \\ -c_2 & c_2 \end{bmatrix} \dot{\mathbf{x}} + \begin{bmatrix} k_1 + k_2 & -k_2 \\ -k_2 & k_2 \end{bmatrix} \mathbf{x} = \mathbf{0} \qquad (4.116)$$

where $\mathbf{x} = [x_1(t) \quad x_2(t)]^T$. Equation (4.116) yields an example of a damping matrix C, defined by

$$C = \begin{bmatrix} c_1 + c_2 & -c_2 \\ -c_2 & c_2 \end{bmatrix} \qquad (4.117)$$

Here c_1 and c_2 refer to the damping coefficients indicated in Figure 4.15. The damping matrix C is symmetric and, in a general n-degree-of-freedom system, will be an

Window 4.7
Modal Damping in the Mode Summation Method

First transform the undamped equations of motion into the $\mathbf{q}$ coordinate system and calculate ω_i and $\mathbf{v}_i$. Choose the modal damping ratios and write

$$\mathbf{q}(t) = \sum_{i=1}^{n} d_i e^{-\zeta_i \omega_i t} \sin(\omega_{di} t + \phi_i) \mathbf{v}_i$$

where $M^{-1/2} K M^{-1/2} \mathbf{v}_i = \omega_i^2 \mathbf{v}_i$, $\omega_{di} = \omega_i \sqrt{1 - \zeta_i^2}$, $d_i = \dfrac{\mathbf{v}_i^T \mathbf{q}(0)}{\sin \phi_i}$

and

$$\phi_i = \tan^{-1} \frac{\omega_{di} \mathbf{v}_i^T \mathbf{q}(0)}{\mathbf{v}_i^T \dot{\mathbf{q}}(0) + \zeta_i \omega_i \mathbf{v}_i^T \mathbf{q}(0)}$$

Recall the initial conditions are found from

$$\mathbf{q}(0) = M^{1/2} \mathbf{x}(0) \qquad \text{and} \qquad \dot{\mathbf{q}}(0) = M^{1/2} \dot{\mathbf{x}}(0)$$

Once $\mathbf{q}(t)$ is computed, transform back to the physical coordinate system by

$$\mathbf{x}(t) = M^{-1/2} \mathbf{q}(t)$$

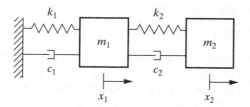

Figure 4.15 A two-degree-of-freedom system with viscous damping.

$n \times n$ matrix. Thus a damped n-degree-of-freedom system is modeled by equations of the form

$$M\ddot{\mathbf{x}} + C\dot{\mathbf{x}} + K\mathbf{x} = 0 \tag{4.118}$$

The difficulty with modeling damping in this fashion is that modal analysis cannot in general be used to solve equation (4.118) unless $CM^{-1}K = KM^{-1}C$ holds. This is true because the damping provides additional coupling between the equations of motion that cannot always be decoupled by the modal transformation S (see Caughey and O'Kelly, 1965). Other methods can be used to solve equation (4.118), as discussed in Sections 4.9 and 4.10.

Modal analysis can be used directly to solve equation (4.118) if the damping matrix C can be written as a linear combination of the mass and stiffness matrix, that is, if

$$C = \alpha M + \beta K \tag{4.119}$$

where α and β are constants. This form of damping is called *proportional damping*. Substitution of equation (4.119) into equation (4.118) yields

$$M\ddot{\mathbf{x}}(t) + (\alpha M + \beta K)\dot{\mathbf{x}}(t) + K\mathbf{x}(t) = \mathbf{0} \tag{4.120}$$

Substitution of $\mathbf{x}(t) = M^{-1/2}\mathbf{q}(t)$ and multiplying by $M^{-1/2}$ yields

$$\ddot{\mathbf{q}}(t) + (\alpha I + \beta\tilde{K})\dot{\mathbf{q}}(t) + \tilde{K}\mathbf{q}(t) = \mathbf{0} \tag{4.121}$$

Continuing to follow the steps of Window 4.5, substituting $\mathbf{q}(t) = P\mathbf{r}(t)$ and premultiplying by P^T where P is the matrix of eigenvectors of $\tilde{K}$, yields

$$\ddot{\mathbf{r}}(t) + (\alpha I + \beta\Lambda)\dot{\mathbf{r}}(t) + \Lambda\mathbf{r}(t) = \mathbf{0} \tag{4.122}$$

This corresponds to the n decoupled modal equations

$$\ddot{r}_i(t) + 2\zeta_i\omega_i\dot{r}_i(t) + \omega_i^2 r_i(t) = 0 \tag{4.123}$$

where $2\zeta_i\omega_i = \alpha + \beta\omega_i^2$ or

$$\zeta_i = \frac{\alpha}{2\omega_i} + \frac{\beta\omega_i}{2} \qquad i = 1, 2, \ldots, n \tag{4.124}$$

Here α and β can be chosen to produce some measured (or desired, in the design case) values of the modal damping ratio ζ_i. On the other hand, if α and β are known, equation (4.124) determines the value of the modal damping ratios ζ_i. The solution of equation (4.123) for the underdamped case ($0 < \zeta_i < 1$) is

$$r_i(t) = A_i e^{-\zeta_i\omega_i t} \sin(\omega_{di}t + \phi_i) \tag{4.125}$$

where A_i and ϕ_i are determined by applying the initial conditions on $\mathbf{r}(t)$. The solution in physical coordinates is then calculated from $\mathbf{x}(t) = S\mathbf{r}(t)$, where $S = M^{-1/2}P$ as before. The most general case of proportional damping is if $CM^{-1}K = KM^{-1}C$ holds. Note that if equation (4.119) holds, then $CM^{-1}K = KM^{-1}C$ is satisfied.

4.6 MODAL ANALYSIS OF THE FORCED RESPONSE

The forced response of a multiple-degree-of-freedom system can also be calculated by use of modal analysis. For example, consider the building system of Figure 4.9 with a force $F_4(t)$ applied to the fourth floor. For example, this force could be the result of an out-of-balance rotating machine on the fourth floor. The equation of motion takes the form

$$M\ddot{\mathbf{x}} + C\dot{\mathbf{x}} + K\mathbf{x} = B\mathbf{F}(t) \tag{4.126}$$

where $\mathbf{F}(t) = \begin{bmatrix} 0 & 0 & 0 & F_4(t) \end{bmatrix}^T$ and the matrix B is given by

$$B = \begin{bmatrix} 0 & 0 & 0 & 0 \\ 0 & 0 & 0 & 0 \\ 0 & 0 & 0 & 0 \\ 0 & 0 & 0 & 1 \end{bmatrix}$$

On the other hand, if the different forces are applied at each degree of freedom, B and $\mathbf{F}(t)$ would take on the form

$$B = \begin{bmatrix} 1 & 0 & 0 & 0 \\ 0 & 1 & 0 & 0 \\ 0 & 0 & 1 & 0 \\ 0 & 0 & 0 & 1 \end{bmatrix}, \quad \mathbf{F}(t) = \begin{bmatrix} F_1(t) \\ F_2(t) \\ F_3(t) \\ F_4(t) \end{bmatrix} \tag{4.127}$$

Alternately, if only a single force is applied at one coordinate, the matrix B may be collapsed to the vector $\mathbf{b}$ and the applied force reduces to the scalar $F(t)$. For example, the single force $F_4(t)$ applied to the fourth coordinate may also be written in equation (4.126) as $\mathbf{b}F_4(t)$, where $\mathbf{b} = \begin{bmatrix} 0 & 0 & 0 & 1 \end{bmatrix}^T$.

The approach of modal analysis again follows Window 4.5 and uses transformations to reduce equation (4.126) to a set of n decoupled modal equations, which in this case will be inhomogeneous. Then the methods of Chapter 3 can be applied to solve for the individual forced response in the modal coordinate system. The modal solution is then transformed back into the physical coordinate system.

To this end, assume that the damping matrix C is proportional of the form given by equation (4.119). Following the procedure in Window 4.5, let $\mathbf{x}(t) = M^{-1/2}\mathbf{q}(t)$ in equation (4.126) and multiply by $M^{-1/2}$. This yields

$$I\ddot{\mathbf{q}}(t) + \tilde{C}\dot{\mathbf{q}}(t) + \tilde{K}\mathbf{q}(t) = M^{-1/2}B\mathbf{F}(t) \tag{4.128}$$

where $\tilde{C} = M^{-1/2}CM^{-1/2}$. Next, calculate the eigenvalue problem for $\tilde{K}$. Let $\mathbf{q}(t) = P\mathbf{r}(t)$, where P is the matrix of eigenvectors of $\tilde{K}$ and multiply by P^T. This yields

$$\ddot{\mathbf{r}}(t) + \text{diag}\,[2\zeta_i\omega_i]\dot{\mathbf{r}}(t) + \Lambda\mathbf{r}(t) = P^T M^{-1/2}B\mathbf{F}(t) \tag{4.129}$$

where the matrix $\text{diag}[2\zeta_i\omega_i]$ follows from equation (4.123). The vector $P^T M^{-1/2} B\mathbf{F}(t)$ has elements $f_i(t)$ that will be linear combinations of the forces F_i applied to each mass. Hence the decoupled modal equations take the form

$$\ddot{r}_i(t) + 2\zeta_i\omega_i\dot{r}_i(t) + \omega_i^2 r_i(t) = f_i(t) \tag{4.130}$$

Referring to Section 3.2, this has the solution (reviewed in Window 4.8 for the underdamped case)

$$r_i(t) = d_i e^{-\zeta_i\omega_i t} \sin(\omega_{di} t + \phi_i) + \frac{1}{\omega_{di}} e^{-\zeta_i\omega_i t} \int_0^t f_i(\tau) e^{\zeta_i\omega_i \tau} \sin\omega_{di}(t - \tau)\,d\tau \tag{4.131}$$

Window 4.8
Forced Response of an Underdamped System from Section 3.2

The response of an underdamped system

$$m\ddot{x}(t) + c\dot{x}(t) + kx(t) = F(t)$$

(with zero initial conditions) is given by (for $0 < \zeta < 1$)

$$x(t) = \frac{1}{m\omega_d} e^{-\zeta\omega_n t} \int_0^t F(\tau) e^{\zeta\omega_n \tau} \sin \omega_d(t - \tau)\, d\tau$$

where $\omega_n = \sqrt{k/m}$, $\zeta = c/(2m\omega_n)$, and $\omega_d = \omega_n\sqrt{1 - \zeta^2}$. With nonzero initial conditions this becomes

$$x(t) = Ae^{-\zeta\omega_n t} \sin(\omega_d t + \phi) + \frac{1}{\omega_d} e^{-\zeta\omega_n t} \int_0^t f(\tau) e^{\zeta\omega_n \tau} \sin \omega_d(t - \tau)\, d\tau$$

where $f = F/m$ and A and ϕ are constants determined by the initial conditions.

where d_i and ϕ_i must be determined by the modal initial conditions and $\omega_{di} = \omega_i\sqrt{1 - \zeta_i^2}$ as before. Note that f_i may represent a sum of forces if more than one force is applied to the system. In addition, if a force is applied to only one mass of the system, this force becomes applied to each of the modal equations (4.131) by the transformation S, as illustrated in the following example.

Example 4.6.1

Consider the simple two-degree-of-freedom system with a harmonic force applied to one mass as indicated in Figure 4.16.

For this example, let $m_1 = 9\,\text{kg}$, $m_2 = 1\,\text{kg}$, $k_1 = 24\,\text{N/m}$, and $k_2 = 3\,\text{N/m}$. Also assume that the damping is proportional with $\alpha = 0$ and $\beta = 0.1$, so that $c_1 = 2.4\,\text{N}\cdot\text{s/m}$ and $c_2 = 0.3\,\text{N}\cdot\text{s/m}$. Calculate the steady-state response.

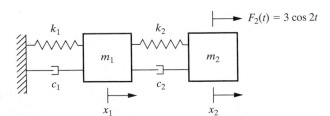

Figure 4.16 A damped two-degree-of-freedom system for Example 4.6.1.

Solution   The equations of motion in matrix form become

$$\begin{bmatrix} 9 & 0 \\ 0 & 1 \end{bmatrix} \ddot{\mathbf{x}} + \begin{bmatrix} 2.7 & -0.3 \\ -0.3 & 0.3 \end{bmatrix} \dot{\mathbf{x}} + \begin{bmatrix} 27 & -3 \\ -3 & 3 \end{bmatrix} \mathbf{x} = \begin{bmatrix} 0 & 0 \\ 0 & 1 \end{bmatrix} \begin{bmatrix} 0 \\ F_2(t) \end{bmatrix}$$

The matrices $M^{1/2}$ and $M^{-1/2}$ become

$$M^{1/2} = \begin{bmatrix} 3 & 0 \\ 0 & 1 \end{bmatrix} \qquad M^{-1/2} = \begin{bmatrix} \dfrac{1}{3} & 0 \\ 0 & 1 \end{bmatrix}$$

so that

$$\tilde{C} = M^{-1/2}CM^{-1/2} = \begin{bmatrix} 0.3 & -0.1 \\ -0.1 & 0.3 \end{bmatrix} \quad \text{and} \quad \tilde{K} = \begin{bmatrix} 3 & -1 \\ -1 & 3 \end{bmatrix}$$

The eigenvalue problem for $\tilde{K}$ yields

$$\lambda_1 = 2 \quad \lambda_2 = 4 \quad P = 0.7071 \begin{bmatrix} 1 & -1 \\ 1 & 1 \end{bmatrix}$$

Hence the natural frequencies of the system are $\omega_1 = \sqrt{2}$ and $\omega_2 = 2$; the matrices $P^T \tilde{C} P$ and $P^T \tilde{K} P$ become

$$P^T \tilde{C} P = \begin{bmatrix} 0.2 & 0 \\ 0 & 0.4 \end{bmatrix} \quad \text{and} \quad P^T \tilde{K} P = \begin{bmatrix} 2 & 0 \\ 0 & 4 \end{bmatrix}$$

The vector $\mathbf{f}(t) = P^T M^{-1/2} B F(t)$ becomes

$$\mathbf{f}(t) = \begin{bmatrix} 0.2357 & 0.7071 \\ -0.2357 & 0.7071 \end{bmatrix} \begin{bmatrix} 0 \\ F_2(t) \end{bmatrix} = 0.7071 \begin{bmatrix} F_2(t) \\ F_2(t) \end{bmatrix}$$

Hence the decoupled modal equations become

$$\ddot{r}_1 + 0.2\dot{r}_1 + 2r_1 = 0.7071(3) \cos 2t = 2.1213 \cos 2t$$

$$\ddot{r}_2 + 0.4\dot{r}_2 + 4r_2 = 0.7071(3) \cos 2t = 2.1213 \cos 2t$$

Comparing the coefficient of $\dot{r}_i$ in each case to $2\zeta_i\omega_i$ yields

$$\zeta_1 = \frac{0.2}{2\sqrt{2}} = 0.0707$$

$$\zeta_2 = \frac{0.4}{2(2)} = 0.1000$$

Thus the damped natural frequencies become

$$\omega_{d1} = \omega_1\sqrt{1 - \zeta_1^2} = 1.4106 \approx 1.41$$

$$\omega_{d2} = \omega_2\sqrt{1 - \zeta_2^2} = 1.9899 \approx 1.99$$

Note that while the force F_2 is applied only to mass m_2, it becomes applied to both coordinates when transformed to modal coordinates. The modal equations for r_1 and r_2 can be solved by equation (4.131), or in this case of a simple harmonic excitation, the particular solution is given directly by equation (2.36) as

$$r_{1p}(t) = \frac{2.1213}{\sqrt{(2-4)^2 + [2(0.0707)\sqrt{2}(2)]^2}} \cos\left(2t - \tan^{-1}\frac{2(0.0707)\sqrt{2}(2)}{\sqrt{2^2} - 2^2}\right)$$

$$= (1.040)\cos(2t + 0.1974) = 1.040\cos(2t - 2.9449)$$

Note that the argument of the arctangent function is negative $\left(\sqrt{2^2} - 2^2 < 0\right)$ so that the fourth quadrant angle must be used (see Window 2.4), yielding 2.9449 radians. The second mode particular solution is

$$r_{2p}(t) = \frac{2.1213}{\sqrt{(4-4)^2 + (2(0.1)(2)(2))^2}} \cos\left(2t - \tan^{-1}\frac{2(0.1)(2)(2)}{2^2 - 2^2}\right)$$

$$= 2.6516\cos\left(2t - \frac{\pi}{2}\right) = 2.6516\sin 2t$$

Here r_{ip} is used to denote the particular solution of the ith modal equation. Note that $r_2(t)$ is excited at its resonance frequency but has high damping, so that the larger but finite amplitude for $r_{2p}(t)$ is not unexpected. If the transient response is ignored [it dies out per equation (2.30)], the preceding solution yields the steady-state response. The solution in the physical coordinate system is

$$\mathbf{x}_{ss}(t) = M^{-1/2}P\mathbf{r}(t) = \begin{bmatrix} 0.2357 & -0.2357 \\ 0.7071 & 0.7071 \end{bmatrix}\begin{bmatrix} 1.040\cos(2t - 2.9442) \\ 2.6516\sin 2t \end{bmatrix}$$

so that in the steady state

$$x_1(t) = 0.2451\cos(2t - 2.9442) - 0.6249\sin 2t$$

$$x_2(t) = 0.7354\cos(2t - 2.9442) + 8749\sin 2t$$

Note that even though there is a fair amount of damping in the resonant mode, the coordinates each have a large component vibrating near the resonant frequency.

□

Resonance

The concept of resonance in multiple-degree-of-freedom systems is similar to that introduced in Section 2.2 for single-degree-of-freedom systems. It is based on the idea that a harmonic driving force is exciting the system at its natural frequency, causing an unbounded oscillation in the undamped case and a response with a maximum amplitude in the damped case. However, in multiple-degree-of-freedom systems, there are n natural frequencies, and the concept of resonance is complicated by the effects of mode shapes. Basically, if a force is applied orthogonally to the mode of the exciting frequency, the system will *not* resonate at any frequency, a fact that can be used in design. The following example illustrates resonance in a two-degree-of-freedom system.

Example 4.6.2

Consider the following system and determine if the driving frequency will cause the system to experience resonance.

$$\begin{bmatrix} 4 & 0 \\ 0 & 9 \end{bmatrix} \ddot{\mathbf{x}} + \begin{bmatrix} 30 & -5 \\ -5 & 5 \end{bmatrix} \mathbf{x}(t) = \begin{bmatrix} 1 \\ 0 \end{bmatrix} \sin(2.757t)$$

If so, which mode experiences resonance? Does this cause both degrees of freedom to experience resonance?

Solution   First, compute the mass-normalized stiffness matrix and then the eigenvalue problem for this system

$$\tilde{K} = M^{-1/2}KM^{-1/2} = \begin{bmatrix} \frac{1}{2} & 0 \\ 0 & \frac{1}{3} \end{bmatrix} \begin{bmatrix} 30 & -5 \\ -5 & 5 \end{bmatrix} \begin{bmatrix} \frac{1}{2} & 0 \\ 0 & \frac{1}{3} \end{bmatrix} = \begin{bmatrix} 7.5 & -0.833 \\ -0.833 & 0.556 \end{bmatrix}$$

Solving the eigenvalue problem for this matrix yields $\lambda_1 = 0.456956$ and $\lambda_2 = 7.5986$ so that $\omega_1 = 0.676$ rad/s and $\omega_2 = 2.757$ rad/s. Note that the second frequency is within round off to the driving frequency so that this is a resonant system. Next, compute the modal equations. From equation (4.129), the modal force vector is computed from

$$P^T M^{-1/2} \mathbf{b} = \begin{bmatrix} 0.118 & 0.993 \\ 0.993 & -0.118 \end{bmatrix} \begin{bmatrix} \frac{1}{2} & 0 \\ 0 & \frac{1}{3} \end{bmatrix} \begin{bmatrix} 1 \\ 0 \end{bmatrix} = \begin{bmatrix} 0.059 \\ 0.497 \end{bmatrix}$$

Thus the modal equations are

$$\ddot{r}_1(t) + (0.676)^2 r_1(t) = \sin(2.57t)$$
$$\ddot{r}_2(t) + (2.57)^2 r_2(t) = \sin(2.57t)$$

Thus the second mode is clearly in resonance. Note however that once transformed back to physical coordinates, each mass will be affected by both modes. That is, both $x_1(t)$ and $x_2(t)$ are a linear combination of $r_1(t)$ and $r_2(t)$. Thus each mass will experience resonance. This is because the transformation back to physical coordinates couples the modal solutions.

□

Forced Response via Mode Summation

First, compute the particular solution of the forced response. Consider

$$M\ddot{\mathbf{x}}(t) + K\mathbf{x}(t) = \mathbf{F}(t) \tag{4.132}$$

Here $\mathbf{F}(t)$ is a general force input. Let $\mathbf{x}_p$ denote the particular solution computed for a given force input. Next, consider the free response using mode summation. First consider the transformation of coordinates $x(t) = M^{-1/2}q(t)$ substituted into equation (4.132), and then premultiplying (4.132) by $M^{-1/2}$ yields

$$\ddot{q}(t) + \tilde{K}q(t) = M^{-1/2}F(t) \tag{4.133}$$

where $\tilde{K} = M^{-1/2}KM^{-1/2}$ as before. From equation (4.92) the homogeneous solution in mode summation form is

$$q_H(t) = \sum_{i=1}^{n} d_i \sin(\omega_i t + \phi_i)v_i \tag{4.134}$$

where v_i are the eigenvectors of the symmetric matrix $\tilde{K}$. Rewriting this last expression in orthogonal form (i.e., sine plus cosine instead of magnitude and phase) and adding in the particular solution yields the total solution:

$$q(t) = \underbrace{\sum_{i=1}^{n} \left[b_i \sin\omega_i t + c_i \cos\omega_i t \right] v_i}_{\text{homogeneous}} + \underbrace{q_p(t)}_{\text{particular}} \tag{4.135}$$

Now it remains to find an expression for q_p and to evaluate the constants of integration b_i and c_i in terms of the given initial conditions. From the coordinate transformation $x(t) = M^{-1/2}q(t)$, q_p is related to x_p by $q_p(t) = M^{1/2}x_p(t)$. Thus equation (4.135) becomes

$$q(t) = \sum_{i=1}^{n} \left(b_i \sin\omega_i t + c_i \cos\omega_i t \right) v_i + M^{1/2}x_p(t) \tag{4.136}$$

The initial conditions can now be used to compute the constants of integration. Setting $t = 0$ in equation (4.136) yields

$$q(0) = q_0 = \sum_{i=1}^{n} \left(b_i \sin\omega_i 0 + c_i \cos\omega_i 0 \right) v_i + M^{1/2}x_p(0) \tag{4.137}$$

Premultiplying this last expression by v_i^T yields

$$v_i^T q_0 = c_i + v_i^T M^{1/2}x_p(0) \tag{4.138}$$

Likewise, differentiating equation (4.136), setting $t = 0$, and multiplying by v_i^T yields

$$v_i^T \dot{q}_0 = \omega_i b_i + v_i^T M^{1/2}\dot{x}_p(0) \tag{4.139}$$

Solving equations (4.138) and (4.139) for the constants of integration yields

$$c_i = v_i^T q_0 - v_i^T M^{1/2}x_p(0) \tag{4.140}$$

$$b_i = \frac{1}{\omega_i} \left(v_i^T \dot{q}_0 - v_i^T M^{1/2}\dot{x}_p(0) \right)$$

Substitution of these values into equation (4.136) yields the expression for $\mathbf{q}(t)$. Premultiplying by $M^{-1/2}$ then yields the displacement in physical coordinates:

$$\mathbf{x}(t) = \sum_{i=1}^{n}\left(b_i \sin \omega_i t + c_i \cos \omega_i t\right)\mathbf{u}_i + \mathbf{x}_p(t) \qquad (4.141)$$

where the constants are given by equation (4.140), $\mathbf{u}_i$ are the mode shapes, ω_i are the natural frequencies, and $\mathbf{x}_p$ is the particular solution.

4.7 LAGRANGE'S EQUATIONS

Lagrange's equation was introduced in Section 1.4 as a follow-on to the energy method of deriving equations of motion. Equations (1.62) and (1.63) along with Example 1.4.7 introduced the method for single-degree-of-freedom systems. Just as in the single-degree-of-freedom case, the Lagrange formulation can be used to model multiple-degree-of-freedom systems as an alternative to using Newton's law (summing forces and moments) for those cases where the free-body diagram is not as obvious. Recall that the Lagrange formulation requires identification of the energy in the system, rather than the identification of forces and moments acting on the system, and requires the use of generalized coordinates. A brief working account of the Lagrange formulation is given here. A more precise and detailed account is given in Meirovitch (1995), for instance.

The procedure begins by assigning a generalized coordinate to each moving part. The standard rectangular coordinate system is an example of a generalized coordinate, but any length, angle, or other coordinate that *uniquely* defines the position of the part at any time forms a generalized coordinate. It is usually desirable to choose coordinates that are independent. It is customary to designate each coordinate by the letter q with a subscript so that a set of n generalized coordinates is written as $q_1, q_2, \ldots, q_n$. Note that we have run out of symbols and that the q_i used here are different than the q_i used to denote mass-normalized coordinates in previous sections.

An example of generalized coordinates is illustrated in Figure 4.17. In the figure, the location of the two masses can be described by the set of four coordinates

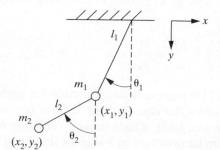

Figure 4.17   An example of generalized coordinates for a double pendulum illustrating an example of constraints.

x_1, y_1, x_2, and y_2 or the two coordinates θ_1 and θ_2. The coordinates θ_1 and θ_2 are taken to be generalized coordinates because they are independent. The Cartesian coordinates (x_1, x_2, y_1, y_2) are not independent and hence would not make a desirable choice of generalized coordinates. Note that

$$x_1^2 + y_1^2 = l_1^2 \quad \text{and} \quad (x_2 - x_1)^2 + (y_2 - y_1)^2 = l_2^2 \tag{4.142}$$

express the dependence of the Cartesian coordinates on each other. The relationships in equation (4.142) are called *equations of constraint*.

A new configuration of the double pendulum of Figure 4.17 can be obtained by changing the generalized coordinates $q_1 = \theta_1$ and $q_2 = \theta_2$ by an amount δq_1 and δq_2, respectively. Here δq_i are referred to as *virtual displacements*, which are defined to be infinitesimal displacements that do not violate constraints and such that there is no significant change in the system's geometry. The *virtual work*, denoted by δW, is the work done in causing the virtual displacement. The principle of virtual work states that if a system at rest (or at equilibrium) under the action of a set of forces is given a virtual displacement, the virtual work done by the forces is zero. The generalized force (or moment) at the ith coordinate, denoted by Q_i, is related to the work done in changing q_i by the amount δq_i and is defined to be

$$Q_i = \frac{\delta W}{\delta q_i} \tag{4.143}$$

The quantity Q_i will be a moment if q_i is a rotational coordinate and a force if it is a translational coordinate.

The Lagrange formulation follows from variational principles and states that the equations of motion of a vibrating system can be derived from

$$\frac{d}{dt}\left(\frac{\partial T}{\partial \dot{q}_i}\right) - \frac{\partial T}{\partial q_i} + \frac{\partial U}{\partial q_i} = Q_i \quad i = 1, 2, \ldots, n \tag{4.144}$$

where $\dot{q}_i = \partial q_i/\partial t$ is the generalized velocity, T is the kinetic energy of the system, U is the potential energy of the system, and Q_i represents all the nonconservative forces corresponding to q_i. Here $\partial/\partial q_i$ denotes the partial derivative with respect to the coordinate q_i. For conservative systems, $Q_i = 0$ and equation (4.144) becomes

$$\frac{d}{dt}\left(\frac{\partial T}{\partial \dot{q}_i}\right) - \frac{\partial T}{\partial q_i} + \frac{\partial U}{\partial q_i} = 0 \quad i = 1, 2, \ldots, n \tag{4.145}$$

Equations (4.144) and (4.145) represent one equation for each generalized coordinate. These expressions allow the equations of motion of complicated systems to be derived without using free-body diagrams and summing forces and moments. The Lagrange equation can be rewritten in a slightly simplified form by defining the

Lagrangian, L, to be $L = (T - U)$, the difference between the kinetic and potential energies. Then if $\partial U/\partial \dot{q}_i = 0$, the Lagrange equation becomes [this is equation (1.63)]

$$\frac{d}{dt}\left(\frac{\partial L}{\partial \dot{q}_i}\right) - \frac{\partial L}{\partial q_i} = 0 \qquad i = 1, 2, \ldots, n \tag{4.146}$$

The following examples illustrate the procedure.

Example 4.7.1

Derive the equations of motion of the system of Figure 4.18 using the Lagrange equation.

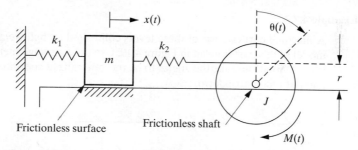

Figure 4.18   A vibration model of a simple machine part. The quantity $M(t)$ denotes an applied moment. The disk rotates without translation.

Solution   The motion of this system can be described by the two coordinates x and θ, so a good choice of generalized coordinates is $q_1(t) = x(t)$ and $q_2(t) = \theta(t)$. The kinetic energy becomes

$$T = \frac{1}{2}m\dot{q}_1^2 + \frac{1}{2}J\dot{q}_2^2$$

The potential energy becomes

$$U = \frac{1}{2}k_1 q_1^2 + \frac{1}{2}k_2(rq_2 - q_1)^2$$

Here $Q_1 = 0$ and $Q_2 = M(t)$. Using equation (4.145) yields, for $i = 1$,

$$\frac{d}{dt}(m\dot{q}_1 + 0) - 0 + k_1 q_1 + k_2(rq_2 - q_1)(-1) = 0$$

or

$$m\ddot{q}_1 + (k_1 + k_2)q_1 - k_2 rq_2 = 0 \tag{4.147}$$

Similarly, for $i = 2$, equation (4.144) yields

$$J\ddot{q}_2 + k_2 r^2 q_2 - k_2 rq_1 = M(t) \tag{4.148}$$

Combining equations (4.147) and (4.148) into matrix form yields

$$\begin{bmatrix} m & 0 \\ 0 & J \end{bmatrix} \ddot{\mathbf{x}}(t) + \begin{bmatrix} k_1 + k_2 & -rk_2 \\ -rk_2 & r^2k_2 \end{bmatrix} \mathbf{x}(t) = \begin{bmatrix} 0 \\ M(t) \end{bmatrix} \qquad (4.149)$$

Here the vector $\mathbf{x}(t)$ is

$$\mathbf{x}(t) = \begin{bmatrix} q_1(t) \\ q_2(t) \end{bmatrix} = \begin{bmatrix} x(t) \\ \theta(t) \end{bmatrix}$$

$\square$

Example 4.7.2

A machine part consists of three levers connected by lightweight linkages. A vibration model of this part is given in Figure 4.19. Use the Lagrange method to obtain the equation of vibration. Take the angles to be the generalized coordinates. Linearize the result and put it in matrix form.

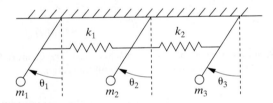

Figure 4.19 A vibration model of three coupled levers. The lengths of the levers are l, and the springs are attached at α units from the pivot points.

Solution   The kinetic energy is

$$T = \frac{1}{2} m_1 l^2 \dot{\theta}_1^2 + \frac{1}{2} m_2 l^2 \dot{\theta}_2^2 + \frac{1}{2} m_3 l^2 \dot{\theta}_3^2$$

The potential energy becomes

$$U = m_1 gl(1 - \cos\theta_1) + m_2 gl(1 - \cos\theta_2) + m_3 gl(1 - \cos\theta_3)$$
$$+ \frac{1}{2} k_1(\alpha\theta_2 - \alpha\theta_1)^2 + \frac{1}{2} k_2(\alpha\theta_3 - \alpha\theta_2)^2$$

Applying the Lagrange equation for $i = 1$ yields

$$m_1 l^2 \ddot{\theta}_1 + m_1 gl \sin\theta_1 - \alpha k_1(\alpha\theta_2 - \alpha\theta_1) = 0$$

For $i = 2$, the Lagrange equation yields

$$m_2 l^2 \ddot{\theta}_2 + m_2 gl \sin\theta_2 + \alpha k_1(\alpha\theta_2 - \alpha\theta_1) - \alpha k_2(\alpha\theta_3 - \alpha\theta_2) = 0$$

and for $i = 3$, the Lagrange equation becomes

$$m_3 l^2 \ddot{\theta}_3 + m_3 gl \sin\theta_3 + \alpha k_2(\alpha\theta_3 - \alpha\theta_2) = 0$$

These three equations can be linearized by assuming θ is small so that $\sin \theta \sim \theta$. Note that this linearization occurs after the equations have been derived. In matrix form this becomes

$$\begin{bmatrix} m_1 l^2 & 0 & 0 \\ 0 & m_2 l^2 & 0 \\ 0 & 0 & m_3 l^2 \end{bmatrix} \ddot{\mathbf{x}}(t)$$

$$+ \begin{bmatrix} m_1 g l + \alpha^2 k_1 & -\alpha^2 k_1 & 0 \\ -\alpha^2 k_1 & m_2 g l + \alpha^2 (k_1 + k_2) & -\alpha^2 k_2 \\ 0 & -\alpha^2 k_2 & m_3 g l + \alpha^2 k_2 \end{bmatrix} \mathbf{x}(t) = \mathbf{0}$$

where $\mathbf{x}(t) = [q_1 \quad q_2 \quad q_3]^T = [\theta_1 \quad \theta_2 \quad \theta_3]^T$ is the generalized set of coordinates.

$\square$

Example 4.7.3

Consider the wing vibration model of Figure 4.20. Using the vertical motion of the point of attachment of the springs, $x(t)$, and the rotation of this point, $\theta(t)$, determine the equations of motion using Lagrange's method. Use the small-angle approximation (recall the pendulum of Example 1.4.6) and write the equations in matrix form. Note that G denotes the center of mass and e denotes the distance between the point of rotation and the center of mass. Ignore the gravitational force.

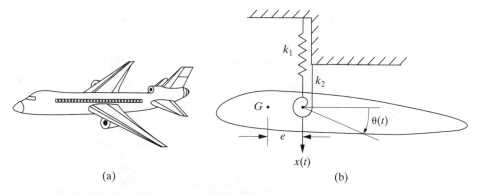

(a)   (b)

Figure 4.20   An airplane in flight (a) presents a number of different vibration models, one of which is given in part (b). In (b) a vibration model of a wing in flight is sketched which accounts for bending and torsional motion by modeling the wing as attached to ground (the aircraft body in this case) through a linear spring k_1 and a torsional spring k_2.

Solution   Let m denote the mass of the wing section and J denote the rotational inertia about point G. The kinetic energy is

$$T = \frac{1}{2} m \dot{x}_G^2 + \frac{1}{2} J \dot{\theta}^2$$

where x_G is the displacement of the point G. This displacement is related to the coordinate, $x(t)$, of the point of attachment of the springs by

$$x_G(t) = x(t) - e \sin \theta(t)$$

which is obtained by examining the geometry of Figure 4.20. Thus $\dot{x}_G(t)$ becomes

$$\dot{x}_G(t) = \dot{x}(t) - e \cos \theta(t) \frac{d\theta}{dt} = \dot{x}(t) - e\dot{\theta} \cos \theta$$

The expression for kinetic energy in terms of the generalized coordinates $q_1 = x$ and $q_2 = \theta$ then becomes

$$T = \frac{1}{2} m[\dot{x} - e\dot{\theta} \cos \theta]^2 + \frac{1}{2} J\dot{\theta}^2$$

The expression for potential energy is

$$U = \frac{1}{2} k_1 x^2 + \frac{1}{2} k_2 \theta^2$$

which is already in terms of the generalized coordinates. The Lagrangian, L, becomes

$$L = T - U = \frac{1}{2} m[\dot{x} - e\dot{\theta} \cos \theta]^2 + \frac{1}{2} J\dot{\theta}^2 - \frac{1}{2} k_1 x^2 - \frac{1}{2} k_2 \theta^2$$

Calculating the derivatives required by equation (4.146) for $i = 1$ yields

$$\frac{\partial L}{\partial \dot{q}_1} = \frac{\partial L}{\partial \dot{x}} = m[\dot{x} - e\dot{\theta} \cos \theta]$$

$$\frac{d}{dt} \left(\frac{\partial L}{\partial \dot{x}} \right) = m\ddot{x} - me\ddot{\theta} \cos \theta + me\dot{\theta}^2 \sin \theta$$

$$\frac{\partial L}{\partial q_1} = \frac{\partial L}{\partial x} = -k_1 x$$

so that equation (4.146) becomes

$$m\ddot{x} - me\ddot{\theta} \cos \theta + me\dot{\theta}^2 \sin \theta + k_1 x = 0$$

Assuming small motions so that the approximations $\cos \theta \to 1$, and $\sin \theta \to \theta$ hold, and assuming that the term $\dot{\theta}^2 \theta$ is small enough to ignore, results in a linear equation in $x(t)$ given by

$$m\ddot{x} - me\ddot{\theta} + k_1 x = 0$$

Calculating the derivatives of the Lagrangian required by equation (4.146) for $i = 2$ yields

$$\frac{\partial L}{\partial \dot{q}_2} = \frac{\partial L}{\partial \dot{\theta}} = m[\dot{x} - e\dot{\theta}\cos\theta](-e\cos\theta) + J\dot{\theta} = -me\cos\theta\dot{x} + me^2\dot{\theta}\cos^2\theta + J\dot{\theta}$$

$$\frac{d}{dt}\left(\frac{\partial L}{\partial \dot{q}_2}\right) = \frac{d}{dt}\left(\frac{\partial L}{\partial \dot{\theta}}\right)$$

$$= -me\cos\theta\ddot{x} + me\dot{x}\sin\theta\dot{\theta} + me^2\ddot{\theta}\cos^2\theta - 2me^2\dot{\theta}^2\sin\theta\cos\theta + J\ddot{\theta}$$

$$\frac{\partial L}{\partial q_2} = \frac{\partial L}{\partial \theta} = m[\dot{x} - e\dot{\theta}\cos\theta](e\dot{\theta}\sin\theta) - k_2\theta$$

$$= me\dot{x}\dot{\theta}\sin\theta - me^2\dot{\theta}^2\sin\theta\cos\theta - k_2\theta$$

so that equation (4.146) becomes

$$J\ddot{\theta} - me\cos\theta\ddot{x} + me^2\cos^2\theta\ddot{\theta} - me^2\dot{\theta}^2\sin\theta\cos\theta + k_2\theta = 0$$

Again if the small-angle, small-motion approximation (i.e., $\sin\theta \to \theta$, $\cos\theta \to 1$, $\dot{\theta}^2\theta \to 0$) is used, a linear equation in $\theta(t)$ results given by

$$(J + me^2)\ddot{\theta} - me\ddot{x} + k_2\theta = 0$$

Combining the expression for $i = 1$ and $i = 2$ into one vector equation in the generalized vector $\mathbf{x} = [q_1(t) \quad q_2(t)]^T = [x(t) \quad \theta(t)]^T$ yields

$$\begin{bmatrix} m & -me \\ -me & me^2 + J \end{bmatrix}\begin{bmatrix} \ddot{x}(t) \\ \ddot{\theta}(t) \end{bmatrix} + \begin{bmatrix} k_1 & 0 \\ 0 & k_2 \end{bmatrix}\begin{bmatrix} x(t) \\ \theta(t) \end{bmatrix} = \begin{bmatrix} 0 \\ 0 \end{bmatrix}$$

Note here that the two equations of motion are coupled, not through stiffness terms, as in Example 4.7.2, but rather through the inertia terms. Such systems are called *dynamically coupled*, meaning that the terms that couple the equation in $\theta(t)$ to the equation in $x(t)$ are in the mass matrix (i.e., meaning that the mass matrix is not diagonal). In all previous examples, the mass matrix is diagonal and the stiffness matrix is not diagonal. Such systems are called *statically coupled*. Dynamically coupled systems have nondiagonal mass matrices, and hence require the use of the Cholesky decomposition for factoring the mass matrix ($M = L^T L$) in the modal analysis steps of Window 4.5 (replacing $M^{-1/2}$ with L). This is discussed in Section 4.9. The programs in the Toolbox and the various codes given in Section 4.9 are capable of solving dynamically coupled systems as easily as those that have a diagonal mass matrix.

□

Example 4.7.3 not only illustrates a dynamically coupled system but also presents a system that is easier to approach using Lagrange's method than by using a force balance to obtain the equation of motion. Several vibration texts have reported an incorrect set of equations of motion for the problem of the preceding example by using the sum of forces and moments rather than taking a Lagrangian approach.

Damping

Viscous damping is a nonconservative force and may be modeled by defining the Rayleigh dissipation function. This function assumes that the damping forces are proportional to the velocities. The Rayleigh dissipation function then takes the form (recall Problem 1.79)

$$F = \frac{1}{2} \sum_{r=1}^{n} \sum_{s=1}^{n} c_{rs} \dot{q}_r \dot{q}_s \tag{4.150}$$

Here the damping coefficients $c_{rs} = c_{sr}$ and n is again the number of generalized coordinates. With this form, the generalized forces for viscous damping can be derived from

$$Q_j = -\frac{\partial F}{\partial \dot{q}_j}, \text{ for each } j = 1, 2, \ldots, n \tag{4.151}$$

Then to derive equations of motion with viscous damping, substitute equation (4.150) into (4.151) and (4.151) into equation (4.144). The following example illustrates the procedure.

Example 4.7.4

Consider again the system of Example 4.7.1 and assume that there is a viscous damper of coefficient c_1, parallel to k_1, and a damper of coefficient c_2, parallel to k_2. Derive the equations of motion for the system using Lagrange's equations.

Solution   The dissipation function given by equation (4.150) becomes

$$F = \frac{1}{2} [c_1 \dot{q}_1^2 + c_2(r\dot{q}_2 - \dot{q}_1)^2]$$

Substitution into equation (4.151) yields the generalized forces

$$Q_1 = -\frac{\partial F}{\partial \dot{q}_1} = -c_1\dot{q}_1 - c_2(r\dot{q}_2 - \dot{q}_1)(-1) = -(c_1 + c_2)\dot{q}_1 + c_2 r\dot{q}_2$$

$$Q_2 = -\frac{\partial F}{\partial \dot{q}_2} = -c_2(r\dot{q}_2 - \dot{q}_1)(r) = -c_2 r^2 \dot{q}_2 + rc_2\dot{q}_1$$

Adding the moment as indicated in Example 4.7.1, the second generalized force becomes

$$Q_2 = M(t) - c_2 r^2 \dot{q}_2 + rc_2\dot{q}_1$$

Next, using T and U as given in Example 4.7.1, recalculate the equations of motion using equation (4.144) to get, for $i = 1$:

$$m\ddot{q}_1 + (k_1 + k_2)q_1 - k_2 r q_2 = Q_1 = -(c_1 + c_2)\dot{q}_1 + c_2 r\dot{q}_2 \text{ or:}$$

$$m\ddot{q}_1 + (c_1 + c_2)\dot{q}_1 - c_2 r\dot{q}_2 + (k_1 + k_2)q_1 - k_2 q_2 = 0$$

and for $i = 2$:

$$J\ddot{q}_2 + k_2 r^2 q_2 - k_2 r q_1 = Q_2 = M(t) - c_2 r^2 \dot{q}_2 + r c_2 \dot{q}_1 \text{ or:}$$

$$J\ddot{q}_2 + c_2 r^2 \dot{q}_2 - r c_2 \dot{q}_1 + k_2 r^2 q_2 - k_2 r q_1 = M(t)$$

Combining the expressions for $i = 1$ and $i = 2$ yields the matrix form of the equations of motion:

$$\begin{bmatrix} m & 0 \\ 0 & J \end{bmatrix} \ddot{\mathbf{x}}(t) + \begin{bmatrix} c_1 + c_2 & -rc_2 \\ -rc_2 & r^2 c_2 \end{bmatrix} \dot{\mathbf{x}}(t) + \begin{bmatrix} k_1 + k_2 & -rk_2 \\ -rk_2 & r^2 k_2 \end{bmatrix} \mathbf{x}(t) = \begin{bmatrix} 0 \\ M(t) \end{bmatrix}$$

□

4.8 EXAMPLES

Several examples of multiple-degree-of-freedom systems, their schematics, and equations of motion are presented in this section. The "art" in vibration analysis and design is often related to choosing an appropriate mathematical model to describe a given structure or machine. The following examples are intended to provide additional "practice" in modeling and analysis.

Example 4.8.1

A drive shaft for a belt-driven machine such as a lathe is illustrated in Figure 4.21(a). The vibration model of this system is indicated in Figure 4.21(b), along with a free-body diagram of the machine. Write the equations of motion in matrix form and solve for the case $J_1 = J_2 = J_3 = 10$ kg m^2/rad, $k_1 = k_2 = 10^3$ N·m/rad, $c = 2$ N·m·s/rad for zero initial conditions, and where the applied moment $M(t)$ is a unit impulse function.

Solution   In Figure 4.21(a) the bearings and shaft lubricant are modeled as lumped viscous damping, and the shafts are modeled as torsional springs. The pulley and machine disks are modeled as rotational inertias. The motor is modeled simply as supplying a moment to the pulley. Figure 4.21(b) illustrates a free-body diagram for each of the three disks, where the damping is assumed to act in proportion to the relative motion of the masses and of the same value at each coordinate (other damping models may be more appropriate, but this choice yields an easy form to solve).

Examining the free-body diagram of Figure 4.21(b) and summing the moments on each of the disks yields

$$J_1\ddot{\theta}_1 = k_1(\theta_2 - \theta_1) + c(\dot{\theta}_2 - \dot{\theta}_1)$$

$$J_2\ddot{\theta}_2 = k_2(\theta_3 - \theta_2) + c(\dot{\theta}_3 - \dot{\theta}_2) - k_1(\theta_2 - \theta_1) - c(\dot{\theta}_2 - \dot{\theta}_1)$$

$$J_3\ddot{\theta}_3 = -k_2(\theta_3 - \theta_2) - c(\dot{\theta}_3 - \dot{\theta}_2) + M(t)$$

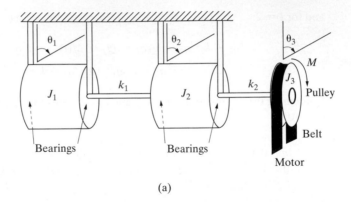

(a)

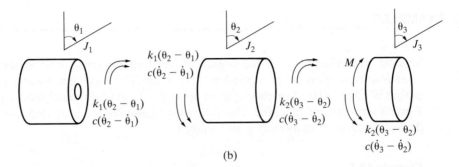

(b)

Figure 4.21 (a) Schematic of the moving parts of a lathe. The bearings that support the rotating shaft are modeled as providing viscous damping while the shafts provide stiffness and the belt drive provides an applied torque. (b) Free-body diagrams of the three inertias in the rotating system of part (a). The shafts are modeled as providing stiffness, or as rotational springs, and the bearings are modeled as rotational dampers.

where θ_1, θ_2, and θ_3 are the rotational coordinates as indicated in Figure 4.21. The unit for θ is radians. Rearranging these equations yields

$$J_1\ddot{\theta}_1 + c\dot{\theta}_1 + k_1\theta_1 - c\dot{\theta}_2 - k_1\theta_2 = 0$$

$$J_2\ddot{\theta}_2 + 2c\dot{\theta}_2 - c\dot{\theta}_1 - c\dot{\theta}_3 + (k_1 + k_2)\theta_2 - k_1\theta_1 - k_2\theta_3 = 0$$

$$J_3\ddot{\theta}_3 + c\dot{\theta}_3 - c\dot{\theta}_2 - k_2\theta_2 + k_2\theta_3 = M(t)$$

In matrix form this becomes

$$\begin{bmatrix} J_1 & 0 & 0 \\ 0 & J_2 & 0 \\ 0 & 0 & J_3 \end{bmatrix}\ddot{\boldsymbol{\theta}} + \begin{bmatrix} c & -c & 0 \\ -c & 2c & -c \\ 0 & -c & c \end{bmatrix}\dot{\boldsymbol{\theta}} + \begin{bmatrix} k_1 & -k_1 & 0 \\ -k_1 & k_1 + k_2 & -k_2 \\ 0 & -k_2 & k_2 \end{bmatrix}\boldsymbol{\theta} = \begin{bmatrix} 0 \\ 0 \\ M(t) \end{bmatrix}$$

where $\mathbf{\theta}(t) = [\theta_1(t) \quad \theta_2(t) \quad \theta_3(t)]^T$. Using the values for the coefficients given previously, this becomes

$$
\begin{bmatrix} 10 & 0 & 0 \\ 0 & 10 & 0 \\ 0 & 0 & 10 \end{bmatrix} \ddot{\mathbf{\theta}} + 2 \begin{bmatrix} 1 & -1 & 0 \\ -1 & 2 & -1 \\ 0 & -1 & 1 \end{bmatrix} \dot{\mathbf{\theta}} + 10^3 \begin{bmatrix} 1 & -1 & 0 \\ -1 & 2 & -1 \\ 0 & -1 & 1 \end{bmatrix} \mathbf{\theta} = \begin{bmatrix} 0 \\ 0 \\ \delta(t) \end{bmatrix}
$$

Note that the damping matrix is proportional to the stiffness matrix so that modal analysis can be used to calculate the solution. Also note that

$$
\tilde{C} = 0.2 \begin{bmatrix} 1 & -1 & 0 \\ -1 & 2 & -1 \\ 0 & -1 & 1 \end{bmatrix} \quad \tilde{K} = 10^2 \begin{bmatrix} 1 & -1 & 0 \\ -1 & 2 & -1 \\ 0 & -1 & 1 \end{bmatrix} \quad M^{-1/2}\mathbf{F} = \frac{1}{\sqrt{10}} \begin{bmatrix} 0 \\ 0 \\ \delta(t) \end{bmatrix}
$$

Following the steps of Example 4.6.1, the eigenvalue problem for $\tilde{K}$ yields

$$\lambda_1 = 0 \quad \lambda_2 = 100 \quad \lambda_3 = 300$$

Note that one of the eigenvalues is zero, thus the matrix $\tilde{K}$ is singular. The physical meaning of this is interpreted in this example. The normalized eigenvectors of $\tilde{K}$ yield

$$
P = \begin{bmatrix} 0.5774 & 0.7071 & 0.4082 \\ 0.5774 & 0 & -0.8165 \\ 0.5774 & -0.7071 & 0.4082 \end{bmatrix} \quad P^T = \begin{bmatrix} 0.5774 & 0.5774 & 0.5774 \\ 0.7071 & 0 & -0.7071 \\ 0.4082 & -0.8165 & 0.4082 \end{bmatrix}
$$

Further computation yields

$$P^T \tilde{C} P = \text{diag}[0 \quad 0.2 \quad 0.6]$$
$$P^T \tilde{K} P = \text{diag}[0 \quad 100 \quad 300]$$

$$
P^T M^{-1/2}\mathbf{F}(t) = \begin{bmatrix} 0.1826 \\ -0.2236 \\ 0.1291 \end{bmatrix} \delta(t)
$$

The decoupled modal equations are

$$\ddot{r}_1(t) = 0.1826\delta(t)$$
$$\ddot{r}_2(t) + 0.2\dot{r}_2(t) + 100r_2(t) = -0.2236\delta(t)$$
$$\ddot{r}_3(t) + 0.6\dot{r}_3(t) + 300r_3(t) = 0.1291\delta(t)$$

Obviously, $\omega_1 = 0$, $\omega_2 = 10$ rad/s, and $\omega_3 = 17.3205$ rad/s.

Comparing coefficients of $\dot{r}_i$ with $2\zeta_i\omega_i$ yields the three modal damping ratios

$$\zeta_1 = 0$$

$$\zeta_2 = \frac{0.2}{2(10)} = 0.01$$

$$\zeta_3 = \frac{0.6}{2(17.3205)} = 0.01732$$

so that the second two modes are underdamped. Hence the two damped natural frequencies become

$$\omega_{d2} = \omega_2\sqrt{1 - \zeta_2^2} = 9.9995 \text{ rad/s}$$

$$\omega_{d3} = \omega_3\sqrt{1 - \zeta_3^2} = 17.3179 \text{ rad/s}$$

As was the case in Example 4.6.1, while the moment is applied to only one physical location, it is applied to each of the three modal coordinates. The modal equations for r_2 and r_3 have solutions given by equation (3.6). The solution corresponding to the zero eigenvalue ($\omega_1 = 0$) can be calculated by direct integration or by using the Laplace transform method. Taking the Laplace transform yields

$$s^2 r(s) = 0.1826 \quad \text{or} \quad r(s) = \frac{0.1826}{s^2}$$

The inverse Laplace transform of this last expression yields (see Appendix B)

$$r_1(t) = (0.1826)t$$

Physically, this is interpreted as the unconstrained motion of the shaft (i.e., the shaft rotates or spins continuously through 360°). This is also called the *rigid-body mode* (see Example 4.6.2) or *zero mode* and results from $\tilde{K}$ being singular (i.e., from the zero eigenvalue). Such systems are also called *semidefinite*, as explained in Appendix C.

Following equation (3.6), the solution for $r_2(t)$ and $r_3(t)$ becomes

$$r_2(t) = \frac{f_2}{\omega_{d2}} e^{-\zeta_2\omega_2 t}\sin\omega_{d2}t = -0.0224e^{-0.1t}\sin 9.9995t$$

$$r_3(t) = \frac{f_3}{\omega_{d3}} e^{-\zeta_3\omega_3 t}\sin\omega_{d3}t = 0.0075e^{-0.2999t}\sin 17.3179t$$

The total solution in physical coordinates is then calculated from $\theta(t) = M^{-1/2}Pr(t) = (1/\sqrt{10})Pr(t)$ or

$$\boldsymbol{\theta}(t) = \begin{bmatrix} 0.0333t - 0.0050e^{-0.1t}\sin(9.9995t) + 0.0010e^{-0.2999t}\sin(17.3179t) \\ 0.0333t - 0.0019e^{-0.2999t}\sin(17.3179t) \\ 0.0333t - 0.0053e^{-0.1t}\sin(9.9995t) + 0.0010e^{-0.2999t}\sin(17.3179t) \end{bmatrix}$$

The three solutions $\theta_1(t)$, $\theta_2(t)$, and $\theta_3(t)$ are plotted in Figure 4.22. Figure 4.23 plots the three solutions without the rigid-body term. This represents the vibrations experienced by each disk as it rotates.

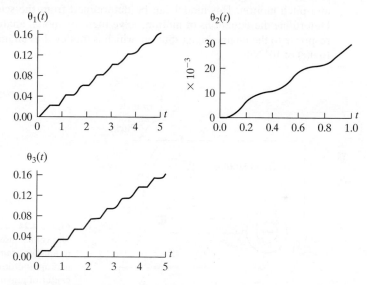

Figure 4.22 The response of each of the disks of Figure 4.21 to an impulse at θ_3, illustrating the effects of a rigid-body rotation.

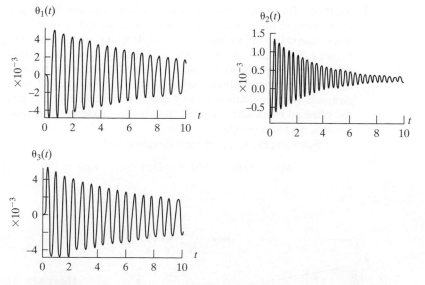

Figure 4.23 The response of each of the disks of Figure 4.21 to an impulse at θ_3 without the rigid-body mode, illustrating the vibration that occurs in each disk.

Example 4.8.2

In Figure 2.17 a vehicle is modeled as a single-degree-of-freedom system. In this example, a two-degree-of-freedom model is used for a vehicle that allows for bounce and pitch motion. This model can be determined from the schematic of Figure 4.24. Determine the equations of motion, solve them by modal analysis, and determine the response to the engine being shut off, which is modeled as an impulse moment applied to $\theta(t)$ of 10^3 Nm.

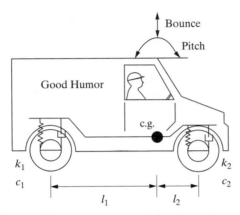

Figure 4.24 A sketch of the side section of a vehicle used to suggest a vibration model for examining its angular (pitch) and up-and-down (bounce) motion. The center of gravity is denoted by c.g.

Solution The sketch of the vehicle of Figure 4.24 can be simplified by modeling the entire mass of the system as concentrated at the center of gravity (c.g.). The tire-and-wheel assembly is approximated as a simple spring–dashpot arrangement as illustrated in Figure 4.25. The rotation of the vehicle in the x–y plane is described by the angle $\theta(t)$, and the up-and-down motion is modeled by $x(t)$. The angle $\theta(t)$ is taken to be positive in the clockwise direction, and the vertical displacement is taken as positive in the downward direction. Rigid translation in the y direction is ignored for the sake of concentrating on the vibration characteristics of the vehicle (e.g., Example 4.8.1 illustrates the concept of ignoring rigid-body motion).

Summing the forces in the x direction yields

$$m\ddot{x} = -c_1(\dot{x} - l_1\dot{\theta}) - c_2(\dot{x} + l_2\dot{\theta}) - k_1(x - l_1\theta) - k_2(x + l_2\theta) \qquad (4.152)$$

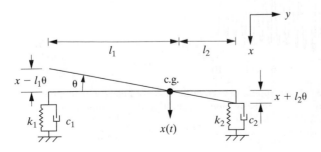

Figure 4.25 The vehicle of Figure 4.24 modeled as having all of its mass at its c.g. and two-degrees-of-freedom, consisting of the pitch, $\theta(t)$, about the c.g. and a translation $x(t)$ of the c.g.

since the spring k_1 experiences a displacement $x - l_1\theta$ and k_2 experiences a displacement $x + l_2\theta$. Similarly, the velocity experienced by the damper c_1 is $\dot{x} - l_1\dot{\theta}$ and that of c_2 is $\dot{x} + l_2\dot{\theta}$. Taking moments about the center of gravity yields

$$J\ddot{\theta} = c_1 l_1(\dot{x} - l_1\dot{\theta}) - c_2 l_2(\dot{x} + l_2\dot{\theta}) + k_1 l_1(x - l_1\theta) - k_2 l_2(x + l_2\theta) \quad (4.153)$$

where $J = mr^2$. Here r is the radius of gyration of the vehicle (recall Example 1.4.6). Equations (4.152) and (4.153) can be rewritten as

$$m\ddot{x} + (c_1 + c_2)\dot{x} + (l_2 c_2 - l_1 c_1)\dot{\theta} + (k_1 + k_2)x + (l_2 k_2 - l_1 k_1)\theta = 0$$

$$mr^2\ddot{\theta} + (c_2 l_2 - c_1 l_1)\dot{x} + (l_2^2 c_2 + l_1^2 c_1)\dot{\theta} + (k_2 l_2 - k_1 l_1)x + (l_1^2 k_1 + l_2^2 k_2)\theta = 0 \quad (4.154)$$

In matrix form, these two coupled equations become

$$m\begin{bmatrix} 1 & 0 \\ 0 & r^2 \end{bmatrix}\ddot{x} + \begin{bmatrix} c_1 + c_2 & l_2 c_2 - l_1 c_1 \\ l_2 c_2 - l_1 c_1 & l_2^2 c_2 + l_1^2 c_1 \end{bmatrix}\dot{x} + \begin{bmatrix} k_1 + k_2 & k_2 l_2 - k_1 l_1 \\ k_2 l_2 - k_1 l_1 & l_1^2 k_1 + l_2^2 k_2 \end{bmatrix}x = 0$$

$$(4.155)$$

where the vector $\mathbf{x}$ is defined by

$$\mathbf{x} = \begin{bmatrix} x(t) \\ \theta(t) \end{bmatrix}$$

Reasonable values for a truck are

$$r^2 = 0.64 \text{ m}^2 \quad m = 4000 \text{ kg} \quad c_1 = c_2 = 2000 \text{ N} \cdot \text{s/m}$$

$$k_1 = k_2 = 20{,}000 \text{ N/m} \quad l_1 = 0.9 \text{ m} \quad l_2 = 1.4 \text{ m}$$

With these values, equation (4.155) becomes

$$\begin{bmatrix} 4000 & 0 \\ 0 & 2560 \end{bmatrix}\ddot{x} + \begin{bmatrix} 4000 & 1000 \\ 1000 & 5540 \end{bmatrix}\dot{x} + \begin{bmatrix} 40{,}000 & 10{,}000 \\ 10{,}000 & 55{,}400 \end{bmatrix}x = \begin{bmatrix} 0 \\ 0 \end{bmatrix} \quad (4.156)$$

Note that $C = (0.1)K$, so that the damping is proportional. If a moment $M(t)$ is applied to the angular coordinate $\theta(t)$ the equations of motion become

$$\begin{bmatrix} 4000 & 0 \\ 0 & 2560 \end{bmatrix}\ddot{x} + \begin{bmatrix} 4000 & 1000 \\ 1000 & 5540 \end{bmatrix}\dot{x} + \begin{bmatrix} 40{,}000 & 10{,}000 \\ 10{,}000 & 55{,}400 \end{bmatrix}x = \begin{bmatrix} 0 \\ \delta(t) \end{bmatrix}10^3$$

Following the usual procedures of modal analysis, calculation of $M^{-1/2}$ yields

$$M^{-1/2} = \begin{bmatrix} 0.0158 & 0 \\ 0 & 0.0198 \end{bmatrix}$$

Thus

$$\tilde{C} = \begin{bmatrix} 1.0000 & 0.3125 \\ 0.3125 & 2.1641 \end{bmatrix} \quad \text{and} \quad \tilde{K} = \begin{bmatrix} 10.000 & 3.1250 \\ 3.1250 & 21.6406 \end{bmatrix}$$

Solving the eigenvalue problem for $\widetilde{K}$ yields

$$P = \begin{bmatrix} 0.9698 & 0.2439 \\ -0.2439 & 0.9698 \end{bmatrix} \quad \text{and} \quad P^T = \begin{bmatrix} 0.9698 & -0.2439 \\ 0.2439 & 0.9698 \end{bmatrix}$$

with eigenvalues $\lambda_1 = 9.2141$ and $\lambda_2 = 22.4265$, so that the natural frequencies are

$$\omega_1 = 3.0355 \text{ rad/s} \quad \text{and} \quad \omega_2 = 4.7357 \text{ rad/s}$$

Thus

$$P^T \widetilde{K} P = \text{diag}\,[9.2141 \quad 22.4265] \quad \text{and} \quad P^T \widetilde{C} P = \text{diag}\,[0.9214 \quad 2.2426]$$

Comparing the elements of $P^T \widetilde{C} P$ to ω_1 and ω_2 yields the modal damping ratios

$$\zeta_1 = \frac{0.9214}{2(3.0355)} = 0.1518 \quad \text{and} \quad \zeta_2 = \frac{2.2426}{2(4.7357)} = 0.2369$$

Using the formula from Window 4.8 for damped natural frequencies yields $\omega_{d1} = 3.0003$ rad/s and $\omega_{d2} = 4.6009$ rad/s. The modal forces are calculated from

$$P^T M^{-1/2} \begin{bmatrix} 0 \\ \delta(t) \end{bmatrix} 10^3 = \begin{bmatrix} 15.3 & -4.8\,\delta \\ 3.9 & 19.2 \end{bmatrix} \begin{bmatrix} 0 \\ \delta(t) \end{bmatrix} = \begin{bmatrix} -4.8\,\delta(t) \\ 19.2\,\delta(t) \end{bmatrix}$$

The decoupled modal equations become

$$\ddot{r}_1(t) + (0.9214)\dot{r}_1(t) + (9.2141)r_1(t) = -4.8\delta(t)$$

$$\ddot{r}_2(t) + (2.2436)\dot{r}_2(t) + (22.4265)r_2(t) = 19.2\delta(t)$$

From equation (3.6) these have solutions

$$r_1(t) = \frac{-4.8}{m\omega_{d1}} e^{-\zeta_1\omega_1 t} \sin\omega_{d1}t = \frac{1}{(1)(3.0003)} e^{-(0.1518)(3.035)t} \sin(3.0003t)$$

$$= -1.6066 e^{-0.4607t} \sin(3.0003t)$$

$$r_2(t) = \frac{19.2}{4.6009} e^{-(0.2369)(4.7357)t} \sin(4.6009t)$$

$$= 4.1659 e^{-1.1219t} \sin(4.6009t)$$

The solution in physical coordinates is obtained from

$$\begin{bmatrix} x(t) \\ \theta(t) \end{bmatrix} = M^{-1/2} P\mathbf{r}(t)$$

which yields

$$x(t) = -2.41 \times 10^{-2} e^{-0.4607t} \sin(3.0003t) + 1.606 \times 10^{-2} e^{-1.1213t} \sin(4.6009t)$$

$$\theta(t) = 7.744 \times 10^{-4} e^{-0.4607t} \sin(3.0003t) + 7.915 \times 10^{-2} e^{-1.1213t} \sin(4.6009t)$$

These coordinates are plotted in Figure 4.26.

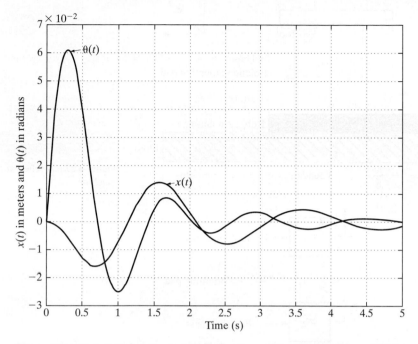

Figure 4.26   Plot of the bounce and pitch vibrations of the vehicle of Figure 4.24 as the result of the engine being shut off.

☐

Example 4.8.3

The punch press of Figure 4.27 can be modeled for vibration analysis in the x direction as indicated by the three-degree-of-freedom system of Figure 4.28. Discuss the solution for the response due to an impact at m_1 using modal analysis.

Solution   The mass and stiffness of the various components can be easily approximated using the static methods suggested in Chapter 1. However, it is very difficult to estimate values for the damping coefficients. Hence, an educated guess is made for the modal damping ratios. Such guesses are often made based on experience or from measurements such as the logarithmic decrement. In this case, the values of various masses and stiffness coefficients are [in mks units and $f(t) = 1000\delta(t)$]

$$m_1 = 400\,\text{kg} \qquad m_2 = 2000\,\text{kg} \qquad m_3 = 8000\,\text{kg}$$

$$k_1 = 300{,}000\,\text{N/m} \qquad k_2 = 80{,}000\,\text{N/m} \qquad k_3 = 800{,}000\,\text{N/m}$$

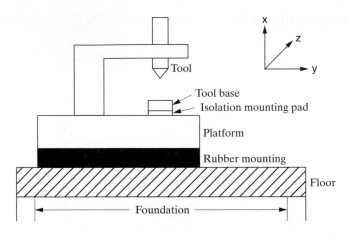

Figure 4.27 A schematic of a punch-press machine.

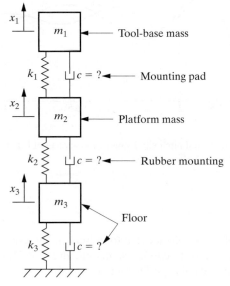

Figure 4.28 A vibration model of the punch press of Figure 4.27.

From free-body diagrams of each mass, the summing of forces in the x direction yields the three coupled equations

$$m_1\ddot{x}_1 = -k_1(x_1 - x_2) + f(t)$$

$$m_2\ddot{x}_2 = k_1(x_1 - x_2) - k_2(x_2 - x_3)$$

$$m_3\ddot{x}_3 = -k_3x_3 + k_2(x_2 - x_3)$$

Rewriting this set of coupled equations in matrix form yields

$$\begin{bmatrix} m_1 & 0 & 0 \\ 0 & m_2 & 0 \\ 0 & 0 & m_3 \end{bmatrix}\ddot{\mathbf{x}} + \begin{bmatrix} k_1 & -k_1 & 0 \\ -k_1 & k_1 + k_2 & -k_2 \\ 0 & -k_2 & k_2 + k_3 \end{bmatrix}\mathbf{x} = \begin{bmatrix} f(t) \\ 0 \\ 0 \end{bmatrix}$$

where $\mathbf{x} = [x_1(t) \quad x_2(t) \quad x_3(t)]^T$. Substituting the numerical values for m_i and k_i yields

$$(10^3)\begin{bmatrix} 0.4 & 0 & 0 \\ 0 & 2 & 0 \\ 0 & 0 & 8 \end{bmatrix} \ddot{\mathbf{x}} + (10^4)\begin{bmatrix} 30 & -30 & 0 \\ -30 & 38 & -8 \\ 0 & -8 & 88 \end{bmatrix}\mathbf{x} = \begin{bmatrix} 1000\delta(t) \\ 0 \\ 0 \end{bmatrix}$$

Following the modal analysis procedure for an undamped system yields

$$M^{1/2} = \begin{bmatrix} 20 & 0 & 0 \\ 0 & 44.7214 & 0 \\ 0 & 0 & 89.4427 \end{bmatrix} \quad M^{-1/2} = \begin{bmatrix} 0.0500 & 0 & 0 \\ 0 & 0.0224 & 0 \\ 0 & 0 & 0.0112 \end{bmatrix}$$

and

$$\widetilde{K} = \begin{bmatrix} 750 & -335.4102 & 0 \\ -335.4102 & 190 & -20 \\ 0 & -20 & 110 \end{bmatrix}$$

Solving the eigenvalue problem for $\widetilde{K}$ yields

$$P = \begin{bmatrix} -0.4116 & -0.1021 & 0.9056 \\ -0.8848 & -0.1935 & -0.4239 \\ -0.2185 & 0.9758 & 0.0106 \end{bmatrix} \quad P^T = \begin{bmatrix} -0.4116 & -0.8848 & -0.2185 \\ -0.1021 & -0.1935 & 0.9758 \\ 0.9056 & -0.4239 & 0.0106 \end{bmatrix}$$

and

$$\lambda_1 = 29.0223 \qquad \omega_1 = 5.3872$$
$$\lambda_2 = 113.9665 \qquad \omega_2 = 10.6755$$
$$\lambda_3 = 907.0112 \qquad \omega_3 = 30.1166$$

The modal force vector becomes

$$P^T M^{-1/2}\begin{bmatrix} 1000\delta(t) \\ 0 \\ 0 \end{bmatrix} = \begin{bmatrix} -20.5805 \\ -5.1026 \\ 45.2814 \end{bmatrix}\delta(t)$$

Hence, the undamped modal equations are

$$\ddot{r}_1(t) + 29.0223\, r_1(t) = -20.5805\delta(t)$$
$$\ddot{r}_2(t) + 113.9665\, r_2(t) = -5.1026\delta(t)$$
$$\ddot{r}_3(t) + 907.0112\, r_3(t) = 45.2814\delta(t)$$

To model the damping, note that each mode shape is dominated by one element. From examining the first column of the matrix P, the second element is larger than the other two elements. Hence if the system were vibrating only in the first mode, the motion of $x_2(t)$ would dominate. This element corresponds to the platform mass, which receives high damping from the rubber support. Hence it is given a large damping ratio of $\zeta_1 = 0.1$ (rubber provides a lot of damping). Similarly, the second mode is dominated by its third element, corresponding to the motion of $x_3(t)$. This is a predominantly metal part, so it is given a low damping ratio of $\zeta_2 = 0.01$. The third mode shape is dominated by the first element, which corresponds to the mounting pad. Hence it is given a medium damping ratio of $\zeta_3 = 0.05$. Recalling that the velocity coefficient in modal coordinates has the form $2\zeta_i\omega_i$, the damped modal coordinates become $2\zeta_1\omega_1 = 2(0.1)(5.3872)$, $2\zeta_2\omega_2 = 2(0.01)(10.6755)$, and $2\zeta_3\omega_3 = 2(0.05)(30.1166)$. Therefore, the damped modal equations become

$$\ddot{r}_1(t) + 1.0774\,\dot{r}_1(t) + 29.0223\,r_1(t) = -20.5805\delta(t)$$
$$\ddot{r}_2(t) + 0.2135\,\dot{r}_2(t) + 113.9665\,r_2(t) = -5.1026\delta(t)$$
$$\ddot{r}_3(t) + 3.0117\,\dot{r}_3(t) + 907.0112\,r_3(t) = 45.2814\delta(t)$$

These have solutions given by equation (3.6) as

$$r_1(t) = -3.8395e^{-0.5387t}\sin(5.3602t)$$
$$r_2(t) = -0.4780e^{-0.1068t}\sin(10.6750t)$$
$$r_3(t) = 1.5054e^{-1.5058t}\sin(30.0789t)$$

Using the transformation $\mathbf{x}(t) = M^{-1/2}P\mathbf{r}(t)$ yields

$$x_1(t) = 0.0790e^{-0.5387t}\sin(5.3602t) + 0.0024e^{-0.1068t}\sin(10.6750t) + 0.0682e^{-1.5058t}\sin(30.0789t)$$
$$x_2(t) = 0.0760e^{-0.5387t}\sin(5.3602t) + 0.0021e^{-0.1068t}\sin(10.6750t) - 0.0143e^{-1.5058t}\sin(30.0789t)$$
$$x_3(t) = 0.0094e^{-0.5387t}\sin(5.3602t) - 0.0052e^{-0.1068t}\sin(10.6750t) + 0.0002e^{-1.5058t}\sin(30.0789t)$$

These solutions are plotted in Figure 4.29.

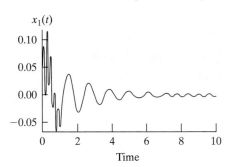

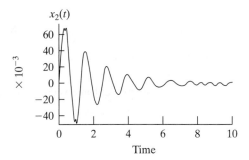

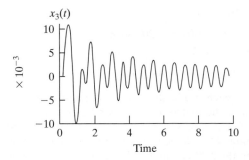

Figure 4.29   A numerical simulation of the vibration of the punch press of Figures 4.27 and 4.28 as the result of the machine tool impacting the tool base.

This example illustrates a method of assigning modal damping to an analytical model. This is a somewhat arbitrary procedure that falls in the category of an educated guess. A more sophisticated method is to measure the modal damping. This is discussed in Chapter 7. Note that the floor, $x_3(t)$, vibrates much longer than the machine parts do. This is something to consider in designing how and where the machine is mounted to the floor of a building.

□

4.9 COMPUTATIONAL EIGENVALUE PROBLEMS FOR VIBRATION

This section examines the various approaches to solving eigenvalue problems and how vibration problems may be solved using these eigenvalue problems. This is presented in the context of the three computer codes Mathematica, MATLAB, and Mathcad. The Engineering Vibration Toolbox also contains a variety of ways of calculating mode shapes, natural frequencies, and damping ratios based on M-files created in MATLAB. The books by Datta (1995), Golub and Van Loan (1996), and Meirovitch (1995) should be consulted for more detail. Additional detail can be found in Appendix C. Each of the matrix computations and manipulations made in the previous sections can be obtained easily by standard functions of most mathematical software packages. Hence the tedious solutions of modal analysis and the eigenvalue problem can be automated and used to solve systems with large numbers of degrees of freedom. Here we introduce the various eigenvalue problems and illustrate how to use various software packages to obtain the needed computation. We refer the reader to Appendix C and references there for details on how the algorithms actually work.

Consider the undamped-vibration problem of equation (4.11) with n degrees of freedom, repeated here:

$$M\ddot{\mathbf{x}}(t) + K\mathbf{x}(t) = \mathbf{0} \tag{4.157}$$

Here the displacement vector $\mathbf{x}$ is $n \times 1$, and the matrices M and K are $n \times n$ and symmetric. There are a number of different ways to relate equation (4.157) to the mathematics of eigenvalue problems and these are presented next along with examples and computer steps for solving them.

Dynamically Coupled Systems

The matrix M is positive definite (see Window 4.9) in most cases and up until Example 4.7.3 was considered to be diagonal, so that factoring M and taking its inverse amounted to scalar arithmetic on the diagonal entries. In Example 4.7.3, the mass matrix was not diagonal, in which case the system is called dynamically coupled, and more sophisticated means are needed to handle the inverse and

Window 4.9
The Definition of Positive Definite

A symmetric matrix M ($M = M^T$) is *positive definite* if, for every nonzero vector $\mathbf{x}$, the scalar $\mathbf{x}^T M \mathbf{x} > 0$. In addition, M is positive definite if and only if all of the eigenvalues of M are positive numbers. The matrix A is said to be positive semidefinite if, for every nonzero vector $\mathbf{x}$, the scalar $\mathbf{x}^T A \mathbf{x} \geq 0$. A matrix is positive semidefinite if and only if all of the eigenvalues of the matrix A are greater than or equal to zero. In particular, A could have one or more zero eigenvalues, as in the case of rigid-body motion.

factoring of the matrix M used in vibration analysis. The method of factoring a positive definite matrix is called Cholesky decomposition, which finds a lower triangular matrix L such that $M = LL^T$. Cholesky decomposition can also be formed from two upper triangular matrices, say U, such that $M = U^T U$. Some codes use upper triangular and others use lower triangular, but the results applied to vibration problems are the same. Both calculating the inverse and computing the factors are simple commands in most computational software programs (and calculators). The algorithms in these codes use the most sophisticated methods. However, one way to compute any function of a matrix is by decomposing the matrix using the eigenvalue problem.

Let M be a symmetric, positive definite matrix and let f be any function defined for positive numbers. Denote the eigenvalues of the matrix M (not the M, K system, just the matrix M) by μ and let R denote the matrix of normalized eigenvectors of M. Then

$$f(M) = R \begin{bmatrix} f(\mu_1) & 0 & \cdots & 0 \\ 0 & f(\mu_2) & \cdots & 0 \\ \vdots & \vdots & \ddots & \vdots \\ 0 & 0 & \cdots & f(\mu_n) \end{bmatrix} R^T \tag{4.158}$$

In particular

$$M^{-1} = R \begin{bmatrix} 1/\mu_1 & 0 & \cdots & 0 \\ 0 & 1/\mu_2 & \cdots & 0 \\ \vdots & \vdots & \ddots & \vdots \\ 0 & 0 & \cdots & 1/\mu_n \end{bmatrix} R^T$$

and

$$M^{-1/2} = R \begin{bmatrix} 1/\sqrt{\mu_1} & 0 & \cdots & 0 \\ 0 & 1/\sqrt{\mu_2} & \cdots & 0 \\ \vdots & \vdots & \ddots & \vdots \\ 0 & 0 & \cdots & 1/\sqrt{\mu_n} \end{bmatrix} R^T$$

which provides one method of computing the matrices used in modal analysis. These decompositions can also be used to prove that the inverse and the square root of a symmetric matrix is symmetric. However, the codes use more numerically sophisticated techniques beyond the scope of this book. In fact, it is better not to compute the inverse directly but to use a modified form of Gaussian elimination to compute the inverse (M\I in MATLAB, for instance). Computing the matrix square root is best not done in the dynamically coupled case, but rather the Cholesky factorization should be used. These are illustrated in the following example.

Example 4.9.1

Consider the nondiagonal mass matrix

$$M = \begin{bmatrix} 5 & 2 & 0 \\ 2 & 4 & 1 \\ 0 & 1 & 3 \end{bmatrix}$$

and compute the inverse and factors.

Solution   The matrix inverse is computed via the following commands:

In MATLAB	inv(M)
In Mathematica	Inverse[M]
In Mathcad	M^{-1} (typed M^-1)

The Cholesky factors of a matrix are computed via the following commands:

In MATLAB	Chol(M) or Chol(M,'lower')
In Mathematica	CholeskyDecomposition[M]
In Mathcad	cholesky(M)

Using any one of these yields the inverse of the mass matrix:

$$M^{-1} = \begin{bmatrix} 0.2558 & -0.1395 & 0.0465 \\ -0.1395 & 0.3488 & -0.1163 \\ 0.0465 & -0.1163 & 0.3721 \end{bmatrix}$$

The factors of the mass matrix are

$$L = \begin{bmatrix} 2.23607 & 0 & 0 \\ 0.89443 & 1.78885 & 0 \\ 0 & 0.55902 & 1.63936 \end{bmatrix} \text{ and } L^T = \begin{bmatrix} 2.23607 & 0.89443 & 0 \\ 0 & 1.78885 & 0.55902 \\ 0 & 0 & 1.63936 \end{bmatrix}$$

such that $M = LL^T$. Also note that $L^{-1}M(L^T)^{-1} = I$, the identity matrix. Note that MATLAB uses an upper triangular Cholesky decomposition. Thus the mass matrix is factored as M = chol(M)'*chol(M) and inv(chol(M)')*M*inv(chol(M)) = I. So when using MATLAB, the following code has to be modified accordingly. Alternately use the command L = chol(M,'lower') which creates the lower triangular Cholesky factor.

□

In order to perform modal analysis on equation (4.157) for a dynamically coupled system, replace $M^{-1/2}$ with the matrix L using Window 4.5 in the following manner:

2. Calculate the mass-normalized stiffness matrix by

$$\tilde{K} = L^{-1}K(L^T)^{-1} \tag{4.159}$$

and note that this is a symmetric matrix.

5. Compute the matrix of mode shapes S from

$$S = (L^T)^{-1}P \quad \text{and} \quad S^{-1} = P^T L^T \tag{4.160}$$

Using Codes

Several examples are given next that illustrate how to use math software to compute the eigenvalues and eigenvectors of a system and then to solve for frequencies and mode shapes. Calculation of the algebraic eigenvalue problem formed the object of intensive study over a 30-year period, resulting in very sophisticated methods. Many of these studies were funded by government agencies and hence are in the public domain. As computer technology advanced, several high-level codes evolved to enable engineers to make eigenvalue calculations simply and accurately. Today almost every code and calculator contains eigensolvers. The Toolbox contains M-files for computing frequencies and mode shapes using MATLAB. The only small difficulty in using math software is that mathematicians always number eigenvalues starting with the largest first and engineers like the frequencies to be numbered with the smallest value first. So in some codes you may want to sort the eigenvalues and eigenvectors accordingly. The following examples illustrate how to use Mathcad, MATLAB, and Mathematica to solve for natural frequencies and mode shapes. Please note that the developers of these codes often update their codes and syntax, so it is wise to check their websites for updates if you have troubles with syntax errors.

Example 4.9.2

Compute the solution of Example 4.2.6 using math software.

Solution   First consider Mathcad. This program enters the elements of a matrix by selecting the appropriate size from the matrix pallet. Note that Mathcad starts counting elements of vectors and matrices with 0 rather than 1. The following illustrates the remaining steps:

Enter the values of M and K:

$$M := \begin{bmatrix} 1 & 0 \\ 0 & 4 \end{bmatrix} \quad K := \begin{bmatrix} 12 & -2 \\ -2 & 12 \end{bmatrix}$$

Compute the root of M, then K tilde:

$$Ms := \begin{bmatrix} 1 & 0 \\ 0 & 2 \end{bmatrix} \quad Kt := Ms^{-1} \cdot K \cdot Ms^{-1} \quad Kt = \begin{bmatrix} 12 & -1 \\ -1 & 3 \end{bmatrix}$$

Compute the eingenvalues:

$$\lambda := \text{eingenvals}(Kt) \qquad \lambda = \begin{bmatrix} 12.109772 \\ 2.890228 \end{bmatrix}$$

Compute the eigenvectors and reorder with the lowest first and compute frequencies:

$$v2 := \text{eigenvec}(Kt, \lambda_0) \quad v1 := \text{eigenvec}(Kt, \lambda_1) \quad \omega1 := \sqrt{\lambda_1} \quad \omega2 := \sqrt{\lambda_0}$$

Display the results:

$$v1 = \begin{bmatrix} 0.109117 \\ 0.994029 \end{bmatrix} \quad v2 = \begin{bmatrix} -0.994029 \\ 0.109117 \end{bmatrix} \quad \omega1 = 1.7 \quad \omega2 = 3.48$$

Check to see the eigenvectors are orthonomal:

$$|v1| = 1 \qquad |v2| = 1 \qquad v1 \cdot v2 = 0$$

Form the matrix P:

$$P := \text{augment }(v1, v2) \quad P = \begin{bmatrix} 0.1091 & -0.994 \\ 0.994 & 0.1091 \end{bmatrix} \quad P^T = \begin{bmatrix} 0.1090.994 \\ -0.9940.109 \end{bmatrix}$$

Show that P is orthoganal and diagonalizes K tilde

$$P^T \cdot P = \begin{bmatrix} 1 & 0 \\ 0 & 1 \end{bmatrix} \quad P^T \cdot Kt \cdot P = \begin{bmatrix} 2.8902 & 0 \\ 0 & 12.1098 \end{bmatrix}$$

Next consider using MATLAB. MATLAB enters matrices using spaces between elements of a row and semicolons to start a new row. Like Mathcad, MATLAB produces the eigenvalues from highest to lowest, so care must be taken to mind the order of eigenvalues and eigenvectors. This is handled here by using the `fliplr(V)` command, which is used to reorder the eigenvectors with the one corresponding to λ_1 first rather than λ_n, as produced from the eigensolver in the code.

```
% enter M and K, compute K tilde
M=[1 0;0 4];K=[12 -2;-2 12];
Mr=sqrtm(M);Kt=inv(Mr)*K*inv(Mr);
% solve eigenvalue problem puts eigenvectors in the
% matrix V, eigenvalues in the diagonal matrix D
[V,D]=eig(Kt);
%check and reorder eigenvalues, smallest first
eignvalues=V'*Kt*V
eignvalues =
   2.8902   -0.0000
   0        12.1098
V'*V % check to see that V1 is orthogonal
ans =
   1.0000   0
   0        1.0000
```

In Mathematica the code is as follows:

Input mass and stiffness matrices:

$$\text{In[1]} := M = \begin{pmatrix} 1 & 0 \\ 0 & 4 \end{pmatrix};$$

$$K = \begin{pmatrix} 12 & -2 \\ -2 & 12 \end{pmatrix};$$

Calculate inverse square root of mass matrix then find $\widetilde{K}$.

```
In[3]:= Mnegsqrt = MatrixPower[M, -0.5];
        Khat = Mnegsqrt.K.Mnegsqrt
        MatrixForm[Mnegsqrt]
        MatrixForm[Khat]
```

Out[5]//Matrix Form=

$$\begin{pmatrix} 1 & 0 \\ 0 & 0.5 \end{pmatrix}$$

Out[6]//MatrixForm=

$$\begin{pmatrix} 12 & -1 \\ -1 & 3 \end{pmatrix}$$

Calculate eingenvalues and eigenvectors. Note that Mathematica returns eigenvectors in rows, not columns as in Mathcad and MATLAB.

```
In[7]:= {λ, v} = Eigensystem[Khat];
        MatrixForm[λ]
        MatrixForm[v]
```

```
Out[8]//MatrixForm=
        ⎛12.1098⎞
        ⎝2.89023⎠

Out[9]//MatrixForm=
        ⎛0.994029   -0.109117⎞
        ⎝0.109117    0.994029⎠
```

Arrange eigenvectors from least to greater and find natural frequencies. Mathematica usually returns eigenvectors normalized to 1.

```
In[10]:= v2 = v[[1, A11]]
         v1 = v[[2, A11]]
         ω1 = √λ[[2]]
         ω2 = √λ[[1]]
Out[10]= {0.994029, -0.109117}
Out[11]= {0.109117, 0.994029}
Out[12]= 1.70007
Out[12]= 3.47991

In[14]:= v1.v1
         v2.v2
         Chop[v1.v2]
Out[14]= 1.

Out[15]= 1.

Out[16]= 0.
```

Since the eigenvectors are in rows in Mathematica, it is easiest to form P^T first, then transpose to get P.

```
In[17]:= PT = {v1, v2}
         P = Transpose[PT]
         MatrixForm[PT]
         MatrixForm[P]

Out[19]//MatrixForm=
        ⎛0.109117    0.994029⎞
        ⎝0.994029   -0.109117⎠

Out[20]//MatrixForm=
        ⎛0.109117    0.994029⎞
        ⎝0.994029   -0.109117⎠
```

Show that P is orthogonal and that P diagonalizes $\tilde{k}$

```
In[21]:= MatrixForm[Chop[PT.P]]
         MatrixForm[Chop[PT.Khat.P]]
```

```
Out[21]//MatrixForm=
```

$$\begin{pmatrix} 1 & 0 \\ 0 & 1 \end{pmatrix}$$

```
Out[22]//MatrixForm=
```

$$\begin{pmatrix} 2.89023 & 0 \\ 0 & 12.1098 \end{pmatrix}$$

Note that Mathematica produces eigenvalues for the highest to the lowest, so an alternative way to reorder them is to use the Reverse command. For example

```
{vals, vecs} = Eigensystem[Khat]
valsr = Reverse[vals]
vecsr = Reverse[vecs]


Transpose[vecsr]
P = Transpose[vecsr]
```

will reorder the eigenvalues and eigenvectors.

                                                                              □

Example 4.9.3

Compute the coefficients for the modal equations for the damped, forced-response problem given in Example 4.6.1.

Solution   The solution in Mathcad is as follows:

```
Enter the matrices M, C, K and the force vector b
```

$$M := \begin{bmatrix} 9 & 0 \\ 0 & 1 \end{bmatrix} \quad C := \begin{bmatrix} 2.7 & -0.3 \\ -0.3 & 0.3 \end{bmatrix} \quad K := \begin{bmatrix} 27 & -3 \\ -3 & 3 \end{bmatrix} \quad b := \begin{bmatrix} 0 \\ 3 \end{bmatrix}$$

```
Note that the damping is proportional, so modal analysis may be used
```

$$C \cdot M^{-1}K - K \cdot M^{-1} \cdot C = \begin{bmatrix} -1.78 \cdot 10^{-15} & 0 \\ 0 & 0 \end{bmatrix} \text{ effectively zero}$$

```
L := cholesky(M)
```

$$Kt := L^{-1} \cdot K \cdot (L^T)^{-1} \quad Kt = \begin{bmatrix} 3 & -1 \\ -1 & 3 \end{bmatrix}$$

$$Ct := L^{-1} \cdot C \cdot (LT)^{-1} \quad Ct = \begin{bmatrix} 0.3 & -0.1 \\ -0.1 & 0.3 \end{bmatrix}$$

$$\lambda := \text{eigenvals (Kt)} \quad v1 := \text{eigenvec(Kt, } \lambda_1) \quad v2 := \text{eigenvec(Kt, } \lambda_0)$$

$$P := \text{augment(v1, v2)} \qquad P = \begin{bmatrix} 0.707 & -0.707 \\ 0.707 & 0.707 \end{bmatrix}$$

$$P^T \cdot P = \begin{bmatrix} 1 & 0 \\ 0 & 1 \end{bmatrix} \quad P^T \cdot Ct \cdot P = \begin{bmatrix} 0.2 & 0 \\ 0 & 0.4 \end{bmatrix} \quad P^T \cdot Kt \cdot P = \begin{bmatrix} 2 & 0 \\ 0 & 4 \end{bmatrix}$$

Next compute the modal force amplitudes:

$$bt := P^T \cdot L^{-1} \cdot b \qquad bt = \begin{bmatrix} 2.121 \\ 2.121 \end{bmatrix}$$

The solution in MATLAB is as follows:

```
% enter M, C, and K
M=[9 0;0 1];K=[27 -3;-3 3];C=K/10;b=[0;3];
%compute L and the mass normalized quantities
L=chol(M);Kt=inv(L)*K*inv(L');Ct=inv(L)*C*inv(L');
% Compute eigensolution, reorder eigenvectors
[V,D]=eig(Kt); P=V
P =
    -0.7071   -0.7071
    -0.7071    0.7071
P'*P
ans =
    1.0000   0
    0        1.0000
P'*Kt*P
ans =
    2.0000   0
    0        4.0000
P'*C*P
ans =
    1.2000   1.2000
    1.2000   1.8000
P'*Ct*P
ans =
    0.2000   0
    0        0.4000
bt=P'*inv(L)*b
bt =
    -2.1213
    2.1213
```

Note that in MATLAB, the first eigenvector is the negative of that produced in Mathcad. This is not a problem as both are correct. In the MATLAB version, note that the mass-normalized input vector also has a sign change. When the modal equations are written out and transformed back to physical coordinates, these signs will recombine to give the same solution as given in Mathcad. Remember that an eigenvector may always be multiplied by a scalar (in this case -1) without changing its direction, and the value 1 does not change its magnitude.

The solution in Mathematica is as follows:

$$\text{In[1]} := M = \begin{pmatrix} 9 & 0 \\ 0 & 1 \end{pmatrix};$$

$$c = \begin{pmatrix} 2.7 & -0.3 \\ -0.3 & 0.3 \end{pmatrix};$$

$$K = \begin{pmatrix} 27 & -3 \\ -3 & 3 \end{pmatrix};$$

$$b = \begin{pmatrix} 0 \\ 3 \end{pmatrix};$$

Note that the damping is proportional, so modal analysis may be used.

```
In[5]:= MatrixForm[Chop[c.Inverse[M].K-K.Inverse[M].c]]
```

```
Out[5]//MatrixForm=
```
$$\begin{pmatrix} 0 & 0 \\ 0 & 0 \end{pmatrix}$$

Calculation of $\tilde{K}$ and $\tilde{C}$

```
In[6]:= L = CholeskyDecomposition[M];
        Khat = Inverse[L].K.Inverse[Transpose[L]];
        Chat = Inverse[L].c.Inverse[Transpose[L]];
        MatrixForm[Khat]
        MatrixForm[Chat]
```

```
Out[10]//MatrixForm=
```
$$\begin{pmatrix} 3 & -1 \\ -1 & 3 \end{pmatrix}$$

```
Out[11]//MatrixForm=
```
$$\begin{pmatrix} 0.3 & -0.1 \\ -0.1 & 0.3 \end{pmatrix}$$

Calculate eigenvalues and eigenvectors. Note that Mathematica returns eigenvectors in rows, not columns as in Mathcad and Matlab. For this particular system, the eigenvalues were found in the correct order and the eigenvectors must be normalized to 1.

```
In[11]:= {λ,v}=Eigensystem[Khat];
         MatrixForm[λ]
         MatrixForm[v]
         v1 = Normalize[v[[1, All]]];
         v2 = Normalize[v[[2, All]]];
         N[MatrixForm][{ v1, v2}]
```

```
Out[12]//MatrixForm=
```
$$\begin{pmatrix} 2 \\ 4 \end{pmatrix}$$

```
Out[13]//MatrixForm=
```
$$\omega2 := \sqrt{\text{Re}(\lambda_1)^2 + \text{Im}(\lambda_1)^2}$$

```
Out[16]//MatrixForm=
```
$$\begin{pmatrix} 0.707107 & 0.707107 \\ -0.707107 & 0.707107 \end{pmatrix}$$

```
In[17]:= PT = {v1, v2};
        P = Transpose[PT];
```

```
In[19]:= MatrixForm[Chop[PT.P]]
        MatrixForm[Chop[PT.Khat.P]]
        MatrixForm[Chop[PT.Chat.P]]
```

```
Out[19]//MatrixForm=
```
$$\begin{pmatrix} 1 & 0 \\ 0 & 1 \end{pmatrix}$$

```
Out[20]//MatrixForm=
```
$$\begin{pmatrix} 2 & 0 \\ 0 & 4 \end{pmatrix}$$

```
Out[21]//MatrixForm=
```
$$\begin{pmatrix} 0.2 & 0 \\ 0 & 0.4 \end{pmatrix}$$

```
Compute modal force amplitudes.
```

```
In[22]:= bt = PT.Inverse[L].b;
        N[MatrixForm[bt]]
```

```
Out[23]//MatrixForm=
```
$$\begin{pmatrix} 2.12132 \\ 2.12132 \end{pmatrix}$$

□

Various Eigenvalue Problems

There are several ways to relate the vibration problem to the eigenvalue problem. The simplest way is unfortunately the worst in terms of computational effort. This is to use the *generalized eigenvalue problem*, formed from equation (4.157) by substitution of $\mathbf{x} = e^{j\omega t}\mathbf{u}$, which results in

$$K\mathbf{u} = \lambda M\mathbf{u} \tag{4.161}$$

Here $\lambda = \omega^2$, and the $\mathbf{u}$ are the mode shapes. This is the most direct approach but suffers from computational burden (four times as much as using the Cholesky approach for a two-degree-of-freedom example). The solution of equation (4.157) then becomes

$$\mathbf{x}(t) = \sum_{i=1}^{n} c_i \sin(\omega_i t + \phi_i)\mathbf{u}_i \tag{4.162}$$

Here c_i and ϕ_i are constants determined by initial conditions.

Next multiply equation (4.161) by the matrix M^{-1}. Again, assuming a solution of the form $\mathbf{x}(t) = e^{j\omega t}\mathbf{u}$ yields

$$-\omega^2 \mathbf{u} + M^{-1}K\mathbf{u} = 0 \tag{4.163}$$

or

$$(M^{-1}K)\mathbf{u} = \lambda \mathbf{u} \tag{4.164}$$

This is the standard *algebraic eigenvalue problem*. The matrix $M^{-1}K$ is neither symmetric nor banded. Again there are n eigenvalues λ_i, which are the squares of the natural frequencies ω_i^2, and n eigenvectors $\mathbf{u}_i$. The solution of equation (4.157), $\mathbf{x}(t)$, is again in the form

$$\mathbf{x}(t) = \sum_{i=1}^{n} c_i \sin(\omega_i t + \phi_i)\mathbf{u}_i \tag{4.165}$$

where c_i and ϕ_i are constants to be determined by the initial conditions. Thus the eigenvectors $\mathbf{u}_i$ are also the mode shapes. Since $M^{-1}K$ is not symmetric, the solution of the algebraic eigenvalue problem could yield complex values for the eigenvalues and eigenvectors. However, they are known to be real valued because of the generalized eigenvalue problem formulation, equation (4.161), which has the same eigenvalues and eigenvectors.

Next, consider the vibration problem (following Window 4.5) obtained by substitution of the coordinate transformation $\mathbf{x}(t) = (L^T)^{-1}\mathbf{q}(t)$ into equation (4.161) and multiplying by L^{-1}. This yields the form

$$\ddot{\mathbf{q}}(t) + \tilde{K}\mathbf{q}(t) = 0 \tag{4.166}$$

where the matrix $\tilde{K}$ is symmetric but not necessarily sparse or banded unless M is diagonal. The solution of equation (4.166) is obtained by assuming a solution of the form $\mathbf{q}(t) = e^{j\omega t}\mathbf{v}$ where $\mathbf{v}$ is a nonzero vector of constants. Substituting this form into equation (4.166) yields

$$-\omega^2 \mathbf{v} + \tilde{K}\mathbf{v} = 0 \tag{4.167}$$

or

$$\tilde{K}\mathbf{v} = \lambda \mathbf{v} \tag{4.168}$$

where again $\lambda = \omega^2$. This is the *symmetric eigenvalue problem* and again results in n eigenvalues λ_i, which are the squares of the natural frequencies ω_i^2, and n eigenvectors $\mathbf{v}_i$. The solution of equation (4.166) becomes

$$\mathbf{q}(t) = \sum_{i=1}^{n} c_i \sin(\omega_i t + \phi_i) \mathbf{v}_i \tag{4.169}$$

where c_i and ϕ_i are again constants of integration. The solution in the original coordinate system $\mathbf{x}$ is obtained from this last expression by multiplying by the matrix $(L^T)^{-1}$:

$$\mathbf{x}(t) = (L^T)^{-1}\mathbf{q}(t) = \sum_{i=1}^{n} c_i \sin(\omega_i t + \phi_i)(L^T)^{-1}\mathbf{v}_i \tag{4.170}$$

Hence the mode shapes are the vectors $(L^T)^{-1}\mathbf{v}_i$, where $\mathbf{v}_i$ are the eigenvectors of the symmetric matrix $\tilde{K}$. Since the eigenvalue problem here is symmetric, it is known that the eigenvalues and eigenvectors are real valued, as are the mode shapes. In addition, the orthogonality of the $\mathbf{v}_i$ allows easy computation of the modal initial conditions. The numerical advantage here is that the eigenvalue problem is symmetric, so more efficient numerical algorithms can be used to solve it. Computationally this is the most efficient method of the four possible formulations presented here.

Again consider the vibration problem defined in equation (4.161), which is

$$\ddot{\mathbf{x}}(t) + M^{-1}K\mathbf{x}(t) = 0 \tag{4.171}$$

This equation can be transformed to a first-order vector differential equation by defining two new $n \times 1$ vectors, $\mathbf{y}_1(t)$ and $\mathbf{y}_2(t)$, by

$$\mathbf{y}_1(t) = \mathbf{x}(t), \qquad \mathbf{y}_2(t) = \dot{\mathbf{x}}(t) \tag{4.172}$$

Note that $\mathbf{y}_1$ is the vector of displacements and $\mathbf{y}_2(t)$ is a vector of velocities. Differentiating these two vectors yields

$$\dot{\mathbf{y}}_1(t) = \dot{\mathbf{x}}(t) = \mathbf{y}_2(t) \tag{4.173}$$

$$\dot{\mathbf{y}}_2(t) = \ddot{\mathbf{x}}(t) = -M^{-1}K\mathbf{y}_1(t)$$

where the equation for $\dot{\mathbf{y}}_2(t)$ has been expanded by solving equation (4.171) for $\ddot{\mathbf{x}}(t)$. Equations (4.173) can be recognized as the first-order vector differential equation

$$\dot{\mathbf{y}}(t) = A\mathbf{y}(t) \tag{4.174}$$

Here

$$A = \begin{bmatrix} 0 & I \\ -M^{-1}K & 0 \end{bmatrix} \tag{4.175}$$

is called the *state matrix*. The 0 denotes an $n \times n$ matrix of zeros, I denotes the $n \times n$ identity matrix, and the *state vector* $\mathbf{y}(t)$ is defined by the $2n \times 1$ vector

$$\mathbf{y}(t) = \begin{bmatrix} \mathbf{y}_1(t) \\ \mathbf{y}_2(t) \end{bmatrix} = \begin{bmatrix} \mathbf{x}(t) \\ \dot{\mathbf{x}}(t) \end{bmatrix} \tag{4.176}$$

The solution of equation (4.174) proceeds by assuming the exponential form $\mathbf{y}(t) = \mathbf{z}e^{\lambda t}$, where $\mathbf{z}$ is a nonzero vector of constants and λ is a scalar. Substitution into equation (4.174) yields $\lambda \mathbf{z} = A\mathbf{z}$ or

$$A\mathbf{z} = \lambda \mathbf{z} \qquad \mathbf{z} \neq \mathbf{0} \tag{4.177}$$

This is again the standard algebraic eigenvalue problem. While the matrix A has many zero elements, it is now a $2n \times 2n$ eigenvalue problem. It can be shown that the $2n$ eigenvalues λ_i again corresponds to the n natural frequencies ω_i by the relation $\lambda_i = \omega_i j$, where $j = \sqrt{-1}$. The extra n eigenvalues are $\lambda_i = -\omega_i j$, so that there are still only n natural frequencies, ω_i. The $2n$ eigenvectors, $\mathbf{z}$ of the matrix A, however, are of the form

$$\mathbf{z}_i = \begin{bmatrix} \mathbf{u}_i \\ \lambda_i \mathbf{u}_i \end{bmatrix} \tag{4.178}$$

where $\mathbf{u}_i$ are the mode shapes of the corresponding vibration problem. The matrix A (see Window 4.10) is not symmetric and the eigenvalues λ_i and eigenvectors $\mathbf{z}_i$ would therefore be complex. In fact, the eigenvalues λ_i in this case are imaginary numbers of the form $\omega_i j$.

Window 4.10
Various Uses of the Symbol A

Do not be confused by the matrix A. The symbol A is used to denote any matrix. Here, A is used to denote

$$A = M^{-1}K$$

$$A = \begin{bmatrix} 0 & I \\ -M^{-1}K & -M^{-1}C \end{bmatrix} \qquad A = \begin{bmatrix} 0 & I \\ M^{-1}K & 0 \end{bmatrix}$$

to name a few. Which matrix A is being discussed should be clear from the context.

Damped Systems

For large-order systems, computing the eigenvalues using equation (4.177) becomes numerically more difficult because it is of order $2n$ rather than n. The main advantage of the state-space form is in numerical simulations and in solving the damped multiple-degree-of-freedom vibration problem, which is discussed next. Now consider the damped vibration problem of the form

$$M\ddot{\mathbf{x}}(t) + C\dot{\mathbf{x}}(t) + K\mathbf{x}(t) = \mathbf{0} \tag{4.179}$$

where C represents the viscous damping in the system (see Section 4.5) and is assumed only to be symmetric and positive semidefinite. The state-matrix approach and related standard eigenvalue problem of equation (4.177) can also be used to describe the nonconservative vibration problem of equation (4.179). Multiplying equation (4.179) by the matrix M^{-1} yields

$$\ddot{\mathbf{x}}(t) + M^{-1}C\dot{\mathbf{x}}(t) + M^{-1}K\mathbf{x}(t) = \mathbf{0} \tag{4.180}$$

Again it is useful to rewrite this expression in a first-order or state-space form by defining the two $n \times 1$ vectors, $\mathbf{y}_1(t) = \mathbf{x}(t)$ and $\mathbf{y}_2(t) = \dot{\mathbf{x}}(t)$, as indicated in equation (4.172). Then equation (4.173) becomes

$$\dot{\mathbf{y}}_1(t) = \dot{\mathbf{x}}(t) = \mathbf{y}_2(t)$$

$$\dot{\mathbf{y}}_2(t) = \ddot{\mathbf{x}}(t) = -M^{-1}K\mathbf{x}(t) - M^{-1}C\dot{\mathbf{x}}(t) \tag{4.181}$$

where the expression for $\ddot{\mathbf{x}}(t)$ in equation (4.181) is taken from equation (4.180) for the damped system by moving the terms $M^{-1}C\dot{\mathbf{x}}(t)$ and $M^{-1}K\mathbf{x}(t)$ to the right of the equal sign. Renaming $\mathbf{x}(t) = \mathbf{y}_1(t)$ and $\dot{\mathbf{x}}(t) = \mathbf{y}_2(t)$ in equation (4.181) and using matrix notation yields

$$\dot{\mathbf{y}}(t) = \begin{bmatrix} \dot{\mathbf{y}}_1(t) \\ \dot{\mathbf{y}}_1(t) \end{bmatrix} = \begin{bmatrix} 0\mathbf{y}_1(t) + I\mathbf{y}_2(t) \\ -M^{-1}K\mathbf{y}_1(t) - M^{-1}C\mathbf{y}_2(t) \end{bmatrix} = \begin{bmatrix} 0 & I \\ -M^{-1}K & -M^{-1}C \end{bmatrix} \begin{bmatrix} \mathbf{y}_1(t) \\ \mathbf{y}_2(t) \end{bmatrix} = A\mathbf{y}(t) \tag{4.182}$$

where the vector $\mathbf{y}(t)$ is defined as the state vector of equation (4.176), and the state matrix A for the damped case is defined as the partitioned form

$$A = \begin{bmatrix} 0 & I \\ -M^{-1}K & -M^{-1}C \end{bmatrix} \tag{4.183}$$

The eigenvalue analysis for the system of equation (4.182) proceeds directly as for the undamped state-matrix system of equation (4.176).

A solution of equation (4.183) is again assumed of the form $\mathbf{y}(t) = \mathbf{z}e^{\lambda t}$ and substituted into equation (4.183) to yield the eigenvalue problem of equation (4.177) (i.e., $A\mathbf{z} = \lambda\mathbf{z}$). This again defines the standard eigenvalue problem of dimension $2n \times 2n$. In this case, the solution again yields $2n$ values λ_i that may be complex valued. The $2n$ eigenvectors $\mathbf{z}_i$ described in equation (4.178) may also be complex valued (if the corresponding λ_i is complex). This, in turn, causes the physical mode shape $\mathbf{u}_i$ to be complex valued as well as the free-response vector $\mathbf{x}(t)$.

Fortunately, there is a rational physical interpretation of the complex eigenvalue, modes and the resulting solution determined by the state-space formulation of the eigenvalue problem given in equation (4.183). The physical time response $\mathbf{x}(t)$ is simply taken to be the real part of the first n coordinates of the vector $\mathbf{y}(t)$ computed from

$$\mathbf{x}(t) = \sum_{i=1}^{2n} c_i \mathbf{u}_i e^{\lambda_i t} \tag{4.184}$$

The time response is discussed in more detail in Section 4.10 and was introduced in equation (4.115). The physical interpretation of the complex eigenvalues λ_i is taken directly from the complex numbers arising from the solution of an underdamped single-degree-of-freedom system given in equations (1.33) and (1.34) of Section 1.3. In particular, the complex eigenvalues λ_i will appear in complex conjugate pairs in the form

$$\lambda_i = -\zeta_i\omega_i - \omega_i\sqrt{1 - \zeta_i^2}j$$
$$\lambda_{i+1} = -\zeta_i\omega_i + \omega_i\sqrt{1 - \zeta_i^2}j \tag{4.185}$$

where $j = \sqrt{-1}$, ω_i is the undamped natural frequency of the ith mode and ζ_i is the *modal damping ratio* associated with the ith mode. The solution of the eigenvalue problem for the state matrix A of equation (4.183) produces a set of complex numbers of the form $\lambda_i = \alpha_i + \beta_i j$, where $\text{Re}(\lambda_i) = \alpha_i$ and $\text{Im}(\lambda_i) = \beta_i$. Comparing these expressions with equations (4.185) yields

$$\omega_i = \sqrt{\alpha_i^2 + \beta_i^2} = \sqrt{\text{Re}(\lambda_i)^2 + \text{Im}(\lambda_i)^2} \tag{4.186}$$

$$\zeta_i = \frac{-\alpha_i}{\sqrt{\alpha_i^2 + \beta_i^2}} = \frac{-\text{Re}(\lambda_i)}{\sqrt{\text{Re}(\lambda_i)^2 + \text{Im}(\lambda_i)^2}} \tag{4.187}$$

which provides a connection to the physical notions of natural frequency and damping ratios for the underdamped case. (See Inman, 2006, for the overdamped and critically damped cases.)

The complex-valued mode-shape vectors $\mathbf{u}_i$ also appear in complex conjugate pairs and are referred to as *complex modes*. The physical interpretation of a complex mode is as follows: each element describes the relative magnitude and phase of the motion of the degree of freedom associated with that element when the system is excited at that mode only. In the undamped real-mode case, the mode-shape vector is real (recall Section 4.1) and indicates the relative positions of each mass at any given instant of time at a single frequency. The difference between the real-mode case and the complex-mode case is that if the mode is complex, the relative position of each mass can also be out of phase by the amount indicated by the complex part of the mode shapes entry (recall that a complex number can be thought of as a magnitude and a phase rather than a real part and an imaginary part).

The state-space formulation of the eigenvalue problem for the matrix A given by equation (4.183) is related to the most general linear vibration problem. It also forms the most difficult computational eigenvalue problem of the five problems discussed previously. The following example illustrates how to compute the natural frequencies and damping ratios using the state-matrix approach.

Example 4.9.4

Consider the following system and compute the natural frequencies and damping ratios. Note that the system will not uncouple into modal equations and has complex modes. The system is given by

$$\begin{bmatrix} 2 & 0 \\ 0 & 1 \end{bmatrix}\ddot{\mathbf{x}}(t) + \begin{bmatrix} 1 & -0.5 \\ -0.5 & 0.5 \end{bmatrix}\dot{\mathbf{x}}(t) + \begin{bmatrix} 3 & -1 \\ -1 & 1 \end{bmatrix}\mathbf{x}(t) = 0$$

Solution In each of the following codes, the *M, C,* and *K* matrices are entered, put into state-space form, and solved. The frequencies and damping ratios are then extracted using equations (4.186) and (4.187).

The solution in Mathcad is as follows:

$$M := \begin{bmatrix} 2 & 0 \\ 0 & 1 \end{bmatrix} \quad C := \begin{bmatrix} 1 & 0.5 \\ 0.5 & 0.5 \end{bmatrix} \quad K := \begin{bmatrix} 3 & -1 \\ -1 & 1 \end{bmatrix}$$

$$0 := \begin{bmatrix} 0 & 0 \\ 0 & 0 \end{bmatrix} \quad I := \begin{bmatrix} 1 & 0 \\ 0 & 1 \end{bmatrix}$$

A := augment(stack(0,-M⁻¹K),stack(I,-M⁻¹ C))

$$A = \begin{bmatrix} 0 & 0 & 1 & 0 \\ 0 & 0 & 0 & 1 \\ -1.5 & 0.5 & -0.5 & 0.25 \\ 1 & -1 & 0.5 & -0.5 \end{bmatrix} \quad \lambda := \text{eigenvals(A)} \quad \lambda = \begin{bmatrix} -0.417 + 1.345i \\ -0.417 - 1.345i \\ -0.083 + 0.705i \\ -0.083 - 0.705i \end{bmatrix}$$

$$\omega 1 := \sqrt{\text{Re}(\lambda_3)^2 + \text{Im}(\lambda_3)^2} \qquad \zeta 1 := \frac{-\text{Re}(\lambda_3)}{\omega 1}$$

$$\omega 2 := \sqrt{\text{Re}(\lambda_1)^2 + \text{Im}(\lambda_1)^2} \qquad \zeta 2 := \frac{-\text{Re}(\lambda_1)}{\omega 2}$$

$$\omega 1 = 0.71 \qquad \omega 2 = 1.408$$

$$\zeta 1 = 0.117 \qquad \zeta 2 = 0.296$$

U := eigenvecs(A)

$$U = \begin{bmatrix} 0.193 + 0.341i & 0.193 - 0.341i & 0.37 + 0.042i & 0.37 - 0.042i \\ -0.126 - 0.407i & -0.126 + 0.407i & 0.725 & 0.725 \\ -0.539 + 0.118i & -0.539 - 0.118i & -0.061 + 0.257i & -0.061 - 0.257i \\ 0.6 & 0.6 & -0.06 + 0.512i & -0.06 - 0.512i \end{bmatrix}$$

The solution in MATLAB requires defining the real and imaginary parts (also see VTB4_3) and is as follows:

```
% enter data
n=2;M=[2 0;0 1];C=[1 -0.5;-0.5 0.5];K=[3 -1;-1 1];
% compute the state matrix
A=[zeros(n) eye(n); -M\K -M\C];
[V,D]=eig(A);% computes eigenvalues and eigenvectors
% compute the real and imaginary parts
ReD=(D+D')/2;ImD=(D'-D)*i/2;
W=(ReD^2+ImD^2).^.5, Zeta=-ReD/W
W =
    1.4078        0        0        0
         0   1.4078        0        0
         0        0   0.7103        0
         0        0        0   0.7103
```

Zeta =

0.2962	0	0	0
0	0.2962	0	0
0	0	0.1169	0
0	0	0	0.1169

The solution in Mathematica is as follows:

$$\text{In[1]:= } M = \begin{pmatrix} 2 & 0 \\ 0 & 1 \end{pmatrix};$$

$$c = \begin{pmatrix} 1 & -.5 \\ -.5 & .5 \end{pmatrix};$$

$$K = \begin{pmatrix} 3 & 1 \\ 1 & 1 \end{pmatrix};$$

$$o = \begin{pmatrix} 0 & 0 \\ 0 & 0 \end{pmatrix};$$

```
i = IdentityMatrix[2];
```

Formation of A matrix.

```
In[6]:= A = ArrayFlatten[{o, i}, {-Inverse[M].K, -Inverse[M].c}}];
        MatrixForm[A]
```

Out[7]//MatrixForm=

$$\begin{pmatrix} 0 & 0 & 1 & 0 \\ 0 & 0 & 0 & 1 \\ -\dfrac{3}{2} & \dfrac{1}{2} & -0.5 & 0.25 \\ 1 & -1 & 0.5 & -0.5 \end{pmatrix}$$

Solution of the eigenvalue problem.

```
In[8]:= {λ, v} = Eigensystem[A];
        MatrixForm[λ]
        MatrixForm[v]
```

Out[9]//MatrixForm=

$$\begin{pmatrix} -0.416934 + 1.34464i \\ -0.416934 - 1.34464i \\ -0.0830665 + 0.705454i \\ -0.0830665 - 0.705454i \end{pmatrix}$$

Out[10]//MatrixForm=

$$\begin{pmatrix} 0.193156+0.341044i & -0.126254-0.40718i & -0.539116+0.117534i & 0.600152+0.0i \\ 0.193156-0.341044i & -0.126254+0.40718i & -0.539116-0.117534i & 0.600152+0.0i \\ 0.369554+0.0422688i & 0.725456+0.i & -0.0605163+0.257193i & -0.060261+0.511776i \\ 0.369554-0.0422688i & 0.725454+0.i & -0.0605163-0.257193i & -0.060261-0.511776i \end{pmatrix}$$

Calculation of natural frequencies and damping ratios.

$$In[11]:= \omega1 = \sqrt{Re[\lambda[[3]]]^2 + Im[\lambda[3]]]^2}$$
$$\omega2 = \sqrt{Re[\lambda[1]]]^2 + Im[\lambda[1]]]^2}$$
$$\zeta1 = \frac{-Re[\lambda[[3]]]}{\omega1}$$
$$\zeta2 = \frac{-Re[\lambda[[1]]]}{\omega2}$$

Out[11]= 0.710328
Out[12]= 1.4078
Out[13]= 0.116941
Out[14]= 0.29616

□

4.10 NUMERICAL SIMULATION OF THE TIME RESPONSE

This section is a simple extension of Sections 1.9, 2.8, and 3.9, which examine the use of numerical integration to simulate and plot the response of a vibration problem. Simulation is a much easier way to obtain the system's response when compared to computing the response by modal decomposition, which was stressed in the last nine sections. However, the modal approach is needed to perform design and to gain insight into the dynamics of the system. Important design criteria are often stated in terms of modal information, not directly available from the time response. In addition, the modal properties may be used to check numerical simulations. Likewise, numerical solutions may also be used to check analytical work.

Consider the forced response of a damped linear system. The most general case can be written as

$$M\ddot{x} + C\dot{x} + Kx = BF(t) \qquad x(0) = x_0, \quad \dot{x}(0) = \dot{x}_0 \qquad (4.188)$$

Following the development of equation (4.181), define $y_1(t) = x(t)$ and $y_2(t) = \dot{x}(t)$ so that $\dot{y}_1(t) = y_2(t)$. Then multiplying equation (4.188) by M^{-1} yields the coupled first-order vector equations

$$\dot{y}_1(t) = y_2(t)$$
$$\dot{y}_2(t) = -M^{-1}Ky_1(t) - M^{-1}Cy_2(t) + M^{-1}BF(t) \qquad (4.189)$$

with initial conditions $y_1(0) = x_0$ and $y_2(0) = \dot{x}_0$. Equation (4.189) can be written as the single first-order equation

$$\dot{y}(t) = Ay(t) + f(t) \qquad y(0) = y_0 \qquad (4.190)$$

where A is the state matrix given by equation (4.183):

$$A = \begin{bmatrix} 0 & I \\ -M^{-1}K & -M^{-1}C \end{bmatrix}$$

and

$$\mathbf{y}(t) = \begin{bmatrix} \mathbf{y}_1(t) \\ \mathbf{y}_2(t) \end{bmatrix} \quad \mathbf{f}(t) = \begin{bmatrix} 0 \\ M^{-1}B\mathbf{F}(t) \end{bmatrix} \quad \mathbf{y}_0 = \begin{bmatrix} \mathbf{y}_1(0) \\ \mathbf{y}_2(0) \end{bmatrix} = \begin{bmatrix} \mathbf{x}_0 \\ \dot{\mathbf{x}}_0 \end{bmatrix}$$

Here $\mathbf{y}(t)$ is the $(2n \times 1)$ state vector, where the first $n \times 1$ elements correspond to the displacement $\mathbf{x}(t)$ and where the second $n \times 1$ elements correspond to the velocities $\dot{\mathbf{x}}(t)$.

The Euler method of the numerical solution given in Section 3.9, equation (3.105) can be applied directly to the vector (state-space) formulation given in equation (4.190), repeated here:

$$\mathbf{y}(t_{i+1}) = \mathbf{y}(t_i) + \Delta t A \mathbf{y}(t_i) \tag{4.191}$$

which defines the Euler formula for integrating the general vibration problem described in equation (4.190) for the zero-force input case. This can be extended to the forced-response case by including the term $\mathbf{f}_i = \mathbf{f}(t_i)$:

$$\mathbf{y}_{i+1} = \mathbf{y}_i + \Delta t A \mathbf{y}_i + \mathbf{f}_i \tag{4.192}$$

where $\mathbf{y}_{i+1}$ denotes $\mathbf{y}(t_{i+1})$, and so on, using $\mathbf{y}(0)$ as the initial value.

As before in Sections 1.9, 2.8, and 3.9, Runge–Kutta integration methods are used in most codes to produce a more accurate approximation to the solution. The following examples illustrate how to use MATLAB, Mathcad, and Mathematica to perform numerical integration to solve vibration problems.

Example 4.10.1

Consider the system given by

$$\begin{bmatrix} 4 & 0 \\ 0 & 9 \end{bmatrix}\begin{bmatrix} \ddot{x}_1(t) \\ \ddot{x}_2(t) \end{bmatrix} + \begin{bmatrix} 30 & -5 \\ -5 & 5 \end{bmatrix}\begin{bmatrix} x_1(t) \\ x_2(t) \end{bmatrix} = \begin{bmatrix} 0.23500 \\ 2.97922 \end{bmatrix}\sin(2.756556t)$$

A quick computation shows that the driving frequency is also the natural frequency of the second mode. However, because of the force distribution vector $\mathbf{b}$ is proportional to the first mode shape, no resonance will occur. The system does not resonate because the force is distributed orthogonal to the second mode. The simulation will verify that this prediction is correct.

Solution   The problem is first put into state-space form and then solved numerically.

In Mathcad the code becomes

$$M := \begin{bmatrix} 4 & 0 \\ 0 & 9 \end{bmatrix} \quad K := \begin{bmatrix} 30 & 5 \\ 5 & 5 \end{bmatrix} \quad B := \begin{bmatrix} 0.23500 \\ 2.97922 \end{bmatrix} \quad \omega := 2.75655$$

$$f := M^{-1} \cdot B \quad M^{-1} \cdot K = K = \begin{bmatrix} 7.5 & -1.25 \\ -0.556 & 0.556 \end{bmatrix} \quad f = \begin{bmatrix} 0.059 \\ 0.331 \end{bmatrix}$$

$$X := \begin{bmatrix} 0 \\ 0 \\ 0 \\ 0 \end{bmatrix} \quad A1 := \begin{bmatrix} 0 & 0 & 1 & 0 \\ 0 & 0 & 0 & 1 \\ -7.5 & 1.25 & 0 & 0 \\ 0.556 & -0.556 & 0 & 0 \end{bmatrix} \quad D(t, X) := A1 \cdot X + \begin{bmatrix} 0 \\ 0 \\ f_0 \cdot \sin(\omega \cdot t) \\ f_1 \cdot \sin(\omega \cdot t) \end{bmatrix}$$

$$Z := rkfixed(X, 0, 20, 3000, D) \quad t := Z^{<0>} \quad x1 := Z^{<1>} \quad x2 := Z^{<2>}$$

The MATLAB code to produce the solution is as follows:

```
clear all

xo=[0; 0; 0; 0];
ts=[0 20];

[t,x]=ode45('f',ts,xo);
plot(t,x(:,1),t,x(:,2),'--')
%-------------------------------------------
function v=f(t,x)

M=[4 0; 0 9];
K=[30 -5; -5 5];
B=[0.23500; 2.97922];
w=2.75655;

A1=[zeros(2) eye(2); -inv(M)*K zeros(2)];

f=inv(M)*B;

v=A1*x+[0;0; f]*sin(w*t);
```

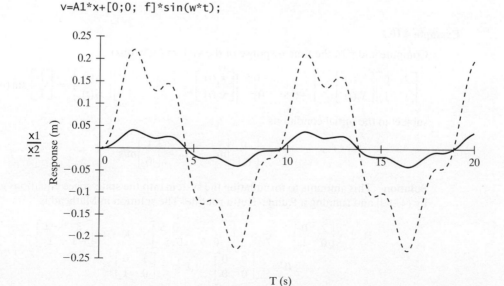

The Mathematica code to produce the solution is as follows:

```
In[1]:= <<PlotLegends`
```

$$
In[2]:= m = \begin{pmatrix} 4 & 0 \\ 0 & 9 \end{pmatrix};
$$

$$
k = \begin{pmatrix} 30 & -5 \\ -5 & 5 \end{pmatrix};
$$

```
ω = 2.756556;
```

$$
f = \begin{pmatrix} .23500*\text{Sin}[ω*t] \\ 2.97922*\text{Sin}[ω*t] \end{pmatrix};
$$

$$
x = \begin{pmatrix} x1[t] \\ x2[t] \end{pmatrix};
$$

$$
xdd = \begin{pmatrix} x1''[t] \\ x2''[t] \end{pmatrix};
$$

```
system = m.xdd + k.x;
```

```
In[9]:= num = NDSolve [{system[[1]] == f[[1]], x1'[0] == 0,
            x1[0] == 0, system[[2]] == f[[2]], x2[0] == 0,
            x2'[0] == 0}, {x1[t], x2[t]}, {t, 0, 20}];
        Plot[{Evaluate[x1[t] /. num], Evaluate[x2[t] /. num]},
            { t, 0, 20},
            PlotStyle → {RGBColor[1, 0, 0], RGBColor[0, 1, 0]},
            PlotLegend → {"x1[t]", "x2[t]"},
            LegendPosition → {1, 0}, LegendSize → {1, .3}]
```

□

Example 4.10.2

Compute and plot the time response of the system (newtons)

$$
\begin{bmatrix} 2 & 0 \\ 1 & 1 \end{bmatrix} \begin{bmatrix} \ddot{x}_1(t) \\ \ddot{x}_2(t) \end{bmatrix} + \begin{bmatrix} 3 & -0.5 \\ -0.5 & 0.5 \end{bmatrix} \begin{bmatrix} \dot{x}_1(t) \\ \dot{x}_2(t) \end{bmatrix} + \begin{bmatrix} 3 & -1 \\ -1 & 1 \end{bmatrix} \begin{bmatrix} x_1(t) \\ x_2(t) \end{bmatrix} = \begin{bmatrix} 1 \\ 1 \end{bmatrix} \sin(\omega t)
$$

subject to the initial conditions

$$
\mathbf{x}_0 = \begin{bmatrix} 0 \\ 0.1 \end{bmatrix} m, \qquad \mathbf{v}_0 = \begin{bmatrix} 1 \\ 0 \end{bmatrix} m/s
$$

Solution This amounts to formulating the system into the state-space equations given by (4.190) and running a Runge–Kutta routine. The solution in Mathcad is

$$
M := \begin{bmatrix} 2 & 0 \\ 0 & 1 \end{bmatrix} \qquad C := \begin{bmatrix} 3 & -0.5 \\ -0.5 & 0.5 \end{bmatrix} \qquad K := \begin{bmatrix} 3 & -1 \\ -1 & 1 \end{bmatrix}
$$

$$
0 := \begin{bmatrix} 0 & 0 \\ 0 & 0 \end{bmatrix} \qquad I := \begin{bmatrix} 1 & 0 \\ 0 & 1 \end{bmatrix}
$$

$$
B := \begin{bmatrix} 1 \\ 1 \end{bmatrix} \qquad \omega := 2
$$

$$A \; := \; \mathrm{augment}(\mathrm{stack}(0,-M^{-1} \cdot K),\mathrm{stack}(I,-M^{-1}\,C)) \qquad X := \begin{bmatrix} 0 \\ 0.1 \\ 1 \\ 0 \end{bmatrix}$$

$$f \; := \; M^{-1} \cdot B \qquad f = \begin{bmatrix} 0.5 \\ 1 \end{bmatrix}$$

$$D(t, \; X) \; := \; A \cdot X + \begin{bmatrix} 0 \\ 0 \\ f_0 \\ f_1 \end{bmatrix} \cdot \sin(\omega \cdot t)$$

$$Z \; := \; \mathrm{rkfixed}\,(X, \; 0, \; 20, \; 3000, \; D) \quad t \; := \; Z^{<0>} \quad x1 \; := \; Z^{<1>} \quad x2 \; := \; Z^{<2>}$$

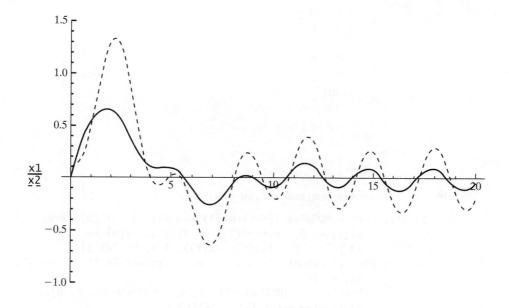

The MATLAB code to produce the same plot is as follows:

```
clear all

xo=[0; 0.1; 1; 0];
ts=[0 20];

[t,x]=ode45('f',ts,xo);
plot(t,x(:,1),t,x(:,2),'--')

%----------------------------------------
function v=f(t,x)

M=[2 0; 0 1];
C=[3 -0.5; -0.5 0.5];
K=[3 -1; -1 1];
B=[1; 1];
```

```
w=2;

A1=[zeros(2) eye(2); -inv(M)*K -inv(M)*C];
f=inv(M)*B;

v=A1*x+[0;0; f]*sin(w*t);
```

The Mathematica code for simulating and plotting the response is as follows:

```
In[1]:= <<PlotLegends'
```

$$In[2]:= \mathbf{m} = \begin{pmatrix} 2 & 0 \\ 0 & 1 \end{pmatrix};$$

$$\mathbf{c} = \begin{pmatrix} 3 & -.05 \\ -.05 & .05 \end{pmatrix};$$

$$\mathbf{k} = \begin{pmatrix} 3 & -1 \\ -1 & 1 \end{pmatrix};$$

$$\omega = 2;$$

$$\mathbf{f} = \begin{pmatrix} Sin[\omega * t] \\ Sin[\omega * t] \end{pmatrix};$$

$$\mathbf{x} = \begin{bmatrix} x1[t] \\ x2[t] \end{bmatrix};$$

$$\mathbf{xd} = \begin{bmatrix} x1'[t] \\ x2'[t] \end{bmatrix};$$

$$\mathbf{xdd} = \begin{bmatrix} x1''[t] \\ x2''[t] \end{bmatrix};$$

```
        system = m.xdd + c.xd + k.x;
```

```
In[11]:= num = NDSolve [{system[[1]] == f[[1]], x1'[0] == 1,
            x1[0] == 0, system[[2]] == f[[2]], x2[0] == .1,
            x2'[0] == 0}, {x1[t], x2[t]}, { t, 0, 20} ];
        Plot [{Evaluate [x1[t] /. num], Evaluate[x2[t] /. num]},
            {t, 0, 20},
            PlotStyle → {RGBColor[1, 0, 0], RGBColor[0, 1, 0]},
            PlotLegend → {"x1[t]", "x2[t]"},
            LegendPosition → {1, 0}, LegendSize → {1, .3}]
```

□

Example 4.10.3

Consider the following system excited by a pulse of duration 0.1 s (units in newtons):

$$\begin{bmatrix} 2 & 0 \\ 0 & 1 \end{bmatrix}\begin{bmatrix} \ddot{x}_1 \\ \ddot{x}_2 \end{bmatrix} + \begin{bmatrix} 0.3 & -0.05 \\ -0.05 & 0.05 \end{bmatrix}\begin{bmatrix} \dot{x}_1 \\ \dot{x}_2 \end{bmatrix} + \begin{bmatrix} 3 & -1 \\ -1 & 1 \end{bmatrix}\begin{bmatrix} x_1 \\ x_2 \end{bmatrix}$$

$$= \begin{bmatrix} 0 \\ 1 \end{bmatrix}[\Phi(t-1) - \Phi(t-1.1)]$$

and subject to the initial conditions

$$\mathbf{x}_0 = \begin{bmatrix} 0 \\ -0.1 \end{bmatrix}\text{m}, \qquad \mathbf{v}_0 = \begin{bmatrix} 0 \\ 0 \end{bmatrix}\text{m/s}$$

Compute and plot the response of the system. Here Φ indicates the Heaviside step function introduced in Section 3.2.

Solution   This again follows the same format as the previous examples of putting the equations of motion into state-matrix form and solving using one of the software programs. In Mathcad the code and solution are

$$M := \begin{bmatrix} 2 & 0 \\ 0 & 1 \end{bmatrix} \quad C := \begin{bmatrix} 0.3 & -0.05 \\ -0.05 & 0.05 \end{bmatrix} \quad K := \begin{bmatrix} 3 & -1 \\ -1 & 1 \end{bmatrix}$$

$$o := \begin{bmatrix} 0 & 0 \\ 0 & 0 \end{bmatrix} \quad I := \begin{bmatrix} 1 & 0 \\ 0 & 1 \end{bmatrix}$$

$$B := \begin{bmatrix} 0 \\ 1 \end{bmatrix} \qquad A := \text{augment (stack } (o, -M^{-1} \cdot K), \text{ stack } (I, -M^{-1} \, C))$$

$$f := M^{-1} \cdot B \qquad f = \begin{bmatrix} 0 \\ 1 \end{bmatrix} \qquad X := \begin{bmatrix} 0 \\ (-0.1) \\ 1 \\ 0 \end{bmatrix}$$

$$D(t, \ X) := A \cdot X + \begin{bmatrix} 0 \\ 0 \\ f_0 \\ f_1 \end{bmatrix} \cdot (\Phi(t - 1) - \Phi(t - 1.1))$$

$$Z := \text{rkfixed } (X, \ 0, \ 30, \ 3000, \ D) \quad t := Z^{<0>} \quad x1 := Z^{<1>} \quad x2 := Z^{<2>}$$

The MATLAB code for producing the same plot is as follows. Note that in this case it is necessary to set the tolerance for ODE45 in order to clearly define the "impulse" as the difference between two step functions. This is done using the options command listed in the following.

```
clear all

xo = [0; -0.1; 0; 0];
ts = [0 30];

options=odeset('RelTol',1e-4);
[t,x] = ode45('f',ts,xo);
plot(t,x(:,1),t,x(:,2),'- -')

%---------------------------------------------
function v = f(t,x)

M = [2 0; 0 1];
C = [0.3 -0.05; -0.05 0.05];
K = [3 -1; -1 1];
B = [0; 1];t1 = 1; t2 = 1.1;
```

```
A1 = [zeros(2) eye(2); -inv(M)*K -inv(M)*C];
f = inv(M)*B;

v = A1*x + [0;0; f]*(stepfun(t,t1)-stepfun(t,t2));
```

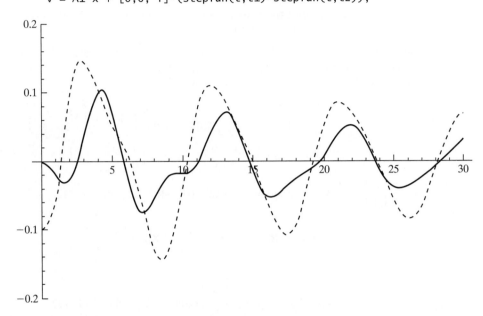

The Mathematica code for simulating and plotting the response is as follows. As in the MATLAB case, the tolerances must be set in order to define the impulse. In this case, the command is `PrecisionGoal->10` as listed below:

$\text{In}[1] := $ **<<PlotLegends'**

$\text{In}[2] := $ **m** $= \begin{pmatrix} 2 & 0 \\ 0 & 1 \end{pmatrix}$;

$\quad$ **c** $= \begin{pmatrix} 0.3 & -0.05 \\ -0.05 & 0.05 \end{pmatrix}$;

$\quad$ **k** $= \begin{pmatrix} 3 & -1 \\ -1 & 1 \end{pmatrix}$;

$\quad$ **f** $= \begin{pmatrix} 0 \\ \text{UnitStep}[t-1] - \text{UnitStep}[t-1.1] \end{pmatrix}$;

$\quad$ **x** $= \begin{bmatrix} \text{x1}[t] \\ \text{x2}[t] \end{bmatrix}$;

$\quad$ **xd** $= \begin{bmatrix} \text{x1}'[t] \\ \text{x2}'[t] \end{bmatrix}$;

$\quad$ **xdd** $= \begin{bmatrix} \text{x1}''[t] \\ \text{x2}''[t] \end{bmatrix}$;

$\quad$ **system = m.xdd + c.xd + k.x;**

```
In[10]:= num = NDSolve [{system[[1]] == f[[1]], x1'[0] == 0,
            x1[0] == 0, system[[2]] == f[[2]], x2[0] == -.1,
            x2'[0] == 0}, {x1[t], x2[t]}, {t, 0, 30},
         PrecisionGoal->10];
         Plot [{Evaluate [x1[t] /. num], Evaluate[x2[t] /. num]},
            {t, 0, 30},
            PlotStyle → {RGBColor[1, 0, 0], RGBColor[0, 1, 0]},
            PlotRange → {-.2, .2},
            PlotLegend → {"x1[t]", "x2[t]"},
            LegendPosition → {1, 0}, LegendSize → {1, .3}]
```

□

The preceding examples illustrate the basic features of using math software to solve vibration problems. These examples are all simple two-degree-of-freedom systems, but the methods and routines work for any number of degrees of freedom, limited only by the array size for a particular code. The Toolbox provides additional solution possibilities.

PROBLEMS

Those problems marked with an asterisk are intended to be solved using computational software.

Section 4.1 (Problems 4.1 through 4.19)

4.1. Consider the system of Figure P4.1. For $c_1 = c_2 = c_3 = 0$, derive the equation of motion and calculate the mass and stiffness matrices. Note that setting $k_3 = 0$ in your solution should result in the stiffness matrix given by equation (4.9).

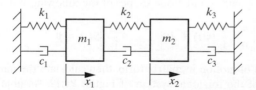

Figure P4.1

4.2. Calculate the characteristic equation from Problem 4.1 for the case

$$m_1 = 8\,\text{kg} \quad m_2 = 2\,\text{kg} \quad k_1 = 24\,\text{N/m} \quad k_2 = 3\,\text{N/m} \quad k_3 = 3\,\text{N/m}$$

and solve for the system's natural frequencies.

4.3. Calculate the vectors $\mathbf{u}_1$ and $\mathbf{u}_2$ for Problem 4.2.

4.4. For initial conditions $\mathbf{x}(0) = \begin{bmatrix} 1 & 0 \end{bmatrix}^T$ and $\dot{\mathbf{x}}(0) = \begin{bmatrix} 0 & 0 \end{bmatrix}^T$ calculate the free response of the system of Problem 4.2. Plot the response $x_1(t)$ and $x_2(t)$.

4.5. Calculate the response of the system

$$\begin{bmatrix} 8 & 0 \\ 0 & 1 \end{bmatrix} \ddot{\mathbf{x}}(t) + \begin{bmatrix} 24 & -2 \\ -2 & 2 \end{bmatrix} \mathbf{x}(t) = 0$$

described in Example 4.1.7, to the initial condition $\mathbf{x}(0) = 0$, $\dot{\mathbf{x}}(0) = \begin{bmatrix} 1 & 0 \end{bmatrix}^T$, plot the response and compare the result to Figure 4.3.

4.6. Write the equations of motion for the system of Figure P4.1 for the case that $k_1 = k_3 = 0$ and identify the mass and stiffness matrix for this case.

4.7. Calculate and solve the characteristic equation for the following system:

$$\begin{bmatrix} 4 & 0 \\ 0 & 1 \end{bmatrix} \ddot{\mathbf{x}}(t) + 10\begin{bmatrix} 1 & -1 \\ -1 & 1 \end{bmatrix} \mathbf{x}(t) = 0$$

4.8. Compute the natural frequencies of the following system:

$$\begin{bmatrix} 6 & 4 \\ 4 & 6 \end{bmatrix} \ddot{\mathbf{x}}(t) + \begin{bmatrix} 3 & -1 \\ -1 & 1 \end{bmatrix} \mathbf{x}(t) = 0.$$

4.9. Calculate the solution

$$\begin{bmatrix} 9 & 0 \\ 0 & 1 \end{bmatrix} \ddot{\mathbf{x}}(t) + \begin{bmatrix} 27 & -3 \\ -3 & 3 \end{bmatrix} \mathbf{x}(t) = 0, \qquad \mathbf{x}(0) = \begin{bmatrix} 1 \\ 3 \\ 1 \end{bmatrix} \qquad \dot{\mathbf{x}}(0) = 0$$

Compare the response with that of Figure 4.3.

4.10. Calculate the solution

$$\begin{bmatrix} 9 & 0 \\ 0 & 1 \end{bmatrix} \ddot{\mathbf{x}}(t) + \begin{bmatrix} 27 & -3 \\ -3 & 3 \end{bmatrix} \mathbf{x}(t) = 0, \qquad \mathbf{x}(0) = \begin{bmatrix} -\frac{1}{3} \\ 3 \\ 1 \end{bmatrix} \qquad \dot{\mathbf{x}}(0) = 0$$

Compare the response with that of Figure 4.3. If you worked Problem 4.9, compare your solution to that response also.

4.11. Compute the natural frequencies and mode shapes of the following system:

$$\begin{bmatrix} 4 & 0 \\ 0 & 1 \end{bmatrix} \ddot{\mathbf{x}}(t) + 10\begin{bmatrix} 4 & -2 \\ -2 & 1 \end{bmatrix} \mathbf{x}(t) = 0$$

4.12. Determine the equation of motion in matrix form, then calculate the natural frequencies and mode shapes of the torsional system of Figure P4.12. Assume that the torsional stiffness values provided by the shaft are equal ($k_1 = k_2$) and that disk 1 has four times the inertia as that of disk 2 ($J_1 = 3J_2$).

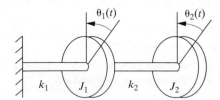

Figure P4.12 Torsional system with two disks and, hence, two degrees of freedom.

4.13. Two subway cars of Fig. P4.13 have 2100 kg mass each and are connected by a coupler. The coupler can be modeled as a spring of stiffness $k = 270{,}000$ N/m. Write the equation of motion and calculate the natural frequencies and (normalized) mode shapes.

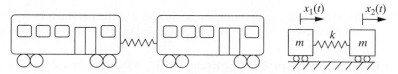

Figure P4.13   Vibration model of two subway cares connected by a coupling device modeled as a massless spring.

4.14. Suppose that the subway cars of Problem 4.13 are given the initial position of $x_{10} = 0$, $x_{20} = 0.05$ m and initial velocities of $v_{10} = v_{20} = 0$. Calculate the response of the cars.

4.15. A slightly more sophisticated model of a vehicle suspension system is given in Figure P4.15. Write the equations of motion in matrix form. Calculate the natural frequencies for $k_1 = 10^3$ N/m, $k_2 = 10^4$ N/m, $m_2 = 60$ kg, and $m_1 = 2400$ kg.

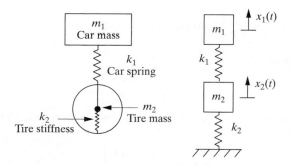

Figure P4.15   A two-degree-of-freedom model of a vehicle suspension system.

4.16. Examine the effect of the initial condition of the system of Figure 4.1(a) on the responses $x_1(t)$ and $x_2(t)$ by repeating the solution of Example 4.1.7 given by

$$\begin{bmatrix} x_1(t) \\ x_2(t) \end{bmatrix} = \begin{bmatrix} \dfrac{1}{3} A_1 \sin\left(\sqrt{2}t + \phi_1\right) - \dfrac{1}{3} A_2 \sin\left(2t + \phi_2\right) \\ A_1 \sin\left(\sqrt{2}t + \phi_1\right) + A_2 \sin\left(2t + \phi_2\right) \end{bmatrix}$$

first for $x_{10} = 0$, $x_{20} = 1$ with $\dot{x}_{10} = \dot{x}_{20} = 0$ and then for $x_{10} = x_{20} = \dot{x}_{10} = 0$ and $\dot{x}_{20} = 1$. Plot the time response in each case and compare your results against Figure 4.3.

4.17. Consider the system defined by

$$\begin{bmatrix} 9 & 0 \\ 0 & 1 \end{bmatrix} \ddot{\mathbf{x}} + \begin{bmatrix} 24 + k_2 & -k_2 \\ -k_2 & k_2 \end{bmatrix} \mathbf{x} = \mathbf{0}$$

Using the initial conditions $x_1(0) = 1$ mm, $x_2(0) = 0$, and $\dot{x}_1(0) = \dot{x}_2(0) = 0$, resolve and plot $x_1(t)$ for the cases that k_2 takes on the values 0.3, 3, 30, and 300. In each case, compare the plots of x_1 and x_2. What can you conclude?

4.18. Consider the system defined by

$$\begin{bmatrix} m_1 & 0 \\ 0 & m_2 \end{bmatrix} \begin{bmatrix} \ddot{x}_1 \\ \ddot{x}_2 \end{bmatrix} + \begin{bmatrix} k_1 + k_2 & -k_2 \\ -k_2 & k_2 \end{bmatrix} \begin{bmatrix} x_1 \\ x_2 \end{bmatrix} = \begin{bmatrix} 0 \\ 0 \end{bmatrix}$$

Determine the natural frequencies in terms of the parameters m_1, m_2, k_1, and k_2. How do these compare to the two single-degree-of-freedom frequencies $\omega_1 = \sqrt{k_1/m_1}$ and $\omega_2 = \sqrt{k_2/m_2}$?

4.19. Consider the problem of Example 4.1.7, which the first degree-of-freedom response given by $x_1(t) = 0.6(\cos\sqrt{2}t + \cos 2t)$. Use a trig identity to show the $x_1(t)$ experiences a beat. Plot the response to show the beat phenomena in the response.

Section 4.2 (Problems 4.20 through 4.35)

4.20. Calculate the square root of the matrix

$$M = \begin{bmatrix} 25 & -32 \\ -32 & 41 \end{bmatrix}$$

$\left[\textit{Hint: } \text{Let } M^{1/2} = \begin{bmatrix} a & -b \\ -b & c \end{bmatrix}; \text{ calculate } (M^{1/2})^2 \text{ and compare to } M. \right]$

4.21. Normalize the vectors

$$\begin{bmatrix} 1 \\ -2 \end{bmatrix}, \begin{bmatrix} 0 \\ 5 \end{bmatrix}, \begin{bmatrix} -0.1 \\ 0.1 \end{bmatrix}$$

first with respect to unity (i.e., $\mathbf{x}^T\mathbf{x} = 1$) and then again with respect to the matrix M (i.e., $\mathbf{x}^T M\mathbf{x} = 1$), where

$$M = \begin{bmatrix} 3 & -0.1 \\ -0.1 & 2 \end{bmatrix}$$

4.22. Consider the vibrating system described by

$$\begin{bmatrix} 4 & 0 \\ 0 & 1 \end{bmatrix} \ddot{\mathbf{x}}(t) + \begin{bmatrix} 4 & -2 \\ -2 & 1 \end{bmatrix} \mathbf{x}(t) = \mathbf{0}$$

Compute the mass-normalized stiffness matrix, the eigenvalues, the normalized eigenvectors, the matrix P, and show that $P^T M P = I$ and $P^T K P$ is the diagonal matrix of eigenvalues Λ.

4.23. Calculate the matrix $\widetilde{K}$ for the system defined by

$$\begin{bmatrix} m_1 & 0 \\ 0 & m_2 \end{bmatrix} \ddot{\mathbf{x}}(t) + \begin{bmatrix} k_1 + k_2 & -k_2 \\ -k_2 & k_2 + k_3 \end{bmatrix} \mathbf{x}(t) = \mathbf{0}$$

and see that it is symmetric.

4.24. Consider the vibrating system described by

$$\begin{bmatrix} 4 & 0 \\ 0 & 1 \end{bmatrix} \ddot{\mathbf{x}}(t) + \begin{bmatrix} 4 & -2 \\ -2 & 1 \end{bmatrix} \mathbf{x}(t) = \mathbf{0}$$

Compute the mass-normalized stiffness matrix, the eigenvalues, the normalized eigenvectors, the matrix P, and show that $P^T M P = I$ and $P^T K P$ is the diagonal matrix of eigenvalues Λ.

4.25. Discuss the relationship or difference between a mode shape of equation (4.54) and an eigenvector of $\tilde{K}$.

4.26. Calculate the units of the elements of matrix $\tilde{K}$.

4.27. Calculate the spectral matrix Λ and the modal matrix P for the vehicle model of Problem 4.15 described by

$$\begin{bmatrix} 2000 & 0 \\ 0 & 50 \end{bmatrix} \ddot{\mathbf{x}}(t) + \begin{bmatrix} 1000 & -1000 \\ -1000 & 11{,}000 \end{bmatrix} \mathbf{x}(t) = \mathbf{0}$$

4.28. Calculate the spectral matrix Λ and the modal matrix P for the system given by

$$\begin{bmatrix} 2000 & 0 \\ 0 & 2000 \end{bmatrix} \ddot{\mathbf{x}}(t) + \begin{bmatrix} 280{,}000 & -280{,}000 \\ -280{,}000 & 280{,}000 \end{bmatrix} \mathbf{x}(t) = 0$$

4.29. Calculate $\tilde{K}$ for the torsional vibration problem given by

$$J_2 \begin{bmatrix} 3 & 0 \\ 0 & 1 \end{bmatrix} \ddot{\boldsymbol{\theta}}(t) + k \begin{bmatrix} 2 & -1 \\ -1 & 1 \end{bmatrix} \boldsymbol{\theta}(t) = \mathbf{0}$$

What are the units of $\tilde{K}$?

4.30. Consider the system in the Figure P4.30 for the case where $m_1 = 1\,\text{kg}$, $m_2 = 9\,kg$, $k_1 = 240\ \text{N/m}$ and $k_2 = 300\ \text{N/m}$. Write the equations of motion in vector form and compute each of the following
  (a) the natural frequencies
  (b) the mode shapes
  (c) the eigenvalues
  (d) the eigenvectors
  (e) show that the mode shapes are not orthogonal
  (f) show that the eigenvectors are orthogonal
  (g) show that the mode shapes and eigenvectors are related by $M^{-1/2}$
  (h) write the equations of motion in modal coordinates

Note the purpose of this problem is to help you see the difference between these various quantities

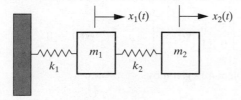

Figure P4.30   A two-degree-of-freedom system.

4.31. Consider the following system:

$$\begin{bmatrix} 1 & 0 \\ 0 & 4 \end{bmatrix} \ddot{\mathbf{x}}(t) + \begin{bmatrix} 3 & -1 \\ -1 & 1 \end{bmatrix} \mathbf{x}(t) = \mathbf{0}$$

where M is in kg and K is in N/m. (a) Calculate the eigenvalues of the system. (b) Calculate the eigenvectors and normalize them.

4.32. The torsional vibration of the wing of an airplane is modeled in Figure P4.32. Write the equation of motion in matrix form and calculate analytical forms of the natural frequencies in terms of the rotational inertia and stiffness of the wing.

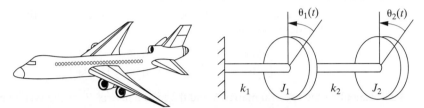

Airplane wing with engines

Wing modeled as two shafts and two disks for torsional vibration

Figure P4.32   A crude model of the torsional vibration of a wing consisting of a two-shaft, two-disk system similar to Problem 4.12 used to estimate the torsional natural frequencies of the wing where the engine inertias are approximated by the disks.

4.33. Calculate the value of the scalar a such that $\mathbf{x}_1 = [a \quad -2 \quad 2]^T$ and $\mathbf{x}_2 = [2 \quad 0 \quad 2]^T$ are orthogonal.

4.34. Normalize the vectors of Problem 4.33. Are they still orthogonal?

4.35. Which of the following vectors are normal? Orthogonal?

$$\mathbf{x}_1 = \begin{bmatrix} \dfrac{1}{\sqrt{2}} \\ 0 \\ \dfrac{1}{\sqrt{2}} \end{bmatrix} \qquad \mathbf{x}_2 = \begin{bmatrix} 0.1 \\ 0.2 \\ 0.3 \end{bmatrix} \qquad \mathbf{x}_3 = \begin{bmatrix} 0.4 \\ 0.3 \\ 0.4 \end{bmatrix}$$

Section 4.3 (Problems 4.36 through 4.46)

4.36. Decouple the following equation of motion into two decoupled equations of motion:

$$\begin{bmatrix} 3 & 0 \\ 0 & 1 \end{bmatrix} \ddot{\mathbf{x}}(t) + \begin{bmatrix} 4 & -2 \\ -2 & 2 \end{bmatrix} \mathbf{x}(t) = 0$$

4.37. Solve the system of Problem 4.12 given by

$$J_2 \begin{bmatrix} 3 & 0 \\ 0 & 1 \end{bmatrix} \ddot{\boldsymbol{\theta}} + k \begin{bmatrix} 2 & -1 \\ -1 & 1 \end{bmatrix} \boldsymbol{\theta} = 0$$

Using modal analysis for the case where the rods have equal stiffness (i.e., $k_1 = k_2$), $J_1 = 3J_2$, and the initial conditions are $\mathbf{x}(0) = [0 \quad 1]^T$ and $\dot{\mathbf{x}}(0) = \mathbf{0}$.

4.38. Consider the system

$$\begin{bmatrix} 9 & 0 \\ 0 & 1 \end{bmatrix} \ddot{\mathbf{x}}(t) + \begin{bmatrix} 27 & -3 \\ -3 & 3 \end{bmatrix} \mathbf{x}(t) = \mathbf{0}$$

of Example 4.3.1. Calculate a value of $\mathbf{x}(0)$ and $\dot{\mathbf{x}}(0)$ such that both masses of the system oscillate with a single frequency of 4 rad/s.

4.39. Consider the system of Figure P4.39 consisting of two pendulums coupled by a spring. Determine the natural frequency and mode shapes. Plot the mode shapes as well as the solution to an initial condition consisting of the first mode shape for $k = 20$ N/m, $l = 0.5$ m and $m_1 = m_2 = 10$ kg, $a = 0.2$ m along the pendulum.

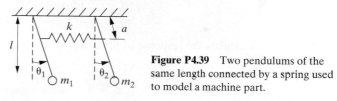

Figure P4.39 Two pendulums of the same length connected by a spring used to model a machine part.

4.40. Compute and plot the response of

$$\begin{bmatrix} 1 & 0 \\ 0 & 10 \end{bmatrix} \ddot{\mathbf{x}}(t) + \begin{bmatrix} 12 & -2 \\ -2 & 12 \end{bmatrix} \mathbf{x}(t) = \mathbf{0}$$

subject to $\mathbf{x}(0) = [1 \quad 1]^T$ and $\dot{\mathbf{x}}(0) = \mathbf{0}$. Compare your result to Example 4.3.2 and Figure 4.6.

4.41. Use modal analysis to calculate the solution of

$$\begin{bmatrix} 2 & 0 \\ 0 & 8 \end{bmatrix} \ddot{\mathbf{x}}(t) + \begin{bmatrix} 3 & -1 \\ -1 & 1 \end{bmatrix} \mathbf{x}(t) = \mathbf{0}$$

for the initial conditions $\mathbf{x}(0) = [0 \quad 1]^T$ (mm) and $\dot{\mathbf{x}}(0) = [0 \quad 0]^T$ (mm/s).

4.42. For the matrices

$$M^{-1/2} = \begin{bmatrix} \dfrac{1}{\sqrt{2}} & 0 \\ 0 & 4 \end{bmatrix} \quad \text{and} \quad P = \frac{1}{\sqrt{2}} \begin{bmatrix} 1 & 1 \\ -1 & 1 \end{bmatrix}$$

calculate $M^{-1/2}P$, $\left(M^{-1/2}P\right)^T$, and $P^T M^{-1/2}$ and hence verify that the computations in equation (4.70) make sense.

4.43. Consider the two-degree-of-freedom system defined by

$$\begin{bmatrix} 9 & 0 \\ 0 & 1 \end{bmatrix} \ddot{\mathbf{x}}(t) + \begin{bmatrix} 27 & -3 \\ -3 & 3 \end{bmatrix} \mathbf{x}(t) = 0$$

Calculate the response of the system to the initial conditions

$$\mathbf{x}_0 = \frac{1}{\sqrt{2}} \begin{bmatrix} 1 \\ \frac{1}{3} \\ 1 \end{bmatrix} \qquad \dot{\mathbf{x}}_0 = 0$$

What is unique about your solution compared to the solution of Example 4.3.1?

4.44. Consider the two-degree-of-freedom system defined by

$$\begin{bmatrix} 9 & 0 \\ 0 & 1 \end{bmatrix} \ddot{\mathbf{x}}(t) + \begin{bmatrix} 27 & -3 \\ -3 & 3 \end{bmatrix} \mathbf{x}(t) = 0$$

Calculate the response of the system to the initial conditions

$$\mathbf{x}_0 = 0 \qquad \text{and} \qquad \dot{\mathbf{x}}_0 = \frac{1}{\sqrt{2}} \begin{bmatrix} 1 \\ \frac{1}{3} \\ -1 \end{bmatrix}$$

What is unique about your solution compared to the solution of Example 4.3.1 and to Problem 4.40, if you also worked that?

4.45. Consider the system defined by

$$\begin{bmatrix} 100 & 0 \\ 0 & 100 \end{bmatrix} \ddot{\mathbf{x}}(t) + \begin{bmatrix} 25{,}000 & -15{,}000 \\ -15{,}000 & 25{,}000 \end{bmatrix} \mathbf{x}(t) = \mathbf{0}$$

Solve for the free response of this system using modal analysis and the initial conditions.

4.46. Consider the model of a vehicle given in Problem 4.15 illustrated in Figure P4.15 defined by

$$\begin{bmatrix} 2000 & 0 \\ 0 & 50 \end{bmatrix} \ddot{\mathbf{x}} + \begin{bmatrix} 1000 & -1000 \\ -1000 & 11{,}000 \end{bmatrix} \mathbf{x} = \mathbf{0}$$

Suppose that the tire rolls over a bump modeled as the initial conditions of $\mathbf{x}(0) = [0 \quad 0.01]^T$ and $\dot{\mathbf{x}}(0) = \mathbf{0}$. Use modal analysis to calculate the response of the car $x_1(t)$. Plot the response for three cycles.

Section 4.4 (Problems 4.47 through 4.59)

4.47. A vibration model of the drive train of a vehicle is illustrated as the three-degree-of-freedom system of Figure P4.47. Calculate the undamped free response [i.e. $M(t) = F(t) = 0$, $c_1 = c_2 = 0$] for the initial condition $\mathbf{x}(0) = \mathbf{0}, \dot{\mathbf{x}}(0) = [0 \quad 0 \quad 1]^T$. Assume that the hub stiffness is 12,000 N/m and that the axle/suspension is 20,000 N/m. Assume the rotational element J is modeled as a translational mass of 75 kg.

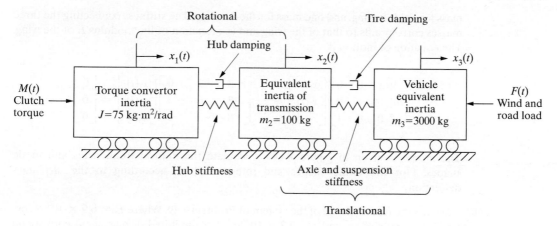

Figure P4.47   A simplified model of an automobile for vibration analysis of the drive train. The parameter values given are representative and should not be considered as exact.

4.48. Calculate the natural frequencies and normalized mode shapes of

$$\begin{bmatrix} 4 & 0 & 0 \\ 0 & 2 & 0 \\ 0 & 0 & 1 \end{bmatrix} \ddot{\mathbf{x}} + \begin{bmatrix} 4 & -1 & 0 \\ -1 & 2 & -1 \\ 0 & -1 & 1 \end{bmatrix} \mathbf{x} = \mathbf{0}$$

4.49. The vibration is the vertical direction of an airplane and its wings can be modeled as a three-degree-of-freedom system, with one mass corresponding to the right wing, one

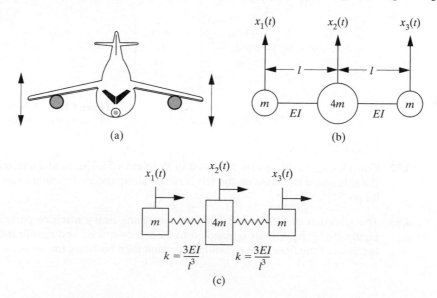

Figure P4.49   A model of the wing vibration of an airplane: (a) vertical wing vibration; (b) lumped mass/beam deflection model; (c) spring–mass model.

mass for the left wing, and one mass for the fuselage. The stiffness connecting the three masses corresponds to that of the wing and is a function of the modulus E of the wing. The equation of motion is

$$m \begin{bmatrix} 1 & 0 & 0 \\ 0 & 4 & 0 \\ 0 & 0 & 1 \end{bmatrix} \begin{bmatrix} \ddot{x}_1(t) \\ \ddot{x}_2(t) \\ \ddot{x}_3(t) \end{bmatrix} + \frac{EI}{l^3} \begin{bmatrix} 3 & -3 & 0 \\ -3 & 6 & -3 \\ 0 & -3 & 3 \end{bmatrix} \begin{bmatrix} x_1(t) \\ x_2(t) \\ x_3(t) \end{bmatrix} = \begin{bmatrix} 0 \\ 0 \\ 0 \end{bmatrix}$$

The model is given in Figure P4.49. Calculate the natural frequencies and mode shapes. Plot the mode shapes and interpret them according to the airplane's deflection.

4.50. Solve for the free response of the system of Problem 4.49. Where $E = 6.9 \times 10^9 \text{ N/m}^2$, $l = 2\,\text{m}$, $m = 4000\,\text{kg}$, and $I = 5.2 \times 10^{-6}\text{m}^4$. Let the initial displacement correspond to a gust of wind that causes an initial condition of $\dot{x}(0) = \mathbf{0}$, $\mathbf{x}(0) = [0.2 \quad 0 \quad 0]^T\text{m}$. Discuss your solution.

4.51. Consider the two-mass system of Figure P4.51. This system is free to move in the $x_1 - x_2$ plane. Hence each mass has two degrees of freedom. Derive the linear equations of motion, write them in matrix form, and calculate the eigenvalues and eigenvectors for $m = 20\,\text{kg}$ and $k = 200 \text{ N/m}$.

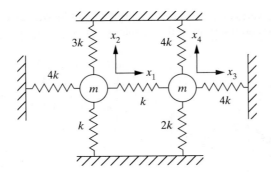

Figure P4.51 A two-mass system free to move in two directions.

4.52. Consider again the system discussed in Problem 4.51. Use modal analysis to calculate the solution if the mass on the left is raised along the x_2 direction exactly 0.02 m and let go.

4.53. The vibration of a floor in a building containing heavy machine parts is modeled in Figure P4.53. Each mass is assumed to be evenly spaced and significantly larger than the mass of the floor. The equation of motion then becomes ($m_1 = m_2 = m_3 = m$).

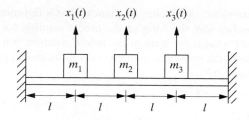

Figure P4.53 A lumped-mass model of boxes loaded on the floor of a building.

$$mI\ddot{x} + \frac{EI}{l^3}\begin{bmatrix} \dfrac{9}{64} & \dfrac{1}{6} & \dfrac{13}{192} \\[2mm] \dfrac{1}{6} & \dfrac{1}{3} & \dfrac{1}{6} \\[2mm] \dfrac{13}{192} & \dfrac{1}{6} & \dfrac{9}{64} \end{bmatrix}\begin{bmatrix} x_1 \\ x_2 \\ x_3 \end{bmatrix} = \mathbf{0}$$

Calculate the natural frequencies and mode shapes. Assume that in placing box m_2 on the floor (slowly) the resulting vibration is calculated by assuming that the initial displacement at m_2 is 0.04 m. If $l = 2\,\text{m}$, $m = 200\,\text{kg}$, $E = 0.6 \times 10^9 \,\text{N/m}^2$, $I = 4.17 \times 10^{-5}\,\text{m}^4$. Calculate the response and plot your results.

4.54. Recalculate the solution to Problem 4.53 for the case that m_2 is increased in mass to 1000 kg. Compare your results to those of Problem 4.53. Do you think it makes a difference where the heavy mass is placed?

4.55. Repeat Problem 4.49 for the case that the airplane body is 10 m instead of 4 m as indicated in the figure. What effect does this have on the response, and which design (4 m or 10 m) do you think is better for reduced vibration?

4.56. Often in the design of a car, certain parts cannot be reduced in mass. For example, consider the drive train model illustrated in Figure P4.47. The mass of the torque converter and transmission are relatively the same from car to car. However, the mass of the car could change as much as 1000 kg (e.g., a two-seater sports car versus a family sedan). With this in mind, resolve Problem 4.47 for the case that the vehicle inertia is reduced to 2000 kg. Which case has the smallest amplitude of vibration?

4.57. Use the *mode summation method* to compute the analytical solution for the response of the two-degree-of-freedom system:

$$\begin{bmatrix} 1 & 0 \\ 0 & 4 \end{bmatrix}\ddot{\mathbf{x}}(t) + \begin{bmatrix} 540 & -300 \\ -300 & 300 \end{bmatrix}\mathbf{x}(t) = 0$$

to the initial conditions of

$$\mathbf{x}_0 = \begin{bmatrix} 0 \\ 0.01 \end{bmatrix}, \quad \dot{\mathbf{x}}_0 = \begin{bmatrix} 0 \\ 0 \end{bmatrix}$$

4.58. For a zero value of an eigenvalue and hence frequency, what is the corresponding time response? Or asked another way, the form of the modal solution for a nonzero frequency is $A \sin(\omega_n t + \phi)$, what is the form of the modal solution that corresponds to a zero frequency? Evaluate the constants of integration if the modal initial conditions are $r_1(0) = 0.1$, and $\dot{r}_1(0) = 0.01$.

4.59. Consider the system described by

$$\begin{bmatrix} 1 & 0 \\ 0 & 4 \end{bmatrix} \ddot{x}(t) + \begin{bmatrix} 400 & -400 \\ -400 & 400 \end{bmatrix} x(t) = 0$$

subject to the initial conditions $x(0) = [1 \quad 0]^T$, $\dot{x}(0) = 0$. Plot the displacements versus time.

Section 4.5 (Problems 4.60 through 4.72)

4.60. Consider the following two-degree-of-freedom system and compute the response assuming modal damping rations of $\zeta_1 = 0.01$ and $\zeta_2 = 0.1$:

$$\begin{bmatrix} 10 & 0 \\ 0 & 1 \end{bmatrix} \ddot{x}(t) + \begin{bmatrix} 20 & -3 \\ -3 & 3 \end{bmatrix} x(t) = 0, \qquad x_0 = \begin{bmatrix} 0.1 \\ 0.05 \end{bmatrix}, \dot{x}_0 = 0$$

Plot the response.

4.61. Consider the example of the automobile drive train system discussed in Problem 4.47, modeled by

$$\begin{bmatrix} 75 & 0 & 0 \\ 0 & 100 & 0 \\ 0 & 0 & 3000 \end{bmatrix} \ddot{x} + 10,000 \begin{bmatrix} 2 & -2 & 0 \\ -2 & 6 & -2 \\ 0 & -2 & 4 \end{bmatrix} x = 0$$

$$x(0) = 0 \text{ and } \dot{x}(0) = [0 \quad 0 \quad 1]^T \text{ m/s}$$

Add 10% modal damping to each coordinate, calculate and plot the system response.

4.62. Consider the following two-degree-of-freedom system and compute the response assuming modal damping ratios of $\zeta_1 = 0.05$ and $\zeta_2 = 0.01$:

$$\begin{bmatrix} 50 & 0 \\ 0 & 1 \end{bmatrix} \ddot{x}(t) + \begin{bmatrix} 20 & -9 \\ -9 & 9 \end{bmatrix} x(t) = 0, \qquad x_0 = \begin{bmatrix} 0.05 \\ 0.09 \end{bmatrix}, \dot{x}_0 = 0$$

Plot the response.

4.63. Consider the model of an airplane discussed in Problem 4.49, Figure P4.49 modeled by

$$\begin{bmatrix} 3000 & 0 & 0 \\ 0 & 12,000 & 0 \\ 0 & 0 & 3000 \end{bmatrix} \ddot{x} + \begin{bmatrix} 13,455 & -13,455 & 0 \\ -13,455 & 26,910 & -13,455 \\ 0 & -13,455 & 13,455 \end{bmatrix} x = 0$$

subject to the initial conditions $x(0) = [0.02 \quad 0 \quad 0]^T$ m and $\dot{x}(0) = 0$. (a) Calculate the response assuming that the damping provided by the wing rotation is $\zeta_i = 0.01$ in each mode. (b) If the aircraft is in flight, the damping forces may increase dramatically to $\zeta_i = 0.1$. Recalculate the response and compare it to the more lightly damped case of part (a).

4.64. Repeat the floor vibration problem of Problem 4.53 given by

$$200\ddot{\mathbf{x}} + 3.197 \times 10^{-4} \begin{bmatrix} \dfrac{9}{64} & \dfrac{1}{6} & \dfrac{13}{192} \\[2mm] \dfrac{1}{6} & \dfrac{1}{3} & \dfrac{1}{6} \\[2mm] \dfrac{13}{192} & \dfrac{1}{6} & \dfrac{9}{64} \end{bmatrix} \mathbf{x} = \mathbf{0}$$

$$\mathbf{x}(0) = [0 \quad 0.05 \quad 0]^T \text{m and } \dot{\mathbf{x}}(0) = \mathbf{0}$$

by assigning modal damping ratios of

$$\zeta_1 = 0.01 \qquad \zeta_2 = 0.1 \qquad \zeta_3 = 0.2$$

4.65. Repeat Problem 4.64 with constant modal damping of $\zeta_1, \zeta_2, \zeta_3 = 0.1$. If you worked the previous problem, compare this solution with the solution of Problem 4.64.

4.66. Consider the damped system of Figure P4.66. Determine the damping matrix and use the formula of equation (4.119) to determine values of the damping coefficient c_I for which this system would be proportionally damped.

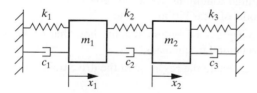

Figure P4.66   A damped two-degree-of-freedom system fixed at each end.

4.67. Let $k_3 = 0$ in Figure P4.66. Also let $m_1 = 1, m_2 = 4, k_1 = 2, k_2 = 1$ and calculate c_1, c_2 and c_3 such that $\zeta_1 = 0.02$ and $\zeta_2 = 0.2$.

4.68. Calculate the constants α and β for the two-degree-of-freedom system given by

$$\begin{bmatrix} 1 & 0 \\ 0 & 4 \end{bmatrix} \ddot{\mathbf{x}} + (\alpha M + \beta K)\dot{\mathbf{x}} + \begin{bmatrix} 3 & -1 \\ -1 & 1 \end{bmatrix} \mathbf{x} = \mathbf{0}$$

such that the system has modal damping of $\zeta_1 = \zeta_2 = 0.3$.

4.69. Equation (4.124) represents n equations in only two unknowns and hence cannot be used to specify all the modal damping ratios for a system with $n > 2$. If the floor vibration system of Problem 4.53 has measured damping of $\zeta_1 = 0.01$ and $\zeta_2 = 0.05$, determine ζ_3.

4.70. If you worked the previous problem, calculate the damping matrix for the system of Problem 4.69. What are the units of the elements of the damping matrix?

4.71. Does the following system decouple? If so, calculate the mode shapes and write the equation in decoupled form.

$$\begin{bmatrix} 1 & 0 \\ 0 & 1 \end{bmatrix} \ddot{\mathbf{x}} + \begin{bmatrix} 5 & -3 \\ -3 & 3 \end{bmatrix} \dot{\mathbf{x}} + \begin{bmatrix} 5 & -1 \\ -1 & 1 \end{bmatrix} \mathbf{x} = \mathbf{0}$$

4.72. Show that if the damping matrix satisfies $C = \alpha M + \beta K$, then the matrix $CM^{-1}K$ is symmetric and hence that $CM^{-1}K = KM^{-1}C$.

Section 4.6 (Problems 4.73 through 4.83)

4.73. Calculate the response of the system of Figure P4.73 discussed in Example 4.6.1 if $F_2(t) = \delta(t)$ and the initial conditions are set to zero. This might correspond to a two-degree-of-freedom model of a car hitting a bump.

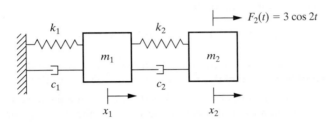

Figure P4.73   A damped two-degree-of-freedom system.

4.74. Calculate the response of the system of Figure P4.73 discussed in Example 4.6.1 if $F_1(t) = \delta(t)$ and the initial conditions are set to zero. This might correspond to a two-degree-of-freedom model of a car hitting a bump.

4.75. For an undamped two-degree-of-freedom system, show that resonance occurs at one or both of the system's natural frequencies.

4.76. Use modal analysis to calculate the response of the drive train system of Problem 4.44 given by

$$\begin{bmatrix} 75 & 0 & 0 \\ 0 & 100 & 0 \\ 0 & 0 & 3000 \end{bmatrix} \ddot{\mathbf{x}} + 10{,}000 \begin{bmatrix} 1 & -1 & 0 \\ -1 & 3 & -2 \\ 0 & -2 & 2 \end{bmatrix} \mathbf{x} = \mathbf{0}$$

to a unit impulse on the car body (i.e., at location x_3). Use the modal damping of 10% in each mode. Calculate the solution in terms of physical coordinates, and after subtracting the rigid-body modes, compare the responses of each part.

4.77. Consider the machine tool of Figure 4.28 with a floor mass of $m = 1000$ kg, subject to a force of $10 \sin t$ (in Newtons) so that the equation of motions is

$$(10^3) \begin{bmatrix} 0.8 & 0 & 0 \\ 0 & 4 & 0 \\ 0 & 0 & 2 \end{bmatrix} \ddot{\mathbf{x}}(t) + (10^4) \begin{bmatrix} 30 & -30 & 0 \\ -30 & 38 & -8 \\ 0 & -8 & 88 \end{bmatrix} \mathbf{x}(t) = \begin{bmatrix} 0 \\ 0 \\ 10 \sin t \end{bmatrix}$$

Calculate the response. How much does this floor vibration affect the machine's toolhead compared to the solution given in Example 4.8.3?

4.78. Consider the airplane of Figure P4.49 with modal damping of 0.1 in each mode. Suppose that the airplane hits a gust of wind, which applies an impulse of $3\delta(t)$ at the end of the left wing and $\delta(t)$ at the end of the right wing. Calculate the resulting vibration of the cabin $[x_2(t)]$.

4.79. Consider again the airplane of Figure P4.49. with 10% modal damping in each mode. Suppose that this is a propeller-driven airplane with an internal combustion engine mounted in the nose. At a cruising speed, the engine mounts transmit an applied force to the cabin mass ($4m$ at x_2) which is harmonic of the form $50\sin 10t$. Calculate the effect of this harmonic disturbance at the nose and on the wing tips after subtracting out the translational or rigid motion.

4.80. Consider the automobile model of Problem 4.15 illustrated in Figure P4.15 with equations of motion:

$$\begin{bmatrix} 2000 & 0 \\ 0 & 50 \end{bmatrix} \ddot{\mathbf{x}} + \begin{bmatrix} 1000 & -1000 \\ -1000 & 11000 \end{bmatrix} \mathbf{x} = \mathbf{0}$$

Add modal damping to this model of $\zeta_1 = 0.01$ and $\zeta_2 = 0.2$ and calculate the response of the body $[x_2(t)]$ to a harmonic input at the second mass of $10\sin 3t$ N.

4.81. Determine the *modal equations* for the following system and comment on whether or not the system will experience resonance.

$$\ddot{\mathbf{x}} + \begin{bmatrix} 2 & -1 \\ -1 & 1 \end{bmatrix} \mathbf{x} = \begin{bmatrix} 1 \\ 0 \end{bmatrix} \sin(0.618t)$$

4.82. Consider the following system and compute the solution using the mode summation method.

$$M = \begin{bmatrix} 9 & 0 \\ 0 & 1 \end{bmatrix}, \quad K = \begin{bmatrix} 27 & -3 \\ -3 & 3 \end{bmatrix}, \quad \mathbf{x}(0) = \begin{bmatrix} 1 \\ 0 \end{bmatrix}, \quad \dot{\mathbf{x}}(0) = \begin{bmatrix} 0 \\ 0 \end{bmatrix}$$

4.83. Consider the following two systems, and in each case determine if a resonance response occurs.

(a) $\begin{bmatrix} m_1 & 0 \\ 0 & m_2 \end{bmatrix} \begin{bmatrix} \ddot{x}_1 \\ \ddot{x}_2 \end{bmatrix} + \begin{bmatrix} k_1 + k_2 & -k_2 \\ -k_2 & k_2 \end{bmatrix} \begin{bmatrix} x_1 \\ x_2 \end{bmatrix} = \begin{bmatrix} 0.642 \\ 0.761 \end{bmatrix} \sin(2t)$

(b) $\begin{bmatrix} m_1 & 0 \\ 0 & m_2 \end{bmatrix} \begin{bmatrix} \ddot{x}_1 \\ \ddot{x}_2 \end{bmatrix} + \begin{bmatrix} k_1 + k_2 & -k_2 \\ -k_2 & k_2 \end{bmatrix} \begin{bmatrix} x_1 \\ x_2 \end{bmatrix} = \begin{bmatrix} 0.23500 \\ 2.97922 \end{bmatrix} \sin(2.756556t)$

where $m_1 = 4$ kg, $k_1 = 25$ N/m, $m_2 = 9$ kg, and $k_2 = 5$ N/m.

Section 4.7 (Problems 4.84 through 4.87)

4.84. Use Lagrange's equation to derive the equations of motion of the lathe of Figure 4.21 for the undamped case.

4.85. Use Lagrange's equations to derive the equations of motion for the automobile of Example 4.8.2 illustrated in Figure 4.25 for the case $c_1 = c_2 = 0$.

4.86. Use Lagrange's equations to derive the equations of motion for the building model presented in Figure 4.9 of Example 4.4.3 for the undamped case.

4.87. Consider again the model of the vibration of an automobile of Figure 4.25. In this case, include the tire dynamics as indicated in Figure P4.87. Derive the equations of motion using Lagrange formulation for the undamped case. Let m_3 denote the mass of the car acting at c.g.

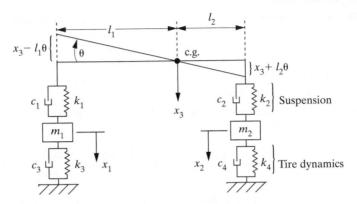

Figure P4.87   A simple car model including tire dynamics.

Section 4.9 (Problems 4.88 through 4.98)

*__4.88.__ Consider the mass matrix

$$M = \begin{bmatrix} 10 & -1 \\ -1 & 1 \end{bmatrix}$$

and calculate $M^{-1}, M^{-1/2}$, and the Cholesky factor of M. Show that

$$LL^T = M$$
$$M^{-1/2} M^{-1/2} = I$$
$$M^{1/2} M^{1/2} = M$$

*__4.89.__ Consider the matrix and vector

$$A = \begin{bmatrix} 1 & -\varepsilon \\ -\varepsilon & \varepsilon \end{bmatrix} \qquad b = \begin{bmatrix} 10 \\ 10 \end{bmatrix}$$

Use a code to solve $Ax = b$ for $\varepsilon = 0.1, 0.01, 0.001, 10^{-4}$, and 1.

*__4.90.__ Calculate the natural frequencies and mode shapes of the system of Example 4.8.3. Use the undamped equation and the form given by equation (4.161).

*__4.91.__ Compute the natural frequencies and mode shapes of the undamped version of the system of Example 4.8.3 using the formulation of equation (4.164) and (4.168). Compare your answers.

*__4.92.__ Use a code to solve for the modal information of the following system

$$\begin{bmatrix} 9 & 0 \\ 0 & 1 \end{bmatrix} \ddot{\mathbf{x}}(t) + \begin{bmatrix} 27 & -3 \\ -3 & 3 \end{bmatrix} \mathbf{x}(t) = \mathbf{0}$$

*4.93. Write a program to normalize the vector $\mathbf{x} = [0.4450 \quad 0.8019 \quad 1]^T$.

*4.94. Use a code to calculate the natural frequencies and mode shapes obtained for the system

$$\begin{bmatrix} 1 & 0 \\ 0 & 4 \end{bmatrix} \ddot{\mathbf{x}}(t) + \begin{bmatrix} 24 & -4 \\ -4 & 24 \end{bmatrix} \mathbf{x}(t) = \mathbf{0}$$

*4.95. Following the modal analysis solution of Window 4.5, write a program to compute the time response of the system of Problem 4.94.

*4.96. Use a code to solve the damped vibration problem

$$\begin{bmatrix} 9 & 0 \\ 0 & 1 \end{bmatrix} \ddot{\mathbf{x}} + \begin{bmatrix} 2.7 & -0.3 \\ -0.3 & 0.3 \end{bmatrix} \dot{\mathbf{x}} + \begin{bmatrix} 27 & -3 \\ -3 & 3 \end{bmatrix} \mathbf{x} = \mathbf{0}$$

by calculating the natural frequencies, damping ratios, and mode shapes.

*4.97. Consider the vibration of the airplane of Problems 4.46 and 4.47 as given in Figure P4.46. The mass and stiffness matrices are given as

$$M = m \begin{bmatrix} 1 & 0 & 0 \\ 0 & 4 & 0 \\ 0 & 0 & 1 \end{bmatrix} \qquad K = \frac{EI}{l^3} \begin{bmatrix} 3 & -3 & 0 \\ -3 & 6 & -3 \\ 0 & -3 & 3 \end{bmatrix}$$

where $m = 4000\,\text{kg}$, $l = 2\,\text{m}$, $I = 5.2 \times 10^{-6}\,\text{m}^4$, $E = 6.9 \times 10^9\,\text{N/m}^2$, and the damping matrix C is taken to be $C = (0.002)K$. Calculate the natural frequencies, normalized mode shapes, and damping ratios.

*4.98. Consider the proportionally damped, dynamically coupled system given by

$$M = \begin{bmatrix} 9 & -1 \\ -1 & 1 \end{bmatrix} \qquad C = \begin{bmatrix} 6 & -2 \\ -2 & 2 \end{bmatrix} \qquad K = \begin{bmatrix} 49 & -2 \\ -2 & 2 \end{bmatrix}$$

and calculate the mode shapes, natural frequencies, and damping ratios.

Section 4.10 (Problems 4.99 through 4.106)

*4.99. Recall the system of Example 1.7.3 for the vertical suspension system of a car modeled by $m\ddot{x}(t) + c\dot{x}(t) + kx(t) = 0$, with $m = 1361$ kg, $k = 2.668 \times 10^5\,\text{N/m}$, and $c = 3.81 \times 10^4$ kg/s subject to the initial conditions of $x(0) = 0$ and $v(0) = 0.01$ m/s². Solve this and plot the solution using numerical integration.

*4.100. Solve for the time response of Example 4.4.3 (i.e., the four-story building of Figure 4.9) modeled by

$$4000 \begin{bmatrix} 1 & 0 & 0 & 0 \\ 0 & 1 & 0 & 0 \\ 0 & 0 & 1 & 0 \\ 0 & 0 & 0 & 1 \end{bmatrix} \ddot{\mathbf{x}}(t) + \begin{bmatrix} 10{,}000 & -5000 & 0 & 0 \\ -5000 & 10{,}000 & -5000 & 0 \\ 0 & -5000 & 10{,}000 & -5000 \\ 0 & 0 & -5000 & 5000 \end{bmatrix} \mathbf{x}(t) = \mathbf{0}$$

subject to an initial displacement of $\mathbf{x}(0) = [0.001 \quad 0.010 \quad 0.020 \quad 0.025]^T$ and zero initial velocity. Compare the solutions obtained with using a modal analysis approach to a solution obtained by numerical integration.

*4.101. Reproduce the plots of Figure 4.13 for the two-degree of freedom system of Example 4.5.1 given by

$$\begin{bmatrix} 9 & 0 \\ 0 & 1 \end{bmatrix} \ddot{\mathbf{x}}(t) + \begin{bmatrix} 27 & -3 \\ -3 & 3 \end{bmatrix} \mathbf{x}(t) = \mathbf{0}, \qquad \mathbf{x}(0) = \begin{bmatrix} 1 \\ 0 \end{bmatrix}, \dot{\mathbf{x}}(t) = \mathbf{0}$$

using an numerical integration code.

*4.102. Consider Example 4.8.3 and (a) using the damping ratios given, compute a damping matrix in physical coordinates, (b) use numerical integration to compute the response and plot it, and (c) use the numerical code to design the system so that all three physical coordinates die out within 5 seconds (i.e., change the damping matrix until the desired response results).

*4.103. Compute and plot the time response of the system (Newtons):

$$\begin{bmatrix} 9 & 0 \\ 0 & 1 \end{bmatrix} \begin{bmatrix} \ddot{x}_1 \\ \ddot{x}_2 \end{bmatrix} + \begin{bmatrix} 3 & -0.5 \\ -0.5 & 0.5 \end{bmatrix} \begin{bmatrix} \dot{x}_1 \\ \dot{x}_2 \end{bmatrix} + \begin{bmatrix} 3 & -1 \\ -1 & 1 \end{bmatrix} \begin{bmatrix} x_1 \\ x_2 \end{bmatrix} = \begin{bmatrix} 1 \\ 1 \end{bmatrix} \sin(4t)$$

subject to the initial conditions:

$$\mathbf{x}_0 = \begin{bmatrix} 0 \\ 0.1 \end{bmatrix} \text{m}, \qquad \mathbf{v}_0 = \begin{bmatrix} 1 \\ 0 \end{bmatrix} \text{m/s}$$

using numerical integration.

*4.104. Consider the following system excited by a pulse of duration 0.1 s (in Newtons):

$$\begin{bmatrix} 4 & 0 \\ 0 & 1 \end{bmatrix} \begin{bmatrix} \ddot{x}_1 \\ \ddot{x}_2 \end{bmatrix} + \begin{bmatrix} 0.3 & -0.05 \\ -0.05 & 0.05 \end{bmatrix} \begin{bmatrix} \dot{x}_1 \\ \dot{x}_2 \end{bmatrix} + \begin{bmatrix} 3 & -1 \\ -1 & 1 \end{bmatrix} \begin{bmatrix} x_1 \\ x_2 \end{bmatrix}$$

$$= \begin{bmatrix} 0 \\ 1 \end{bmatrix} [\Phi(t - 1) - \Phi(t - 3)]$$

and subject to the initial conditions:

$$\mathbf{x}_0 = \begin{bmatrix} 0 \\ -0.1 \end{bmatrix} \text{m}, \qquad \mathbf{v}_0 = \begin{bmatrix} 0 \\ 0 \end{bmatrix} \text{m/s}$$

Compute and plot the response of the system using numerical integration. Here Φ indicates the Heaviside Step Function introduced in Section 3.2.

*4.105. Use numerical integration to compute and plot the time response of the system (newtons)

$$\begin{bmatrix} 5 & 0 \\ 0 & 1 \end{bmatrix} \begin{bmatrix} \ddot{x}_1 \\ \ddot{x}_2 \end{bmatrix} + \begin{bmatrix} 3 & -0.5 \\ -0.5 & 0.5 \end{bmatrix} \begin{bmatrix} \dot{x}_1 \\ \dot{x}_2 \end{bmatrix} + \begin{bmatrix} 30 & -1 \\ -1 & 1 \end{bmatrix} \begin{bmatrix} x_1 \\ x_2 \end{bmatrix} = \begin{bmatrix} 1 \\ 1 \end{bmatrix} \sin(4t)$$

subject to the initial conditions

$$\mathbf{x}_0 = \begin{bmatrix} 0 \\ 0.1 \end{bmatrix} \text{m}, \qquad \mathbf{v}_0 = \begin{bmatrix} 1 \\ 0 \end{bmatrix} \text{m/s}$$

*4.106. Use numerical integration to compute and plot the time response of the system (newtons)

$$\begin{bmatrix} 4 & 0 & 0 & 0 \\ 0 & 3 & 0 & 0 \\ 0 & 0 & 2.5 & 0 \\ 0 & 0 & 0 & 6 \end{bmatrix} \begin{bmatrix} \ddot{x}_1 \\ \ddot{x}_2 \\ \ddot{x}_3 \\ \ddot{x}_4 \end{bmatrix} + \begin{bmatrix} 4 & -1 & 0 & 0 \\ -1 & 2 & -1 & 0 \\ 0 & -1 & 2 & -1 \\ 0 & 0 & -1 & 1 \end{bmatrix} \begin{bmatrix} \dot{x}_1 \\ \dot{x}_2 \\ \dot{x}_3 \\ \dot{x}_4 \end{bmatrix}$$

$$+ \begin{bmatrix} 500 & -100 & 0 & 0 \\ -100 & 200 & -100 & 0 \\ 0 & -100 & 200 & -100 \\ 0 & 0 & -100 & 100 \end{bmatrix} \begin{bmatrix} x_1 \\ x_2 \\ x_3 \\ x_4 \end{bmatrix} = \begin{bmatrix} 0 \\ 0 \\ 0 \\ 1 \end{bmatrix} \sin(4t)$$

subject to the initial conditions

$$\mathbf{x}_0 = \begin{bmatrix} 0 \\ 0 \\ 0 \\ 0.01 \end{bmatrix} \text{m}, \qquad \mathbf{v}_0 = \begin{bmatrix} 1 \\ 0 \\ 0 \\ 0 \end{bmatrix} \text{m/s}$$

MATLAB ENGINEERING VIBRATION TOOLBOX

If you have not yet used the *Engineering Vibration Toolbox* program, return to the end of Chapter 1 or Appendix G for a brief introduction to using MATLAB files.

TOOLBOX PROBLEMS

TB4.1. Calculate the natural frequencies and mode shapes of the system of Example 4.1.5 using file VTB4_1.

TB4.2. Recalculate Example 4.2.6 using file VTB4_1 and compare your answer with that of the example obtained with a calculator. Verify that $P^T \widetilde{K} P = \Lambda$ and $P^T P = I$.

TB4.3. Consider Example 4.3.1 and investigate the effect of the initial condition on the response by using file VTB4_2 and plotting the responses to the following initial displacements (initial velocities all zero):

$$\mathbf{x}_0 = \begin{bmatrix} 1 \\ 1 \end{bmatrix} \quad \mathbf{x}_0 = \begin{bmatrix} 0 \\ 1 \end{bmatrix} \quad \mathbf{x}_0 = \begin{bmatrix} 1 \\ -1 \end{bmatrix} \quad \mathbf{x}_0 = \begin{bmatrix} 0.1 \\ 0.9 \end{bmatrix} \quad \mathbf{x}_0 = \begin{bmatrix} 0.2 \\ 0.8 \end{bmatrix} \quad \text{etc.}$$

Discuss the results by formulating a short paragraph summarizing what you observed.

TB4.4. Check the calculation of Example 4.4.2.

TB4.5. Using file VTB4_2, examine the effect of increasing the mass m_4 in the building vibration problem of Example 4.4.3. Do this by doubling m_4 and recalculating the solution. Notice what happens to the various responses. Try doubling m_4 until the response does not change or the program fails. Discuss your observations.

TB4.6. Consider the system of Example 4.7.3. Choose the values $m = 10, J = 5, e = 1$, and $k_1 = 1000$ and calculate the eigenvalues as k_2 varies from 10 to 10,000, in increments of 100. What can you conclude?

5

Design for Vibration Suppression

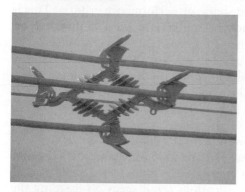

This chapter introduces the techniques useful in designing structures and machines so they vibrate as little as possible. Often this happens after a product is designed, prototyped, and tested. In many cases, vibration problems are found late in the process so redesign is often needed. In this chapter the important concepts of *shock and vibration isolation* and *vibration absorption* are introduced, as they are key methods in vibration design. Optimization as a design tool, use of adding damping and the concept of critical speeds in rotating machines are also introduced.

The cable spacer-damper used on power lines and pictured at the top is used to cut down the cable whistling caused by vibration induced by moderate winds and stop the conductors from hitting one another in strong winds (or resonance) as the cables vibrate. The ideas introduced in the chapter are fundamental to the design of such devices. The cables themselves are modeled in Chapter 6.

The suspension system of any ground vehicle is an example a common design problem. A spring and damper system for an off-road vehicle is shown in the picture at the bottom. A vehicle's size, mass, and road conditions all affect the ability of the suspension system to perform its function of isolating passengers from shock and vibration induced by road conditions and speed.

In this chapter, it is assumed that vibration is undesirable and is to be suppressed. The topics of the previous chapters present a number of techniques and methods for analyzing the vibration response of various systems subject to various inputs. Here the focus is to use the skills developed in the preceding chapters to determine ways of adjusting the physical parameters of a system or device in such a way that the vibration response meets some specified shape or performance criteria. This is called *design;* design was introduced in Section 1.7, and it is the focus of this chapter.

Vibration can often lead to a number of undesirable circumstances. For example, vibration of an automobile or truck can lead to driver discomfort and, eventually, fatigue. Structural or mechanical failure can often result from sustained vibration (e.g., cracks in airplane wings). Electronic components used in airplanes, automobiles, machines, and so on may also fail because of vibration, shock, and/or sustained vibration input.

The "fragility level" of devices, how much vibration a given device can withstand, is addressed by the International Organization of Standards (ISO) as well as by some national agencies. Almost every device manufactured for use by the military must meet certain military specifications ("mil specs") regarding the amount of vibration it can withstand. In addition to government and international agencies, manufacturers also set desired vibration performance standards for some products. If a given device does not meet these regulations, it must be redesigned so that it does. This chapter presents several formulations that are useful for designing and redesigning various devices and structures to meet desired vibration standards.

Design is a difficult subject that does not always lend itself to simple formulations. Design problems typically do not have a unique solution. Many different designs may all give acceptable results. Sometimes a design may simply consist of putting together a number of existing (off-the-shelf) devices to create a new device with desired properties. Here we focus on design as it refers to adjusting a system's physical parameters to cause its vibration response to behave in a desired fashion.

5.1 ACCEPTABLE LEVELS OF VIBRATION

To design a device in terms of its vibration response, the desired response must be clearly stated. Many different methods of measuring and describing acceptable levels of vibration have been proposed. Whether or not the criteria should be established in terms of displacement, velocity, or acceleration, and exactly how these should be measured needs to be clarified before a design can begin. These choices often depend on the specific application. For instance, in practice, it is generally accepted that the best indication of potential structural damage is the amplitude of the structure's velocity, while acceleration amplitude is the most perceptible by humans. Some common ranges of vibration frequency and displacement are given in Table 5.1.

TABLE 5.1   RANGES OF FREQUENCY AND DISPLACEMENT OF VIBRATION

	Frequency (Hz)	Displacement amplitude (mm)
Atomic vibration	10^{12}	10^{-7}
Threshold of human perception	1–8	10^{-2}
Machinery and building vibration	10–100	10^{-2}–1
Swaying of tall buildings	1–5	10–1000

The ISO (International Organization for Standardization, www.iso.org) provides a published standard of acceptable levels of vibration that has the intent of providing a mechanism to facilitate communications between manufacturers and consumers. The standards are tested in terms of root mean-square (rms) values of displacement, velocity, and acceleration. Recall that the value (defined in Section 1.2) is the square root of the time average of the square of a quantity. For the displacement $x(t)$, the rms value is given in equation (1.21) as

$$x_{\text{rms}} = \left[\lim_{T \to \infty} \frac{1}{T} \int_0^T x^2(t) dt \right]^{1/2}$$

A convenient way to express the acceptable values of vibration allowed under ISO standards is to plot them on a nomograph, as illustrated in Figure 5.1. Several examples and further details of nomographs can be found in Niemkiewicz, J. (2002).

The nomograph of Figure 5.1 is a graphical representation of the relationship among displacement, velocity, acceleration, and frequency for an undamped single-degree-of-freedom system. Figure 5.1 is representative and is based on vibration induced around machines as an example of how ISO attempts to classify vibration levels. Acceptable vibration levels are then stated in terms of all three physical responses: displacement, velocity, and acceleration, as well as the frequency. The solution for the displacement is given by equation (1.19) as

$$x(t) = A \sin \omega_n t$$

(for zero phase), which has amplitude A. Differentiating the displacement solution yields the velocity

$$v(t) = \dot{x}(t) = A\omega_n \cos \omega_n t$$

which has amplitude $\omega_n A$. Differentiating again yields the acceleration

$$a(t) = \ddot{x}(t) = -A\omega_n^2 \sin \omega_n t$$

which has amplitude $A\omega_n^2$. These three expressions for the magnitude, along with the definition of rms value, allow the nomograph of Figure 5.1 to be constructed.

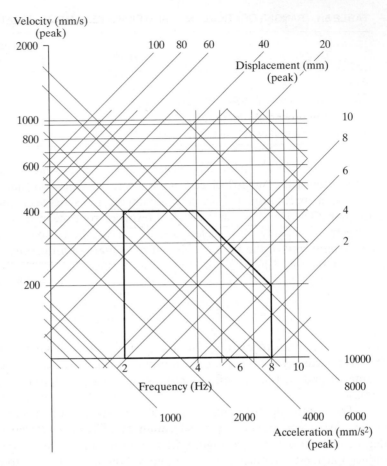

Figure 5.1   An example of a nomograph for specifying acceptable levels of vibration.

From equation (1.21) the average value of $x(t) = A \sin\omega_n t$ is

$$\bar{x}^2 = \lim_{T \to \infty} \frac{1}{T} \int_0^T A^2 \sin^2 \omega_n t \, dt = \lim_{T \to \infty} \left[\frac{A^2}{2T\omega_n} \omega_n T - \frac{1}{2} \frac{A^2}{T\omega_n} (\sin \omega_n T \cos \omega_n T) \right] = \frac{A^2}{2}$$

so that $\bar{x} = A/\sqrt{2}$. Likewise, $\bar{v} = A\omega_n/\sqrt{2}$ and $\bar{\alpha} = A\omega_n^2/\sqrt{2}$.

Example 5.1.1

A machine part is subject to an rms amplitude of vibration of 6 mm. The mass and stiffness are measured to be 5 kg and 20,000 Nm, respectively. Use Figure 5.1 to determine if there is any concern with these values based on the acceptable region marked on the figure. Determine the velocity and acceleration experienced by the part. If the standard represented by Figure 5.1 is not met, then suggest a means of redesigning the bearing cap so that the response does meet the standard.

Solution   The natural frequency of a simple spring–mass system is

$$\omega_n = \sqrt{\frac{k}{m}} = \sqrt{\frac{20000 \text{ N/m}}{5 \text{ kg}}} = 63.246 \text{ rad/s}, \quad f_n = \frac{\omega_n}{2\pi} = 10.066 \text{ Hz}$$

This frequency is outside the acceptable range and must be adjusted. Thus, the standard is not met and some redesign is required.

In order to bring the frequency to an acceptable range, the mass or stiffness of the bearing must be changed to reduce the frequency. For instance, if a 45-kg mass is added to the bearing, and a 2500-N/m spring is added in series with the bearing stiffness (recall Figure 1.31), the resulting frequency is

$$\omega_n = \sqrt{\frac{\frac{1}{1/k + 1/200}}{m + 2}} = \sqrt{\frac{198.02 \text{ N/m}}{7\,kg}} = 37.231 \text{ rad/s}, f_n = \frac{\omega_n}{2\pi} \approx 6 \text{ Hz}$$

This brings the design in line with the acceptable frequency range, but is quite a reduction in stiffness, which may not be acceptable for other reasons. Looking at the chart, the 6 mm line crosses 6 Hz inside the box, at an acceleration of about 8000 mm/s² and a velocity of about 200 mm/s. Using the formulas to get a more precise value yields

$$\bar{v} = \omega_n \bar{x} = (37.231 \text{ rad/s})(6 \text{ mm}) = 223.4 \text{ mm/s}$$

$$\bar{a} = \omega_n^2 \bar{x} = (37.231 \text{ rad/s})^2(6 \text{ mm}) = 8.317 \times 10^3 \text{ mm/s}^2$$

While the redesign achieves the goal of reducing the vibration levels of the bearing, it is an order-of-magnitude change in the bearing stiffness. This could have negative effects on other parts of the machine (see critical speeds in Section 5.7, for instance). This example is what makes design so difficult and challenging. Changing the design to meet one specification could cause another specification to be violated.

□

The design procedure just suggested is oversimplified but helps introduce the ad hoc nature of many design problems. Analysis is used as a tool. Here the desired vibration criteria are provided by an ISO standard represented in a nomograph. This plot, together with the formula $\omega_n = \sqrt{k/m}$ and the series spring formula, provides the analysis tools.

In using any thought process to perform a design, it is important to think through potential flaws in the procedure. In the preceding example there are several possible points of error. One important issue is how well the simple single-degree-of-freedom spring–mass model captures the dynamics of the part. It might be that a more sophisticated multiple-degree-of-freedom model is required (recall Example 4.8.2). Another possible problem with a proposed design change is that the stiffness of the part may not be allowed to be lower than a certain value because of load requirements (static deflection) or other design constraints. In this case, the mass might be changed, but that too may have other constraints on it. In some cases it just may not be possible to design a system to have this desired vibration response. That is, not all design problems have a solution.

The range of vibrations with which an engineer is concerned is usually from about 10^{-4} mm at between 0.1 and 1 Hz for objects such as optical benches or sophisticated medical equipment, to a meter displacement for tall buildings in the range of 0.1 to 5 Hz. Machine vibrations can range between 10 and 1000 Hz, with deflections between fractions of a millimeter and several centimeters. As technology advances, limitations and acceptable levels of vibration change. Thus these numbers should be considered as rough indications of common values.

Example 5.1.2

Calculate and compare the natural frequency, damping ratio, and damped natural frequency of the single-degree-of-freedom model of a stereo turntable and of the automobile given in Figure 5.2. Also plot and compare their frequency response functions and their impulse response functions. Discuss the similarities and differences of these two devices.

Solution   To calculate the undamped natural frequency, damping ratio, and damped natural frequency of the car is simple. From the definitions

$$\omega_n = \sqrt{\frac{k}{m}} = \sqrt{\frac{400{,}000}{1000}} = 20 \text{ rad/s}$$

$$\zeta = \frac{c}{2m\omega_n} = \frac{8000}{2(1000)(20)} = 0.2$$

$$\omega_d = \omega_n\sqrt{1 - \zeta^2} = 20\sqrt{1 - (0.2)^2} = 19.5959 \text{ rad/s}$$

The same calculations for the stereo turntable are

$$\omega_n = \sqrt{\frac{k}{m}} = \sqrt{\frac{400}{1}} = 20 \text{ rad/s}$$

$$\zeta = \frac{c}{2m\omega_n} = \frac{8}{2(1)(20)} = 0.2$$

$$\omega_d = \omega_n\sqrt{1 - \zeta^2} = 20\sqrt{1 - (0.2)^2} = 19.5959 \text{ rad/s}$$

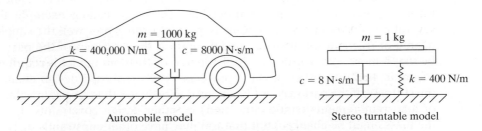

Automobile model   Stereo turntable model

Figure 5.2   A single-degree-of-freedom model of an automobile and a stereo turntable, each with the same frequency.

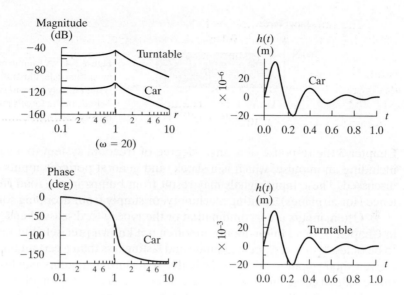

Figure 5.3   The frequency response function and the impulse response of that for the car and stereo turntable of Figure 5.2.

This illustrates that two objects of very different size can have the same natural frequencies and damping ratios.

In Figure 5.3 the transfer function of each device is plotted as well as the impulse response function for each. Note that these plots do indicate a difference in devices. The phase plots of the transfer function of both the car and the stereo are identical, while the magnitude plots differ by a constant. The impulse response function of the car has a smaller amplitude, although the responses both die out at the same time since the decay rate (ζ) and hence log decrement are the same for each device. The acceptable levels of vibration for the car will be much larger than those of the stereo. For instance, a displacement amplitude of 10 mm for the stereo would cause its needle to skip out of the groove in a record and, hence, not perform properly (a reason why phonograph records never made it in automobiles—thank goodness for Bluetooth technology and mp3 players). On the other hand, a similar amplitude of vibration for the car is below the perception level on a nomograph.

☐

An important consideration in specifying vibration response is to specify the nature of the input or driving force that causes the response. Disturbances, or inputs, are normally classified as either *shock* or as vibration, depending on how long the input lasts. An input is considered to be a shock if the disturbance is a sharp, aperiodic one lasting a relatively short time. In contrast, an input is considered to be a vibration if it lasts for a long time and has some oscillatory features.

The distinction between shock and vibration is not always clear as the sources of shock and vibration disturbances are numerous and very difficult to place into categories. In Chapter 2 only vibration inputs (e.g., harmonic inputs) are discussed. In

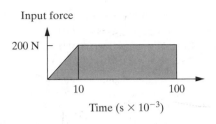

Figure 5.4 A sample of a shock input to an automobile, illustrating that the form is not entirely known—that only the bounds of the force's magnitude and time history are known.

Chapter 3 the response of a single-degree-of-freedom system to a variety of inputs, including an impulse, which is a shock, and general periodic inputs (vibration) are discussed. These input signals may result from bumps in the road (for cars), turbulence (for airplanes), rotating machinery, or simply from dropping something.

Often, inputs are a combination of the types just discussed and those discussed in Chapter 3. In addition, inputs are often not known precisely but rather are known to be of less than a certain magnitude and lasting less than a certain time. For instance, a given shock input to an automobile due to its hitting a bump may take the form of a single-valued curve falling somewhere inside the shaded region of Figure 5.4.

Vibrations that are not harmful exist in many devices. For instance, automobiles continually experience vibration without being damaged or causing harm to passengers. However, some vibrations are extremely damaging, such as severe vibration from an earthquake or a badly out-of-balance tire on a car. The difficult issue for design engineers is deciding between acceptable levels of vibration and those that will cause damage or become so annoying that consumers will not use the device. Once acceptable levels are established, several techniques can be used to limit and alter the shock and vibration response of mechanical systems and structures. These are discussed in the following section.

5.2 VIBRATION ISOLATION

The most effective way to reduce unwanted vibration is to stop or modify the source of the vibration. If this cannot be done, it is sometimes possible to design a *vibration isolation system* to isolate the source of vibration from the system of interest or to isolate the device from the source of vibration. This can be done by using highly damped, compliant materials, such as rubber, to change the stiffness and damping between the source of vibration and the device that is to be protected from the vibrations. The problem of isolating a device from a source of vibration is analyzed in terms of reducing vibration *displacement* transmitted through base motion, as discussed in Section 2.3 and summarized in Window 5.1. The problem of isolating a source of vibration from its surroundings is analyzed in terms of reducing the *force* transmitted by the source through its mounting points, as discussed in Section 2.4 and summarized in Window 5.1. Both force transmissibility and displacement transmissibility are called *isolation problems.*

Window 5.1
Summary of Vibration Isolation Formulas for Both Force and Displacement Transmissibility

The moving-base model on the left is used in designing isolation to protect the device from motion of its point of attachment (base). The model on the right is used to protect the point of attachment (ground) from vibration of the mass.

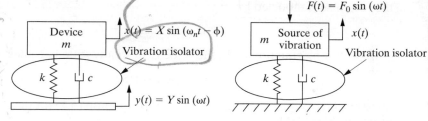

Displacement Transmissibility   Force Transmissibility

$$F(t) = F_0 \sin(\omega t)$$

Device m — $x(t) = X \sin(\omega_n t - \phi)$ Vibration isolator

Source of vibration m — $x(t)$ Vibration isolator

$y(t) = Y \sin(\omega t)$

Moving base (source of vibration)   Fixed base

Vibration source modeled as base motion   Vibration source mounted on isolator

Here $y(t) = Y \sin \omega t$ is the disturbance and from equation (2.71)

$$\frac{X}{Y} = \left[\frac{1 + (2\zeta r)^2}{(1 - r^2)^2 + (2\zeta r)^2} \right]^{1/2}$$

defines the displacement transmissibility and is plotted in Figure 2.13. From equation (2.77),

$$\frac{F_T}{kY} = r^2 \left[\frac{1 + (2\zeta r)^2}{(1 - r^2)^2 + (2\zeta r)^2} \right]^{1/2}$$

defines the related force transmissibility and is plotted in Figure 2.15.

Here $F(t) = F_0 \sin \omega t$ is the disturbance and

$$\frac{F_T}{F_0} = \left[\frac{1 + (2\zeta r)^2}{(1 - r^2)^2 + (2\zeta r)^2} \right]^{1/2}$$

defines the force transmissibility for isolating the source of vibration as derived in Section 5.2.

The analysis tool used to design isolators is the concept of force and/or displacement transmissibility introduced in Section 2.4. By way of review, consider the problem of calculating the *transmissibility ratio,* denoted T.R., defined as the ratio of the magnitude of the force transmitted through the spring and dashpot to the fixed base to the sinusoidal force applied (see Window 5.1) by the machine (modeled as a mass). Symbolically, T.R. $= F_T/F_0$. To calculate the value of T.R., first consider the force transmitted. The force transmitted to the fixed base in

Window 5.1 is denoted by $F_T(t)$ and is the force applied to the base acting through the spring and dashpot, that is,

$$F_T(t) = kx(t) + c\dot{x}(t) \qquad (5.1)$$

The solution for the case that the driving force is harmonic of the form $F_0 \cos \omega t$ is given in Section 2.2, equation (2.37), to be of the form

$$x(t) = Ae^{-\zeta\omega_n t} \sin(\omega_d t + \theta) + X \cos(\omega t - \phi)$$

In steady state (i.e., after some time has elapsed), the first term decays to zero and the response is modeled by

$$x(t) = X \cos(\omega t - \phi) \qquad (5.2)$$

Differentiating equation (5.2), the velocity in steady state becomes

$$\dot{x}(t) = -\omega X \sin(\omega t - \phi) \qquad (5.3)$$

Substitution of equations (5.2) and (5.3) into equation (5.1) for the force transmitted at steady state yields

$$F_T(t) = kX \cos(\omega t - \phi) - c\omega X \sin(\omega t - \phi) \qquad (5.4)$$
$$= kX \cos(\omega t - \phi) + c\omega X \cos(\omega t - \phi + \pi/2)$$

The magnitude of $F_T(t)$, denoted by F_T, can be calculated from equation (5.4) by noting that the two terms are $90°(\pi/2)$ out of phase with each other and hence can be thought of as two perpendicular vectors (recall Figure 2.11 of Section 2.3). Hence the magnitude of $F_T(t)$ is calculated by taking the vector sum of the two terms on the right of equation (5.4). This yields that the magnitude of the transmitted force is

$$F_T = \sqrt{(kX)^2 + (c\omega X)^2} = X\sqrt{k^2 + c^2\omega^2} \qquad (5.5)$$

Window 5.2
Review of the Steady-State Response of an Underdamped System Subject to Harmonic Excitation as Discussed in Section 2.2

The steady-state response of

$$\ddot{x} + 2\zeta\omega_n\dot{x} + \omega_n^2 x = f_0 \cos \omega t$$

where $\omega_n = \sqrt{k/m}$, $\zeta = c/(2m\omega_n)$ and $f_0 = F_0/m$, is $x(t) = X \cos(\omega t - \phi)$.
Here

$$X = \frac{f_0}{\sqrt{(\omega_n^2 - \omega^2)^2 + (2\zeta\omega_n\omega^2)}}, \qquad \phi = \tan^{-1}\frac{2\zeta\omega_n\omega}{\omega_n^2 - \omega^2}$$

The value of X, the amplitude of vibration at steady state, is given in Window 5.2 to be

$$X = \frac{f_0}{[(\omega_n^2 - \omega^2)^2 + (2\zeta\omega_n\omega)^2]^{1/2}} = \frac{F_0/k}{[(1 - r^2)^2 + (2\zeta r)^2]^{1/2}}$$

where $r = \omega/\omega_n$, as before. Substituting this value of X into equation (5.5) yields

$$F_T = \frac{F_0/k}{[(1 - r^2)^2 + (2\zeta r)^2]^{1/2}}\sqrt{k^2 + c^2\omega^2}$$

$$= F_0\frac{\sqrt{1 + c^2\omega^2/k^2}}{[(1 - r^2)^2 + (2\zeta r)^2]^{1/2}} = F_0\sqrt{\frac{1 + (2\zeta r)^2}{(1 - r^2)^2 + (2\zeta r)^2}} \tag{5.6}$$

where $c^2\omega^2/k^2 = (2m\omega_n\zeta)^2\omega^2/k^2 = (2\zeta r)^2$. The *transmissibility ratio*, or *transmission ratio*, denoted T.R., is defined as the ratio of the magnitude of the transmitted force to the magnitude of the applied force. By a simple manipulation of equations (5.6), this becomes

$$\text{T.R.} = \frac{F_T}{F_0} = \sqrt{\frac{1 + (2\zeta r)^2}{(1 - r^2)^2 + (2\zeta r)^2}} \tag{5.7}$$

A comparison of this force transmissibility expression with the displacement transmissibility given in Window 5.1 indicates that they are identical. It is important to note, however, that even though they have the same value, they come from different isolation problems and hence describe different phenomena.

The displacement transmissibility ratio given in the left column of Window 5.1 describes how a steady-state displacement (Y) of the base of a device mounted on an isolator is transmitted into motion of the device (X). Figure 5.5 is a plot of the T.R. for various values of the damping ratio ζ and frequency ratio r. The larger the value of T.R., the larger the amplitude of motion of the mass. These curves are useful for designing the isolators. In particular, the design process consists of choosing ζ and r, within the available isolator's material, such that T.R. is small.

Note from Figure 5.5 that if the frequency ratio r is greater than $\sqrt{2}$, the magnitude of vibration of the device is smaller than the disturbance magnitude Y and vibration isolation occurs. For r less than $\sqrt{2}$, the amplitude X increases (i.e., X is larger than Y). The value of the damping ratio (each curve in Figure 5.5 corresponds to a different ζ) determines how much smaller the amplitude of vibration is for a given frequency ratio. Near resonance, the T.R. is determined completely by the value of ζ (i.e., by the damping in the isolator). In the isolation region, the smaller the value of ζ, the smaller the value of T.R. and the better the isolation. Also note that as r is increased for a fixed ω, the value of T.R. decreases. This corresponds to increasing the mass or decreasing the stiffness of the isolator.

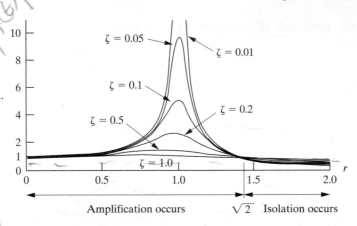

Amplification occurs $\sqrt{2}$ Isolation occurs

Magnification of the isolation area

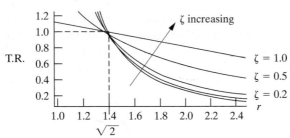

Figure 5.5 Plot of the transmissibility ratio, T.R., indicating the value of T.R. for a variety of choices of the damping ratio ζ and the frequency ratio r. This is a repeat of Figure 2.14 and is a plot of equation (5.7).

Example 5.2.1

An electronic control system for an automobile engine is to be mounted on top of the fender inside the engine compartment of the automobile as illustrated in Figure 5.6. The control module electronically computes and controls the engine timing, fuel/air mixture, and so on, and completely controls the engine. To protect it from fatigue and breakage, it is desirable to isolate the module from the vibration induced in the car body by road and engine vibration, hence the module is mounted on an isolator. Design the isolator (i.e., pick c and k) if the mass of the module is 3 kg and the dominant vibration of the fender is approximated by $y(t) = (0.01)(\sin 35t)$ m. Here it is desired to keep the displacement of the module less than 0.005 m at all times. Once the design values for isolators are chosen, calculate the magnitude of the force transmitted to the module through the isolator.

Solution Since it is desired to keep the vibration of the module, $x(t)$, less than 0.005 m, the response amplitude becomes $X = 0.005$ m. The amplitude of $y(t)$ is $Y = 0.01$ m; hence the desired displacement transmissibility ratio becomes

$$\text{T.R.} = \frac{X}{Y} = \frac{0.005}{0.01} = 0.5$$

Examining the transmissibility curves of Figure 5.5 yields several possible solutions for ζ and ω_n. A straight horizontal line through T.R. = 0.5, illustrated in Figure 5.7, crosses at several values of ζ and r. For instance, the $\zeta = 0.02$ curve intersects the

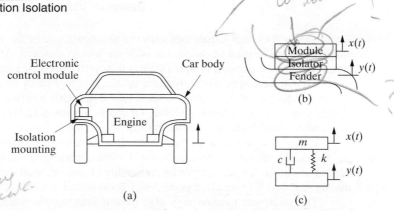

Figure 5.6   (a) A cutaway sketch of the engine compartment of an automobile illustrating the location of the car's electronic control module. (b) A close-up of the control module mounted on the inside fender on an isolator. (c) A vibration model of the module isolator system.

T.R. = 0.5 line at $r = 1.73$. Thus $r = 1.73$, $\zeta = 0.02$ provides one possible design solution.

Recalling that $r = \omega/\omega_n = 1.73$ and $\omega = 35$ rad/s, the isolation system's natural frequency is about $\omega_n = 35/1.73 = 20.231$ rad/s. Since the mass of the module is $m = 3$ kg, the stiffness is (recall $\omega_n = \sqrt{k/m}$)

$$k = m\omega_n^2 = (3 \text{ kg})(20.231 \text{ rad/s})^2 = 1228 \text{ N/m}$$

Thus the isolation mount must be made of a material with this stiffness (or add a stiffener). The damping ratio ζ is related to the damping constant c by equation (1.30):

$$c = 2\zeta m\omega_n = 2(0.02)(3 \text{ kg})(20.231 \text{ rad/s})$$
$$= 2.428 \text{ kg/s}$$

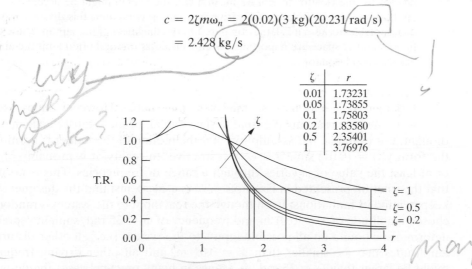

ζ	r
0.01	1.73231
0.05	1.73855
0.1	1.75803
0.2	1.83580
0.5	2.35401
1	3.76976

Figure 5.7   The transmissibility curve of Figure 5.6, repeated here, indicating possible design solutions for Example 5.2.1. Each point of intersection with one of the curves of constant ζ yields the desired T.R.

These values for c and k, together with the geometric size of the module and fender shape, can now be used to choose the isolation mount material. At this point, the designer would look through vendor catalogs to search for existing isolation mounts and materials that have approximately these values. If none exactly meet these values, the curves in Figure 5.7 are consulted to see if one of the other solutions corresponds more closely to an existing isolation material. Of course, equation (2.71) or (5.7) can be used to calculate solutions lying in between those illustrated in Figure 5.7.

If many solutions are still available after a search of existing products is made, the choice of a mount can be "optimized" by considering other functions, such as cost, ease of assembly, temperature range, reliability of vendor, availability of product, and quality required. Eventually, a good design must consider all of these aspects.

The electronic module may also have a limit on the amount of force it can withstand. One way to estimate the amount of force is to use the theory developed in Section 2.4, in particular in equation (2.77), which is reproduced in Window 5.1. This expression relates the force transmitted to the module by the motion of the fender through the isolator. Using the values just calculated in equation (2.77) yields

$$F_T = kYr^2 \left[\frac{1 + (2\zeta r)^2}{(1 - r^2)^2 + (2\zeta r)^2} \right]^{1/2} = kYr^2 \text{(T.R.)}$$

$$= (1228 \text{ N/m})(0.01 \text{ m})(1.73)^2(0.5) = 18.375 \text{ N}$$

If this force happens to be too large, the design must be redone. With the maximum force transmitted as an additional design consideration, the curves of Figure 2.15 must also be consulted when choosing the values of r and ζ to meet the required design specifications. The static deflection caused by this design is $\delta = mg/k = 0.024$ m. The static deflection and the ratio X/Y define the *rattle space* or physical dimensions needed for the isolator to vibrate in. In a car, the 2.4 cm might be acceptable. If the application were to isolate a CD drive in a laptop computer, this distance would be unacceptable because it is large compared to the thickness of the laptop. Static deflection and rattle space are important design considerations and often limit the ability to design a good isolator.

□

Example 5.2.1 may have seemed very reasonable. However, many assumptions were made in reaching the final design, and all of these must be given careful thought. For example, the assumption that the motion of the fender is harmonic of the form $y(t) = (0.01) \sin 35t$ is very restrictive. In reality, it is probably random, or at least the value of ω varies through a range of frequencies. This is not to say that the solution presented in Example 5.2.1 is useless, just that the designer should keep in mind its limitations. Even though the real input to the system is random, the chosen amplitude of $Y = 0.01$ m and frequency of $\omega = 35$ rad/s might represent a deterministic bound on all possible inputs to the system (i.e., all other disturbance amplitudes may be smaller than $Y = 0.01$ m, and all other driving frequencies might be larger than $\omega = 35$ rad/s). Hence in many practical cases the designer is faced with choosing an isolator that will protect the part from, say, 5 g between 20 and 200 Hz, or the designer will be given a plot of PSD versus ω_r (recall Section 3.5) and try to design the isolator to service these types of inputs. Section 5.9 examines the isolation problem from the practical aspect of working with manufacturers of isolation products.

The design of shock isolation systems is performed by examining the shock spectrum, as introduced in Section 3.6. To make the comparison to vibration isolation clear, the shock spectrum is reconsidered here as a plot of the ratio of the maximum motion of the response acceleration amplitude (i.e., $\omega^2 X$) to the disturbance acceleration amplitude [i.e., the amplitude of $\ddot{y}(t)$] versus the product of the natural frequency and the time duration of the pulse, t_1. Here the disturbance $y(t)$ is modeled as a half sinusoid of the form

$$y(t) = \begin{cases} Y \sin \omega_p t & 0 \le t \le t_1 = \dfrac{\pi}{\omega_p} \\ 0 & t > t_1 = \dfrac{\pi}{\omega_p} \end{cases} \tag{5.8}$$

as indicated in Figure 5.8. This type of disturbance is often called a *shock pulse*. The frequency ω_p and the corresponding time $t_1 = \pi/\omega_p$ determine how long the shock pulse lasts. The product $\omega_n t_1$ is used for plotting shock transmissibility rather than plotting the frequency ratio used to design vibration isolators.

A plot of the acceleration amplitude ratio versus the product ωt_1 is given in Figure 5.8. This figure is determined by calculating the acceleration amplitude of the response and comparing it to the magnitude of the acceleration of the input disturbance. Note that as the abscissa increases, corresponding to a longer pulse width, the acceleration experienced by the module is larger than the input acceleration. By examining the plots in Figure 5.8, it can be seen that reduction of the acceleration

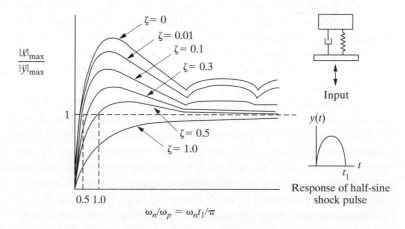

Figure 5.8   A plot of the ratio of output acceleration magnitude to input acceleration magnitude versus a frequency ratio (ω_n/ω_p) for a single-degree-of-freedom system and a base excitation consisting of a shock pulse for different values of damping ratios. Note that a large ω_p value corresponds to a short pulse width.

through isolation occurs only if the transmissibility falls below the horizontal line passing through the number 1. So, for instance, consider the line defined by $\zeta = 0.5$. For shock isolation to occur, this would require

$$\frac{\omega_n t_1}{\pi} < 1.0, \quad \text{or} \quad k < \frac{m\pi^2}{t_1^2}$$

Thus shock isolation enforces a bound on the stiffness of the isolation material.

Next consider the effects of damping on shock isolation. Examination of the plots in Figure 5.8 shows that increasing the damping greatly reduces the maximum acceleration to the point that for critical damping ($\zeta = 1.0$) isolation occurs for any pulse duration (t_1). Thus, good shock isolators require high damping.

Next consider the problem of isolating a source of harmonic vibration from its surroundings. This is the fixed-base isolation problem illustrated in Window 5.1, where the right side is concerned with reducing the force transmitted through the isolator due to harmonic excitation at the mass. The common example is a rotating machine generating a harmonic force at a constant frequency (recall Figure 2.18). Examples of such machines are electric motors, steam turbines, internal combustion engines, generators, washing machines, and disk drive motors.

Examination of the transmissibility curve of Figure 5.5 indicates that vibration isolation begins for isolation stiffness values such that $\omega/\omega_n > \sqrt{2}$. This gives transmissibility ratios of less than 1, so that the force transmitted to ground is less than the force generated by the rotating machine. Since the mass is usually fixed by the nature of the machine, the isolation mounts are generally chosen based on their stiffness, so that $r > \sqrt{2}$ is satisfied. If this does not give an acceptable solution (for low-frequency excitation), mass can sometimes be added to the machine $\left(\omega_n = \sqrt{k/m}\right)$. Since $r = \omega/\sqrt{k/m}$, lower stiffness values correspond to larger values of r, which yields better isolation (lower T.R. values).

As damping is increased for a fixed r, the value of T.R. increases, so that low damping is often used. However, some damping is desirable, since when the machine starts up and causes a harmonic disturbance through a range of frequencies, it generally passes through resonance ($r = 1$) and the presence of damping is required to reduce the transmissibility at resonance. Examination of the transmissibility curve indicates that for a large enough frequency ratio (about $r > 3$) and small enough damping ($\zeta < 0.2$) the T.R. value is not affected by damping. Since most springs have very small internal damping (e.g., less than 0.01), the term $(2\zeta r)^2$ is very small [e.g., for $r = 3$, $(2\zeta r)^2 = 0.0036$]. Hence it is common to design a vibration isolation system by neglecting the damping in equation (5.7). In this case T.R. becomes (taking the negative square root for positive values of T.R.)

$$\text{T.R.} = \frac{1}{r^2 - 1} \quad (r > 3) \tag{5.9}$$

Equation (5.9) can be used to construct design charts for use in choosing vibration isolation pads for mounting rotating machinery.

The driving frequency of a machine is usually specified in terms of its speed of rotation, or revolutions per minute (rpm). If n is the motor speed in rpm,

$$\omega = \frac{2\pi n}{60} \tag{5.10}$$

In addition, springs are often classified in terms of their static deflection defined by $\Delta = W/k = mg/k$, where m is the mass of the machine and g the acceleration due to gravity. It has become very common to design isolators in terms of the machine's rotating speed n and the static deflection Δ. A third quantity, R, defined as the *reduction* in transmissibility,

$$R = 1 - \text{T.R.} \tag{5.11}$$

is commonly used to quantify the success of the vibration isolator.

Substitution of the (undamped) value of T.R. into equation (5.11) and solving for r yields

$$r = \frac{\omega}{\sqrt{k/m}} = \sqrt{\frac{2 - R}{1 - R}} \tag{5.12}$$

Substitution of (5.10) for ω and replacing $k = mg/\Delta$ yields

$$n = \frac{30}{\pi}\sqrt{\frac{g(2 - R)}{\Delta(1 - R)}} = 29.9093\sqrt{\frac{2 - R}{\Delta(1 - R)}} \tag{5.13}$$

which relates the motor speed to the reduction factor and the static deflection of the spring. Equation (5.13) can be used to generate design curves, by taking the log of the expression. This yields

$$\log n = -\frac{1}{2}\log\Delta + \log\left(29.9093\sqrt{\frac{2 - R}{1 - R}}\right) \tag{5.14}$$

which is a straight line on a log–log plot for each value of R. This expression is then used to provide the design chart of Figure 5.9, consisting of plots of motor speed versus static deflection.

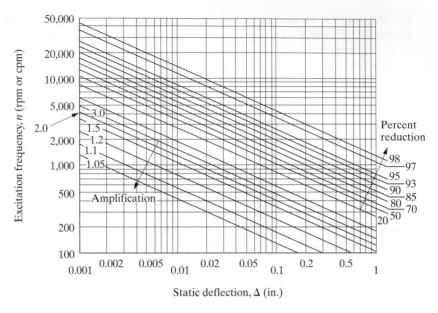

Figure 5.9   Design curves consisting of plots of running speed versus static deflection (or stiffness) for various values of percent reduction in transmitted force.

Example 5.2.2

Consider the computer disk drive of Figure 5.10. The disk drive motor is mounted to the computer chassis through an isolation pad (spring). The motor has a mass of 3 kg and operates at 5000 rpm. Calculate the value of the stiffness of the isolator needed to provide a 95% reduction in force transmitted to the chassis (considered as ground). How much clearance is needed between the motor and the chassis?

Solution   From the chart of Figure 5.9, the line corresponding to a speed of $n = 5000$ rpm hits the curve corresponding to 95% reduction at a static deflection of 0.03 in. or 0.0762 cm. This corresponds to a spring stiffness of

$$k = \frac{mg}{\Delta} = \frac{(3 \text{ kg})(9.8 \text{ m/s}^2)}{0.000762 \text{ m}} = 38{,}582 \text{ N/m}$$

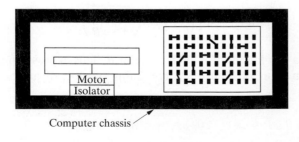

Figure 5.10   A schematic model of a personal computer illustrating the motor running the disk drive system. A small amount of out-of-balance in the motor can transmit harmonic forces to the chassis and onto circuit boards and other components if not properly isolated.

The choice of clearance (i.e., distance needed between the motor and the chassis) should be more than twice the static deflection so that the spring has room to extend and compress, providing isolation.

□

The issue of vibration isolation against harmonic forces for large (heavy) machines quickly becomes one of static deflection. For machines requiring extreme isolation, coiled springs often must be used to provide the large static deflection required at low frequency. This can be seen by examining the 98% curves in Figure 5.9 for low values of n. In some cases, the static deflection required may be too large to be physically obtainable even in small devices. An example of a similar design constraint is the miniaturization of computers. Manufacturers of laptop computers believe sales are tied to how compact and, in particular, how thin the chassis can be. One constraint could be the isolation system required for the disk drive motor or other components.

In designing isolators, it is often difficult to design an isolator that works effectively against both shock and vibration (harmonic) excitations. One reason for this can be seen by examining Figure 5.7 for vibration isolation and Figure 5.8 for shock isolation. In Figure 5.7, isolation occurs in the region $r > \sqrt{2}$, and in this region it is clear that increased damping reduces the effective isolation. Thus, low damping is required for vibration isolation. However, examining the shock isolation plots in Figure 5.8 shows that large damping is required for effective shock isolation. These two requirements are often at odds, as the following example points out.

Example 5.2.3

In this example the design goal is to develop an effective isolator for base-induced shock motion that will also provide an acceptable level of vibration isolation in terms of force transmissibility from the vibrating equipment supported by the isolation system to the base. In actual isolation applications it is often difficult to design an effective combined shock and vibration isolation system. As mentioned earlier and is clear from Figures 5.7 and 5.8, this is due to the fact that an effective vibration isolator must be very lightly damped, whereas an effective shock isolator tends to require large damping forces. Typically in design problems, a main constraint is that parts such as isolators are available only in discrete values of stiffness and damping. Here we suppose that a set of three off-the-shelf isolators (mounts) are available for use, that their natural frequencies are 5 Hz, 6 Hz, and 7 Hz, and all have 8% damping. The shock input that is being isolated is a 15-g, half-sine pulse as shown in Figure 5.8 for the case $t_1 = 40$ ms. With this as input, it is desired to limit the mount response to 15 g's and the mount deflection to 3 in. (76.2 mm). The vibration isolation goal is 20 dB of isolation from an above-mount vibration source of 15 Hz.

Figures 5.11 and 5.12 show the time simulation results to shock input in terms of mount deflection and above-mount response for the three mount possibilities. From Figures 5.11 and 5.12, it is clear that only the 7-Hz mount satisfies all the shock isolation goals, that is, <15 g's above mount and <3 in. of deflection.

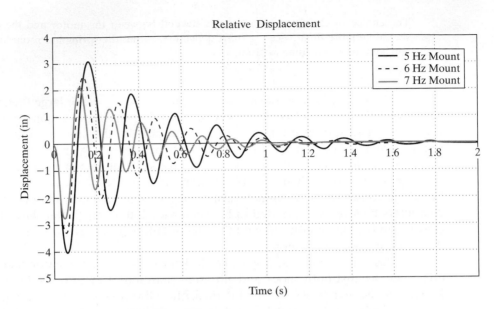

Figure 5.11   A simulated response [relative displacement, $z(t)$].

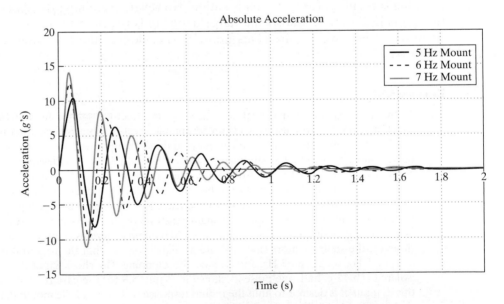

Figure 5.12   A simulated response [absolute acceleration, $\ddot{x}(t)$].

Now consider the vibration isolation performance using the same mount. A minimum of 20 dB of vibration isolation to the 15-Hz above-mount vibration source is desired. Recall that the mount has an inherent damping that can adequately be

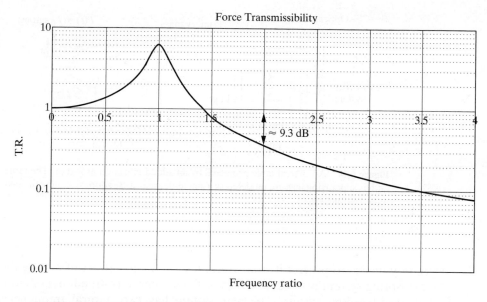

Figure 5.13   Force transmissibility (above-mount source to base).

approximated as 8% viscous damping. To determine the vibration isolation performance, the force transmissibility is plotted in Figure 5.13 for an 8% damped isolator.

The frequency ratio of interest is $r = \omega/\omega_n = 2 \cdot \pi \cdot 15/2 \cdot \pi \cdot 7 \approx 2.1$ for this
example. Recall that a minimum of 20 dB of isolation is required, but from Figure 5.13
only about 9.4 dB of isolation is achieved. To achieve the required vibration isolation
would require a lower damping ratio (see Figure 5.7) than is inherent to the isolator.
It is not possible to lower the damping ratio, as damping is a fixed property of the isolation material. Damping can be raised by adding external dampers, adding layers of
damping material, and so forth, but it cannot be decreased without significantly modifying the mount design. As such, these shock and vibration isolation design parameters
cannot simultaneously be met using the devices at hand, a typical situation in design.

$\square$

In designing isolation mounts, two factors are key. The first is deciding whether
to design for vibration or to design for shock. The next is to check the static deflection. The following section examines the use of absorbers to reduce vibrations from
harmonic disturbances.

5.3   VIBRATION ABSORBERS

Another approach to protecting a device from steady-state harmonic disturbance
at a constant frequency is a *vibration absorber.* Unlike the isolator of the previous
sections, an absorber consists of a second mass–spring combination added to the

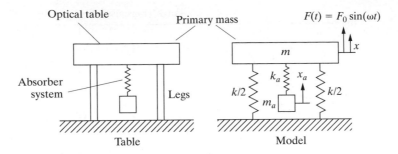

Figure 5.14 An optical table protected by an added vibration absorber. The table and its supporting legs are modeled as a single-degree-of-freedom system with mass m and stiffness k.

primary device to protect it from vibrating. The major effect of adding the second mass–spring system is to change from a single-degree-of-freedom system to a two-degree-of-freedom system. The new system has two natural frequencies (recall Section 4.1). The added spring–mass system is called the absorber. The values of the absorber mass and stiffness are chosen such that the motion of the original mass is a minimum. This is accompanied by substantial motion of the added absorber system, as illustrated in the following:

Absorbers are often used on machines that run at constant speed, such as sanders, compactors, reciprocating tools, and electric razors. Probably the most visible vibration absorbers can be seen on transmission lines and telephone lines. A dumbbell-shaped vibration absorber is often used on such wires to provide vibration suppression against wind blowing, which can cause the wire to oscillate at its natural frequency. The presence of the absorber prevents the wire from vibrating so much at resonance that it breaks (or fatigues). Figure 5.14 illustrates a simple vibration absorber attached to a spring–mass system. The equations of motion [summing forces in the vertical direction (refer to Chapter 4)] are

$$\begin{bmatrix} m & 0 \\ 0 & m_a \end{bmatrix}\begin{bmatrix} \ddot{x} \\ \ddot{x}_a \end{bmatrix} + \begin{bmatrix} k + k_a & -k_a \\ -k_a & k_a \end{bmatrix}\begin{bmatrix} x \\ x_a \end{bmatrix} = \begin{bmatrix} F_0 \sin \omega t \\ 0 \end{bmatrix} \qquad (5.15)$$

where $x = x(t)$ is the displacement of the table modeled as having mass m and stiffness k, x_a is the displacement of the absorber mass (of mass m_a and stiffness k_a), and the harmonic force $F_0 \sin \omega t$ is the disturbance applied to the table mass. It is desired to design the absorber (i.e., choose m_a and k_a) such that the displacement of the primary system is as small as possible in steady state. Here it is desired to reduce the vibration of the table, which is the primary mass.

Window 5.3

Recall that the inverse of a 2×2 matrix A given by

$$A = \begin{bmatrix} a & b \\ c & d \end{bmatrix}$$

is defined to be

$$A^{-1} = \frac{1}{\det A} \begin{bmatrix} d & -b \\ -c & a \end{bmatrix}$$

where

$$\det A = ad - bc$$

In contrast to the solution technique of modal analysis used in Chapter 4, here it is desired to obtain a solution in terms of parameters (m, k, m_a, and k_a) that can then be solved for as part of a design process. To this end, let the steady-state solution of $x(t)$ and $x_a(t)$ be of the form

$$x(t) = X \sin \omega t \tag{5.16}$$

$$x_a(t) = X_a \sin \omega t$$

Substitution of these steady-state forms into equation (5.15) yields (after some manipulation)

$$\begin{bmatrix} k + k_a - m\omega^2 & -k_a \\ -k_a & k_a - m_a\omega^2 \end{bmatrix} \begin{bmatrix} X \\ X_a \end{bmatrix} \sin \omega t = \begin{bmatrix} F_0 \\ 0 \end{bmatrix} \sin \omega t \tag{5.17}$$

which is an equation in the vector $[X \quad X_a]^T$. Dividing by $\sin \omega t$, taking the inverse of the matrix coefficient of $[X \quad X_a]^T$ (see Window 5.3), and multiplying from the left yields

$$\begin{bmatrix} X \\ X_a \end{bmatrix} = \frac{1}{\left(k + k_a - m\omega^2\right)\left(k_a - m_a\omega^2\right) - k_a^2} \begin{bmatrix} k_a - m_a\omega^2 & k_a \\ k_a & k + k_a - m\omega^2 \end{bmatrix} \begin{bmatrix} F_0 \\ 0 \end{bmatrix}$$

$$= \frac{1}{\left(k + k_a - m\omega^2\right)\left(k_a - m_a\omega^2\right) - k_a^2} \begin{bmatrix} \left(k_a - m_a\omega^2\right)F_0 \\ k_a F_0 \end{bmatrix} \tag{5.18}$$

Equating elements of the vector equality given by equation (5.18) yields the result that the magnitude of the steady-state vibration of the device (table) becomes

$$X = \frac{\left(k_a - m_a\omega^2\right)F_0}{\left(k + k_a - m\omega^2\right)\left(k_a - m_a\omega^2\right) - k_a^2} \tag{5.19}$$

while the magnitude of vibration of the absorber mass becomes

$$X_a = \frac{k_a F_0}{\left(k + k_a - m\omega^2 \right)\left(k_a - m_a \omega^2 \right) - k_a^2} \tag{5.20}$$

Note from equation (5.19) that the absorber parameters k_a and m_a can be chosen such that the magnitude of steady-state vibration, X, is exactly zero. This is accomplished by equating the coefficient of F_0 in equation (5.19) to zero:

$$\omega^2 = \frac{k_a}{m_a} \tag{5.21}$$

Hence if the absorber parameters are chosen to satisfy the tuning condition of equation (5.21), the steady-state motion of the primary mass is zero (i.e., $X = 0$). In this event the steady-state motion of the absorber mass is calculated from equations (5.20) and (5.16) with $k_a = m_a \omega^2$ to be

$$x_a(t) = -\frac{F_0}{k_a} \sin \omega t \tag{5.22}$$

Thus the absorber mass oscillates at the driving frequency with amplitude $X_a = F_0/k_a$.

Note that the magnitude of the force acting on the absorber mass is just $k_a x_a = k_a(-F_0/k_a) = -F_0$. Hence when the absorber system is tuned to the driving frequency and has reached steady state, the force provided by the absorber mass is equal in magnitude and opposite in direction to the disturbance force. With zero net force acting on the primary mass, it does not move and the motion is "absorbed" by motion of the absorber mass. Note that while the applied force is completely absorbed by the motion of the absorber mass, the system is not experiencing resonance because $\sqrt{k_a/m_a}$ is not a natural frequency of the two-mass system.

The success of the vibration absorber discussed previously depends on several factors. First, the harmonic excitation must be well known and not deviate much from its constant value. If the driving frequency drifts much, the tuning condition will not be satisfied, and the primary mass will experience some oscillation. There is also some danger that the driving frequency could shift to one of the combined systems' resonant frequencies, in which case one or the other of the system's coordinates would be driven to resonance and potentially fail. The analysis used to design the system assumes that it can be constructed without introducing any appreciable damping. If damping is introduced, the equations cannot necessarily be decoupled and the magnitude of the displacement of the primary mass will not be zero. In fact, damping defeats the purpose of a tuned vibration absorber and is desirable only if the frequency range of the driving force is too wide for effective operation of the absorber system. This is discussed in the next section. Another key factor in absorber design is that the absorber spring stiffness k_a must be capable of withstanding the full force of the excitation and hence must be capable of the corresponding deflections. The issue of spring size and deflection, as well as the value of the absorber mass, places a geometric limitation on the design of a vibration absorber system.

The issue of avoiding resonance in absorber design in case the driving frequency shifts can be quantified by examining the mass ratio μ, defined as the ratio of the absorber mass to the primary mass:

$$\mu = \frac{m_a}{m}$$

In addition, it is convenient to define the frequencies

$\omega_p = \sqrt{\dfrac{k}{m}}$ the original natural frequency of the primary system without the absorber attached

$\omega_a = \sqrt{\dfrac{k_a}{m_a}}$ the natural frequency of the absorber system before it is attached to the primary system

With these definitions, also note that

$$\frac{k_a}{k} = \mu \frac{\omega_a^2}{\omega_p^2} = \mu \beta^2 \tag{5.23}$$

where the frequency ratio β is $\beta = \omega_a/\omega_p$. Substitution of the values for μ, ω_p, and ω_a into equation (5.19) for the amplitude of vibration of the primary mass yields (after some manipulation)

$$\frac{Xk}{F_0} = \frac{1 - \omega^2/\omega_n^2}{\left[1 + \mu\left(\omega_a/\omega_p\right)^2 - \left(\omega/\omega_p\right)^2\right]\left[1 - \left(\omega/\omega_a\right)^2 - \mu\left(\omega_a/\omega_p\right)^2\right]} \tag{5.24}$$

The absolute value of this expression is plotted in Figure 5.15 for the case $\mu = 0.25$. Such plots can be used to illustrate how much drift in driving frequency can be tolerated by the absorber design. Note that if ω should drift to either $0.781\,\omega_a$ or $1.28\,\omega_a$, the combined system would experience resonance and fail, since these are the natural frequencies of the combined system. In fact, if the driving frequency shifts such that $|Xk/F_0| > 1$, the force transmitted to the primary system is amplified and the absorber system is not an improvement over the original design of the primary system. The shaded area of Figure 5.15 illustrates the values of ω/ω_a such that $|Xk/F_0| \le 1$. This illustrates the useful operating range of the absorber design (i.e., $0.908\,\omega_a < \omega < 1.118\,\omega_a$). Hence if the driving frequency drifts within this range, the absorber design still offers some protection to the primary system by reducing its steady-state vibration magnitude.

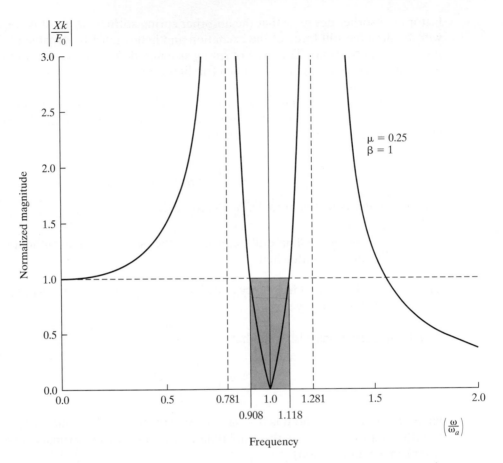

Figure 5.15 A plot of normalized magnitude of the primary mass versus the normalized driving frequency for the case $\mu = 0.25$. The two natural frequencies of the system occur at 0.781 and 1.281.

The design of an absorber can be further illuminated by examining the mass ratio μ and the frequency ratio β. These two dimensionless quantities indirectly specify both the mass and stiffness of the absorber system. The frequency equation (characteristic equation) for the two-mass system is obtained by setting the determinant of the matrix coefficient in equation (5.17) [i.e., the denominator of equation (5.18)] to zero and interpreting ω as the system natural frequency ω_n. Substitution for the value of β and rearranging yields

$$\beta^2 \left(\frac{\omega_n^2}{\omega_a^2}\right)^2 - \left[1 + \beta^2(1 + \mu)\right] \frac{\omega_n^2}{\omega_a^2} + 1 = 0 \qquad (5.25)$$

which is a quadratic equation in $\left(\omega_n^2/\omega_a^2\right)$. Solving this yields

$$\left(\frac{\omega_n}{\omega_a}\right)^2 = \frac{1 + \beta^2(1 + \mu)}{2\beta^2} \pm \frac{1}{2\beta^2}\sqrt{\beta^4(1 + \mu)^2 - 2\beta^2(1 - \mu) + 1} \quad (5.26)$$

which illustrates how the system's natural frequencies vary with the mass ratio μ and the frequency ratio β. This is plotted for $\beta = 1$ in Figure 5.16. Note that as μ is increased, the natural frequencies split farther apart, and farther from the operating point $\omega = \omega_a$ of the absorber. Therefore, if μ is too small, the combined system will not tolerate much fluctuation in the driving frequency before it fails. As a rule of thumb, μ is usually taken to be between 0.05 and 0.25 (i.e., $0.05 \le \mu \le 0.25$), as larger values of μ tend to indicate a poor design. Vibration absorbers can also fail because of fatigue if $x_a(t)$ and the stresses associated with this motion of the absorber are large. Hence limits are often placed on the maximum value of X_a by the designer. The following example illustrates an absorber design.

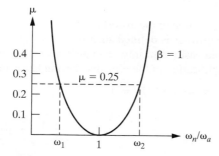

Figure 5.16   A plot of mass ratio versus system natural frequency (normalized to the frequency of the absorber system), illustrating that increasing the mass ratio increases the useful frequency range of a vibration absorber. Here ω_1 and ω_2 indicate the normalized value of the system's natural frequencies.

Example 5.3.1

A radial saw base has a mass of 73.16 kg and is driven harmonically by a motor that turns the saw's blade as illustrated in Figure 5.17. The motor runs at constant speed and produces a 13-N force at 180 cycles/min because of a small unbalance in the motor. The resulting forced vibration was not detected until after the saw had been manufactured. The manufacturer wants a vibration absorber designed to drive the table oscillation to zero simply by retrofitting an absorber onto the base. Design the absorber assuming that the effective stiffness provided by the table legs is 2600 N/m. In addition, the absorber must fit inside the table base and hence has a maximum deflection of 0.2 cm.

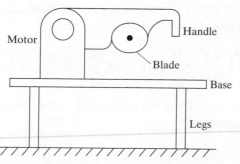

Figure 5.17   A schematic of a radial saw system in need of a vibration absorber.

Solution To meet the deflection requirement, the absorber stiffness is chosen first. This is calculated by assuming that $X = 0$, so that $|X_a k_a| = |F_0|$ [i.e., so that the mass m_a absorbs all of the applied force, see equation (5.20) with $k_a = m_a \omega^2$]. Hence

$$k_a = \frac{F_0}{X_a} = \frac{13 \text{ N}}{0.2 \text{ cm}} = \frac{13 \text{ N}}{0.002 \text{ m}} = 6500 \text{ N/m}$$

Since the absorber is designed such that $\omega = \omega_a$,

$$m_a = \frac{k_a}{\omega^2} = \frac{6500 \text{ N/m}}{[(180/60)^{2\pi}]^2} = 18.29 \text{ kg}$$

Note in this case that $\mu = 18.29/73.16 = 0.25$.

$\square$

Example 5.3.2

Calculate the bandwidth of operation of the absorber design of Example 5.3.1. Assume that the useful range of an absorber is defined such that $|Xk/F_0| < 1$. For values of $Xk/F_0 > 1$, the machine could easily drift into resonance and the amplitude of vibration actually becomes an amplification of the effective driving force amplitude.

Solution From equation (5.24) with $Xk/F_0 = 1$,

$$1 - \left(\frac{\omega}{\omega_a}\right)^2 = \left[1 + \mu\left(\frac{\omega_a}{\omega_p}\right)^2 - \left(\frac{\omega}{\omega_p}\right)^2\right]\left[1 - \left(\frac{\omega}{\omega_p}\right)^2\right] - \mu\left(\frac{\omega_a}{\omega_p}\right)^2$$

Solving this for ω/ω_a yields the two solutions

$$\frac{\omega}{\omega_a} = \pm\sqrt{1 + \mu}$$

For the system of Example 5.3.1, $\mu = 0.25$, so that the second solution becomes

$$\frac{\omega}{\omega_a} = 1.1180$$

The condition that $|Xk/F_0| = 1$ is also satisfied for $Xk/F_0 = -1$. Substitution of this into equation (5.24) followed by some manipulation yields

$$\left(\frac{\omega_a}{\omega_p}\right)^2\left(\frac{\omega}{\omega_a}\right)^4 = \left[2 + (\mu + 1)\left(\frac{\omega_a}{\omega_p}\right)^2\right]\left(\frac{\omega}{\omega_a}\right)^2 + 2 = 0$$

which is quadratic in $(\omega/\omega_a)^2$. Using the values of $\omega_a^2 = 6500/18.29$, $\omega_p^2 = 2600/73.16$, and $\mu = 0.25$, this simplifies to

$$10\left(\frac{\omega}{\omega_a}\right)^4 - 14.5\left(\frac{\omega}{\omega_a}\right)^2 + 2 = 0$$

Solving for ω/ω_a yields

$$\left(\frac{\omega}{\omega_a}\right)^2 = 0.1544,\ 1.2956 \quad \text{or} \quad \frac{\omega}{\omega_a} = 0.3929,\ 1.1382$$

Hence the three roots satisfying $|Xk/F_0| = 1$ are 0.3929, 1.1180, and 1.1382. Following the example of Figure 5.15 indicates that the driving frequency may vary between $0.3929\omega_a$ and $1.1180\omega_a$, or, since $\omega_a = 18.857$,

$$7.4089 < \omega < 21.0821 \ (\text{rad/s})$$

before the response of the primary mass is amplified or the system is in danger of experiencing resonance.

□

The preceding discussion and examples illustrate the concept of *performance robustness;* that is, the examples illustrate how the design holds up as the parameter values (k, k_a, etc.) drift from the values used in the original design. Example 5.3.2 illustrates that the mass ratio greatly affects the robustness of absorber designs. This is stated in the caption of Figure 5.16; up to a certain point, increasing μ increases the robustness of the absorber design. The effects of damping on absorber design are examined in the next section.

5.4 DAMPING IN VIBRATION ABSORPTION

As mentioned in Section 5.3, damping is often present in devices and has the potential for destroying the ability of a vibration absorber to protect the primary system fully by driving X to zero. In addition, damping is sometimes added to vibration absorbers to prevent resonance or to improve the effective bandwidth of operation of a vibration absorber. Also, a damper by itself is often used as a vibration absorber by dissipating the energy supplied by an applied force. Such devices are called *vibration dampers* rather than absorbers.

First consider the effect of modeling damping in the standard vibration absorber problem. A vibration absorber with damping in both the primary and absorber system is illustrated in Figure 5.18. This system is dynamically equal to the system of Figure 4.15 of Section 4.5. The equations of motion are given in matrix form by equation (4.116) as

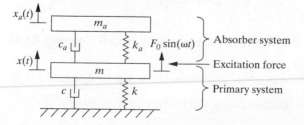

Figure 5.18   A schematic of a vibration absorber with damping in both the primary and absorber system.

$$\begin{bmatrix} m & 0 \\ 0 & m_a \end{bmatrix} \begin{bmatrix} \ddot{x}(t) \\ \ddot{x}_a(t) \end{bmatrix} + \begin{bmatrix} c + c_a & -c_a \\ -c_a & c_a \end{bmatrix} \begin{bmatrix} \dot{x}(t) \\ \dot{x}_a(t) \end{bmatrix}$$
$$+ \begin{bmatrix} k + k_a & -k_a \\ -k_a & k_a \end{bmatrix} \begin{bmatrix} x(t) \\ x_a(t) \end{bmatrix} = \begin{bmatrix} F_0 \\ 0 \end{bmatrix} \sin \omega t \qquad (5.27)$$

Note, as was mentioned in Section 4.5, that these equations cannot necessarily be solved by using the modal analysis technique of Chapter 4 because the equations do not decouple ($KM^{-1}C \ne CM^{-1}K$). The steady-state solution can be calculated, however, by using a combination of the exponential approach discussed in Section 2.3 and the matrix inverse used in previous sections for the undamped case.

To this end, let $F_0 \sin \omega t$ be represented in exponential form by $F_0 e^{j\omega t}$ in equation (5.27) and assume that the steady-state solution is of the form

$$\mathbf{x}(t) = \mathbf{X} e^{j\omega t} = \begin{bmatrix} X \\ X_a \end{bmatrix} e^{j\omega t} \qquad (5.28)$$

where X is the amplitude of vibration of the primary mass and X_a is the amplitude of vibration of the absorber mass. Substitution into equation (5.27) yields

$$\begin{bmatrix} (k + k_a - m\omega^2) + (c + c_a)\omega j & -k_a - c_a\omega j \\ -k_a - c_a\omega j & (k_a - m_a\omega^2) + c_a\omega j \end{bmatrix} \begin{bmatrix} X \\ X_a \end{bmatrix} e^{j\omega t} = \begin{bmatrix} F_0 \\ 0 \end{bmatrix} e^{j\omega t} \qquad (5.29)$$

Note that the coefficient matrix of the vector $\mathbf{X}$ has complex elements. Dividing equation (5.29) by the nonzero scalar $e^{j\omega t}$ yields a complex matrix equation in the amplitudes X and X_a. Calculating the matrix inverse using the formula of Example 4.1.4, reviewed in Window 5.3, and multiplying equation (5.29) by the inverse from the right yields

$$\begin{bmatrix} X \\ X_a \end{bmatrix} = \frac{\begin{bmatrix} (k_a - m_a\omega^2) + c_a\omega j & k_a + c_a\omega j \\ k_a + c_a\omega j & k + k_a - m\omega^2 + (c + c_a)\omega j \end{bmatrix} \begin{bmatrix} F_0 \\ 0 \end{bmatrix}}{\det(K - \omega^2 M + \omega j C)} \qquad (5.30)$$

Here the determinant in the denominator is given by (recall Example 4.1.4)

$$\det(K - \omega^2 M + \omega j C) = mm_a\omega^4 + (c_a c + m_a(k_a + k_a) + k_a m)\omega^2 + k_a k$$
$$+ [(kc_a + ck_a)\omega - (c_a(m + m_a) + cm_a)\omega^3]j \qquad (5.31)$$

and the system coefficient matrices M, C, and K are given by

$$M = \begin{bmatrix} m & 0 \\ 0 & m_a \end{bmatrix} \qquad C = \begin{bmatrix} c + c_a & -c_a \\ -c_a & c_a \end{bmatrix} \qquad K = \begin{bmatrix} k + k_a & -k_a \\ -k_a & k_a \end{bmatrix}$$

Simplifying the matrix vector product yields

$$X = \frac{\left[(k_a - m_a\omega^2) + c_a\omega j\right]F_0}{\det\left(K - \omega^2 M + \omega jC\right)} \tag{5.32}$$

$$X_a = \frac{(k_a + c_a\omega j)F_0}{\det\left(K - \omega^2 M + \omega jC\right)} \tag{5.33}$$

which expresses the magnitude of the response of the primary mass and absorber mass, respectively. Note that these values are now complex numbers and are multiplied by the complex value $e^{j\omega t}$ to get the time responses.

Equations (5.32) and (5.33) are the two-degree-of-freedom version of the frequency response function given for a single-degree-of-freedom system in equation (2.52). The complex nature of these values reflects a magnitude and phase. The magnitude is calculated following the rules of complex numbers and is best done with a symbolic computer code, or after substitution of numerical values for the various physical constants. It is important to note from equation (5.32) that unlike the tuned undamped absorber, the response of the primary system cannot be exactly zero even if the tuning condition is satisfied. Hence the presence of damping ruins the ability of the absorber system to exactly cancel the motion of the primary system.

Equations (5.32) and (5.33) can be analyzed for several specific cases. First, consider the case for which the internal damping of the primary system is neglected ($c = 0$). If the primary system is made of metal, the internal damping is likely to be very low and it is reasonable to neglect it in many circumstances. In this case, the determinant of equation (5.31) reduces to the complex number

$$\det(K - \omega^2 M + \omega Cj) \tag{5.34}$$

$$= \left[(-m\omega^2 + k)(-m_a\omega^2 + k_a) - m_a k_a \omega^2\right] + \left[(k - (m + m_a)\omega^2)c_a\omega\right]j$$

The maximum deflection of the primary mass is given by equation (5.32) with the determinant in the denominator evaluated as given in equation (5.34). This is the ratio of two complex numbers and hence is a complex number representing the phase and the amplitude of the response of the primary mass. Using complex arithmetic (see Window 5.4) the amplitude of the motion of the primary mass can be written as the real number

$$\frac{X^2}{F_0^2} = \frac{(k_a - m_a\omega^2)^2 + \omega^2 c_a^2}{\left[(k - m\omega^2)(k_a - m_a\omega^2) - m_a k_a \omega^2\right]^2 + \left[k - (m + m_a)\omega^2\right]^2 c_a^2 \omega^2} \tag{5.35}$$

Window 5.4
Reminder of Complex Arithmetic

The response magnitude given by equation (5.32) can be written as the ratio of two complex numbers:

$$\frac{X}{F_0} = \frac{A_1 + B_1 j}{A_2 + B_2 j}$$

where $A_1, A_2, B_1,$ and B_2 are real numbers and $j = \sqrt{-1}$. Multiplying this by the conjugate of the denominator divided by itself yields

$$\frac{X}{F_0} = \frac{(A_1 + B_1 j)(A_2 - B_2 j)}{(A_2 + B_2 j)(A_2 - B_2 j)} = \frac{(A_1 A_2 + B_1 B_2)}{A_2^2 + B_2^2} + \frac{B_1 A_2 - A_1 B_2}{A_2^2 + B_2^2} j$$

which indicates how X/F_0 is written as a single complex number of the form $X/F_0 = a + bj$. This is interpreted, as indicated, that the response magnitude has two components: one in phase with the applied force and one out of phase. The magnitude of X/F_0 is the length of the preceding complex number (i.e., $|X/F_0| = \sqrt{a^2 + b^2}$). This yields

$$\left| \frac{X}{F_0} \right| = \frac{A_1^2 + B_1^2}{A_2^2 + B_2^2}$$

which corresponds to the expression given in equation (5.35). (Also see Appendix A.)

It is instructive to examine this amplitude in terms of the dimensionless ratios introduced in Section 5.3 for the undamped vibration absorber. The amplitude X is written in terms of the static deflection $\Delta = F_0/k$ of the primary system. In addition, consider the mixed "damping ratio" defined by

$$\zeta = \frac{c_a}{2 m_a \omega_p} \tag{5.36}$$

where $\omega_p = \sqrt{k/m}$ is the original natural frequency of the primary system without the absorber attached. Using the standard frequency ratio $r = \omega/\omega_p$, the ratio of natural frequencies $\beta = \omega_a/\omega_p$ (where $\omega_a = \sqrt{k_a/m_a}$), and the mass ratio $\mu = m_a/m$, equation (5.35) can be rewritten as

$$\frac{X}{\Delta} = \frac{Xk}{F_0} = \sqrt{\frac{(2\zeta r)^2 + (r^2 - \beta^2)^2}{(2\zeta r)^2 (r^2 - 1 + \mu r^2)^2 + [\mu r^2 \beta^2 - (r^2 - 1)(r^2 - \beta^2)]^2}} \tag{5.37}$$

which expresses the dimensionless amplitude of the primary system. Note from examining equation (5.37) that the amplitude of the primary system response is determined by four physical parameter values:

μ the ratio of the absorber mass to the primary mass

β the ratio of the decoupled natural frequencies

r the ratio of the driving frequency to the primary natural frequency

ζ the ratio of the absorber damping and $2m_a\omega_p$

These four numbers can be considered as design variables and are chosen to give the smallest possible value of the primary mass's response, X, for a given application. Figure 5.19 illustrates how the damping value, as reflected in ζ, affects the response for a fixed value of $\mu = 0.25$ and $\beta = 1$, as r varies.

As mentioned at the beginning of this section, damping is often added to the absorber to improve the bandwidth of operation. This effect is illustrated in Figure 5.19. Recall that if there is no damping in the absorber ($\zeta = 0$), the magnitude of the response of the primary mass as a function of the frequency ratio r is as illustrated in Figure 5.15 (i.e., zero at $r = 1$ but infinite at $r = 0.781$ and $r = 1.281$). Thus the completely undamped absorber has poor bandwidth (i.e., if r changes by a small amount, the amplitude grows). In fact, as noted in Section 5.3, the bandwidth, or useful range of operation of that undamped absorber, is $0.908 \leq r \leq 1.118$. For these values of r, $|Xk/F_0| \leq 1$. However, if damping is added to the absorber ($\zeta \neq 0$), Figure 5.19 results, and the bandwidth, or useful range of operation, is extended. The price for this increased operating region is that $|Xk/F_0|$ is never zero in the damped case (see Figure 5.19).

Examination of Figure 5.19 shows that as ζ is varied, the amplification of $|Xk/F_0|$ over the range of r can be reduced. The design question now becomes: For

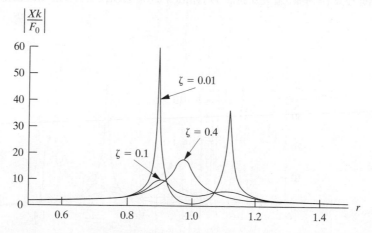

Figure 5.19   The normalized amplitude of vibration of the primary mass as a function of the frequency ratio for several values of the damping in the absorber system for the case of negligible damping in the primary system [i.e., a plot of equation (5.37)].

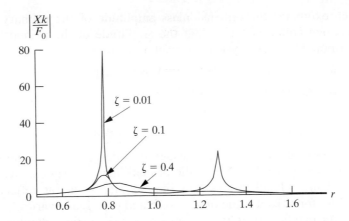

Figure 5.20   Repeat of the plot of Figure 5.19 with $\mu = 0.25$ and $\beta = 0.9$ for several values of ζ. Note that in this case, $\zeta = 0.4$ yields a lower magnitude than does $\zeta = 0.1$.

what values of the mass ratio μ, the absorber damping ratio ζ, and the frequency ratio β is the magnitude $|Xk/F_0|$ smallest over the region $0 \le r \le 2$? Just increasing the damping with μ and β fixed does not necessarily yield the lowest amplitude. Note from Figure 5.19 that $\zeta = 0.1$ produces a smaller amplification over a larger region of r than does the higher ratio, $\zeta = 0.4$. Figures 5.20 and 5.21 yield some hint of how the various parameters affect the magnitude by providing plots of $|Xk/F_0|$ for various combinations of ζ, μ, and β.

A solution of the best choice of μ and ζ is discussed again in Section 5.5. Note from Figure 5.21 that $\mu = 0.25$, $\beta = 0.8$, and $\zeta = 0.27$ yield a minimum value of $|Xk/F_0|$ over a large range of values of r. However, amplification of the response

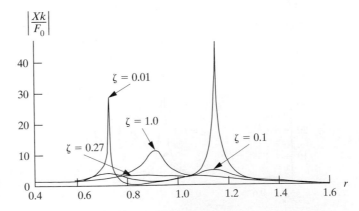

Figure 5.21   Repeat of the plots of Figure 5.19 with $\mu = 0.25$, $\beta = 0.8$ for several values of ζ. In this case, $\zeta = 0.27$ yields the lowest amplification over the largest bandwidth.

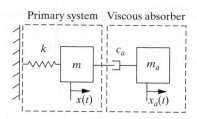

Figure 5.22   Damper–mass system added to a primary mass (with no damping) to form a viscous vibration absorber.

X still occurs (i.e., $|Xk/F_0| > 1$ for values of $r < \sqrt{2}$), but no order-of-magnitude increase in $|X|$ occurs as in the case of the undamped absorber.

Next consider the case of an appended absorber mass connected to an undamped primary mass only by a dashpot, an arrangement illustrated in Figure 5.22. Systems of this form arise in the design of vibration reduction devices for rotating systems such as engines, where the operating speed (and hence the driving frequency) varies over a wide range. In such cases a viscous damper is added to the end of the crankshaft (or other rotating device) as indicated in Figure 5.23. The shaft spins through an angle θ_1 with torsional stiffness k and inertia J_1. The damping inertia J_2 spins through an angle θ_2 in a viscous film providing a damping force $c_a(\theta_1 - \theta_2)$. If an external harmonic torque is applied of the form $M_0 e^{\omega t j}$, the equation of motion of this system becomes

$$\begin{bmatrix} J_1 & 0 \\ 0 & J_2 \end{bmatrix}\begin{bmatrix} \ddot{\theta}_1 \\ \ddot{\theta}_2 \end{bmatrix} + \begin{bmatrix} c_a & -c_a \\ -c_a & c_a \end{bmatrix}\begin{bmatrix} \dot{\theta}_1 \\ \dot{\theta}_2 \end{bmatrix} + \begin{bmatrix} k & 0 \\ 0 & 0 \end{bmatrix}\begin{bmatrix} \theta_1 \\ \theta_2 \end{bmatrix} = \begin{bmatrix} M_0 \\ 0 \end{bmatrix} e^{\omega t j} \qquad (5.38)$$

This is a rotational equivalent to the translational model given in Figure 5.22. It is easy to calculate the undamped natural frequencies of this two-degree-of-freedom system. They are

$$\omega_p = \sqrt{\frac{k}{J_1}} \text{ and } \omega_a = 0$$

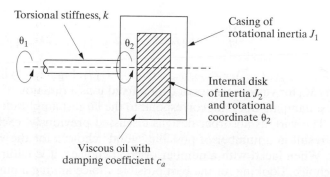

Figure 5.23   A viscous damper and mass added to a rotating shaft for broadband vibration absorption. Often called a *Houdaille damper*.

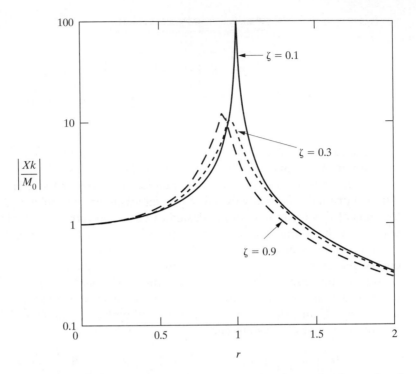

Figure 5.24   The amplitude curves for a system with a viscous absorber, a plot of equation (5.39), for the case $\mu = 0.25$ and for three different values of ζ.

The solution of this set of equations is given by equations (5.32) and (5.33) with m and m_a replaced by J_1 and J_2, respectively, $c = 0$, $k_a = 0$, and F_0 replaced by M_0. Equation (5.32) is given in nondimensional form as equation (5.37). Hence letting $\beta = \omega_a/\omega_p = 0$ in equation (5.37) yields that amplitude of vibration of the primary inertia J_1 [i.e., the amplitude of $\theta_1(t)$] is described by

$$\frac{Xk}{M_0} = \sqrt{\frac{4\zeta^2 + r^2}{4\zeta^2(r^2 + \mu r^2 - 1)^2 + (r^2 - 1)^2 r^2}} \qquad (5.39)$$

where $\zeta = c/(2J_2\omega_p)$, $r = \omega/\omega_p$, and $\mu = J_2/J_1$. Figure 5.24 illustrates several plots of Xk/M_0 for various values of ζ for a fixed μ as a function of r. Note again that the highest damping does not correspond to the largest amplitude reduction.

The various absorber designs discussed previously, excluding the undamped case, result in a number of possible "good" choices for the various design parameters. When faced with a number of good choices, it is natural to ask which is the best choice. Looking for the best possible choice among a number of acceptable or good choices can be made systematic by using methods of optimization introduced in the next section.

5.5 OPTIMIZATION

In the design of vibration systems, the best selection of system parameters is often sought. In the case of the undamped vibration absorber of Section 5.3 the best selection for values of mass and stiffness of the absorber system is obvious from examining the expression for the amplitude of vibration of the primary system. In this case, the amplitude could be driven to zero by tuning the absorber mass and stiffness to the driving frequency. In the other cases, especially when damping is included, the choice of parameters to produce the best response is not obvious. In such cases, optimization methods can often be used to help select the best performance. Optimization techniques often produce results that are not obvious. An example is in the case of the undamped primary system or the damped absorber system discussed in the preceding section. In this case Figures 5.19 to 5.21 indicate that the best selection of parameters does not correspond to the highest value of the damping in the system, as intuition might dictate. These figures essentially represent an optimization by trial and error. In this section a more systematic approach to optimization is suggested by taking advantage of calculus.

Recall from elementary calculus that minimums and maximums of particular functions can be obtained by examining certain derivatives. Namely, if the first derivative vanishes and the second derivative of the function is positive, the function has obtained a minimum value. This section presents a few examples where optimization procedures are used to obtain the best possible vibration reduction for various isolator and absorber systems. A major task of optimization is first deciding what quantity should be minimized to best describe the problem under study. The next question of interest is to decide which variables to allow to vary during the optimization. Optimization methods have developed over the years that allow the parameters during the optimization to satisfy constraints, for example. This approach is often used in design for vibration suppression.

Recall from calculus that a function $f(x)$ experiences a maximum (or minimum) at value of $x = x_m$ given by the solution of

$$f'(x_m) = \frac{d}{dx}\big[f(x_m)\big] = 0 \tag{5.40}$$

If this value of x causes the second derivative, $f''(x_m)$, to be less than zero, the value of $f(x)$ at $x = x_m$ is the maximum value that $f(x)$ takes on in the region near $x = x_m$. Similarly, if $f''(x_m)$ is greater than zero, the value of $f(x_m)$ is the smallest or minimum value that $f(x)$ obtains in the interval near x_m. Note that if $f''(x) = 0$, at $x = x_m$, the value $f(x_m)$ is neither a minimum or maximum for $f(x)$. The points where $f'(x)$ vanish are called *critical points*.

These simple rules were used in Section 2.2, Example 2.2.5, for computing the value (r_{peak}) where the maximum value of normalized magnitude of the steady-state response of a harmonically driven single-degree-of-freedom system occurs. The second derivative test was not checked because several plots of the function clearly indicated that the curve contains a global maximum value rather than a minimum. In both absorber and isolator design, plots of the magnitude of the response can be used to avoid having to calculate the second derivative (second derivatives are often unpleasant to calculate).

If the function f to be minimized (or maximized) is a function of two variables [i.e., $f = f(x, y)$], the preceding derivative tests become slightly more complicated and involve examining the various partial derivatives of the function $f(x, y)$. In this case, the critical points are determined from the equations

$$f_x(x, y) = \frac{\partial f(x, y)}{\partial x} = 0$$

$$f_y(x, y) = \frac{\partial f(x, y)}{\partial y} = 0 \tag{5.41}$$

Whether or not these critical points (x, y) are a maximum of the value $f(x, y)$ or a minimum depends on the following:

1. If $f_{xx}(x, y) > 0$ and $f_{xx}(x, y)f_{yy}(x, y) > f_{xy}^2(x, y)$, then $f(x, y)$ has a relative minimum value at x, y.
2. If $f_{xx}(x, y) < 0$ and $f_{xx}(x, y)f_{yy}(x, y) > f_{xy}^2(x, y)$, then $f(x, y)$ has a relative maximum value at x, y.
3. If $f_{xy}^2(x, y) > f_{xx}(x, y)f_{yy}(x, y)$, then $f(x, y)$ is neither a maximum nor a minimum value; the point x, y is a *saddle* point.
4. If $f_{xy}^2(x, y) = f_{xx}(x, y)f_{yy}(x, y)$, the test fails and the point x, y could be any or none of the above.

Plots of $f(x, y)$ can also be used to determine whether or not a given critical point is a maximum, minimum, saddle point, or none of these. These rules can be used to help solve vibration design problems in some circumstances. As an example of using these optimization formulations for designing a vibration suppression system, recall the damped absorber system of Section 5.4. In this case, the magnitude of the primary mass-normalized displacement with respect to the input force (moment) magnitude is given in equation (5.39) to be

$$\frac{Xk}{M_0} = \sqrt{\frac{4\zeta^2 + r^2}{4\zeta^2(r^2 + \mu r^2 - 1)^2 + (r^2 - 1)^2 r^2}} = f(r, \zeta) \tag{5.42}$$

which is considered to be a function of the mixed damping ratio ζ and the frequency ratio r for a fixed mass ratio μ.

In Section 5.4, values of $f(r)$ are plotted versus r for several values of ζ in an attempt to find the value of ζ that yields the smallest maximum value of $f(r, \zeta)$. This is illustrated in Figure 5.24. Figure 5.25 illustrates the magnitude as a function of both ζ and r. From the figure it can be concluded that the derivative $\partial f/\partial r = 0$ yields the maximum value of the magnitude for each fixed ζ.

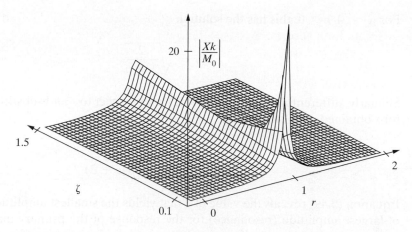

Figure 5.25   A plot of the normalized magnitude of the primary system versus both ζ and r [i.e., a two-dimensional plot of equation (5.42) for $\mu = 0.25$]. This illustrates that the most desirable response is obtained at the saddle point.

Looking along the ζ axis, the partial derivative $\partial f/\partial \zeta = 0$ yields the minimum value of $f(r, \zeta)$ for each fixed value of r. The best design, corresponding to the smallest of the largest amplitudes, is thus illustrated in Figure 5.25. This point corresponds to a saddle point and can be calculated by evaluating the appropriate first partial derivatives.

First consider $\partial(Xk/M_0)/\partial \zeta$. From equation (5.42), the function to be differentiated is of the form

$$f = \frac{A^{1/2}}{B^{1/2}} \tag{5.43}$$

where $A = 4\zeta^2 + r^2$ and $B = 4\zeta^2(r^2 + \mu r^2 - 1)^2 + (r^2 - 1)^2 r^2$. Differentiating and equating the resulting derivatives to zero yields

$$\frac{\partial f}{\partial \zeta} = \frac{1}{2}\frac{A^{-1/2}dA}{B^{1/2}} - \frac{1}{2}A^{1/2}\frac{dB}{B^{3/2}} = 0 \tag{5.44}$$

Solving this yields the form $[B\,dA - A\,dB]/2B^{3/2} = 0$ or

$$B dA = A dB \tag{5.45}$$

where A and B are as defined previously and

$$dA = 8\zeta \qquad \text{and} \qquad dB = 8\zeta\left(r^2 + \mu r^2 - 1\right)^2 \tag{5.46}$$

Substitution of these values of A, dA, B, and dB into equation (5.45) yields

$$\left(1 - r^2\right)^2 = \left(1 - r^2 - \mu r^2\right)^2 \tag{5.47}$$

For $\mu \neq 0, r > 0$, this has the solution

$$r = \sqrt{\frac{2}{2 + \mu}} \tag{5.48}$$

Similarly, differentiating equation (5.42) with respect to r and substituting the value for r obtained previously yields

$$\zeta_{op} = \frac{1}{\sqrt{2(\mu + 1)(\mu + 2)}} \tag{5.49}$$

Equation (5.49) reveals the value of ζ that yields the smallest amplitude at the point of largest amplitude (resonance) for the response of the primary mass. The maximum value of the displacement for the optimal damping is given by

$$\left(\frac{Xk}{M_0}\right)_{max} = 1 + \frac{2}{\mu} \tag{5.50}$$

which is obtained by substitution of equations (5.48) and (5.49) into equation (5.42). This last expression suggests that μ should be as large as possible. However, the practical consideration that the absorber mass should be smaller than the primary mass requires $\mu \leq 1$. The value $\mu = 0.25$ is fairly common.

The second derivative conditions for the function f to have a saddle point (condition 3 in the preceding list) are too cumbersome to calculate. However, the plot of Figure 5.25 clearly illustrates that these conditions are satisfied. Furthermore, the plot indicates that f as a function of ζ is convex and f as a function of r is concave so that the saddle point condition is also the solution of minimizing the maximum value $f(r, \zeta)$, called the *min–max problem* in applied mathematics and optimization.

Example 5.5.1

A viscous damper–mass absorber is added to the shaft of an engine. The mass moment of inertia of the shaft system is 1.5 kg·m²/rad and has a torsional stiffness of 6 × 10³ N·m/rad. The nominal running speed of the engine is 2000 rpm. Calculate the values of the added damper and mass moment of inertia such that the primary system has a magnification (Xk/M_0) of less than 5 for all speeds and is as small as possible at the running speed.

Solution   Since $\omega_p = \sqrt{k/J}$, the natural frequency of the engine system is

$$\omega_p = \sqrt{\frac{6.0 \times 10^3 \text{ N·m/rad}}{1.5 \text{ kg·m}^2/\text{rad}}} = 63.24 \text{ rad/s}$$

The running speed of the engine is 2000 rpm or 209.4 rad/s, which is assumed to be the driving frequency (actually, it is a function of the number of cylinders). Hence the frequency ratio is

$$r = \frac{\omega}{\omega_p} = \frac{209.4}{63.24} = 3.31$$

so that the running speed is well away from the maximum amplification as illustrated in Figures 5.24 and 5.25 and the absorber is not needed to protect the shaft at its running speed. However, the engine spends some time getting to the running speed and often runs at lower speeds. The peak response occurs at

$$r_{peak} = \frac{\omega}{\omega_p} = \sqrt{\frac{2}{2 + \mu}}$$

as given by equation (5.48), and has a value of

$$\left(\frac{Xk}{M_0}\right)_{max} = 1 + \frac{2}{\mu}$$

as given by equation (5.50). The magnification is restricted to be 5, so that

$$1 + \frac{2}{\mu} \le 5 \quad \text{or} \quad \mu \ge 0.5$$

Thus $\mu = 0.5$ is chosen for the design. Since the mass of the primary system is $J_1 = 1.5$ kg · m²/rad and $\mu = J_2/J_1$, the mass of the absorber is

$$J_2 = \mu J_1 = \frac{1}{2}(1.5) \text{ kg} \cdot \text{m}^2/\text{rad} = 0.75 \text{ kg} \cdot \text{m}^2 \cdot \text{rad}$$

The damping value required for equation (5.50) to hold is given by equation (5.49) or

$$\zeta_{op} = \frac{1}{\sqrt{2(\mu + 1)(\mu + 2)}} = \frac{1}{\sqrt{2(1.5)(2.5)}} = 0.3651$$

Recall from Section 5.4 [just following equation (5.39)], that $\zeta = c/(2J_2\omega_p)$, so that the optimal damping constant becomes

$$c_{op} = 2\zeta_{op}J_2\omega_p = 2(0.3651)(0.75)(63.24) = 34.638 \text{ N} \cdot \text{m} \cdot \text{s/rad}$$

The two values of J_2 and c given here form an optimal solution to the problem of designing a viscous damper–mass absorber system so that the maximum deflection of the primary shaft is satisfied $|Xk/M_0| < 5$. This solution is optimal in terms of a choice of ζ, which corresponds to the saddle point of Figure 5.25 and yields a minimum value of all maximum amplifications.

□

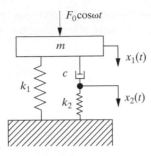

Figure 5.26 Model of a machine mounted on an elastic foundation through an elastic damper to provide vibration isolation.

Optimization methods can also be useful in the design of certain types of vibration isolation systems. For example, consider the model of a machine mounted on an elastic damper and spring system as illustrated in Figure 5.26. The equations of motion of the system of Figure 5.26 are

$$m\ddot{x}_1 + c(\dot{x}_1 - \dot{x}_2) + k_1 x_1 = F_0 \cos \omega t$$

$$c(\dot{x}_1 - \dot{x}_2) = k_2 x_2 \tag{5.51}$$

Because no mass term appears in the second equation, the system given by equation (5.51) is of third order. Equation (5.51) can be solved by assuming periodic motions of the form

$$x_1(t) = X_1 e^{j\omega t} \quad \text{and} \quad x_2(t) = X_2 e^{j\omega t} \tag{5.52}$$

and considering the exponential representation of the harmonic driving force. Substitution of equation (5.52) into (5.51) yields

$$(k_1 - m\omega^2 + jc\omega)X_1 - jc\omega X_2 = F_0$$

$$jc\omega X_1 - (k_2 + jc\omega)X_2 = 0 \tag{5.53}$$

Solving for the amplitudes X_1 and X_2 yields

$$X_1 = \frac{F_0(k_2 + jc\omega)}{k_2\left(k_1 + m\omega^2\right) + jc\omega\left(k_1 + k_2 - m\omega^2\right)} \tag{5.54}$$

and

$$X_2 = \frac{c\omega F_0 j}{k_2\left(k_1 + m\omega^2\right) + c\omega\left(k_1 + k_2 - m\omega^2\right)j} \tag{5.55}$$

These two amplitude expressions can be simplified further by substituting the non-dimensional quantities $r = \omega/\sqrt{k_1/m}$, $\gamma = k_1/k_2$, and $\zeta = c/\left(2\sqrt{k_1 m}\right)$. The force transmitted to the base is the vector sum of the two forces $k_1 x_1$ and $k_2 x_2$. Using complex arithmetic and a vector sum (recall Sections 2.3 and 2.4) the force transmitted can be written as

$$\text{T.R.} = \frac{F_T}{F_0} = \frac{\sqrt{1 + 4(1 + \gamma)^2 \zeta^2 r^2}}{\sqrt{(1 - r^2) + 4\zeta^2 r^2 (1 + \gamma - r^2 \gamma)^2}} \tag{5.56}$$

which describes the transmissibility ratio for the system of Figure 5.26.

The force transmissibility ratio can be optimized by viewing the ratio F_T/F_0 as a function of r and ζ. Figure 5.27 yields a plot of F_T/F_0 versus r for $\gamma = 0.333$ and for several values of ζ. This illustrates that the value of the damping ratio greatly affects the transmissibility at resonance. A three-dimensional plot of F_T/F_0 versus r and ζ is given in Figure 5.28, which illustrates that the saddle point value of ζ and r yields the best design for the minimum transmissibility of the maximum force transmitted.

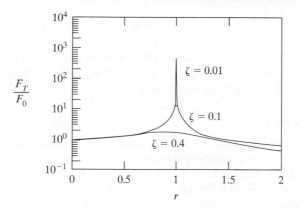

Figure 5.27 Plot of equation (5.56) illustrating the effect of damping on the magnification of force transmitted to ground.

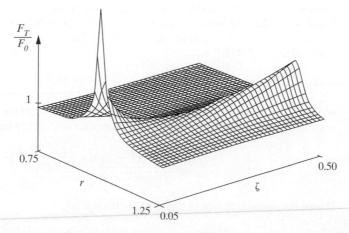

Figure 5.28 Plot of equation (5.56) illustrating F_T/F_0 versus ζ versus r. The plot shows the point where damping minimizes the maximum transmissibility.

The saddle point illustrated in Figure 5.28 can be found from the derivative of T.R. as given in equation (5.56). These partial derivatives are

$$\frac{\partial(\text{T.R.})}{\partial \zeta} = 0 \quad \text{yields} \quad r_{\max} = \frac{\sqrt{2(1 + \gamma)}}{\sqrt{1 + 2\gamma}} \tag{5.57}$$

and

$$\frac{\partial(\text{T.R.})}{\partial r} = 0 \quad \text{yields} \quad \zeta_{\text{op}} = \frac{\sqrt{2(1 + 2\gamma)/\gamma}}{4(1 + \gamma)} \tag{5.58}$$

These values of r correspond to an optimal design of this type of isolation device. At the saddle point, the value of T.R. becomes

$$(\text{T.R.})_{\max} = 1 + 2\gamma \tag{5.59}$$

which results from substitution of equations (5.57) and (5.58) into equation (5.56). This illustrates that as long as $\gamma < 1$, T.R. < 3 and the isolation system will not cause much difficulty at resonance.

Example 5.5.2

An isolation system is to be designed for a machine modeled by the system of Figure 5.26 (i.e., an elasticity coupled viscous damper). The mass of the machine is $m = 100$ kg and the stiffness $k_1 = 400$ N/m. The driving frequency is 10 rad/s at nominal operating conditions. Design this system (i.e., choose k_2 and c) such that the maximum transmissibility ratio at any speed is 2 (i.e., design the system for "startup" or "run through"). What is the T.R. at the normal operating condition of a driving frequency of 10 rad/s?

Solution   For $m = 100$ kg   and $k_1 = 400$ N/m, $\omega_n = \sqrt{400/100} = 2$ rad/s, 2 rad/s, so that the normal operating condition is well away from resonance (i.e., $r = \omega/\omega_n = 10/2 = 5$ at running conditions). Equation (5.59) yields that the maximum value for T.R. is

$$(\text{T.R.})_{\max} = 1 + 2\gamma \leq 2$$

so that $\gamma = 0.5$ and $k_2 = (0.5)(k_1) = (0.5)(400 \text{ N/m}) = 200$ N/m. With $\gamma = 0.5$, the optimal choice of damping ratio is given by equation (5.58) to be

$$\zeta_{\text{op}} = \frac{\sqrt{2(1 + 2\gamma)/\gamma}}{4(1 + \gamma)} = 0.4714$$

Hence the optimal choice of damping coefficient is

$$c_{\text{op}} = 2\zeta_{\text{op}}\omega_n m = 2(0.4714)(2)(100) = 188.56 \text{ kg/s}$$

The T.R. value at nominal operating frequency of $\omega = 10$ rad/s is given by equation (5.56) to be ($r = 10/2 = 5$)

$$\text{T.R.} = \frac{\sqrt{1 + 4(1 + 0.5)^2(0.4714)^2(5)^2}}{\sqrt{(1 - 5^2)^2 + 4(0.4714)^2(5)^2[1 + 0.5 - 5^2(0.5)]^2}} = 0.12$$

Hence the design $k_2 = 200$ N/m and $c = 188.56$ kg/s will protect the surroundings by a T.R. of 0.12 (i.e., only 12% of the applied force is transmitted to ground) and limits the force transmitted near resonance to a factor of 2.

$\square$

5.6 VISCOELASTIC DAMPING TREATMENTS

A common and very effective way to reduce transient and steady-state vibration is to increase the amount of damping in the system so there is greater energy dissipation. This is especially useful in aerospace structures applications, where the added mass of an absorber system may not be practical. While a rigorous derivation of the equations of vibration for structures with damping treatments is beyond the scope of this book, formulas are presented that provide a sample of design calculations for using damping treatments.

A damping treatment consists of adding a layer of viscoelastic material, such as rubber, to an existing structure. The combined system often has a higher damping level and thus reduces unwanted vibration. This is standard in the auto industry to reduce vibration-induced noise in the car's interior and can be found under the flooring carpet. This procedure is described by using the *complex stiffness* notation. The concept of complex stiffness results from considering the harmonic response of a damped system of the form

$$m\ddot{x} + c\dot{x} + kx = F_0 e^{j\omega t} \tag{5.60}$$

Recall from Section 2.3 that the solution to equation (5.60) can be calculated by assuming the form of the solution to be $x(t) = Xe^{j\omega t}$, where X is a constant and $j = \sqrt{-1}$. Substitution of the assumed form into equation (5.60) and dividing by the nonzero function $e^{j\omega t}$ yields

$$[-m\omega^2 + (k + j\omega c)]X = F_0 \tag{5.61}$$

This can be written as

$$\left[-m\omega^2 + k\left(1 + \frac{\omega c}{k}j\right)\right]X = F_0 \tag{5.62}$$

or

$$[-m\omega^2 + k^*]X = F_0 \tag{5.63}$$

where $k^* = k(1 + \overline{\eta}j)$. Here $\overline{\eta} = \omega c/k$ is called the *loss factor* and k^* is called the *complex stiffness*. This illustrates that in steady state, the viscous damping in a system can be represented as an "undamped" system with a complex-valued stiffness. The imaginary part of the stiffness, $\overline{\eta}$, corresponds to the energy dissipation in the system. Since the loss factor has the form

$$\overline{\eta} = \frac{c}{k}\omega \tag{5.64}$$

the loss factor depends on the driving frequency and hence is said to be frequency dependent. Thus the value of the energy dissipation term depends on the value of the driving frequency of the external force exciting the structure.

The concept of complex stiffness just developed is called the *Kelvin–Voigt model* of a material. This corresponds to the standard spring–dashpot configuration as sketched in Figure 5.29 and used extensively in the first four chapters. The difference between the Kelvin–Voigt model used here and the viscous-damping model of the previous chapters is that the Kelvin–Voigt model used here is valid only in steady-state harmonic motion. The complex stiffness and the corresponding frequency-dependent loss factor, $\overline{\eta} = \omega c/k$, model the energy dissipation at steady state during harmonic excitation of frequency ω only. The viscous dashpot representation introduced in Section 1.3 models energy dissipation in free decay as well as in other transient and forced-response excitation. However, the Kelvin–Voigt representation is a more accurate, though limited, model of the internal damping in materials.

The complex stiffness formulation can also be derived from the stress–strain relationship for a linear viscoelastic material. Such materials are called *viscoelastic* because they exhibit both *elastic* behavior and *viscous* behavior, as captured in the Kelvin–Voigt model described in Figure 5.29. Other viscoelastic models exist in addition to this one, but such models are beyond the scope of this book [see Snowden (1968)]. An alternative viscoelastic model is given in Figure 5.26, for instance.

The stress–strain relationship for viscoelastic material can be summarized by extending the modulus of a material, denoted by E, to a complex modulus, denoted E^*, by the relation

$$E^* = E(1 + \eta j) \tag{5.65}$$

where $j = \sqrt{-1}$ as before and η is the loss factor of the viscoelastic material. The complex modulus of a material, as defined in equation (5.65), can be measured,

Material with viscoelastic properties

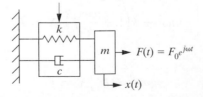

$$F(t) = F_0 e^{j\omega t}$$

$x(t)$

Figure 5.29 Kelvin–Voigt damping model gives rise to the complex stiffness concept of representing damping in steady-state vibration.

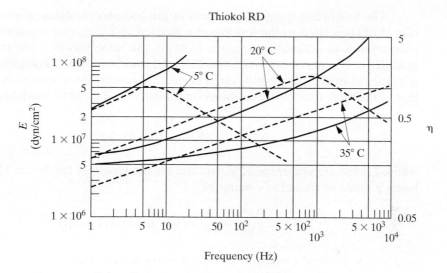

Figure 5.30 A sample plot of elastic modulus (solid lines) and loss factor (dashed lines) versus frequency for several fixed temperatures.

and in general is both frequency and temperature dependent over a broad range of values. Some sample values for frequency dependence are given in Figure 5.30 for fixed temperatures.

Materials that exhibit viscoelastic behavior are rubber and rubber-like substances (e.g., butyl rubber, neoprene, polyurethane) as well as plexiglass, vinyl, and nylon. A common use of these viscoelastic materials in design is as an additive damping treatment to increase the combined structure's damping or as an isolator. Layers of viscoelastic material are often added to structures composed of lightly damped material such as aluminum or steel to form a new structure that has sufficient stiffness for static loading and sufficient damping for controlling vibration. Table 5.2 lists some values of E and η for a viscoelastic material at two different temperatures and several frequencies.

TABLE 5.2 SOME COMPLEX MODULUS DATA (I.E., E AND η) FOR PARACRIL-BJ WITH 50 PHRC[a]

E (psi)	η	T (°F)	ω (Hz)	ω (rad/s)	E (N/m^2)
3×10^3	0.21	75	10	62.8	2.068×10^7
4×10^3	0.28	75	100	628.3	2.758×10^7
7×10^3	0.55	75	1000	6283.2	4.826×10^7
4×10^3	0.25	50	10	62.8	2.758×10^7
6×10^3	0.5	50	100	628.3	4.137×10^7
13×10^3	1	50	1000	6283.2	8.963×10^7

[a]Nitrile rubber elastomeric material made by U.S. Rubber Company.

Source: Nashif, Jones, and Henderson, 1985, Data Sheet 27.

The loss factor η defined in terms of the complex modulus as given in equation (5.65) is related to the loss factor η defined by examining the notion of complex stiffness as defined in equation (5.64) in the same way that the stiffness and modulus of a material are related in Table 1.1 and Section 1.5. For example, if the specimen of interest is a cantilevered beam, the stiffness associated with the deflection of the tip in the transverse direction is related to the elastic modulus by

$$k = \frac{3EI}{l^3} \tag{5.66}$$

where I is the area moment of inertia and l is the length of the beam. Hence if the beam is made of viscoelastic material,

$$k^* = \frac{3E^*I}{l^3} = \frac{3I}{l^3}E(1 + \eta j) = k(1 + \bar{\eta}j)$$

so that $\eta = \bar{\eta}$ and the two notions of loss factor are identical.

The notion of loss factor η is related to the definition of a damping ratio ζ only at resonance (i.e., $\omega = \omega_n = \sqrt{k/m}$). When the driving frequency is the same as the system's natural frequency, $\eta = 2\zeta$. This simple relationship is often used to describe the free decay of a viscoelastic material (an approximation). The design of structures for reduced vibration magnitude often consists of adding a viscoelastic damping treatment to an existing structure. Many structures are made of metals and alloys that have relatively little internal damping. A viscoelastic damping material (such as rubber) is often added as a layer to the outside surface of a structure (called *free-layer damping* treatment or *unconstrained-layer damping*). A much more effective approach is to cover the free layer with another layer of metal to form a *constrained-layer damping* treatment. In the constrained-layer damping treatment, the damping layer is covered with a (usually thin) layer of metal (stiff) to produce shear deformation in the viscoelastic layer. The constrained-layer approach produces higher loss factors and generally costs more. These damping treatments are manufactured as sheets, tapes, and adhesives for ease of application.

A free-layer damping treatment for a pinned–pinned beam (see Table 6.4) in transverse or bending vibration is illustrated in Figure 5.31. Material 1, the bottom layer, is usually a metal providing the appropriate stiffness. The second layer, denoted as having modulus E_2 and thickness H_2, is the damping treatment. Using

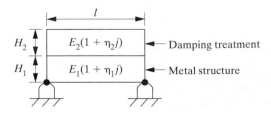

Figure 5.31 A simple supported beam with an unconstrained damping treatment illustrating the geometry and physical parameters.

the notation of Figure 5.31, the combined stiffness EI is related to the original stiffness E_1I_1 by

$$\frac{EI}{E_1I_1} = 1 + e_2h_2^3 + 3(1 + h_2)^2 \frac{e_2h_2}{1 + e_2h_2} \tag{5.67}$$

where $e_2 = E_2/E_1$ and $h_2 = H_2/H_1$ are dimensionless. Note that since all the quantities on the left side of equation (5.67) are positive, the damping treatment increases the stiffness of the system a small amount ($h_2 < 1$). In addition, the combined system's loss factor, η, is given by [assuming that $(e_2h_2)^2 << e_2h_2$]

$$\eta = \frac{e_2h_2\left(3 + 6h_2 + 4h_2^2 + 2e_2h_2^3 + e_2^2h_2^4\right)}{\left(1 + e_2h_2\right)1 + 4e_2h_2 + 6e_2h_2^2 + 4e_2h_2^3 + e_2^2h_2^4} \eta_2 \tag{5.68}$$

Equation (5.68) yields a formula that can be used in the design of add-on damping treatments, as illustrated in the following example.

Example 5.6.1

An electric motor that drives a cooling fan is mounted on an aluminum shelf (1 cm thick) in a cabinet holding electronic parts (perhaps a mainframe computer) as illustrated in Figure 5.32. The vibration of the motor causes the mounting platform, and hence the surrounding cabinet, to shake. The motor rotates at an effective frequency of 100 Hz. The temperature in the cabinet remains at 75°F. A damping treatment is added to reduce the vibration of the shelf.

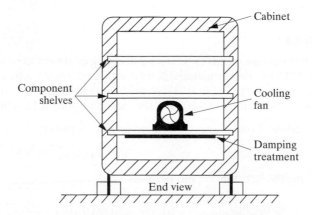

Figure 5.32   Electronic cabinet with cooling fan illustrating the use of a damping treatment.

Solution   The shelf is modeled as a simply supported beam so that equation (5.68) can be used to design the damping treatment. If nitrile rubber is used as the damping treatment, calculate the loss factor of the combined system at 75°F if $H_2 = 1$ cm. Referring to Table 5.2, the modulus of the rubber at 100 Hz and 75°F is

$$E_2 = 2.758 \times 10^7 \, \text{Pa}$$

The modulus of aluminum is $E_1 = 7.1 \times 10^{10}$ Pa, so that

$$e_2 = \frac{2.758 \times 10^7}{7.1 \times 10^{10}} = 0.00039 = 3.885 \times 10^{-4}$$

The thickness of both the shelf and the damping treatment are taken to be the same, so that $h_2 = 1$. From equation (5.68) the combined loss factor becomes

$$\eta = \frac{(0.00039)\left[3 + 6 + 4 + 2(0.00039) + (0.00039)^2\right]}{(1.00039)\left[1 + 4(0.00039) + 6(0.00039) + 4(0.00039) + (0.00039)^2\right]}\eta_2$$

$$= 5.021 \times 10^{-3}\eta_2$$

From Table 5.2, $\eta_2 = 0.28$ at 100 Hz and 75°C, so that

$$\eta = 0.00141$$

which is about 50% higher than the loss factor given by pure aluminum.

     The formula given in equation (5.68) is a bit cumbersome for design work. Often it is approximated by

$$\eta = 14\left(e_2 h_2^2\right)\eta_2 \tag{5.69}$$

which is reasonable for many situations. The values of e_2 and η_2 are fixed by the choice of materials and the operating temperature. Once these parameters are fixed, the parameter $h_2 = H_2/H_1$ is the only remaining design choice. Since H_1 is usually determined by stiffness considerations, the remaining design choice is the thickness of the damping layer, H_2.

□

Example 5.6.2

An aluminum shelf is to be given a damping treatment to raise the system loss factor to $\eta = 0.03$. A rubber material is used with modulus at room temperature of 1% of that of aluminum (i.e., $e_2 = 0.01$). What should the thickness of the damping material be if its loss factor is $\eta_2 = 0.261$ and the aluminum shelf is 1 cm thick?

Solution   From the approximation given by equation (5.69),

$$\eta = 14\eta_2 e_2 h_2^2$$

Using the values given, this becomes

$$0.03 = 14(0.261)(0.01)\frac{H_2^2}{(1 \text{ cm})^2}$$

Solving this for H_2^2 yields

$$H_2^2 = \frac{0.03}{14(0.01)(0.261)}(\text{cm})^2 = 0.82 \text{ cm}^2$$

so that $H_2 = 0.91$ cm will provide the desired loss factor.

□

5.7 CRITICAL SPEEDS OF ROTATING DISKS

Of primary concern in the design of rotating machinery is the vibration phenomenon of *critical speeds*. This phenomenon occurs when a rotating shaft with a disk, such as a jet engine turbine blade rotating about its shaft mounted between two bearings, rotates at a speed that excites the natural bending frequency of the shaft–disk system. This defines a resonance condition that causes large deflection of the shaft, which in turn causes the system to fail violently (i.e., the engine blows apart). The nature of the resonance and the factors that control the resonance values need to be known and calculated by designers so that they can ensure that a given design is safe for production. The analytical formulation of the critical speed problem also provides some insight into how to avoid such resonance, or critical speeds.

If the rotating mass modeled by the disk is not quite homogeneous or symmetric due to some imperfection, its geometric center and center of gravity will be some distance apart (say, *a*). This is illustrated in Figure 5.33, which presents a simplified model of a large electric motor's shaft-and-rotor system (or a bladed turbine engine, for example). The shaft is constrained from moving in the radial direction by two bearings. As the shaft rotates about its long axis with angular velocity ω, the offset center of gravity pulls the shaft away from the centerline, causing it to bow as it rotates. This is called *whirling*.

The forces acting on the center of mass are the inertial force, any damping force (internal or external), and the elastic force of the shaft. In vector form, the force balance yields

$$m\ddot{\mathbf{r}} = -kx\hat{\mathbf{i}} - ky\hat{\mathbf{j}} - c\dot{x}\hat{\mathbf{i}} - c\dot{y}\hat{\mathbf{j}} \qquad (5.70)$$

where $\hat{\mathbf{i}}$ and $\hat{\mathbf{j}}$ are unit vectors, **r** the position vector defined by the line *OG*, *m* the mass of the disk, *c* the damping coefficient of the shaft system, and *k* the stiffness

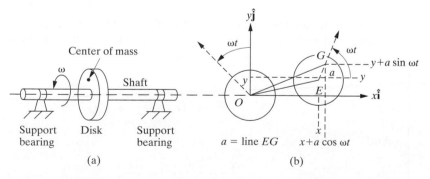

Figure 5.33   A schematic of a model of a disk rotating on a shaft and the corresponding geometry of the center of mass, *G*, of the disk relative to the neutral axis of the shaft, *O*, and the center of the rotating shaft, *E*: (a) side view; (b) end view. This diagram is useful in modeling the whirling of rotating machines (engines, turbine compressors, etc.), which are not perfectly balanced (i.e., $a \neq 0$).

coefficient provided by the shaft system. From examining the end view of Figure 5.33, the vector **r** can also be written in terms of the unit vectors $\hat{\mathbf{i}}$ and $\hat{\mathbf{j}}$ as

$$\mathbf{r} = (x + a\cos\omega t)\hat{\mathbf{i}} + (y + a\sin\omega t)\hat{\mathbf{j}} \tag{5.71}$$

Taking two derivatives yields that the acceleration vector of the center of mass is

$$\ddot{\mathbf{r}} = (\ddot{x} - a\omega^2\cos\omega t)\hat{\mathbf{i}} + (\ddot{y} - a\omega^2\sin\omega t)\hat{\mathbf{j}} \tag{5.72}$$

Substituting equation (5.72) into equation (5.70) yields

$$\left(m\ddot{x} - ma\omega^2\cos\omega t + c\dot{x} + kx\right)\hat{\mathbf{i}} + \left(m\ddot{y} - ma\omega^2\sin\omega t + c\dot{y} + ky\right)\hat{\mathbf{j}} = 0 \tag{5.73}$$

Since this is a vector equation, it is equivalent to the two scalar equations

$$m\ddot{x} + c\dot{x} + kx = ma\omega^2\cos\omega t \tag{5.74}$$

$$m\ddot{y} + c\dot{y} + ky = ma\omega^2\sin\omega t \tag{5.75}$$

These two equations are exactly the form of equation (2.82) for the response of a spring–mass system to a rotating unbalance discussed in Section 2.5. In this case, the x and y motion corresponds to the bending vibration of the shaft instead of the translational motion of a machine in the vertical direction discussed in Section 2.5.

Window 5.5
Solution of the Rotating Unbalance Equation from Section 2.5

The steady-state solution to

$$m\ddot{x} + c\dot{x} + kx = m_0 e\omega^2\sin\omega t$$

where ω is the driving frequency of the unbalanced mass, m_0 the mass of the unbalance, and e the distance from m_0 to the center of rotation, is $X\sin(\omega t - \phi)$. Here

$$X = \frac{m_0 e}{m}\frac{r^2}{\sqrt{(1 - r^2)^2 + (2\zeta r)^2}} \tag{2.84}$$

And

$$\phi = \tan^{-1}\frac{2\zeta r}{1 - r^2} \tag{2.85}$$

where $r = \omega/\sqrt{k/m}$ and $\zeta = c/(2m\omega_n)$.

Referring to Window 5.5, equation (5.75) has steady-state response magnitude given by equation (2.84); that is, equation (5.75) has the steady-state solution (since $m = m_0$ and $e = a$ in this case)

$$y(t) = \frac{ar^2}{\sqrt{\left(1 - r^2\right)^2 + (2\zeta r)^2}} \sin\left(\omega t - \tan^{-1}\frac{2\zeta r}{1 - r^2}\right) \qquad (5.76)$$

Similarly, equation (5.74) has steady-state solution of the form

$$x(t) = \frac{ar^2}{\sqrt{\left(1 - r^2\right)^2 + (2\zeta r)^2}} \cos\left(\omega t - \tan^{-1}\frac{2\zeta r}{1 - r^2}\right) \qquad (5.77)$$

since the solution given by equations (2.84) and (2.85) is 90° out of phase and the phase angle ϕ does not depend on the phase of the exciting force. The angle ϕ given by equation (2.85) becomes the angle between the lines OE and EG. From Figure 5.33, the angle θ made between the x axis and the line OE is

$$\tan\theta = \frac{y}{x} = \frac{\sin(\omega t - \phi)}{\cos(\omega t - \phi)} = \tan(\omega t - \phi) \qquad (5.78)$$

or

$$\theta = \omega t - \phi \qquad (5.79)$$

Differentiating equation (5.79) with respect to t yields $\dot{\theta} = \omega$.

The velocity $\dot{\theta}$ is the velocity of whirling. Whirling is the angular motion of the deflected shaft rotating about the neutral axis of the shaft. The calculation leading to equation (5.79) and its derivative shows that the whirling velocity is the same as the speed with which the disk rotates about the shaft (i.e., $\dot{\theta} = \omega$). This is called *synchronous whirl*.

The amplitude of motion of the center of the shaft about its neutral axis is the line $\mathbf{r} = OE$ in the end view of Figure 5.33. Note that vector $OE = \mathbf{r} = x$: $OE = r = x\hat{\mathbf{i}} + y\hat{\mathbf{j}}$. The magnitude of this vector is just

$$|\mathbf{r}(t)| = \sqrt{x^2 + y^2} = X\sqrt{\sin^2(\omega t - \phi) + \cos^2(\omega t - \phi)} = X \qquad (5.80)$$

where X is the magnitude of $x(t)$. Note that $X = Y$, where Y is the magnitude of $y(t)$ as given in equation (5.76). This calculation indicates that the distance between the shaft and its neutral axis is constant and has magnitude

$$X = \frac{ar^2}{\sqrt{\left(1 - r^2\right)^2 + (2\zeta r)^2}} \qquad (5.81)$$

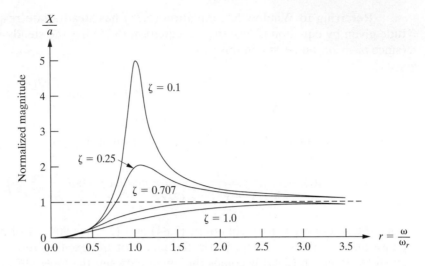

Figure 5.34 A plot of the ratio of radius of deflection (OE) to the distance to the center of mass of the disk (a) versus the frequency ratio for four different values of the damping ratio for the disk and shaft system of Figure 5.33.

This, of course, is exactly the same form as the magnitude plot of equation (2.84) given in Figure 2.21 for a spring–mass–damper system driven by a rotating out-of-balance mass. This plot is repeated for the rotational amplitude case of interest in Figure 5.34. Note that a resonance phenomenon occurs near $r = 1$, as expected. For lightly damped shafts this corresponds to unacceptably high amplitudes of rotation. The special case of $r = 1$ (i.e., $\omega_r = \sqrt{k/m}$) is called the rotor system's *critical speed*. If a rotor system runs at its critical speed, the large deflection will cause a large force to be transmitted to the bearings and eventually lead to failure. From the design point of view, the running speed, mass, and stiffness are examined for a given rotor and redesigned until $r > 3$, so that the deflections are limited to the size of the distance to the center of the mass of the disk. However, when the rotor system is started up, it must pass through the region near $r = 1$. If this startup procedure occurs too slowly, the resonance phenomena could damage the rotor bearings. Hence some damping in the system is desirable to avoid excessive amplitude at resonance. Note from Figure 5.34 that as ζ increases, X/a at resonance becomes substantially smaller.

Example 5.7.1

Consider a 55-kg compressor rotor with a shaft stiffness of 1.4×10^7 N/m, with an operation speed of 6000 rpm, and a measured internal damping providing a damping ratio of $\zeta = 0.05$. The rotor is assumed to have a worst-case eccentricity of 1000 μm ($a = 0.001$ m). Calculate (a) the rotor's critical speed, (b) the radial amplitude at operating speed, and (c) the whirl amplitude at the system's critical speed.

Solution

(a) The critical speed of the rotor is just the rotor's natural frequency, so that

$$\omega_c = \sqrt{k/m} = \sqrt{\frac{1.4 \times 10^7 \, \text{N/m}}{55 \, \text{kg}}} = 504.5 \, \text{rad/s}$$

which corresponds to a rotor speed of

$$504.5 \, \frac{\text{rad}}{\text{s}} \times \frac{60 \, \text{s}}{\text{min}} \times \frac{\text{cycle}}{2\pi \, \text{rad}} = 4817.6 \, \text{rpm}$$

(b) The value of r at running speed is just

$$r = \frac{\omega}{\sqrt{k/m}} = \frac{(2\pi/60)}{(2\pi/60)\sqrt{k/m}} = \frac{6000}{4817.6} = 1.2454$$

or about 1.25. The value of the radial amplitude of whirl at the operating speed is then given by equation (5.81) with this value of r:

$$X = |\mathbf{r}(t)| = \frac{ar^2}{\sqrt{(1-r^2)^2 + (2\zeta r)^2}} = \frac{(0.001)(1.25)^2}{\sqrt{[1-(1.25)^2]^2 + [2(0.05)(1.25)]^2}}$$
$$= 0.0027116 \, \text{m}$$

or about 2.7 mm. Here $r = 1.25$, $\zeta = 0.05$, and $a = 0.001$. Note that if $r = 1.2454$ is used, $X = 0.0027455$ m results. This gives some feeling for the sensitivity of the value of X to knowing exact values of r. Minor speed variation of, say, 10% in the running speed would cause r to vary between 1.12 and 1.37.

(c) At critical speed, $r = 1$ and X becomes

$$X = \frac{a}{2\zeta} = \frac{0.001}{2(0.05)} = 0.01 \, \text{m}$$

or 1 cm, an order of magnitude larger than the whirl amplitude at running speed.

□

Example 5.7.2

In designing a rotor system, there are many factors besides the deflection calculation indicated previously that determine the damping, stiffness, mass, and operating speed of the rotor system. Hence the designer concerned about dynamic deflections and critical speeds is often only allowed to change the design a little. Otherwise, an entire redesign must be performed, which may become very costly. With this in mind, again consider the rotor of Example 5.7.1. The clearance specification for the rotor inside the compressor housing limits the whirl amplitude at resonance to be 2 mm. Since the whirl amplitude at operating speed is greater than the allowable clearance, what percent of change in mass is required to redesign this system? What percent change in stiffness would result in the same design? Discuss the feasibility of such a change.

Solution   The required mass for a 2-mm deflection can be calculated from equation (5.81) by first determining a value of r corresponding to 2 mm. This yields

$$X = 0.002 \text{ m} = \frac{(0.001)r^2}{\sqrt{(1 - r)^2 + [2(0.05)r^2]^2}}$$

or r must satisfy $r^4 - 2.653r^2 + 1.3332 = 0$. This is a quadratic equation in r^2, which has solutions $r^2 = 0.6737, 1.979$ or $r = 0.8207, 1.406$, since the values of frequency ratio must be positive and real. Examination of the plot in Figure 5.34 of the magnitude yields that the value of r of interest is $r = 1.406$. At running speed,

$$r = \frac{6000 \text{ rpm} \dfrac{\text{min}}{60 \text{ s}} \cdot \dfrac{2\pi \, \text{rad}}{\text{rev}}}{\sqrt{k/m} \, \text{rad/s}} = \frac{628.12}{\sqrt{1.4 \times 10^7/m}} = 1.406$$

Solving for the mass m yields

$$m = 70.15 \text{ kg}$$

Since the original design value of the mass of the disk is 55 kg, the mass must be increased by 27.5% to produce a design that has its running speed deflection limited to 2 mm.

If the compressor is to be used in an application fixed to ground (such as a building), then adding 15 kg of mass to the disk may be a perfectly reasonable solution, provided that the bearings are capable of the increased force. However, if the compressor is to be used in a vehicle where weight is a consideration, such as an airplane, a 27% increase in mass may not be an acceptable design. In this case equation (5.81) can be used to examine a possible redesign by making a change in stiffness. Equation (5.81) with the appropriate parameter values yields

$$r = 1.406 = \frac{628.12 \text{ rad/s}}{\sqrt{k/55}}$$

Solving for k yields

$$k = 1.0977 \times 10^7 \text{ N/m}$$

This amounts to about a 27% change in the stiffness. Unfortunately, the stiffness of the shaft cannot be changed very easily. It is determined by geometric and material properties. The material is often determined by temperature and cost considerations as well as toughness. It can be difficult to change the stiffness by 27%.

□

Note from Example 5.7.2 that the amplitude of whirling is sensitive to changes in mass and changes in stiffness. Also note from Figure 5.34 that the damping value is of little concern when choosing the design for whirl amplitude if chosen far enough from resonance (i.e., $r > 2$). Rather, damping is chosen to limit the amplitude near resonance, which should occur only during startup and run down (i.e., $X = a/2\zeta$ at resonance). The analysis of critical speeds and rotor dynamics presented here provides a quick introduction to the topic with simplifying assumptions. The topic of rotor dynamics constitutes a separate field of study, and a text on rotor dynamics should be consulted for complete details [see, for example, Ehrich (1992), or Childs (1993)].

PROBLEMS

Section 5.1 (Problems 5.1 through 5.5)

5.1. Using the nomograph of Figure 5.1, determine the frequency range of vibration for which a machine oscillation remains at a satisfactory level under rms acceleration of 1.2 g.

5.2. Using the nomograph of Figure 5.1, determine the frequency range of vibration for which a structure's rms acceleration will not cause wall damage if vibrating with an rms displacement of 0.8 mm or less.

5.3. What natural frequency must a hand drill have if its vibration must be limited to a minimum rms displacement of 20μ m and rms acceleration of $0.1 \, \text{m/s}^2$? What rms velocity will the drill have?

5.4. A machine of mass 600 kg is mounted on support of stiffness $2.2 \times 10^8 \, \text{N/m}$. Is the vibration of this machine acceptable (Figure 5.1) for an rms amplitude of $10\mu m$? If not, suggest a way to make it acceptable.

5.5. Using the expression for the amplitude of the displacement, velocity, and acceleration of an undamped single-degree-of-freedom system, calculate the velocity and acceleration amplitude of a system with a maximum displacement of 8 cm and a natural frequency of 12 Hz. If this corresponds to the vibration of the wall of a building under a wind load, is it an acceptable level?

Section 5.2 (Problems 5.6 through 5.26)

5.6. A 120-kg machine is supported on an isolator of stiffness $700 \times 10^3 \, \text{N/m}$. The machine causes a vertical disturbance force of 350 N at a revolution of 3600 rpm. The damping ratio of the isolator is $\zeta = 0.2$. Calculate (a) the amplitude of motion caused by the unbalanced force, (b) the transmissibility ratio, and (c) the magnitude of the force transmitted to ground through the isolator.

5.7. Plot the T.R. of Problem 5.6 for the cases $\zeta = 0.001, \zeta = 0.025$, and $\zeta = 1.2$.

5.8. A simplified model of a washing machine is illustrated in Figure P5.8. A bundle of wet clothes forms a mass of 10 kg (m_b) in the machine and causes a rotating unbalance. The rotating mass is 25 kg (including m_b) and the diameter of the washer basket ($2e$) is 60 cm. Assume that the spin cycle rotates at 240 rpm. Let k be 1000 N/m and $\zeta = 0.01$. Calculate the force transmitted to the sides of the washing machine. Discuss the assumptions made in your analysis in view of what you might know about washing machines.

Top view

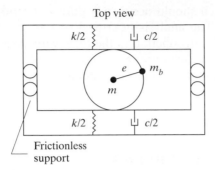

Figure P5.8   A simple model of the vibration of a washing machine induced by a rotating imbalance such as commonly caused by an uneven distribution of wet clothes during a rinse cycle.

5.9. Referring to Problem 5.8, let the spring constant and damping rate become variable. The quantities m, m_b, e, and ω are all fixed by the previous design of the washing machine. Design the isolation system (i.e., decide on which value of k and c to use) so that the force transmitted to the side of the washing machine (considered as ground) is less than 50 N.

5.10. A harmonic force of maximum value 25 N and frequency of 240 cycles $>$ min acts on a machine of 25 kg mass. Design a support system for the machine (i.e., choose c, k) so that only 8% of the force applied to the machine is transmitted to the base supporting the machine.

5.11. Consider a machine of mass 80 kg mounted to ground through an isolation system of total stiffness 30,000 N/m, with a measured damping ratio of 0.2. The machine produces a harmonic force of 500 N at 13 rad/s during steady-state operating conditions. Determine (a) the amplitude of motion of the machine, (b) the phase shift of the motion (with respect to a zero phase exciting force), (c) the transmissibility ratio, (d) the maximum dynamic force transmitted to the floor, and (e) the maximum velocity of the machine.

5.12. A small compressor weighs about 30 kg and runs at 1200 rpm. The compressor is mounted on four supports made of metal with negligible damping.
   (a) Design the stiffness of these supports so that only 10% of the harmonic force produced by the compressor is transmitted to the foundation.
   (b) Design a metal spring that provides the appropriate stiffness using Section 1.5 (refer to Table 1.2 for material properties).

5.13. Typically, in designing an isolation system, one cannot choose any continuous value of k and c but rather works from a parts catalog wherein manufacturers list isolators available and their properties (and costs, details of which are ignored here). Table 5.3 lists several made-up examples of available parts. Using this table, design an isolator for a 500-kg compressor running in steady state at 1500 rev/min. Keep in mind that as a rule of thumb compressors usually require a frequency ratio of $r = 3$.

TABLE 5.3   CATALOG VALUES OF STIFFNESS AND DAMPING PROPERTIES OF VARIOUS OFF-THE-SHELF ISOLATORS

Part No.[a]	R-1	R-2	R-3	R-4	R-5	M-1	M-2	M-3	M-4	M-5
k (10^3 N/m)	250	500	1000	1800	2500	75	150	250	500	750
c (N·s/m)	2000	1800	1500	1000	500	110	115	140	160	200

[a]The "R" in the part number designates that the isolator is made of rubber, and the "M" designates metal. In general, metal isolators are more expensive than rubber isolators.

5.14. An electric motor of mass 12 kg is mounted on four identical springs as indicated in Figure P5.14. The motor operates at a steady-state speed of 1750 rpm. The radius of gyration (see Example 1.4.6 for a definition) is 100 mm. Assume that the springs are undamped and choose a design (i.e., pick k) such that the transmissibility ratio in the vertical direction is 0.0194. With this value of k, determine the transmissibility ratio for the torsional vibration (i.e., using θ rather than x as the displacement coordinates).

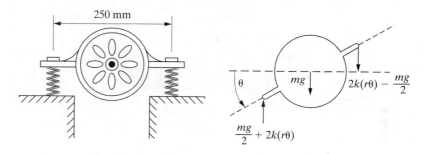

Figure P5.14   A vibration model of an electric motor mount.

5.15. A large industrial exhaust fan is mounted on a steel frame in a factory. The plant manager has decided to mount a storage bin on the same platform. Adding mass to a system can change its dynamics substantially and the plant manager wants to know if this is a safe change to make. The original design of the fan support system is not available. Hence measurements of the floor amplitude (horizontal motion) are made at several different motor speeds in an attempt to measure the system dynamics. No resonance is observed in running the fan from zero to 600 rpm. Deflection measurements are made and it is found that the amplitude is 10 mm at 600 rpm and 4.5 mm at 400 rpm. The mass of the fan is 50 kg, and the plant manager would like to store up to 50 kg on the same platform. The best operating speed for the exhaust fan is between 400 and 600 rpm depending on environmental conditions in the plant.

5.16. A 300-kg rotating machine operates at 900 cycles/min. It is desired to reduce the transmissibility ratio by one-fourth of its current value by adding a rubber vibration isolation pad. How much static deflection must the pad be able to withstand?

5.17. A 70-kg electric motor is mounted on an isolator of mass 1200 kg. The natural frequency of the entire system is 160 cycles/min and has a measured damping ratio of $\zeta = 1$. Determine the amplitude of vibration and the force transmitted to the floor if the out-of-balance force produced by the motor is $F(t) = 100\sin(31.4t)$ in Newtons.

5.18. The force exerted by an eccentric ($e = 0.22$ mm) flywheel of 1200 kg, is 600 cos (52.4t) in Newtons. Design a mounting to reduce the amplitude of the force exerted on the floor to 1% of the force generated. Use this choice of damping to ensure that the maximum force transmitted is never greater than twice the generated force.

5.19. A rotating machine weighing 1800-kg has an operating speed of 2000 rpm. It is desired to reduce the amplitude of the transmitted force by 80% using isolation pads. Calculate the stiffness required of the isolation pads to accomplish this design goal.

5.20. The mass of a system may be changed to improve the vibration isolation characteristics. Such isolation systems often occur when mounting heavy compressors on factory floors. This is illustrated in Figure P5.20. In this case the soil provides the stiffness of the isolation system (damping is neglected) and the design problem becomes that of choosing the value of the mass of the concrete block/compressor system. Assume that the stiffness of the soil is about $k = 2.0 \times 10^7$ N/m and design the size of the concrete block (i.e., choose m) such that the isolation system reduces the transmitted force by 75%. Assume that the density of concrete is $\rho = 23{,}000$ N/m^3. The surface area of the cement block is 4 m^2. The steady-state operating speed of the compressor is 2400 rpm.

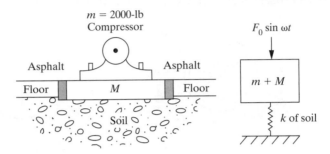

Figure P5.20   A model of a floor-mounted compressor illustrating the use of added mass to design a vibration-isolation system.

5.21. The instrument board of an aircraft is mounted on an isolation pad to protect the panel from vibration of the aircraft frame. The dominant vibration in the aircraft is measured to be at 2100 rpm. Because of size limitation in the aircraft's cabin, the isolators are only allowed to deflect 2-mm. Find the percent of motion transmitted to the instrument panel if it weighs 25-kg.

5.22. Design a base isolation system for an electronic module of mass 8 kg so that only 10% of the displacement of the base is transmitted into displacement of the module at 50 Hz. What will the transmissibility be if the frequency of the base motion changes to 100 Hz? What if it reduces to 25 Hz?

5.23. Redesign the system of Problem 5.22 such that the smallest transmissibility ratio possible is obtained over the range 50 to 75 Hz.

5.24. A 1.8-kg printed circuit board for a computer is to be isolated from external vibration of frequency 3 rad/s at a maximum amplitude of 1 mm, as illustrated in Figure P5.24. Design an undamped isolator such that the transmitted displacement is 10% of the base motion. Also calculate the range of transmitted force.

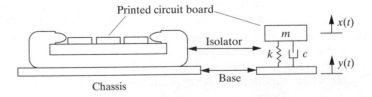

Figure P5.24   Isolation system for a printed circuit board.

5.25. Change the design of the isolator of Problem 5.24 by using a damping material with damping value ζ chosen such that the maximum T.R. at resonance is 2.

5.26. Calculate the damping ratio required to limit the displacement transmissibility to 6 at resonance for any damped isolation system.

Section 5.3 (Problems 5.27 through 5.36)

5.27. A motor is mounted on a platform that is observed to vibrate excessively at an operating speed of 5400 rpm, producing a 240-N force. Design a vibration absorber (undamped) to add to the platform. Note that in this case, the absorber mass will only be allowed to move 1 mm because of geometric and size constraints.

5.28. Consider an undamped vibration absorber with $\beta = 1$ and $\mu = 0.2$. Determine the operating range of frequencies for which $|Xk/F_0| \leq 0.5$.

5.29. Consider an internal combustion engine that is modeled as a lumped inertia attached to ground through a spring. Assuming that the system has a measured resonance of 100 rad/s, design an absorber so that the amplitude is 0.02 m for a (measured) force input of 120 N.

5.30. A small rotating machine weighing 24-kg runs at a constant speed of 6000 rpm. The machine was installed in a building, and it was discovered that the system was operating at resonance. Design a retrofit undamped absorber such that the nearest resonance is at least 20% away from the driving frequency.

5.31. A 3000-kg machine tool exhibits a large resonance at 120 Hz. The plant manager attaches an absorber to the machine of 600 kg tuned to 120 Hz. Calculate the range of frequencies at which the amplitude of the machine vibration is less with the absorber fitted than without the absorber.

5.32. A motor-generator set is designed with steady-state operating speed between 2000 and 4000 rpm. Unfortunately, due to an imbalance in the machine, a large violent vibration occurs at around 3000 rpm. An initial absorber design is implemented with a mass of 2 kg tuned to 3000 rpm. This, however, causes the combined system natural frequencies to occur at 2400 and 3000 rpm. Redesign the absorber so that $\omega_1 = 62000$ rpm and $\omega_2 = 74000$ rpm, rendering the system safe for operation.

5.33. A rotating machine is mounted on the floor of a building. Together, the mass of the machine and the floor is 1000-kg. The machine operates in steady state at 600 rpm and causes the floor of the building to shake. The floor–machine system can be modeled as a spring–mass system similar to the optical table of Figure 5.14. Design an undamped absorber system to correct this problem. Make sure you consider the bandwidth.

5.34. A pipe carrying steam through a section of a factory vibrates violently when the driving pump hits a speed of 300 rpm. (See Figure P5.34.) In an attempt to design an absorber, a trial 8-kg absorber tuned to 300 rpm was attached. By changing the pump speed, it was found that the pipe–absorber system has a resonance at 207 rpm. Redesign the absorber so that the natural frequencies are 40% away from the driving frequency.

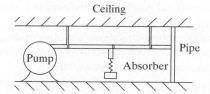

Figure P5.34 A schematic of a steam-pipe system with an absorber attached.

5.35. A machine sorts bolts according to their size by moving a screen back and forth using a primary system of 3000 kg with a natural frequency of 400 cycles/min. Design a vibration absorber so that the machine–absorber system has natural frequencies below 160 cycles/min and above 320 rpm. The machine is illustrated in Figure P5.35.

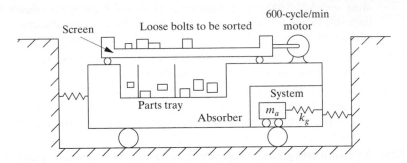

Figure P5.35   Model of a parts sorting machine. The parts (bolts here) are placed on a screen that shakes. Parts that are small enough fall through the screen into the tray below. The larger ones remain on the screen.

5.36. A dynamic absorber is designed with $\mu = 0.2$ and $\omega_a = \omega_p$. Calculate the frequency range for which the ratio $|Xk/F_0| < 1$.

Section 5.4 (Problems 5.37 through 5.52)

5.37. A machine, largely made of aluminum, is modeled as a simple mass (of 100 kg) attached to ground through a spring of 2000 N/m. The machine is subjected to a 100-N harmonic force at 20 rad/s. Design an undamped tuned absorber system (i.e., calculate m_a and k_a) so that the machine is stationary at steady state. Aluminum, of course, is not completely undamped and has internal damping that gives rise to a damping ratio of about $\zeta = 0.001$. Similarly, the steel spring for the absorber gives rise to internal damping of about $\zeta_a = 0.0015$. Calculate how much this spoils the absorber design by determining the magnitude X using equation (5.32).

5.38. Plot the magnitude of the primary system calculated in Problem 5.37 with and without the internal damping. Discuss how the damping affects the bandwidth and performance of the absorber designed without knowledge of internal damping.

5.39. Derive equation (5.35) for the damped absorber from equations (5.34) and (5.32) along with Window 5.4. Also derive the nondimensional form of equation (5.37) from equation (5.35). Note that the definition of ζ given in equation (5.36) is not the same as the ζ values used in Problems 5.37 and 5.38.

*5.40. (Project) If you have a three-dimensional graphics routine available, plot equation (5.37) [i.e., plot (X/Δ) versus both r and ζ for $0 < \zeta < 1$ and $0 < r < 3$, and a fixed μ and β]. Discuss the nature of your results. Does this plot indicate any obvious design choices? How does it compare to the information obtained by the series of plots given in Figures 5.19 to 5.21? (Three-dimensional plots such as these are commonplace.)

*5.41. (Project) Repeat Problem 5.40 by plotting $|X/\Delta|$ versus r and β for a fixed ζ and μ.

*5.42. (Project) The full damped vibration absorber equations (5.32) and (5.33) have not historically been used in absorber design because of the complicated nature of the complex arithmetic involved. However, if you have a symbolic manipulation code available to you, calculate an expression for the magnitude X by using the code to calculate the magnitude and phase of equation (5.32). Apply your results to the absorber design indicated in Problem 5.37 by using m_a, k_a, and ζ_a as design variables (i.e., design the absorber).

5.43. A machine of mass 180 kg is driven harmonically by a 100-N force at 10 rad/s. The stiffness of the machine is 18,000 N/m. Design a broadband vibration absorber [i.e., equation (5.37)] to limit the machine's motion as much as possible over the frequency range 8 to 12 rad/s. Note that other physical constraints limit the added absorber mass to be, at most, 50 kg.

5.44. Often, absorber designs are afterthoughts, such as indicated in Example 5.3.1. Add a damper to the absorber design of Figure 5.17 to increase the useful bandwidth of operation of the absorber system in the event the driving frequency drifts beyond the range indicated in Example 5.3.2 (Recall that $m = 73.16$ kg, $k = 2600$ N/m, $m_a = 18.29$ kg, $k_a = 6500$ N/m, and $7.4059 < \omega < 21.0821$ rad/s).

5.45. Again consider the absorber design of Example 5.3.1 ($m = 73.16$ kg, $k = 2600$ N/m subject to a force of 13 N at 180 cycles/min constrained to a maximum deflection of 0.2 cm). If the absorber spring is made of aluminum and introduces a damping ratio of $\zeta = 0.002$, calculate the effect of this on the deflection of the saw (primary system) with the design given in Example 5.3.1.

5.46. Consider the undamped primary system with a viscous absorber as modeled in Figure 5.22 and the rotational counterpart of Figure 5.23 repeated in Figure P5.46. Calculate the magnification factor $|Xk/M_0|$ for a 400-kg compressor having a natural frequency of 16.2 Hz if driven at resonance, for an absorber system defined by $\mu = 0.133$ and $\zeta = 0.025$.

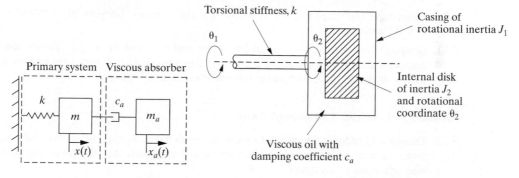

Figure P5.46   A repeat of Figures 5.22 and 5.23.

5.47. Recalculate the magnification factor $|Xk/M_0|$ for the compressor of Problem 5.46 if the damping factor is changed to $\zeta = 0.05$. Which absorber design produces the smallest displacement of the primary system $\zeta = 0.025$ or $\zeta = 0.05$?

5.48. Consider a one-degree-of-freedom model of the nose of an aircraft (A-10) as illustrated in Figure P5.48. The nose cracked under fatigue during battle conditions. This problem has been fixed by adding a viscoelastic material to the inside of the skin to act as a damped vibration absorber as illustrated in Figure P5.48. This fixed the problem and the vibration fatigue cracking disappeared in the A-10s after they were retrofitted with viscoelastic damping treatments. While the actual values remain classified, use the following data to calculate the required damping ratio given the following: $M = 100$ kg, $f = 31$ Hz, $F_0 = 100$ N, and $k = 3.533 \times 10^6$ N/m such that the maximum response is less than 0.25 mm. Since mass always needs to be limited in an aircraft, use $\mu = 0.1$ in your design.

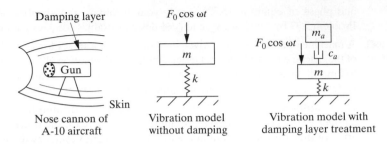

Figure P5.48 Simplified model of the A-10 nose cannon vibration problem. The nose cannon of the A-10 aircraft can be modeled as applying a harmonic force of $F_0 \cos \omega t$ to the skin of the aircraft nose. The skin can be modeled as a spring–mass system based on the stiffness model of Figure 1.26 (i.e., $k = 3EI/l^3$).

5.49. Plot an amplification curve (such as Figure 5.24) by using equation (5.39) for $\zeta = 0.02$ after several values of μ ($\mu = 0.1, 0.25, 0.5$, and 1). Can you form any conclusions about the effect of the mass ratio on the response of the primary system? Note that as μ gets large, $|(Xk/M_0)|$ gets very small. What is wrong with using very large μ in an absorber design?

5.50. A Houdaille damper is to be designed for an automobile engine. Choose a value for ζ and μ if the magnification $|(Xk/M_0)|$ is to be limited to 4 at resonance. (One solution is $\mu = 1, \zeta = 0.129$.)

5.51. Determine the amplitude of vibration for the various dampers of Problem 5.46 if $\zeta = 0.1$ and $F_0 = 120\,\text{N}$.

5.52. (Project) Use your knowledge of absorbers and isolation to design a device that will protect a mass from both shock inputs and harmonic inputs. It may help to have a particular device in mind such as the module discussed in Figure 5.6.

Section 5.5 (Problems 5.53 through 5.66)

5.53. Design a Houdaille damper for an engine modeled as having an inertia of $1.5\ \text{kg} \cdot \text{m}^2$ and a natural frequency of 33 Hz. Choose a design such that the maximum dynamic magnification is less than 6:

$$\left| \frac{Xk}{M_0} \right| < 6$$

The design consists of choosing J_2 and c_a, the required optimal damping.

5.54. Recall the optimal vibration absorber of Problem 5.53. This design is based on a steady-state response. Calculate the response of the primary system to an impulse of magnitude M_0 applied to the primary inertia J_1. How does the maximum amplitude of the transient compare to that in steady state?

5.55. Consider the damped vibration absorber of equation (5.37) with β fixed at $\beta = 1/2$ and μ fixed at $\mu = 0.25$. Calculate the value of ζ that minimizes $|X/\Delta|$. Plot this function for several values of $0 < \zeta < 1$ to check your design. If you cannot solve this analytically, consider using a three-dimensional plot of $|X/\Delta|$ versus r and ζ to determine your design.

5.56. For a Houdaille damper with mass ratio $\mu = 0.24$, calculate the optimum damping ratio and the frequency at which the damper is most effective at reducing the amplitude of vibration of the primary system.

5.57. Consider again the system of Problem 5.53. If the damping ratio is changed to $\zeta = 0.12$, what happens to $|Xk/M_0|$?

5.58. Derive equation (5.42) from equation (5.35) and derive equation (5.49) for the optimal damping ratio.

5.59. Consider the design suggested in Example 5.5.1 (mass moment of inertia of 1.5 kg m²/rad, torsional stiffness of 6×10^3 N m/rad, and a running speed of 2000 rpm). Calculate the percent change in the maximum deflection if the damping constant changes 10% from its optimal value. If the optimal damping is fixed but the mass of the absorber changes by 10%, what percent change in $|Xk/M_0|_{max}$ results? Is the optimal absorber design more sensitive to changes in c_a or m_a?

5.60. Consider the elastic isolation problem described in Figure 5.26 and repeated in Figure P5.60. Derive equations (5.54) and (5.55) from equation (5.53).

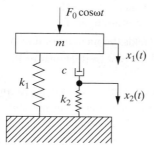

Figure P5.60   A repeat of Figure 5.26 for reference in the following problems.

5.61. Use the derivative calculation for finding maximum and minimum to derive equations (5.57) and (5.58) for the elastic damper system.

5.62. A 1200-kg mass is suspended from ground by a 48,000-N/m spring. A viscoelastic damper is added, as indicated in Figure P5.60. Design the isolator (choose k_2 and c) such that when a 60-N sinusoidal force is applied to the mass, no more than 80 N is transmitted to ground.

5.63. Consider the isolation design of Example 5.5.2 ($c = 188.56$ kg/s and $k_2 = 200$ N/m with $r = 5$ and $\gamma = 0.5$). If the value of the damping coefficient changes 10% from the optimal value (of 188.56 kg/s), what percent change occurs in $(T.R.)_{max}$? If c remains at its optimal value and k_2 changes by 10%, what percent change occurs in $(T.R.)_{max}$? Is the design of this type of isolation more sensitive to changes in damping or stiffness?

5.64. A 3200-kg machine is mounted on an isolator with an elastically coupled viscous damper such as indicated in Figure P5.60. The machine stiffness ($k1$) is 2.943×10^6 N/m, $\gamma = 0.5$, and $c = 56.4 \times 10^3$ N·s/m. The machine, a large compressor, develops a harmonic force of 1000 N at 7 Hz. Determine the amplitude of vibration of the machine.

5.65. Again consider the compressor isolation design given in Problem 5.64. If the isolation material is changed so that the damping in the isolator is changed to $\zeta = 0.16$, what is the force transmitted? Next determine the optimal value for the damping ratio and calculate the resulting transmitted force.

5.66. Consider the optimal vibration isolation design of Problem 5.65. Calculate the optimal design if the compressor's steady-state driving frequency changes to 24 Hz. If the wrong optimal point is used (i.e., if the optimal damping for the 7-Hz driving frequency is used), what happens to the transmissibility ratio?

Section 5.6 (Problems 5.67 through 5.73)

5.67. Compare the resonant amplitude at steady state (assume a driving frequency of 100 Hz) of a piece of nitrile rubber at 50°F versus the value at 75°F. Use the values for η from Table 5.2.

5.68. Using equation (5.67), calculate the new modulus of a $0.05 \times 0.01 \times 1$ m piece of pinned–pinned aluminum covered with a 1-cm-thick piece of nitrile rubber at 75°F driven at 100 Hz.

5.69. Calculate Problem 5.68 again at 50°F. What percent effect does this change in temperature have on the modulus of the layered material?

5.70. Repeat the design of Example 5.6.1 (recall $E_1 = 7.1 \times 10^{10}$ N/m^2 and $h_2 = 1$) by

    (a) changing the operating frequency to 1000 Hz, and
    (b) changing the operating temperature to 50°F.

Discuss which of these designs yields the most favorable system.

5.71. Reconsider Example 5.6.2. Make a plot of thickness of the damping treatment versus loss factor.

5.72. Calculate the maximum transmissibility coefficient of the center of the shelf of Example 5.6.1. Make a plot of the maximum transmissibility ratio for this system frequency using Table 5.1 for each temperature.

5.73. The damping ratio associated with steel is about $\zeta = 0.001$. Does it make any difference whether the shelf in Example 5.6.1 is made out of aluminum or steel? What percent improvement in damping ratio at resonance does the rubber layer provide the steel shelf?

Section 5.7 (Problems 5.74 through 5.80)

5.74. A 100-kg compressor rotor has a shaft stiffness of 1.5×10^7 N/m. The compressor is designed to operate at a speed of 6000 rpm. The internal damping of the rotor shaft system is measured to be $\zeta = 0.02$.

    (a) If the rotor has an eccentric radius of 1 cm, what is the rotor system's critical speed?
    (b) Calculate the whirl amplitude at critical speed. Compare your results to those of Example 5.7.1.

5.75. Redesign the rotor system of Problem 5.74 such that the whirl amplitude at critical speed is less than 5mm by changing the mass of the rotor.

5.76. Determine the effect of the rotor system's damping ratio on the design of the whirl amplitude at critical speed for the system of Example 5.7.1 ($r = 1$ and $\alpha = 0.001$ m) by plotting X at $r = 1$ for ζ between $0 < \zeta < 1$.

5.77. Consider the design of the compressor rotor system of Example 5.7.1 ($r = 1$ and $\alpha = 0.001$ m). The amplitude of the whirling motion depends on the parameters a, ζ, m, k, and the driving frequency. Which parameter has the greatest effect on the amplitude? Discuss your results.

5.78. The flywheel of an automobile engine has a mass of about 50 kg and an eccentricity of about 2 cm. The operating speed ranges from 1200 rpm (idle) to 5000 rpm (red line). Choose the remaining parameters so that the whirling amplitude is never more than 1 mm.

5.79. At critical speed the amplitude is determined entirely by the damping ratio and the eccentricity. If a rotor has an eccentricity of 2 cm, what value of damping ratio is required to limit the deflection to 2 cm?

5.80. A rotor system has damping limited by $\zeta < 0.05$. What is the maximum value of eccentricity allowable in the rotor design if the maximum amplitude at critical speed must be less than 0.5 cm?

MATLAB ENGINEERING VIBRATION TOOLBOX

If you have not yet used the *Engineering Vibration Toolbox* program, return to the end of Chapter 1 or Appendix G for a brief introduction to using MATLAB files.

The files contained in folder VTB5 may be used to help solve the preceding problems. The M-files from earlier chapters (VTBX.X.M, etc.) may also be useful. The following Toolbox problems are intended to help you gain some experience with the concepts introduced in this chapter for designing vibrating systems and to build experience with the various formulas. These problems may also be solved using any of the codes introduced in Sections 1.9, 1.10, 2.8, 2.9, 3.8, 4.9, and 4.10.

TOOLBOX PROBLEMS

TB5.1. Use file VTB5_1 to reproduce the plots of Figure 5.5 by inputting various values of m, c, and k. First fix m and k and vary c. Then fix m and c and vary k.

TB5.2. Use file VTB5_2 to verify the solution of Example 5.2.1 for the magnitude of the force transmitted to the electronic module.

TB5.3. Use file VTB5_3 to examine what happens to the shaded region in Figure 5.15 as μ is varied. Do this by increasing the absorber mass ($\mu = m_a/m$) so that μ varies in increments of 0.1 from 0.1 to 1 for a fixed value of β.

TB5.4. Examine the effect of β on Figure 5.16 by using file VTB5_4 to plot Figure 5.16 over again for several different values of β ($\beta = 0.1, 0.5, 1, 1.5$). What do you notice? What does changing β correspond to in terms of choosing the values of the absorber design?

TB5.5. Consider the amplitude plot of the damped absorber given in Figure 5.19. Use file VTB5_5 to see the effect changing the primary mass m has on the design. Choose $m_a = 10, c_a = 1, k = 1000, k_a = 1000$. Plot $|Xk/F_0|$ for various values of the primary mass m (i.e., $m = 1, 10, 100, 1000$) versus r. Can you draw any conclusions?

TB5.6. File VTB5_6 plots the three-dimensional mesh used to generate Figure 5.25, which illustrates the optimal damped absorber design. Use this program to note the effects of changing the spring–mass on the values of ζ corresponding to the absolute minimum of the maximum response.

6 Distributed-Parameter Systems

So far this book has focused on the vibration of rigid bodies. This chapter introduces the analysis needed to describe the vibration of systems that have flexible components. Flexibility of structural components arises when the mass and stiffness properties are modeled as being distributed throughout the spatial definition of the component rather than at lumped positions, as done in Chapter 4. Examples of such systems are the wings and panels of aircraft such as the Reaper pictured on the left. The vibrations of the wing of a commercial aircraft can usually be seen during takeoff and landing or during turbulence. The blades of the wind turbine in the bottom photo form another example of a distributed-parameter system. Increased reliance on wind energy has promoted larger, and hence more flexible, wind turbine blades. Many structures, such as wings, blades, and other components, can be modeled by the simple string, beam, and plate models discussed in this chapter. Many systems, such as truck chassis, buildings, dance floors, and computer disk drives, can be modeled and analyzed by the methods presented in the chapter. The major concept presented here is that distributed-parameter systems have an infinite number of natural frequencies. The concepts of mode shapes and modal analysis used in Chapter 4 are extended here to treat the vibration of distributed-parameter systems.

In previous chapters, all systems considered are modeled as lumped-parameter systems; that is, the motion of each point in the system under consideration is modeled as if the mass were concentrated at that point. Multiple-degree-of-freedom systems are considered as arrangements of various lumped masses separated by springs and dampers. In this sense, the parameters of the system are discrete sets of finite numbers. Hence, such systems are also called discrete systems or finite dimensional systems. In this chapter, the flexibility of structures is considered. Here the mass of an object is considered to be distributed throughout the structure as a series of infinitely small elements. When a structure vibrates, each of these infinite number of elements move relative to each other in a continuous fashion. Hence these systems are called *infinite-dimensional systems, continuous systems*, or *distributed-parameter systems*. The choice of modeling a given mechanical system as a lumped-parameter system or a distributed-parameter system depends on the purpose at hand as well as the nature of the object. There are only a few distributed-parameter models that have closed-form solutions. However, these solutions provide insight into a large number of problems that cannot be solved in closed form.

The time response of a distributed-parameter system is described spatially by a continuous function of the relative position along the system. In contrast, the time response of a lumped-parameter system is described spatially by labeling a discrete number of points throughout the system in the form of a vector. Here the terms *lumped parameter* and *distributed parameter* are used rather than discrete and continuous to avoid confusion with discrete-time systems (used in numerical integration and measurement). The specific cases considered here are the vibrations of strings, rods, beams, membranes, and plates. Common examples of such structures are a vibrating guitar string and the swaying motion of a bridge or tall building. In addition, systems having both lumped parameters and distributed parameters are considered.

The single-degree-of-freedom systems discussed in Chapters 1 through 3 have only one natural frequency $\omega_n = \sqrt{k/m}$. In Chapter 4, multiple-degree-of-freedom systems introduce the concept of multiple natural frequencies, denoted by ω_i, one frequency for each degree of freedom. Design to avoid resonance becomes more difficult with multiple natural frequencies because of the increased chance that a harmonic driving frequency will correspond to one of the natural frequencies, causing resonance. The distributed-parameter systems considered in this chapter have an infinite number of degrees of freedom and hence an infinite number of natural frequencies, again increasing concern about resonance in design.

The numerous frequencies of a distributed-parameter system are also denoted by ω_n for one-dimensional structures and ω_{nm} for structures defined in a plane. There is thus a slight notational inconsistency that persists in the vibration literature by using ω_n for both the natural frequency of a single-degree-of-freedom system and to mean the nth natural frequency of a distributed system. The distinction is ultimately clear from the context of usage.

6.1 VIBRATION OF A STRING OR CABLE

String instruments (guitars, violins, etc.) provide an excellent and intuitive example of the vibration of a distributed-parameter object. Strings are also the easiest system to solve and provide a systematic way to approach other distributed-parameter structures, much like the simple spring–mass system formed the basics for analysis for the lumped-parameter system analysis. Consider the string of Figure 6.1 with mass density ρ and cross-sectional area A, fixed at both ends and under a tension denoted by τ. The string moves up and down in the y direction. The motion at any point on the string must be a function of both the time, t, and the position along the string, x. The deflection of the string is thus denoted by $w(x, t)$. Let $f(x, t)$ be an external force per unit length also distributed along the string, and consider the infinitesimal element (Δx long) of the displaced string indicated in Figure 6.1.

The net force acting on the infinitesimal element in the y direction must be equal to the inertial force in the y direction, $\rho A \Delta x (\partial^2 w / \partial t^2)$ so that

$$-\tau_1 \sin \theta_1 + \tau_2 \sin \theta_2 + f(x,t)\Delta x = \rho A \Delta x \frac{\partial^2 w(x,t)}{\partial t^2} \tag{6.1}$$

Note that the acceleration is stated in terms of partial derivatives $(\partial^2 / \partial t^2)$ because w is a function of two variables. The expressions in equation (6.1) can be approximated in the case of small deflections so that θ_1 and θ_2 are small. In this case, τ_1 and τ_2 can easily be related to the initial tension in the string τ by noting that the horizontal component of the deflected string tension is $\tau_1 \cos \theta_1$ at end 1, and $\tau_2 \cos \theta_2$ at

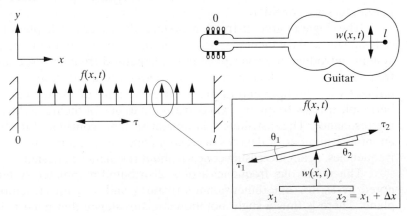

Infinitesimal element of string
displaced from rest

Figure 6.1 The geometry of a vibrating string with applied force $f(x, t)$ and displacement $w(x, t)$.

end 2. In the small-angle approximation, $\cos\theta_1 \simeq 1$ and $\cos\theta_2 \simeq 1$, so it is reasonable to set $\tau_1 = \tau_2 = \tau$. Also, for small θ_1,

$$\sin\theta_1 \simeq \tan\theta_1 = \left.\frac{\partial w(x,t)}{\partial x}\right|_{x_1} \tag{6.2}$$

and

$$\sin\theta_2 \simeq \tan\theta_2 = \left.\frac{\partial w(x,t)}{\partial x}\right|_{x_2} \tag{6.3}$$

where $(\partial w/\partial x)|_{x_1}$ is the slope of the string at point x_1 and $(\partial w/\partial x)|_{x_2}$ is the slope of the string at point $x_2 = x_1 + \Delta x$. The notation for partial and total derivatives is reviewed in Window 6.1.

With these approximations, equation (6.1) now becomes

$$\left.\left(\tau\frac{\partial w(x,t)}{\partial x}\right)\right|_{x_2} - \left.\left(\tau\frac{\partial w(x,t)}{\partial x}\right)\right|_{x_1} + f(x,t)\Delta x = \rho A \frac{\partial^2 w(x,t)}{\partial t^2}\Delta x \tag{6.4}$$

The slopes can be evaluated further by recalling the Taylor series expansion for the function $\tau(\partial w/\partial x)$ around the point x_1 from calculus. This yields

$$\left.\left(\tau\frac{\partial w}{\partial x}\right)\right|_{x_2} = \left.\left(\tau\frac{\partial w}{\partial x}\right)\right|_{x_1} + \Delta x\frac{\partial}{\partial x}\left.\left(\tau\frac{\partial w}{\partial x}\right)\right|_{x_1} + O(\Delta x^2) \tag{6.5}$$

Window 6.1
Notation for Vibrations Described by Distributed Mass and Stiffness

Derivatives of a function of multiple variables, such as $f(x,y)$, deal with partial derivatives with the following notation:

$$f_x(x,y) = \frac{\partial f(x,y)}{\partial x} = \lim_{\Delta x \to 0}\left(\frac{f(x+\Delta x, y) - f(x,y)}{\Delta x}\right)$$

In contrast, the total derivative of a function of a single variable, say $T(t)$, is denoted by

$$\frac{dT(t)}{dt} = \ddot{T}(t) = \lim_{\Delta t \to 0}\frac{T(t+\Delta t) - T(t)}{\Delta t}$$

The exact differential is

$$df = \frac{\partial f}{\partial x}dx + \frac{\partial f}{\partial y}dy$$

where $O(\Delta x^2)$ denotes the rest of the Taylor series, which consists of terms of order Δx^2 and higher. Since Δx is small, $O(\Delta x^2)$ is even smaller and hence neglected. Substitution of expression (6.5) into equation (6.4) yields

$$\frac{\partial}{\partial x}\left(\tau\,\frac{\partial w}{\partial x}\right)\Bigg|_{x_1}\Delta x + f(x,t)\Delta x = \rho A\,\frac{\partial^2 w(x,t)}{\partial t^2}\Delta x \tag{6.6}$$

Dividing by Δx and realizing that since Δx is infinitesimal, the designation of point 1 becomes unnecessary, the equation of motion for the string becomes

$$\frac{\partial}{\partial x}\left(\tau\,\frac{\partial w(x,t)}{\partial x}\right) + f(x,t) = \rho A\,\frac{\partial^2 w(x,t)}{\partial t^2} \tag{6.7}$$

Since the tension τ is constant, and if the external force is zero, this becomes

$$c^2\,\frac{\partial^2 w(x,t)}{\partial x^2} = \frac{\partial^2 w(x,t)}{\partial t^2} \quad\text{or}\quad \frac{\partial^2 w(x,t)}{\partial x^2} = \frac{1}{c^2}\,\frac{\partial^2 w(x,t)}{\partial t^2} \tag{6.8}$$

where $c = \sqrt{\tau/\rho A}$ depends only on the physical properties of the string (called the *wave speed* and is not to be confused with the symbol used for the damping coefficient in earlier chapters). Equation (6.8) is the one-dimensional wave equation, also called the *string equation*, and is subject to two initial conditions in time because of the dependence on the second time derivative. These are written as $w(x, 0) = w_0(x)$ and $w_t(x,0) = \dot{w}_0(x)$, where the subscript t is an alternative notation for the partial derivative $\partial/\partial t$, and where $w_0(x)$ and $\dot{w}_0(x)$ are the initial displacement and velocity distributions of the string, respectively. The second spatial derivative in equation (6.8) implies that two other conditions must be applied to the solution $w(x, t)$ in order to calculate the two constants of integration arising from integrating these spatial derivatives. These conditions come from examining the boundaries of the string. In the configuration of Figure 6.1 the string is fixed at both ends (i.e., at $x = 0$ and $x = l$). This means that the deflection $w(x, t)$ must be zero at these points so that

$$w(0, t) = w(l, t) = 0 \quad t > 0 \tag{6.9}$$

These two conditions at the boundary provide the other two constants of integration resulting from the two spatial derivatives of $w(x, t)$. Because these conditions occur at the boundaries, the problem described by equations (6.8) and (6.9) and the initial conditions is called a *boundary-value problem*.

Using the subscript notation for partial differentiation, the various derivatives of the deflection, $w(x, t)$, have the following physical interpretations. The quantity $w_x(x, t)$ denotes the slope of the string, while $\tau w_{xx}(x, t)$ corresponds to the restoring force of the string (i.e., the string's stiffness or elastic property). The quantity $w_t(x, t)$ is the velocity and $w_{tt}(x, t)$ is the acceleration of the string at any point x and time t.

The string vibration problem just described forms a convenient and simple model to study the vibration of distributed-parameter systems. This is analogous to the spring–mass model of Section 1.1, which provided a building block for the study of lumped-parameter systems. To that end, the string with fixed endpoints is used in the next section to develop general techniques of solving for the vibration response of distributed-parameter systems. More about the string equation and its use in wave propagation can be found in introductory physics texts. Note that these developments apply to cables as well as strings.

Example 6.1.1

Consider the cable of Figure 6.2, which is pinned at one end and attached to a spring at the other end held in a frictionless slider such that the cable remains in constant tension. Determine the governing equation for the vibration of the system.

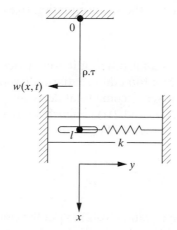

Figure 6.2 A cable fixed at one end and attached to a spring at the other end with a frictionless slide. Note that the motion $w(x, t)$ is in the y direction.

Solution The equations of motion for the spring and the cable are the same and are given by equation (6.8). The initial conditions are also unaffected. However, the boundary condition at $x = l$ changes. Writing a force balance in the y direction at point $x = l$ yields

$$\sum_y F|_{x=l} = \tau \sin \theta + kw(l, t) = 0$$

where k is the stiffness of the (lumped) spring. Again, enforcing the small-angle approximation, this becomes

$$\tau \frac{\partial w(x, t)}{\partial x}\bigg|_{x=l} = -kw(x, t)|_{x=l}$$

The boundary condition at $x = 0$ remains unchanged.

$\square$

6.2 MODES AND NATURAL FREQUENCIES

In this section the string equation is solved for the case of fixed–fixed boundary conditions using the technique of *separation of variables*. This method leads in a natural way to modal analysis and the concepts of mode shapes and natural frequencies for distributed-parameter systems that are used extensively for lumped-parameter systems in Chapter 4. The solution procedures are described in detail in introductory differential equations (see Boyce and DiPrima, 2009, for instance) and reviewed here. First, it is assumed that the displacement $w(x, t)$ can be written as the product of two functions, one depending only on x and the other depending only on t (hence separation of variables). Thus

$$w(x, t) = X(x)T(t) \tag{6.10}$$

Substitution of this separated form into the string equation (6.8) yields

$$c^2 X''(x)T(t) = X(x)\ddot{T}(t) \tag{6.11}$$

where the primes on $X''(x)$ denote total differentiation (twice in this case) with respect to x, and the overdots indicate total differentiation (twice in this case) with respect to the time, t. These derivatives become total derivatives, instead of partial derivatives, because the functions $X(x)$ and $T(t)$ are each a function of a single variable. A simple rearrangement of equation (6.11) yields

$$\frac{X''(x)}{X(x)} = \frac{\ddot{T}(t)}{c^2 T(t)} \tag{6.12}$$

Note that the choice of which side in equation (6.12) to put the constant c^2 is arbitrary. Some choose to place it on the left side and some, as done here, on the right side. The final solution remains the same.

Since each side of the equation is a function of a different variable, it is argued that each side must be constant. To see this, differentiate with respect to x. This yields

$$\frac{d}{dx}\left(\frac{X''}{X}\right) = 0 \tag{6.13}$$

which becomes, upon integration,

$$\frac{X''}{X} = \text{constant} = -\sigma^2 \tag{6.14}$$

In this case $-\sigma^2$ is the constant chosen to ensure that the quantity on the right side of equation (6.14) is negative. Actually, all possible choices (negative, positive, and zero) for this constant need to be considered. The other two choices (positive and zero) lead to physically unacceptable results as discussed in Example 6.2.1 to follow. Equation (6.14) also requires that

$$\frac{\ddot{T}(t)}{c^2 T(t)} = -\sigma^2 \tag{6.15}$$

in order to satisfy equation (6.12).

Rearranging equation (6.14) yields the result that the function $X(x)$ must satisfy

$$X''(x) + \sigma^2 X(x) = 0 \tag{6.16}$$

Equation (6.16) has the solution (see Example 6.2.1)

$$X(x) = a_1 \sin \sigma x + a_2 \cos \sigma x \tag{6.17}$$

where a_1 and a_2 are constants of integration. To solve for these constants of integration, consider the boundary conditions of equation (6.9) in the separated form implied by equation (6.10). They become

$$X(0)T(t) = 0 \quad \text{and} \quad X(l)T(t) = 0 \quad t > 0 \tag{6.18}$$

Since it is assumed that $T(t)$ cannot be zero for all t [this would yield only the uninteresting solution $w(x, t) = 0$ for all time], equation (6.18) reduces to

$$X(0) = 0 \quad X(l) = 0 \tag{6.19}$$

Applying these two conditions to the solution, equation (6.17), yields the two simultaneous equations

$$X(0) = a_2 = 0$$

$$X(l) = a_1 \sin \sigma l = 0 \tag{6.20}$$

The first of these two expressions eliminates the cosine term in the solution, and the last of these two expressions yields values of σ by requiring that $\sin \sigma l = 0$. This is called the *characteristic equation*, which has solutions $\sigma l = n\pi$. Since the sine function vanishes when its argument is 0 and any integer multiple of π, there is one solution of the characteristic equation for each value of $n = 0, 1, 2, 3, \ldots$ Hence, there exists an infinite number of values of σ that satisfy the condition $\sigma l = n\pi$. Thus, σ is indexed to be σ_n and takes on the following values ($n = 0$ implies a zero solution):

$$\sigma_n = \frac{n\pi}{l} \quad n = 1, 2, 3, \ldots \tag{6.21}$$

The two simultaneous equations resulting from the boundary conditions given by equation (6.20) can also be written in matrix form as

$$\begin{bmatrix} \sin \sigma l & 0 \\ 0 & 1 \end{bmatrix} \begin{bmatrix} a_1 \\ a_2 \end{bmatrix} = \begin{bmatrix} 0 \\ 0 \end{bmatrix}$$

Recall from Section 4.1, equation (4.19), that this vector equation has a nonzero solution (for a_1 and a_2) as long as the coefficient matrix has a zero determinant. Thus this alternative formulation yields the characteristic equation

$$\det \begin{bmatrix} \sin \sigma l & 0 \\ 0 & 1 \end{bmatrix} = \sin \sigma l = 0$$

This provides a more systematic calculation of the characteristic equation from the statement of the boundary conditions.

Since there are an infinite number of values of σ_n, the solution (6.17) then becomes the infinite number of solutions

$$X_n(x) = a_n \sin \left(\frac{n\pi}{l} x \right) \quad n = 1, 2, \ldots \tag{6.22}$$

Here X is now indexed by n because of its dependence on σ_n, and the a_n are still-to-be-determined arbitrary constants, potentially depending on the index n as well.

Equation (6.22) forms the spatial solution of the vibrating string problem. The functions $X_n(x)$ satisfy the boundary value problem

$$\frac{\partial^2}{\partial x^2}(X_n) = \lambda_n X_n$$

$$X_n(0) = X_n(l) = 0 \tag{6.23}$$

where λ_n are constants $\left(\lambda_n = \sigma_n^2 \right)$ and where the function X_n is never identically zero over all values of x. Comparison of this with the definition of the matrix eigenvalue and eigenvector problem of Section 4.2 yields some very strong similarities. Instead of an eigenvector consisting of a column of constants, equation (6.23) defines the *eigenfunctions, $X_n(x)$,* which are nonzero functions of x satisfying boundary conditions as well as a differential equation. The constants λ_n are called *eigenvalues* just as in the matrix case. The differential operator $-\partial^2/\partial x^2$ takes the place of the matrix in this eigenproblem. Similar to the eigenvector of Chapter 4, the eigenfunction is only known to within a constant. That is, if $X_n(x)$ is an eigenfunction, so is $aX_n(x)$, where a is any constant. In fact, the concept of eigenvector and that of eigenfunction are mathematically identical. As is illustrated in the following, the eigenvalues, as determined by the characteristic equation given in this case by expression (6.20), and the eigenfunctions, described in (6.22), will determine the natural frequencies and mode shapes of the vibrating string.

To this end, consider next the temporal equation given by (6.15) with the quantities of (6.21) substituted for σ_n:

$$\ddot{T}_n(t) + \sigma_n^2 c^2 T_n(t) = 0 \quad n = 1, 2, \ldots \tag{6.24}$$

where $T(t)$ is now indexed because there is one solution for each value of σ_n. The coefficient of $T_n(t)$ in the temporal equation defines the natural frequency by noting that $\omega_n = c\sigma_n$ and hence

$$\omega_n = c\sigma_n = \frac{n\pi}{l}\sqrt{\frac{\tau}{\rho A}} \text{ rad/s}$$

The general form of the solution of (6.24) is given in Window 1.4 as

$$T_n(t) = A_n \sin \omega_n t + B_n \cos \omega_n t \tag{6.25}$$

where A_n and B_n are constants of integration. Since both of the functions $X_n(x)$ and $T_n(t)$ are found to be dependent on n, the solution $w(x, t) = X_n(x)T_n(t)$ must also be a function of n, so that

$$w_n(x, t) = c_n \sin\left(\frac{n\pi}{l}x\right)\sin\left(\frac{n\pi c}{l}t\right) + d_n \sin\left(\frac{n\pi}{l}x\right)\cos\left(\frac{n\pi c}{l}t\right) \quad n = 1, 2, \dots \tag{6.26}$$

where c_n and d_n are new constants to be determined. Note that an unknown constant a_n times another unknown constant A_n is the unknown constant c_n (similarly, $d_n = a_n B_n$). Since the string equation is linear, any linear combination of solutions is a solution. Hence the general solution is of the form

$$w(x, t) = \sum_{n=1}^{\infty}(c_n \sin \sigma_n x \sin \sigma_n ct + d_n \sin \sigma_n x \cos \sigma_n ct) \tag{6.27}$$

The set of constants $\{c_n\}$ and $\{d_n\}$ can be determined by applying the initial conditions on $w(x, t)$ and the orthogonality of the set of functions $\sin(n\pi x/l)$. The orthogonality of the set of functions $\sin(n\pi x/l)$ states that

$$\int_0^l \sin\frac{n\pi x}{l} \sin\frac{m\pi x}{l}\, dx = \begin{cases} \dfrac{l}{2} & n = m \\ 0 & n \neq m \end{cases} \tag{6.28}$$

which is identical to the orthogonality of mode shape vectors discussed in Section 4.2. This orthogonality condition can be derived by simple trigonometric identities and integration (as suggested in Section 3.3).

Consider the initial condition on the displacement applied to equation (6.27):

$$w(x, 0) = w_0(x) = \sum_{n=1}^{\infty} d_n \sin\frac{n\pi x}{l}\cos(0) \tag{6.29}$$

Multiplying both sides of this equality by sin $m\pi x/l$ and integrating over the length of the string yields

$$\int_0^l w_0(x) \sin\frac{m\pi x}{l}\, dx = \sum_{n=1}^{\infty} d_n \int_0^l \sin\frac{n\pi x}{l} \sin\frac{m\pi x}{l}\, dx = d_m\left(\frac{l}{2}\right) \qquad (6.30)$$

where each term in the summation on the right-hand side of equation (6.29) is zero except for the mth term, by direct application of the orthogonality condition of (6.28). Equation (6.30) must hold for each value of m, so that

$$d_m = \frac{2}{l}\int_0^l w_0(x) \sin\frac{m\pi x}{l}\, dx \quad m = 1, 2, 3, \ldots \qquad (6.31)$$

It is customary to replace m by n in the preceding, since the index runs over all positive values [i.e., the index in equation (6.31) is a free index and it does not matter what it is named]. It is most convenient to rename it d_n.

A similar expression for the constants $\{c_n\}$ is obtained by using the initial velocity condition. Time differentiation of the summation of equation (6.27) yields

$$\dot{w}_0(x) = w_t(x, 0) = \sum_{n=1}^{\infty} c_n \sigma_n c \sin\frac{n\pi x}{l} \cos(0) \qquad (6.32)$$

Again, multiplying by sin $m\pi x/l$, integrating over the length of the string, and applying the orthogonality condition of (6.28) yields

$$c_n = \frac{2}{n\pi c}\int_0^l \dot{w}_0(x) \sin\frac{n\pi x}{l}\, dx \quad n = 1, 2, 3, \ldots \qquad (6.33)$$

where the index has been renamed n. Equations (6.31) and (6.33) combined with equation (6.27) form the complete solution for the vibrating string (i.e., they describe the vibration response of the string at any point x and any time t).

Example 6.2.1

The solution of a second-order ordinary differential equation with constant coefficients subject to boundary conditions is used throughout this section. This example clarifies the choice of a negative constant for the separation-of-variables procedure and reviews the solution technique for second-order differential equations with constant coefficients (see, e.g., Boyce and DiPrima, 2009). Consider again equation (6.14), where this time the separation constant is chosen to be $-\beta$, where β is of arbitrary sign. This yields

$$X'' + \beta X = 0 \qquad (6.34)$$

Assume that the solution of (6.34) is of the form $X(x) = e^{\lambda x}$, where λ is to be determined. Substitution into (6.34) yields

$$(\lambda^2 + \beta)e^{\lambda x} = 0 \qquad (6.35)$$

Since $e^{\lambda x}$ is never zero, this requires that

$$\lambda = \pm\sqrt{-\beta} \qquad (6.36)$$

Here λ will be purely imaginary or real, depending on the sign of the separation constant β. Thus there are two solutions and the general solution is the sum

$$X(x) = Ae^{-\sqrt{-\beta}x} + Be^{+\sqrt{-\beta}x} \qquad (6.37)$$

where A and B are constants of integration to be determined by the boundary conditions. Applying the boundary conditions to (6.37) yields

$$X(0) = A + B = 0$$

$$X(l) = Ae^{-\sqrt{-\beta}l} + Be^{\sqrt{-\beta}l} = 0 \qquad (6.38)$$

This system of equations can be written in the matrix form

$$\begin{bmatrix} 1 & 1 \\ e^{-\sqrt{-\beta}l} & e^{\sqrt{-\beta}l} \end{bmatrix} \begin{bmatrix} A \\ B \end{bmatrix} = \begin{bmatrix} 0 \\ 0 \end{bmatrix}$$

which has a nonzero solution for A and B if and only if (recall Section 4.1) the determinant of the matrix coefficient is zero. This yields

$$e^{+\sqrt{-\beta}l} - e^{-\sqrt{-\beta}l} = 0 \qquad (6.39)$$

which must be satisfied. For negative real values of β, say $\beta = -\sigma^2$, this becomes [recall the definition $\sinh u = (e^u - e^{-u})/2$]

$$e^{\sigma l} - e^{-\sigma l} = 2\sinh \sigma l = 0 \qquad (6.40)$$

which has only the trivial solution and hence $\beta \neq -\sigma^2$. This means that the separation constant σ in the development of equation (6.14) cannot be positive. Thus $\sqrt{-\beta}$ must be complex (i.e., $\beta = \sigma^2$), so that equation (6.39) becomes

$$e^{\sigma jl} - e^{-\sigma jl} = 0 \qquad (6.41)$$

where $j = \sqrt{-1}$. Euler's formula for the sine function is $\sin u = (e^{uj} - e^{-uj})/2j$, so that (6.41) becomes

$$\sin \sigma l = 0 \qquad (6.42)$$

which has the solution $\sigma = n\pi/l$ as used in equation (6.20).

The only possibility that remains to check is the case $\beta = 0$, which yields $X'' = 0$, or, upon integrating twice,

$$X(x) = a + bx \qquad (6.43)$$

Applying the boundary conditions results in

$$X(0) = a = 0$$

$$X(l) = bl = 0 \qquad (6.44)$$

which yields only the trivial solution $a = b = 0$. Hence the choice of separation constant in equation (6.14) as $-\sigma^2$ is fully justified, and the solution of the spatial equation for the string fixed at both ends is of the form

$$X_n(x) = a_n \sin \frac{n\pi x}{l} \qquad n = 1, 2, 3, \ldots$$

Here the subscript n has been added to $X(x)$ to denote its dependence on the index n and to indicate that more than one solution results. In this case, an infinite number of solutions result, one for each integer n.

$\square$

Note that the solution of this spatial equation is very much like the solution of the equation of the single-degree-of-freedom oscillator in Section 1.1 and reviewed in Window 6.2. There are, however, two main differences. First, the sign of the coefficient in the single-degree-of-freedom oscillator is determined on physical grounds (i.e., the ratio of stiffness to mass is positive), and second, the constants of integration were both evaluated at the beginning of the interval instead of one at each end. The spatial string equation is a boundary-value problem, while the single-degree-of-freedom oscillator equation is an initial-value problem. Note that the temporal equation for the string, however, is identical to the equation of a single-degree-of-freedom oscillator. The temporal equation is also an initial-value problem.

Window 6.2
Review of the Solution of a Single-Degree-of-Freedom System

The solution to $m\ddot{x} + kx = 0$, $x(0) = x_0$, $\dot{x}(0) = v_0$ for $m, k > 0$ is

$$x(t) = \frac{\sqrt{\omega_n^2 x_0^2 + v_0^2}}{\omega_n} \sin\left(\omega_n t + \tan^{-1}\frac{\omega_n x_0}{v_0}\right) \qquad (1.10)$$

where $\omega_n = \sqrt{k/m}$.

Now that the mathematical solution of the vibrating string is established, consider the physical interpretation of the various terms in equation (6.27). Consider giving the string an initial displacement of the form

$$w_0(x) = \sin\frac{\pi x}{l} \qquad (6.45)$$

and an initial velocity of $\dot{w}_0(x) = 0$. With these values of the initial conditions, the computation of the coefficients in (6.27) by using equations (6.31) and (6.33) yields $c_n = 0$ for $n = 1, 2, \ldots$, $d_n = 0$ for $n = 2, 3, \ldots$, and $d_1 = 1$. Substitution of these coefficients back into equation (6.27) yields the solution

$$w(x, t) = \sin\frac{\pi x}{l} \cos\frac{\pi c}{l} t \qquad (6.46)$$

which is the first term of the series. Comparing the second factor of equation (6.46) to $\cos \omega t$ states that the string is oscillating in time at a frequency of

$$\omega_1 = \frac{\pi c}{l} = \frac{\pi}{l}\sqrt{\frac{\tau}{\rho A}} \text{ rad/s or } f_1 = \frac{\omega_1}{2\pi} \text{ Hz} \tag{6.47}$$

in the spatial shape of $\sin (\pi x/l)$. Hence for a fixed time t, the string will be deformed in the shape of a sinusoid. Each point of the string is moving up and down in time (i.e., vibrating at frequency $\pi c/l$). Borrowing the jargon of Chapter 4, the function $X_1(x) = \sin (\pi x/l)$ is called a *mode shape*, or *mode*, of the string and the quantity $(\pi c/l)$ is called a *natural frequency* of the string.

This procedure can be repeated for each value of the index n by choosing the initial displacement to be $\sin (n\pi x/l)$ and the velocity to be zero. This gives rise to an infinite number of mode shapes, $\sin (n\pi x/l)$, and natural frequencies, $(n\pi c/l)$, and is the reason distributed-parameter systems are called *infinite-dimensional systems*. Figure 6.3 illustrates the first two mode shapes of the fixed-endpoint string. If the string is excited in the second mode by using the initial conditions $w_0 = \sin (2\pi x/l)$, $\dot{w}_0 = 0$, and the resulting motion is viewed by a stroboscope flashing with frequency $2\pi c/l$, the curve labeled $n = 2$ in Figure 6.3 would be visible.

An interesting property of the modes of a string are the points where the modes $\sin (n\pi x/l)$ are zero. These points are called *nodes*. Note from Figure 6.3 that the node of the second mode is at the point $l/2$. If the string is given an initial displacement equal to the second mode, there will be no motion at the point $l/2$ for any value of t. This node, as in the case of lumped-parameter systems, is a point on the string that does not move if the structure is excited only in that mode.

The modes and natural frequencies for a string and those for a multiple-degree-of-freedom system are very similar. For a multiple-degree-of-freedom system, the mode shape is a vector, the elements of which yield the relative amplitude of vibration of each coordinate of the system. The same is true for an eigenfunction of the string, the difference being that instead of elements of a vector, the eigenfunction gives the modal response magnitude at each value of the position x along the string.

Equation (6.27) is the distributed-parameter equivalent of the modal expansion given by equation (4.103) for lumped-parameter systems. Instead of an expansion of the solution vector $\mathbf{x}(t)$ in terms of the modal vectors $\mathbf{u}_i$ indicated in equation (4.103)

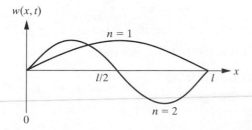

Figure 6.3   A plot of the deflection $w(x, t)$ versus the position x for a fixed time illustrating the first two mode shapes of a vibrating string fixed at both ends.

Window 6.3
***A Review of the Modal Expansion Solution
of a Multiple-Degree-of-Freedom System***

The solution of $M\ddot{\mathbf{x}} + K\mathbf{x} = \mathbf{0}$, $\mathbf{x}(0) = \mathbf{x}_0$, $\dot{\mathbf{x}}(0) = \dot{\mathbf{x}}_0$ is

$$\mathbf{x}(t) = \sum_{i=1}^{n} c_i \sin(\omega_i t + \phi_i)\mathbf{u}_i \qquad (4.103)$$

where $\mathbf{u}_i$ are the eigenvectors of the matrix $M^{-1/2}KM^{-1/2}$ and ω_i^2 are the associated eigenvalues of $M^{-1/2}KM^{-1/2}$. The constants c_i and ϕ_i are determined from the initial conditions by using the orthogonality of $\mathbf{u}_i$ to obtain

$$c_i = \frac{\mathbf{u}_i^T M \mathbf{x}(0)}{\sin \phi_i}$$

$$\phi_i = \tan^{-1} \frac{\omega_i \mathbf{u}_i^T M \mathbf{x}(0)}{\mathbf{u}_i^T M \dot{\mathbf{x}}(0)}$$

and reviewed in Window 6.3, the distributed-parameter equivalent consists of an expansion in terms of the mode shape functions $\sin \sigma_n x$. Just as the mode shape vectors determine the relative magnitude of the motion of various masses of the lumped system, the mode shape functions determine the magnitude of the motion of the mass distribution of the distributed-parameter system. The procedures outlined in this section constitute the modal analysis of a distributed-parameter system. This basic procedure is repeated in the following sections to determine the vibration response of a variety of different distributed-parameter systems. Modal analysis for damped systems is given in Section 6.7 and for the forced response in Section 6.8.

Example 6.2.2

For the sake of obtaining a feeling for the units of a vibrating string, consider a piano wire. A reasonable model of a piano wire is the string fixed at both ends of Figure 6.1. A piano wire is 1.4- m long, has a mass of about 110 g, and has a tension of about $\tau = 11.1 \times 10^4$ N. Its first natural frequency is then

$$\omega_1 = \frac{\pi c}{l} = \frac{\pi}{1.4 \text{ m}} \sqrt{\frac{\tau}{\rho A}} \text{ rad/s}$$

Since $\rho A = 110$ g per 1.4 m $= 0.0786$ kg/m, this yields

$$\omega_1 = \frac{\pi}{1.4 \text{ m}} \left(\frac{11.1 \times 10^4 \text{ N}}{0.0786 \text{ kg/m}} \right)^{1/2} = 2666.69 \text{ rad/s} \quad \text{or} \quad f_1 = 424 \text{ Hz}$$

□

Example 6.2.3

Calculate the mode shapes and natural frequencies of the spring–cable system of Example 6.1.1. The solution of the string equation will be the same as that presented for the string fixed at both ends except for the evaluation of the constants by using the boundary conditions.

Solution   Referring to equation (6.17) and applying the boundary conditions given in Example 6.1.1 yields

$$X(0) = a_2 = 0 \tag{6.48}$$

at one end. At the other end, substitution of $w(x, t) = X(x)T(t)$ into the boundary condition yields

$$\tau X'(l)T(t) = -kX(l)T(t) \tag{6.49}$$

Upon further substitution of $X = a_1 \sin \sigma x$ this becomes

$$\tau \sigma \cos \sigma l = -k \sin \sigma l \tag{6.50}$$

so that

$$\tan \sigma l = -\frac{\tau \sigma}{k} \tag{6.51}$$

As in the case of the string fixed at both ends, satisfying the second boundary condition yields an infinite number of values of σ. Hence σ becomes σ_n, where σ_n are the solutions of equation (6.51). These values of σ_n can be visualized (and computed) by plotting $\tan \sigma l$ and $-\tau \sigma / k$ versus σ on the same plot, as illustrated in Figure 6.4. Thus the eigenfunctions or mode shapes of a cable connected to a spring are

$$a_n \sin \sigma_n x, n = 1, 2, 3, \ldots \infty \tag{6.52}$$

where the σ_n must be calculated numerically for given values of k, τ, and l from equation (6.51). Note that by examining the curves of Figure 6.4, the values of σ_n (the point where the two curves cross) for large values of n become very close to $(2n - 1)\pi/(2l)$.

□

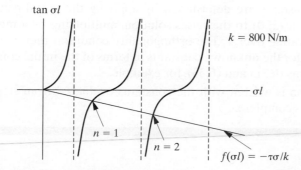

Figure 6.4   A graphical solution of the transcendental equation $\tan \sigma l = -\tau \sigma / k$.

The separation-of-variables modal analysis method illustrated in this section is summarized in Window 6.4. Note that the procedure is analogous to that summarized in Window 6.3 for lumped-parameter systems.

Window 6.4
Summary of the Separation-of-Variables Solutions Method

1. A solution to a partial differential equation is assumed to be of the form $w(x, t) = X(x)T(t)$ (i.e., separated).

2. The separated form is next substituted into the partial differential equation of motion and separated by trying to position all of the terms containing $X(x)$ on one side of the equality, and all of the terms containing $T(t)$ on the other side. (Note that this cannot always be done.)

3. Once written in separated form, equation (6.12), for example, each side of the equality is set equal to the same constant to obtain two new equations: a spatial equation in the single-valued variable $X(x)$ and a temporal equation in the single-valued variable $T(t)$. This constant is called the separation constant and is denoted by σ^2.

4. The boundary conditions are applied to the spatial equation which results in an algebraic eigenvector problem that yields an infinite number of values of the separation constant, σ_n, and the mode shapes $X_n(x)$.

5. The separation constants obtained in step 4 are substituted into the temporal equation, and the solutions $T_n(t)$ are obtained in terms of two arbitrary constants.

6. The infinite number of solutions $w_n(x, t) = X_n(x)T_n(t)$ are then formed and summed to produce the series solution

$$w(x, t) = \sum_{n=1}^{\infty} X_n(x)T_n(t)$$

which contains two sets of undetermined constants.

7. The unknown constants are determined by applying the initial conditions $w(x, 0)$ and $w_t(x, 0)$ to the series solution, multiplying by a mode $X_m(x)$, and integrating over x. The orthogonality condition then yields simple equations for the unknown constants in terms of the initial conditions. See equations (6.31) and (6.33), for example.

8. The series solution is written with the constants evaluated from step 7, and the solution is complete.

6.3 VIBRATION OF RODS AND BARS

Next consider the vibration of an elastic bar (or rod) of length l and of varying cross-sectional area in the direction indicated in Figure 6.5. The density of the bar is denoted by ρ (not to be confused with mass per unit length used for the string) and the cross-sectional area by $A(x)$. Using the coordinate system indicated in the figure, the forces on the infinitesimal element summed in the x direction are

$$F + dF - F = \rho A(x)dx \frac{\partial^2 w(x, t)}{\partial t^2} \tag{6.53}$$

where $w(x, t)$ is the deflection of the rod in the x direction, F denotes the force acting on the infinitesimal element to the left, and $F + dF$ denotes the force to the right, as the element is displaced. From introductory strength of materials, the force F is given by $F = \sigma_s A$, where σ_s is the unit stress in the x direction and has the value $Ew_x(x, t)$, where E is the Young's modulus (or modulus of elasticity), and w_x is the unit strain. This yields

$$F = EA(x) \frac{\partial w(x, t)}{\partial x} \tag{6.54}$$

The differential force becomes $dF = (\partial F/\partial x)\,dx$, from the chain rule for partial derivatives. Substitution of these quantities into equation (6.53) and dividing by dx yields

$$\frac{\partial}{\partial x}\left(EA(x)\frac{\partial w(x, t)}{\partial x}\right) = \rho A(x)\frac{\partial^2 w(x, t)}{\partial t^2} \tag{6.55}$$

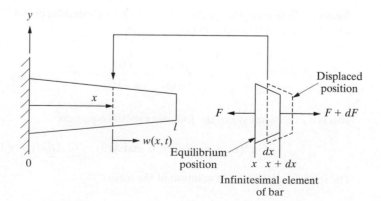

Figure 6.5   A cantilevered bar in longitudinal vibration along x.

In those cases where $A(x)$ is a constant this becomes

$$\left(\frac{E}{\rho}\right)\frac{\partial^2 w(x,t)}{\partial x^2} = \frac{\partial^2 w(x,t)}{\partial t^2} \tag{6.56}$$

which has exactly the same form as the string equation of Section 6.2. The quantity $c = \sqrt{E/\rho}$ defines the velocity of propagation of the displacement (or stress wave) in the bar. Hence the solution technique used in Section 6.2 is exactly applicable here.

Because the bar can support its own weight, a variety of boundary conditions are possible. Various springs, masses, or dashpots can be attached to one end of the bar or the other, to model a variety of situations. Consider the boundary conditions for the clamped–free configuration (called cantilevered) of Figure 6.5. At the clamped end, $x = 0$, the displacement must be zero, so that

$$w(0,t) = 0 \quad t > 0 \tag{6.57}$$

At the free end, $x = l$, the force in the bar must be zero or

$$EA\left.\frac{\partial w(x,t)}{\partial x}\right|_{x=l} = 0 \tag{6.58}$$

These are the simplest boundary conditions.

Example 6.3.1

Calculate the mode shapes and natural frequencies of the cantilevered (clamped–free) bar of Figure 6.5 for the case of constant cross section.

Solution   Following the method of separation of variables outlined in Window 6.4, let $w(x,t) = X(x)T(t)$ in equation (6.56) to get

$$\frac{X''(x)}{X(x)} = \frac{\ddot{T}(t)}{c^2 T(t)} = -\sigma^2 \tag{6.59}$$

This last expression yields the following spatial equation:

$$X''(x) + \sigma^2 X(x) = 0, \text{ and } X(0) = 0,\, AEX'(l) = 0$$

The spatial equation has a solution of the form

$$X(x) = a\sin\sigma x + b\cos\sigma x$$

Applying the condition at the fixed end, $x = 0$, yields the result that $b = 0$, so that $X(x)$ has the form

$$X(x) = a \sin \sigma x$$

At $x = l$, the boundary condition yields

$$AEX'(l) = 0 = AE\sigma a \cos \sigma l \tag{6.60}$$

Since A, E, a, and σ are nonzero, this requires that $\cos \sigma l = 0$, or that

$$\sigma_n = \frac{2n - 1}{2l} \pi \quad n = 1, 2, 3, \ldots \tag{6.61}$$

The *mode shapes* are thus of the form

$$a_n \sin \frac{(2n - 1)\pi x}{2l} \tag{6.62}$$

Now consider the temporal equation resulting from equation (6.59):

$$\ddot{T}(t) + c^2\sigma^2 T(t) = 0 \Rightarrow T(t) = A \sin \sigma ct + B \cos \sigma ct$$

This suggests an oscillation with frequency σc, or natural frequencies of the form

$$\omega_n = \frac{(2n - 1)\pi c}{2l} = \frac{(2n - 1)\pi}{2l} \sqrt{\frac{E}{\rho}} \, \text{rad/s} \quad n = 1, 2, 3, \ldots \tag{6.63}$$

The solution is given by equation (6.27) with $c = \sqrt{E/\rho}$ and σ_n as given by expression (6.61).

□

In distributed-parameter systems, the form of the spatial differential equation is often determined by the stiffness term. The differential form of the stiffness term along with the boundary conditions determine whether or not the eigenfunctions, and hence the mode shapes, are orthogonal. The eigenfunction orthogonality condition for distributed-parameter systems is an exact replica of the eigenvector and mode shape orthogonality condition for lumped-parameter systems. Precise orthogonality conditions are beyond the scope of this chapter and can be found in Inman (2006), for instance. Note in Example 6.3.1 that the mode shapes given by expression (6.62) are orthogonal [recall equation (6.28)]. Just as in the lumped-parameter case, the orthogonality of mode shapes for a distributed-parameter system is extremely useful in calculating the vibration response of the system. The following example illustrates the use of modes to calculate the response to a given set of initial conditions.

Example 6.3.2

Calculate the response of the bar of Example 6.3.1 to an initial velocity of 3 cm/s at the free end and a zero initial displacement. Assume that the bar is 5 m long, has a density of $\rho = 8 \times 10^3$ kg/m^3, and has a modulus of $E = 20 \times 10^{10}$ N/m^2. Plot the response at $x = l$ and $x = l/2$.

Solution   The solution $w(x, t)$ for the longitudinal vibration of a clamped–free bar is given by equation (6.27) as

$$w(x, t) = \sum_{n=1}^{\infty} (c_n \sin \sigma_n ct + d_n \cos \sigma_n ct) \sin \frac{(2n - 1)\pi x}{2l}$$

where σ_n is calculated in Example 6.3.1 and given in equation (6.61). To calculate the coefficients c_n and d_n, note that $w(x, 0) = 0$ and $w_t(l, 0) = 0.03$ m/s. Thus from equation (6.31)

$$d_n = \frac{2}{l} \int_0^l w_0(x) \sin \frac{n\pi x}{l} \, dx = \frac{2}{l} \int_0^l (0) \sin \frac{n\pi x}{l} \, dx = 0$$

The coefficients c_n are computed from differentiating the series $w(x, t)$ and setting $t = 0$:

$$w_t(x, 0) = 0.03\delta(x - l) = \sum_{n=1}^{\infty} c_n \sigma_n c \sin \sigma_n x \cos (0)$$

where δ is the Dirac delta function. Multiplying by $\sin \sigma_m x$ and integrating yields

$$0.03 \int_0^l \delta(x - l) \sin(\sigma_m x) dx = \sum_{n=1}^{\infty} c_n \sigma_n c \int_0^l \sin (\sigma_n x) \sin (\sigma_m x) dx$$

Because $\{\sin \sigma_m x\}$ is an orthogonal set of functions, each term on the right side in the summation is zero except for the one term when $n = m$:

$$\int_0^l \sin\left(\frac{2n - 1}{2l} \pi x\right) \sin\left(\frac{2m - 1}{2l} \pi x\right) dx = \begin{cases} l/2, & n = m \\ 0, & n \neq m \end{cases}$$

and the integral on the left side can be evaluated using the filtering property of the Dirac delta function from Section 3.1, $\int f(x)\delta(x - a)dx = f(a)$. Thus the preceding equation becomes

$$0.03 \sin\left(\frac{(2m - 1)\pi}{2}\right) = \frac{c\sigma_m l}{2} c_m$$

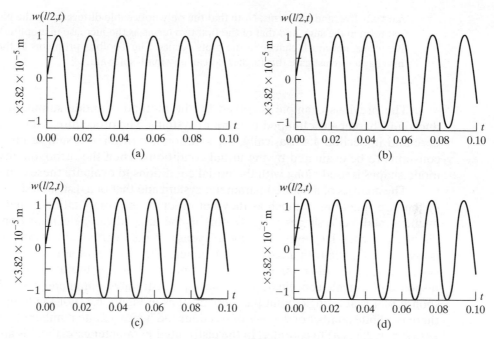

Figure 6.6   Response of the bar of Example 6.3.2 plotted for $x = l/2$ for (a) one, (b) two, (c) five, and (d) ten terms of the series solution. Note the slight amplitude changes.

The numerator on the left-hand side can be written as ± 1 depending on the value of the sine function. Solving for c_m, evaluating the physical constants, and renaming m by n yields the equation

$$c_n = \frac{0.06(-1)^{n+1}}{c\sigma_n l} = \frac{0.6l(-1)^{n+1}}{c\pi(2n-1)} = 3.82 \times 10^{-6} \frac{(-1)^{n+1}}{(2n-1)} \quad m,n = 1, 2, 3, \ldots$$

Thus the total solution becomes

$$w(x, t) = 3.82 \times 10^{-6} \sum_{n=1}^{\infty} \frac{(-1)^{n+1}}{(2n-1)} \sin \frac{(2n-1)\pi x}{2l} \sin[512.35(2n-1)\pi t] \text{m}$$

At $x = l$ this becomes

$$w(l, t) = 3.82 \times 10^{-6} \sum_{n=1}^{\infty} \frac{(-1)^{n+1}}{(2n-1)} \sin[512.35(2n-1)\pi t] \text{m}$$

and at $x = l/2$ the response is given by

$$w(l/2, t) = 3.82 \times 10^{-5} \sum_{n=1}^{N} \frac{(-1)^{n+1}}{(2n-1)^2} \sin(512.35)(2n-1)\pi t \text{ m}$$

where $N = \infty$. To plot these, the infinite series must be approximated by truncating the series taking finite values of N. In Figure 6.6, the response of $w(l/2, t)$ is plotted for one, two, five, and ten terms. Note that the only noticeable difference in the plot of the first term in the sum, and that of the first two terms, is the increase in amplitude. After five terms, not much change in the plots occurs. The Toolbox problems at the end of this chapter investigate this further, as can any math software.

□

The previous example is solved by the method of mode summation introduced in Chapter 4 for lumped-parameter systems. The procedure is the same both here and in Section 4.4. Basically, the solution is written as a sum of modes with constants to be evaluated by the initial conditions. Then the orthogonality of the mode shapes is used along with the initial conditions to evaluate these constants.

The analysis of a lumped-parameter system and that of a distributed-parameter system are similar in as much as they both require the computation of natural frequencies and mode shapes (eigenvectors in the lumped case, eigenfunctions in the distributed case). However, the time response of the distributed-parameter system is distributed spatially in a continuous manner, whereas that of a lumped-parameter system is spatially discrete. Thus for a lumped system it is useful to plot the time response of a single coordinate $x_i(t)$, and for a distributed-parameter system the response plot is at a single point [i.e., $w(l/2, t)$] as in Figure 6.6. For the lumped case, there is a finite number of degrees of freedom and hence a finite number of responses $x_i(t)(i = 1, 2, \ldots n)$ to consider. In the distributed-parameter case, there is an infinite number of responses $w(x, t)$ to consider, as the spatial variable x can take on any value between 0 and l. Boundary conditions for various configurations of the bar are listed in Table 6.1. Table 6.2 lists the various frequencies and mode shapes using these boundary conditions.

TABLE 6.1 A SUMMARY OF VARIOUS BOUNDARY CONDITIONS FOR THE LONGITUDINAL VIBRATION OF THE BAR OF FIGURE 6.5

Fixed at left end: $w(x, t)\big|_{x=0} = 0$

Fixed at right end: $w(x, t)\big|_{x=l} = 0$

Free at left end: $EA \dfrac{\partial w(x, t)}{\partial x}\bigg|_{x=0} = 0$

Free at right end: $EA \dfrac{\partial w(x, t)}{\partial x}\bigg|_{x=l} = 0$

Attached to a mass of mass m at left end: $AE \dfrac{\partial w(x, t)}{\partial x}\bigg|_{x=0} = m \dfrac{\partial^2 w(x, t)}{\partial t^2}\bigg|_{x=0}$

Attached to a mass of mass m at right end: $AE \dfrac{\partial w(x, t)}{\partial x}\bigg|_{x=l} = -m \dfrac{\partial^2 w(x, t)}{\partial t^2}\bigg|_{x=l}$

Attached to a spring of stiffness k at left end: $AE \dfrac{\partial w(x, t)}{\partial x}\bigg|_{x=0} = kw(x, t)\big|_{x=0}$

Attached to a spring of stiffness k at right end: $AE \dfrac{\partial w(x, t)}{\partial x}\bigg|_{x=l} = -kw(x, t)\big|_{x=l}$

TABLE 6.2   VARIOUS CONFIGURATIONS OF A UNIFORM BAR OF LENGTH *l* IN LONGITUDINAL VIBRATION ILLUSTRATING THE NATURAL FREQUENCIES AND MODE SHAPES[a]

Configuration	Frequency (rad/s) or characteristic equation	Mode shape
Free–free	$\omega_n = \dfrac{n\pi c}{l}$, $n = 0, 1, 2, \ldots$	$\cos \dfrac{n\pi x}{l}$[b]
Fixed–free	$\omega_n = \dfrac{(2n-1)\pi c}{2l}$, $n = 1, 2, \ldots$	$\sin \dfrac{(2n-1)\pi x}{2l}$
Fixed–fixed	$\omega_n = \dfrac{n\pi c}{l}$, $n = 1, 2, \ldots$	$\sin \dfrac{n\pi x}{l}$
Fixed–spring	$\lambda_n \cot \lambda_n = -\left(\dfrac{kl}{EA}\right)$ $\omega_n = \dfrac{\lambda_n c}{l}$	$\sin \dfrac{\lambda_n x}{l}$
Fixed–mass	$\cot \lambda_n = \left(\dfrac{m}{\rho A l}\right)\lambda_n$ $\omega_n = \dfrac{\lambda_n c}{l}$	$\sin \dfrac{\lambda_n x}{l}$

[a]Note that the last two examples require a numerical procedure (suggested in the Toolbox problems at the end of the chapter) in order to calculate the values of the natural frequencies. Here $c = \sqrt{E/\rho}$.
[b]The free mode shape is a constant.

6.4 TORSIONAL VIBRATION

The bar of Section 6.3 may also vibrate in the torsional direction as indicated by the circular shaft of Figure 6.7. In this case, the vibration occurs in an angular direction around the center axis of the shaft in a plane perpendicular to the cross section of the shaft or rod. The rotation of the shaft, θ, about the center axis is a function of both the position along the length of the rod, x, and the time, t. Thus θ is a function of two variables denoted $\theta(x, t)$. The equation of motion can be determined by considering a moment balance of an infinitesimal element of the rod of length dx illustrated in

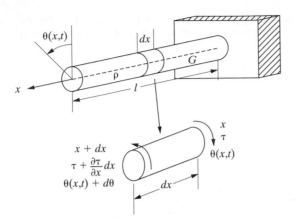

Figure 6.7　A circular shaft illustrating an angular motion, $\theta(x, t)$, as the result of a moment acting on a differential element dx of the shaft of density ρ, length l, and given modulus G. The function $\theta(x, t)$ denotes the angle of twist.

the figure. Referring to Figure 6.7, the torque at the right face of the element dx (i.e., at position x) is τ (here τ is used as a torque, not as tension as in Section 6.1), while that at the left end, at position $x + dx$, is $\tau + \dfrac{\partial \tau}{\partial x}\, dx$. From solid mechanics, the applied torque is related to the torsional deflection by (Shames, 1989)

$$\tau = GJ \frac{\partial \theta(x, t)}{\partial x} \tag{6.64}$$

where GJ is the torsional stiffness composed of the shear modulus G and the polar moment of area J of the cross section. Note that J could be a function of x as well, but is considered constant here.

The total torque acting on dx becomes

$$\tau + \frac{\partial \tau}{\partial x}\, dx - \tau = J_0 \frac{\partial^2 \theta}{\partial t^2}\, dx \tag{6.65}$$

where J_0 is the polar moment of inertia of the shaft per unit length and $\partial^2 \theta/\partial t^2$ is the angular acceleration. If the shaft is of uniform circular cross section, J_0 becomes simply $J_0 = \rho J$, where ρ is the shaft's material mass density. Substitution of the expression for the torque given by equation (6.64) into equation (6.65) yields

$$\frac{\partial}{\partial x}\left(GJ \frac{\partial \theta}{\partial x} \right) = \rho J \frac{\partial^2 \theta}{\partial t^2}$$

Simplifying for the case of constant stiffness GJ yields

$$\frac{\partial^2 \theta(x, t)}{\partial t^2} = \left(\frac{G}{\rho} \right) \frac{\partial^2 \theta(x, t)}{\partial x^2} \tag{6.66}$$

for the equation of twisting vibration of a rod. This is, again, identical in mathematical form to the formula of equation (6.8) for the string or cable and the formula of equation (6.56) for the longitudinal vibrations of a rod or bar.

For other types of cross sections, the torsional equation of a shaft can still be used to approximate the torsional motion by replacing J in equation (6.64) with a *torsional constant* γ defined to be the moment required to produce a torsional rotation of 1 rad on a unit length of shaft divided by the shear modulus. Thus a shaft with noncircular cross section can be approximated by the equation

$$\frac{\partial^2 \theta(x, t)}{\partial t^2} = \left(\frac{G\gamma}{\rho J}\right) \frac{\partial^2 \theta(x, t)}{\partial x^2} \tag{6.67}$$

Some values of γ are presented in Table 6.3 for several common cross sections. Keep in mind that equation (6.67) is approximate and assumes in particular that the

TABLE 6.3   SOME VALUES OF THE TORSIONAL CONSTANT FOR VARIOUS CROSS-SECTIONAL SHAPES USED IN APPROXIMATING TORSIONAL VIBRATION FOR NONCIRCULAR CROSS SECTIONS

Cross section	Torsional constant, γ
R (circle)	Circular shaft $\dfrac{\pi R^4}{2}$
R_2, R_1 (hollow circle)	Hollow circular shaft $\dfrac{\pi}{2}(R_2^4 - R_1^4)$
$a \times a$ (square)	Square shaft $0.1406a^4$
B, A, b, a (hollow rectangle)	Hollow rectangular shaft $\dfrac{2AB(a - A)^2 (b - B)^2}{aA + bB - A^2 - B^2}$

center of mass and center of rotation coincide so that torsional and flexural vibrations do not couple.

The solution of equation (6.66) depends on two initial conditions in time [i.e., $\theta(x, 0)$ and $\theta_t(x, 0)$] and two boundary conditions, one at each end of the rod. The possible choices of boundary conditions are similar to those for the string and the bar (i.e., either the deflection is zero if the rod is fixed at a boundary or the torque is zero if the rod is free at a boundary). For example, the clamped–free rod of Figure 6.7 has boundary conditions

$$\text{(deflection at 0)} \quad \theta(0, t) = 0 \tag{6.68}$$

$$\text{(torque at } l) \quad \gamma G \theta_x(l, t) = 0 \tag{6.69}$$

Table 6.4 lists some common boundary conditions for the torsional vibration of shafts.

TABLE 6.4   A SUMMARY OF VARIOUS BOUNDARY CONDITIONS FOR THE TORSIONAL VIBRATION OF THE SHAFT OF FIGURE 6.7

Fixed at left end: $\theta(x, t)\big|_{x=0} = 0$

Fixed at right end: $\theta(x, t)\big|_{x=l} = 0$

Free at left end: $\gamma G \left. \dfrac{\partial \theta(x, t)}{\partial x} \right|_{x=0} = 0$

Free at right end: $\gamma G \left. \dfrac{\partial \theta(x, t)}{\partial x} \right|_{x=l} = 0$

Attached to an inertial mass J_1 at left end: $\gamma G \left. \dfrac{\partial \theta(x, t)}{\partial x} \right|_{x=0} = J_1 \left. \dfrac{\partial^2 \theta(x, t)}{\partial t^2} \right|_{x=0}$

Attached to an inertial mass J_1 at right end: $\gamma G \left. \dfrac{\partial \theta(x, t)}{\partial x} \right|_{x=l} = -J_1 \left. \dfrac{\partial^2 \theta(x, t)}{\partial t^2} \right|_{x=l}$

Attached to a rotational spring of stiffness k at left end: $\gamma G \left. \dfrac{\partial \theta(x, t)}{\partial x} \right|_{x=0} = k\theta(x, t) \big|_{x=0}$

Attached to a rotational spring of stiffness k at right end: $\gamma G \left. \dfrac{\partial \theta(x, t)}{\partial x} \right|_{x=l} = -k\theta(x, t) \big|_{x=l}$

Example 6.4.1

The vibration of a large grinding machine can be modeled as a shaft or rod with a disk at one end as illustrated in Figure 6.8. The top end of the shaft, at $x = 0$, is connected to a pulley. The effects of the drive belt and motor are accounted for by including their collective inertia in with the mass moment of inertia of the pulley, denoted by J_1. Determine the natural frequencies of the system.

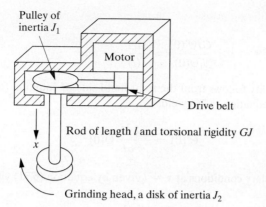

Pulley of
inertia J_1

Motor

Drive belt

Rod of length l and torsional rigidity GJ

x

Grinding head, a disk of inertia J_2

Figure 6.8   A simple model of a large grinding machine.

Solution   The frequencies of vibration are determined by equation (6.66) subject to the appropriate boundary conditions. The boundary conditions are determined by examining either the deflections or torques at the boundaries. In this example, the deflection is unspecified at $x = 0$, but the torque must match that supplied by the pulley or

$$GJ \frac{\partial \theta(0,t)}{\partial x} = J_1 \frac{\partial^2 \theta(0,t)}{\partial t^2} \tag{6.70}$$

Similarly, at $x = l$ a balance of torques yields

$$GJ \frac{\partial \theta(l,t)}{\partial x} = -J_2 \frac{\partial^2 \theta(l,t)}{\partial t^2} \tag{6.71}$$

where the minus sign arises from the right-hand rule. Following the method of Section 6.2, the solution for $\theta(x, t)$ is assumed to separate and be of the form $\theta(x, t) = \Theta(x)T(t)$. Substitution into equation (6.66) and rearranging yields

$$\frac{\Theta''(x)}{\Theta(x)} = \left(\frac{\rho}{G}\right) \frac{\ddot{T}(t)}{T(t)} = -\sigma^2 \tag{6.72}$$

Defining $(\rho/G) = 1/c^2$ and realizing that equation (6.72) splits into a temporal equation and a spatial equation, the spatial equation becomes

$$\Theta''(x) + \sigma^2 \Theta(x) = 0 \tag{6.73}$$

where σ is the separation constant and is related to the natural frequencies of the system by

$$\omega = \sigma c = \sigma \sqrt{\frac{G}{\rho}} \tag{6.74}$$

From equation (6.70) the boundary condition at $x = 0$ becomes

$$GJ\Theta'(0)T(t) = J_1\Theta(0)\ddot{T}(t) \tag{6.75}$$

or

$$\frac{GJ\Theta'(0)}{J_1\Theta(0)} = \frac{\ddot{T}(t)}{T(t)} = -c^2\sigma^2 \qquad (6.76)$$

where the last equality follows from the right-hand side of equation (6.72). Upon further manipulation, equation (6.76) yields

$$\Theta'(0) = -\frac{\sigma^2 J_1}{\rho J}\Theta(0) \qquad (6.77)$$

Similarly, the boundary conditions at $x = l$ given by equation (6.71) yield

$$\Theta'(l) = \frac{\sigma^2 J_2}{\rho J}\Theta(l) \qquad (6.78)$$

The general solution of equation (6.73) is

$$\Theta(x) = a_1\sin\sigma x + a_2\cos\sigma x \qquad (6.79)$$

so that

$$\Theta'(x) = a_1\sigma\cos\sigma x - a_2\sigma\sin\sigma x \qquad (6.80)$$

Substitution of these expressions into equation (6.77) for the boundary condition at $x = 0$ yields

$$a_1 = -\frac{\sigma J_1}{\rho J}a_2 \qquad (6.81)$$

Application of the boundary conditions at $x = l$ by substitution of equations (6.79), (6.80), and (6.81) into equation (6.78) yields the characteristic equation given by

$$\tan\sigma l = \frac{\rho J l(J_1 + J_2)(\sigma l)}{J_1 J_2(\sigma l)^2 - \rho^2 J^2 l^2} \qquad (6.82)$$

This expression is a transcendental equation in the quantity σl, which must be solved numerically, similar to Example 6.2.3. Because of the tangent term, equation (6.82) has an infinite number of solutions, which can be denoted $\sigma_n l$, $n = 1, 2, 3, \ldots \infty$. The numerical solutions σ_n determine the natural frequencies of vibrations according to equation (6.74):

$$\omega_n = \sigma_n\sqrt{\frac{G}{\rho}} \qquad (6.83)$$

Note that the first solution of equation (6.82) corresponding to the first eigenvalue yields

$$\omega_1 = 0$$

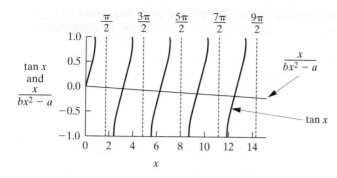

Figure 6.9 The plot of $\tan x$ versus x and $x/(bx^2 - a)$ versus x. These points of intersection are determined using the `fzero` command in MATLAB, as discussed at the end of the problem section. This yields the roots of equation (6.84).

Since $\omega_1 = \sigma_1 = 0$, the time equation (6.72) becomes $\ddot{T}_1(t) = 0$, or $T_1(t) = a + bt$, where a and b are constants determined by the initial conditions. This solution corresponds to the rigid-body mode of the system, which in this case is a constant shaft rotation. The rigid-body mode shape is calculated from equation (6.73) with $\sigma = 0$. This yields $\Theta_1''(x) = 0$, so that $\Theta_1(x) = a_1 + b_1 x$. Applying the boundary condition at $x = 0$ yields $GJ(b_1) = 0$, so that $b_1 = 0$. The boundary condition at $x = l$ also results in $b_1 = 0$, so that the first mode shape, or first eigenfunction, is simply $\Theta_1(x) = a_1$, a nonzero constant. This defines the rigid-body mode shape.

For very large values of σl, equation (6.82) reduces to $\tan \sigma l = 0$, so that the high frequencies of vibration approach $\omega_n = n\pi c$. Examining the form of the characteristic equation indicates that it can be written as

$$(bx^2 - a)\tan x = x \tag{6.84}$$

where $x = \sigma l$, $a = \rho\, Jl/(J_1 + J_2)$, and $b = J_1 J_2/[(J_1 + J_2)\,\rho\, Jl]$. The solution of this expression for the case $J_1 = 10$ kg·m^2/rad, $J_2 = 10$ kg·m^2/rad, $\rho = 7870$ kg/m^3, $J = 5$ m^4, $l = 0.425$ m is illustrated by the points of intersection of the two curves in Figure 6.9. These points of intersection are determined by using MATLAB to solve for the roots of equation (6.84). For $G = 80 \times 10^9$ Pa, this results in

$$f_1 = 0 \qquad\qquad f_2 = 3{,}980 \text{ Hz} \qquad f_3 = 7{,}961 \text{ Hz}$$

$$f_4 = 11{,}940 \text{ Hz} \qquad f_5 = 15{,}920 \text{ Hz} \qquad f_6 = 19{,}900 \text{ Hz}$$

with the higher frequencies approaching $n\pi c$ in rad/s.

$\square$

The natural frequencies for various configurations of torsional systems are listed in Table 6.5. Note the similarity to Table 6.2 for longitudinal vibration. The only difference is the physical interpretation of the motion. Table 6.5 can be combined with Table 6.3 to approximate the natural frequencies of torsional vibration for noncircular cross sections as well. A more complete tabulation can be found in Blevins (1987).

TABLE 6.5 SAMPLE OF VARIOUS CONFIGURATIONS OF A UNIFORM SHAFT IN TORSIONAL VIBRATION, OF LENGTH l, ILLUSTRATING THE NATURAL FREQUENCIES AND MODE SHAPES[a]

Configuration	Frequency (rad/s) or characteristic equation	Mode shape
Free–free	$\omega_n = \dfrac{n\pi c}{l}, \quad n = 0, 1, 2, \ldots$	$\cos \dfrac{n\pi x}{l}$ [b]
Fixed–free	$\omega_n = \dfrac{(2n-1)\pi c}{2l}, \quad n = 1, 2, \ldots$	$\sin \dfrac{(2n-1)\pi x}{2l}$
Fixed–fixed	$\omega_n = \dfrac{n\pi c}{l}, \quad n = 1, 2, \ldots$	$\sin \dfrac{n\pi x}{l}$
Fixed–spring (rotational)	$\lambda_n \cot \lambda_n = -\dfrac{kl}{G\gamma}$ $\omega_n = \dfrac{\lambda_n c}{l}$	$\sin \dfrac{\lambda_n x}{l}$
Fixed–inertia	$\cot \lambda_n = \dfrac{J_0}{\rho l \gamma} \lambda_n$ $\omega_n = \dfrac{\lambda_n c}{l}$	$\sin \dfrac{\lambda_n x}{l}$

[a] Here $c = \sqrt{G\gamma/\rho J}$. The values of γ can be found in Table 6.3. Note that a numerical procedure is required for the last two cases, as illustrated in Example 6.4.1.
[b] The free mode shape is a constant.

6.5 BENDING VIBRATION OF A BEAM

This section again considers the vibration of the bar or beam of Figure 6.5. In this case, however, vibration of the beam in the direction perpendicular to its length is considered. Such vibrations are often called *transverse vibrations*, or *flexural vibrations*, because they move across the length of the beam. Transverse vibration is easily felt by humans when walking over bridges, for example.

Euler–Bournoulli Beam Theory

Figure 6.10 illustrates a cantilevered beam with the transverse direction of vibration indicated [i.e., the deflection, $w(x, t)$, is in the y direction]. The beam is of rectangular cross section $A(x)$ with width h_y, thickness h_z, and length l. Also associated with the beam is a flexural (bending) stiffness $EI(x)$, where E is Young's elastic modulus for the beam and $I(x)$ is the cross-sectional area moment of inertia about the "z axis." From mechanics of materials, the beam sustains a bending moment $M(x, t)$, which is related to the beam deflection, or bending deformation, $w(x, t)$, by

$$M(x, t) = EI(x) \frac{\partial^2 w(x, t)}{\partial x^2} \tag{6.85}$$

A model of bending vibration may be derived from examining the force diagram of an infinitesimal element of the beam as indicated in Figure 6.10. Assuming the deformation to be small enough such that the shear deformation is much smaller than

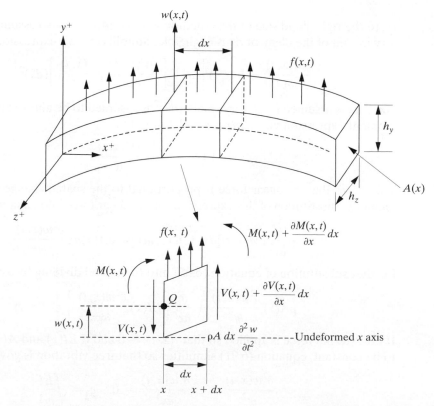

Figure 6.10   A simple Euler–Bernoulli beam of length (l) in transverse vibration and a free-body diagram of a small element of the beam as it is deformed by a distributed force-per-unit length, denoted by $f(x, t)$.

$w(x, t)$ (i.e., so that the sides of the element dx do not bend), a summation of forces in the y direction yields

$$\left(V(x, t) + \frac{\partial V(x, t)}{\partial x} dx\right) - V(x, t) + f(x, t)dx = \rho A(x) dx \frac{\partial^2 w(x, t)}{\partial t^2} \qquad (6.86)$$

Here $V(x, t)$ is the shear force at the left end of the element dx, $V(x, t) + V_x(x, t) dx$ is the shear force at the right end of the element dx, $f(x, t)$ is the total external force applied to the element per unit length, and the term on the right side of the equality is the inertial force of the element. The assumption of small shear deformation used in the force balance of equation (6.86) is true if $l/h_z \geq 10$ and $l/h_y \geq 10$ (i.e., for long slender beams or Euler–Bernoulli beams).

Next, the moments acting on the element dx about the z axis through point Q are summed. This yields

$$\left[M(x, t) + \frac{\partial M(x, t)}{\partial x} dx\right] - M(x, t) + \left[V(x, t) + \frac{\partial V(x, t)}{\partial x} dx\right] dx + [f(x, t)dx] \frac{dx}{2} = 0 \qquad (6.87)$$

Here the right-hand side of the equation is zero since it is also assumed that the rotary inertia of the element dx is negligible. Simplifying this expression yields

$$\left[\frac{\partial M(x, t)}{\partial x} + V(x, t)\right] dx + \left[\frac{\partial V(x, t)}{\partial x} + \frac{f(x, t)}{2}\right](dx)^2 = 0 \qquad (6.88)$$

Since dx is assumed to be very small, $(dx)^2$ is assumed to be almost zero, so that this moment expression yields (dx is small, but not zero)

$$V(x, t) = -\frac{\partial M(x, t)}{\partial x} \qquad (6.89)$$

This states that the shear force is proportional to the spatial change in the bending moment. Substitution of this expression for the shear force into equation (6.86) yields

$$-\frac{\partial^2}{\partial x^2} [M(x, t)]dx + f(x, t)dx = \rho A(x)dx \frac{\partial^2 w(x, t)}{\partial t^2} \qquad (6.90)$$

Further substitution of equation (6.85) into (6.90) and dividing by dx yields

$$\rho A(x) \frac{\partial^2 w(x, t)}{\partial t^2} + \frac{\partial^2}{\partial x^2}\left[EI(x) \frac{\partial^2 w(x, t)}{\partial x^2}\right] = f(x, t) \qquad (6.91)$$

If no external force is applied so that $f(x, t) = 0$ and if $EI(x)$ and $A(x)$ are assumed to be constant, equation (6.91) simplifies so that free vibration is governed by

$$\frac{\partial^2 w(x, t)}{\partial t^2} + c^2 \frac{\partial^4 w(x, t)}{\partial x^4} = 0, \quad c = \sqrt{\frac{EI}{\rho A}} \qquad (6.92)$$

Note that unlike the previous equations, the free vibration equation (6.92) contains four spatial derivatives and hence requires four (instead of two) boundary conditions in calculating a solution. The presence of the two time derivatives again requires that two initial conditions, one for the displacement and one for the velocity, be specified.

The boundary conditions required to solve the spatial equation in a separation-of-variables solution of equation (6.92) are obtained by examining the deflection $w(x, t)$, the slope of the deflection $\partial w(x, t)/\partial x$, the bending moment $EI\partial^2 w(x, t)/\partial x^2$, and the shear force $\partial[EI\partial^2 w(x, t)/\partial x^2]/\partial x$ at each end of the beam. A common configuration is *clamped–free* or *cantilevered* as illustrated in Figure 6.10. In addition to a boundary being clamped or free, the end of a beam could be resting on a support restrained from bending or deflecting. The situation is called *simply supported* or *pinned*. A *sliding* boundary is one in which displacement is allowed but rotation is not. The shear load at a sliding boundary is zero.

If a beam in transverse vibration is free at one end, the deflection and slope at that end are unrestricted, but the bending moment and shear force must vanish:

$$\text{bending moment} = EI\frac{\partial^2 w}{\partial x^2} = 0$$

$$\text{shear force} = \frac{\partial}{\partial x}\left[EI\frac{\partial^2 w}{\partial x^2}\right] = 0 \tag{6.93}$$

If, on the other hand, the end of a beam is clamped (or fixed), the bending moment and shear force are unrestricted, but the deflection and slope must vanish at that end:

$$\text{deflection} = w = 0$$

$$\text{slope} = \frac{\partial w}{\partial x} = 0 \tag{6.94}$$

At a simply supported or pinned end, the slope and shear force are unrestricted and the deflection and bending moment must vanish:

$$\text{deflection} = w = 0$$

$$\text{bending moment} = EI\frac{\partial^2 w}{\partial x^2} = 0 \tag{6.95}$$

At a sliding end, the slope or rotation is zero and no shear force is allowed. On the other hand, the deflection and bending moment are unrestricted. Hence, at a sliding boundary,

$$\text{slope} = \frac{\partial w}{\partial x} = 0$$

$$\text{shear force} = \frac{\partial}{\partial x}\left(EI\frac{\partial^2 w}{\partial x^2}\right) = 0 \tag{6.96}$$

Other boundary conditions are possible by connecting the ends of a beam to a variety of devices such as lumped masses, springs, and so on. These boundary conditions can be determined by force and moment balances.

In addition to satisfying four boundary conditions, the solution of equation (6.92) for free vibration can be calculated only if two initial conditions (in time) are specified. As in the case of the rod, string, and bar, these initial conditions are specified initial deflection and velocity profiles:

$$w(x, 0) = w_0(x) \quad \text{and} \quad w_t(x, 0) = \dot{w}_0(x)$$

assuming that $t = 0$ is the initial time. Note that if w_0 and $\dot{w}_0$ were both zero, no motion would result.

The solution of equation (6.92) subject to four boundary conditions and two initial conditions proceeds following exactly the same steps used in previous sections. A separation-of-variables solution of the form $w(x, t) = X(x)T(t)$ is assumed. This is substituted into the equation of motion, equation (6.92), to yield (after rearrangement)

$$c^2 \frac{X''''(x)}{X(x)} = -\frac{\ddot{T}(t)}{T(t)} = \omega^2 \tag{6.97}$$

where the partial derivatives have been replaced with total derivatives as before (*note:* $X'''' = d^4X/dx^4$, $\ddot{T} = d^2T/dt^2$). Here the choice of separation constant, ω^2, is made, based on experience with the systems of Section 6.4 that the natural frequency comes from the temporal equation

$$\ddot{T}(t) + \omega^2 T(t) = 0 \tag{6.98}$$

which is the right side of equation (6.97). This temporal equation has a solution of the form

$$T(t) = A \sin \omega t + B \cos \omega t \tag{6.99}$$

where the constants A and B will eventually be determined by the specified initial conditions after being combined with the spatial solution.

The spatial equation comes from rearranging equation (6.97), which yields

$$X''''(x) - \left(\frac{\omega}{c}\right)^2 X(x) = 0 \tag{6.100}$$

By defining [recall equation (6.92)]

$$\beta^4 = \frac{\omega^2}{c^2} = \frac{\rho A \omega^2}{EI} \left(\text{so that } \omega = \beta^2 \sqrt{\frac{EI}{\rho A}} \text{ rad/s} \right) \tag{6.101}$$

and assuming a solution to equation (6.100) of the form $Ae^{\sigma x}$, the general solution of equation (6.100) can be calculated to be of the form (see Problem 6.44)

$$X(x) = a_1 \sin \beta x + a_2 \cos \beta x + a_3 \sinh \beta x + a_4 \cosh \beta x \tag{102}$$

Here the value for β and three of the four constants of integration $a_1, a_2, a_3,$ and a_4 will be determined from the four boundary conditions. The fourth constant becomes combined with the constants A and B from the temporal equation, which are then determined from the initial conditions. The following example illustrates the solution procedure for a beam fixed at one end and simply supported at the other end.

Example 6.5.1

Calculate the natural frequencies and mode shapes for the transverse vibration of a beam of length l that is fixed at one end and pinned at the other end.

Solution   The boundary conditions in this case are given by equation (6.94) at the fixed end and equation (6.95) at the pinned end. Substitution of the general solution given by equation (6.102) into equation (6.94) at $x = 0$ yields

$$X(0) = 0 \Rightarrow a_2 + a_4 = 0 \tag{a}$$

$$X'(0) = 0 \Rightarrow \beta(a_1 + a_3) = 0 \tag{b}$$

Similarly, at $x = l$ the boundary conditions result in

$$X(l) = 0 \Rightarrow a_1 \sin \beta l + a_2 \cos \beta l + a_3 \sinh \beta l + a_4 \cosh \beta l = 0 \tag{c}$$

$$EIX''(l) = 0 \Rightarrow \beta^2(-a_1 \sin \beta l - a_2 \cos \beta l + a_3 \sinh \beta l + a_4 \cosh \beta l) = 0 \tag{d}$$

These four boundary conditions thus yield four equations [(a) through (d)] in the four unknown coefficients a_1, a_2, a_3, and a_4. These can be written as the single vector equation

$$\begin{bmatrix} 1 & 1 & 0 & 1 \\ \beta & 0 & \beta & 0 \\ \sin \beta l & \cos \beta l & \sinh \beta l & \cosh \beta l \\ -\beta^2 \sin \beta l & -\beta^2 \cos \beta l & \beta^2 \sin \beta l & \beta^2 \cosh \beta l \end{bmatrix} \begin{bmatrix} a_1 \\ a_2 \\ a_3 \\ a_4 \end{bmatrix} = \begin{bmatrix} 0 \\ 0 \\ 0 \\ 0 \end{bmatrix}$$

Recall from Chapter 4 that this vector equation can have a nonzero solution for the vector $\mathbf{a} = [a_1 \ a_2 \ a_3 \ a_4]^T$ only if the determinant of the coefficient matrix vanishes (i.e., if the coefficient matrix is singular). Furthermore, recall that since the coefficient matrix is singular, not all of the elements of the vector $\mathbf{a}$ can be calculated.

Setting the determinant above equal to zero yields the characteristic equation

$$\tan \beta l = \tanh \beta l$$

This equality is satisfied for an infinite number of choices for β, denoted β_n. The solution can be visualized by plotting both $\tan \beta l$ and $\tanh \beta l$ versus βl on the same plot. This is similar to the solution technique used in Example 6.4.1 and illustrated in Figure 6.9. The first five solutions are

$$\beta_1 l = 3.926602 \qquad \beta_2 l = 7.068583 \qquad \beta_3 l = 10.210176$$

$$\beta_4 l = 13.351768 \qquad \beta_5 l = 16.49336143$$

For the rest of the modes (i.e., for values of the index $n > 5$), the solutions to the characteristic equation are well approximated by

$$\beta_n l = \frac{(4n + 1)\pi}{4}$$

The weighted frequencies determine the system natural frequencies by

$$\omega_n = \beta_n^2 \sqrt{\frac{EI}{\rho A}} \text{ rad/s,} \quad \text{and} \quad f_n = \frac{\beta_n^2}{2\pi} \sqrt{\frac{EI}{\rho A}} \text{ Hz}$$

With these values of the weighted frequencies $\beta_n l$, the individual modes of vibration can be calculated. Solving the preceding matrix equation for the individual coefficients a_i yields $a_1 = -a_3$, $a_2 = -a_4$, and

$$(\sinh \beta_n l - \sin \beta_n l)a_3 + (\cosh \beta_n l - \cos \beta_n l)a_4 = 0$$

Thus

$$a_3 = -\frac{\cosh \beta_n l - \cos \beta_n l}{\sinh \beta_n l - \sin \beta_n l} a_4$$

for each n. The fourth coefficient a_4 cannot be determined by this set of equations, because the coefficient matrix is singular (otherwise, each a_i would be zero). This remaining coefficient becomes the arbitrary magnitude of the eigenfunction. As this constant depends on n, denote it by $(a_4)_n$. Substitution of these values of a_i in the expression $X(x)$ for the spatial solution yields the result that the eigenfunctions or mode shapes have the form

$$X_n(x) = (a_4)_n \left[\frac{\cosh \beta_n l - \cos \beta_n l}{\sinh \beta_n l - \sin \beta_n l} (\sinh \beta_n x - \sin \beta_n x) - \cosh \beta_n x + \cos \beta_n x \right],$$

$$n = 1, 2, 3, \ldots$$

The first three mode shapes are plotted in Figure 6.11 for $(a_2)_n = 1$ and $n = 1, 2, 3$.

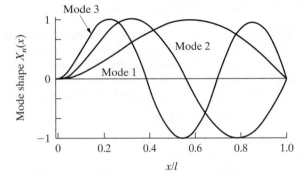

Figure 6.11 A plot of the first three mode shapes of the clamped–pinned beam of Example 6.5.1, arbitrarily normalized to unity.

These mode shapes can be shown to be orthogonal, so that

$$\int_0^l X_n(x)X_m(x)dx = 0$$

for $n \neq m$ (see Problem 6.47). As in Example 6.3.2, this orthogonality, along with the initial conditions, can be used to calculate the constants A_n and B_n in the series solution for the displacement

$$w(x, t) = \sum_{n=1}^{\infty} (A_n \sin \omega_n t + B_n \cos \omega_n t)X_n(x)$$

□

Table 6.6 summarizes a number of different boundary configurations for the slender beam. The slender-beam model given in equation (6.91) is often referred to

TABLE 6.6 SAMPLE OF VARIOUS BOUNDARY CONFIGURATIONS OF A SLENDER BEAM IN TRANSVERSE VIBRATION OF LENGTH *l* ILLUSTRATING WEIGHTED NATURAL FREQUENCIES AND MODE SHAPES[a]

Configuration	Weighted frequencies $\beta_n l$ and characteristic equation	Mode shape	σ_n
Free–free	0 (rigid-body mode) 4.73004074 7.85320462 10.9956078 14.1371655 17.2787597 $\dfrac{(2n+1)\pi}{2}$ for $n > 5$ $\cos \beta l \cosh \beta l = 1$	$\cosh \beta_n x + \cos \beta_n x$ $-\sigma_n (\sinh \beta_n x + \sin \beta_n x)$[b]	 0.9825 1.0008 0.9999 1.0000 0.9999 1 for $n > 5$
Clamped–free	1.87510407 4.69409113 7.85475744 10.99554073 14.13716839 $\dfrac{(2n-1)\pi}{2}$ for $n > 5$ $\cos \beta l \cosh \beta l = -1$	$\cosh \beta_n x - \cos \beta_n x$ $-\sigma_n (\sinh \beta_n x - \sin \beta_n x)$	0.7341 1.0185 0.9992 1.0000 1.0000 1 for $n > 5$
Clamped–pinned	3.92660231 7.06858275 10.21017612 13.35176878 16.49336143 $\dfrac{(4n+1)\pi}{4}$ for $n > 5$ $\tan \beta l = \tanh \beta l$	$\cosh \beta_n x - \cos \beta_n x$ $-\sigma_n (\sinh \beta_n x - \sin \beta_n x)$	1.0008 1 for $n > 1$
Clamped–sliding	2.36502037 5.49780392 8.63937983 11.78097245 14.92256510 $\dfrac{(4n-1)\pi}{4}$ for $n > 5$ $\tan \beta l + \tanh \beta l = 0$	$\cosh \beta_n x - \cos \beta_n x$ $-\sigma_n (\sinh \beta_n x - \sin \beta_n x)$	0.9825 1 for $n > 1$
Clamped–clamped	4.73004074 7.85320462 10.9956079 14.1371655 17.2787597 $\dfrac{(2n+1)\pi}{2}$ for $n > 5$ $\cos \beta l \cosh \beta l = 1$	$\cosh \beta_n x - \cos \beta_n x$ $-\sigma_n (\sinh \beta_n x - \sin \beta_n x)$	0.982502 1.00078 0.999966 1.0000 1.0000 1 for $n > 5$
Pinned–pinned	$n\pi$ $\sin \beta l = 0$	$\sin \dfrac{n\pi x}{l}$	none

[a]Here the weighted natural frequencies $\beta_n l$ are related to the natural frequencies by equation (6.101) or $\omega_n = \beta_n^2 \sqrt{EI/\rho A}$, as used in Example 6.5.1. The values of σ_i for the mode shapes are computed from the formulas given in Table 6.5.

[b]There are two free–free mode shapes: $X_0 = $ constant and $X_0 = A(x - l/2)$; the first is translational, the second rotational.

as the *Euler–Bernoulli* or Bernoulli–Euler beam equation. The assumptions used in formulating this model are that the beam be

- Uniform along its span, or length, and slender ($l > 10h$)
- Composed of a linear, homogeneous, isotropic elastic material without axial loads
- Such that plane sections remain plane
- Such that the plane of symmetry of the beam is also the plane of vibration so that rotation and translation are decoupled
- Such that rotary inertia and shear deformation can be neglected

The key to solving for the time response of distributed-parameter systems is the orthogonality of the mode shapes. Note from Table 6.7 that the mode shapes are quite complicated in many configurations. This does not mean that orthogonality is necessarily violated, just that evaluating the integrals in the modal analysis procedure becomes more difficult.

TABLE 6.7 EQUATIONS FOR THE MODE SHAPE
COEFFICIENTS σ_n FOR USE IN TABLE 6.4[a]

Boundary condition	Formula for σ_n
Free–free	$\sigma_n = \dfrac{\cosh \beta_n l - \cos \beta_n l}{\sinh \beta_n l - \sin \beta_n l}$
Clamped–free	$\sigma_n = \dfrac{\sinh \beta_n l - \sin \beta_n l}{\cosh \beta_n l + \cos \beta_n l}$
Clamped–pinned	$\sigma_n = \dfrac{\cosh \beta_n l - \cos \beta_n l}{\sinh \beta_n l - \sin \beta_n l}$
Clamped–sliding	$\sigma_n = \dfrac{\sinh \beta_n l - \sin \beta_n l}{\cosh \beta_n l + \cos \beta_n l}$
Clamped–clamped	Same as free–free

[a]These coefficients are used in the calculations for the mode shapes, as illustrated in Example 6.5.1.

Timoshenko Beam Theory

The model of the transverse vibration of the beam presented in equation (6.91) ignores the effects of shear deformation and rotary inertia. Beams modeled including the effects of rotary inertia and shear deformation are called *Timoshenko beams*. These effects are considered next. As mentioned previously, it is safe to ignore the shear deformation as long as the h_z and h_y illustrated in Figure 6.10 are small compared with the length of the beam. As the beam becomes shorter, the effect of shear deformation becomes evident. This is illustrated in Figure 6.12, which is a repeat of the element dx of Figure 6.10 with shear deformation included.

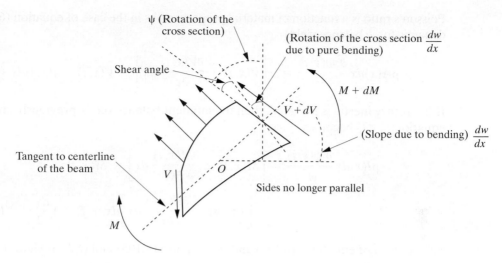

Figure 6.12   The effects of shear deformation on an element of a bending beam.

Referring to the figure, the line OA is a line through the center of the element dx perpendicular to the face at the right side. The line OB, on the other hand, is the line through the center tangent to the centerline of the beam, while the line OC is the centerline of the beam while at rest. As the beam bends, the shear angle appears as the length l is decreased relative to the beam width. For the case of a long beam, the lines OB and OA coincide, as in Figure 6.10. The presence of significant shear causes the rectangular infinitesimal element of Figure 6.10 to deform into the distorted almost-diamond shape of Figure 6.12. Referring to Figure 6.12, note that the shear angle given by $\psi - dw/dx$ (i.e., the difference between the total angle due to bending, ψ, and the slope of the centerline of the beam, dw/dx), represents the effect of shear deformation. From elastic considerations [see, e.g., Reismann and Pawlik (1974)] the bending moment equation becomes

$$EI\frac{d\psi(x, t)}{dx} = M(x, t) \tag{6.103}$$

and the shear force equation becomes

$$\kappa^2 AG\left[\psi(x, t) - \frac{dw(x, t)}{dx}\right] = V(x, t) \tag{6.104}$$

where E, I, A, ψ, V, and M are as defined previously, G is the *shear modulus*, and κ^2 is a dimensionless factor that depends on the shape of the cross-sectional area. Note that some texts use κ instead of κ^2 for the shear coefficient, also sometimes referred to as the *Timoshenko shear coefficient*. The constant κ^2 is called a *shear coefficient* and has been tabulated by Cowper (1966). The shear modulus G and the elastic modulus E are related by the Poisson's ratio v, according the relationship $E = 2(1 + v)G$. The

Poisson's ratio is a function of material properties. As in the case of equation (6.86), a dynamic force balance yields

$$\rho A(x)dx \frac{\partial^2 w(x, t)}{\partial t^2} = -\left[V(x, t) + \frac{\partial V(x, t)}{\partial x} dx \right] + V(x, t) + f(x, t)dx \quad (6.105)$$

If the rotary inertia is included, then the moment balance on dx previously given by equation (6.87) becomes

$$\rho I(x)dx \frac{\partial^2 \psi(x, t)}{\partial t^2} = \left[M(x, t) + \frac{\partial M(x, t)}{\partial x} dx \right] - M(x, t)$$

$$+ \left[V(x, t) + \frac{\partial V(x, t)}{\partial x} dx \right]dx + f(x, t) \frac{dx^2}{2} \quad (6.106)$$

Substitution of equations (6.103) and (6.104) into (6.105) and (6.106) yields the two coupled equations

$$\frac{\partial}{\partial x}\left[EI \frac{\partial \psi}{\partial x} \right] + \kappa^2 AG\left(\frac{\partial w}{\partial x} - \psi \right) = \rho I \frac{\partial^2 \psi}{\partial t^2} \quad (6.107)$$

and

$$\frac{\partial}{\partial x}\left[\kappa^2 AG\left(\frac{\partial w}{\partial x} - \psi \right) \right] + f(x, t) = \rho A \frac{\partial^2 w}{\partial t^2} \quad (6.108)$$

governing the vibration of a beam, including the effects of rotary inertia and shear deformation. Assuming that the coefficients are all constant and that no external force is applied, $\psi(x, t)$ can be eliminated and the coupled equations can be reduced to one single equation for free vibration of uniform beams. This is

$$EI \frac{\partial^4 w}{\partial x^4} + \rho A \frac{\partial^2 w}{\partial t^2} - \rho I\left(1 + \frac{E}{\kappa^2 G} \right)\frac{\partial^4 w}{\partial x^2 \partial t^2} + \frac{\rho^2 I}{\kappa^2 G}\frac{\partial^4 w}{\partial t^4} = 0 \quad (6.109)$$

Equation (6.109) is now subject to four initial conditions and four boundary conditions. For a clamped end, the boundary conditions become (at $x = 0$, say)

$$\psi(0, t) = w(0, t) = 0 \quad (6.110)$$

At a simply supported end, they become

$$EI \frac{\partial \psi(0, t)}{\partial x} = w(0, t) = 0 \quad (6.111)$$

and at a free end they become

$$\kappa^2 AG\left(\frac{\partial w}{\partial x} - \psi \right) = EI \frac{\partial \psi}{\partial x} = 0 \quad (6.112)$$

These equations can be solved by the methods suggested for the beam model given by equation (6.91). Equation (6.91) is called the *Euler–Bernoulli beam model* or classical beam model and equation (6.109) is called the *Timoshenko beam model*.

Which of these two beam models to use is largely dependent on the beam geometry, which modes are of interest, and how many modes are important. A steel beam 12- m long, 15- cm wide, and 0.6- m deep shows a difference of only 0.4% between the first natural frequency of the Euler–Bernoulli model and that of the Timoshenko model. This grows to a 10% difference in the fifth natural frequency. Hence if only the first mode is of interest, the Euler–Bernoulli model for this system would be good enough. On the other hand, if the fifth mode is of interest, the Timoshenko model might be worth the extra complexity. For a beam of typical metal with rectangular cross section, the shear deformation is about three times more significant than rotary inertia effects. This is examined in the next example.

Example 6.5.2

To obtain a feeling for the differences between the Euler–Bernoulli beam model and the more complicated Timoshenko model, calculate the natural frequencies of a pinned–pinned beam using the Timoshenko equation and compare them to the natural frequencies predicted by the beam without shear deformation or rotary inertia as given in Table 6.6.

Solution   Equation (6.109) cannot easily be solved by separation of variables as suggested in Window 6.4. This results because substitution of $w(x, t) = X(x)T(t)$ into equation (6.3) yields an equation of the form (a_i are constants)

$$a_1 X'''' T + (a_2 X + a_3 X'')\ddot{T} + a_4 X \dddot{T} = 0$$

which does not readily separate because of the middle term and the fourth time derivative. This expression can be solved by assuming that the mode shapes of pinned–pinned Euler–Bernoulli beam and that of the Timoshenko beam are the same (a reasonable assumption, as one is a more complete account of the other) and that the temporal response is periodic (also reasonable, as the system is undamped). Proceeding with these assumptions, a solution of equation (6.109) is *assumed* to be of the specific separated form

$$w_n(x, t) = \sin\frac{n\pi x}{l}\cos\omega_n t$$

where $\sin(n\pi x/l)$ is the assumed nth mode shape of the pinned–pinned configuration and ω_n is, at this point, the unknown natural frequency. Substitution of this form into equation (6.109) yields

$$EI\left(\frac{n\pi}{l}\right)^4 \sin\frac{n\pi x}{l}\cos\omega_n t - \rho I\left(1 + \frac{E}{\kappa^2 G}\right)\left(\frac{n\pi}{l}\right)^2 \omega_n^2 \sin\frac{n\pi}{l}\cos\omega_n t$$

$$= -\frac{\rho^2 I}{\kappa^2 G}\omega_n^4 \sin\frac{n\pi x}{l}\cos\omega_n t + \rho A \omega_n^2 \sin\frac{n\pi x}{l}\cos\omega_n t$$

Each term contains the factor $\sin(n\pi x/l)\cos(\omega_n t)$, which can be factored out to reveal the characteristic equation

$$\omega_n^4 \frac{\rho r^2}{\kappa^2 G} - \left(1 + \frac{n^2\pi^2 r^2}{l^2} + \frac{n^2\pi^2 r^2}{l^2}\frac{E}{\kappa^2 G}\right)\omega_n^2 + \frac{\alpha^2 n^4\pi^4}{l^4} = 0$$

where r and α are defined by

$$\alpha^2 = \frac{EI}{\rho A} \quad r^2 = \frac{I}{A}$$

The characteristic equation for the frequencies ω_n is quadratic in ω_n^2, and hence easily solved. This expression for ω_n provides a mechanism for observing the effects of shear deformation and rotary inertia on the natural frequencies of a pinned–pinned beam. Note the following comparisons:

1. Of the two roots for each value of n determined by the frequency equation, the smaller value is associated with bending deformation, and the larger root is associated with shear deformation.

2. The natural frequencies for just the Euler–Bernoulli beam are

$$\omega_n^2 = \frac{\alpha^2 n^4 \pi^4}{l^4}$$

3. The natural frequencies for including just rotary inertia (i.e., no shear, so that terms involving κ are eliminated) are

$$\omega_n^2 = \frac{\alpha^2 n^4 \pi^4}{l^4(1 + n^2 \pi^2 r^2 / l^2)}$$

4. The natural frequencies for neglecting the rotary inertia and including the shear deformation are

$$\omega_n^2 = \frac{\alpha^2 n^4 \pi^4}{l^4[1 + (n^2 \pi^2 r^2 / l^2)\, E/\kappa G]}$$

These expressions can be used to investigate the various effects on the natural frequencies of prismatic beams in the pinned–pinned configuration. These expressions can also be used to gain insight into the effects of rotary inertia and shear deformation for other configurations. By comparing notes 2, 3, and 4, the general effect of shear deformation and rotary inertia is to reduce the value of the natural frequencies. Also note that for high frequencies (large n) the effects of shear deformation and rotary inertia are more pronounced because of the $1 + (n\pi r/l)^2$ term in the denominator. This is investigated further in Problem TB.6.4 at the end of this chapter.   □

6.6 VIBRATION OF MEMBRANES AND PLATES

The string, rod, and beam models considered in the previous five sections have displacements that are a function of a single direction, x, along the mass. In this sense they are one-dimensional problems. In this section membranes and plates are considered having displacements which are functions of two dimensions (i.e., they are

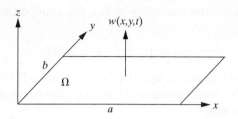

Figure 6.13   A schematic of a rectangular membrane illustrating the vibration perpendicular to its surface. The boundary conditions are specified around the edge of the membrane (i.e., along the lines $x = a$, $x = 0$, $y = b$, and $y = 0$).

defined in a plane region in space as illustrated in Figure 6.13). A membrane is essentially a two-dimensional string, and a plate is essentially a two-dimensional beam. The equation for a membrane and plate are not derived here but follow similar arguments used in developing the string and Euler–Bernoulli beam equations, respectively.

First, consider the vibrational equations for a membrane. A membrane is basically a two-dimensional system that lies in a plane when in equilibrium. A drumhead and a soap film are physical examples of objects that may be modeled as a membrane. The structure itself provides no resistance to bending, so the restoring force is due only to the tension in the membrane. Thus a membrane is similar to a string. The reader is referred to Weaver, Timoshenko, and Young (1990) for the derivation of the membrane equation.

Let $w(x, y, t)$ represent the displacement in the z direction of a membrane, lying in the xy plane at the point (x, y) and time t. The displacement is assumed to be small, with small slopes, and is perpendicular to the xy plane. Let τ be the constant tension per unit length of the membrane and ρ be the mass per unit area of the membrane. Then the equation for free vibration is given by

$$\tau \nabla^2 w(x,y,t) = \rho w_{tt}(x,y,t) \tag{6.113}$$

where x and y lie inside the region, Ω, occupied by the membrane as indicated in Figure 6.13. Here ∇^2 is the *Laplace operator*. In rectangular coordinates this operator has the form

$$\nabla^2 = \frac{\partial^2}{\partial x^2} + \frac{\partial^2}{\partial y^2} \tag{6.114}$$

The Laplace operator takes on other forms if a different coordinate system is used. For example, a drumhead is best written in a circular coordinate system. For the rectangular system considered here, equation (6.113) becomes

$$\frac{\partial^2 w(x, y, t)}{\partial x^2} + \frac{\partial^2 w(x, y, t)}{\partial y^2} = \frac{1}{c^2} \frac{\partial^2 w(x, y, t)}{\partial t^2} \tag{6.115}$$

where $c = \sqrt{\tau/\rho}$. The boundary conditions for the membrane must be specified along the shape of the boundary, not just at points, as in the case of the string. If the membrane is fixed or clamped at a segment of the boundary, the deflection must be zero along that segment. If $\partial\Omega$ is the curve in the xy plane corresponding to the edge of the membrane (i.e., the boundary of Ω), the clamped boundary condition is denoted by

$$w(x, y, t) = 0 \quad \text{for} \quad x, y \in \delta\Omega \tag{6.116}$$

If for some segment of $\delta\Omega$, denoted by $\delta\Omega_1$, the membrane is free to deflect transversely, there can be no force component in the transverse direction, and the boundary condition becomes

$$\frac{\partial w(x, y, t)}{\partial n} = 0 \quad \text{for} \quad x, y \in \delta\Omega_1 \tag{6.117}$$

Here, $\partial/\partial n$ denotes the derivative of $w(x, y, t)$ normal to the boundary in the reference plane of the membrane. The following example illustrates the procedure for calculating the solution for a vibrating membrane.

Example 6.6.1

Consider the vibration of a square membrane, as indicated in Figure 6.13. The membrane is clamped at all the edges. The equation of motion is given by equation (6.115). Compute the natural frequencies and mode shapes for the case $a = b = 1$.

Solution   Assuming that the solution separates [i.e., that $w(x, y, t) = X(x)Y(y)T(t)$], equation (6.115) becomes

$$\frac{1}{c^2}\frac{\ddot{T}}{T} = \frac{X''}{X} + \frac{Y''}{Y} \tag{6.118}$$

This implies that $\ddot{T}/(Tc^2)$ is a constant (recall the argument used in Example 6.2.1). Denote the constant by ω^2, so that

$$\frac{\ddot{T}}{Tc^2} = -\omega^2 \tag{6.119}$$

This assumption leads to the appropriate time solution as before. Then equation (6.118) implies that

$$\frac{X''}{X} = -\omega^2 - \frac{Y''}{Y} \tag{6.120}$$

By the same argument as that used before, both X''/X and Y''/Y must be constant (i.e., independent of t and x or y). Hence

$$\frac{X''}{X} = -\alpha^2 \tag{6.121}$$

and

$$\frac{Y''}{Y} = -\gamma^2 \tag{6.122}$$

where α^2 and γ^2 are constants. Equation (6.120) then yields

$$\omega^2 = \alpha^2 + \gamma^2 \tag{6.123}$$

This results in two spatial equations to be solved,

$$X'' + \alpha^2 X = 0 \tag{6.124}$$

which has a solution (A and B constants of integration) of the form

$$X(x) = A \sin \alpha x + B \cos \alpha x \tag{6.125}$$

and

$$Y'' + \gamma^2 Y = 0 \tag{6.126}$$

which yields a solution of the form (C and D are constants of integration)

$$Y(y) = C \sin \gamma y + D \cos \gamma y \tag{6.127}$$

The total spatial solution is the product $X(x)Y(y)$ or

$$\begin{aligned} X(x)Y(y) = &A_1 \sin \alpha x \sin \gamma y + A_2 \sin \alpha x \cos \gamma y \\ &+ A_3 \cos \alpha x \sin \gamma y + A_4 \cos \alpha x \cos \gamma y \end{aligned} \tag{6.128}$$

Here the constants A_i consist of the products of the constants in equations (6.125) and (6.127) and are to be determined by the boundary and initial conditions.

Equation (6.128) can now be used with the boundary conditions to calculate the eigenvalues and eigenfunctions of the system. The clamped boundary condition along $x = 0$ in Figure 6.13 yields

$$T(t)X(0)Y(y) = T(t)(A_3 \sin \gamma y + A_4 \cos \gamma y) = 0$$

or

$$A_3 \sin \gamma y + A_4 \cos \gamma y = 0 \tag{6.129}$$

Now, equation (6.129) must hold for any value of y. Thus, as long as γ is not zero (a reasonable assumption, since if it is zero the system has a rigid-body motion), A_3 and A_4 must be zero. Hence the solution must have the form

$$X(x)Y(y) = A_1 \sin \alpha x \sin \gamma y + A_2 \sin \alpha x \sin \gamma y \tag{6.130}$$

Next, application of the boundary condition $w = 0$ along the line $x = 1$ yields

$$A_1 \sin \alpha \sin \gamma y + A_2 \sin \alpha \cos \gamma y = 0 \tag{6.131}$$

Factoring this expression yields

$$\sin \alpha (A_1 \sin \gamma y + A_2 \cos \gamma y) = 0 \tag{6.132}$$

Now either $\sin \alpha = 0$ or, by the preceding argument, A_1 and A_2 must be zero. However, if A_1 and A_2 are both zero, the solution is trivial. Hence for a nontrivial solution to exist, $\sin \alpha = 0$. This yields

$$\alpha = n\pi \quad n = 1, 2, \ldots, \infty \tag{6.133}$$

Using the boundary condition $w = 0$ along the lines $y = 1$ and $y = 0$ results in a similar procedure and yields

$$\gamma = m\pi \quad m = 1, 2, \ldots, \infty \tag{6.134}$$

Note that the possibility of $\gamma = \alpha = 0$ is not used because it was necessary to assume that $\gamma \neq 0$ in order to derive equation (6.130). Equation (6.123) yields that the constant ω in the temporal equation must have the form

$$\omega_{mn} = \sqrt{\alpha_n^2 + \gamma_m^2}$$
$$= \pi\sqrt{m^2 + n^2} \quad m, n = 1, 2, 3, \ldots, \infty$$

Thus the eigenvalues and eigenfunctions for the clamped membrane are, respectively, $\pi\sqrt{m^2 + n^2}$ and $\{\sin n\pi x \sin m\pi y\}$ (because $A_2 = a_3 = a_4 = 0$). The solution of equation (6.115) becomes

$$w(x, y, t) = \sum_{m=1}^{\infty}\sum_{n=1}^{\infty}(\sin m\pi x \sin n\pi y)\left\{A_{mn}\sin\sqrt{n^2 + m^2}c\pi t\right.$$
$$\left. + B_{mn}\cos\sqrt{n^2 + m^2}c\pi t\right\} \tag{6.135}$$

where A_{mn} and B_{mn} are determined by the initial conditions. To see this, multiply equation (6.135) by the mode shape $(\sin m\pi x \sin n\pi y)$ and integrate the resulting equation over dx and dy along the edges of the membrane. Because the set of functions $\{\sin m\pi x \sin n\pi y\}$ is orthogonal, the summations are reduced to the single term

$$\int_0^1\int_0^1 w(x, y, t)\sin m\pi x \sin n\pi y \, dx \, dy = \frac{1}{4}\left(A_{mn}\cos\omega_{mn}ct + B_{mn}\sin\omega_{mn}ct\right) \tag{6.136}$$

Setting $t = 0$ to obtain the initial conditions in equation (6.136) yields

$$A_{mn} = 4\int_0^1\int_0^1 w(x, y, 0)\sin m\pi x \sin n\pi y \, dx \, dy$$

Differentiating equation (6.136) and setting $t = 0$ yields

$$B_{mn} = \frac{4}{\omega_{nm}c}\int_0^1\int_0^1 w_t(x, y, 0)\sin m\pi x \sin n\pi y \, dx \, dy$$

These last two expressions yield the expansion coefficients in terms of the initial displacement $w(x, y, 0)$ and the initial velocity $w_t(x, y, 0)$.

The mode shapes of the membrane are the set of functions

$$w_{nm}(x, y) = \sin m\pi x \sin n\pi y, \quad m = 1, 2, \ldots, \quad n = 1, 2, \ldots$$

The first mode corresponds to the index $m = n = 1$. If the initial velocity is zero [i.e., $w_t(x, y, 0) = 0$], the coefficients B_{mn} are all zero. If, in addition, the initial displacement $w(x, y, 0)$ is chosen such that all the coefficients A_{mn} are zero except for A_{11}, the solution becomes the single term

$$w_{11}(x, y, t) = (A_{11}\sin\omega_{11}ct)\sin\pi x \sin\pi y$$

where $\omega_{11} = \pi\sqrt{2}$. This last expression describes the fundamental mode shape of the membrane vibrating at the single frequency $\pi\sqrt{2}$ rad/s.

Note from the expression for w_{nm} that the frequency corresponding to $m = 1$, $n = 2$ will be the same as that corresponding to $m = 2, n = 1$ (i.e., $\omega_{12} = \omega_{21} = \pi\sqrt{5}$). However, the mode shapes are different:

$$w_{12}(x, y, t) = (A_{12} \cos c\pi\sqrt{5}t) \sin \pi x \sin 2\pi y$$

$$w_{21}(x, y, t) = (A_{21} \cos c\pi\sqrt{5}t) \sin 2\pi x \sin \pi y$$

Thus the membrane can vibrate the single frequency $c\pi\sqrt{5}$ in two different ways, exhibiting different nodal lines.

□

Note from this example that the mode shapes are functions of both x and y, so that the nodes of the modes of a membrane form a line of no motion along the membrane when excited at a particular natural frequency. This turns out to be relatively simple to verify experimentally, adding credibility to the analysis presented here.

In progressing from the vibration of a string to considering the transverse vibration of a beam, the beam equation allowed for bending stiffness. In somewhat the same manner, a plate differs from a membrane because plates have bending stiffness. The reader is referred to Reismann (1988) for a more detailed explanation and for a precise derivation of the plate equation. Basically, the plate, like the membrane, is defined in a plane (xy) with the deflection $w(x, y, t)$ taking place along the z axis perpendicular to the xy plane. The basic assumption is again small deflections with respect to the thickness h. Thus the plane running through the middle of the plate is assumed not to deform during bending (called a *neutral plane* or *surface*). In addition, normal stresses in the direction transverse to the plate are assumed to be negligible. Again there is no thickness stretch. The displacement equation of motion for the free vibration of the plate is

$$-D_E\nabla^4 w(x, y, t) = \rho w_{tt}(x, y, t) \tag{6.137}$$

where E again denotes the elastic modulus, ρ is the mass density, and the constant D_E, the plate flexural rigidity, is defined in terms of Poisson's ratio ν and the plate thickness h as

$$D_E = \frac{Eh^3}{12(1 - \nu^2)} \tag{6.138}$$

The operator ∇^4, called the *biharmonic operator*, is a fourth-order operator, the exact form of which depends on the choice of coordinate systems. In rectangular coordinates the biharmonic operator becomes

$$\nabla^4 = \frac{\partial^4}{\partial x^4} + 2\frac{\partial^4}{\partial x^2 \partial y^2} + \frac{\partial^4}{\partial y^4} \tag{6.139}$$

The boundary conditions for a plate are a little more difficult to write, as their form, in some cases, also depends on the coordinate system in use.

For a clamped edge the deflection and normal derivative, $\partial/\partial n$, are both zero along the edge:

$$w(x, y, t) = 0 \quad \text{and} \quad \frac{\partial w(x, y, t)}{\partial n} = 0 \quad \text{for} \quad x, y \text{ along } \partial\Omega \qquad (6.140)$$

Here the normal derivative is the derivative of w normal to the plate boundary and in the neutral plane. For a rectangular plate, the simply supported boundary conditions become

$$w(x, y, t) = 0 \quad \text{along all edges} \qquad (6.141)$$

$$\frac{\partial^2 w(x, y, t)}{\partial x^2} = 0 \quad \text{along the edges } x = 0, x = l_1 \qquad (6.142)$$

$$\frac{\partial^2 w(x, y, t)}{\partial y^2} = 0 \quad \text{along the edges } y = 0, y = l_2$$

where l_1 and l_2 are the lengths of the plate edges and the second partial derivatives reflect the normal strains along these edges.

This plate model is basically a two-dimensional analog of the Euler–Bernoulli beam and is referred to as *thin plate theory*. Hence it is limited to cases where shear deformation and rotary inertia are negligible. The plate equation can be improved by adding shear deformation and rotary inertia to produce a two-dimensional analog of the Timoshenko beam. This is called Mindlin–Timoshenko theory and is not discussed here (see, e.g., Magrab, 1979).

6.7 MODELS OF DAMPING

The models discussed in the preceding six sections do not account for energy dissipation. As in Section 4.5 for lumped-mass systems, damping can be introduced in two ways: either as modal damping or as a physical damping model. In modeling single-degree-of-freedom systems, viscous damping was used as much for mathematical convenience as for physical truth. This is the case here as well.

A simple procedure for including damping is to add it to the temporal equation after separation of variables. For example, consider the temporal equation for the string as given by equation (6.24):

$$\ddot{T}_n(t) + \sigma_n^2 c^2 T_n(t) = 0 \quad n = 1, 2, \ldots$$

These expressions yield the distributed parameter analog of equations (4.85) for a lumped-mass system and can be called *modal equations*. Modal damping can be added to equation (6.24) by including the term

$$2\zeta_n \omega_n \dot{T}_n(t) \quad n = 1, 2, 3, \ldots \tag{6.143}$$

where ω_n is the nth natural frequency and ζ_n is the nth modal damping ratio. The damping ratios ζ_n are chosen, like those of equation (4.123), based on experience or on experimental measurements. Usually, ζ_n is a small positive number between 0 and 1, with most common values of $\zeta_n \leq 0.05$.

Once the modal damping ratios are assigned, the damping term of equation (6.143) is added to equation (6.24) to yield

$$\ddot{T}(t) + 2\zeta_n \omega_n \dot{T}(t) + \omega_n^2 T_n(t) = 0 \quad n = 1, 2, 3, \ldots \tag{6.144}$$

where $\omega_n = \sigma_n c$. The solution, for an underdamped mode, becomes (see Window 6.5)

$$T_n(t) = A_n e^{-\zeta_n \omega_n t} \sin(\omega_{dn} t + \phi_n) \quad n = 1, 2, 3, \ldots \tag{6.145}$$

where $\omega_{dn} = \omega_n \sqrt{1 - \zeta_n^2}$ and where A_n and ϕ_n are constants to be determined by the initial condition. Once the temporal coefficients are determined, the rest of the solution procedure is the same as given in Section 6.2.

Window 6.5
Review of a Damped Single-Degree-of-Freedom System

The solution of $m\ddot{x} + c\dot{x} + kx = 0$, $x(0) = x_0$, $\dot{x}(0) = \dot{x}_0$, or $\ddot{x} + 2\zeta\omega_n\dot{x} + \omega^2 x = 0$ is (for the underdamped case $0 < \zeta < 1$)

$$x(t) = Ae^{-\zeta\omega_n t} \sin(\omega_d t + \phi)$$

where $\omega_n = \sqrt{k/m}$, $\zeta = c/2m\omega$, and

$$A = \left[\frac{(\dot{x}_0 + \zeta\omega_n x_0)^2 + (x_0\omega_d)^2}{\omega_d^2} \right]^{1/2} \qquad \phi = \tan^{-1} \frac{x_0\omega_d}{\dot{x}_0 + \zeta\omega_n x_0}$$

from equations (1.36), (1.37), and (1.38).

Example 6.7.1

Calculate the response of the bar of Example 6.3.1 to an initial displacement of $w(x, 0) = 2(x/l)$ cm and an initial velocity of $w_t(x, 0) = 0$. Assume that the bar exhibits modal damping of $\zeta_n = 0.01$ in each mode.

Solution From Example 6.3.1, the mode shapes are $a_n \sin[(2n - 1)\pi x/2l]$ and the undamped natural frequencies are

$$\omega_n = \sigma_n \sqrt{\frac{E}{\rho}} = \frac{(2n - 1)\pi}{2l} \sqrt{\frac{E}{\rho}} \text{ rad/s}$$

Since the modal damping ratio is chosen to be 0.01, the damped natural frequencies become

$$\omega_{dn} = \omega_n \sqrt{1 - \zeta_n^2} = 0.9999 \frac{2n - 1}{2l} \pi \sqrt{\frac{E}{\rho}} \text{ rad/s}$$

From equation (6.145), the temporal solution becomes

$$T_n(t) = A_n e^{-0.01\omega_n t} \sin(\omega_{dn} t + \phi_n)$$

The total solution is then of the form

$$w(x, t) = \sum_{n=1}^{\infty} A_n e^{-0.01\omega_n t} \sin(\omega_{dn} t + \phi_n) \sin \frac{2n - 1}{2l} \pi x \qquad (6.146)$$

Applying the initial condition yields (changing 2 cm to 0.02 m)

$$0.02\left(\frac{x}{l}\right) = \sum_{n=1}^{\infty} A_n \sin \phi_n \sin \sigma_n x$$

Multiplying the last expression by $\sin \sigma_m x$ and integrating over the length of the bar yields

$$\frac{0.02}{l} \int_0^l x \sin \sigma_m x \, dx = \frac{0.02}{l\sigma_m^2}(-1)^{m+1} = \sum_{n=1}^{\infty} A_n \sin \phi_n \int_0^l \sin \sigma_n x \sin \sigma_m x \, dx \qquad (6.147)$$

The integral on the right side is the orthogonality condition for the modes:

$$\int_0^l \sin \sigma_n x \sin \sigma_m x \, dx = \left(\frac{l}{2}\right)\delta_{mn} \qquad (6.148)$$

Substitution of the orthogonality conditions into the summation of (6.147) yields

$$\frac{0.02}{l\sigma_m^2}(-1)^{m+1} = (A_m \sin \phi_m)\left(\frac{l}{2}\right) \quad m = 1, 2, \ldots \qquad (6.149)$$

which provides one equation in the two sets of unknown coefficients A_m and ϕ_m. A second equation for the unknown coefficients A_m and ϕ_m is obtained from the second initial condition. This yields

$$w_t(x, t) = 0$$

$$= \sum_{n=1}^{\infty} A_n \left[-0.01\omega_n e^{-0.01\omega_n t} \sin\left(\omega_{dn} t + \phi_n\right) + e^{-0.01\omega_n t}\omega_{dn} \cos\left(\omega_{dn} t + \phi_n\right)\right] \sin \sigma_n x$$

Again multiplying by $\sin \sigma_m x$, integrating over the length of the beam, and using the orthogonality condition results in

$$0 = A_m(-0.01\omega_n \sin \phi_m + \omega_{dm} \cos \phi_m)\frac{l}{2} \quad m = 1, 2, \ldots$$

Since $A_m \neq 0$, the term in parentheses must be zero so that

$$\tan \phi_m = \frac{\sqrt{1 - \zeta^2}}{0.01} = 99.9949 \quad m = 1, 2, \ldots$$

and hence $\phi_m = 1.5607$, which is almost $\pi/2$ radians or $90°$. Substitution of this value back into equation (6.149) yields

$$A_m = \frac{0.04}{l^2 \sigma_m^2} (-1)^{m+1} \quad m = 1, 2, \ldots$$

Hence the general solution, from equation (6.146), becomes

$$w(x, t) = \sum_{n=1}^{\infty} \left(\frac{0.04}{l^2 \sigma_n^2} (-1)^{n+1} \right) e^{-0.01 \omega_n t} \cos \omega_{dn} t \sin \sigma_n x \qquad (6.150)$$

where $\sigma_n = (2n - 1)\pi/2l$, $\omega_n = \sigma_n \sqrt{E/\rho}$, and $\omega_{dn} = 0.9999 \omega_n$.

□

Next consider some physical models of damping. Again, as was the case for single- and lumped-parameter models, physical damping mechanisms are elusive and difficult to derive. Hence some common examples are presented here. First, consider the effects of an external damping mechanism, such as air. As a string, membrane, or beam in transverse vibration oscillates it pushes air around, causing an energy loss from a nonconservative force proportional to velocity [see Blevins (1977) for a more complete explanation].

In some circumstances, this energy dissipation can be approximated as linear-viscous damping of the form $\gamma w_t(x, t)$, where γ is a constant damping parameter. In this case, the equation of a clamped–clamped string becomes

$$\rho A w_{tt}(x, t) + \gamma w_t(x, t) - \tau w_{xx}(x, t) = 0$$

$$w(0, t) = 0 = w(l, t) \qquad (6.151)$$

where ρA and τ are the density and tension as defined in equation (6.4) and l is the length of the string. The equation for a square membrane moving in a fluid and clamped around its edges becomes

$$\rho w_{tt}(x, y, t) + \gamma w_t(x, y, t) - \tau \left[w_{xx}(x, y, t) + w_{yy}(x, y, t) \right] = 0$$

$$w(0, y, t) = w(l, y, t) = w(x, 0, t) = w(x, l, t) = 0 \qquad (6.152)$$

where ρ and τ are as defined in equation (6.113) and l is the length of the sides of the membrane.

Internal damping can be modeled by examining the various forces and moments involved in deriving the equations of motion. For example, consider the Euler–Bernoulli bending vibration model given by equation (6.91), developed from Figure 6.10. One possible choice for internal damping is to assign a viscous damping proportional to the rate of strain in the beam. The equation of motion for a beam with viscous air damping (external) and strain-rate damping (internal) is

$$\rho A w_{tt}(x, t) + \gamma w_t(x, t) + \beta \frac{\partial^2}{\partial x^2}\left[I \frac{\partial^3 w(x, t)}{\partial x^2 \partial t}\right] + \frac{\partial^2}{\partial x^2}\left[EI \frac{\partial^2 w(x, t)}{\partial x^2}\right] = 0 \qquad (6.153)$$

For the clamped–free case, the boundary conditions become

$$w(0, t) = w_x(0, t) = 0$$

$$EIw_{xx}(l, t) + \beta Iw_{xxt}(l, t) = 0 \qquad (6.154)$$

$$\frac{\partial}{\partial x}\left[EIw_{xx}(l, t) + \beta Iw_{xxt}(l, t)\right] = 0$$

Here E, I, ρ, and A are as defined in equation (6.91) and γ and β are constant damping parameters. If $I(x)$ is constant, the boundary conditions and equation of motion can be somewhat simplified. Note that the inclusion of strain-rate damping changes the boundary conditions as well as the equation of motion. For constant $I(x)$, the change in the boundary condition does not affect the solution. See Cudney and Inman (1989) for an experimental verification of these equations of motion. Strain-rate damping is also called *Kelvin–Voigt damping*.

The solution technique for systems with the preceding physical damping models remains the same for constant parameters (i.e., E, I, ρ constant) as for the undamped case. This is so because these damping terms are all proportional to the effective mass and stiffness terms as in the lumped-parameter case. Caughey and O'Kelly (1965) present a more detailed explanation. The following example illustrates the use of modal analysis to solve the proportionally damped case.

Example 6.7.2

Calculate the solution of the damped string, equation (6.151), by modal analysis. Equation (6.151) is first restated by using separation of variables [i.e., by substituting $w(x, t) = T(t)X(x)$]. This yields

$$\frac{\rho A \ddot{T} + \gamma \dot{T}}{\tau T} = \frac{X''}{X} = \text{constant} = -\sigma^2$$

This results in two equations: one in time and one in space. The spatial equation $X''(x) + \sigma^2 X(x) = 0$ is subject to the two boundary conditions. This problem was solved in Section 6.2, which resulted in $X_n(x) = a_n \sin(n\pi x/l)$ and $\sigma_n = n\pi/l$, for $n = 1, 2, \ldots$. Substitution of these values into the preceding temporal equation yields

$$\ddot{T}_n(t) + \frac{\gamma}{\rho A}\dot{T}_n(t) + \frac{\tau}{\rho A}\left(\frac{n\pi}{l}\right)^2 T_n(t) = 0 \qquad (6.155)$$

Comparing the coefficient of $\dot{T}_n(t)$ with $2\zeta_n\omega_n$ yields

$$\zeta_n = \frac{1}{2\omega_n}\frac{\gamma}{\rho A} = \frac{\gamma l}{2n\pi\sqrt{\tau\rho A}} \tag{6.156}$$

The solution of the temporal equation (6.155) yields (see Window 6.4)

$$T_n(t) = A_n e^{-\zeta_n\omega_n t}\sin(\omega_{dn}t + \phi_n)$$

where $\omega_{dn} = \omega_n\sqrt{1 - \zeta_n^2}$. The total solution becomes

$$w(x, t) = \sum_{n=1}^{\infty} A_n e^{-\zeta_n\omega_n t}\sin\left(\omega_{dn}t + \phi_n\right)\sin\frac{n\pi x}{l}$$

where A_n and ϕ_n are constants to be determined by the initial conditions using the orthogonality relationship.

$\square$

Damping can also be modeled at the boundary of a structure. In fact, in many cases more energy is dissipated at joints or points of connection than through internal mechanisms such as strain-rate damping. For example, the longitudinal vibration of the bar of Section 6.5 could be modeled as being attached to a lumped spring–damper system as indicated in Figure 6.14.

The equation of motion remains as given in equation (6.56). However, summing forces in the x direction at the boundary yields

$$AE\frac{\partial w(0, t)}{\partial x} = kw(0, t) + c\frac{\partial w(0, t)}{\partial t}$$

$$AE\frac{\partial w(l, t)}{\partial x} = -kw(l, t) - c\frac{\partial w(l, t)}{\partial t} \tag{6.157}$$

These new boundary conditions affect both the orthogonality conditions and the system's temporal solution.

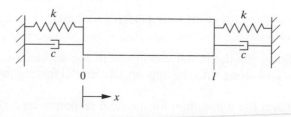

Figure 6.14 A clamped–clamped bar in longitudinal vibration with the point of attachment to ground modeled as providing viscous damping c and stiffness k.

6.8 MODAL ANALYSIS OF THE FORCED RESPONSE

The forced response of a distributed-parameter system can be calculated using modal analysis just as in the lumped-parameter case of Section 4.6. The approach again uses the orthogonality condition of the unforced system's eigenfunctions to reduce the calculation of the response to a system of decoupled modal equations for the time response. The procedure is summarized in Window 6.6 and illustrated by a simple example.

Window 6.6
Modal Analysis of the Total Forced Response for Distributed-Parameter Systems

1. Compute the normalized mode shapes and natural frequencies of the undamped, homogenous system; $X_n(x)$ and ω_n.

2. Assume separation of variables of the form $w(x, t) = X_n(x)T_n(t)$ and substitute this in to the equation of motion.

3. Multiply the result of step 2 by $X_n(x)$ and integrate over the structure (length of the structure in case of a string, rod, bar, or beam) which results in an equation for $T_n(t)$ driven by the modal forcing function, $f_n(t)$ given by

$$f_n(t) = \int_0^l F(x, t)X_n(x)dx$$

4. Solve the resulting second order, ordinary differential equation in time for the form of $T_n(t)$ using the methods of Chapters 2 and 3, for single-degree-of-freedom systems.

5. Compute the modal initial conditions for each $T_n(t)$ from the given initial conditions $w(x, 0)$ and $w_t(x, 0)$ by substituting $w(x, t) = X_n(x)T_n(t)$ and integrating across x:

$$w(x, 0) = X_n(x)T_n(0) \Rightarrow T_n(0) = \int_0^l X_n(x)w(x, 0)\,dx$$

$$w_t(x, 0) = X_n(x)\dot{T}_n(0) \Rightarrow \dot{T}_n(0) = \int_0^l X_n(x)w_t(x, 0)\,dx$$

6. Solve for the initial conditions in the expression for $T_n(t)$ obtained in step 4 along with the appropriate modal forcing function.

7. Form the summation for the total response: $w(x, t) = \sum_{n=1}^{\infty} X_n(x)T_n(t)$.

Example 6.8.1

Calculate the forced response of the string fixed at both ends (discussed in Section 6.2) with external damping coefficient γ, subject to unit impulse applied at $l/4$, where l is the length of the string. Assume that the initial conditions are both zero.

Solution   The equation of motion is

$$\rho A w_{tt}(x, t) + \gamma w_t(x, t) - \tau w_{xx}(x, t) = f(x, t) \qquad (6.158)$$

where $f(x, t) = \delta(x - l/4)\delta(t)$. Here $\delta(x - l/4)$ is a Dirac delta function indicating that the unit force is applied at $x = l/4$ and $\delta(t)$ indicates that the force is applied at time $t = 0$. The boundary conditions are $w(0, t) = w(l, t) = 0$. The eigenfunctions of the undamped, unforced clamped–clamped string are given by equation (6.22) to be of the form

$$X_n(x) = a_n \sin \frac{n\pi x}{l}$$

obtained by using the method of separation of variables.

The modal analysis procedure continues by assuming that the solution is of the form $w_n(x, t) = T_n(t)X_n(x)$, substituting this form into equation (6.158), multiplying by $X_n(x)$, integrating over the length of the string, and solving for $T_n(t)$. Following this procedure, equation (6.158) becomes

$$\left\{ \rho A \ddot{T}_n(t) + \gamma \dot{T}_n(t) - \tau \left[-\left(\frac{n\pi}{l} \right)^2 \right] T_n(t) \right\} \sin \frac{n\pi x}{l} = \delta\left(x - \frac{1}{4} \right) \delta(t)$$

where the constant coefficient a_n of X_n has been arbitrarily set equal to unity. Multiplying by $\sin(n\pi x/l)$ and integrating yields

$$\left[\rho A \ddot{T}_n(t) + \gamma \dot{T}_n(t) + \tau \left(\frac{n\pi}{l} \right)^2 T_n(t) \right] \frac{l}{2}$$

$$= \delta(t) \int_0^l \delta(x - l/4) \sin \frac{n\pi x}{l}\, dx = \delta(t) \left(\sin \frac{n\pi}{4} \right)$$

Rewriting this expression in the form of single-degree-of-freedom oscillator yields

$$\ddot{T}_n(t) + \frac{\gamma}{\rho A} \dot{T}_n(t) + \left(\frac{cn\pi}{l} \right)^2 T_n(t) = \left(\frac{2}{l\rho A} \sin \frac{n\pi}{4} \right) \delta(t) \quad n = 1, 2, \dots \quad (6.159)$$

where $c = \sqrt{\tau/\rho A}$ as before. Equation (6.159) can be written in terms of the modal damping ratio, natural frequency, and input magnitude by comparing the coefficients to

$$\ddot{T}_n(t) + 2\zeta_n \omega_n \dot{T}_n(t) + \omega_n^2 T_n(t) = \hat{F}_n \delta(t) \quad n = 1, 2, \dots$$

which has solution given by equations (3.7) and (3.8) and repeated in Window 6.7. Comparing coefficients between the expressions in Window 6.5 and equation (6.159) yields (for $n = 1, 2, 3, \ldots$)

$$\omega_n = \frac{cn\pi}{l} = \frac{n\pi}{l}\sqrt{\frac{\tau}{\rho A}}$$

$$\hat{F}_n = \frac{2}{l\rho A}\sin\frac{n\pi}{4}$$

$$\zeta_n = \frac{\gamma l}{2cn\pi\rho A} = \frac{\gamma l}{2n\pi\sqrt{\tau\rho A}}$$

Similarly, the nth damped natural frequency becomes

$$\omega_{dn} = \omega_n\sqrt{1 - \zeta_n^2} = \frac{cn\pi}{l}\sqrt{1 - \frac{\gamma^2 l^2}{4c^2 n^2 \pi^2 \rho^2 A^2}} \quad n = 1, 2, 3, \ldots$$

$$= \frac{1}{2\rho Al}\sqrt{(2cn\pi\rho A)^2 - \gamma^2 l^2} \quad n = 1, 2, 3, \ldots$$

Substitution of these values for ω_n, ω_{dn}, ζ_n, and $\hat{F}_n$ into the expression of Window 6.7 yields that the solution for the nth temporal equation becomes

$$T_n(t) = \frac{\hat{F}_n}{\omega_{dn}}e^{-\zeta_n\omega_n t}\sin\omega_{dn}t$$

$$= \frac{4\sin(n\pi/4)}{\sqrt{(2\rho Acn\pi)^2 - (\gamma l)^2}}e^{-\gamma t/2\rho}A\sin\left[\frac{1}{2\rho lA}\sqrt{(2\rho Acn\pi)^2 - (\gamma l)^2}\,t\right]$$

Window 6.7
Solution of the Single-Degree-of-Freedom Impulse Response

The response of an underdamped single-degree-of-freedom system to an impulse modeled by

$$m\ddot{x} + c\dot{x} + kx = \hat{F}\delta(t)$$

where $\delta(t)$ is a unit impulse, is given by equations (3.7) and (3.8) as

$$x(t) = \frac{\hat{F}}{m\omega_d}e^{-\zeta\omega_n t}\sin\omega_d t$$

where $\omega_n = \sqrt{k/m}$, $\zeta = c/2m\omega_n$, $\omega_d = \omega_n\sqrt{1 - \zeta^2}$, and $0 < \zeta < 1$ must hold.

Combining this with $X_n = \sin(n\pi x/l)$ and forming the summation over all modes, the total solution becomes

$$w(x, t) = \sum_{n=1}^{\infty} \frac{4 \sin(n\pi/4)}{\sqrt{(2\rho A c n\pi)^2 - (\gamma l)^2}} e^{-\gamma t/2\rho A} \sin\left[\left(\frac{1}{2\rho l A} \sqrt{(2\rho A c n\pi)^2 - (\gamma l)^2}\right)t\right] \sin\frac{n\pi x}{l}$$

(6.160)

This represents the solution of the damped string subject to a unit impulse applied at $l/4$ units from one end. Similar to the series solutions for the free response, the series must converge, and hence not all terms need to be calculated to obtain a reasonable solution. In fact, usually only the first few terms need be calculated, as illustrated in Figure 6.6 for a similar example.

□

The response of a single-degree-of-freedom system to an impulse can be used to calculate the response of any general force input by use of the impulse response functions. This was illustrated in Section 3.2. The response of a distributed-parameter system to an arbitrary force input can also be calculated using the concept of an impulse response function by defining a modal impulse response function.

The solution for the general forced response of an underdamped single-degree-of-freedom system as detailed in Section 3.2 is summarized in Window 6.8. Following the reasoning used in Example 6.8.1 for the impulse response and referring to Window 6.8, the response of a damped distributed-parameter system to any force can be calculated. As in the case of the impulse response, the method is best illustrated by example.

Window 6.8
Response of an Underdamped System to an Arbitrary Excitation from Section 3.2

The response of an underdamped system

$$m\ddot{x} + c\dot{x} + kx = F(t)$$

(with zero initial conditions) is given by (for $0 < \zeta < 1$)

$$x(t) = \frac{1}{m\omega_d} e^{-\zeta\omega_n t} \int_0^t F(\tau) e^{\zeta\omega_n \tau} \sin\omega_d(t - \tau)\, d\tau$$

where $\omega_n = \sqrt{k/m}$, $\zeta = c/2m\omega_n$, and $\omega_d = \omega_n\sqrt{1 - \zeta^2}$. With nonzero initial conditions this becomes

$$x(t) = A e^{-\zeta\omega_n t} \sin(\omega_d t + \phi) + \frac{1}{\omega_d} e^{-\zeta\omega_n t} \int_0^t f(\tau) e^{\zeta\omega_n \tau} \sin\omega_d(t - \tau)\, d\tau$$

where $f = F/m$ and A and ϕ are constants determined by the initial conditions.

Example 6.8.2

A rotating machine is mounted on the second floor of a building as illustrated in Figure 6.15. The machine excites the floor support beam with a force of 100 N at 3 rad/s. Model the floor support beam as an undamped simply-supported Euler–Bernoulli beam and calculate the forced response.

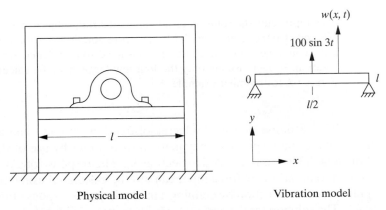

Physical model   Vibration model

Figure 6.15 A model of a building with an out-of-balance rotating machine mounted in the middle of the second floor over a support beam. The vibration model is simplified to be that of a harmonic force applied to the center of a simply-supported beam.

Solution The sketch on the right side of Figure 6.15 suggests that a reasonable model for the vibration is given by equation (6.92) and boundary conditions given by equation (6.95) with a driving force of 100 sin 3t. Hence the problem is to solve

$$w_{tt}(x, t) + c^2 w_{xxxx}(x, t) = 100 \sin 3t \, \delta\left(x - \frac{l}{2}\right) \tag{6.161}$$

subject to $w(0, t) = w(l, t) = w_{xx}(0, t) = w_{xx}(l, t) = 0$, where $c = \sqrt{EI/\rho A}$. Note that the harmonic time dependence of the input force suggests that equation (2.8) will be used to solve the temporal equation. The Dirac delta function in equation (6.161) is used to denote that the force is applied only at the point $l/2$ and not everywhere along x. First, separation of variables is used to calculate the spatial mode shapes using the homogeneous version of equation (6.161). To this end, let $w(x, t) = T(t)X(x)$ in equation (6.161) so that

$$\frac{\ddot{T}(t)}{T(t)} = -c^2 \frac{X''''(x)}{X(x)} = -\omega^2$$

following equation (6.97). This leads to equations (6.98) and (6.102), so that the solution of the spatial equation is given by equation (6.102):

$$X(x) = \alpha_1 \sin \beta x + \alpha_2 \cos \beta x + \alpha_3 \sinh \beta x + \alpha_4 \cosh \beta x \tag{6.162}$$

Here $\beta^4 = \rho A \omega^2 / EI$. Applying the four boundary conditions given in equations (6.95) yields the desired eigenfunctions. The deflection must be zero at $x = 0$ [i.e., $X(0) = 0$] so that $a_2 + a_4 = 0$. Similarly, the bending moment must vanish at $x = 0$ [i.e., $X''(0) = 0$] so that $-a_2 + a_4 = 0$. Hence $a_2 = a_4 = 0$ is required to satisfy the boundary conditions at $x = 0$. Thus equation (6.162) for the spatial solution reduces to

$$X(x) = \alpha_1 \sin \beta x + \alpha_3 \sinh \beta x \tag{6.163}$$

Applying the boundary condition at $x = l$ yields

$$\alpha_1 \sin \beta l + \alpha_3 \sinh \beta l = 0 \tag{6.164}$$

$$-\alpha_1 \sin \beta l + \alpha_3 \sinh \beta l = 0 \tag{6.165}$$

which has the solution $a_3 = 0$ and $\sin \beta l = 0$. Hence the characteristic equation is $\sin \beta l = 0$, which yields $\beta l = n\pi$, and the eigenfunction becomes

$$X_n(x) = A_n \sin \frac{n\pi x}{l} \quad n = 1, 2, \ldots \tag{6.166}$$

Recalling that $\beta^4 = \rho A \omega^2 / EI$ yields

$$\omega = \omega_n = \sqrt{\frac{EI}{\rho A}} \left(\frac{n\pi}{l}\right)^2 \quad n = 1, 2, \ldots \tag{6.167}$$

It is convenient at this point to normalize the eigenfunction given by equation (6.166). Following the definition of Chapter 4 for vectors, a set of eigenfunctions $X_n(x)$ is said to be *normal* if for each value of n

$$\int_0^l X_n(x)X_n(x)dx = 1 \tag{6.168}$$

As in the case of eigenvectors, the normalization condition fixes the arbitrary constant associated with the eigenfunction. The constant A_n in equation (6.166) can be determined by substitution of the eigenfunction $X_n(x) = A_n \sin(n\pi x/l)$ into equation (6.168). This yields

$$A_n^2 \int_0^l \sin^2 \frac{n\pi x}{l} dx = 1$$

Performing the indicated integration yields

$$A_n^2 \frac{l}{2} = 1 \quad \text{or} \quad A_n = \sqrt{2/l}$$

so that the normalized eigenfunction becomes

$$X_n(x) = \sqrt{\frac{2}{l}} \sin \frac{n\pi x}{l} \quad n = 1, 2, \ldots \tag{6.169}$$

These are also referred to as normalized mode shapes. These functions also have the property that

$$\int_0^l X_n(x)X_m(x)dx = 0 \quad m \neq n \tag{6.170}$$

which is an orthogonality condition similar to that for eigenvectors. If a set of functions $X_n(x)$ satisfied both equations (6.170) and (6.168) for all combinations of the index n, it is said to be an *orthonormal set.*

Continuing with the modal analysis solution of equation (6.161), substitution of the form $w(x, t) = T_n(t)X_n(x)$ into the equation of motion yields

$$\ddot{T}_n(t)X_n(x) + c^2 T_n(t)X_n''''(x) = 100 \sin 3t \, \delta\left(x - \frac{l}{2}\right) \tag{6.171}$$

From the equation that follows (6.161), $X_n'''' = (\omega_n^2/c^2)X_n(x)$. Substitution of this into (6.171) yields

$$\left[\ddot{T}_n(t) + \omega_n^2 T_n(t)\right]X_n(x) = (100 \sin 3t)\delta\left(x - \frac{l}{2}\right) \tag{6.172}$$

Multiplying equation (6.172) by $X_n(x)$ and integrating over the length of the beam yields

$$\ddot{T}_n(t) + \omega_n^2 T_n(t) = 100 \sin 3t \sqrt{\frac{2}{l}} \int_0^l \delta\left(x - \frac{l}{2}\right)\sin\frac{n\pi x}{l} \, dx \tag{6.173}$$

where the normalization condition is used to evaluate the integral on the left side of equation (6.173). Evaluating the integral on the left side yields

$$\ddot{T}_n(t) + \omega_n^2 T_n(t) = \sqrt{\frac{2}{l}} 100 \sin 3t \sin\frac{n\pi}{2} \quad n = 1, 2, 3, \ldots \tag{6.174}$$

or

$$\ddot{T}_n(t) + \frac{EI}{\rho A}\left(\frac{n\pi}{l}\right)^4 T_n(t) = 0, \quad n = 2, 4, 6, \ldots \tag{6.175}$$

$$\ddot{T}_n(t) + \frac{EI}{\rho A}\left(\frac{n\pi}{l}\right)^4 T_n(t) = \sqrt{\frac{2}{l}} 100 \sin 3t \quad n = 1, 5, 9, \ldots \tag{6.176}$$

$$\ddot{T}_n(t) + \frac{EI}{\rho A}\left(\frac{n\pi}{l}\right)^4 T_n(t) = -\sqrt{\frac{2}{l}} 100 \sin 3t \quad n = 3, 7, \ldots \tag{6.177}$$

where equation (6.167) has been used to evaluate ω_n^2.

Since the forced response is of interest here, the solution for $T_n(t)$ for even values of the index n determined by equation (6.175) is zero (zero force input and zero initial conditions). The solution to equations (6.176) and (6.177) is given by equation (2.7) as

$$T_n(t) = \frac{100\sqrt{2/l}}{(EI/\rho A)(n\pi/l)^4 - 9} \sin 3t \quad n = 1, 5, 9, \ldots \tag{6.178}$$

and

$$T_n(t) = \frac{-100\sqrt{2/l}}{(EI/\rho A)(n\pi/l)^4 - 9} \sin 3t \quad n = 3, 7, 11, \ldots \tag{6.179}$$

The forced response of the floor beam is then given by the series

$$w(x, t) = \sum_{n=1}^{\infty} T_n(t) X_n(x) \tag{6.180}$$

with T_n and X_n as indicated in equations (6.169), (6.178), and (6.179). Writing out the first few nonzero terms of this solution yields

$$w(x, t) =$$

$$\frac{200}{l} \left[\frac{\sin - (\pi x/l)}{(\pi^4 EI/l^4 \rho A) - 9} - \frac{\sin - (3\pi x/l)}{(81\pi^4 EI/l^4 \rho A) - 9} + \frac{\sin - (5\pi x/l)}{(625\pi^4 EI/l^4 \rho A) - 9} \cdots \right] \sin 3t \tag{6.181}$$

Given values of the material parameters ρ, E, and the dimension of the support beam l, I, and A, equation (6.181) describes the forced response of the beam in the model of the vibration of a building floor due to a rotating machine mounted above it, as sketched in Figure 6.15.

□

Example 6.8.3

Consider the camera mounted on a square shaft in Figure 6.16. Use Tables 6.3 and 6.5 to compute the first three natural frequencies and then compute the magnitude of the first mode forced response to a wind load providing a torque of $M_0 = 15 \times L$ Nm at

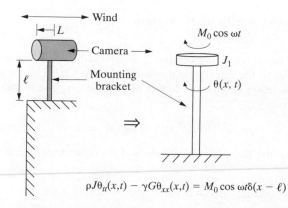

$$\rho J \theta_{tt}(x,t) - \gamma G \theta_{xx}(x,t) = M_0 \cos \omega t \delta(x - \ell)$$

Figure 6.16 A distributed model of a mounting bracket for a security camera.

a frequency of $\omega = 10$ Hz. Model the camera in torsional vibration as suggested in the figure. The mounting bracket is a solid piece of aluminum 0.02×0.02 m in cross section, with $J = 2.667 \times 10^{-8}$ m^4, density 2.7×10^3 kg/m^3, shear modulus 2.67×10^{10} N, and shaft length $\ell = 0.55$ m. The camera has length $L = 0.2$ m, with mass $m = 3$ kg. Approximate the rotational inertia of the camera by $J_1 = mL^2$.

Solution   Following Example 6.8.2, first compute the solution of the eigenvalue problem for the free response. The equation of motion for the free response is given in equation (6.67) by

$$\theta_{tt}(x, t) = \frac{G\gamma}{\rho J} \theta_{xx}(x, t), \quad \theta(0, t) = 0 \quad \text{and} \quad \gamma G\theta_x(\ell, t) = -J_1\theta_{tt}(\ell, t)$$

Here γ is the torsional constant of Table 6.3. Using separation of variables to obtain the frequency equation yields

$$\varphi(x)\ddot{T}(t) = \frac{G\gamma}{\rho J} \varphi''(x)T(t), \quad \varphi(0)T(t) = 0, \quad \text{and} \quad \gamma G\varphi'(\ell)T(t) = -J_1\varphi(\ell)\ddot{T}(t)$$

Rearranging and simplifying reveals the eigenvalue problem

$$\frac{\varphi''(x)}{\varphi(x)} = \frac{\rho J}{G\gamma}\frac{\ddot{T}(t)}{T(t)} = -\sigma^2, \quad \varphi(0) = 0, \quad \frac{\gamma G}{J_1}\frac{\varphi'(\ell)}{\varphi(\ell)} = -\frac{\ddot{T}(t)}{T(t)} = \frac{\sigma^2 G\gamma}{\rho J}$$

The spatial equation is then summarized as

$$\varphi''(x) + \sigma^2\varphi(x) = 0 \Rightarrow \varphi(x) = a\sin\sigma x + b\cos\sigma x$$

$$\varphi(0) = 0$$

$$\gamma G\rho J\varphi'(\ell) = \gamma\sigma^2 GJ_1\varphi(\ell)$$

Substitution of the solution form into the first boundary condition yields that $b = 0$, so that the eigenfunctions will have the form $\varphi(x) = a\sin\sigma x$. Substitution of this form into the boundary condition at $x = \ell$ yields the characteristic equation

$$\rho Ja\sigma\cos\sigma\ell = \sigma^2 J_1 a\sin\sigma\ell \Rightarrow \tan\sigma l = \left(\frac{\rho\ell J}{J_1}\right)\frac{1}{\sigma\ell}$$

This characteristic equation is transcendental and must be solved numerically for the values of $\sigma\ell$, similar to the calculation made in Example 6.2.3. Thus, σ will depend on the index n as there are an infinite number of solutions, σ_n. The inertia of the camera is computed from

$$J_1 = mL^2 = 3(0.2)^2 = 0.12 \text{ kg} \cdot \text{m}^2$$

With the values given, the first three numerical values of the solutions of the transcendental equation yield

$$\sigma_1\ell = 0.0182, \quad \sigma_2\ell = 3.1417, \quad \sigma_3\ell = 6.2832.$$

Note that the values of $\sigma_n\ell$ quickly approach $n\pi$ as n increases. Forming the temporal equation

$$\ddot{T}_n(t) + \frac{\gamma G}{\rho J}\sigma_n^2 T_n(t) = 0$$

the natural frequency is related to $\sigma_n\ell$ by

$$\omega_n = \frac{\sigma_n\ell}{\ell}\sqrt{\frac{\gamma G}{\rho J}}\,\text{rad/s}$$

Using the numerically computed values of $\sigma_n\ell$ and the given data,

$$\omega_1 = 95.392\,\text{rad/s} \quad \omega_2 = 1.6497 \times 10^4\,\text{rad/s} \quad \omega_3 = 3.2994 \times 10^4\,\text{rad/s}$$

Next. consider the first mode solution ($n = 1$) for the force response. Note that after the second index, the driving frequency is well away from the modal frequency. The modal solution then takes on the form

$$(\rho J\ddot{T}_1(t) + \gamma G\sigma_1^2 T_1(t))a_1\sin\sigma_1 x = M_0\cos\omega t\,\delta(x - \ell)$$

Normalizing the arbitrary constant $a_n = 1$ and multiplying the first mode equation by $\sin\sigma_1 x$ yields (using the normality of $\sin\sigma_1 x$)

$$(\rho J\ddot{T}_1(t) + \gamma G\sigma_1^2 T_1(t))\frac{\ell}{2} = M_0\cos\omega t\int_0^\ell \sin\sigma_1 x\,\delta(x - \ell)dx$$

$$\Rightarrow \rho J\ddot{T}_1(t) + \gamma G\sigma_1^2 T_1(t) = \frac{2M_0}{\ell}\sin\sigma_1\ell\cos\omega t$$

This last expression can be solved for the particular solution using the method of Section 2.1. Dividing by the leading coefficient yields

$$\ddot{T}_1(t) + \omega_1^2 T_1(t) = \frac{2M_0}{\ell\rho J}(0.0182)\cos\omega t$$

From Example 2.1.2, the magnitude of the force response is

$$\left|\frac{2f_0}{\omega_n^2 - \omega^2}\right| = \left|\frac{2(0.0182)\,M_0/\ell\rho J}{\omega_n^2 - \omega^2}\right|$$

$$= \left|\frac{2(0.0182)(15\cdot 0.2)/(0.55\cdot 2.7\times 10^3\cdot 2.667\times 10^{-8})}{95.392^2 - 62.832^2}\right| = 0.3764$$

□

The modal analysis calculations used in Examples 6.8.1, 6.8.2, and 6.8.3 as well as the example in Section 6.7 can be outlined for a general case just as for lumped-parameter systems in Chapter 4. This is summarized in Window 6.4 and again here. Modal analysis proceeds by substitution of the separation-of-variables form

$$w(x, t) = X_n(x)T_n(t)$$

into the equation of motion. This leads to two equations: one a boundary-value problem in $X_n(x)$ and the other an initial-value problem in $T_n(t)$. The boundary conditions are applied to the solution of the spatial equation for $X_n(x)$. This yields the eigenvalues (natural frequencies) and eigenfunctions (mode shapes) of the system. This step yields the same type of information as solving the eigenvalue problems for the matrix $M^{-1/2}KM^{-1/2}$ for lumped-mass systems. Next the eigenfunctions are normalized by using equation (6.168).

With the functions $X_n(x)$ completely determined and the natural frequencies known, the temporal equation for the function $T_n(t)$ can be solved by substitution of $w(x, t) = T_n(t)X_n(x)$ into the equation of motion and the initial conditions. With $T_n(t)$ known for all n, the total solution is assembled as the sum

$$w(x, t) = \sum_{n=1}^{\infty} T_n(t)X_n(x) \tag{6.182}$$

This equation is also called the *expansion theorem*. Certain mathematical arguments need to be made to check the convergence of this infinite series [see, e.g., Inman (2006)], but these are beyond the scope of presentation here. In many cases it suffices to use just a few of the first terms in the series to compute a meaningful approximation to the solution. Chapter 8 discuss more about approximating the solution given in equation (6.182).

As mentioned in Window 6.4 and confirmed in Example 6.5.2, the separation of variables/modal-analysis approach does not always work. That is, it is often difficult or even impossible to separate the governing partial differential equations into a separate spatial equation and a separate temporal equation. This happens, for instance, in attempting to solve the Timoshenko beam equation because of the cross term

$$\frac{\partial^4 w(x, t)}{\partial x^2 \partial t^2}$$

which involves derivatives of both variables, and because of the existence of a fourth time derivative. However, because the Euler–Bernoulli beam mode shapes are assumed to be satisfactory for use with the Timoshenko method, a separated solution is assumed by writing $w(x, t)$ as a product of $T(t)$ and the assumed mode shape. This works in Example 6.5.2. Many other distributed-parameter vibration problems cannot be directly solved by the analytical approach of separation of variables. Hence, a number of approximate approaches have been developed, such as assuming the mode shape (called the *assumed mode method*), as is done in Example 6.5.2. Other approximate methods are developed in detail in Chapter 8. For distributed-parameter structures complicated beyond the simple configurations considered in this chapter, the finite element of Chapter 8 or similar approximation method must be used for vibration analysis.

PROBLEMS

Section 6.2 (Problems 6.1 through 6.8)

6.1. Compute the first two natural frequencies of a fixed-fixed string with density $\rho A = 0.8\,\text{g/m}$, tension of 120 N, and length $l = 300\,\text{mm}$.

6.2. Prove the orthogonality condition of equation (6.28).

6.3. Calculate the orthogonality of the modes in Example 6.2.3.

6.4. Plot the first four modes of Example 6.2.3 for the case $l = 1.2\,\text{m}$, $k = 1000\,\text{N/m}$, and $\tau = 880\,\text{N/m}$.

6.5. Consider a cable that has one end fixed and one end free. The free end cannot support a transverse force, so that $w_x(l, t) = 0$. Calculate the natural frequencies and mode shapes.

6.6. Calculate the coefficients c_n and d_n of equation (6.27) for the system of a clamped–clamped string to the initial displacement given in Figure P6.6 and an initial velocity of $w_t(x, 0) = 0$.

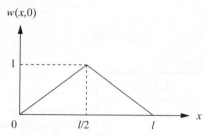

Figure P6.6   The initial displacement (mm) for the string of Problem 6.6.

6.7. Plot the response of the string in Problem 6.6 for the piano string of Example 6.2.2 ($l = 1.2\,\text{m}$, $m = 100\,\text{g}$, $\tau = 11.1 \times 10^4\,\text{N}$) at $x = l/4$ and $x = l/2$, using 3, 5, and 10 terms in the solution.

***6.8.** Consider the clamped string of Problem 6.6. Calculate the response of the string to the initial condition

$$w(x, 0) = \sin\frac{3\pi x}{l} \quad w_t(x, 0) = 0$$

Plot the response at $x = l/2$ and $x = l/4$, for the parameters of Example 6.2.2.

Section 6.3 (Problems 6.9 through 6.31)

6.9. Calculate the natural frequencies and mode shapes for a free–free bar. Calculate the temporal solution of the first mode.

6.10. Calculate the natural frequencies and mode shapes of a clamped–clamped bar.

6.11. It is desired to design a 4.6-m clamped–free bar such that its first natural frequency is 1800 Hz. Of what material should it be made?

6.12. Suppose that for a particular bar, $A(x)$ is not constant. Show that the temporal solution of

$$\frac{\partial}{\partial x}\left(EA(x)\frac{\partial w(x, t)}{\partial x}\right) = \rho A(x)\frac{\partial^2 w(x, t)}{\partial t^2}$$

is still of the form $T(t) = A\sin\omega_n t + B\cos\omega_n t$, where ω_n is proportional to $\sqrt{E/\rho}$.

6.13. Compare the natural frequencies of a clamped–free 1.2-m aluminum bar to that of a 1.2-m bar made of steel, a carbon composite, and a piece of wood.

6.14. Derive the boundary conditions for a clamped–free bar with a solid lumped mass of mass M attached to the free end.

6.15. Calculate the mode shapes and natural frequencies of the bar of Problem 6.12. State how the lumped mass affects the natural frequencies and the mode shapes.

***6.16.** Calculate and plot the first three mode shapes of a clamped–free bar.

*6.17. Calculate and plot the first three mode shapes of a clamped–clamped bar and compare them to the plots of Problem 6.16.

*6.18. Calculate and compare the eigenvalues of the free–free, clamped–free, and clamped–clamped bar. Are they related? What does this state about the system's natural frequencies?

6.19. Consider the nonuniform bar of Figure P6.19, which changes cross-sectional area as indicated in the figure. In the figure, A_1, E_1, ρ_1, and l_1 are the cross-sectional area, modulus, density, and length of the first segment, respectively, and A_2, E_2, ρ_2, and l_2 are the corresponding physical parameters of the second segment. Determine the characteristic equation.

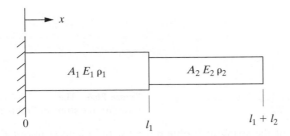

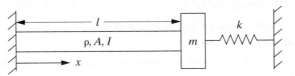

Figure P6.19 A bar with two separate cross-sectional areas made of two different materials.

6.20. Show that the solution obtained to Problem 6.19 is consistent with that of a uniform bar.

6.21. Calculate the first three natural frequencies for the cable and spring system of Example 6.2.3 for $l = 1.2, k = 80, \tau = 120$ (SI units).

6.22. Calculate the first three natural frequencies of a clamped–free cable with a mass of value m attached to the free end. Compare these to the frequencies obtained in Problem 6.19 if you also worked that problem.

6.23. Calculate the boundary conditions of a bar fixed at $x = 0$ and connected to ground through a mass and a spring as illustrated in Figure P6.23.

Figure P6.23 A beam with a tip mass connected to ground via a spring.

6.24. Calculate the natural frequency equation for the system of Problem 6.23.

6.25. Estimate the natural frequencies of an automobile frame for vibration in its longitudinal direction (i.e., along the length of the car) by modeling the frame as a (one-dimensional) steel bar.

6.26. Consider the first natural frequency of the bar of Problem 6.23 with $k = 0$ and Table 6.2, which is fixed at one end and has a lumped-mass, M, attached at the free end. Compare this to the natural frequency of the same system modeled as a single-degree-of-freedom spring–mass system given in Figure 1.23. What happens to the comparison as M becomes small and goes to zero?

6.27. Following the line of thought suggested in Problem 6.26, model the system of Problem 6.23 as a lumped-mass single-degree-of-freedom system and compare this frequency to the first natural frequency obtained in Problem 6.24 if you solved that.

6.28. Calculate the response of a clamped–free bar to an initial displacement of 1.2 cm at the free end and a zero initial velocity. Assume that density $= 7.8 \times 10^3$ kg/m^3, $A = 0.002\,m^2$, $E = 2.1 \times 10^{10}$ N/m^2, and $l = 0.6$ m. Plot the response at $x = l$ and $x = l/2$ using the first three modes.

6.29. Repeat the plots of Problem 6.28 for 5 modes, 10 modes, 15 modes, and so on, to answer the question of how many modes are needed in the summation of equation (6.27) in order to yield an accurate plot of the response for this system.

6.30. A moving bar is traveling along the x axis with constant velocity and is suddenly stopped at the end at $x = 0$, so that the initial conditions are $w(x, 0) = 0$ and $w_t(x, 0) = v$. Calculate the vibration response.

6.31. Calculate the response of the clamped–clamped string of Section 6.2 to a zero initial velocity and an initial displacement of $w_0(x) = \sin(2\pi x/l)$. Plot the response at $x = l/2$.

Section 6.4 (Problems 6.32 through 6.41)

6.32. Calculate the first three natural frequencies of torsional vibration of a shaft of Figure 6.7 clamped at $x = 0$, if a disk of inertia $J_0 = 12$ kg $\cdot$ m^2/rad is attached to the end of the shaft at $x = l$. Assume that $l = 0.6$ m, $J = 5$ m^4, $G = 2.5 \times 10^9$ Pa, density $\rho = 2700$ kg/m^3.

6.33. If you worked Problem 6.32, compare the frequencies calculated in that problem to the frequencies of the lumped-mass single-degree-of-freedom approximation of the same system.

6.34. Calculate the natural frequencies and mode shapes of a shaft in torsion of shear modulus G, length l, polar inertia J, and density ρ that is free at $x = 0$ and connected to a disk of inertia J_0 at $x = l$.

6.35. Consider the lumped-mass model of Figure 4.21 and the corresponding three-degree-of-freedom model of Example 4.8.1. Let $J_1 = k_1 = 0$ in this model and collapse it to a two-degree-of-freedom model. Comparing this to Example 6.4.1, it is seen that they are a lumped-mass model and a distributed-mass model of the same physical device. Referring to Chapter 1 for the effects on lumped stiffness of a rod in torsion (k_2), compare the frequencies of the lumped-mass two-degree-of-freedom model with those of Example 6.4.1.

6.36. The modulus and density of a 1.2-m aluminum rod are $E = 7.2 \times 10^{10}$ N/m^2, $G = 2.7 \times 10^{10}$ N/m^2, and $\rho = 2.7 \times 10^3$ kg/m^3. Compare the torsional natural frequencies with the longitudinal natural frequencies for a free-clamped rod.

6.37. Consider the aluminum shaft of Problem 6.32. Add a disk of inertia J_0 to the free end of the shaft. Plot the torsional natural frequencies versus increasing the tip inertia J_0 of a single-degree-of-freedom model and for the first natural frequency of the distributed-parameter model in the same plot. Are there any values of J_0 for which the single-degree-of-freedom model gives the same frequency as the fully distributed model?

6.38. Calculate the mode shapes and natural frequencies of a bar with circular cross section in torsional vibration with free–free boundary conditions. Express your answer in terms of G, l, and ρ.

6.39. Calculate the mode shapes and natural frequencies of a bar with circular cross section in torsional vibration with fixed boundary conditions. Express your answer in terms of G, l, and ρ.

6.40. Consider the torsional vibrations of a fixed–free shaft and compare the torsional frequencies to its longitudinal vibrations.

6.41. Compute the mode shapes and natural frequencies for a fixed-spring shaft of Table 6.5.

Section 6.5 (Problems 6.42 through 6.49)

6.42. Calculate the natural frequencies and mode shapes of a clamped–free beam. Express your solution in terms of E, I, ρ, and l. This is called the cantilevered beam problem.

*6.43. Plot the first three mode shapes calculated in Problem 6.40. Next calculate the strain-mode shape [i.e., $X'(x)$], and plot these next to the displacement-mode shapes $X(x)$. Where is the strain the largest?

6.44. Derive the general solution to a fourth-order ordinary differential equation with constant coefficients of equation (6.100) given by equation (6.102).

6.45. Derive the natural frequencies and mode shapes of a pinned–pinned beam in transverse vibration. Calculate the solution for $w_0(x) = \sin 2\pi x/l$ and $\dot{w}_0(x) = 0$.

6.46. Derive the natural frequencies and mode shapes of a fixed–fixed beam in transverse vibration.

6.47. Show that the eigenfunctions or mode shapes of Example 6.5.1 are orthogonal. Make them normal.

6.48. Derive equation (6.109) from equations (6.107) and (6.108).

6.49. Show that if shear deformation and rotary inertia are neglected, the Timoshenko equation reduces to the Euler–Bernoulli equation and the boundary conditions for each model become the same.

Section 6.6 (Problems 6.50 through 6.54)

6.50. Calculate the natural frequencies of the membrane of Example 6.6.1 for the case that one edge $x = 1$ is free.

6.51. Repeat Example 6.6.1 for a rectangular membrane of size a by b. What is the effect of a and b on the natural frequencies?

6.52. Plot the first three mode shapes of Example 6.6.1.

6.53. The lateral vibrations of a circular membrane are given by

$$\frac{\partial^2 \omega(r, \phi, t)}{\partial r^2} + \frac{1}{r}\frac{\partial \omega(r, \phi, t)}{\partial r} + \frac{1}{r^2}\frac{\partial^2 \omega(r, \phi, t)}{\partial \phi \partial r} = \frac{\rho}{\tau}\frac{\partial^2 \omega(r, \phi, t)}{\partial t^2}$$

where r is the distance from the center point of the membrane along a radius and ϕ is the angle around the center. Calculate the natural frequencies if the membrane is clamped around its boundary at $r = R$.

6.54. Discuss the orthogonality condition for Example 6.6.1.

Section 6.7 (Problems 6.55 through 6.65)

6.55. Calculate the response of bar of length $l = 1.2\,\mathrm{m}$, $E = 2.8 \times 10^{10}\,\mathrm{N/m^2}$, and $\rho = 8.4 \times 10^3\,\mathrm{kg/m^3}$. Assume modal damping of 0.08 in each mode and initial conditions of $w(x,0) = 2(x/l)$ cm and an initial velocity of $w_t(x,0) = 0$. Plot the response using the first three modes at $x = l/2, l/4$, and $3l/4$. How many modes are needed to represent accurately the response at the point $x = l/2$?

6.56. Compute the solution for a bar with modal damping ratio of $\zeta_n = 0.01$, subject to initial conditions $w(x, 0) = 2(x/l)$ cm, and an initial velocity of $w_t(x, 0) = 0$.

6.57. Repeat Problem 6.55 for the case of Problem 6.56. Does it take more or fewer modes to accurately represent the response at $l/2$?

6.58. Calculate the form of modal damping ratios for the clamped string of equations (6.151) and the clamped membrane of equation (6.152).

6.59. Calculate the units on γ and β in equation (6.153).

6.60. Assume that E, I, and ρ are constant in equations (6.153) and (6.154) and calculate the form of the modal damping ratio ζ_n.

6.61. Calculate the form of the solution $w(x, t)$ for the system of Problem 6.58.

6.62. For a given cantilevered composite beam, the following values have been measured for bending vibration:

$$E = 2.71 \times 10^{10}\,\text{N/m}^2 \qquad \rho = 1710\,\text{kg/m}^3$$
$$A = 0.597 \times 10^{-3}\,\text{m}^2 \qquad l = 1\,\text{m}$$
$$I = 1.64 \times 10^{-9}\,\text{m}^4 \qquad \gamma = 1.75\,\text{N}\cdot\text{s/m}^2$$
$$\beta = 20{,}500\,\text{N}\cdot\text{s/m}^2$$

Calculate the solution for the beam to an initial displacement of $w_t(x,0) = 0$ and $w(x,0) = 4\sin\pi x$.

*6.63. Plot the solution of Example 6.7.2 for the case $w_t(x,0) = 0$, $w(x,0) = \sin(\pi x/l)$, $= 10\,\text{N}\cdot\text{s/m}^2, \tau = 10^4\,\text{N}, l = 1.2\,\text{m}$, and $\rho = 0.012\,\text{kg/m}^3$.

6.64. Calculate the orthogonality condition for the system of Example 6.7.2. Then calculate the form of the temporal solution.

6.65. Calculate the form of modal damping for the longitudinal vibration of the beam of Figure 6.14 with boundary conditions specified by equation (6.157).

Section 6.8 (Problems 6.66 through 6.70)

6.66. Calculate the response of the damped string of Example 6.8.1 to a disturbance force of $f(x, t) = (\sin\pi x/l)\sin 10t$. That is, solve the following equation:

$$\rho A w_{tt}(x, t) + \gamma w_t(x, t) - \tau w_{xx}(x, t) = (\sin\pi x/l)\sin 10t$$

6.67. Consider the clamped–free bar of Example 6.3.2. The bar can be used to model a truck bed frame. If the truck hits an object (at the free end) causing an impulsive force of 120 N, calculate the resulting vibration of the frame. Note here that the truck cab is so massive compared to the bed frame that the end with the cab is modeled as clamped. This is illustrated in Figure P6.67.

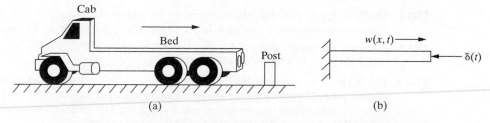

Figure P6.67 (a) Model of a truck hitting an object; (b) simplified vibration model.

6.68. A rotating machine sits on the second floor of a building just above a support column as indicated in Figure P6.68. Calculate the response of the column in terms of E, A, and ρ of the column modeled as a bar.

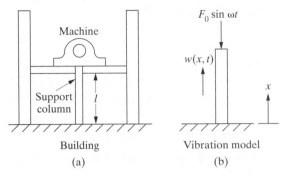

Figure P6.68   (a) Model of a rotating out-of-balance machine mounted on top of a column on the second floor of a building; (b) vibration model.

6.69. Recall Example 6.8.2, which models the vibration of a building due to a rotating machine imbalance on the second floor. Suppose that the floor is constructed so that the beam is clamped at one end and pinned at the other, and recalculate the response (recall Example 6.5.1). Compare your solution and that of Example 6.8.2, and discuss the difference.

6.70. Use the modal analysis procedure suggested at the end of Section 6.8 to calculate the response of a clamped–free beam with a sinusoidal loading $F_0 \sin \omega t$ at its free end.

MATLAB ENGINEERING VIBRATION TOOLBOX

If you have not yet used the *Engineering Vibration Toolbox*, return to the end of Chapter 1 or Appendix G for a brief introduction. The files contained in the folder VTB6 may be used to help solve the following Toolbox problems. The M-files from earlier chapters may also be useful. Alternately, the codes may be used directly following the introductions given in Sections 1.9, 1.10, 2.8, 2.9, 3.8, 3.9, 4.9, and 4.10 and Appendix F.

TOOLBOX PROBLEMS

TB6.1. Use file VTB6_1 to investigate the effects of changing the length, density, and modulus (or tension) on the frequency and mode shapes of bars and shafts. For example, consider the clamped–free bar of Example 6.3.1. Study the effect of changing the length l on the frequencies by calculating ω_n for fixed value of E and ρ (say, for aluminum). What happens to the mode shapes?

TB6.2. Use VTB6_2 to calculate the frequencies for torsional vibration for the first three boundary conditions of Table 6.5. In particular, check the results of Example 6.4.1.

TB6.3. Compare the mode shapes of a cantilevered beam with those of the fixed–pinned beam of Example 6.5.1 by using file VTB6_3 to plot the mode shapes.

TB6.4. Use VTB6_4 to compare the effects of rotary inertia and shear deformation on a pinned–pinned beam by trying several different values of the various physical parameters. Try to conclude circumstances under which the Timoshenko theory gives drastically different frequencies than the Euler–Bernoulli theory.

7 Vibration Testing and Experimental Modal Analysis

This chapter presents methods of testing and measurement useful for obtaining experimental models of a variety of devices and structures. The analyzer pictured here receives analog signals from transducers (accelerometers and hammer) mounted on a structure, digitizes the signals, and transforms them into the frequency domain for analysis. The entire process is controlled by a personal computer. A schematic of such a test setup is given later in Figure 7.1. The picture at the bottom exhibits the use of a laser vibrometer to measure velocity without physically being mounted on the test object (satellite component). Section 7.1 describes the measurement hardware, and the remainder of the chapter is devoted to methods of analyzing the data. In particular, the method of modal analysis is discussed.

This chapter discusses vibration measurement and focuses in particular on techniques associated with experimental modal analysis. Vibration measurements are made for a variety of reasons. As pointed out in previous chapters, especially in Chapter 5 on design, the natural frequencies of a structure or machine are extremely important in predicting and understanding a system's dynamic behavior. Hence, a major reason for performing a vibration test of a system is to determine its natural frequencies. Another reason for vibration testing is to verify an analytical model proposed for the test system. For example, the analytical models proposed for the various examples of Section 4.8 and those of Chapter 6 yield a specific set of frequencies and mode shapes. A vibration test can then be performed on the same system. If the measured frequencies and mode shapes agree with those predicted by the analytical model, the model is verified and can be used in design and response prediction with some confidence.

Another important use of vibration testing is to determine experimentally the dynamic durability of a particular device. In this case, a test article is driven or forced to vibrate by specified inputs for a specific length of time. When the test is over, the test piece must still perform its original task. The purpose of this type of testing is to provide experimental evidence that a machine part or structure can survive a specific dynamic environment.

Vibration testing is also used in machinery diagnostics for maintenance. The idea here is continuous monitoring of the natural frequencies of a structure or machine. A shift in frequency or some other vibration parameter may indicate a pending failure or a need for maintenance. This use of vibration measurement is part of the general topic of condition monitoring of machinery and of structural health monitoring. It is similar in concept to observing the oil pressure in an automobile engine to determine if engine failure may occur or if maintenance is required.

The primary requirement of each of the aforementioned uses of vibration tests is a determination of a system's natural frequencies. This chapter focuses on *experimental modal analysis* (EMA), which is the determination of natural frequencies, mode shapes, and damping ratios from experimental vibration measurements. The fundamental idea behind modal testing is that of resonance introduced in Section 2.2. If a structure is excited at resonance, its response exhibits two distinct phenomena, as indicated in Figure 2.8. As the driving frequency approaches the natural frequency of the structure, the magnitude at resonance rapidly approaches a sharp maximum value, provided that the damping ratio is less than about 0.5. The second, often neglected phenomenon of resonance is that the phase of the response shifts by 180° as the frequency sweeps through resonance, with the value of the phase at resonance being 90°. This physical phenomenon is used to determine the natural frequency of a structure from measurements of the magnitude and phase of the structure's forced response as the driving frequency is swept through a wide range of values.

The vibration test methods presented in this chapter depend on several assumptions. First, it is assumed that the structure or machine being tested can be described adequately by a lumped-parameter model. There are several other assumptions commonly made but not stated (or understated) in vibration testing. The most obvious of these is that the system under test is linear and is driven by the test input only in its linear range. This assumption is essential and should not be neglected.

Vibration testing and measurement for modeling purposes has grown into a large industry. This field is referred to as *modal testing, modal analysis*, or *experimental modal analysis*. Understanding modal testing requires knowledge of several areas. These include instrumentation, signal processing, parameter estimation, and vibration analysis. These topics are presented in the following sections. The first few sections of this chapter deal with the hardware considerations and digital signal analysis necessary for making a vibration measurement for any purpose.

7.1 MEASUREMENT HARDWARE

The data acquisition and signal processing hardware has changed considerably over the past decades and continues to change rapidly as the result of advances in solid-state and computer technology. Hence specific hardware capabilities change very quickly, and only generic hardware is discussed here. A vibration measurement generally requires several hardware components. The basic hardware elements required consist of a source of excitation, called an *exciter*, for providing a known or controlled input force to the structure, a *transducer* to convert the mechanical motion of the structure into an electrical signal, a signal conditioning amplifier to match the characteristics of the transducer to the input electronics of the digital data acquisition system, and an analysis system (or analyzer) in which signal processing and modal analysis computer programs reside. This arrangement is illustrated in Figure 7.1; it includes a power amplifier and a signal generator for the exciter, as well as a transducer to measure, and possibly control, the driving force or other input. Each of these devices and their functions are discussed briefly in this section.

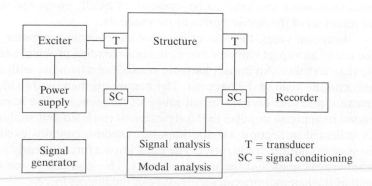

Figure 7.1   A schematic of hardware used in performing a vibration test.

First consider the excitation system. This system provides an input motion or, more commonly, a driving force $F_i(t)$, as in equation (4.130). The physical device may take several forms, depending on the desired input and the physical properties of the test structure. The two most commonly used exciters in modal testing are the *shaker* (electromagnetic or electrohydraulic) and the *impulse hammer*. The preferred device is often the electromagnetic exciter, which has the ability, when properly sized, to provide inputs large enough to result in easily measured responses. Also, the output is easily controlled electronically, sometimes using force feedback. The excitation signal, which can be tailored to match the requirements of the structure being tested, can be a swept sinusoidal, random, or other appropriate signal. A swept sine input consists of applying a harmonic force of constant magnitude f_i at a variety of different frequencies, ranging from a small value to larger values covering a frequency range of interest. At each incremental value of the driving frequency, the structure is allowed to reach steady state before the response magnitude and phase are measured. The electromagnetic shaker is basically a linear electric motor consisting of coils of wire surrounding a shaft in a magnetic field. An alternating current applied to the coil causes a force to be applied to the shaft, which, in turn, transfers force to the structure. The input electrical signal to the shaker is usually a voltage that causes a proportional force to be applied to the test structure, thus a signal generator can be used to impart a variety of different input signals to the structure.

Since shakers are attached to the test structure and since they have significant mass, care should be taken by the experimenter in choosing the size of shaker and method of attachment to minimize the effect of the shaker on the structure. The shaker and its attachment can add mass to the structure under test (called *mass loading*) as well as otherwise constraining the structure. Mass loading will lower the apparent measured frequency since $\omega_n = \sqrt{k/m}$. Mass loading and other effects can be minimized by attaching the shaker to the structure through a *stinger*. A stinger consists of a short thin rod (usually made of steel or nylon) running from the driving point of the shaker to a force transducer mounted directly on the structure. The stinger serves to isolate the shaker from the structure, reduces the added mass, and causes the force to be transmitted axially along the stinger, controlling the direction of the applied force more precisely.

In recent years, the *impact hammer* has become a popular excitation device. The use of an impact hammer avoids the mass loading problem and is much faster to use than a shaker. An impact hammer consists of a hammer with a force transducer built into the head of the hammer. The hammer is then used to hit (impact) the test structure and thus excite a broad range of frequencies. The impact hammer is intended to apply an impulse to the structure as modeled and analyzed in Section 3.1. As indicated in Section 4.6, the impulse response contains excitations at each of the system's natural frequencies. The peak impact force is nearly proportional to the hammerhead mass and the impact velocity. The load cell (force transducer) in the head of the hammer provides a measure of the impact force.

Figure 7.2 illustrates both time history and the corresponding frequency response of a typical hammer hit. Note that the time history is not a perfect delta function (as in Figure 3.1) but rather has a finite time duration, T. Hence the frequency response is not a flat straight line as indicated by the transform of an exact impulse, but rather has the periodic form given in Figure 7.2. The duration of the pulse and hence the shape of the frequency response is controlled by the mass and stiffness of both the hammer and the structure. In the case of a small hammer mass used on a hard structure (such as metal), the stiffness of the hammer tip determines the shape of the spectrum and, in particular, the cutoff frequencies ω_c. The *cutoff frequency* is the largest value of frequency reasonably well excited by the hammer hit. As illustrated in the figure, ω_c corresponds roughly to the point where the magnitude of the frequency response falls more than 10 to 20 dB from its maximum value. This means that at frequency higher than ω_c, the test structure does not receive enough energy to excite modes above ω_c. Thus ω_c determines the useful range of frequency excitation.

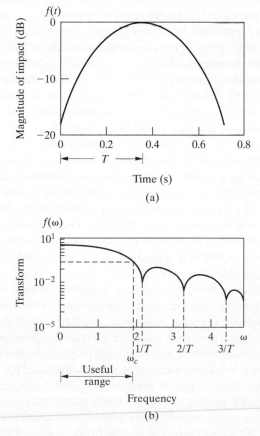

Figure 7.2 Time (a) and frequency response (b) of a hammer hit, indicating the useful range of excitation and its dependence on the pulse duration, T.

The upper frequency limit excited by the hammer is decreased by increasing the hammerhead mass and is increased with increasing stiffness of the tip of the hammer. The hammer hit is less effective in exciting the modes of the structure with frequencies larger than ω_c than it is for those less than ω_c. The built-in force transducer in impact hammers should be dynamically calibrated for each tip used, as this will affect the sensitivity. Although the impact hammer is simple and does not add mass loading to the structure, it is often incapable of transforming sufficient energy to the structure to obtain adequate response signals in the frequency range of interest. Also, peak impact loads are potentially damaging, and the direction of the applied load is difficult to control. Nonetheless, impact hammers remain a popular and useful excitation device, as they generally are much faster to use than shakers, are portable, and are relatively inexpensive.

Next consider the transducers required to measure the response of the structure as well as the impact force. The most popular and widely used transducers are made from piezoelectric crystals. Piezoelectric materials generate electrical charge when strained. By various designs, transducers incorporating these materials can be built to produce signals proportional to either force or local acceleration. *Accelerometers*, as they are called, actually consist of two masses, one of which is attached to the structure, separated by a piezoelectric material. The piezoelectric material acts like a very stiff spring. This causes the transducer to have a resonant frequency. The maximum measurable frequency is usually a fraction of the accelerometer's resonance frequency (recall Figure 2.26). In fact, the upper frequency limit is usually determined by the so-called mounted resonance, since the connection of the transducer to the structure is always somewhat compliant. The dynamics of accelerometers are discussed in some detail in Section 2.6.

Strain gauges can also be used to pick up vibrational responses. A *strain gauge* is a metallic or semiconductor material that exhibits a change in electrical resistance when subjected to a strain. Strain gauges are constructed by bending a conducting wire back and forth in a serpentine fashion over a very small surface, which is then bonded to the device or structure to be measured. As the structure strains, the resistance of the serpentine wire changes. The gauge is made part of a Wheatstone bridge circuit, which is used to measure the resistance change of the gauge and hence the strain in the test specimen [see Figliola and Beasley (1991)]. Strain gauges are also used to form load cells.

The output impedance of most transducers is not well suited for direct input into signal analysis equipment. Hence *signal conditioners*, which may be charge amplifiers or voltage amplifiers, match and often amplify signals prior to analyzing the signal. It is very important that each set of transducers along with signal conditioning are properly calibrated in terms of both magnitude and phase over the frequency range of interest. While accelerometers are convenient for many applications, they provide weak signals if one is interested in lower-frequency vibrations incurred in terms of velocity or displacement. Even substantial low-frequency vibration displacements may result in only small accelerations, since a harmonic displacement of amplitude X has acceleration of amplitude $-\omega^2 X$. Strain gauges and potentiometers as well as various optical, capacitive, and inductive transducers are often more suitable than accelerometers for low-frequency motion measurement.

Once the response signal has been properly conditioned, it is routed to an analyzer for signal processing. There are several types of analyzers in use. The type that has become the standard is called a digital Fourier analyzer, also called the fast Fourier transform (often abbreviated FFT) analyzer; it is introduced briefly here. Basically, the analyzer accepts analog voltage signals that represent the acceleration (force, velocity, displacement, or strain) from a signal conditioning amplifier. This signal is filtered and digitized for computation. Discrete frequency spectra of individual signals and cross-spectra between the input and various outputs are computed. The analyzed signals can then be manipulated in a variety of ways to produce such information as natural frequencies, damping ratios, and mode shapes in numerical or graphic displays.

While almost all of the commercially available analyzers are marketed as turnkey devices, it is important to understand a few details of the signal processing performed by these analysis units in order to carry out valid experiments. This forms the topic of the next two sections.

A relatively new instrument for obtaining high spatial density vibration measurements is the scanning laser Doppler vibrometer. The SLDV uses the Doppler shift in a reflected laser beam to determine the velocity of the test object. The laser sensor is noncontacting, but often structures must be painted with reflective paint to produce a sufficient intensity of the reflected signal. The SLDV can measure vibration up to frequencies of 250 KHz and at amplitudes in the nanometer range. The system software and a video camera allow generating and viewing a mesh pattern of points on the structure. The laser is programmed to measure the vibration at all mesh points and display the vibration as time responses, Fourier transforms, frequency response functions (FRFs), spectral densities, coherence functions, or operational deflection shapes (ODSs). When used with a high-bandwidth actuator such as a piezoceramic patch, highly accurate vibration patterns or ODSs of a structure can be generated. The ODS can help in redesign of structures and detecting faults or damage in structures.

7.2 DIGITAL SIGNAL PROCESSING

Much of the analysis done in modal testing is performed in the frequency domain, inside the analyzer. The analyzer's task is to convert analog time-domain signals into digital frequency-domain information compatible with digital computing and then to perform the required computations with these signals. The method used to change an analog signal, $x(t)$, into frequency-domain information is the Fourier transform [defined by equations (3.45) and (3.46)], or a *Fourier series* as defined by equations (3.20) to (3.23). The Fourier series is used here to introduce the digital Fourier transform (DFT).

As pointed out in Section 3.3, a periodic time signal of period T can be represented by a Fourier series in time of the form given by equation (3.20) with Fourier coefficients, or spectral coefficients as defined by equations (3.21) to (3.23). Essentially, the spectral coefficients represent frequency-domain information about a given time signal. These equations are repeated in Window 7.1.

Window 7.1
Review of the Fourier Series of a Periodic Signal F(t) of Period T

$$F(t) = \frac{a_0}{2} + \sum_{n=1}^{\infty} \left(a_n \cos n\omega_T t + b_n \sin n\omega_T t \right) \tag{3.20}$$

where

$$\omega_T = \frac{2\pi}{T}$$

$$a_0 = \frac{2\pi}{T} \int_0^T F(t)\,dt \tag{3.21}$$

$$a_n = \frac{2}{T} \int_0^T F(t)\cos n\omega_T t\,dt \qquad n = 1, 2 \ldots \tag{3.22}$$

$$b_n = \frac{2}{T} \int_0^T F(t)\sin n\omega_T t\,dt \qquad n = 1, 2 \ldots \tag{3.23}$$

The Fourier coefficients a_n and b_n given by equations (3.21) to (3.23) also represent the connection between Fourier analysis and vibration experiments. The analog output signals of accelerometers and force transducers, represented by $x(t)$, are inputs to the analyzer. The analyzer, in turn, calculates the spectral coefficients of these signals, thus setting the stage for a frequency-domain analysis of the signals. Some signals and their Fourier spectra are illustrated in Figure 7.3. The analyzer first converts the analog signals into digital records. It samples the signals $x(t)$ at many different equally spaced values and produces a digital record, or version, of the signal in the form of a set of numbers $\{x(t_k)\}$. Here $k = 1, 2, \ldots, N$, where the digit N denotes the number of samples and t_k indicates a discrete value of the time.

This process is performed by an analog-to-digital (A/D) converter. This conversion from an analog to a digital signal can be thought of in two ways. First, one can imagine a gate that samples the signal every Δt seconds and passes through the signal $x(t_k)$. The process of A/D conversion can also be considered as multiplying the signal $x(t)$ by a square-wave function, which is zero over alternate values of t_k

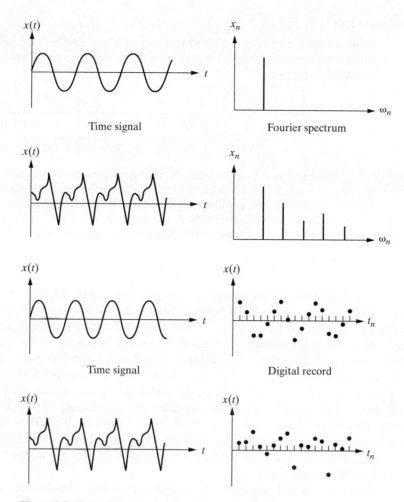

Figure 7.3 Several signals, their Fourier representations, and their digital representations.

and has the value of 1 at each t_k for a short time. Some signals and their digital representation are illustrated in Figure 7.3.

In calculating digital Fourier transforms, care must be taken in choosing the sampling time (i.e., the time elapsed between successive t_k's). A common error introduced in digital signal analysis caused by improper sampling time is called *aliasing*. Aliasing results from A/D conversion and refers to the misrepresentation of the analog signal by the digital record. Basically, if the sampling rate is too slow to catch the details of the analog signal, the digital representation will cause high frequencies to appear as low frequencies. The following example illustrates two periodic analog signals of different frequency and phase that produce the same digital record.

Example 7.2.1

Consider the signals $x_1(t) = \sin[(\pi/4)t)]$ and $x_2(t) = -\sin[(7\pi/4)t)]$, and suppose that these signals are both sampled at 1-s intervals. The digital record of each signal is given in the following table:

t_k	0	1	2	3	4	5	6	7	8	ω_i
x_1	0	0.707	1	0.707	0	−0.707	−1	−0.707	0	$\frac{1}{8}\pi$
x_2	0	0.707	1	0.707	0	−0.707	−1	−0.707	0	$\frac{7}{8}\pi$

As is easily seen from the table, the digital sample records of x_1 and x_2 are the same [i.e., $x_1(t_k) = x_2(t_k)$ for each value of k]. Hence no matter what analysis is performed on the digital record, x_1 and x_2 will appear the same. Here the sampling frequency, $\Delta\omega$, is 1. Note that the difference between the frequency of the first signal, $x_1(t)$, and the sampling frequency is $\frac{1}{8} - 1 = -\frac{7}{8}$, which is the frequency of the second signal $x_2(t)$.

□

To avoid aliasing, the sampling interval, denoted by Δt, must be chosen small enough to provide at least two samples per cycle of the highest frequency to be calculated. That is, to recover a signal from its digital samples, the signal must be sampled at a rate at least twice the highest frequency in the signal. In fact, experience [see Otnes and Encochson (1972)] indicates that 2.5 samples per cycle is a better choice. This is referred to as the *sampling theorem*, or *Shannon's sampling theorem*.

Aliasing can be avoided in signals containing many frequencies by subjecting the analog signal $x(t)$ to an antialiasing filter. An antialiasing filter is a low-pass (i.e., only allows low frequencies through) sharp–cutoff filter. The filter effectively cuts off frequencies higher than about half the maximum frequency of interest, denoted by ω_{max}, and also called the *Nyquist frequency*. Most modern digital analyzers provide built-in antialiasing filters.

Once the digital record of the signal is available, the discrete version of the Fourier transform is performed. This transform provides a series representation of a discrete-time history value. This is accomplished by a digital Fourier transform or series defined by

$$x_k = x(t_k) = \frac{a_0}{2} + \sum_{i=1}^{N/2}\left(a_i \cos \frac{2\pi t_k}{T} + b_i \sin \frac{2\pi t_k}{T}\right) \qquad k = 1, 2, \ldots N \quad (7.1)$$

where the *digital spectral coefficients* are given by

$$a_0 = \frac{1}{N}\sum_{k=1}^{N} x_k \qquad (7.2)$$

$$a_i = \frac{1}{N}\sum_{k=1}^{N} x_k \cos \frac{2\pi ik}{N} \qquad (7.3)$$

$$b_i = \frac{1}{N}\sum_{k=1}^{N} x_k \sin \frac{2\pi ik}{N} \qquad (7.4)$$

These are the digital versions of equations (3.21), (3.22), and (3.23), respectively. The task of the analyzer is to calculate equations (7.2) to (7.4) given the digital record $x(t_k)$, also denoted by x_k, for the measured signals. The transform size or number of samples, N, is usually fixed for a given analyzer and is a power of 2. Some common sizes are 512 and 1024.

Writing out equations (7.2) to (7.4) for each of the N samples yields N linear equations in the N spectral coefficients $\left(a_0, \ldots, a_{N/2}, b_0, \ldots, b_{N/2} \right)$. These equations can also be written in the form of matrix equations. In matrix form they become

$$\mathbf{x} = C\mathbf{a} \tag{7.5}$$

where $\mathbf{x}$ is the vector of samples with elements x_k and $\mathbf{a}$ is the vector of the spectral coefficients, a_0, a_i, and b_i. The matrix C consists of elements containing the coefficients $\cos(2\pi i t_k / T)$ and $\sin(2\pi i t_k / T)$, as indicated in equation (7.1). The solution of equation (7.5) for the spectral coefficients is then given simply by

$$\mathbf{a} = C^{-1}\mathbf{x} \tag{7.6}$$

The task of the analyzer is to compute the matrix C^{-1} and hence the coefficient $\mathbf{a}$. The most widely used method of computing the inverse of this matrix C is called the fast Fourier transform (FFT), developed by Cooley and Tukey (1965). Note that while $\mathbf{x}$ represents the digital version, the spectral coefficient $\mathbf{a}$ represents the frequency content of the response (or input) signal.

To make digital analysis feasible, the periodic signal must be sampled over a finite time (N must obviously be finite). This can give rise to another problem referred to as *leakage*. To make the signal finite, one could simply cut the signal at any integral multiple of its period. Unfortunately, there is no convenient way to do this for complicated signals containing many different frequencies. Hence, if no further steps are taken, the signal may be cut off midperiod. This causes erroneous frequencies to appear in the digital representation because the digital Fourier transform of the finite-length signal assumes that the signal is periodic *within* the sample record length. Thus the actual frequency will "leak" into a number of fictitious frequencies. This is illustrated in Figure 7.4.

Leakage can be corrected to some degree by the use of a *window function*. Windowing, as it is called, involves multiplying the original analog signal by a weighting function, or window function, $w(t)$, which forces the signal to be zero outside the sampling period. A common window function, called the *Hanning window*, is illustrated in Figure 7.5, along with the effect it has on a periodic signal. A properly windowed signal will yield a spectral plot with much less leakage. This is also illustrated in the figures.

As noted in this section, if the signal's properties are precisely known (i.e., the frequency content), the choice of sampling rate and N would be obvious and correct. However, the reason a signal is being measured in the first place is to determine its frequency content; hence part of the art of modal analysis is choosing the sampling frequency and data size, N.

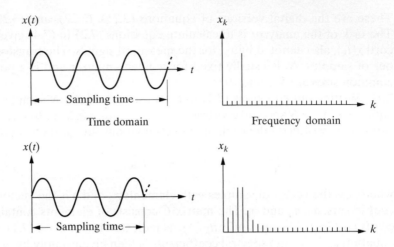

Figure 7.4   An example of leakage (i.e., frequencies caused by not sampling over an integer multiple of frequencies).

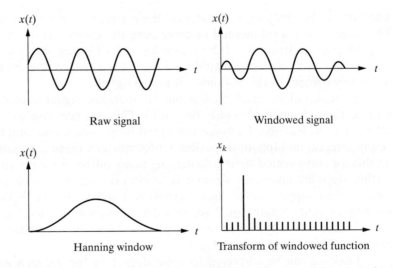

Figure 7.5   Use of a window function, in this case a Hanning window, to reduce leakage in calculating the frequency content of a signal.

7.3 RANDOM SIGNAL ANALYSIS IN TESTING

The transducer used to measure both the input and output during a vibration test usually contains noise (i.e., random components that make it difficult to analyze the measured data in a deterministic fashion). In addition, confidence in a measured quantity is increased by performing a number of identical tests and averaging the results. This is fairly common practice when measuring almost anything. In fact, the stiffness of a single structure is determined by multiple measurements, not just one, as indicated in Figure 1.3 of Section 1.1. Thus, it is important to consider the random input vibration response developed in Section 3.5.

Recall the definition of the autocorrelation function of a signal and the associated power spectral density (PSD). These are reviewed in Window 7.2. Also recall that the PSD of the input or driving force can be related to the PSD of the response and the frequency response function of the system by

$$S_{xx}(\omega) = |H(\omega)|^2 S_{ff}(\omega) \tag{7.7}$$

Window 7.2
Review of Some of the Definitions Used in Random Vibration Analysis

The autocorrelation function of the random signal $x(t)$ is given by

$$R_{xx}(\tau) = \lim_{T \to \infty} \frac{1}{T} \int_0^T x(t)x(t + \tau)dt \tag{3.50}$$

The power spectral density (PSD) of a signal is the Fourier transform of the signal's autocorrelation:

$$S_{xx}(\omega) = \frac{1}{2\pi} \int_{-\infty}^{\infty} R_{xx}(\tau)e^{-j\omega\tau}d\tau \tag{3.51}$$

as indicated by equation (3.62) and reviewed in Window 7.3. Equation (7.7) relates the dynamics of the test structure contained in $H(\omega)$ to measurable quantities (i.e., the PSDs). As pointed out at the end of Section 3.7, the common approach to measuring the frequency response function is to average several matched sets of input force time histories and output response time histories. These averages are used to produce correlation functions that are transformed to yield the corresponding PSDs. Equation (7.7) is then used to calculate the magnitude of the frequency response function $|H(\omega)|$. The experimental vibration data are then taken from the plot of $|H(\omega)|$ as indicated in Figure 3.20, or by means to be discussed in Section 7.4.

The frequency response function can also be related to the cross-correlation between the two signals $x(t)$ and $f(t)$. The *cross-correlation function*, denoted $R_{xf}(\tau)$, for the two signals $x(t)$ and $f(t)$, is defined by

$$R_{xf}(\tau) = \lim_{T \to \infty} \frac{1}{T} \int_0^T x(t)f(t + \tau)dt \tag{7.8}$$

Window 7.3
Comparison between Calculation for the Response of a
Spring–Mass–Damper System to Deterministic and Random Excitations

$$\text{transfer function} = G(s) = \frac{1}{ms^2 + cs + k}$$

Frequency response function:
$$G(j\omega) = H(\omega) = \frac{1}{k - m\omega^2 + c\omega j}$$

Impulse response function:
$$h(t) = \frac{1}{m\omega_d} e^{-\zeta\omega_n t} \sin \omega_d t$$

which has Laplace transform
$$L[h(t)] = \frac{1}{ms^2 + cs + k} = G(s)$$

and the Fourier transform of the impulse response function is just the frequency response function $H(\omega)$. These quantities relate to the input and response by

For deterministic $f(t)$:   For random $f(t)$:

$$X(s) = G(s) F(s) \quad \longleftrightarrow \quad S_{xx}(\omega) = |H(\omega)|^2 S_{ff}(\omega)$$

$$x(t) = \int_0^t h(t - \tau) f(\tau) \, d\tau \quad \longleftrightarrow \quad E[\bar{x}^2] = \int_{-\infty}^{\infty} |H(\omega)|^2 S_{ff}(\omega) d\omega$$

Here $x(t)$ is considered to be the response of the structure to the driving force $f(t)$. Similarly, the *cross-spectral density* is defined as the Fourier transform of the cross correlation:

$$S_{xf}(\omega) = \frac{1}{2\pi} \int_{-\infty}^{\infty} R_{xf}(\tau) e^{-j\omega\tau} dt \tag{7.9}$$

These correlation and density functions also allow calculation of the transfer functions of test structures. The frequency response function, $H(j\omega)$, can be shown [see, e.g., Ewins (2000)] to be related to the spectral density functions by the two equations

$$S_{fx}(\omega) = H(j\omega) S_{ff}(\omega) \tag{7.10}$$

and

$$S_{xx}(\omega) = H(j\omega)S_{xf}(\omega) \tag{7.11}$$

These hold if the structure is excited by a random input $f(t)$ resulting in the response $x(t)$. Note that the cross-correlation functions include information about the phase and magnitude of the structure's transfer function and not just the magnitude, as in the case of the correlation function of equation (7.7), repeated in the bottom right corner of Window 7.3.

The spectrum analyzer calculates (or estimates) the various spectral density functions from the transducer outputs. Then, using equation (7.10) or (7.11), the analyzer can calculate the desired frequency response function $H(j\omega)$. Note that equations (7.10) and (7.11) use different power spectral densities to calculate the same quantity. This fact can be used to check the consistency of $H(j\omega)$. The *coherence function*, denoted by γ^2, is defined to be the ratio of the two values of $H(j\omega)$ calculated from equations (7.10) and (7.11). In particular, the coherence function is defined to be

$$\gamma^2 = \frac{|S_{xf}(\omega)|^2}{S_{xx}(\omega)S_{ff}(\omega)} \tag{7.12}$$

which always lies between 0 and 1. In fact, if the measurements are consistent, $H(j\omega)$ should be the same value, independent of how it is calculated, and the coherence should be 1 ($\gamma^2 = 1$). The coherence is a measurement of the noise in the signal. If it is zero, the measurement is of a pure noise; if the value of the coherence is 1, the signals x and f are not contaminated with noise. In practice, coherence is plotted versus frequency (see Figure 7.6) and is taken as an indication of how accurate the measurement process is over a given range of frequencies. Generally, the values of $\gamma^2 = 1$ should occur at values of ω far from the structure's resonant frequencies. Near resonance, the signals are large and magnify the noise. In practice, data with a coherence of less than 0.75 are not used and indicate that the test should be done again or the data should be examined over a smaller frequency range.

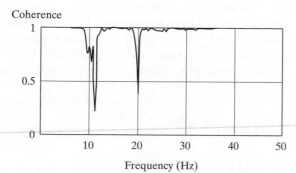

Figure 7.6   A sample plot of a coherence function.

7.4 MODAL DATA EXTRACTION

Once the frequency response of a test structure is calculated from equation (7.10) or (7.11), the analyzer is used to construct various vibration parameter information from the processed measurements. This is what is referred to as *experimental modal analysis*. In what follows, it is assumed that the frequency response function $H(j\omega)$ has been measured via equation (7.10) or (7.11) or their equivalents.

The task of interest is to compute the natural frequencies, damping ratios, and modal amplitudes associated with each resonant peak of the measured frequency response function. There are several ways to examine the measured frequency response function to extract these data. To examine all of them is beyond the scope of this book and the interested reader should consult Ewins (2000). To illustrate the basic method, consider the somewhat idealized (compliance) frequency response function record of Figure 7.7, resulting from measurements taken between two points on a simple structure. Here it is assumed that a sinusoidal force of adjustable frequency is applied to one point on the structure and that the displacement response is measured at a second point. The response is measured for many values of the driving frequency to produce the plot of Figure 7.7. The procedure is examined for a single-degree-of-freedom system as illustrated in Section 3.7 and Figure 3.20.

One of the gray areas in modal testing is deciding on the number of degrees of freedom to assign to a test structure. In many cases, simply counting the number of clearly defined peaks or resonances, three in Figure 7.7, determines the order, and the procedure continues with a three-mode model. However, this procedure is not accurate if the structure has closely spaced natural frequencies or repeated natural frequencies.

The easiest method to use on these data is the so-called *single-degree-of-freedom curve fit* (often called the SDOF method) approach. In this method, the frequency response function for the compliance is sectioned off into frequency ranges bracketing each successive peak. Each peak is then analyzed by assuming that it is the frequency response of a single-degree-of-freedom system. This assumes that in the vicinity of the resonance, the frequency response function is dominated by that single mode.

In other words, in the frequency range around the first resonant peak, it is assumed that the plot is due to the response of a damped single-degree-of-freedom system due to a harmonic input at, and near, the first natural frequency. Recall from Section 3.7 that the point of resonance corresponds to that value of the frequency for which the magnification curve has its maximum or peak value and the phase shift is 90°. Hence each of the frequencies ω_1, ω_2, and ω_3 of the plot of Figure 7.7 are determined simply by

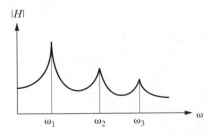

Figure 7.7   A sample magnitude plot of the frequency response function from a test article excited at one point and measured at another point.

noting where the three peaks lie on the horizontal (frequency) axis and is confirmed by examining the value of the phase at each of these frequencies (should be 90°).

The damping ratio associated with each peak is assumed to be the modal damping ratio, ζ_i, defined in Sections 4.5 and 6.7 in the modal coordinate system. To obtain the modal damping ratios, consider the frequency response functions (compliance) magnitude plot illustrated in Figure 7.8. For systems with light enough damping, so that the peak $|H(\omega)|$ at resonance is well defined, the modal damping ratio ζ is related to the frequencies corresponding to the two points on the magnitude plot where

$$|H(\omega_a)| = |H(\omega_b)| = \frac{|H(\omega_d)|}{\sqrt{2}} \qquad (7.13)$$

by $\omega_b - \omega_a = 2\zeta\omega_d$, so that

$$\zeta = \frac{\omega_b - \omega_a}{2\omega_d} \qquad (7.14)$$

Here ω_d is the damped natural frequency at resonance and ω_a and ω_b satisfy condition (7.13). The condition of equation (7.13) is also called the *3-dB down point*, since $H(\omega_d)/\sqrt{2}$ corresponds to $H(\omega_d)$ at 3 dB below its peak value when the magnitude is plotted on a log scale.

Equation (7.14) can be shown to apply for inertance (acceleration) transfer functions as well. The peak of the inertance frequency response magnitude plot also occurs at ω_d. Often, because the damping is small, ω_n, the natural frequency, and ω_d, the damped natural frequency, are taken to be the same. In fact, for $\zeta = 0.01$, $\omega_d = \omega_n\sqrt{1 - \zeta^2} = 0.999949998\omega_n$ and for $\zeta = 0.1$, $\omega_d = 0.99498\omega_n$, so that they are very nearly the same for an order-of-magnitude spread of damping ratios. Both the natural frequency and the damping ratio can be determined directly for accelerometer

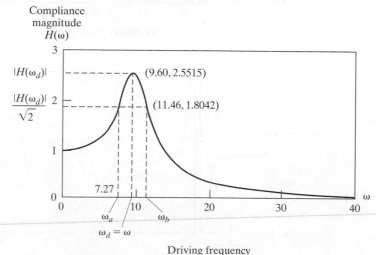

Figure 7.8   The magnitude of the compliance frequency response function of a single-degree-of-freedom system, illustrating the calculation of the modal damping ratio by using the *quadrature peak picking* method for lightly damped systems.

measurements and force input measurements by plotting the magnitude of the inertance frequency response function and applying the method of Figure 7.8.

In the case of a multiple-degree-of-freedom system, as indicated by the three peaks of Figure 7.7, the number of peaks indicates the number of degrees of freedom. Each peak is then treated as if it resulted from a single-degree-of-freedom system. For example, the three natural frequencies of Figure 7.7 are determined by the positions of the peaks on the frequency axis, and the three damping ratios are determined by treating each peak as if it were from a single-degree-of-freedom system, computing the 3-db down frequencies, and using equation (7.14) three times. This yields the three modal damping ratios ζ_1, ζ_2, and ζ_3.

Example 7.4.1

Consider the experimentally determined compliance transfer function plotted in Figure 7.9 and calculate the number of degrees of freedom, modal damping ratios, and natural frequencies.

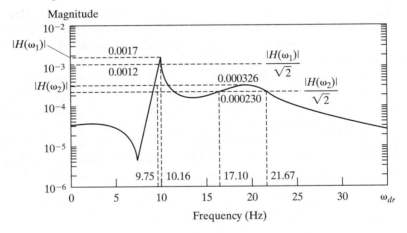

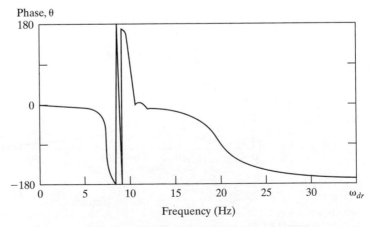

Figure 7.9 A plot of the magnitude and phase versus driving frequency of a test specimen, illustrating the peak amplitude method of determining modal damping ratios and natural frequencies.

Solution   Since the magnitude plot indicates two distinct peaks, the test system is assumed to have two degrees of freedom. This is confirmed by examining the phase plot at the peaks. Since the phase is $\pm 90°$ at each peak, each of the peaks corresponds to a natural frequency. Reading the vertical axis yields

$$\omega_1 = 10 \text{ Hz} \qquad \omega_2 = 20 \text{ Hz}$$

Next, since $|H(\omega_1)| = 0.0017$, the 3-dB down points are those two values of ω where $H(\omega_a) = H(\omega_b) = 0.0017/\sqrt{2} = 0.0012$. From the plot, these values yield $\omega_a = 9.95$ Hz and $\omega_b = 10.16$ Hz. Using equation (7.14) yields

$$\zeta_1 = \frac{\omega_b - \omega_a}{2\omega_1} = \frac{10.16 - 9.75}{20} = 0.02$$

Repeating this procedure for the second peak yields

$$\zeta_2 = \frac{\omega_b - \omega_a}{2\omega_2} = \frac{21.67 - 17.10}{40} = 0.11$$

$\square$

7.5 MODAL PARAMETERS BY CIRCLE FITTING

In Section 7.4, the frequency and damping ratios are essentially determined by visually examining the frequency response function. In this section a more systematic method is examined that can be programmed so that an analyzer can calculate the frequencies and damping ratios in a more automated fashion. This method also assumes that a single mode dominates the behavior of the mobility transfer function in a frequency range around the natural frequency. If the real part of the mobility frequency response function is plotted versus the imaginary part of the frequency response function for a range of frequencies, a circle results. Plots of $\text{Re}[(H(\omega))]$ versus $\text{Im}[H(\omega)]$ are called *Nyquist plots*, or *Nyquist circles*, or *Argand plane plots*.

The mobility transfer function and corresponding frequency response function are presented in Section 7.3 and reviewed in Window 7.4. The real and imaginary parts of the mobility frequency response function can be calculated to be

$$\text{Re}(\alpha) = \frac{\omega^2 c}{\left(k - \omega^2 m \right)^2 + (\omega c)^2} \tag{7.15}$$

and

$$\text{Im}(\alpha) = \frac{\omega \left(k - \omega^2 m \right)}{\left(k - \omega^2 m \right)^2 + (\omega c)^2} \tag{7.16}$$

respectively. The imaginary part is plotted versus the real part in Figure 7.10 for increasing values of ω. The plots of Figure 7.10 are formed by computing values for the pairs $[\text{Im}(\alpha), \text{Re}(\alpha)]$ for each value of ω. This triple of values$-\omega$, $\text{Im}(\alpha)$, $\text{Re}(\alpha)-$correspond to information available in digital forms in the analyzer used to manipulate measured data.

Window 7.4
Mobility Frequency Response Function

Recall from Table 3.2 that the mobility transfer function is the ratio of the Laplace transform of the response velocity to the force input. From equation (3.87) this is

$$\frac{sX(s)}{F(s)} = sH(s) = \frac{s}{ms^2 + cs + k}$$

for a single-degree-of-freedom system. Substitution of $s = j\omega$ yields the mobility frequency response function defined by

$$\alpha(\omega) = j\omega H(\omega) = \frac{j\omega}{\left(k - \omega^2 m\right) + j\omega c}$$

which is a complex-valued function usually denoted by $\alpha(\omega)$. However, $\alpha(\omega)$ is used to denote the frequency response function associated with any transfer function.

The imaginary part is plotted versus the real part for equally spaced increments of frequency. That is, the frequency range of interest is divided up into equally spaced values of the driving frequency ω (say, every 1 Hz or every 0.1 Hz). At each of the values of ω, the quantities Re(α) and Im (α) are calculated from the measured data in the analyzer, and these are plotted as indicated in Figure 7.10(a).

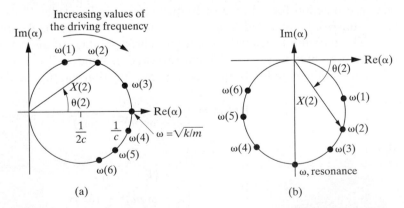

(a)   (b)

Figure 7.10   (a) A plot of the imaginary part of the mobility frequency response function versus the real part for increasing values of ω, starting at $\omega = 0$. (b) Receptance transfer function. The values in parentheses correspond to numbered data points. These are called Nyquist plots. The plots here are for underdamped systems.

For a single-degree-of-freedom system, equations (7.15) and (7.16) predict that these points will all lie on a circle tangent to the origin centered on the $\text{Re}(\alpha)$ axis. The equally spaced driving frequencies are labeled $\omega(1)$, $\omega(2)$, and so on in the figure. Note that these points are not equally spaced around the circle, but are at equal increments of frequency. Note also on this plot that the input force magnitude is lined up along the $\text{Re}(\alpha)$ axis and that the response of magnitude X lags the force by the phase angle θ. These quantities are also marked on the plot of Figure 7.10 for the second value of the driving frequency [these are labeled $X(2)$, $\theta(2)$]. Thus, the distance from the origin to any point on the circle drawn through the data points is the magnitude of the response. Since resonance is defined as the driving frequency at which the response magnitude X is the largest, the point on the circle farthest from the origin corresponds to the resonant condition. Resonance is also defined, for small damping, as the condition when the driving frequency and the system's natural frequency coincide. Hence, the point labeled ω corresponds to the system's natural frequency, which is calculated from equations (7.15) and (7.16) evaluated at the point of intersection of the circle and the $\text{Re}(\alpha)$ axis.

The fact that the Nyquist plot of the mobility frequency response function is a circle can be seen by defining $A = (\alpha) - (1/2c)$ and $B = \text{Im}(\alpha)$ and by using equations (7.15) and (7.16), so that

$$A^2 + B^2 = \left[\text{Re}(\alpha) - \frac{1}{2c} \right]^2 + ((\text{Im}(\alpha))^2 = \left[\frac{1}{2c} \right]^2$$

which is the equation of a circle of radius $(1/2c)$ centered at the point [$\text{Im}(\alpha) = 0$, $\text{Re}(\alpha) = 1/2c$]. Note from equation (7.16) that at the point where the circle intersects the $\text{Re}(\alpha)$ axis, $\text{Im}(\alpha) = 0$, so that $\omega = \sqrt{k/m}$, which is the condition for resonance. If the frequency at the point where the circle crosses the axis is not necessarily a point that was measured or plotted, the value of the resonant frequency can be determined by fitting a circle to the points $\omega(i)$ and numerically determining the value of ω corresponding to the intersection of the axis and the circle. This also gives the value of the damping coefficient from the simple relationship $\text{Re}(\alpha) = 1/c$ at the value of ω corresponding to resonance.

Figure 7.10(b) shows the Nyquist circle for the receptance transfer function (i.e., displacement measurements) for the same system. Most analyzers allow the user to plot any of the transfer functions given in Table 3.1 and their corresponding Nyquist plots. The point corresponding to resonance on the receptance plot of a single-degree-of-freedom system given in Figure 7.10(b) can be characterized in four different ways. The point corresponding to the natural frequency can be thought of as

1. The point on the circle corresponding to a maximum distance from the origin.
2. The point lying halfway between the two adjacent frequencies forming the largest arc length on the circle. The arc between $\omega(3)$ and $\omega(4)$ is the largest in the example given in Figure 7.10(b).

3. The point on the circle a maximum distance away from the $\text{Re}(\alpha)$ axis.

4. The point on the circle intersecting the line $\text{Im}(\alpha)$ axis.

For single-degree-of-freedom systems, each of these points is, of course, the same. However, for multiple-degree-of-freedom systems several changes occur. First, the Nyquist circle becomes many circles (roughly one for each mode) and the circles drift from being tangent to the origin centered on the $\text{Im}(\alpha)$ axis. As this happens, the four points just listed no longer coincide. This is illustrated in Figure 7.11. The single point labeled ω in Figure 7.10(b) becomes the four points labeled 1, 2, 3, and 4 in Figure 7.11(a) as the circle moves away from the origin. These labels correspond numerically to the preceding list, characterizing the resonance point. Any one of these four points could be used to define the natural frequency for the mode described by the circle. The most common choice, and the choice least affected by the presence of the other modes, is to use point 2 (i.e., the point halfway between adjacent frequencies defining the largest arc length).

Referring to Figure 7.11(b), the procedure for using the circle is as follows: First, the analyzer computes the points marked x in the figure from $\text{Re}(\alpha)$ and $\text{Im}(\alpha)$ evaluated at equal intervals of the driving frequency. A numerical curve-fit procedure is then used to calculate the best circle through these points, the center of the circle (O), and the arc lengths between each point x. This determines point 2 as defined by the largest arc length. The equations for $\text{Re}(\alpha)$ and $\text{Im}(\alpha)$ are then used to calculate the value for the natural frequency, denoted by ω_3 here, since it corresponds to the third mode. The points ω_a and ω_b and O are also calculated and used to derive the angle α.

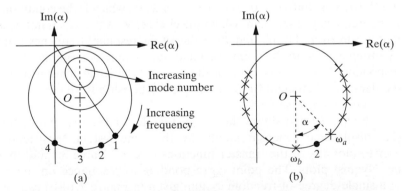

(a)　　　　　　　　　　　　　　(b)

Figure 7.11 Nyquist plot (receptance or compliance) for a three-degree-of-freedom lightly damped system. (a) Four points that could be used to define the frequency of the third mode, defined by the circle centered at O. (b) Third mode plotted without the other modes and the geometry label that is used to determine the modal damping ratio. Point 2 corresponds to the natural frequency for the third mode. The points labeled ω_a and ω_b are adjacent frequencies that form the largest arc length, and α is the angle between the radii defined by ω_a and ω_b. The x's denote measured points of equally spaced frequencies.

As long as the angle $a/2 < 45°$, which it will be for any reasonable amount of data, the angle α is related to the modal damping ratio ζ_3 and natural frequency ω_3 by

$$\tan\frac{\alpha}{2} = \frac{1 - (\omega/\omega_3)^2}{2\zeta_3\omega/\omega_3} \tag{7.17}$$

Applying equation (7.17) to ω_a yields

$$\tan\frac{\alpha}{2} = \frac{(\omega_a/\omega_3)^2 - 1}{2\zeta_3\omega_a/\omega_3} \tag{7.18}$$

and applying equation (7.17) to ω_b yields

$$\tan\frac{\alpha}{2} = \frac{1 - (\omega_b/\omega_3)^2}{2\zeta_3\omega_b/\omega_3} \tag{7.19}$$

Equations (7.18) and (7.19) can be added to yield (after some manipulation)

$$\zeta_3 = \frac{\omega_b^2 - \omega_a^2}{2\omega_3\left[\omega_a \tan(\alpha/2) + \omega_b \tan(\alpha/2)\right]} \tag{7.20}$$

which yields an expression for the damping ratio for the mode under study.

To check this result, note that if the half-power points used in Section 7.4 are taken to be those frequencies ω_a and ω_b, where $\alpha = 90°$, equation (7.20) reduces to

$$\zeta_i = \frac{\omega_b - \omega_a}{2\omega_i} \tag{7.21}$$

which matches the quadrature formula given to equation (7.14) of Section 7.4.

The method of using the Nyquist circle for determining natural frequencies and damping ratios is first to divide the frequency response function into segments by looking at the magnitude plot to determine regions of ω for which the frequency response looks approximately like that of a single-degree-of-freedom system (i.e., take a frequency range around each peak). The data points comprising each peak are then chosen to use in generating a Nyquist circle for that mode. Each frequency range must contain at least three points.

The circle generated by these points will contain noise, and so on, and will not be perfect circles. To rectify this, the data points are curve fit to a circle using a simple least squares method. This yields the equation of the ("best") circle through the data points. This circle is then used to calculate ω_i and ζ_i from the preceding formulas. Since a curve-fit procedure is used, this method is called the *circle fit* method of extracting modal parameters. The method was formulated by Kennedy and Pancu (1947).

7.6 MODE SHAPE MEASUREMENT

Determining the mode shapes from experimentally measured transfer functions is slightly more complicated and involves the measurement of several transfer functions. First, the concept of a transfer function matrix, or *receptance matrix*, needs to be established. To this end, consider the response of the multiple-degree-of-freedom system as described by equation (4.122) to a harmonic force input represented in a complex form by $\mathbf{f}e^{j\omega t}$. The equation of motion becomes

$$M\ddot{\mathbf{x}} + C\dot{\mathbf{x}} + K\mathbf{x} = \mathbf{f}e^{j\omega t} \tag{7.22}$$

The forced response can be constructed by assuming that the solution $\mathbf{x}(t)$ is harmonic, of the form $\mathbf{x}(t) = \mathbf{u}e^{j\omega t}$. Substitution of this form into equation (7.22) yields

$$\left(K - \omega^2 M + j\omega C\right)\mathbf{u} = \mathbf{f} \tag{7.23}$$

after rearranging terms and factoring out the nonzero scalar $e^{j\omega t}$. Equation (7.23) relates the magnitude of the response vector (i.e., the vector $\mathbf{u}$) to the magnitude of the input vector $\mathbf{f}$, both of which are constants. Solving equation (7.23) yields

$$\mathbf{u} = \left(K - \omega^2 M + j\omega C\right)^{-1}\mathbf{f} \tag{7.24}$$

The inverse of the complex matrix coefficient is the *receptance matrix*, denoted $\alpha(\omega)$ and defined by

$$\alpha(\omega) = \left(K - \omega^2 M + j\omega C\right)^{-1} \tag{7.25}$$

so that equation (7.24) becomes simply $\mathbf{u} = \alpha(\omega)\mathbf{f}$. A two-degree-of-freedom example of the receptance matrix is used in Section 5.4 on damped absorbers to calculate equation (5.35). An undamped version of $\alpha(\omega)$ for a two-degree-of-freedom system is given by equation (5.18).

The receptance matrix can be further analyzed by recalling the transformation used in Chapter 4 to derive modal coordinates. In particular, recall that the modal stiffness matrix can be represented in diagonal form by

$$\Lambda_K = \text{diag}\left[\omega_i^2\right] = P^T M^{-1/2} K M^{-1/2} P \tag{7.26}$$

where P is the matrix of normalized eigenvectors of the matrix $M^{-1/2}KM^{-1/2}$. Similarly, the modal damping matrix can be written as

$$\Lambda_c = \text{diag}\left[2\zeta_i\omega_i\right] = P^T M^{-1/2} C M^{-1/2} P \tag{7.27}$$

if the damping is assumed to be proportional. Multiplying equation (7.26) by $M^{1/2}P$ from the left and $P^T M^{1/2}$ from the right yields

$$K = M^{1/2}P\Lambda_K P^T M^{1/2} \tag{7.28}$$

since $P^T P = I$. Similarly, the damping matrix can be written from equation (7.27) as

$$C = M^{1/2} P \Lambda_C P^T M^{1/2} \tag{7.29}$$

Substitution of equations (7.28) and (7.29) for C and K into equation (7.25) for the receptance matrix yields

$$\alpha(\omega) = \left[M^{1/2} P \left(\Lambda_K - \omega^2 I + j\omega \Lambda_C \right) P^T M^{1/2} \right]^{-1} \tag{7.30}$$

$$= \left[S \left(\Lambda_K - \omega^2 I + j\omega \Lambda_C \right) S^T \right]^{-1} \tag{7.31}$$

$$= \left[S \, \text{diag} \left(\omega_i^2 - \omega^2 + 2\zeta_i \omega_i \omega_j \right) S^T \right]^{-1} \tag{7.32}$$

where $S = M^{1/2} P$. Here the inside matrix is a combination of diagonal matrices and hence is diagonal. Recall from matrix theory that $(AB)^{-1} = B^{-1} A^{-1}$ (see Appendix C), so that equation (7.32) can be written as

$$\alpha(\omega) = S^{-T} \text{diag} \left[\frac{1}{\omega_i^2 - \omega + 2\zeta_i \omega_i \omega_j} \right] S^{-1} \tag{7.33}$$

since the inverse of a diagonal matrix is obtained simply by inverting its nonzero diagonal elements. Note that this formulation assumes proportional damping.

Equation (7.33) for the receptance matrix can be expressed as a summation of n matrices rather than the product of three matrices by realizing that the columns of S^{-T} are the mode-shape vectors of the undamped system, denoted $\mathbf{u}_i$, of equation (4.164). Equation (7.33) can thus be written as

$$\alpha(\omega) = \sum_{i=1}^{n} \left[\frac{\mathbf{u}_i \mathbf{u}_i^T}{\left(\omega_i^2 - \omega^2 \right) + \left(2\zeta_i \omega_i \omega \right) j} \right] \tag{7.34}$$

where $\mathbf{u}_i \mathbf{u}_i^T$ is the outer product of two $n \times 1$ mode-shape vectors. This outer product results in an $n \times n$ matrix. This representation provides a connection between the receptance matrix and the system's mode shapes, which can be exploited in testing to provide a measurement of the test article's mode shapes.

The element of the receptance matrix located at the intersection of the sth row and r^{th} column $\alpha(\omega)$ is essentially the transfer function between the response at point s, u_s, and the input at point r, f_r, when all other inputs are held at zero. The sr^{th} element of $\alpha(\omega)$ is

$$\alpha_{sr}(\omega) = \sum_{i=1}^{n} \frac{\left[\mathbf{u}_i \mathbf{u}_i^T \right]_{sr}}{\left(\omega_i^2 - \omega^2 \right) + \left(2\zeta_i \omega_i \omega \right) j} \tag{7.35}$$

which relates the transfer function between a given input and output, $\alpha_{sr}(\omega)$, to elements of the mode shapes $\mathbf{u}_i$. This interpretation of $\alpha_{sr}(\omega)$ is a generalization of the single-degree-of-freedom concept of a transfer function. Since $\alpha(\omega)$ is a matrix, it cannot be written as the ratio of an output to an input. However, each element of $\alpha(\omega)$ is a transfer function:

$$\frac{u_s}{f_r} = \left[\alpha(\omega)\right]_{sr} = H_{sr}(\omega) \tag{7.36}$$

where $H_{sr}(\omega)$ is the transfer function between an input at point r and an output at point s. An example of equation (7.36) is the ratio x_1/F_0k used in Sections 5.3 and 5.4 to discuss absorbers.

If it is assumed that the modes, or peaks, of the system are well spaced, the summation in equation (7.35) evaluated at a natural frequency will be dominated by one term, the term corresponding to that frequency. This can be seen by substituting $\omega = \omega_i$ into equation (7.35) and taking the magnitude. This yields the approximation

$$\left|\alpha_{sr}(\omega_i)\right| = \frac{\left|\mathbf{u}_i\mathbf{u}_i^T\right|_{sr}}{\left|(\omega_i^2 - \omega_i^2) - 2\zeta_i\omega_i\omega_i j\right|} = \frac{\left|\mathbf{u}_i\mathbf{u}_i^T\right|_{sr}}{2\zeta_i\omega_r^2} \tag{7.37}$$

where it is assumed that the contributions from the other terms in the summation are all much smaller because of the nonzero term $\omega_i^2 - \omega^2$ in their denominators. Equation (7.37) can be rearranged to yield

$$\left|\mathbf{u}_i\mathbf{u}_i^T\right|_{sr} = \left|2\zeta_i\omega_i^2\right|\left|H_{sr}(\omega_i)\right| \tag{7.38}$$

where $\left|H_{sr}(\omega_i)\right| = \left|\alpha_{sr}(\omega_i)\right|$ is the magnitude of the frequency response function measured between points s and r and evaluated at the ith natural frequency. Equation (7.38) holds for proportionally damped systems with underdamped, widely spaced modes. This equation relates the measured damping ratio, ζ_i, measured natural frequency, ω_i, and the measured magnitude of the transfer function, $\left|H_{sr}(\omega_i)\right|$, to the ith mode shape, ω_i, and hence provides a measure of the mode shape of the test structure.

Equation (7.38) only provides a measurement of the magnitude of one element of the matrix $\left[\mathbf{u}_i\mathbf{u}_i^T\right]_{sr}$. The phase plot of $H(\omega_i)$ is used to determine the sign of the element $\left[\mathbf{u}_i\mathbf{u}_i^T\right]_{sr}$. Equation (7.37) is a mathematical statement that the ith peak in the transfer function plot of Figure 7.8 results from only a single-degree-of-freedom system. Note that the matrix $\mathbf{u}_i\mathbf{u}_i^T$ has n^2 elements but that only n of them are unique where n is the number of measured natural frequencies. Hence n measurements of $\left|H_{sr}(\omega_i)\right|$ must be made. This is accomplished by stepping through the n elements of the vector $\mathbf{f}$ one at a time so that r ranges from 1 to n. This amounts to measuring the response at point s with first an input at point 1, then at point 2, and so on, until all n input positions have been used. This provides a measurement of one row of the matrix $\mathbf{u}_i\mathbf{u}_i^T$. From this one row, the entire vector $\mathbf{u}_i$ can be determined as the following example illustrates. Note that this process could be interchanged so that the driving point is fixed and the measurement point is moved. The following example illustrates this procedure.

Example 7.6.1

Consider a simple beam of Figure 7.12. A transfer function measurement made by applying a force at point 1 and measuring the response at point 1 (called the *driving point* frequency response) yields three distinct peaks, indicating that the system has three natural frequencies and can be modeled by a three-degree-of-freedom system. This initial measurement suggests that the beam be measured at two other points in order to establish enough data to determine the mode shapes. These other two points are marked on the beam of Figure 7.12. Since a shaker is used, it is easier to move the accelerometer to obtain the required two additional transfer functions. Alternatively, a multichannel frequency analyzer can be used with two additional accelerometers to obtain simultaneously the required three transfer functions: $H_{11}(\omega), H_{21}(\omega)$, and $H_{31}(\omega)$. This is the procedure illustrated in Figure 7.12. Plots of the three transfer functions are given in Figure 7.13.

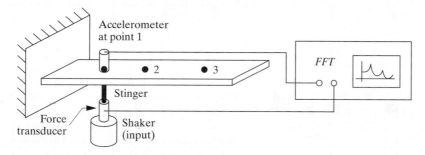

Figure 7.12   A cantilevered beam labeled with three measurement and driving points.

From the driving point transfer function $H_{11}(\omega)$, the values of the modal damping ratio and natural frequencies are obtained using the peak method illustrated in Example 7.4.1. They are

$$\omega_1 = 10 \text{ rad/s} \qquad \zeta_1 = 0.01$$
$$\omega_2 = 20 \text{ rad/s} \qquad \zeta_2 = 0.01 \qquad (7.39)$$
$$\omega_3 = 32 \text{ rad/s} \qquad \zeta_3 = 0.05$$

These values can be checked against similar calculations of the remaining two transfer functions $H_{21}(\omega)$ and $H_{31}(\omega)$. To determine the mode-shape vectors, the value of $|H_{11}(\omega_1)|$ is measured to be $|H_{11}(\omega_1)| = 0.423$ and the phase of $H_{11}(\omega_1)$ is noted to be phase $[H_{11}(\omega_1)] = -90°$. In addition, the magnitude and phase of the remaining two transfer functions yield

$$|H_{21}(\omega_1)| = 0.917 \qquad \text{phase}[H_{21}(\omega_1)] = -90° \qquad (7.40)$$
$$|H_{31}(\omega_1)| = 0.317 \qquad \text{phase}[H_{31}(\omega_1)] = -90°$$

From equation (7.38) and the measured values of $\zeta_1, \omega_1, H_{11}(\omega_1), H_{21}(\omega_1)$, and $H_{31}(\omega_1)$, the first row of the matrix $\mathbf{u}_1\mathbf{u}_1^T$ is known:

$$|\mathbf{u}_1\mathbf{u}_1^T|_{11} = 0.846 \qquad |\mathbf{u}_1\mathbf{u}_1^T|_{21} = 1.834 \qquad |\mathbf{u}_1\mathbf{u}_1^T|_{31} = 4.633 \qquad (7.41)$$

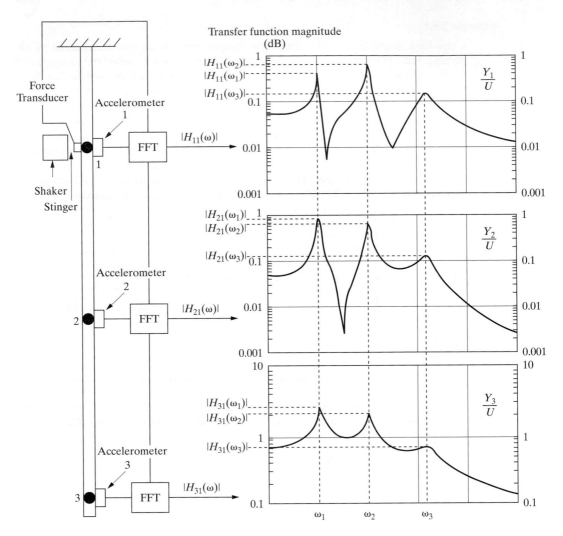

Figure 7.13 (cont'd on next page) Instrumentation required to construct the necessary frequency response curves to allow computation of the mode shapes for a three-degree-of-freedom model of the test specimen. The peaks of the three transfer functions determine the values of the receptance matrix and hence the system mode shapes.

This, along with the phase information, allows determination of the vector $\mathbf{u}_1$. To see this, let $\mathbf{u}_1 = [a_1 \quad a_2 \quad a_3]^T$, so that

$$
\mathbf{u}_i \mathbf{u}_i^T = \begin{bmatrix} a_1^2 & a_1 a_2 & a_1 a_3 \\ a_2 a_1 & a_2^2 & a_2 a_3 \\ a_3 a_1 & a_3 a_2 & a_3^2 \end{bmatrix}
$$

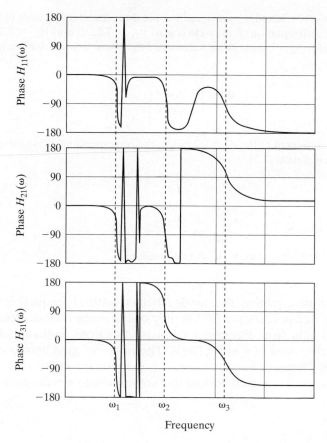

Frequency

Figure 7.13   (cont'd)

Note that this matrix is symmetric. Hence, from the values given in equation (7.41), the elements of the matrix $|\mathbf{u}_i\mathbf{u}_i^T|$ must satisfy

$$a_1^2 = 0.846 \qquad a_1a_2 = 1.834 \qquad a_1a_3 = 4.633 \qquad (7.42)$$

This forms a set of three equations in three unknowns that is readily solved to yield

$$a_1 = 0.920 \qquad a_2 = 1.993 \qquad a_3 = 5.036 \qquad (7.43)$$

Using the phase as either a $+$ or $-$ sign [i.e., $H(\omega_1)$ is either in phase or out of phase], the phase is either $-90°$ or $+90°$ at resonance. If phase $[H_{ij}(\omega_1)] = +90°$, the element associated with $[u_1u_1^T]_{ij}$ is assigned a positive value. If the phase is $-90°$, the element is assigned a negative value. Examining the phase plots of Figure 7.13 and the numerical values given in equation (7.43), the vector $\mathbf{u}_1$ becomes

$$\mathbf{u}_1 = \begin{bmatrix} a_1 \\ a_2 \\ a_3 \end{bmatrix} = \begin{bmatrix} -0.920 \\ -1.993 \\ -5.036 \end{bmatrix}$$

Next, consider the $\omega_2 = 20$ rad/s peak in each of the three transfer functions of Figure 7.13. These along with equation (7.38) yield $|(\mathbf{u}_2\mathbf{u}_2^T)|_{11} = 7.12$, $|(\mathbf{u}_2\mathbf{u}_2^T)|_{21} = 7.72$, and $|(\mathbf{u}_2\mathbf{u}_2^T)|_{31} = 15.681$. Again, using these values and the phase information yields

$$\mathbf{u}_2 = \begin{bmatrix} -2.67 \\ -2.89 \\ 5.873 \end{bmatrix}$$

Similarly, the measurement of the $\omega_3 = 32$ rad/s, the corresponding phase values, and the modal data in equation (7.38) yield

$$\mathbf{u}_3 = \begin{bmatrix} -4.22 \\ 2.99 \\ -15.311 \end{bmatrix}$$

Hence, the three mode shapes $\mathbf{u}_1$, $\mathbf{u}_2$, $\mathbf{u}_3$ are determined.

□

The method of determining the mode shapes, natural frequencies, and damping ratios illustrated in Example 7.6.1 is only one of many methods available for extracting modal data from frequency response functions constructed from test data. These are discussed in Ewins (2000). The notation used here is consistent with that used in Chapters 4 and 5. However, the modal analysis and testing community has begun to try to standardize the notation, and it will likely differ from that used here.

Example 7.6.2

This example is of a laboratory experiment using a hammer as the force excitation device and a laser vibrometer as the measurement sensor. The laser, of course, is non-contacting and the hammer does not provide mass loading, so there should be little effect of the measurement hardware on the dynamics of the beam being measured. A top view of the experiment along with measurement points is shown in Figure 7.14.

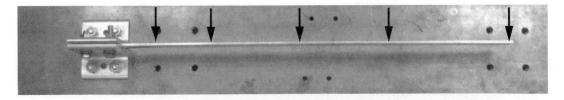

Figure 7.14   The top view of the cantilever beam used in the experiment, showing the measurement points. The points are numbered starting with 1 on the left and ending with 5 at the tip. Point 2 is the driving point.

The data taken is formed into magnitude plots and these are given in Figures 7.15(a) through 7.15(e) with the last number indicating where the transfer function was measured. The transfer functions in Figures 7.15 are mobility transfer functions (velocity out, force in) and the units on the magnitude are volts. The laser sensor constant is 125 mm/s/v. The experiment was conducted on a clamped–free beam made of aluminum. The beam dimensions were 0.5128 m in length, 25.5 mm in height, and 3.2 mm in width. The modulus of elasticity was 6.9×10^{10} N/m^2 and the density was 2715 kg/m^3. The cross-sectional area is $A = 8.16$ m^2 and $I = 6.96 \times 10^{-11}$ m^4. The excitation to the structure was provided with an impact hammer (Kistler type: 9722A500, s/n 014019). The impulse signal was fed to a signal conditioner (Dytran model: 4114B1) and then passed to the DSP SigLab unit. SigLab is a signal processing plug-in compatible with MATLAB. The velocity readings were made by a laser vibrometer composed of a sensor head and a vibrometer controller (model OFV 303 and OFV 3001, respectively, with resolution of 125 mm/s/V), for all five points. The data collected by the laser vibrometer was also fed into a SigLab DSP board.

Compute the natural frequencies using Table 6.6 and the theory presented in Section 6.5, and compare the results to the measured natural frequencies.

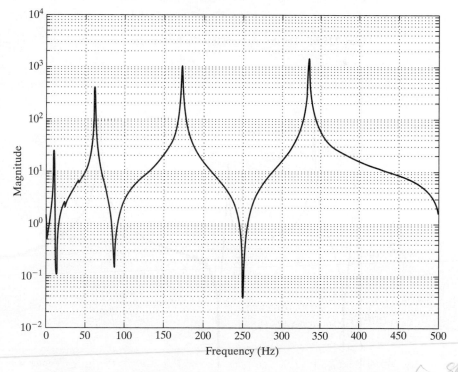

Figure 7.15(a)   The transfer function between points 1 and 2 in Figure 7.14.

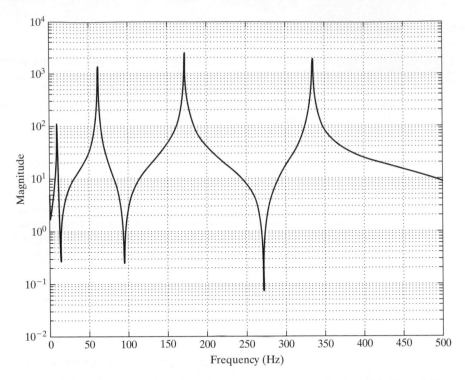

Figure 7.15(b) The driving point transfer function between points 2 and 2 in Figure 7.14.

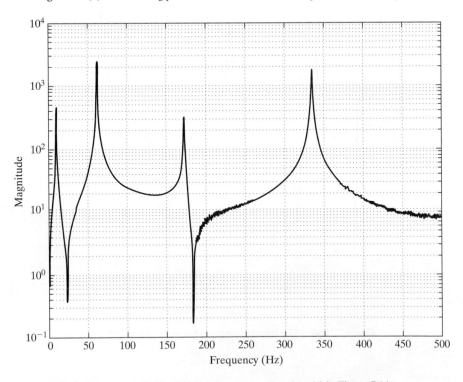

Figure 7.15(c) The transfer function between points 2 and 3 in Figure 7.14.

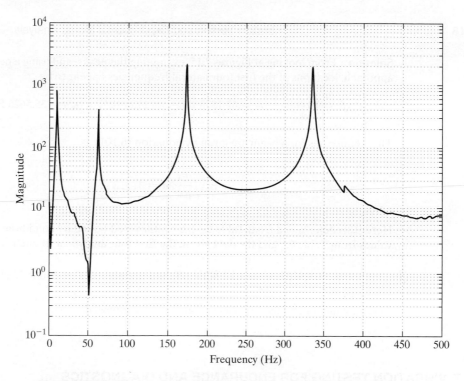

Figure 7.15(d) The transfer function between points 2 and 4 in Figure 7.14.

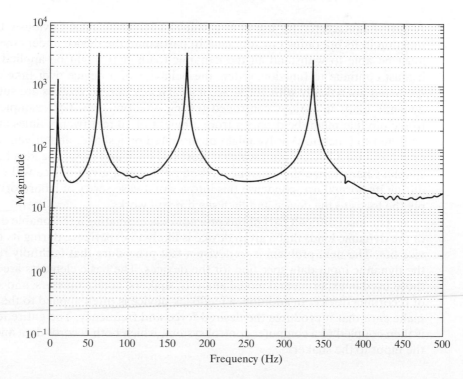

Figure 7.15(e) The transfer function between points 2 and 5 in Figure 7.14.

Solution   From looking at Figures 7.15, examining the peaks, and using a peak-picking approach, locations of the first four natural frequencies appear to be

$$f_1 = 10.000 \qquad f_2 = 62.1875 \qquad f_3 = 172.8125 \qquad f_4 = 335.5626 \text{ Hz}$$

The frequencies for a beam are given by equation (6.101) as

$$\omega_n = \beta_n^2 \sqrt{\frac{EI}{\rho A}}, n = 1, 2, 3 \ldots$$

Here β_n is determined from Table 6.6 for cantilevered boundary conditions. Using the values from Table 6.6 and the data given, the first four analytical frequencies are

$$f_1 = 9.9105 \quad f_2 = 62.1082 \qquad f_3 = 173.0946 \qquad f_4 = 340.7837 \text{ Hz}$$

Thus, the theory and the experiment agree reasonably well.

$\square$

7.7 VIBRATION TESTING FOR ENDURANCE AND DIAGNOSTICS

A part manufactured for use in a machine or structure must obviously be able to function in its operating environment. In particular, the device under consideration must be able to withstand all the dynamic loads that might be applied to it and it must continue to function. Often, the behavior of a device over time cannot be predicted analytically with 100% accuracy. Hence sample devices are subjected to dynamic loads in various controlled testing environments. For example, an electronic module might be dropped from a height of 10 feet twenty times and still be expected to function. A valve might be mounted on a shaker and driven with a random input for 12 hours, after which it must still open and close. The idea here is very simple: Characterize the dynamic environment that a given device will experience during its normal use; condense this into a worst-case series of laboratory tests; apply the loads to the device and then see if it still works.

The most difficult part of this procedure is estimating a reasonable description of the dynamic loads that a given device is likely to experience during its useful service life. The next problem is to devise a test procedure that faithfully reproduces the dynamic input data specified for the devices. The basic elements are similar to those used in experimental modal analysis: shakers, accelerometers, and some kind of recording device. In addition, a feedback device is often applied to the shaker to make sure that it produces the desired force and frequency. The entire test system is then coupled to a computer control system, which both records data and controls the input to the shaker.

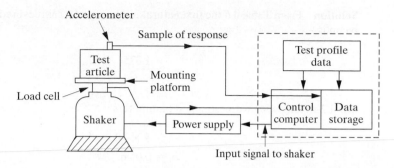

Figure 7.16   A schematic of a computer-controlled vibration endurance test.

The control computer can be used to program the shaker to perform long hours of testing at a variety of loads and frequencies and signal inputs. For example, a load may call for a device to be driven at 10 G for 3 hours at 10 Hz, then 3 hours at 50 Hz followed by 3 hours of random vibration of a specified strength. A schematic of a computer-controlled vibration testing device is given in Figure 7.16. In the figure, the test profile data containing timing and signal information is fed to the control computer, which assigns the appropriate signal to the power supply, which in turn drives the shaker. The force output of the shaker is monitored by a load cell. This signal is returned to the computer which matches the signal to the required test profile. If there are some differences, the computer adjusts its output to the shaker accordingly. The acceleration of the test article is also measured if transmissibility values are needed and to provide a record of the test. All signals are stored in the computer data storage section to provide evidence that the tests were performed and exactly what the responses and inputs were.

Another use of vibration testing is diagnostics or machine health monitoring. The basic idea here is that periodic measurements of frequency and damping may yield information regarding changes in the integrity of a structure or predict the pending failure of a machine. If a system's frequencies are measured and monitored over a period of time and a change is observed, then since $\omega_n = \sqrt{k/m}$, some change in the system's mass or stiffness must be the cause. A change in stiffness could imply a cracked or malfunctioning part and a change in mass might reflect excessive wear.

Example 7.7.1

A static deflection test is performed on a cantilevered aluminum beam, both with and without a small cut in the aluminum. The results indicate that with the cut, the modulus is measured to be 10% lower than its nominal value of 71×10^{10} N/m². Based on this information, the fundamental frequency of aircraft wings is measured after each flight to see if any fatigue cracks are present. What sort of frequency shift would be expected to detect a crack?

Solution From Table 6.6 the first natural frequency of a cantilevered Euler–Bernoulli beam is

$$\omega_1 = \frac{1.8751}{l^2}\sqrt{\frac{EI}{\rho A}}$$

For a 2-m long wing with estimated parameters

$$I = 5 \times 10^{-5}\ \mathrm{m}^4$$
$$A = 0.05\ \mathrm{m}^2$$

the nominal value of the wing frequency will be ($\rho = 2.7 \times 10^3\ \mathrm{kg/m^3}$)

$$\omega_1 = \frac{1.871}{(2)^2}\sqrt{\frac{\left(71 \times 10^{10}\right)\left(5 \times 10^{-5}\right)}{\left(2.7 \times 10^3\right)(0.05)}} \approx 240\ \mathrm{rad/s}$$

If E changes by 10% (i.e., is reduced to 63.9×10^{10}) the new frequency becomes

$$\omega_1 = \frac{1.871}{(2)^2}\sqrt{\frac{\left(63.9 \times 10^{10}\right)\left(5 \times 10^{-5}\right)}{\left(2.7 \times 10^3\right)(0.05)}} \approx 228\ \mathrm{rad/s}$$

Thus the change in frequency is both noticeable and possible to measure, forming a reasonable diagnostic.

□

In some cases the vibration response is examined as a signature of the device. If the time history of the response changes over time, it is possible that the change has been caused by some deterioration of the part. The use of these ideas is an emerging technology and engineers are involved in trying to make strong connections between certain types of changes (such as frequency) and the condition or health of the device.

As an example of health monitoring, consider the plot of Figure 7.17. The plot consists of a record of displacement measurements of a bearing housing of a rotating shaft on a machine made over several months. The measurements over time indicate a trend. The increase (changes in normal operating deflections) in later months is thought to show that something is changing in the bearing system so that maintenance or repair is required. The other way to examine this is to stop the machine and physically look for damage or wear. If the machine is required for production, stopping the machine to dismantle it could be very expensive. Using vibration monitoring techniques, such as indicated in Figure 7.17, the routine of stopping the machine periodically to check it or waiting for it to fail can be avoided. A more complete discussion of machine health monitoring can be found in Wowk (1991).

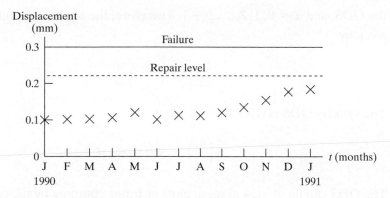

Figure 7.17 A record of average displacement of a housing for the bearing of a rotating shaft over a period of months.

7.8 OPERATIONAL DEFLECTION SHAPE MEASUREMENT

An operational deflection shape (ODS) is a vibration response pattern of a structure that is excited by sinusoidal forces. Here we will restrict our study to the simplest case of forces applied at a single excitation frequency and where all forces have the same or opposite phase. The ODSs can be studied to decide design changes to reduce vibration and noise of a structure, and also to detect damage in structures. The mathematical description of the ODS and the differences between ODS and mode shapes are described next.

Using equation (7.25), the steady-state forced displacement of a structure can be written as

$$\mathbf{x}(t) = \mathrm{Re}\left(\alpha(\omega)\mathbf{f}e^{j\omega t}\right) \tag{7.44}$$

The velocity response is

$$\mathbf{v}(t) = \mathrm{Re}\left(j\omega\alpha(\omega)\mathbf{f}e^{j\omega t}\right) \tag{7.45}$$

The velocity ODS will be considered here because they can be measured using the SLDV. The ODS is defined by evaluating equation (7.45) at different angles or times during a steady-state sinusoidal response. The angles are defined to be

$$\theta_a = \omega t_a \tag{7.46}$$

where ω is one excitation frequency. The ODS can be evaluated at specified angles, $\theta_a = \dfrac{2\pi a}{b}$, where b is the number of points in one cycle of vibration to evaluate the ODS, and $a = 0, 1, 2, \ldots, b - 1$. Therefore, the times to evaluate the ODS are given by

$$t_a(\omega) = \frac{2\pi a}{b\omega} \tag{7.47}$$

The velocity ODS is given by

$$\mathbf{v}\left(\frac{\theta_a}{\omega}\right) = \text{Re}\left(j\omega\alpha(\omega)\mathbf{f}e^{j\theta_a}\right) \tag{7.48}$$

The ODS can be plotted as mesh plots or fringe contours by the software supplied with the laser measurement system, or by MATLAB algorithms. Equation (7.48) can be used to study the ODS in terms of the force vector, excitation frequency, and angle of the response.

Another way to understand the ODS is to write it in terms of modal parameters, that is, mode shapes and damping ratios. To do this the damping must be modal or proportional so that the modes are real and the equations can decouple. By substituting equation (7.34) into equation (7.48), the velocity ODS becomes

$$\mathbf{v}\left(\frac{\theta_a}{\omega}\right) = \omega \sum_{i=1}^{n} \frac{\mathbf{u}_i\left(\mathbf{u}_i^T\mathbf{f}\right)\left((\omega\omega_i 2\zeta_i)\cos\theta_a + (\omega_i^2 - \omega^2)\sin\theta_a\right)}{\left((\omega_i^2 - \omega^2)^2 + (\omega\omega_i 2\zeta_i)^2\right)} \tag{7.49}$$

Equation (7.49) can be used to study the ODS in terms of the mode shapes, force vectors, excitation frequencies, damping ratios, and the angle of the response. From (7.49) it is obvious that there are a large number of ODSs that can occur. Note that the velocity ODS depends on all of the mode shapes of the structure. If the damping ratios are small and the structure is excited at its ith natural frequency, that is, $\omega = \omega_i$, then the denominator of equation (7.49) will be small and the ith term (modal response) in the series may be the largest contribution to the response. The response in each mode, however, also depends on the degree of orthogonality between that mode and the forcing vector. If $\left(\mathbf{u}_i^T\mathbf{f}\right) = 0$, then there will be no response in the ith mode. The effect of the force is the major difference between mode shapes and operational deflection shapes. The mode shapes are computed from the FRF matrix and do not depend on the forcing function magnitude or location. Conversely, the ODS depends on the location and relative magnitudes of the forces acting at different locations on the structure. In practice, if a single input is used and the structure is excited at a resonance, and damping is small, then the mode shapes and ODS are similar. However, other cases can occur where (1) the ODSs have nonstationary nodal points and (2) all parts of the ODS do not reach their maximum displacements and do not pass through equilibrium at the same time. This nonmodal behavior may be observed when two or more excitations are used or when multiple mode shapes that are close in frequency participate in the response.

The ODSs can be more difficult to understand than mode shapes, but they represent the actual structural response from all modes combined and damping is included. No assumptions or approximation other than linearity and the FFT operation are used. Because of their sensitivity, a recent application is to measure ODSs to detect cracks or other damage to structures. Changes in the ODS over time can indicate damage, and if higher-frequency ODSs are used, the damage site can often be located. A research topic is how to apply forces to a structure to obtain the maximum sensitivity of ODS to damage.

PROBLEMS

Section 7.2 Problems (7.1 through 7.5)

7.1. A low-frequency signal is to be measured by using an accelerometer. The signal is physically a displacement of the form $6 \sin (0.2t)$. The noise floor of the accelerometer (i.e. the smallest magnitude signal it can detect) is 0.4 volt/g. The accelerometer is calibrated at 1 volt/g. Can the accelerometer measure this signal?

7.2. Referring to Chapter 2, calculate the response of a single-degree-of-freedom system to a unit impulse and then to a unit triangle input lasting T second. Compare the two responses. The differences correspond to the differences between a "perfect" hammer hit and a more realistic hammer hit, as indicated in Figure 7.2. Use $\zeta = 0.01$ and $\omega = 5 \, \text{rad/s}$ for your model.

7.3. Compare the Laplace transform of $\delta(t)$ with the Laplace transform of the triangle input of Figure 7.2 and Problem 7.2.

7.4. Plot the error in measuring the natural frequency of a single-degree-of-freedom system of mass 12 kg and stiffness 300 N/m if the mass of the excitation device (shaker) is included and varies from 0.5 to 5 kg.

7.5. Calculate the Fourier transform of $f(t) = 4 \sin 2t - 2 \sin t - \cos t$ and plot the spectral coefficients.

Section 7.3 (Problems 7.6 through 7.9)

7.6. Represent $6 \sin 3t$ as a digital signal by sampling the signal at $\pi/3$, $\pi/6$ and $\pi/12$ seconds. Compare these three digital representations.

7.7. Compute the Fourier coefficient of the signal $|240 \sin (120 \pi t)|$.

7.8. Consider the periodic function

$$x(t) = \begin{cases} -5 & 0 < t < \pi \\ 5 & \pi < t < 2\pi \end{cases}$$

and $x(t) = (t + 2\pi)$. Calculate the Fourier coefficients. Next plot $x(t)$: $x(t)$ represented by the first term in the Fourier series, $x(t)$ represented by the first two terms of the series, and $x(t)$ represented by the first three terms of the series. Discuss your results.

7.9. Consider a signal $x(t)$ with maximum frequency of 600 Hz. Discuss the choice of record length and sampling interval.

Section 7.4 (Problems 7.10 through 7.19)

7.10. Consider the magnitude plot of Figure P7.10. How many natural frequencies does this system have and what are their approximate values?

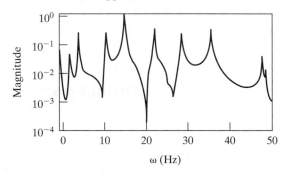

Figure P7.10   A sample magnitude plot of $H(\omega)$ for a simple structure.

7.11. Consider the experimental transfer function plot of Figure P7.11. Use the methods of Example 7.4.1 to determine ζ_i and ω_i.

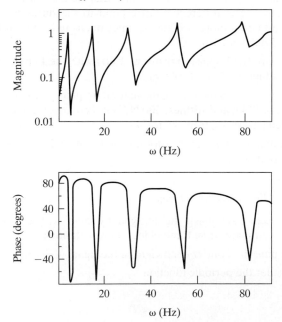

Figure P7.11   Experimental plot of phase and magnitude of a simple laboratory structure.

7.12. Consider a two-degree-of-freedom system with frequencies $\omega_1 = 12\,\mathrm{rad/s}, \omega_2 = 18\,\mathrm{rad/s}$, and damping ratios $\zeta_1 = \zeta_2 = 0.01$. With modal $S = \dfrac{1}{\sqrt{2}}\begin{bmatrix} 1 & -1 \\ 1 & 1 \end{bmatrix}$, calculate the transfer function of this system for an input at x_1 and a response measurement at x_2.

7.13. Plot the magnitude and phase of the transfer function of Problem 7.12 and see if you can reconstruct the modal data ($\omega_1, \omega_2, \zeta_1$, and ζ_2) from your plot.

7.14. Consider equation (7.14) for determining the damping ratio of a single mode. If the measurement in frequency varies by 0.5 %, how much will the value of ζ change?

7.15. Discuss the problems of using equation (7.14) if the natural frequencies of the structure are very close together.

7.16. Discuss the limitation of using equation (7.15) if ζ is very small. What happens if ζ is very large?

7.17. Consider the two-degree-of-freedom system described by

$$\begin{bmatrix} 1 & 0 \\ 0 & 1 \end{bmatrix}\begin{bmatrix} \ddot{x}_1 \\ \ddot{x}_2 \end{bmatrix} + \begin{bmatrix} 0 & 0 \\ 0 & c \end{bmatrix}\begin{bmatrix} \dot{x}_1 \\ \dot{x}_2 \end{bmatrix} + \begin{bmatrix} 2 & -1 \\ -1 & 2 \end{bmatrix}\begin{bmatrix} x_1 \\ x_2 \end{bmatrix} = \begin{bmatrix} f_0 \sin \omega t \\ 0 \end{bmatrix}$$

and calculate the transfer function $|X/F|$ as a function of the damping parameter c.

7.18. Plot the transfer function of Problem 7.17 for the four cases: $c = 0.01$, $c = 0.2$, $c = 1$, and $c = 10$. Discuss the difficulty in using these plots to measure ζ_i and ω_i for each value of c.

7.19. Use a numerical procedure to calculate the natural frequencies and damping ratios of the system of Problem 7.18. Label these on your plots from Problem 7.18 and discuss the possibility of measuring these values using the methods of Section 7.4.

Section 7.5 (Problems 7.20 through 7.24)

7.20. Using the definition of the mobility transfer function of Window 7.4, calculate the Re and Im parts of the frequency response function and hence verify equations (7.15) and (7.16).

7.21. Using equations (7.15) and (7.16), verify that the Nyquist plot of the mobility frequency response function does in fact form a circle.

7.22. Consider a single-degree-of-freedom system of mass 8kg, stiffness 1200 N/m, and damping ratio of 0.01. Pick five values of ω between 0 and 20 rad/s and plot five points of the Nyquist circle using equations (7.15) and (7.16). Do these form a circle?

7.23. Derive equation (7.20) for the damping ratio from equations (7.18) and (7.19). Then verify that equation (7.20) reduces to equation (7.21) at the half-power points.

7.24. Consider the experimental curve-fit Nyquist circle of Figure P7.24. Determine the modal damping ratio for this mode.

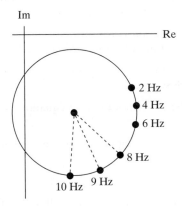

Figure P7.24   Experimentally determined Nyquist circle consisting of five data points. The 9-Hz point is constructed as halfway along the longest arc.

Section 7.6 (Problems 7.25 through 7.31)

7.25. Referring to Section 5.4 and Window 5.3, calculate the receptance matrix of equation (7.25) for the following two-degree-of-freedom system without using the system's mode shapes.

$$\begin{bmatrix} 2 & 0 \\ 0 & 1 \end{bmatrix} \begin{bmatrix} \ddot{x}_1 \\ \ddot{x}_2 \end{bmatrix} + \begin{bmatrix} 3 & -1 \\ -1 & 1 \end{bmatrix} \begin{bmatrix} \dot{x}_1 \\ \dot{x}_2 \end{bmatrix} + \begin{bmatrix} 6 & -2 \\ -2 & 2 \end{bmatrix} \begin{bmatrix} x_1 \\ x_2 \end{bmatrix} = \begin{bmatrix} f_0 \\ 0 \end{bmatrix} \sin \omega t$$

7.26. Repeat Problem 7.25 using the undamped mode shapes. Note that the system has proportional damping since $C = \alpha M + \beta K$, where $\alpha = 0, \alpha = 1/2$. Use this result and the result of Problem 7.25 to verify equation (7.33).

7.27. Compare equation (7.36) to equations (5.19) and (5.20) for the undamped vibration absorber problem and with equation (5.29) for the damped vibration absorber.

7.28. Consider the damped vibration absorber equation given by (5.29) and write out the four terms of the matrix $H_{sr}(\omega)$ given in equation (7.36). Physically interpret each term of H_{sr} by relating the input and output points to Figure 5.19.

7.29. Consider the transfer function of Figure P7.29 and determine the natural frequencies.

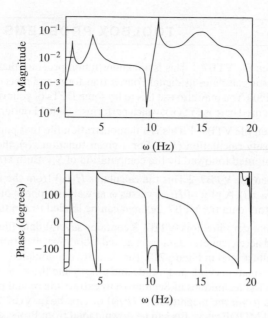

Figure P7.29   A sample magnitude and phase plot of a simple structure.

7.30. Try to determine the modal damping ratios from the plot of Figure P7.29.

7.31. The first row of the matrix $\left[\mathbf{u}_i \mathbf{u}_i^T \right]$ is measured to be

$$\begin{bmatrix} 1 & -1 & 3 & -0.25 & 4 \end{bmatrix}$$

Construct the entire matrix.

MATLAB ENGINEERING VIBRATION TOOLBOX

If you have not yet used the *Engineering Vibration Toolbox*, return to the end of Chapter 1 or Appendix G for a brief introduction. The files contained in the folder VTB7 may be used to help solve the following Toolbox problems. Some of these files contain real experimental data for those who do not have access to labs. The file VTB7_3.m contains experimental data, and updates of data are occasionally added to the Toolbox. The M-files from earlier chapters may also be useful. Alternately, the codes may be used directly following the introductions given in Sections 1.9, 1.10, 2.8, 2.9, 3.8, 3.9, 4.9, and 4.10 and Appendix F.

TOOLBOX PROBLEMS

TB7.1. Open file VTB7_1. This is a demonstration program that inputs a periodic signal, digitizes it, calculates its digital Fourier transform, and plots out both the digital record and its PSD. You may also use this to try some DFTs of your own. The basic MATLAB function used here is `fft(x)`, which performs a digital Fourier transform of the data vector x.

TB7.2. Open file VTB7_1. This is a demonstration file that performs a crude power spectral density calculation (S_{xx}). For a given function $x(t)$, the digital Fourier transform is computed followed by the computation of S_{xx}. Both $x(t)$ and $S_{xx}(\omega)$ are plotted.

TB7.3. Open file VTB7_2. This file calculates $H(j\omega)$ from the input data $f(t)$ and the output data $x(t)$. A plot of $H(\omega)$ versus ω as well as a phase plot are given. The sample time history data file VTB7_2ex.mat can be loaded to run this program.

TB7.4. Subfolder/directory VTB7_3 contains several data files containing force transducer and accelerometer data from actual laboratory experiments on a clamped–free beam, as illustrated in Figure 7.12. The data are manipulated to produce frequency response information (both magnitude and phase) like those of Figure 7.13. Plot these data and use the techniques of Section 7.6 to extract the modal parameters. Use the command abs to get the magnitude of $H(j\omega)$ or type `help vtb7` for help. (*Note:* FRF curve fits and MDOF curve fits can be downloaded from Professor Slater's website.)

TB7.5. Subfolder/directory VTB7_3 contains experimental data presented as Nyquist plots. Plot these data and use the techniques of Section 7.5 to determine the natural frequencies and damping ratios. Type `help vtb7` for help.

8

Finite Element Method

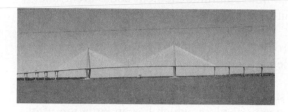

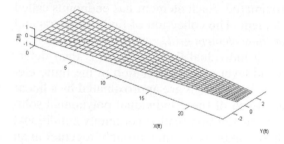

This chapter introduces the finite element method for vibration analysis. The finite element method is extremely useful for modeling complicated structures and machines such as the bridge and wing pictured here. In finite element analysis, the vibration of the structure is approximated by the motion of the points of connection of the lines illustrated in the finite element model of the airplane wing illustrated here. Such animations are valuable in locating troublesome vibration, in design, and in predicting the response of such devices before they are actually constructed and tested. The finite element methods presented here are often used in conjunction with the modal analysis test methods of Chapter 7 and the analysis methods of Chapter 4 for multiple-degree-of-freedom systems.

The finite element method is a powerful numerical technique that uses variational and interpolation methods for modeling and solving boundary-value problems such as described in Chapter 6, associated with distributed-parameter vibration problems (e.g., bars, beams, and plates). The method is also extremely useful for complicated devices and structures with unusual geometric shapes (e.g., trusses, frames, and machine parts). The finite element method is very systematic and modular. Hence the finite element method may easily be implemented on a digital computer to solve a wide range of practical vibration problems simply by changing the input to a computer program. In fact, several large commercial finite element codes are available. These commercial codes can be run on almost every type of computer, ranging from laptops to large super computers.

The finite element method approximates a structure in two distinct ways. The first approximation made in finite element modeling is to divide the structure up into a number of small, simple parts. These small parts are called *finite elements*, and the procedure of dividing up the structure is called *discretization*. Each element is usually very simple, such as a bar, beam, or plate, which has an equation of motion that can easily be solved or approximated. Each element has endpoints called *nodes*, which connect it to the next element. The collection of finite elements and nodes is called a *finite element mesh* or *finite element grid*.

The equation of vibration for each individual finite element is then determined and solved. This forms the second level of approximation in the finite element method. The solutions of the element equations are approximated by a linear combination of low-order polynomials. Each of these individual polynomial solutions is made compatible with the adjacent solution (called continuity conditions) at nodes common to two elements. These solutions are then brought together in an assembly procedure, resulting in global mass and stiffness matrices, which describe the vibration of the structure as a whole. This global mass and stiffness model represents a lumped-parameter approximation of the structure that can be analyzed and solved using the methods of Chapter 4. The vector $\mathbf{x}(t)$ of displacements associated with the solution of the global finite element model corresponds to the motion of the nodes of the finite element mesh.

The finite element procedure is best illustrated by examining some of the simple beams discussed in Chapter 6. Because simple structures have closed-form solutions, developing the finite element approximation on such structures provides an easy comparison with a more exact solution. However, the power and usefulness of the finite element method is not found in examining simple structures with closed-form solutions, but rather in modeling and solving complicated parts and structures that do not have closed-form solutions.

As a final introductory comment, note that the word *node* in finite element analysis means something completely different from a *node* in vibration analysis. This is an unfortunate situation that must be kept in mind. A *node* in vibration analysis refers to a node of a mode shape (i.e., a place where no motion occurs). In finite element analysis, a node is a point on the structure representing the boundary between two elements, corresponding to the coordinate or point on the structure that represents the motion of the structure as a whole. The nodes in finite element methods are used to capture the global motion of the structure as it vibrates.

The phrase *finite element method* is often abbreviated "FEM." This abbreviation can also denote the phrase *finite element model*. Another often used abbreviation is "FEA," which denotes *finite element analysis*. Sometimes the abbreviation "FE" is used to denote *finite elements*.

8.1 EXAMPLE: THE BAR

The longitudinal vibration of a bar provides a simple example of how a finite element model is constructed as well as how it is used to approximate the vibration of a distributed-parameter system with that of a lumped-parameter system (FEM). Recall the longitudinal vibration of a bar introduced in Section 6.3, which is reviewed in Window 8.1. Two finite element models of such a bar in a cantilevered configuration are illustrated in Figure 8.1. Note that the figure illustrates two *different* finite element grids of the *same* bar. Consider first the single-element model of Figure 8.1(a). The static (time-independent) displacement of this bar element must satisfy the equation

$$EA \, \frac{d^2 u(x)}{dx^2} = 0 \qquad\qquad (8.1)$$

for each value of x in the interval from 0 to l. Equation (8.1) can be integrated directly to yield

$$u(x) = c_1 x + c_2 \qquad\qquad (8.2)$$

where c_1 and c_2 are constants of integration with respect to x. Hence, although c_1 and c_2 are called constants, they could be functions of another variable, such as t. Equation (8.2) for the time-dependent displacement follows from the static deflection equation and is known because the structure being modeled is so simple. For more complicated structures, the functional form of expression (8.2) must be guessed, usually as some low-order polynomial. As pointed out in the introduction, the finite element method proceeds with two levels of approximation. One approximation involves deciding which model of Figure 8.1 to use (i.e., which mesh and size of mesh, where to put elements and nodes, etc.). The second level of approximation is the choice of polynomials to use in equation (8.1).

At each node, the value of u is allowed to be time dependent, hence the use of the labels $u_1(t)$ and $u_2(t)$ in Figure 8.1(a). The time-variable functions $u_1(t)$ and $u_2(t)$ are called the *nodal* displacements of the model and will eventually be solved for and used to describe the longitudinal vibration of the bar. The spatial function

Window 8.1
***Review of the Vibration of an Undamped Cantilevered
Bar in Longitudinal Vibration from Example 6.3.1***

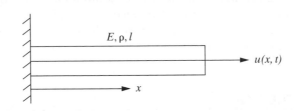

Displacement:	$u(x,t)$	Velocity:	$u_t(x,t)$
Acceleration:	$u_{tt}(x,t)$	Elastic modulus:	E
Cross-sectional area:	A	Length:	l

Equation of motion: $\rho A u_{tt}(x,t) - EA u_{xx}(x,t) = 0$

Boundary conditions: $u(0,t) = 0$ at $x = 0$, $u_x(l,t) = 0$ at $x = l$

Initial conditions: $u(x,0) = u_0(x)$ at $t = 0$, $u_t(x,0) = \dot{u}_0(x)$ at $t = 0$

Solution: $$u(x,t) = \sum_{n=1}^{\infty} A_n \sin(\omega_n t + \phi_n) \sin \frac{n\pi}{l} x$$

where A_n and ϕ_n are constants determined by
the initial conditions: $u(x,0)$ and $u_t(x,0)$

Natural frequencies: $$\omega_n = \frac{(2n-1)\pi}{2l} \sqrt{\frac{E}{\rho}}, \qquad n = 1, 2, 3, \ldots$$

Mode shapes: $$\sin\left(\frac{2n-1}{2l}\pi x\right), \qquad n = 1, 2, 3, \ldots$$

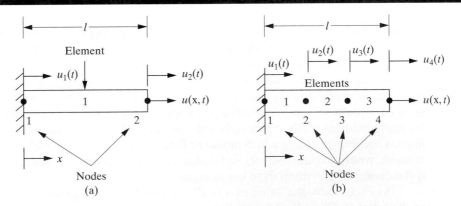

Figure 8.1 Two different finite element grids of the same cantilevered bar of length l
in longitudinal vibration. (a) Single-element, two-node mesh. (b) Three-element, four-
node mesh.

$u(x)$ and the temporal functions $u_1(t)$ and $u_2(t)$ are related through using the nodes as boundaries to evaluate the spatial constants in equation (8.2). At $x = 0$, equation (8.2) becomes

$$u(0) = u_1(t) = c_1(0) + c_2 \qquad t \geq 0$$

so that $c_2 = u_1(t)$. Similarly, at $x = l$, equation (8.2) yields

$$u(l) = u_2(t) = c_1 l + c_2$$

so that $c_1 = [u_2(t) - u_1(t)]/l$. Substitution of these (time-dependent) values of c_1 and c_2 into the expression for $u(x)$ given by equation (8.2) yields the approximation of the displacement $u(x, t)$, given by

$$u(x, t) = \left(1 - \frac{x}{l}\right)u_1(t) + \frac{x}{l}u_2(t) \tag{8.3}$$

If $u_1(t)$ and $u_2(t)$ were known at this stage, equation (8.3) would provide an approximate solution to the bar equation. The coefficient functions $(1 - x/l)$ and (x/l) are called *shape functions*, because they determine the spatial distribution or shape of the solution $u(x, t)$.

Next, consider the energy associated with the approximation given by equation (8.3). The strain energy of a bar is given by the integral [see, e.g., Shames (1989)]

$$V(t) = \frac{1}{2} \int_0^l EA \left[\frac{\partial u(x, t)}{\partial x}\right]^2 dx \tag{8.4}$$

Substituting the approximate solution for $u(x, t)$ given by equation (8.3) into equation (8.4) yields

$$V(t) = \frac{1}{2} \int_0^l \frac{EA}{l^2}[-u_1(t) + u_2(t)]^2 dx = \frac{EA}{2l}(u_1^2 - 2u_1u_2 + u_2^2) \tag{8.5}$$

The last expression can be recognized as proportional to the product of the transpose of the vector $\mathbf{u}(t)$ defined by

$$\mathbf{u}(t) = \begin{bmatrix} u_1(t) \\ u_2(t) \end{bmatrix} \tag{8.6}$$

with the matrix K and the vector $\mathbf{u}(t)$, where

$$K = \frac{EA}{l} \begin{bmatrix} 1 & -1 \\ -1 & 1 \end{bmatrix} \tag{8.7}$$

that is, $V(t) = \frac{1}{2}\mathbf{u}^T K \mathbf{u}$. Equation (8.7) defines the stiffness matrix associated with the single element of Figure 8.1(a).

The kinetic energy of the element can be calculated from the integral

$$T(t) = \frac{1}{2}\int_0^l A\rho(x)\left[\frac{\partial u(x, t)}{\partial t}\right]^2 dx \tag{8.8}$$

where $\rho(x)$ is the density of the bar as discussed in Section 6.3 and reviewed in Window 8.1. Using the approximation for $u(x, t)$ given by equation (8.3), the approximate velocity becomes

$$\frac{\partial u(x, t)}{\partial t} = \left(1 - \frac{x}{l}\right)\dot{u}_1(t) + \frac{x}{l}\dot{u}_2(t) \tag{8.9}$$

Assuming a constant density [i.e., $\rho(x) = \rho$] and substituting equation (8.9) into equation (8.8) yields

$$T(t) = \frac{1}{2}\frac{\rho A l}{3}\left(\dot{u}_1^2 + \dot{u}_1\dot{u}_2 + \dot{u}_2^2\right) \tag{8.10}$$

Looking at this expression as a matrix-based quadratic form yields that equation (8.10) can be factored into the velocity vector $\dot{\mathbf{u}}(t) = [\dot{u}_1(t) \quad \dot{u}_2(t)]^T$, the matrix M defined by

$$M = \frac{\rho A l}{6}\begin{bmatrix} 2 & 1 \\ 1 & 2 \end{bmatrix} \tag{8.11}$$

and the vector $\dot{\mathbf{u}}^T$ as

$$T(t) = \frac{1}{2}\dot{\mathbf{u}}^T M \dot{\mathbf{u}} \tag{8.12}$$

The matrix M defined by equation (8.11) is the mass matrix associated with the single finite element of 1(a).

The equations of vibration can be obtained from the preceding expressions for the kinetic energy $T(t)$ and strain energy $V(t)$ by using the variational or Lagrangian approach introduced in Section 4.7. Recall that the equations of motion can be calculated from the energy in the structure from

$$\frac{\partial}{\partial t}\left(\frac{\partial T}{\partial \dot{u}_i}\right) - \frac{\partial T}{\partial u_i} + \frac{\partial V}{\partial u_i} = f_i(t) \qquad i = 1, 2, \ldots, n \tag{8.13}$$

where u_i is the ith coordinate of the system, which is assumed to have n degrees of freedom, and where f_i is the external force applied at coordinates u_i (for the problem at hand, $f_i = 0$). Here the energies T and V are the total kinetic energy and strain energy, respectively, in the structure.

Examination of the boundary conditions in Figure 8.1(a) indicates that the time response at the clamped end must be zero. Hence the total kinetic energy becomes

$$T(t) = \frac{1}{2} \frac{\rho A l}{3} \left(\dot{u}_2^2 \right)$$

and the total strain energy becomes

$$V(t) = \frac{1}{2} \frac{EA}{l} u_2^2$$

Substitution of these two energy expressions into the Lagrange equation given in (8.13) yields

$$\frac{\rho A l}{3} \ddot{u}_2(t) + \frac{EA}{l} u_2(t) = 0 \tag{8.14}$$

This becomes

$$\ddot{u}_2(t) + \frac{3E}{\rho l^2} u_2(t) = 0 \tag{8.15}$$

which constitutes a simple finite element model of the cantilevered bar using only one element.

This finite element model of the bar can now be solved, given a set of initial conditions, for the nodal displacement $u_2(t)$. The solution to equation (8.15) is (see Window 8.2)

$$u_2(t) = \sqrt{u_0^2 + \left(\frac{\dot{u}_0}{\omega_n} \right)^2} \sin\left(\omega_n t + \tan^{-1} \frac{\omega_n u_0}{\dot{u}_0} \right) \tag{8.16}$$

Window 8.2
Review of the Solution for the Free Response
of an Undamped Single-Degree-of-Freedom System

From Section 1.1, equation (1.10), recall that the solution of $\ddot{x}(t) + \omega_n^2 x(t) = 0$ subject to initial conditions $x(0) = x_0$, $\dot{x}(0) = v_0$ is

$$x(t) = \sqrt{x_0^2 + \left(\frac{v_0}{\omega_n} \right)^2} \sin\left(\omega_n t + \tan^{-1} \frac{\omega_n x_0}{v_0} \right)$$

where u_0 is the initial nodal displacement, $\dot{u}_0$ is the initial nodal velocity, and, from the coefficient of $u_2(t)$ in equation (8.15), the natural frequency ω_n is

$$\omega_n = \frac{1}{l}\sqrt{\frac{3E}{\rho}}$$

This solution for $u_2(t)$ can be combined with equation (8.3) to yield the approximate solution for the transient displacement of the bar. The bar displacement becomes

$$u(x, t) = \sqrt{u_0^2 + \left(\frac{\dot{u}_0}{\omega_n}\right)^2}\, \frac{x}{l} \sin\left[\left(\frac{1}{l}\sqrt{\frac{3E}{\rho}}\right)t + \tan^{-1}\frac{\omega_n u_0}{\dot{u}_0}\right] \qquad (8.17)$$

This describes a vibration of frequency $(1/l)\sqrt{3E/\rho}$, in contrast to the exact solution given in Window 8.1, which describes vibration at an infinite number of frequencies.

Example 8.1.1

Compare the solution of the clamped–free bar of Window 8.1 obtained by the methods of Chapter 6 with the solution obtained by the finite element approach as given by equation (8.17).

Solution   The solution given by the finite element approach contains a time oscillation at only one frequency, oscillating in only one spatial mode shape, whereas the exact solution consists of an infinite number of mode shapes oscillating at an infinite number of frequencies superimposed on one another (depending, of course, on the initial conditions). The single undamped natural frequency of the finite element model is

$$\omega_{\text{FEM}} = \frac{1}{l}\sqrt{\frac{3E}{\rho}} = \frac{1.732}{l}\sqrt{\frac{E}{\rho}}$$

whereas the first natural frequency of the exact solution is $\omega_1 = (\pi/2l)\sqrt{E/\rho}$ or approximately

$$\omega_1 = \frac{1.57}{l}\sqrt{\frac{E}{\rho}}$$

This is smaller than the frequency predicted by the FEM. The second (exact) natural frequency is approximately

$$\omega_2 = \frac{4.712}{l}\sqrt{\frac{E}{\rho}}$$

which is larger than ω_{FEM}. In addition, the exact first mode shape is different, yet has some of the same characteristics as the FEM shape function. This is illustrated in Figure 8.2.

   If the bar is excited only in its first mode, the shape of vibration for the finite element model is a fairly reasonable approximation to the exact mode shape in the sense

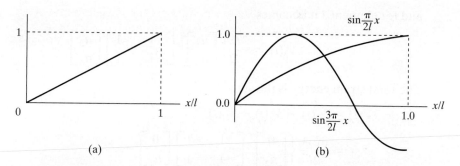

Figure 8.2   A plot of the normalized magnitude versus the normalized length of (a) the FEM shape function and (b) the first mode shape $[\sin(\pi x/2l)]$ of the exact solution, both for a clamped–free bar. Part (b) also shows the second mode shape $[\sin(3\pi x/2l)]$.

that it captures the basic behavior of the first mode. However, examination of the second exact mode shape in Figure 8.2 illustrates that the FEM is a terrible representation of the exact response if the bar is excited in the second mode.

□

Example 8.1.1 indicates that the finite element approximation is useful only in a certain context. However, the next section illustrates that the FEM can often be increased in size to produce a more accurate description of the structure under consideration.

8.2 THREE-ELEMENT BAR

Consider increasing the size of the finite element model of the bar in Section 8.1 to four nodes and three elements, as indicated in Figure 8.1(b). Each element of the bar will again have a strain energy relationship as calculated in equation (8.5), with two differences. The first difference is that the length of the element becomes $l/3$, instead of l, and the second is that the node coordinates are different in each of the three elements. Taking these changes into consideration and using matrix notation, the strain energy for element 1 is

$$V_1(t) = \frac{3EA}{2l} \begin{bmatrix} 0 \\ u_2 \end{bmatrix}^T \begin{bmatrix} 1 & -1 \\ -1 & 1 \end{bmatrix} \begin{bmatrix} 0 \\ u_2 \end{bmatrix} \tag{8.18}$$

For element 2 the strain energy becomes

$$V_2(t) = \frac{3EA}{2l} \begin{bmatrix} u_2 \\ u_3 \end{bmatrix}^T \begin{bmatrix} 1 & -1 \\ -1 & 1 \end{bmatrix} \begin{bmatrix} u_2 \\ u_3 \end{bmatrix} \tag{8.19}$$

and for element 3 it becomes

$$V_3(t) = \frac{3EA}{2l} \begin{bmatrix} u_3 \\ u_4 \end{bmatrix}^T \begin{bmatrix} 1 & -1 \\ -1 & 1 \end{bmatrix} \begin{bmatrix} u_3 \\ u_4 \end{bmatrix} \tag{8.20}$$

The total strain energy is the sum

$$V(t) = V_1(t) + V_2(t) + V_3(t)$$

$$= \frac{3EA}{2l} \left\{ \begin{bmatrix} 0 \\ u_2 \end{bmatrix}^T \begin{bmatrix} 1 & -1 \\ -1 & 1 \end{bmatrix} \begin{bmatrix} 0 \\ u_2 \end{bmatrix} \right.$$

$$+ \begin{bmatrix} u_2 \\ u_3 \end{bmatrix}^T \begin{bmatrix} 1 & -1 \\ -1 & 1 \end{bmatrix} \begin{bmatrix} u_2 \\ u_3 \end{bmatrix} + \begin{bmatrix} u_3 \\ u_4 \end{bmatrix}^T \begin{bmatrix} 1 & -1 \\ -1 & 1 \end{bmatrix} \begin{bmatrix} u_3 \\ u_4 \end{bmatrix} \right\}$$

$$= \frac{3EA}{2l} \left(2u_2^2 - 2u_2u_3 - 2u_3^2 - 2u_3u_4 + u_4^2 \right) \tag{8.21}$$

The vector of derivatives of this total strain energy indicated in the Lagrangian given by equation (8.13) becomes

$$\begin{bmatrix} \dfrac{\partial V}{\partial u_2} \\[2mm] \dfrac{\partial V}{\partial u_3} \\[2mm] \dfrac{\partial V}{\partial u_4} \end{bmatrix} = \frac{3EA}{l} \begin{bmatrix} 2u_2 - u_3 \\ -u_2 + 2u_3 - u_4 \\ -u_3 + u_4 \end{bmatrix} = \frac{3EA}{l} \begin{bmatrix} 2 & -1 & 0 \\ -1 & 2 & -1 \\ 0 & -1 & 1 \end{bmatrix} \begin{bmatrix} u_2(t) \\ u_3(t) \\ u_4(t) \end{bmatrix} \tag{8.22}$$

In a similar fashion, the total kinetic energy can be written as

$$T(t) = \frac{\rho A l}{36} \left\{ \begin{bmatrix} 0 \\ \dot{u}_2 \end{bmatrix}^T \begin{bmatrix} 2 & 1 \\ 1 & 2 \end{bmatrix} \begin{bmatrix} 0 \\ \dot{u}_2 \end{bmatrix} + \begin{bmatrix} \dot{u}_2 \\ \dot{u}_3 \end{bmatrix}^T \begin{bmatrix} 2 & 1 \\ 1 & 2 \end{bmatrix} \begin{bmatrix} \dot{u}_2 \\ \dot{u}_3 \end{bmatrix} + \begin{bmatrix} \dot{u}_3 \\ \dot{u}_4 \end{bmatrix}^T \begin{bmatrix} 2 & 1 \\ 1 & 2 \end{bmatrix} \begin{bmatrix} \dot{u}_3 \\ \dot{u}_4 \end{bmatrix} \right\} \tag{8.23}$$

The first term in the Lagrange equation (8.13) then becomes

$$\frac{d}{dt} \begin{bmatrix} \dfrac{\partial T}{\partial \dot{u}_2} \\[2mm] \dfrac{\partial T}{\partial \dot{u}_3} \\[2mm] \dfrac{\partial T}{\partial \dot{u}_4} \end{bmatrix} = \frac{\rho A l}{18} \begin{bmatrix} 4 & 1 & 0 \\ 1 & 4 & 1 \\ 0 & 1 & 2 \end{bmatrix} \begin{bmatrix} \ddot{u}_2(t) \\ \ddot{u}_3(t) \\ \ddot{u}_4(t) \end{bmatrix} \tag{8.24}$$

Combining equations (8.24) and (8.22) with equation (8.13) yields the familiar form

$$M\ddot{\mathbf{u}}(t) + K\mathbf{u}(t) = 0 \tag{8.25}$$

where $\mathbf{u}(t) = [u_2 \quad u_3 \quad u_4]^T$ is the vector of nodal displacements. Here the coefficient

$$M = \frac{\rho A l}{18} \begin{bmatrix} 4 & 1 & 0 \\ 1 & 4 & 1 \\ 0 & 1 & 2 \end{bmatrix} \tag{8.26}$$

is the *global mass matrix* and the coefficient

$$K = \frac{3EA}{l} \begin{bmatrix} 2 & -1 & 0 \\ -1 & 2 & -1 \\ 0 & -1 & 1 \end{bmatrix} \tag{8.27}$$

is the *global stiffness matrix* defining the dynamic finite element model of the bar. Equation (8.25) can be solved and analyzed by the methods of Section 4.4. Note that equation (8.25) is both dynamically and statically coupled.

Example 8.2.1

Compare the natural frequencies of the finite element model of Figure 8.1(b) with those of the distributed-mass model given in Window 8.1.

Solution The natural frequencies of the three-element finite element model of the clamped–free bar are determined by substituting the global stiffness matrix of equation (8.27) and the global mass matrix of equation (8.26) into equation (8.25). Following the procedures of Section 4.4 yields the characteristic equation

$$\det\left\{ \frac{3EA}{l} \begin{bmatrix} 2 & -1 & 0 \\ -1 & 2 & -1 \\ 0 & -1 & 1 \end{bmatrix} - \omega^2 \frac{\rho A l}{18} \begin{bmatrix} 4 & 1 & 0 \\ 1 & 4 & 1 \\ 0 & 1 & 2 \end{bmatrix} \right\} = 0 \tag{8.28}$$

Assuming that the beam is made of aluminum and is 1 m in length, then $l = 1$ m, $\rho = 2700$ kg/m^3, and $E = 7.0 \times 10^{10}$ N/m^2. Note that the value of A is not required. Equation (8.28) can be solved for the values of ω^2 using MATLAB to yield the natural frequencies

$$\omega_1 = 8092 \text{ rad/s}$$

$$\omega_2 = 26{,}458 \text{ rad/s}$$

$$\omega_3 = 47{,}997 \text{ rad/s}$$

On the other hand, the first three natural frequencies for the beam from the distributed-parameter solution given in Window 8.1 are

$$\omega_1 = 7998 \text{ rad/s} \ (0.55\%)$$

$$\omega_2 = 23{,}994 \text{ rad/s} \ (9.64\%)$$

$$\omega_3 = 39{,}900 \text{ rad/s} \ (19.3\%) \tag{8.29}$$

The numbers in parentheses in equations (8.29) are percents of the difference between the actual natural frequencies of the aluminum beam, as calculated from the distributed model, and the three corresponding natural frequencies of the three-element finite element model of the same beam. Note that the first natural frequency of the three-element model calculated here is much closer to the actual value than the natural frequency calculated in Example 8.1.1 from a one-element model. The value for ω from a one-element bar is 8819 rad/s. It is a rule of thumb in finite element analysis that many more (usually at least twice as many) elements must be used than the number of accurate frequencies required.

□

In the preceding analysis, each element is of the same length. However, the finite element method does not require the sizes of the various elements to be uniform. In fact, for complicated structures it is often necessary to choose a nonuniform size. The following example illustrates the use of a nonuniform element size.

Example 8.2.2

Consider the longitudinal vibration of a clamped bar and calculate two natural frequencies using the two-element mesh arrangement suggested in Figure 8.3. Compare these frequencies to the exact frequencies given in Window 8.1.

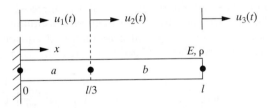

Figure 8.3   A bar of length l, area A, modulus E, and density ρ divided into two elements of dissimilar length.

Solution   The energy expressions for the first element are identical to the energy expressions calculated for the three-element model of a cantilevered bar analyzed previously. In particular, equation (8.18) yields that the potential energy in element 1 is

$$V_1(t) = \frac{3EA}{2l} u_2^2(t) \tag{8.30}$$

and equation (8.23) yields that the kinetic energy in element 1 is

$$T_1(t) = \frac{\rho A l}{18} \dot{u}_2^2(t) \tag{8.31}$$

Following the example of Section 8.1, the energy in the second element can be calculated from the assumed form of the solution given by

$$u(x, t) = c_3(t)x + c_4(t) \tag{8.32}$$

This displacement equation must satisfy $u(l/3, t) = u_2(t)$ and $u(l, t) = u_3(t)$, which yields the coupled algebraic equations

$$u_2(t) = \frac{l}{3} c_3(t) + c_4(t)$$

$$u_3(t) = lc_3(t) + c_4(t) \tag{8.33}$$

Equations (8.33) can be solved for $c_3(t)$ and $c_4(t)$ in terms of $u_2(t)$ and $u_3(t)$ to yield

$$c_3(t) = -\frac{3}{2l}\left[u_2(t) - u_3(t)\right] \quad\text{and}\quad c_4(t) = \frac{1}{2}\left[3u_2(t) - u_3(t)\right] \tag{8.34}$$

Substitution of these values of $c_3(t)$ and $c_4(t)$ into equation (8.32) yields

$$u(x, t) = \frac{3}{2}\left(1 - \frac{x}{l}\right)u_2(t) + \frac{1}{2}\left(\frac{3x}{l} - 1\right)u_3(t) \tag{8.35}$$

which can be used to calculate the energy expressions for the second element.
The potential energy in the second element becomes

$$V_2(t) = \frac{EA}{2}\int_{l/3}^{l}\left[u_x(x, t)\right]^2 dx = \frac{EA}{2}\int_{l/3}^{l}\left(\frac{3}{2l}\right)^2\left[u_3 - u_2\right]^2 dx$$

$$= \frac{EA}{2l}\left(\frac{3}{2}\right)\left[u_2^2(t) - 2u_2(t)u_3(t) + u_3^2(t)\right] \tag{8.36}$$

Adding the potential energy in each element as given by equations (8.30) and (8.36) yields the total potential energy in the bar:

$$V(t) = \frac{EA}{2l}\left(\frac{3}{2}\right)\left[3u_2^2(t) - 2u_2(t)u_3(t) + u_3^2(t)\right] \tag{8.37}$$

This can be written in matrix form as

$$V(t) = \frac{EA}{2l}\left(\frac{3}{2}\right)\begin{bmatrix} u_2 \\ u_3 \end{bmatrix}^T \begin{bmatrix} 3 & -1 \\ -1 & 1 \end{bmatrix}\begin{bmatrix} u_2 \\ u_3 \end{bmatrix} \tag{8.38}$$

which implies that the stiffness matrix has the value

$$K = \frac{EA}{l}\left(\frac{3}{2}\right)\begin{bmatrix} 3 & -1 \\ -1 & 1 \end{bmatrix} \tag{8.39}$$

The kinetic energy in the second element becomes

$$T_2(t) = \frac{A\rho}{2}\int_{l/3}^{l}\left[u_t(x, t)\right]^2 dx$$

$$= \frac{A\rho}{8l^2}\int_{l/3}^{l}\left[9(l - x)^2\dot{u}_2^2 + 6(3x - l)(l - x)\dot{u}_2\dot{u}_3 + (3x - l)^2\dot{u}_3^2\right] dx$$

$$= \frac{A\rho l}{9}\left[\dot{u}_2^2 + \dot{u}_2\dot{u}_3 + \dot{u}_3^2\right] \tag{8.40}$$

where $u_t(x, t)$ is calculated from equation (8.35). Adding equations (8.31) and (8.40), the total kinetic energy in the bar becomes

$$T(t) = T_1(t) + T_2(t) = \frac{A\rho l}{18} \left[3\dot{u}_2^2(t) + 2\dot{u}_2(t)\dot{u}_3(t) + 2\dot{u}_3^2(t) \right] \tag{8.41}$$

In matrix form this becomes simply

$$T(t) = \frac{1}{2} \frac{A\rho l}{18} \begin{bmatrix} \dot{u}_2 \\ \dot{u}_3 \end{bmatrix}^T \begin{bmatrix} 6 & 2 \\ 2 & 4 \end{bmatrix} \begin{bmatrix} \dot{u}_2 \\ \dot{u}_3 \end{bmatrix} \tag{8.42}$$

so that the system mass matrix is identified as

$$M = \frac{A\rho l}{18} \begin{bmatrix} 6 & 2 \\ 2 & 4 \end{bmatrix} \tag{8.43}$$

Using the kinetic energy and potential energy in the Euler–Lagrange equation yields that the equations of motion are

$$\frac{\rho l}{p} \begin{bmatrix} 3 & 1 \\ 1 & 2 \end{bmatrix} \begin{bmatrix} \ddot{u}_2 \\ \ddot{u}_3 \end{bmatrix} + \frac{3E}{2l} \begin{bmatrix} 3 & -1 \\ -1 & 1 \end{bmatrix} \begin{bmatrix} u_2 \\ u_3 \end{bmatrix} = \begin{bmatrix} 0 \\ 0 \end{bmatrix} \tag{8.44}$$

Using the methods of Section 4.3, the natural frequencies of equation (8.44) are

$$\omega_1 = (1.6432) \frac{1}{l} \sqrt{\frac{E}{\rho}} \quad \left(1.5708 \frac{1}{l} \sqrt{\frac{E}{\rho}} \right)$$

$$\omega_2 = (5.1962) \frac{1}{l} \sqrt{\frac{E}{\rho}} \quad \left(4.7124 \frac{1}{l} \sqrt{\frac{E}{\rho}} \right)$$

where the number in parentheses indicates the exact value from the solution given in Window 8.1. Although not illustrated here, the best use of nonuniform elements is for those applications where the geometry or physical parameters are variable along x.

$\square$

8.3 BEAM ELEMENTS

Many parts and structures cannot be modeled by axial vibration only. Hence a finite element is needed to describe transverse vibration as well. Window 8.3 summarizes the vibration analysis for an Euler–Bernoulli beam as discussed in Section 6.5. Figure 8.4 indicates the coordinate system and variables used in the finite element analysis of a free–free beam for transverse vibration. The coordinates used in the finite element model of the beam are the two linear coordinates $u_1(t)$ and $u_3(t)$ and the two rotational coordinates $u_2(t)$ and $u_4(t)$. One of each type is required to describe the motion of each node. That is, each node is modeled as having two degrees

Window 8.3
Review of the Transverse Vibration Analysis of an Undamped Pinned–Pinned Beam

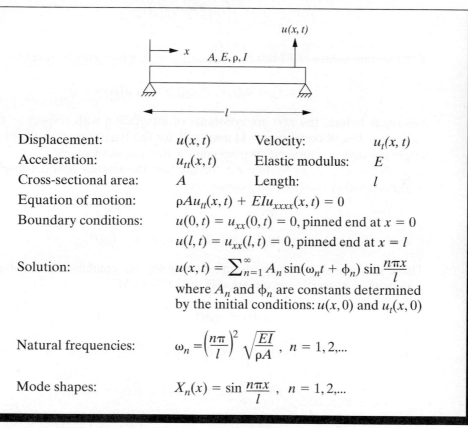

Displacement: $u(x,t)$ Velocity: $u_t(x,t)$

Acceleration: $u_{tt}(x,t)$ Elastic modulus: E

Cross-sectional area: A Length: l

Equation of motion: $\rho A u_{tt}(x,t) + E I u_{xxxx}(x,t) = 0$

Boundary conditions: $u(0,t) = u_{xx}(0,t) = 0,$ pinned end at $x = 0$

$u(l,t) = u_{xx}(l,t) = 0,$ pinned end at $x = l$

Solution: $u(x,t) = \sum_{n=1}^{\infty} A_n \sin(\omega_n t + \phi_n) \sin \dfrac{n\pi x}{l}$

where A_n and ϕ_n are constants determined by the initial conditions: $u(x,0)$ and $u_t(x,0)$

Natural frequencies: $\omega_n = \left(\dfrac{n\pi}{l}\right)^2 \sqrt{\dfrac{EI}{\rho A}}$, $n = 1, 2,...$

Mode shapes: $X_n(x) = \sin \dfrac{n\pi x}{l}$, $n = 1, 2,...$

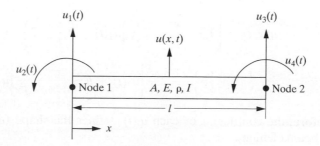

Figure 8.4   A single finite element model of a beam illustrating the two transverse coordinates $u_1(t)$ and $u_3(t)$ and the two rotational coordinates $u_2(t)$ and $u_4(t)$ required to describe the vibration of this element.

of freedom. One of these accounts for the slope and the other for the lateral motion. The transverse static displacement must satisfy

$$\frac{\partial^2}{\partial x^2}\left[EI\frac{\partial^2 u(x,t)}{\partial x^2}\right] = 0 \tag{8.45}$$

For constant values of EI this becomes $u_{xxxx} = 0$, which may be integrated to yield

$$u(x,t) = c_1(t)x^3 + c_2(t)x^2 + c_3(t)x + c_4(t) \tag{8.46}$$

where, as before, the $c_i(t)$ are constants of integration with respect to the spatial variable x [recall equations (8.1) and (8.2) for the bar]. Equation (8.46) is used to approximate the transverse displacement within the element.

Proceeding as in Section 8.1 for the bar, the unknown nodal displacements $u_i(t)$ must satisfy the boundary conditions

$$u(0,t) = u_1(t) \qquad u_x(0,t) = u_2(t)$$
$$u(l,t) = u_3(t) \qquad u_x(l,t) = u_4(t) \tag{8.47}$$

These relationships along with equation (8.46) are combined and solved for the constants of integration in equation (8.46). This yields

$$c_4(t) = u_1(t) \qquad c_3(t) = u_2(t)$$

$$c_2(t) = \frac{1}{l^2}\left[3(u_3 - u_1) - l(2u_2 + u_4)\right] \tag{8.48}$$

$$c_1(t) = \frac{1}{l^3}\left[2(u_1 - u_3) + l(u_2 + u_4)\right]$$

Substitution of equations (8.48) into equation (8.46), and rearranging terms as co-efficients of the unknown nodal displacement (again recall Section 8.1), yields the result that the approximate displacement $u(x,t)$ for the element can be expressed as

$$u(x,t) = \left[1 - 3\frac{x^2}{l^2} + 2\frac{x^3}{l^3}\right]u_1(t) + l\left[\frac{x}{l} - 2\frac{x^2}{l^2} + \frac{x^3}{l^3}\right]u_2(t)$$

$$+ \left[3\frac{x^2}{l^2} - 2\frac{x^3}{l^3}\right]u_3(t) + l\left[-\frac{x^2}{l^2} + \frac{x^3}{l^3}\right]u_4(t) \tag{8.49}$$

As before, the coefficient of each $u_i(t)$ defines the shape functions for the transverse beam element.

The mass and stiffness matrices can be calculated following exactly the same procedure as followed for the bar. In the case of the beam, the mass matrix calculation results from substituting equation (8.49) into the formula for kinetic energy:

$$T(t) = \tfrac{1}{2} \int_0^l \rho A \left[u_t(x, t) \right]^2 dx \tag{8.50}$$

and recognizing that this can be written in the form

$$T(t) = \tfrac{1}{2} \dot{\mathbf{u}}^T M \dot{\mathbf{u}} \tag{8.51}$$

where M is the desired mass matrix. The vector $\dot{\mathbf{u}}$ is the time derivative of the 4×1 vector $\mathbf{u}(t)$ of nodal displacements defined by

$$\mathbf{u}(t) = \begin{bmatrix} u_1(t) \\ u_2(t) \\ u_3(t) \\ u_4(t) \end{bmatrix} \tag{8.52}$$

After performing the integrations and factoring out the nodal displacement vector, the mass matrix for the beam of Figure 8.4 becomes

$$M = \frac{\rho A l}{420} \begin{bmatrix} 156 & 22l & 54 & -13l \\ 22l & 4l^2 & 13l & -3l^2 \\ 54 & 13l & 156 & -22l \\ -13l & -3l^2 & -22l & 4l^2 \end{bmatrix} \tag{8.53}$$

The stiffness matrix calculation proceeds in a similar fashion, again following the example of the bar in Section 8.1. The strain energy for the beam results from substituting the assumed solution form given by equation (8.49) into the formula for the strain energy given by

$$V(t) = \tfrac{1}{2} \int_0^l EI \left[u_{xx}(x, t) \right]^2 dx \tag{8.54}$$

The result can be factored into the form

$$V(t) = \tfrac{1}{2} \mathbf{u}^T K \mathbf{u} \tag{8.55}$$

where $\mathbf{u}$ is as defined previously. This defines the stiffness matrix K to be

$$K = \frac{EI}{l^3} \begin{bmatrix} 12 & 6l & -12 & 6l \\ 6l & 4l^2 & -6l & 2l^2 \\ -12 & -6l & 12 & -6l \\ 6l & 2l^2 & -6l & 4l^2 \end{bmatrix} \tag{8.56}$$

The mass and stiffness matrices of equations (8.53) and (8.56) define the beam finite element for transverse vibration.

Example 8.3.1

Use the beam finite element mass and stiffness matrices defined previously to calculate the natural frequencies of a simply supported beam. Use a single element. Compare these with the frequencies obtained for a simply supported beam in Chapter 6 and reviewed in Window 8.3.

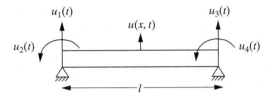

Figure 8.5   The coordinates of a simply supported beam modeled as a single finite element.

Solution   The simply supported boundary condition allows the slope at the boundary to vary but fixes the displacement to be zero at the boundary. Examining Figure 8.5, the pinned boundary condition requires that both $u_1(t)$ and $u_3(t)$ be zero. This is accomplished by setting these terms to zero in the kinetic and potential energy expression. Setting $u_1(t)$ and $u_3(t)$ to zero has the effect of deleting the rows and columns of the mass and stiffness matrices corresponding to $u_1(t)$ and $u_3(t)$. Deleting the first and third row and column results in the dynamic equation

$$\frac{\rho Al}{420}\begin{bmatrix} 4l^2 & -3l^2 \\ -3l^2 & 4l^2 \end{bmatrix}\begin{bmatrix} \ddot{u}_2(t) \\ \ddot{u}_4(t) \end{bmatrix} + \frac{EI}{l^3}\begin{bmatrix} 4l^2 & 2l^2 \\ 2l^2 & 4l^2 \end{bmatrix}\begin{bmatrix} u_2(t) \\ u_4(t) \end{bmatrix} = \begin{bmatrix} 0 \\ 0 \end{bmatrix} \tag{8.57}$$

The preceding can be written as

$$\begin{bmatrix} 4 & -3 \\ -3 & 4 \end{bmatrix}\ddot{u} + \frac{840EI}{\rho Al^4}\begin{bmatrix} 2 & 1 \\ 1 & 2 \end{bmatrix}u = 0 \tag{8.58}$$

where **u** is now interpreted as $\mathbf{u} = [u_2(t) \quad u_4(t)]^T$. Following the procedures of Section 4.2 (i.e., let $\mathbf{u} = \mathbf{x}e^{\omega t}$ and solve for ω), equation (8.58) yields the two natural frequencies

$$\omega_1 = 10.95\sqrt{\frac{EI}{\rho Al^4}} \quad \left(9.87\sqrt{\frac{EI}{\rho Al^4}}\right)$$

$$\omega_2 = 50.20\sqrt{\frac{EI}{\rho Al^4}} \quad \left(39.48\sqrt{\frac{EI}{\rho Al^4}}\right) \tag{8.59}$$

The numbers in parentheses are the actual values calculated using the distributed-mass models and methods of Chapter 6. The first frequency is within 11% of the actual value, while the second frequency is only within 28% of the actual value. As stated previously, it is generally thought that it is required to take at least twice as many degrees of freedom in the finite element model as the number of frequencies required.

□

The next step in applying the finite element method is to incorporate more than one element in the analysis of the beam. In the case of the bar, Section 8.2, the equations of motion for multiple-element structures were obtained by returning to the energy computation and the Euler–Lagrange formula. Such equations become too lengthy for the more complicated beam element, so an alternative but equivalent method of superimposing individual element matrices is used to obtain the global mass and stiffness matrices. The procedure is presented by repeating the three-element-bar example of Section 8.2.

Example 8.3.2

Consider again the three-element model of the longitudinal vibration of the bar of Figure 8.1(b) repeated in Figure 8.6. Here the local mass and stiffness matrix for each element is obtained from the basic element matrices given by equations (8.7) and (8.11) with l replaced by $l/3$. The resulting mass and the stiffness matrices are

$$M_1 = \frac{\rho A l}{18} \begin{bmatrix} 2 & 1 \\ 1 & 2 \end{bmatrix} \qquad K_1 = \frac{3EA}{l} \begin{bmatrix} 1 & -1 \\ -1 & 1 \end{bmatrix} \tag{8.60}$$

corresponding to local coordinates $u_1(t)$ and $u_2(t)$. The local equation of motion then becomes

$$\frac{\rho A l}{18} \begin{bmatrix} 2 & 1 \\ 1 & 2 \end{bmatrix} \begin{bmatrix} \ddot{u}_1 \\ \ddot{u}_2 \end{bmatrix} + \frac{3EA}{l} \begin{bmatrix} 1 & -1 \\ -1 & 1 \end{bmatrix} \begin{bmatrix} u_1 \\ u_2 \end{bmatrix} = \begin{bmatrix} 0 \\ 0 \end{bmatrix} \tag{8.61}$$

Similarly, for element 2,

$$M_2 = \frac{\rho A l}{18} \begin{bmatrix} 2 & 1 \\ 1 & 2 \end{bmatrix} \qquad K_2 = \frac{3EA}{l} \begin{bmatrix} 1 & -1 \\ -1 & 1 \end{bmatrix} \tag{8.62}$$

with local coordinates u_2 and u_3. Element 3 has mass and stiffness matrices

$$M_3 = \frac{\rho A l}{18} \begin{bmatrix} 2 & 1 \\ 1 & 2 \end{bmatrix} \qquad K_3 = \frac{3EA}{l} \begin{bmatrix} 1 & -1 \\ -1 & 1 \end{bmatrix} \tag{8.63}$$

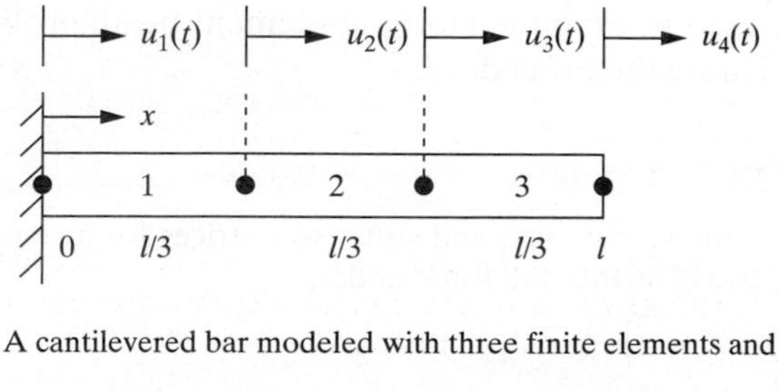

Figure 8.6   A cantilevered bar modeled with three finite elements and four nodes.

and local coordinates u_3 and u_4. These three sets of matrices and their corresponding equations [i.e., three sets of equations identical to equation (8.61) with different sets of modal displacements $u_i(t)$] can be assembled together by superimposing them to yield

$$\frac{\rho Al}{18}\begin{bmatrix} 2 & 1 & 0 & 0 \\ 1 & 2+2 & 1 & 0 \\ 0 & 1 & 2+2 & 1 \\ 0 & 0 & 1 & 2 \end{bmatrix}\begin{bmatrix} \ddot{u}_1 \\ \ddot{u}_2 \\ \ddot{u}_3 \\ \ddot{u}_4 \end{bmatrix} + \frac{3EA}{l}\begin{bmatrix} 1 & -1 & 0 & 0 \\ -1 & 1+1 & -1 & 0 \\ 0 & -1 & 1+1 & -1 \\ 0 & 0 & -1 & 1 \end{bmatrix}\begin{bmatrix} u_1 \\ u_2 \\ u_3 \\ u_4 \end{bmatrix} = \begin{bmatrix} 0 \\ 0 \\ 0 \\ 0 \end{bmatrix}$$

$$(8.64)$$

Here each of the local matrices from each of the individual elements is overlapped with that of adjacent elements. Equation (8.64) represents the global finite element model of the bar with global coordinates vector $\mathbf{u}(t) = [u_1(t) \quad u_2(t) \quad u_3(t) \quad u_4(t)]^T$ and defines the global mass and stiffness matrices, whereas M_1, M_2, M_3, K_1, K_2, and K_3 are called local mass and stiffness matrices. The coordinates $u_i(t)$ taken separately in pairs are called local coordinates, while the vector $\mathbf{u}(t)$ defines the global coordinate system.

To finish this modeling exercise, the boundary conditions must be accounted for. Since the bar is clamped at $u_1(t)$, this coordinate is eliminated in equation (8.64) by striking out the row and column associated with it, resulting in

$$M = \frac{\rho Al}{18}\begin{bmatrix} 4 & 1 & 0 \\ 1 & 4 & 1 \\ 0 & 1 & 2 \end{bmatrix} \qquad K = \frac{3EA}{l}\begin{bmatrix} 2 & -1 & 0 \\ -1 & 2 & -1 \\ 0 & -1 & 1 \end{bmatrix} \qquad (8.65)$$

These are in perfect agreement with the global mass and stiffness matrices derived using a complete energy calculation for a three-element model of the clamped bar in Section 8.2 as given in equations (8.26) and (8.27), respectively.

$\square$

It is important to note that the method of superposition used in the example to assemble the global mass and stiffness matrices yields the same results as the more rigorous variational approach used in Section 8.2. Since the superposition method is much easier to use, this approach of assembling global mass and stiffness matrices will be used to examine multiple-element beam problems. An example is again used to illustrate the procedure.

Example 8.3.3

Derive the global mass and stiffness matrices for a clamped–free beam of Figure 8.7 using two elements and three nodes.

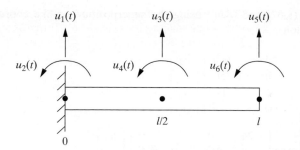

Figure 8.7 A two-element, three-node mesh of a cantilevered beam illustrating the nodal coordinates.

Solution The equations for the first element are obtained directly from equations (8.53) and (8.56) for a general beam element of length l, by replacing l with $l/2$ in these equations.

Thus the equation for the finite element in Figure 8.7 becomes

$$
\frac{\rho A l}{840}
\begin{bmatrix}
156 & 11l & 54 & -\frac{13}{2}l \\
11l & l^2 & \frac{13}{2}l & -\frac{3}{4}l^2 \\
54 & \frac{13}{2}l & 156 & -11l \\
-\frac{13}{2}l & -\frac{3}{4}l^2 & -11l & l^2
\end{bmatrix}
\begin{bmatrix}
\ddot{u}_1 \\ \ddot{u}_2 \\ \ddot{u}_3 \\ \ddot{u}_4
\end{bmatrix}
$$

$$
+ \frac{8EI}{l^3}
\begin{bmatrix}
12 & 3l & -12 & 3l \\
3l & l^2 & -3l & 0.5l^2 \\
-12 & -3l & 12 & -3l \\
3l & 0.5l^2 & -3l & l^2
\end{bmatrix}
\begin{bmatrix}
u_1 \\ u_2 \\ u_3 \\ u_4
\end{bmatrix}
=
\begin{bmatrix}
0 \\ 0 \\ 0 \\ 0
\end{bmatrix}
\tag{8.66}
$$

At this point it is possible to apply the clamped boundary condition, as it affects only this first element. The clamped end requires that both the deflection and slope at $x = 0$ must vanish so that $u_1 = u_2 = 0$. Striking out the rows and columns associated with these two coordinates yields

$$
\frac{\rho A l}{840}
\begin{bmatrix}
156 & -11l \\
-11l & l^2
\end{bmatrix}
\begin{bmatrix}
\ddot{u}_3 \\ \ddot{u}_4
\end{bmatrix}
+ \frac{8EL}{l^3}
\begin{bmatrix}
12 & -3l \\
-3l & l^2
\end{bmatrix}
\begin{bmatrix}
u_3 \\ u_4
\end{bmatrix}
=
\begin{bmatrix}
0 \\ 0
\end{bmatrix}
\tag{8.67}
$$

The equations for the second element are identical to equation (8.66) with the vector $[u_1 \; u_2 \; u_3 \; u_4]^T$ replaced with $[u_3 \; u_4 \; u_5 \; u_6]^T$ or

$$
\frac{\rho A l}{840}
\begin{bmatrix}
156 & 11l & 54 & -\frac{13}{2}l \\
11l & l^2 & \frac{13}{2}l & -\frac{3}{4}l^2 \\
54 & \frac{13}{2}l & 156 & -11l \\
-\frac{13}{2}l & -\frac{3}{4}l^2 & -11l & l^2
\end{bmatrix}
\begin{bmatrix}
\ddot{u}_3 \\ \ddot{u}_4 \\ \ddot{u}_5 \\ \ddot{u}_6
\end{bmatrix}
+ \frac{8EI}{l^3}
\begin{bmatrix}
12 & 3l & -12 & 3l \\
3l & l^2 & -3l & 0.5l^2 \\
-12 & -3l & 12 & -3l \\
3l & 0.5l^2 & -3l & l^2
\end{bmatrix}
\begin{bmatrix}
u_3 \\ u_4 \\ u_5 \\ u_6
\end{bmatrix}
= \mathbf{0}
\tag{8.68}
$$

Combining equations (8.67) and (8.68) using the superposition of like coordinates yields the global equation

$$
\frac{\rho A l}{840}
\begin{bmatrix}
312 & 0 & 54 & -6.5l \\
0 & 2l^2 & 6.5l & -0.75l^2 \\
54 & 6.5l & 156 & -11l \\
-6.5l & -0.75l^2 & -11l & l^2
\end{bmatrix}
\begin{bmatrix}
\ddot{u}_3 \\
\ddot{u}_4 \\
\ddot{u}_5 \\
\ddot{u}_6
\end{bmatrix}
$$

$$
+ \frac{8EI}{l^3}
\begin{bmatrix}
24 & 0 & -12 & 3l \\
0 & 2l^2 & -3l & \frac{1}{2}l^2 \\
-12 & -3l & 12 & -3l \\
3l & \frac{1}{2}l^2 & -3l & l^2
\end{bmatrix}
\begin{bmatrix}
u_3 \\
u_4 \\
u_5 \\
u_6
\end{bmatrix}
= \mathbf{0} \qquad (8.69)
$$

which constitutes the two-element finite element model of a cantilevered beam. Note that the matrix element in the 1–1 position in equation (8.69) is the sum of the element in the 1–1 position of equation (8.67) and the element in the 1–1 position of equation (8.68). This is also true for the elements in the mass and stiffness matrices in the 2–1, 1–2, and 2–2 positions. These four positions correspond to the common coordinates, u_3 and u_4, between the two finite elements.

$\square$

8.4 LUMPED-MASS MATRICES

In this section an alternative procedure for constructing the mass matrix is considered. In Section 8.3, the mass matrix was constructed by using the shape functions derived from the static displacement of a given element along with the definition of kinetic energy. Mass matrices constructed in this fashion are called *consistent-mass matrices*, because they are derived from a set of shape functions and displacement functions consistent with the stiffness matrix calculation.

Recall from the exercises and examples that the integrations required for the mass matrix involve higher-order polynomials than those required for the stiffness matrix. Hence an alternative to performing these calculations is to use a lumped-mass approximation. This involves a simple lumping of the mass of the structure at the nodes of the finite element model in proportion to the number of elements in the model. Such mass matrices are called *inconsistent-mass matrices*.

The lumped-mass approach has an advantage in that it generally produces lower-frequency estimates and is very easy to calculate. The lumped-mass matrices are diagonal, making computation easier. However, the lumped-mass method has several disadvantages. First, it can cause errors through a loss of accuracy. If the element under consideration has a rotational coordinate, such as the slope coordinates of the beam element, this coordinate has no mass assigned to it and the resulting mass matrix becomes singular (i.e., M^{-1} does not exist). Such systems require special methods to solve.

   The lumped-mass matrix is obtained by simply placing a lumped mass at each node equal to the appropriate proportions of the total mass of the system. For example, consider the bar element of Section 8.1. The total mass of the bar element of length l is ρAl. Placing one-half of this at each of the two nodes yields

$$M = \frac{\rho Al}{2} \begin{bmatrix} 1 & 0 \\ 0 & 1 \end{bmatrix} \tag{8.70}$$

This is the lumped-mass matrix for the bar element.

   Next consider the beam element of Section 8.3. The mass of an element of length l is ρAl. If this mass is divided evenly among the two transverse coordinates (u_1 and u_3), the mass matrix becomes

$$M = \frac{\rho Al}{2} \begin{bmatrix} 1 & 0 & 0 & 0 \\ 0 & 0 & 0 & 0 \\ 0 & 0 & 1 & 0 \\ 0 & 0 & 0 & 0 \end{bmatrix} \tag{8.71}$$

Note that since the rotational coordinates (u_2 and u_4) are not assigned any mass, the diagonal-mass matrix has two zeros along its diagonal and hence is singular. The singularity of the mass matrix can cause great difficulties in computing and interpreting the eigenvalues and hence the corresponding natural frequencies. The singular nature of the beam mass matrix can be removed by assigning some inertia to the rotational coordinates u_2 and u_4. This is done by computing the mass moment of inertia of half of the beam element about each of its ends. For a uniform beam this becomes

$$I = \frac{1}{3}\left(\frac{\rho Al}{2}\right)\left(\frac{l}{2}\right)^2 = \frac{\rho Al^3}{24} \tag{8.72}$$

Assigning this inertia to u_2 and u_4, the beam element lumped-mass matrix becomes

$$M = \frac{\rho Al}{2} \operatorname{diag}\left[(1) \quad \left(\frac{l^2}{12}\right) \quad (1) \quad \left(\frac{l^2}{12}\right) \right] \tag{8.73}$$

This diagonal lumped-mass matrix is nonsingular and, when combined with the beam stiffness matrix, can easily be solved for the system's natural frequencies using the methods of Chapter 4.

Example 8.4.1

   Compute the frequencies of a clamped–clamped bar of length l, modulus E, cross section A, and density ρ using two elements and both a consistent-mass and lumped-mass matrix. Compare the natural frequencies between the two systems and to the exact frequencies obtained by the methods of Chapter 6. The bar is illustrated in Figure 8.8.

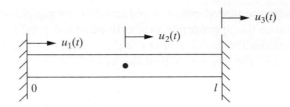

Figure 8.8 The coordinates for a two-element model of a clamped–clamped bar.

Solution The stiffness and consistent-mass matrices for a bar element are given by equations (8.7) and (8.11), respectively. Substituting $l/2$ in for l yields the following equation for each element:

element 1:
$$\frac{\rho A l}{12}\begin{bmatrix} 2 & 1 \\ 1 & 2 \end{bmatrix}\begin{bmatrix} \ddot{u}_1 \\ \ddot{u}_2 \end{bmatrix} + \frac{2EA}{l}\begin{bmatrix} 1 & -1 \\ -1 & 1 \end{bmatrix}\begin{bmatrix} u_1 \\ u_2 \end{bmatrix} = \begin{bmatrix} 0 \\ 0 \end{bmatrix} \qquad (8.74)$$

element 2:
$$\frac{\rho A l}{12}\begin{bmatrix} 2 & 1 \\ 1 & 2 \end{bmatrix}\begin{bmatrix} \ddot{u}_2 \\ \ddot{u}_3 \end{bmatrix} + \frac{2EA}{l}\begin{bmatrix} 1 & -1 \\ -1 & 1 \end{bmatrix}\begin{bmatrix} u_2 \\ u_3 \end{bmatrix} = \begin{bmatrix} 0 \\ 0 \end{bmatrix} \qquad (8.75)$$

These two equations are combined to form the global equation by striking out the first column and row of (8.74) because of the clamped boundary (see Figure 8.8) at $u_1(t)$, and the last row and column of equation (8.75) because of the clamped boundary at $u_3(t)$ (i.e., at $x = 1$). This yields the single-degree-of-freedom system

$$\frac{\rho A l}{12}[2 + 2]\ddot{u}_2 + \frac{2EA}{l}[1 + 1]u_2 = 0 \qquad (8.76)$$

Solving for ω^2 yields

$$\omega = 2\sqrt{3}\sqrt{\frac{E}{\rho l^2}} \cong 3.464\sqrt{\frac{E}{\rho l^2}} \qquad (8.77)$$

Next consider making the same calculation with the lumped-mass matrix of equation (8.70). Substitution of $l/2$ for l in equation (8.70) and repeating equations (8.74) and (8.75) yields

element 1:
$$\frac{\rho A l}{4}\begin{bmatrix} 1 & 0 \\ 0 & 1 \end{bmatrix}\begin{bmatrix} \ddot{u}_1 \\ \ddot{u}_2 \end{bmatrix} + \frac{2EA}{l}\begin{bmatrix} 1 & -1 \\ -1 & 1 \end{bmatrix}\begin{bmatrix} u_1 \\ u_2 \end{bmatrix} = \begin{bmatrix} 0 \\ 0 \end{bmatrix} \qquad (8.78)$$

element 2:
$$\frac{\rho A l}{4}\begin{bmatrix} 1 & 0 \\ 0 & 1 \end{bmatrix}\begin{bmatrix} \ddot{u}_2 \\ \ddot{u}_3 \end{bmatrix} + \frac{2EA}{l}\begin{bmatrix} 1 & -1 \\ -1 & 1 \end{bmatrix}\begin{bmatrix} u_2 \\ u_3 \end{bmatrix} = \begin{bmatrix} 0 \\ 0 \end{bmatrix} \qquad (8.79)$$

Here the lumped-mass matrix is equation (8.70) with l replaced by $l/2$.

Again using the assembly procedure and applying the boundary conditions yields the single-degree-of-freedom model

$$\frac{\rho A l}{4}[1 + 1]\ddot{u}_2 + \frac{2EA}{l}[1 + 1]u_2 = 0 \qquad (8.80)$$

Solving this single-degree-of-freedom system for ω yields

$$\omega = 2\sqrt{2}\sqrt{\frac{E}{\rho l^2}} \cong 2.8284\sqrt{\frac{E}{\rho l^2}} \tag{8.81}$$

The first natural frequency of a clamped–clamped bar is given in Chapter 6 as

$$\omega_1 = \pi\sqrt{\frac{E}{\rho l^2}} \cong 3.14159\sqrt{\frac{E}{\rho l^2}}$$

Note that the frequency estimates with the two different mass models are each about 10% away from the actual value as determined by the distributed-parameter model. The inconsistent-mass matrix yields an approximate value that is 10% lower, and the consistent-mass matrix yields an estimate that is about 10% higher than the actual value.

$\square$

8.5 TRUSSES

The power of finite element analysis is its ability to model complicated structures of odd geometry using simple elements such as bar, beam, and torsional rod elements. While the analysis of such structures is well beyond the scope of this introduction, an analysis of a truss structure is presented here to illustrate the main features of using finite element analysis on a more complicated structure.

Consider the simple truss structure of Figure 8.9. Note in particular that the coordinate system for each of the two elements (u_1, u_2, u_3, and u_4) is pointing in different directions. The truss element model describes vibration only along its axis, while the combined structure can vibrate in both the X and Y directions. To accommodate this situation, a *global coordinate system* aligned with the X–Y coordinate direction is defined. The final model for the full structure will be defined relative to the global X–Y coordinate. This is accomplished by defining global coordinates for each of the structure's nodes. These are denoted by capital U_i in the figure and are called global joint displacements.

The geometric configuration of the frame can be used to establish a relationship between the *local* nodal displacement u_i and the global joint displacements U_i. From the figure, u_3 and u_4 can be related to U_3, U_4, U_5, and U_6 by examining the projections of the global coordinates along the *local coordinate* direction (e.g., u_3 and u_4). This yields the relationships

$$u_3(t) = U_3\cos\theta + U_4\sin\theta$$
$$u_4(t) = U_5\cos\theta + U_6\sin\theta \tag{8.82}$$

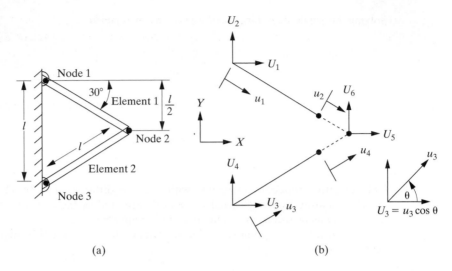

Figure 8.9 (a) A two-member framed structure mounted to a wall through pinned connections. Each of the two members is modeled as a bar. The two bars are pinned together. (b) A coordinate system.

where θ is the angle between the global system, X–Y, and the local coordinate system that is aligned along each of the two bars. Equation (8.82) can be written as the product of a matrix and vector:

$$
\begin{bmatrix} u_3(t) \\ u_4(t) \end{bmatrix} = \begin{bmatrix} \cos\theta & \sin\theta & 0 & 0 \\ 0 & 0 & \cos\theta & \sin\theta \end{bmatrix} \begin{bmatrix} U_3(t) \\ U_4(t) \\ U_5(t) \\ U_6(t) \end{bmatrix} \tag{8.83}
$$

or written symbolically as

$$
\mathbf{u}_2(t) = \Gamma \mathbf{U}_2(t) \tag{8.84}
$$

where Γ denotes the coordinate transformation matrix between the local and global coordinate systems and $\mathbf{U}_2$ is that part of the global coordinate vector $\mathbf{U}$ containing those coordinates associated with the second element (i.e., $\mathbf{U}_2 = [U_3 \ U_4 \ U_5 \ U_6]^T$). The vector $\mathbf{u}_2$ is the collection of local coordinates (i.e., $\mathbf{u}_2 = [u_3 \ u_4]^T$).

The kinetic and potential energy of element 2 in the figure can now be written in two ways, which must be equivalent. That is, the energy written in terms of either coordinate system must be the same. Equating the strain energy in the local coordinate system with that of the global coordinate system yields

$$
V(t) = \tfrac{1}{2} \mathbf{u}^T K_e \mathbf{u} = \tfrac{1}{2} \mathbf{U}^T \Gamma^T K_e \Gamma \mathbf{U} \tag{8.85}
$$

where K_e is the element stiffness matrix of the local coordinate system. In this case the element stiffness matrix in the local coordinate system is that of a bar. Hence the stiffness matrix in the global coordinate system for element 2 becomes

$$K_{(2)} = \Gamma^T K_e \Gamma \tag{8.86}$$

Note that the stiffness matrix in the global coordinate system is the product of the three matrices: the element stiffness matrix, the transformation matrix, and its transpose.

For the example of element 2 of Figure 8.9, the element stiffness matrix in global coordinates becomes

$$K_{(2)} = \frac{EA}{l} \begin{bmatrix} \cos\theta & 0 \\ \sin\theta & 0 \\ 0 & \cos\theta \\ 0 & \sin\theta \end{bmatrix} \begin{bmatrix} 1 & -1 \\ -1 & 1 \end{bmatrix} \begin{bmatrix} \cos\theta & \sin\theta & 0 & 0 \\ 0 & 0 & \cos\theta & \sin\theta \end{bmatrix} \tag{8.87}$$

where K_e, the general bar element stiffness matrix in the local coordinate system, is given by equation (8.7). Performing the indicated matrix products yields

$$K_{(2)} = \frac{EA}{l} \begin{bmatrix} \cos^2\theta & \sin\theta\cos\theta & -\cos^2\theta & -\sin\theta\cos\theta \\ \sin\theta\cos\theta & \sin^2\theta & -\sin\theta\cos\theta & -\sin^2\theta \\ -\cos^2\theta & -\sin\theta\cos\theta & \cos^2\theta & \sin\theta\cos\theta \\ -\sin\theta\cos\theta & -\sin^2\theta & \sin\theta\cos\theta & \sin^2\theta \end{bmatrix} \tag{8.88}$$

which is a 4×4 matrix corresponding to the global coordinates $\mathbf{U}_2$. Following this procedure for the other member of the truss, the global stiffness matrix for element 1 becomes

$$K_{(1)} = \frac{EA}{l} \begin{bmatrix} \cos^2\theta & \sin\theta\cos\theta & -\cos^2\theta & \sin\theta\cos\theta \\ -\sin\theta\cos\theta & \sin^2\theta & \sin\theta\cos\theta & -\sin^2\theta \\ -\cos^2\theta & \sin\theta\cos\theta & \cos^2\theta & -\sin\theta\cos\theta \\ \sin\theta\cos\theta & -\sin^2\theta & -\sin\theta\cos\theta & \sin^2\theta \end{bmatrix} \tag{8.89}$$

which corresponds to the global coordinate vector.

$$\mathbf{U}_{(1)} = \begin{bmatrix} U_1(t) \\ U_2(t) \\ U_5(t) \\ U_6(t) \end{bmatrix} \tag{8.90}$$

To combine the two-element matrices in the global coordinates [i.e., $K_{(1)}$ and $K_{(2)}$] into a global matrix in the full global coordinate $\mathbf{U} = [U_1 \quad U_2 \quad U_3 \quad U_4 \quad U_5 \quad U_6]^T$, $K_{(1)}$ is expanded to

$$K'_{(1)} = \frac{EA}{l} \begin{bmatrix} \cos^2\theta & -\sin\theta\cos\theta & 0 & 0 & -\cos^2\theta & \sin\theta\cos\theta \\ \sin\theta\cos\theta & \sin^2\theta & 0 & 0 & \sin\theta\cos\theta & -\sin^2\theta \\ 0 & 0 & 0 & 0 & 0 & 0 \\ 0 & 0 & 0 & 0 & 0 & 0 \\ -\cos^2\theta & \sin\theta\cos\theta & 0 & 0 & \cos^2\theta & -\sin\theta\cos\theta \\ \sin\theta\cos\theta & -\sin^2\theta & 0 & 0 & \sin\theta\cos\theta & \sin^2\theta \end{bmatrix} \tag{8.91}$$

Here zeros have been added to those positions corresponding to the missing coordinates U_3 and U_4. Similarly, $K_{(2)}$ is expanded to become compatible with the size of the full global vector $\mathbf{U}$. This yields

$$K'_{(2)} = \frac{EA}{l} \begin{bmatrix} 0 & 0 & 0 & 0 & 0 & 0 \\ 0 & 0 & 0 & 0 & 0 & 0 \\ 0 & 0 & \cos^2\theta & \sin\theta\cos\theta & -\cos^2\theta & -\sin\theta\cos\theta \\ 0 & 0 & \sin\theta\cos\theta & \sin^2\theta & -\sin\theta\cos\theta & -\sin^2\theta \\ 0 & 0 & -\cos^2\theta & -\sin\theta\cos\theta & \cos^2\theta & \sin\theta\cos\theta \\ 0 & 0 & -\sin\theta\cos\theta & -\sin^2\theta & \sin\theta\cos\theta & \sin^2\theta \end{bmatrix} \tag{8.92}$$

The two terms $K'_{(1)}\mathbf{U}$ and $K'_{(2)}\mathbf{U}$ can be added to yield the full global stiffness matrix in the full global coordinate system $\mathbf{U}$.

For the example of Figure 8.9, the sum of equations (8.91) and (8.92) yields the global stiffness matrix (for the case that each rod of the frame has the same physical parameters, $\theta = 30°$) defined by

$$K\mathbf{U} = (K'_{(1)} + K'_{(2)})\mathbf{U}$$

$$= \frac{EA}{l} \begin{bmatrix} 0.75 & -0.4330 & 0 & 0 & -0.75 & 0.4330 \\ -0.4330 & 0.25 & 0 & 0 & 0.4330 & -0.25 \\ 0 & 0 & 0.75 & 0.4330 & -0.75 & -0.4330 \\ 0 & 0 & 0.4330 & 0.25 & -0.4330 & -0.25 \\ -0.75 & 0.4330 & -0.75 & -0.4330 & 1.5 & 0 \\ 0.4330 & -0.25 & -0.4330 & -0.25 & 0 & 0.5 \end{bmatrix} \begin{bmatrix} U_1 \\ U_2 \\ U_3 \\ U_4 \\ U_5 \\ U_6 \end{bmatrix}$$
$$\tag{8.93}$$

Note that the effective stiffness corresponding to coordinates U_5 and U_6 has increased. This corresponds to the point where the two beams join together at a common node. Before equation (8.93) can be used as part of a vibration analysis, the boundary conditions must be applied. Examining Figure 8.9, it is clear that the pinned boundary condition at the connection of the two elements to ground requires that $U_1 = U_2 = U_3 = U_4 = 0$. Hence after applying the boundary conditions, the global stiffness matrix reduces to

$$K = \frac{EA}{l}\begin{bmatrix} 1.5 & 0 \\ 0 & 0.5 \end{bmatrix} \tag{8.94}$$

which is obtained by deleting those rows and columns of the coefficient matrix in equation (8.93) corresponding to U_1, U_2, U_3, and U_4. Similarly, the global displacement vector reduces to $\mathbf{U} = [U_5 \quad U_6]^T$.

Next consider assembling a consistent-mass matrix for this truss from the local mass matrix of the bar element given by equation (8.11) and from the frame geometry captured by the transformation Γ. Following the steps used in determining the global stiffness matrix, the kinetic energy of bar element number i (here $l = 1, 2$) as stated in global coordinates is equated to the kinetic energy of the same element stated in the local coordinate system. Using the notation defined previously, this yields

$$T_{(i)} = \tfrac{1}{2}\mathbf{u}^T M_i \mathbf{u} = \tfrac{1}{2}\mathbf{U}^T M_{(i)}\mathbf{U}$$

where M_i is the element mass matrix given by equation (8.11) and Γ is the coordinate transformation between the local coordinate $\mathbf{u}$ and the global coordinates $\mathbf{U}$, as given by equation (8.83). Thus the mass matrix in the global coordinates for the elements is of the form

$$M_{(i)} = \Gamma^T M_i \Gamma$$

Each of these matrices, $M_{(1)}$ and $M_{(2)}$ in the case of the example Figure 8.9, is then expanded by adding zeros as indicated in equations (8.91) and (8.92) for the corresponding stiffness matrices. The expanded matrices are combined (added) as in equation (8.93) to produce a 6×6 matrix corresponding to the full global coordinate system. This 6×6 mass matrix is then collapsed by applying the boundary conditions to the global coordinates to produce a 2×2 mass matrix compatible with the stiffness matrix of equation (8.94).

Alternatively, a lumped-mass matrix, as defined in Section 8.4, can be defined. In this case a reasonable choice for a lumped-mass matrix is to assign the mass of each element to each of the two remaining coordinates. Since the mass of each bar (of length l) is ρAl, a reasonable lumped-mass matrix is

$$M = \rho Al \begin{bmatrix} 1 & 0 \\ 0 & 1 \end{bmatrix} \tag{8.95}$$

Problem TB8.6 at the end of the chapter addresses the difference between the lumped-mass matrix as defined by equation (8.95) and the consistent-mass matrix suggested above.

The vibration problem using the finite element approach for the simple truss of Figure 8.9 becomes

$$\rho Al \begin{bmatrix} 1 & 0 \\ 0 & 1 \end{bmatrix} \begin{bmatrix} \ddot{U}_5 \\ \ddot{U}_6 \end{bmatrix} + \frac{EA}{l} \begin{bmatrix} 1.5 & 0 \\ 0 & 0.5 \end{bmatrix} \begin{bmatrix} U_5 \\ U_6 \end{bmatrix} = \begin{bmatrix} 0 \\ 0 \end{bmatrix} \tag{8.96}$$

Thus in this case with a lumped-mass matrix, the vibration of the truss system of Figure 8.9 is modeled as the independent motion of the tip in the x and y directions of frequency.

$$\omega_1 = \frac{1}{l} \sqrt{\frac{0.5E}{\rho}}$$

$$\omega_2 = \frac{1}{l} \sqrt{\frac{1.5E}{\rho}}$$

The model in equation (8.96) provides a crude result in the sense that the motions of U_6 and U_5 are decoupled. The model can be improved by choosing to model each of the two bars with more elements. This is addressed in the Toolbox problems at the end of the chapter.

8.6 MODEL REDUCTION

A difficulty with many design and analysis methods is that they work best for systems with a small number of degrees of freedom. Unfortunately, many interesting problems have a large number of degrees of freedom. In fact, to obtain accurate results with finite element models, the number of elements and hence the order of the vibration is increased. Thus finite element models of practical structures and machines are often very large. One approach to this dilemma is to reduce the size of the original model by essentially removing those parts of the model that affect its dynamic response the least. This process is called *model reduction* or *reduced-order modeling*. It is an attempt to reduce the size of an FEM but still retain the dynamic character of the system.

Quite often the mass matrix of a system may be singular or nearly singular, due to some elements being much smaller than others. In fact, in the case of finite element modeling, the mass matrix may contain zeros along a portion of the diagonal. Coordinates associated with zero or relatively small mass are likely candidates for being removed from the model. Another set of coordinates that are likely choices for removal from the model are those that do not respond when the structure is excited. Stated another way, some coordinates may have more significant responses than others. The distinction between significant and insignificant coordinates leads to a convenient formulation of the model reduction problem due to Guyan (1965).

Consider the undamped, forced-vibration, finite element model and partition the mass and stiffness matrices according to significant displacements denoted by $\mathbf{u}_1$ and insignificant displacements $\mathbf{u}_2$. This yields

$$
\begin{bmatrix} M_{11} & M_{12} \\ M_{21} & M_{22} \end{bmatrix} \begin{bmatrix} \ddot{\mathbf{u}}_1 \\ \ddot{\mathbf{u}}_2 \end{bmatrix} + \begin{bmatrix} K_{11} & K_{12} \\ K_{21} & K_{22} \end{bmatrix} \begin{bmatrix} \mathbf{u}_1 \\ \mathbf{u}_2 \end{bmatrix} = \begin{bmatrix} \mathbf{f}_1 \\ \mathbf{f}_2 \end{bmatrix}
\tag{8.97}
$$

Note here that the coordinates $\mathbf{u}(t)$ have been rearranged so that those having the least significant displacements associated with them appear last in the displacement vector, $\mathbf{u} = \begin{bmatrix} \mathbf{u}_1^T & \mathbf{u}_2^T \end{bmatrix}$. Next consider the potential energy of the system defined by the scalar $V = \frac{1}{2}\mathbf{u}^T K \mathbf{u}$ or, in partitioned form,

$$
V = \frac{1}{2}\begin{bmatrix} \mathbf{u}_1 \\ \mathbf{u}_2 \end{bmatrix}^T \begin{bmatrix} K_{11} & K_{12} \\ K_{21} & K_{22} \end{bmatrix} \begin{bmatrix} \mathbf{u}_1 \\ \mathbf{u}_2 \end{bmatrix}
\tag{8.98}
$$

Similarly, the kinetic energy of the system can be written as the scalar $T = \frac{1}{2}\dot{\mathbf{u}}^T M \dot{\mathbf{u}}$, which becomes

$$
T = \frac{1}{2}\begin{bmatrix} \dot{\mathbf{u}}_1 \\ \dot{\mathbf{u}}_2 \end{bmatrix}^T \begin{bmatrix} M_{11} & M_{12} \\ M_{21} & M_{22} \end{bmatrix} \begin{bmatrix} \dot{\mathbf{u}}_1 \\ \dot{\mathbf{u}}_2 \end{bmatrix}
\tag{8.99}
$$

in partitioned form. Since each coordinate u_i is acted on by a force f_i, the condition that there is no force in the direction of the insignificant coordinates $\mathbf{u}_2$ requires that $\mathbf{f}_2 = 0$ and that $\partial V/\partial \mathbf{u}_2 = 0$. This yields

$$
\frac{\partial}{\partial \mathbf{u}_2}\left(\mathbf{u}_1^T K_{11}\mathbf{u}_1 + \mathbf{u}_1^T K_{12}\mathbf{u}_2 + \mathbf{u}_2^T K_{21}\mathbf{u}_1 + \mathbf{u}_2^T K_{22}\mathbf{u}_2 \right) = 0
\tag{8.100}
$$

Hence the constraint relation between $\mathbf{u}_1$ and $\mathbf{u}_2$ must be (since $K_{12} = K_{21}^T$)

$$
\mathbf{u}_2 = -K_{22}^{-1}K_{21}\mathbf{u}_1
\tag{8.101}
$$

Expression (8.101) suggests a coordinate transformation (which is not a similarity transformation) from the full coordinate system to the reduced coordinate system $\mathbf{u}_1$. If the transformation matrix Q is defined by

$$
Q = \begin{bmatrix} I \\ -K_{22}^{-1}K_{21} \end{bmatrix}
\tag{8.102}
$$

then if $\mathbf{u} = Q\mathbf{u}_1$ is substituted into equation (8.97) and this expression is premultiplied by Q^T, a new reduced-order system of the form

$$
Q^T M Q \ddot{\mathbf{u}}_1 + Q^T K Q \mathbf{u}_1 = Q^T \mathbf{f}
\tag{8.103}
$$

results. The vector $Q^T\mathbf{f}$ now has the dimension of $\mathbf{u}_1$. Equation (8.103) represents the reduced-order form of equation (8.97), where

$$Q^T M Q = M_{11} - K_{21}^T K_{22}^{-1} M_{21} - M_{12} K_{22}^{-1} K_{21} + K_{21}^T K_{22}^{-1} M_{22} K_{22}^{-1} K_{21} \quad (8.104)$$

and

$$Q^T K Q = K_{11} - K_{12} K_{22}^{-1} K_{21} \quad (8.105)$$

These last expressions are frequently used to reduce the order of finite element vibration models in a systematic and consistent manner. Such model reduction schemes are used when a finite element model has coordinates (represented by $\mathbf{u}_2$) that do not contribute substantially to the response of the system. Model reduction can greatly simplify design and analysis problems under certain circumstances.

   If some of the masses in the system are negligible or zero, the preceding formulas can be used to reduce the order of the vibration problem simply by setting $M_{22} = 0$ in equation (8.104). This is essentially the model reduction technique referred to as *mass condensation*.

Example 8.6.1

   Consider a four-degree-of-freedom system finite element model with mass matrix

$$M = \frac{1}{420} \begin{bmatrix} 312 & 54 & 0 & -13 \\ 54 & 156 & 13 & -22 \\ 0 & 13 & 8 & -3 \\ -13 & -22 & -3 & 4 \end{bmatrix}$$

and stiffness matrix

$$K = \begin{bmatrix} 24 & -6 & 0 & 6 \\ -6 & 12 & -6 & -6 \\ 0 & -6 & 10 & 4 \\ 6 & -6 & 4 & 4 \end{bmatrix}$$

Note that this system is both dynamically and statically coupled. To remove the effect of the last two coordinates, the submatrices of equation (8.97) are easily identified as

$$M_{11} = \frac{1}{420} \begin{bmatrix} 312 & 54 \\ 54 & 156 \end{bmatrix} \qquad M_{12} = \frac{1}{420} \begin{bmatrix} 0 & -13 \\ 13 & -22 \end{bmatrix} = M_{21}^T$$

$$M_{22} = \frac{1}{420} \begin{bmatrix} 8 & -3 \\ -3 & 4 \end{bmatrix} \qquad K_{22} = \begin{bmatrix} 10 & 4 \\ 4 & 4 \end{bmatrix}$$

$$K_{11} = \begin{bmatrix} 24 & -6 \\ -6 & 12 \end{bmatrix} \qquad K_{12} = \begin{bmatrix} 0 & 6 \\ -6 & -6 \end{bmatrix} = K_{21}^T$$

Using equations (8.104) and (8.105) yields

$$Q^T M Q = \begin{bmatrix} 1.012 & 0.198 \\ 0.198 & 0.236 \end{bmatrix}$$

$$Q^T K Q = \begin{bmatrix} 9 & 3 \\ 3 & 3 \end{bmatrix}$$

These matrices form the resulting reduced-order model of the structure.

□

PROBLEMS

Section 8.1 (Problems 8.1 Through 8.7)

8.1. Consider the one-element model of a bar discussed in Section 8.1. Calculate the finite element of the bar for the case that it is free at both ends rather than clamped.

8.2. Calculate the natural frequencies of the free–free bar of Problem 8.1. To what does the first natural frequency correspond? How do these values compare with the exact values obtained from the methods of Chapter 6?

8.3. Consider the system of Figure P8.3, consisting of a spring connected to a clamped–free bar. Calculate the finite element model and discuss the accuracy of the frequency prediction of this model by comparing it with the method of Chapter 6.

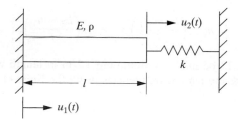

Figure P8.3   A one-element model of a cantilevered bar connected to a spring.

8.4. Consider a clamped–free bar with a force $f(t)$ applied in the axial direction at the free end as illustrated in Figure P8.4. Calculate the equations of motion using a single-element finite element model.

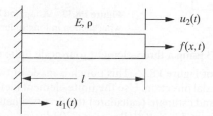

Figure P8.4   A cantilevered bar with an externally applied axial force.

8.5. Compare the solution of a cantilevered bar modeled as a single finite element with that of the distributed-parameter method summarized in Figure 8.1 truncated at three modes by calculating (a) $u(x, t)$ and (b) $u(l/2, t)$ for a 1-m aluminum beam at $t = 0.1, 1,$ and 10 s using both methods. Use the initial condition $u(x, 0) = 0.1x$ m and $u_t(x, 0) = 0$.

8.6. Repeat Problem 8.5 using a five-mode model. Can you draw any conclusions?

8.7. Repeat Problem 8.5 using only the first mode in the series solution and the initial condition $u(x, 0) = 0.1 \sin(\pi x/2l), u_t(x, 0) = 0$. For this initial condition, the first mode is exact. Why?

Section 8.2 (Problems 8.8 Through 8.20)

8.8. Consider the bar of Figure P8.3 and model the bar with two elements. Calculate the frequencies and compare them with the solution obtained in Problem 8.3. Assume material properties of aluminum, a cross-sectional area of 1 m, and a spring stiffness of 1×10^6 N/m.

8.9. Repeat Problem 8.8 with a three-element model. Calculate the frequencies and compare them with those of Problem 8.8.

8.10. Consider Example 8.2.2. Repeat this example with node 2 moved to $l/2$ so that the mesh is uniform. Calculate the natural frequencies and compare them to those obtained in the example. What happens to the mass matrix?

8.11. Compare the frequencies obtained in Problem 8.10 with those obtained in Section 8.2 using three elements.

8.12. As mentioned in the text, the usefulness of the finite element method rests in problems that cannot readily be solved in closed form. To this end, consider a section of an air frame sketched in Figure P8.13 and calculate a two-element finite model of this structure (i.e., find M and K) for a bar with

$$A(x) = \frac{\pi}{4}\left[h_1^2 + \left(\frac{h_2 - h_1}{l}\right)^2 x^2 + 2h_1\left(\frac{h_2 - h_1}{l}\right)x \right]$$

8.13. Let the bar in Figure P8.13 be made of aluminum 1.2 m in length with $h_1 = 20$ cm and $h_2 = 10$ cm. Calculate the natural frequencies using the finite element model of Problem 8.12.

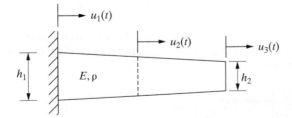

Figure P8.13 A tapered bar model of a wing section in longitudinal vibration.

8.14. Repeat Problems 8.12 and 8.13 using a three-element four-node finite element model.

8.15. Consider the machine punch of Figure P8.15. This punch is made of two materials and is subject to an impact in the axial direction. Use the finite element method with two elements to model this system and estimate (calculate) the first two natural frequencies. Assume $E_1 = 8 \times 10^{10}$ Pa, $E_2 = 2.0 \times 10^{11}$ Pa, $\rho_1 = 7200$ kg/m³, $\rho_2 = 7800$ kg/m³, $l = 0.22$ m, $A_1 = 0.008$ m², and $A_2 = 0.0008$ m².

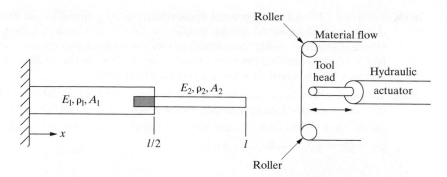

Figure P8.15   A bar made of two materials of two sizes used for a punch. Material passes around the rollers. At appropriate points on the material, the roller stops and the toolhead is actuated in the x direction and impacts the bar, causing a hole to be punched in the material. The punch itself (E_2) is made of hardened steel while the base is made of cast iron (E_1).

8.16. Recalculate the frequencies of Problem 8.15 assuming that it is made entirely of one material and size (i.e., $E_1 = E_2$, $\rho_1 = \rho_2$, and $A_1 = A_2$), say steel, and compare your results to those of Problem 8.15.

8.17. A bridge support column is illustrated in Figure P8.17. The column is made of concrete with a cross-sectioned area defined by $A(x) = A_0 e^{-x/l}$, where A_0 is the area of the column at ground. Consider this pillar to be cantilevered (i.e., fixed) at ground level and to be excited sinusoidally at its tip in the longitudinal direction due to traffic over the bridge. Calculate a single-element finite element model of this system and compute its approximate natural frequency.

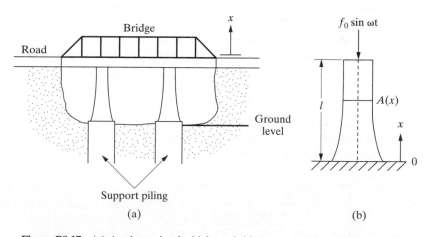

Figure P8.17   (a) A schematic of a highway bridge over a ravine. Traffic over the bridge causes a harmonic motion in the up-and-down direction (labeled x here). (b) The vibration of the column holding up the bridge. The pillars are modeled as bars of odd cross section.

8.18. Redo Problem 8.17 using two elements. What would happen if the "traffic" frequency corresponds with one of the natural frequencies of the support column?

8.19. Problems 8.17 and 8.18 represent approximations. As pointed out in Problem 8.18, it is important to know the natural frequencies of this column as precisely as possible. Hence consider modeling this column as a uniform bar of average cross section, calculate the first few natural frequencies, and compare them to the results in Problems 8.17 and 8.18. Which model do you think is closest to reality?

8.20. Torsional vibration can also be modeled by finite elements. Referring to Figure P8.20, calculate a single-element mass and stiffness matrix for the torsional vibration following the steps of Section 8.1. (*Hint:* $\theta(x, t) = c_1(t)\theta + c_2(t)$, $T(t) = \frac{1}{2} \int_0^1 \rho I_p \big[\theta_t(x, t) \big]^2 dx$, and $V(t) = \frac{1}{2} \int_0^1 GI_p \big[\theta_x(x, t) \big]^2 dx$.)

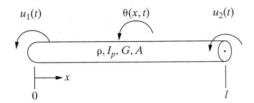

$u_1(t)$ $\theta(x, t)$ $u_2(t)$

ρ, I_p, G, A

x

0 l

Figure P8.20 Coordinate system used for finite element analysis of torsional vibration.

Section 8.3 (Problems 8.21 Through 8.33)

8.21. Use equations (8.47) and (8.46) to derive equation (8.48) and hence make sure that the author and reviewer have not cheated you.

8.22. It is instructive, though tedious, to derive the beam element deflection given by equation (8.49). Hence derive the beam shape functions.

8.23. Using the shape functions of Problem 8.22, calculate the mass and stiffness matrices given by equations (8.53) and (8.56). Although tedious, this involves only simple integration of polynomials in x.

8.24. Calculate the natural frequencies of the cantilevered beam given in equation (8.69) using $l = 1$ m and compare your results with those listed in Table 6.6.

8.25. Calculate the finite element model of a cantilevered beam 1.2 m in length using three elements. Calculate the natural frequencies and compare them to those obtained in Problem 8.23 and with the exact values listed in Table 6.6.

8.26. Consider the cantilevered beam of Figure P8.26 attached to a lumped spring–mass system. Model this system using a single finite element and calculate the natural frequencies. Assume $m = (\rho A l)/420$.

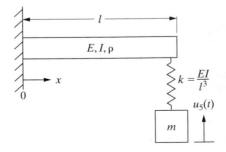

l

E, I, ρ

x

0

$k = \dfrac{EI}{l^3}$

$u_5(t)$

m

Figure P8.26 A cantilevered beam with a spring–mass system attached to its end.

8.27. Repeat Problem 8.26 using two finite elements for the beam and compare the frequencies.

8.28. Calculate the natural frequencies of a clamped–clamped beam for the physical parameters $l = 1.2$ m, $E = 2.1 \times 10^{11}$ N/m^2, $\rho = 7800$ kg/m^3, $I = 10 - 6$ m^4, and $A = 10^{-2}$ m^2, using the beam theory of Chapter 6 and a four-element finite element model of the beam.

8.29. Repeat Problem 8.28 with two elements and compare the frequencies with the four-element model. Calculate the frequencies of a clamped–clamped beam using one element. Any comment?

8.30. Estimate the first natural frequency of a clamped–simply supported beam. Use a single finite element.

8.31. Consider the stepped bar of Figure P8.31 clamped at each end. Both pieces are made of aluminum. Use two elements, one for each step, and calculate the natural frequencies.

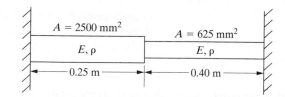

Figure P8.31   A clamped two-step aluminum beam.

8.32. Use a two-element model of nonuniform length to estimate the first few natural frequencies of a clamped–clamped beam. Use the spacing indicated in Figure P8.32. Compare the result to the actual frequencies and to those of Problems 8.28 and 8.29.

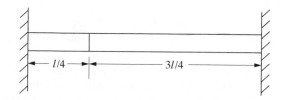

Figure P8.32   A clamped beam modeled with two nonuniform finite elements.

8.33. Calculate the first natural frequency of a clamped–pinned beam using first one, then two elements.

Section 8.4 (Problems 8.34 Through 8.43)

8.34. Refer to the tapered bar of Figure P8.13. Calculate a lumped-mass matrix for this system and compare it to the solution of Problem 8.13. Since the beam is tapered, be careful how you divide up the mass.

8.35. Calculate and compare the natural frequencies obtained for a tapered bar by using first, the consistent-mass matrix (Problem 8.12), and second, the lumped-mass matrix (Problem 8.34).

8.36. Consider again the machine punch of Problem 8.16 and Figure P8.15. Calculate the natural frequencies of this system using a lumped-mass matrix and compare the results to those obtained with the consistent-mass matrix.

8.37. Consider again the bridge support of Figure P8.17 discussed in connection with Problem 8.17. Develop a four-element finite element model of this structure using a lumped-mass approximation and calculate the natural frequencies. Use constant area elements.

8.38. Consider the torsional vibration problem illustrated in Figure P8.20 and discussed in Problem 8.20. Calculate a lumped-mass matrix for the single element.

8.39. Estimate the first three natural frequencies of a clamped–free bar of length l in torsional vibration by using a lumped-mass model and four elements.

8.40. Calculate the natural frequencies of a pinned–pinned beam of length l using one element and the consistent-mass matrix of equation (8.73).

8.41. Calculate the natural frequencies of a pinned–pinned beam of length l using one element and the lumped-mass matrix of equation (8.73). Compare your results to those obtained with a consistent-mass matrix of Problem 8.40.

8.42. Calculate a three-element finite element model of a cantilevered beam (see Problem 8.25) using a lumped mass that includes rotational inertia. Also calculate the system's natural frequencies and compare them with those obtained with a consistent-mass matrix of Problem 8.25 and with the values obtained by the methods of Chapter 6.

8.43. Repeat Problem 8.42 using a lumped-mass matrix that neglects the rotational degree of freedom. Discuss any problems you encounter when trying to solve the related eigenvalue problem.

Section 8.5 (Problems 8.44 Through 8.49)

8.44. Derive a consistent-mass matrix for the system of Figure 8.9. Compare the natural frequencies of this system with those calculated with the lumped-mass matrix computed in Section 8.5.

*8.45. Consider the two-beam system of Figure P8.45. Use VTB8_1 to create a two-element, rod/beam element model and compute the first three natural frequencies. Use $A = 0.0005\,\text{m}^2$, $I = 1.33 \times 10^{-8}\,\text{m}^4$, and the properties of aluminum. Assume that nodes 1 and 3 are clamped.

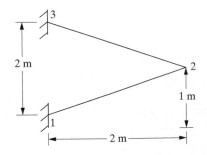

Figure P8.45   A clamped two-element beam system.

*8.46. Follow the procedure of Problem 8.45 using two elements for each beam. Compare the natural frequencies and mode shapes of the four-element model produced here to those of the two-element model of Problem 8.45. State which model is better and why.

8.47. Determine a finite element model of the three-bar truss of Figure P8.47 using a lumped-mass matrix.

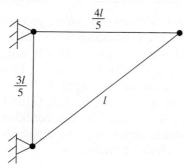

Figure P8.47   A pinned three-element truss.

8.48. Determine a finite element model for the three-bar truss of Figure P8.47 using a consistent-mass matrix.

8.49. Compare the frequencies obtained for the system of Problem 8.48 with those of Figure P8.47.

Section 8.6 (Problems 8.50 Through 8.54)

8.50. Consider the machine punch of Figure P8.15. Recalculate the fundamental natural frequency by reducing the model obtained in Problem 8.16 to a single-degree-of-freedom using Guyan reduction.

8.51. Compute a reduced-order model of the three-element model of a cantilevered bar given in Example 8.3.2 by eliminating $\mathbf{u}_2$ and $\mathbf{u}_3$ using Guyan reduction. Compare the frequencies of each model to those of the distributed model given in Window 8.1.

8.52. Consider the system defined by the matrices

$$M = \begin{bmatrix} 2 & 0 & 0 & 0 \\ 0 & 0 & 0 & 0 \\ 0 & 0 & 2 & 0 \\ 0 & 0 & 0 & 0 \end{bmatrix} \qquad K = \begin{bmatrix} 20 & -1 & 0 & 0 \\ -1 & 20 & -3 & 0 \\ 0 & -3 & 20 & -17 \\ 0 & 0 & -17 & 17 \end{bmatrix}$$

Use mass condensation to reduce this to a two-degree-of-freedom system with a nonsingular-mass matrix.

8.53. Recall the punch-press problem modeled in Figure 4.28 and treated in Example 4.8.3. The mass and stiffness matrices are given by

$$M = \begin{bmatrix} 0.4 \times 10^3 & 0 & 0 \\ 0 & 2.0 \times 10^3 & 0 \\ 0 & 0 & 8.0 \times 10^3 \end{bmatrix} \quad K = \begin{bmatrix} 30 \times 10^4 & -30 \times 10^4 & 0 \\ -30 \times 10^4 & 38 \times 10^4 & -8 \times 10^4 \\ 0 & -8 \times 10^4 & 88 \times 10^4 \end{bmatrix}$$

Recalling that the only external force acting on the machine is at the $x_1(t)$ coordinate, reduce this to a single-degree-of-freedom system using Guyan reduction to remove x_2 and x_3. Compare this single frequency with those of Example 4.8.3.

8.54. Consider the beam example given in Example 7.6.2. Using the values given there (an aluminum beam, 0.5128 m × 25.5 mm × 3.2 mm, $E = 6.9 \times 10^{10}$ N/m^2, $\rho = 2715$ kg/m^3, $A = 8.16$ m^2, and $I = 6.96 \times 10^{-11}$ m^4), compute the first four natural frequencies as accurately as possible and compare them to both the analytical values and the measured values.

MATLAB ENGINEERING VIBRATION TOOLBOX

If you have not yet used the *Engineering Vibration Toolbox* program, return to the end of Chapter 1 or Appendix G for a brief introduction to MATLAB. The finite element modeling method introduced in this chapter is ideally suited for computer implementation. Toolbox folder/directory VTB_8 contains a finite element program. This interactive program calls for the user to input node locations, followed by the physical parameters between the nodes (i.e., E, A, I, G, and ρ). The code is based on a two-dimensional Timoshenko beam. The file contains an example of how to use the program for an aluminum beam with five elements. The following problems should help clarify the finite element procedure.

TOOLBOX PROBLEMS

TB8.1. Use the file VTB_8 to solve for the frequencies of a cantilevered aluminum bar using five elements. Compare the frequencies to those of the analytical solution of the distributed-parameter model.

TB8.2. Use file VTB_8 to solve for the natural frequencies of a pinned–pinned beam of Window 8.3 using 10 elements. Compare your results with the analytical solution.

TB8.3. Recalculate the frequencies of the machine punch of Figure P8.15 using four elements for each bar. How does your result compare with those obtained in Problem 8.16?

TB8.4. Write a MATLAB file to perform mass condensation.

TB8.5. Use file VTB_8 to calculate the consistent-mass matrix of the two-bar truss system of Section 8.5. Compute the frequencies of this system and compare them to those calculated in Section 8.5.

TB8.6. Use VTB_8 to create a finite element model of the bridge in Figure PTB8.6. The bridge is made of steel, and the beam cross sections are 0.1 m wide and 0.15 m high. The x and y displacements of nodes 1 and 5 are constrained to be zero. Find the first seven natural frequencies and mode shapes. Removing the center diagonal members, find the first seven natural frequencies and mode shapes again. What is the most significant result of removing the center diagonal members? Can you explain why the seventh natural frequency and mode shape do not change?

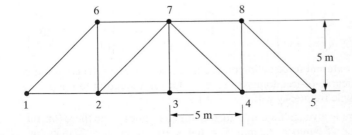

Figure PTB8.6   A 13-element bridge model.

A Complex Numbers and Functions

Complex numbers occur naturally in vibration analysis from the solution of differential equations through their algebraic characteristic equations. In particular, solution of the damped single-degree-of-freedom system given by equation (1.36) is dependent on the values of λ satisfying the algebraic equation

$$m\lambda^2 + c\lambda + k = 0 \tag{A.1}$$

This is the familiar quadratic equation that has the solution

$$\lambda = -\frac{c}{2m} \pm \frac{1}{2m} \sqrt{c^2 - 4mk} \tag{A.2}$$

The roots given in equation (A.2) are complex valued if $c^2 - 4mk < 0$ (the underdamped case). In this case the formula given in equation (A.2) calls for the square root of a negative number. Stated algebraically, equation (A.1) in the underdamped case calls for a number j (sometimes denoted i) such that

$$j^2 = -1 \tag{A.3}$$

or symbolically, the "imaginary number" j is defined to be

$$j = \sqrt{-1} \tag{A.4}$$

This representation allows the expression of the two roots of equation (A.1) as the two pairs of real numbers

$$\left(-\frac{c}{2m}, -\frac{1}{2m}\sqrt{4mk - c^2}\right) \quad \text{and} \quad \left(-\frac{c}{2m}, \frac{1}{2m}\sqrt{4mk - c^2}\right) \tag{A.5}$$

which are written as

$$-\frac{c}{2m} - \frac{1}{2m}\sqrt{4mk - c^2}\,j \quad \text{and} \quad -\frac{c}{2m} + \frac{1}{2m}\sqrt{4mk - c^2}\,j \tag{A.6}$$

where $4mk - c^2 > 0$.

With the preceding as motivation, a general complex number, x, is written as $x = a + bj$. The real number a is referred to as the *real part* of the number x and the real number b is referred to as the *imaginary part* of the number x. Such complex-valued numbers are represented in a plane, called the complex plane, as illustrated in Figure A.1. The notation Re x, is used to denote the real part of the number x (i.e., Re $x = a$) and Im x is used to denote the value of the imaginary part of x (i.e., Im $x = b$).

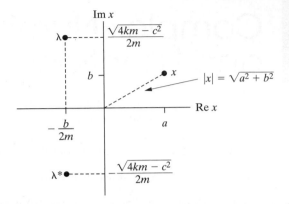

Figure A.1   A complex plane used to represent a complex
number x and the roots of equation (A.1): $-(b/2m) \pm (1/2m)$
$\sqrt{4km - c^2}j$. This plot is called an Argand diagram.

The complex numbers $a + bj$ and $a - bj$ are called *conjugates* of each other.
The notation $\bar{x}$, or x^*, is used to denote the conjugate of the complex number x.
That is, if $x = a + bj$, then $\bar{x} = x^* = a - bi$. The roots of equation (A.1) appear
as a complex conjugate pair in the underdamped case. Another useful property of
complex numbers is their *absolute value* or *modulus*, denoted $|x|$. The modulus of a
complex number is the distance from the origin in Figure A.1 to the point x:

$$|x| = |a + bj| = \sqrt{a^2 + b^2} \tag{A.7}$$

The modulus is illustrated in Figure A.1, as is the conjugate of the root λ. Note that
conjugate pairs of numbers fall on the same vertical lines as they have the same real
part. Also notice that a complex number and its conjugate both have the same mod-
ulus (i.e., $|x| = |x^*|$).

Complex numbers may be manipulated using real arithmetic following rules
similar to those for vectors. Addition of two complex numbers is defined simply by
adding the real parts and imaginary parts as separate entities. In particular, if $x = a + bj$ and $y = c + dj$, then

$$x + y = (a + c) + (b + d)j \tag{A.8}$$

In a consistent fashion, if β is a real number, then

$$\beta x = \beta a + \beta bj \tag{A.9}$$

expresses the product of a real number and a complex number.

Multiplication of two complex numbers is defined by starting with two basic
definitions:

$$(a)(j) = aj$$

where a is real, and

$$(j)(j) = -1$$

Then the product of two general complex numbers x and y becomes

$$(x)(y) = (a + bj)(c + dj) = (ac - bd) + (ad + bc)j \qquad \text{(A.10)}$$

so that $\text{Re}(xy) = (ac - bd)$ and $\text{Im}(xy) = (ad + bc)$. The product of x and its conjugate $\bar{x}$ becomes a real number, that is,

$$x\bar{x} = (a + bj)(a - bj) = a^2 + b^2$$

This is consistent with definitions of the modulus of x defined by equation (A.7) with

$$|x|^2 = a^2 + b^2 = xx^* \qquad \text{(A.11)}$$

which is a real number.

 With addition defined (hence subtraction) and multiplication defined, it is important to define the division of one complex number by another. First note that the multiplicative identity of a complex number is simply the real number 1 (i.e., $1x = x$ for any complex number x). The inverse of the complex number $x = a + bj$ is given by

$$(a + bj)^{-1} = \frac{1}{a + bj} = \frac{1}{a^2 + b^2}(a - bj) \qquad \text{(A.12)}$$

provided that $x = a + bj \neq 0$. Note that

$$(a + bj)^{-1}(a + bj) = \frac{1}{a^2 + b^2}(a - bj)(a + bj) = 1$$

the multiplicative inverse. Equation (A.12) allows the formulation of division of two complex numbers. To see this, consider the division of y by x:

$$\frac{y}{x} = \frac{c + dj}{a + bj} = (c + dj)(a + bj)^{-1} \qquad \text{(A.13)}$$

Invoking equation (A.12) for the inverse of x yields

$$\frac{y}{x} = (c + dj)\frac{1}{a^2 + b^2}(a - bj) = \frac{1}{a^2 + b^2}[(ac + bd) + (ad - cb)j] \quad \text{(A.14)}$$

for $x \neq 0$. Note that $x = 0$ if and only if both a and b are zero.

 Note that the imaginary number j was used in equation (A.6) to manipulate the square root of a negative number. Specifically, for $c^2 - 4mk < 0$, the $4mkx - c^2 > 0$ and the discriminant in equation (A.2) becomes

$$\sqrt{c^2 - 4mk} = \sqrt{(-1)(4mk - c^2)} = (\sqrt{4mk - c^2})(\sqrt{-1})$$

$$= (\sqrt{4mk - c^2})j \qquad \text{(A.15)}$$

which yields the complex number representation of the roots of the characteristic equation for a single-degree-of-freedom underdamped system. Note also from the definition of addition that the real and imaginary parts of a complex number are determined by simple arithmetic. If $x = a + bj$, then

$$\text{Re } x = a = \frac{x + \bar{x}}{2}$$

$$\text{Im } x = b = \frac{x - \bar{x}}{2j} \tag{A.16}$$

Applying these formulas to the roots λ of the characteristic equation for an underdamped single-degree-of-freedom system yields

$$\text{Re } \lambda = -\frac{c}{2m}$$

$$\text{Im } \lambda = -\frac{\sqrt{4mk - c^2}}{2m} \tag{A.17}$$

where λ satisfies equation (A.1).

A complex number can also be represented in terms of a polar coordinate system as illustrated in Figure A.2. Here θ is the angle between the line between the origin and the point (a, b). Note that $r = |x| = \sqrt{a^2 + b^2}$ and $\theta = \tan^{-1}(b/a)$. With this in mind, x can be written as

$$x = a + bj = r(\cos\theta + j\sin\theta) \tag{A.18}$$

This polar representation can be used to write the exponential as

$$e^{jt} = \cos t + j\sin t \tag{A.19}$$

equation (A.19) can be manipulated to yield

$$\cos t = \frac{1}{2}(e^{jt} + e^{-jt})$$

$$\sin t = \frac{1}{2j}(e^{jt} - e^{-jt}) \tag{A.20}$$

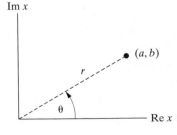

Figure A.2 An Argand plot illustrating the polar representation of a complex number $x = a + bj$.

These equations are referred to as Euler formulas and can be derived by writing the exponential e^{jt} as a power series and recalling the series expressions for $\sin t$ and $\cos t$. The hyperbolic functions can be written as

$$\cosh t = \frac{1}{2}(e^t + e^{-t})$$

$$\sinh t = \frac{1}{2}(e^t - e^{-t}) \tag{A.21}$$

which are comparable to the Euler formulas for the trigonometric functions.

The foregoing algebraic formulation for complex numbers can be extended to functions of a complex variable. An example of a function of a complex variable is given by equation (A.6) [i.e., $f(x) = |x| = (a^2 + b^2)^{1/2}$]. This is a real-valued function of a complex variable since x is complex and $f(x)$ is real. In general, however, a function of a complex number will also be complex. A theory of limits, differentiation, and integration can be defined, with a few cautions, following that of functions of a real variable. One major difference is that the real and imaginary parts of a complex function may have continuous derivatives of all orders at a point, yet the function itself may not be differentiable. Hence one needs to proceed with caution in examining the calculus of complex functions.

There are several graphical representations of a complex function that are useful in vibration analysis. First note that if x is a complex variable, then $f(x)$ is also potentially complex and will be of the general form

$$f(x) = u(x) + v(x)j \tag{A.22}$$

where $u(x)$ and $v(x)$ are real-valued functions. Multiplication of complex functions follows that of complex numbers as given by equation (A.10). The conjugate function is just

$$\bar{f}(x) = u(x) - v(x)j \tag{A.23}$$

and all of the formulas for arithmetic of complex numbers developed previously apply to the arithmetic of complex functions. In particular,

$$|f(x)| = \sqrt{f\bar{f}} = \sqrt{u^2(x) + v^2(x)}$$

and for values of x such that $f(x) \neq 0$,

$$\frac{1}{f(x)} = \frac{1}{u^2(x) + v^2(x)}[u(x) - v(x)j] \tag{A.24}$$

which satisfies the relation $f(x)[1/f(x)] = 1$.

The graphical representation of a complex function becomes difficult because both the argument and the function require two dimensions to represent graphically. One approach is to plot $u(x)$ versus $v(x)$, as indicated in Figure A.3 for several

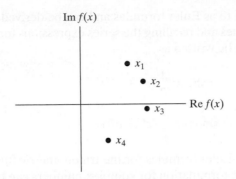

Figure A.3 The imaginary part of a complex function plotted versus the real part for several values of the variable x. Such plots are called Nyquist diagrams.

different values of the complex variable x. These are called Nyquist plots and are used extensively in analyzing vibration measurement data. Another method of plotting complex functions is to examine the magnitude and phase separately using the polar form suggested in Figure A.2. In the case that $|f(x)|$ is plotted versus b, the imaginary part of x, the plot is called a Bode magnitude plot. Similarly, a plot of $\theta(x)$ is $\tan^{-1}[v(x)/u(x)]$ versus the imaginary part of x, called a Bode phase plot. These plots are also used extensively in analyzing vibration test data as discussed in Sections 1.6 and 7.4.

B Laplace Transforms

An integral transform is the procedure of integrating the time dependence of a function into becoming a function of an alternative variable, or parameter, which can be manipulated algebraically. A common integral transform is the Laplace transform. Laplace transforms are viewed here as a method of solving differential equations of motion by reducing the computation to that of integration and algebraic manipulation. Transforms provide both an alternative solution technique for vibration problems and an important analytical tool in the analysis and measurement of vibrating systems.

The definition of a Laplace transform of the function $f(t)$ is

$$L[f(t)] = F(s) = \int_0^\infty f(t)e^{-st}dt \tag{B.1}$$

for an integrable function $f(t)$ such that $f(t) = 0$ for $t < 0$. The variable s is complex valued. The Laplace transform changes the domain of the function from the positive real number line (t) to the complex number plane (s). The integration in the Laplace transform changes differentiation into multiplication. From the definition of the Laplace transform, it is a simple matter to see that the procedure is linear. Thus the transform of a linear combination of two functions is the same linear combination of the transform of these functions. The Laplace transform of various functions can be calculated in closed form by using equation (B.1). In addition, the Laplace transform of a derivative of an arbitrary function can easily be calculated in symbolic form. In particular, the Laplace transform of $\dot{x}(t)$ is just

$$L[\dot{x}(t)] = sX(s) - x(0) \tag{B.2}$$

where the capital X denotes a transformed version of $x(t)$. Similarly,

$$L[\ddot{x}(t)] = s^2X(s) - sx(0) - \dot{x}(0) \tag{B.3}$$

Here $x(0)$ and $\dot{x}(0)$ are the initial values of the function $x(t)$. Note that in the transformed domain, often called the s-domain, differentiating a variable $X(s)$ corresponds to simple multiplication [i.e., $sX(s)$], and integration in the time domain corresponds to dividing by s in the transform domain.

Table B.1 lists some common functions of time preceded by their Laplace transforms as calculated using equation (B.1) for the case that all of the initial conditions are set to zero. The table provides a method of quickly finding the Laplace transforms given a particular function of time by reading from right to left. However, reading the table from left to right provides the ability to determine a

TABLE B.1 PARTIAL LIST OF FUNCTIONS AND THEIR LAPLACE TRANSFORMS
WITH ZERO INITIAL CONDITIONS AND $t > 0$

$F(s)$	$f(t)$
(1) 1	$\delta(t_0)$ unit impulse at t_0
(2) $\dfrac{1}{s}$	1, unit step
(3) $\dfrac{1}{s+a}\left(\dfrac{1}{s-a}\right)$	$e^{-at}\ (e^{at})$
(4) $\dfrac{1}{(s+a)(s+b)}$	$\dfrac{1}{b-a}(e^{-at}-e^{-bt})$
(5) $\dfrac{\omega}{s^2+\omega^2}$	$\sin \omega t$
(6) $\dfrac{s}{s^2+\omega^2}$	$\cos \omega t$
(7) $\dfrac{1}{s(s^2+\omega^2)}$	$\dfrac{1}{\omega^2}(1-\cos \omega t)$
(8) $\dfrac{1}{s^2+2\zeta\omega s+\omega^2}$	$\dfrac{1}{\omega_d}e^{-\zeta\omega t}\sin \omega_d t,\ \zeta<1,\ \omega_d=\omega\sqrt{1-\zeta^2}$
(9) $\dfrac{\omega^2}{s(s^2+2\zeta\omega s+\omega^2)}$	$1-\dfrac{\omega}{\omega_d}e^{-\zeta\omega t}\sin(\omega_d t+\phi),\ \phi=\cos^{-1}\zeta,\ \zeta<1$
(10) $\dfrac{1}{s^n}$	$\dfrac{t^{n-1}}{(n-1)!},\ n=1,2\ldots$
(11) $\dfrac{n!}{(s-\omega)^{n+1}}$	$t^n e^{\omega t},\ n=1,2\ldots$
(12) $\dfrac{1}{s(s+\omega)}$	$\dfrac{1}{\omega}(1-e^{-\omega t})$
(13) $\dfrac{1}{s^2(s+\omega)}$	$\dfrac{1}{\omega^2}(e^{-\omega t}+\omega t-1)$
(14) $\dfrac{\omega}{s^2-\omega^2}$	$\sinh \omega t$
(15) $\dfrac{s}{s^2-\omega^2}$	$\cosh \omega t$
(16) $\dfrac{1}{s^2(s^2+\omega^2)}$	$\dfrac{1}{\omega^3}(\omega t-\sin \omega t)$
(17) $\dfrac{1}{(s^2+\omega^2)^2}$	$\dfrac{1}{2\omega^3}(\sin \omega t-\omega t\cos \omega t)$
(18) $\dfrac{s}{(s^2+\omega^2)^2}$	$\dfrac{t}{2\omega}\sin \omega t$
(19) $\dfrac{s^2-\omega^2}{(s^2+\omega^2)^2}$	$t\cos \omega t$

TABLE B.1 CONTINUED

$F(s)$	$f(t)$
(20) $\dfrac{\omega_1^2 - \omega_2^2}{(s^2 + \omega_1^2)(s^2 + \omega_2^2)}$	$\dfrac{1}{\omega_2} \sin \omega_2 t - \dfrac{1}{\omega_1} \sin \omega_1 t$
(21) $\dfrac{(\omega_1^2 - \omega_2^2)s}{(s^2 + \omega_1^2)(s^2 + \omega_2^2)}$	$\cos \omega_2 t - \cos \omega_1 t$
(22) $\dfrac{\omega}{(s + a)^2 + \omega^2}$	$e^{-at} \sin \omega t$
(23) $\dfrac{s + a}{(s + a)^2 + \omega^2}$	$e^{-at} \cos \omega t$
(24) $F(s - a)$	$e^{at} f(t)$
(25) $e^{-as} F(s)$	$f(t - a) \, \Phi(t - a)$

function, $x(t)$, from its Laplace transform $X(s)$. This gives rise to the concept of an inverse Laplace transform, denoted L^{-1}, which is formally defined by

$$L^{-1}[X(s)] = x(t) \tag{B.4}$$

which can be computed by using Table B.1.

The procedure for using Laplace transforms to solve equations of motion expressed as an inhomogeneous ordinary differential equation is to take the Laplace transform of both sides of the equation, treating the time derivatives symbolically using equations (B.2) and (B.3) and using Table B.1 to compute the Laplace transform of the driving force. This renders an algebraic equation in the variable $X(s)$, which is easily solved by simple manipulation. The inverse Laplace transform is applied to the resulting expression for $X(s)$ by using Table B.1 again, resulting in the time response.

As an example of using Laplace transforms to solve a homogeneous differential equation, consider the undamped single-degree-of-freedom system described by

$$\ddot{x}(t) + \omega_n^2 x(t) = 0, \quad x(0) = x_0, \quad \dot{x}(0) = v_0 \tag{B.5}$$

Taking the Laplace transform of $\ddot{x} + \omega_n^2 x = 0$ results in

$$s^2 X(s) - s x_0 - v_0 + \omega_n^2 X(s) = 0 \tag{B.6}$$

by direct application of equation (B.3) and the linear nature of the Laplace transform. Algebraically solving equation (B.6) for $X(s)$ yields

$$X(s) = \frac{x_0 + s v_0}{s^2 + \omega_n^2} \tag{B.7}$$

Using $L^{-1}[X(s)] = x(t)$ and entries (6) and (5) of Table B.1 yields that the solution is

$$x(t) = x_0 \cos \omega_n t + \frac{v_0}{\omega_n} \sin \omega_n t \tag{B.8}$$

The same procedure works for calculating the forced response. However, in the forced response, calculating the algebraic solution for $X(s)$ often results in quotients of polynomials in s. These polynomial ratios may not be found in tables directly, but such quotients may be resolved into simple terms by using the method of *partial fractions*.

The method of partial fractions is one of finding unknown coefficients of terms used in combining simple fractions by computing the lowest common denominator. For instance, consider a function $X(s)$ given by

$$X(s) = \frac{s + 1}{s(s + 2)} \tag{B.9}$$

which does not appear in Table B.1, rendering it difficult to invert. This quotient of polynomials can be written as

$$\frac{s + 1}{s(s + 2)} = \frac{A}{s} + \frac{B}{s + 2} \tag{B.10}$$

where A and B are unknown constant factors. Clearing the fractions in equation (B.10) yields

$$s + 1 = (A + B)s + 2A \tag{B.11}$$

after a little manipulation. Since the coefficients on each side of the equality must match, equation (B.11) implies that

$$A + B = 1 \quad \text{(from the coefficient of } s)$$

$$2A = 1 \quad \text{(from the coefficient of } s^0)$$

so that $A = B = \dfrac{1}{2}$. Thus $X(s)$ may be written as

$$X(s) = \frac{1}{2s} + \frac{1}{2} \frac{1}{s + 2} \tag{B.12}$$

Inverting $X(s)$ is easily performed by using entries (2) and (3) in Table B.1. This yields that

$$x(t) = \frac{1}{2} \left(1 + e^{-2t} \right) \tag{B.13}$$

providing the desired time response.

The partial fraction method requires that repeated linear factors and quadratic factors have additional coefficients. For example, if the polynomial in the denominator has a repeated linear factor such as $(s + \omega)^2$, an additional term is needed. To see this, suppose that $X(s) = \omega^2/[s(s + \omega)^2]$; then the partial fraction expansion becomes

$$\frac{\omega^2}{s(s + \omega)^2} = \frac{A}{s} + \frac{B}{s + \omega} + \frac{C}{(s + \omega)^2} \tag{B.14}$$

Multiplying both sides by $s(s + \omega)^2$ and solving for the coefficients A, B, and C by comparing coefficients of powers of s yields $A = 1$, $B = -1$, and $C = -\omega$, so that

$$\frac{\omega^2}{s(s + \omega)^2} = \frac{1}{s} - \frac{1}{s + \omega} - \frac{\omega}{(s + \omega)^2} \tag{B.15}$$

For quadratic factors such as $(s^2 + 2s + 5)$ the numerator of the expansion must contain a first-order polynomial element of the form $As + B$. For example, consider $X(s) = (s + 3)/[(s + 1)(s^2 + 2s + 5)]$. Its partial fraction expansion is

$$\frac{s + 3}{(s + 1)(s^2 + 2s + 5)} = \frac{A}{s + 1} + \frac{Bs + C}{s^2 + 2s + 5} \tag{B.16}$$

Multiplying by $(s + 1)(s^2 + 2s + 5)$ yields

$$s + 3 = (A + B)s^2 + (2A + B + C)s + (5A + C)$$

after grouping terms as coefficients of the powers of s. Comparing coefficients of s yields

$$A + B = 0, \quad 2A + B + C = 1, \quad 5A + C = 3$$

which has the solution $A = \dfrac{1}{2}, B = -\dfrac{1}{2}, C = \dfrac{1}{2}$. Thus the partial fraction expansion in equation (B.16) becomes

$$X(s) = \frac{1}{2(s + 1)} + \frac{1 - s}{2(s^2 + 2s + 5)} = \frac{1}{2(s + 1)} - \frac{s + 1 - 2}{2(s^2 + 2s + 5)}$$

$$= \frac{1}{2(s + 1)} - \frac{1}{2}\frac{s + 1}{(s + 1)^2 + 4} + \frac{2}{2(s^2 + 2s + 5)} \tag{B.17}$$

This can easily be inverted by using entries (3), (23), and (8) of Table B.1, respectively, to yield

$$x(t) = \frac{1}{2}e^{-t} - \frac{1}{2}e^{-t}\cos 2t + e^{-t}\sin 2t \tag{B.18}$$

since the denominator in the last fraction in equation (B.17) implies that $2\zeta\omega_n = 2$ and $\omega_n^2 = 5$. Hence $\omega_n = \sqrt{5}, \zeta = 1/\sqrt{5}$, and $\omega_d = 2$.

Matrix Basics

Some matrix definitions and manipulations useful in vibration analysis are summarized in this appendix. A matrix is an array of numbers (real or complex) arranged in rows and columns according to the following scheme:

$$
A = \begin{bmatrix}
a_{11} & a_{12} & \cdots & a_{1n} \\
a_{21} & a_{22} & \cdots & a_{2n} \\
\vdots & & & \\
a_{m1} & a_{m2} & & a_{mn}
\end{bmatrix}
$$

where A denotes the matrix as a single entity and a_{ij} denotes the element in the ij position (i.e., the element at the intersection of the ith row and jth column). Such a matrix A is said to be of order $m \times n$ and to have m rows and n columns. A majority of the matrices used in vibration analysis are square (i.e., $m = n$), having the same number of rows and columns, or are rectangular, consisting of a single row ($1 \times n$) or a single column ($n \times 1$), which is then called a row vector or column vector, respectively. However, in the analysis of vibration measurements, other size rectangular matrices occur.

Matrix arithmetic can be defined only if the matrices to be combined are of compatible dimensions. The sum of two matrices, A and B (i.e., $C = A + B$), is defined if A and B are of the same size (say $m \times n$), by

$$c_{ij} = a_{ij} + b_{ij} \tag{C.1}$$

for each value of i between 1 and m and each value of j between 1 and n. This definition of matrix addition is simply to create a new matrix C of the same size with elements formed by the sum of the corresponding elements of the two matrices A and B. Multiplication of a matrix by a scalar α is defined on a per-element basis (i.e., the product of the scalar α and the matrix A, denoted αA, has elements αa_{ij}).

The product of two matrices is only defined in a specific order for matrices of a compatible size. In particular, the matrix product $C = AB$ is defined in terms of the elements of the matrix C by

$$C_{ij} = \sum_{k=1}^{p} a_{ik} b_{kj} \tag{C.2}$$

where p is the common size of each matrix. Here, if the matrix A is $m \times p$, the matrix B must be $p \times n$ in order for the definition to be consistent. Thus not all matrices can be multiplied together. The matrix C resulting from equation (C.2) will be of the size $m \times n$.

One extremely common matrix product in vibration analysis is that of a square matrix times a column vector. A matrix is square if it has the same number of rows and columns (i.e., the special case $m = n$). If A is an $n \times n$ matrix and $\mathbf{x}$ is an $n \times 1$ vector, the product $\mathbf{y} = A\mathbf{x}$ is the $n \times 1$ vector with ith element given by

$$y_i = \sum_{j=1}^{n} a_{ij} x_j \tag{C.3}$$

here $i = 1, 2, \ldots, n$. Another useful matrix manipulation in vibration analysis is the concept of the transpose of a matrix. The transpose of the matrix A is formed from A by interchanging the rows and columns of the matrix A. In particular, if A has elements a_{ij}, the transpose of A, denoted A^T, has elements a_{ji}. If A is $n \times m$, then A^T will be $m \times n$. If $\mathbf{x}$ is a column vector ($n \times 1$), then $\mathbf{x}^T$ is a $1 \times n$ row vector. The transpose operation on a matrix (or a vector) satisfies the following rules:

$$(A + B)^T = A^T + B^T \tag{C.4}$$

$$(\alpha A)^T = \alpha A^T \tag{C.5}$$

$$(AB)^T = B^T A^T \tag{C.6}$$

$$(A)^T = A \tag{C.7}$$

where A and B are any two matrices for which the indicated operations can be defined and α is a scalar.

The product of two vectors can be defined in two different ways, both of which are useful in vibration analysis. First consider the product $\mathbf{y}^T\mathbf{x}$, where both $\mathbf{x}$ and $\mathbf{y}$ are $n \times 1$ column vectors. Following the definition given in equation (C.2), this product becomes

$$\mathbf{y}^T\mathbf{x} = \sum_{i=1}^{n} y_i x_i \tag{C.8}$$

This corresponds to the familiar "dot product" and is also called the *inner product* or *scalar product*. The phrase *scalar product* arises because the product $\mathbf{y}^T\mathbf{x}$ results in a scalar. The scalar product is useful in defining the magnitude of mode shapes in vibration analysis. A second product of two column vectors can be defined by taking the product of a column vector times a row vector. This product, called an *outer product*, produces an $n \times n$ matrix:

$$\mathbf{y}\mathbf{x}^T = \begin{bmatrix} y_1 x_1 & y_1 x_2 & \cdots & y_1 x_n \\ y_2 x_1 & y_2 x_2 & \cdots & y_2 x_n \\ \vdots & \vdots & \cdots & \vdots \\ y_n x_1 & y_n x_2 & \cdots & y_n x_n \end{bmatrix} \tag{C.9}$$

The outer product is extremely useful for analyzing vibration measurements.

Returning to the inner product, if the inner product of a vector **x** and itself is formed, the result leads to the interpretation of the length of a vector. The length of a vector is an example of a vector *norm*. In particular, if **x** is an $n \times 1$ vector, the *norm* of **x** is defined by

$$\|\mathbf{x}\| = (\mathbf{x}^T\mathbf{x})^{1/2} \tag{C.10}$$

If the norm of a vector is the number 1, the vector is called a *unit vector* and is said to be *normalized*. If the inner product of two vectors is zero, they are said to be *orthogonal*.

Often vectors occur in sets. If a set of $n \times 1$ vectors $\mathbf{x}_i$ satisfies the relation

$$\mathbf{x}_i^T\mathbf{x}_j = \begin{cases} 0 & i \neq j \\ 1 & i = j \end{cases} \tag{C.11}$$

for all vectors in the set, the unit vectors are each orthogonal to each other. Such a set of vectors is called *orthonormal*. Another useful property of sets of vectors is the concept of linear independence. A set of n, $n \times 1$ vectors $\mathbf{x}_i$ is *linearly dependent* if there exist n scalars $\alpha_1, \alpha_2, ..., \alpha_n$ that are not all zero, such that

$$\alpha_1\mathbf{x}_1 + \alpha_2\mathbf{x}_2 + ... + \alpha_n\mathbf{x}_n = \mathbf{0} \tag{C.12}$$

Essentially, this statement means that there is at least one vector in the set that can be written as a linear combination of some of the other vectors in the set. If such a set of scalars cannot be found, the set of vectors $\{\mathbf{x}_i\}$ is said to be *linearly independent*. Linearly independent vectors are very useful for expressing the solution of multiple-degree-of-freedom vibration problems. A familiar example of an orthonormal, linearly independent set of vectors are the three unit vectors $(\hat{\mathbf{i}}, \hat{\mathbf{j}}, \hat{\mathbf{k}})$ used in statics and dynamics to analyze motion and forces in three dimensions.

Vectors can be differentiated by defining the derivatives in terms of each element. In this way the derivatives of a vector are simply

$$\frac{d}{dt}(\mathbf{x}) = \begin{bmatrix} \dot{x}_1 \\ \dot{x}_2 \\ \vdots \\ \dot{x}_n \end{bmatrix} \tag{C.13}$$

where the overdot denotes the usual time derivatives of each element. Such derivatives are used in representing the equations of motion for multiple-degree-of-freedom systems.

A majority of the matrices used in vibration analysis are square and have the same number of rows as columns. Some special square matrices are the *identity matrix*, I, which has its element δ_{ij}, and the zero matrix, which has a zero as each of its elements. These special matrices satisfy the following conditions for any square matrix A:

$$AI = A$$

$$IA = A$$

$$0A = 0$$

$$A0 = 0 \qquad \text{(C.14)}$$

The identity matrix I is, of course, the multiplicative identity for the set of square matrices, and the zero matrix is the additive identity. A matrix that has zeros as all of its elements except those along the diagonal (i.e., a_{ii}) is called a diagonal matrix. Diagonal matrices can be manipulated almost like scalars and are often used in modal analysis.

The determinant of a matrix A is defined by the formula

$$\det(A) = \sum_{j=1}^{n} (-1)^{1+j} a_{1j} \det(A_{1j}) \qquad \text{(C.15)}$$

where a_{1j} is the element in the $(1-j)$th position of the matrix A, and A_{1j} is the $(n-1) \times (n-1)$th matrix formed from the matrix A by deleting the first row and jth column of the matrix A. For a 1×1 matrix (i.e., a scalar), the determinant of A is just

$$\det(A) = a$$

the value of the scalar. For a 2×2 matrix the determinant becomes

$$\det(A) = a_{11}a_{22} - a_{12}a_{21} \qquad \text{(C.16)}$$

The expression given by equation (C.15) can be used to calculate the determinant of a 3×3 matrix in terms of the determinant formula for a 2×2 matrix given previously, and so on. Note that the determinant is a scalar value.

The matrix A_{ij} formed by deleting the ith row and jth column of the matrix A is called a minor of A. Let α_{ij} denote the scalar found by taking the det (A_{ij}) with a particular sign:

$$\sigma_{ij} = (-1)^{i+j} \det(A_{ij}) \qquad \text{(C.17)}$$

Then the matrix defined by forming the elements α_{ij} into a matrix is called the *adjoint* of A, denoted adj A. The matrix A^{-1} such that $A^{-1}A = I$ is called the *inverse* of the matrix A and can be calculated from the adjoint by

$$A^{-1} = \frac{\text{adj } A}{\det(A)} \qquad \text{(C.18)}$$

Note from this calculation that the matrix A does not have an inverse if det $(A) = 0$. In this case, A is said to be *singular*. If A^{-1} does exist [i.e., if det $(A) \neq 0$], the matrix

A is called *nonsingular*. The following is a list of properties of the matrix inverse and determinant:

$$\left(A^{-1}\right)^T = \left(A^T\right)^{-1}$$

$$\left(AB\right)^{-1} = B^{-1}A^{-1}$$

$$\det\left(AB\right) = \det\left(A\right)\det\left(B\right)$$

$$\det\left(A^T\right) = \det\left(A\right)$$

$$\det\left(\alpha A\right) = \alpha \det\left(A\right) \qquad \text{where } \alpha \text{ is a scalar}$$

Each of these expressions can be derived from the definitions given previously.

Some special and important types of matrices in vibration analysis are summarized in the following list; that is, a matrix A is

symmetric if $A = A^T$

skew symmetric if $A = -A^T$

positive definite if $\mathbf{x}^T A \mathbf{x} > 0$ for all $\mathbf{x} \neq \mathbf{0}$

nonnegative definite if $\mathbf{x}^T A \mathbf{x} \geq 0$ for all $\mathbf{x} \neq \mathbf{0}$ (also called *semidefinite*)

indefinite if $(\mathbf{x}^T A \mathbf{x})(\mathbf{y}^T A \mathbf{y}) < 0$ for some $\mathbf{x}$ and $\mathbf{y}$

orthogonal if $A^T A = I$

If a matrix A is not symmetric, it can be written as the sum of a symmetric matrix, A_s, and a skew symmetric matrix, A_{ss}, defined by

$$A_s = \frac{A^T + A}{2} \qquad A_{ss} = \frac{A - A^T}{2} \tag{C.19}$$

Another useful matrix definition is that of the *trace* of a matrix. The trace of the matrix A, denoted $\text{tr}(A)$, is simply the sum of the diagonal elements of A:

$$\text{tr}\left(A\right) = \sum_{i=1}^{n} a_{ij} \tag{C.20}$$

which is a scalar.

The frequency response methods used in steady-state vibration analysis and modal testing methods often result in complex-valued matrices. In this situation the matrix transpose used previously is replaced with a conjugate transpose. In particular, let $\bar{a}_{ij}$ denote the complex conjugate of the number a_{ij}, then define A^* as the conjugate transpose by

$$A^* = \overline{A}^T \tag{C.21}$$

which has elements $\bar{a}_{ij}$. In the complex-valued case, a matrix A is said to be *Hermitian* if

$$A^* = A$$

and *unitary* if

$$A^* A = I$$

In the case of a complex-valued vector, the inner product becomes

$$\mathbf{x}*\mathbf{y} = \sum_{i=1}^{n} \bar{x}_i y_i \tag{C.22}$$

and a complex-valued matrix A is positive definite if

$$\mathbf{x}*A\mathbf{x} > 0 \tag{C.23}$$

for all nonzero complex vectors $\mathbf{x}$.

Next consider the matrix eigenvalue problem. Let A be a square matrix ($n \times n$). A scalar λ is an *eigenvalue* of the matrix A, with *eigenvector* $\mathbf{x}, \mathbf{x} \neq \mathbf{0}$, if

$$A\mathbf{x} = \lambda\mathbf{x} \tag{C.24}$$

is satisfied. Note that λ can be zero but $\mathbf{x}$ cannot. Also note that if $\mathbf{x}$ is an eigenvector, so is $\alpha\mathbf{x}$, where α is any scalar. As was developed in Chapter 4, the natural frequencies and mode shapes of an undamped system are calculated by solving a matrix eigenvalue problem. If the matrix A is known to be symmetric (i.e., $A = A^T$), the eigenvalues of A are real numbers and the eigenvectors are real valued. If, in addition, A is positive definite, the eigenvalues must be positive numbers. If A is size $n \times n$, then A will have n eigenvalues and n eigenvectors. In particular, if A is symmetric (and real), the eigenvectors $\mathbf{x}_i$ form a linearly independent set. Furthermore, these eigenvectors can be normalized to form an orthonormal linearly independent set. These normalized eigenvectors can be used to define an orthogonal matrix P by

$$P = [\mathbf{x}_1 \quad \mathbf{x}_2 \quad \ldots \quad \mathbf{x}_n] \tag{C.25}$$

such that

$$P^T P = I \quad \text{and} \quad P^T A P = \text{diag}(\lambda_i)I \tag{C.26}$$

This last expression is the foundation of the modal analysis method used so extensively in vibration theory.

The eigenvector defined in equation (C.24) is called a *right eigenvector*. A *left eigenvector* can also be defined by

$$\mathbf{y}^T A = \lambda\mathbf{y}^T \quad \mathbf{y} \neq \mathbf{0} \tag{C.27}$$

For symmetric matrices, the left and right eigenvectors are the same. For matrices that are not symmetric, these vectors may or may not be the same. In fact the eigenvalues, and hence the eigenvectors, of a nonsymmetric matrix may or may not be complex valued. The concept of an eigenvalue and eigenvector can be generalized by introducing a *lambda matrix*, or *matrix polynomial*. In particular, the solution of the equations of motion for a lumped-parameter multiple-degree-of-freedom system with damping generates the matrix polynomial problem of calculating a scalar λ and a nonzero vector $\mathbf{x}$ satisfying

$$(M\lambda^2 + C\lambda + K)\mathbf{x} = \mathbf{0} \tag{C.28}$$

The matrix $(M\lambda^2 + C\lambda + K)$ is called a matrix polynomial (it is viewed as a polynomial in λ, with matrix coefficients). If M, C, and K are $n \times n$ matrices, there are $2n$ values of λ and $2n$ values of $\mathbf{x}$ satisfying equation (C.28). A particularly simple case to solve is that where M, C, and K are symmetric, real-valued matrices satisfying $CM^{-1}K = KM^{-1}C$ (which in general is not true). In this special case, the eigenvectors of the case $C = 0$ will also satisfy (C.28). For a more complex discussion of matrix methods in vibration analysis, see Inman (2006) and Golub and Van Loan (1996).

The vibration problem for a multiple-degree-of-freedom system, or a finite element model of a system, can be expressed as an eigenvalue problem in several ways. In Section 4.2 the undamped vibration problem is related to the symmetric eigenvalue problem by calculating the inverse-square root of the positive definite mass matrix M. In particular, if the equations of motion are stated in matrix/vector form as

$$M\ddot{\mathbf{x}} + K\mathbf{x} = \mathbf{0} \tag{C.29}$$

where M is the mass matrix and K is the $n \times n$, symmetric positive semidefinite stiffness matrix, then there are four different eigenvalue problems that can be used to solve for the natural frequencies and mode shapes. As before, $\mathbf{x}$ is an $n \times 1$ vector of displacements and $\ddot{\mathbf{x}}$ is the second time derivative of $\mathbf{x}$.

The first approach considered is the standard approach of multiplying equation (C.29) from the left by the inverse of the matrix M. This yields

$$\ddot{\mathbf{x}} + M^{-1}K\mathbf{x} = \mathbf{0} \tag{C.30}$$

First the matrix M^{-1} must be calculated. In many cases M is a diagonal matrix (i.e., $M = \text{diag}[m_1\, m_2 \ldots m_n]$ and the inverse becomes simply

$$M^{-1} = \begin{bmatrix} \dfrac{1}{m} & 0 & 0 & \cdots & \cdots & 0 \\ 0 & \dfrac{1}{m_2} & 0 & \cdots & \cdots & 0 \\ \cdots & 0 & \dfrac{1}{m_3} & 0 & \cdots & \cdots \\ \cdots & \cdots & \cdots & \cdots & \cdots & \cdots \\ \cdots & \cdots & \cdots & \cdots & \cdots & \cdots \\ 0 & \cdots & \cdots & \cdots & \cdots & \dfrac{1}{m_n} \end{bmatrix} \tag{C.31}$$

In this case, calculating the matrix $M^{-1}K$ becomes a simple matrix multiplication. If K is a banded matrix, $M^{-1}K$ will also be banded. However, the product is usually not symmetric, even though both K and M are symmetric matrices, and computing the eigenvalues of a symmetric matrix is more efficient than calculating the eigenvalue problem of an asymmetric matrix.

If the matrix M is not diagonal (i.e., if the system is dynamically coupled), the inverse can be computed using the eigenvalue problem for the matrix M. Let $\mathbf{v}_i$ denote the eigenvector of the matrix M and let μ_i denote the associated eigenvalues. Then the statement of the eigenvalue problem for the matrix M becomes

$$M\mathbf{v}_i = \mu_i \mathbf{v}_i \quad \mathbf{v}_i \neq \mathbf{0} \tag{C.32}$$

Since M is symmetric and positive definite, each of the μ_i is greater than zero and each $\mathbf{v}_i$ can be normalized such that they form an orthonormal set. The $n \times n$ matrix V formed by taking the vectors $\mathbf{v}_i$ as its columns (similar to equation (4.49), i.e., $V = [\mathbf{v}_1 \quad \mathbf{v}_2 \quad \cdots \quad \mathbf{v}_n]$) is called orthogonal because $V^T = V^{-1}$, so that $V^T V = I$, the $n \times n$ identity. The matrix V also diagonalizes the matrix M, so that

$$V^T M V = \mathrm{diag}\left(\mu_1, \mu_2, \ldots, \mu_n\right)$$

Thus the matrix M can be decomposed into

$$M = V \mathrm{diag}\left(\mu_1, \mu_2, \ldots, \mu_n\right) V^T \tag{C.32}$$

This decomposition can be used to calculate the inverse of the matrix M using the identity $(AB)^{-1} = B^{-1} A^{-1}$, via

$$M^{-1} = V \mathrm{diag}\left(\frac{1}{\mu_1}, \frac{1}{\mu_2}, \ldots, \frac{1}{\mu_n}\right) V^T \tag{C.34}$$

Hence one way to compute the inverse of the matrix M is to solve the eigenvalue problem associated with M by the power method and use equation (C.34) to construct the inverse.

It is also useful to note that for any symmetric matrix, the formula for a function of a matrix is given in the same form as equation (C.34). That is, if $f(\cdot)$ is any real function for which $f(\mu_i)$ is defined,

$$f(M) = V \mathrm{diag}\left[f(\mu_1), f(\mu_2), \ldots, f(\mu_n)\right] V^T \tag{C.35}$$

defines the function f of the symmetric matrix M. In particular, the matrix $M^{-1/2}$ was used extensively in Chapter 4. Using this function of a matrix definition, the matrices $M^{-1/2}$ and $M^{1/2}$ can be calculated from the eigenvalues and eigenvectors of M by

$$M^{1/2} = V \mathrm{diag}\left(\mu_1^{1/2}, \mu_2^{1/2}, \ldots, \mu_n^{1/2}\right) V^T \tag{C.36}$$

and

$$M^{-1/2} = V \mathrm{diag}\left(\mu_1^{-1/2}, \mu_2^{-1/2}, \ldots, \mu_n^{-1/2}\right) V^T \tag{C.37}$$

This formulation allows the extension of the material of Chapter 4 to dynamically coupled systems. Again, in the case of only static coupling, the matrix M is diagonal and computation of $M^{1/2}$ or $M^{1/2}$ is trivial. For instance, if $M = \mathrm{diag}(\mu_1, \mu_2, \ldots, \mu_n)$, $M^{-1/2} = \mathrm{diag}(\mu_1^{-1/2}, \mu_2^{-1/2}, \ldots, \mu_n^{-1/2})$ as was used in Chapter 4.

The use of equation (C.34) to calculate the inverse of a matrix is not a very numerically efficient method. A better way to calculate M^{-1} is to view the calculation as the solution of the set of linear equations

$$MA = K \tag{C.38}$$

where A and K are vectors or matrices, K is known, and A is unknown. This can be solved by a Gaussian elimination (think of the case where A and K are vectors) and back substitution. Gaussian elimination is essentially a systematic version of the method of elimination often taught in high school algebra classes and involves a series of steps, each of which eliminates one variable from the system of equations. To examine this method, consider the set of linear algebraic equations of the form

$$A\mathbf{x} = \mathbf{b} \qquad (C.39)$$

where A is a known $n \times n$ matrix, $\mathbf{x}$ is an unknown $n \times 1$ vector of the form $\mathbf{x} = [x_1 \quad x_2 \quad \ldots \quad x_n]^T$, and $\mathbf{b}$ is an $n \times 1$ vector of known elements. The method of Gaussian elimination uses the fact that if P is any $n \times n$ nonsingular matrix, equation (C.39) and

$$PA\mathbf{x} = P\mathbf{b} \qquad (C.40)$$

have the same solution $\mathbf{x}$. The approach of the Gaussian elimination algorithm is to find a matrix P such that the matrix PA is upper triangular (i.e., PA has zeros as every element below the diagonal elements). If such a matrix P can be determined, the last equation in the relation specified by equation (C.40) will be of the form

$$x_n = (P\mathbf{b})_n \qquad (C.41)$$

where $(P\mathbf{b})_n$ denotes the nth element of the vector $P\mathbf{b}$. The second-to-last equation will be only in x_{n-1} and x_n. Hence each element x_i is calculated from the last equation backward (back substitution) until the entire vector is known.

The required matrix P in equations (C.41) and (C.40) can be determined from a series of elementary matrices. Elementary matrices are nonsingular matrices that when postmultiplied by a given matrix (A, in this case) result in the subtraction of multiples of one column of the matrix A from each of the other columns of A. An example of an elementary matrix is the matrix P_i defined to be the $n \times n$ identity matrix with the ith column replaced by

$$\begin{bmatrix} 0 \\ 0 \\ \vdots \\ 1 \\ -p_{i,i+1} \\ \vdots \\ -p_{i,n} \end{bmatrix} \qquad (C.42)$$

where the element p_{ij} are to be determined by the elements of the matrix A in such a way as to reduce A to a triangular form.

The algorithm for Gaussian elimination then proceeds in a step-by-step manner by writing the initial set of linear equations given in equation (C.39) as

$$A_0\mathbf{x} = \mathbf{b}_0 \qquad (C.43)$$

Multiplying this expression by P_1 to get $P_1A_0\mathbf{x} = P_1\mathbf{b}_0$ produces the next step. The last expression is renamed $A_1\mathbf{x} = \mathbf{b}_1$, where $A_1 = P_1A_0$ and $\mathbf{b}_1 = P_1\mathbf{b}_0$. Then P_2 is multiplied times $A_1\mathbf{x} = \mathbf{b}_1$ to yield $A_2\mathbf{x} = \mathbf{b}_2$, where $A_2 = P_2A_1$ and $\mathbf{b}_2 = P_2\mathbf{b}_1$. This is repeated so that the rth step is

$$A_r\mathbf{x} = \mathbf{b}_r \qquad (C.44)$$

Here multiplying by P_r has made A_{r-1} upper triangular in the first r column. For example, for $n = 6$ and $r = 3$, equation (C.44) has the form

$$\begin{bmatrix} \times & \times & \times & \times & \times & \times \\ 0 & \times & \times & \times & \times & \times \\ 0 & 0 & \times & \times & \times & \times \\ 0 & 0 & 0 & \times & \times & \times \\ 0 & 0 & 0 & \times & \times & \times \\ 0 & 0 & 0 & \times & \times & \times \end{bmatrix} \begin{bmatrix} x_1 \\ x_2 \\ x_3 \\ x_4 \\ x_5 \\ x_6 \end{bmatrix} = \begin{bmatrix} \times \\ \times \\ \times \\ \times \\ \times \\ \times \end{bmatrix} \qquad (C.45)$$

where $\times$ represents any nonzero number. Equation (C.45) illustrates how the equation $A\mathbf{x} = \mathbf{b}$ becomes more "triangular" at each step.

Examination of each step illustrates that the matrix P_rA_{r-1} has as its rth column,

$$P_r \begin{bmatrix} a_{1r}^{(r-1)} \\ a_{2r}^{(r-1)} \\ \vdots \\ a_{nr}^{(r-1)} \end{bmatrix} \qquad (C.46)$$

where $a_{ij}^{(r-1)}$ denotes the ijth element of the matrix A_{r-1}. The choice of the matrix P_r that performs the desired triangularization is then given by equation (C.42) with

$$p_{ir} = \frac{a_{ir}^{(r-1)}}{a_{rr}^{(r-1)}} \qquad (C.47)$$

This formula results in a matrix P_r such that P_rA_{r-1} has one more upper triangular column, eliminating x_r from the equation in the $r-1$ position. Successive multiplication of the matrices P_r defined by equations (C.46) and (C.47) yields a triangular system of equations that are easily solved for the elements x_r of the vector $\mathbf{x}$.

The solution of the linear system of equations $A\mathbf{x} = \mathbf{b}$ is then solved by computing $P_nP_{n-1}P_{n-2}\ldots P_1 A\mathbf{x} = P_nP_{n-1}P_{n-2}\ldots P_1\mathbf{x}$. The method works well and is numerically superior to computing the solution via $\mathbf{x} = A^{-1}\mathbf{b}$. However, the Gaussian elimination method fails if any of the pivot elements $a_{rr} = 0$. The matrix equation $MA = K$ given in equation (C.38) can also be solved by a Gaussian elimination procedure (applied to each column of A) or by more sophisticated triangularization procedures. This produces a numerically more reliable calculation of the matrix $A = M^{-1}K$ than is obtained by first calculating the matrix M^{-1} via equation (C.34) and computing the matrix product of M^{-1} and K. The procedure of computing the matrix A from the equation $M A = K$ to form the matrix $M^{-1} K$ without actually computing an inverse is called *matrix division*.

The ability to effectively compute the inverse matrix M^{-1} and the inverse-square root matrix $M^{-1/2}$ allows the formulation of five eigenvalue problems associated with vibration analysis. Each of these eigenvalue problems has computational advantages and disadvantages depending on the nature of both the structures and values of the elements of both the mass and stiffness matrices.

In Section 4.9 several approaches to solving for the natural frequencies and mode shapes of the undamped system given of equation (C.30) were given. There are several approaches available for solving the natural frequencies and mode shapes of the undamped system given of equation (C.30). The question arises as to which one is best. One way to answer this question is to use MATLAB'S ability to count the number of floating-point operations it takes to solve the related eigenvalue problem. This is done by using the `flops` command. Here we show the results for the problem of Example 4.1.5 using MATLAB.

```
EDU>M=[9 0;0 1];K=[27 -3;-3 3];
EDU>[V,D]=eig(K,M) %these vectors are not orthogonal
V =
     0.3162   -0.3162
     0.9487    0.9487
D =
     2.0000         0
          0    4.0000
EDU>flops
ans =
     417
```

```
EDU>M=[9 0;0 1];K=[27 -3;-3 3];
EDU>L=chol(M);
EDU>S=inv(L);
EDU>Kh=S*K*S;
EDU>[V,D]=eig(Kh) %these vectors are orthogonal
V =
    -0.7071   -0.7071
     0.7071   -0.7071
D =
     4   0
     0   2
EDU>P=L*V
P =
    -2.1213   -2.1213
     0.7071   -0.7071
EDU>flops
ans =
     118
```

Similar runs show the following:

Using `inv(L)*K*inv(L')` requires 118 flops.

Using `inv(M)*K` requires 191 flops.

Using `inv(sqrtm(M))*K* inv(sqrtm(M))` requires 228 flops.

Using the generalized eigenvalue problem $\lambda M\mathbf{u} = K\mathbf{u}$ requires 417 flops.

While these do not include normalization steps to get the eigenvectors into normalized mode shapes, they do get the mode shapes all in the same coordinate system. As the number of degrees of freedom increases, these differences in flop count increase. Note that the best approach using this measure is to use the Cholesky decomposition approach suggested in Chapter 4 and keep all the eigenvalue problems symmetric. The standard approach $(M^{-1} K)$ as taught in most books is almost twice as expensive for a two-degree-of-freedom system. Such differences usually increase exponentially with the number of degrees of freedom.

The Vibration Literature

There are several outstanding texts and reference books on vibration that merit consultation for those who wish a second explanation for the sake of understanding or for those who wish to pursue these topics further. In addition, there are several publications devoted to presenting research results and case studies in vibration. A few such general references are listed here. A computer search in your local library should turn up the rest.

Introductory Texts

Mechanical Vibration, J. P. Den Hartog, Dover, New York, 1985.

Theory of Vibration with Applications, 5th ed., W. T. Thomson, and M. D. Dahleh, Prentice Hall, Upper Saddle River, N.J., 1998.

Structural Dynamics: An Introduction to Computer Methods, R. R. Craig, Jr., Wiley, New York, 1981.

Mechanical Vibrations, 5th ed., S. S. Rao, Prentice Hall, Upper Saddle River, N.J., 2010.

Mechanical Vibrations: Theory and Applications, S.G. Kelly, Cengage Learning, Stamford, Conn., 2012.

Elements of Vibration Analysis, 2nd ed., L. Meirovitch, McGraw-Hill, New York, 1986.

Shock and Vibration Handbook, 5th ed., C. M. Harris and A. G. Piersol, editors, McGraw-Hill, New York, 2002.

Vibrations, 2nd ed., B. Balachandran and E. B. Magrab, Cengage Learning, Stamford, Conn., 2009.

Principles of Vibration, B. H. Tongue, 2nd ed., Oxford Press, New York, 2002.

Vibration, Fundamentals and Practice, 2nd ed., C. W. de Silva, CRC Press, Boca Raton, Fla., 2006.

Advanced Texts

Vibration Problems in Engineering, 5th ed., W. Weaver, Jr., S. P. Timoshenko, and D. H. Young, Wiley, New York, 1990.

Analytical Methods in Vibration, L. Meirovitch, Macmillan, New York, 1967.

Vibration with Control Measurement and Stability, D. J. Inman, Prentice Hall, Englewood Cliffs, N.J., 1989.

Mechanical Vibration, H. Benaroya, 2nd ed., Marcel Decker, New York, 2004.

Principles and Techniques of Vibrations, L. Meirovitch, Prentice Hall, Upper Saddle River, N.J., 1997.

Fundamentals of Vibration, L. Meirovitch, McGraw-Hill, New York, 2001.

Mechanical and Structural Vibrations: Theory and Applications, J. H. Ginsberg, Wiley, New York, 2001.

Vibration and Control, D. J. Inman, Wiley, Chichester, U.K., 2006.

Periodicals

Sound and Vibration, Acoustical Publications, Bay Village, Ohio.

Journals

Shock and Vibration, IOS Press, Amsterdam.
Journal of Sound and Vibration, Elsevier, Oxford.
Journal of Vibration and Acoustics, American Society of Mechanical Engineering, New York.
Journal of Mechanical Systems and Signals, Academic Press, New York.
AIAA Journal, American Institute of Aeronautics and Astronautics, Washington, D. C.

List of Symbols

This appendix lists the symbols used in the book. There are not enough symbols to go around so some symbols are used to denote more than one quantity. In addition, the choice of a symbol to denote a specific quantity has evolved over time often by different groups of people. Hence it is not uncommon for a particular symbol to represent several different quantities. Here, symbols have been chosen consistent with the most common used in the vibration literature.

In general, a lowercase italic symbol is a scalar, an uppercase italic symbol is a matrix, and a bold lowercase letter is used to denote a vector. However, when working with transforms, it is common to denote the Laplace transform of a scalar by its uppercase symbol. Hence, it is not possible to read an uppercase italic letter and know whether the quantity is matrix or a transformed scalar. Symbols must always be clarified by the context in which they appear. Greek symbols are almost always scalars. Letters that are not italic or bold are usually units (such as N for newtons or mm for millimeters). Units are summarized on the inside front cover.

Several attempts have been made to establish an international standard of symbols for modal analysis and hence, vibration. The basic problem remains, however, there are just not enough symbols to go around.

CHAPTER 1

f_k = elastic restoring force of a spring
m = mass
g = acceleration of gravity
x_0 = an initial displacement
$x, x(t)$ = displacement
k = spring constant, spring stiffness
N = normal force
x, y = coordinates of a two dimensional space
$\theta, \theta(t)$ = angular displacement
t = time
ω_n = $\sqrt{k/m}$ undamped natural frequency, angular frequency
ϕ = phase angle
A = amplitude of vibration, also a constant of integration
f = $\omega/2\pi$ frequency in hertz, not to be confused with a force
T = $1/f = 2\pi/\omega$, period of oscillation
v_0 = initial velocity
a, a_1, a_2 = unknown constants of integration

A_1, A_2 = unknown constants of integration
λ = constant used in solving differential equations
j = $\sqrt{-1}$ the imaginary unit
$\bar{x}, \bar{x}^2$ = average displacement, root mean-square value, respectively
$\dot{x}(t), \ddot{x}(t)$ = time derivatives of displacement, i.e., velocity and acceleration, respectively
f_c = $c\dot{x}$, a damping force
C = damping coefficient
ζ = $c/2\sqrt{km}$ damping ratio
ω_d = $\omega\sqrt{1 - \zeta^2}$, for $0 < \zeta < 1$, called the damped natural frequency
c_{cr} = $2\sqrt{km}$, critical damping coefficient so that $\zeta = c/c_{cr}$
f_{xi} = ith force acting in the x direction
M_{0i} = ith torque acting about the point 0
I_0, J, J_0 = moment of inertia about point 0
U_1, U_2 = potential energy at time t_1 and t_2, respectively

T_1, T_2	= kinetic energy at time t_1 and t_2, respectively	k_0	= radius of gyration ($k_0^2 = q_0 r$)
T_{max}, U_{max}	= maximum kinetic energy, maximum potential energy, respectively	E	= Young's modulus (elastic modulus)
r, l	= radius (capital R is often used as well), length, respectively	A	= cross-sectional area when used with E
$\dfrac{d}{dt}(\ ), (\dot{\ })$	= always refers to a time derivative of ( )	Jp	= area moment of inertia
		G	= shear modulus
m_s	= mass of a spring	ρ	= mass density
γ	= a density (specific weight)	W	= weight
q_0	= radius of the center of percussion	δ	= logarithmic decrement (see Window 4.2 for other uses of this symbol)
		Δ	= static deflection

CHAPTER 2

$F(t)$	= an external or driving force	$H(j\omega)$	= complex frequency response function
F_0	= constant magnitude of a harmonic driving force	s	= transform variable
ω	= input frequency or driving frequency	$X(s)$	= the Laplace transform of $x(t)$
		θ	= used to denote phase shifts
f_0	= F_0/m	$H(s)$	= $X(s)/F(s)$, a transfer function
x_p	= the particular solution	$z(t)$	= $x(t) - y(t)$, relative displacement
X	= amplitude of the particular solution	Z	= amplitude of relative displacement
r	= frequency ratio = ω_{dr}/ω (not to be confused with a radius)	ψ	= phase of relative displacement
$y(t)$	= displacement of the base of spring–mass–damper system	z_p	= particular solution for relative displacement
ω_b	= frequency of base motion, i.e., the frequency of an applied harmonic displacement	F_d	= damping force
		μ	= coefficient of sliding friction
X	= magnitude of the harmonic response	A_1, B_1	= constants of integration
		ΔE	= energy dissipated
Y	= magnitude of the applied harmonic displacement	c_{eq}	= equivalent viscous-damping coefficient
F_T	= magnitude of the force transmitted to a mass through a spring and dashpot	U	= potential energy
		U_{max}	= peak potential energy
		η	= loss factor
e	= radius of an out-of-balance mass (not to be confused with the exponential)	β	= hysteretic damping constant
		C	= drag coefficient
		α	= $C\rho A/2$

CHAPTER 3

τ	= a specific value of time, or a dummy (time) variable of integration	$\delta(t)$	= the Dirac delta function (not a decrement)
		$\hat{F}$	= $F\Delta t$, impulse
ε	= a very small positive value of time	$h(t - \tau)$	= impulse response function

a_0, a_n, b_n	= Fourier coefficients
n, m, I	= integers (be careful not to confuse m, the mass, with m, the index)
ω_T	= $2\pi/T$
$x_{cn}(t)$	= the nth solution if the driving force is $\cos n\omega_n t$
$x_{sn}(t)$	= the nth solution if the driving force is $\sin n\omega_n t$
$L[\ ]$	= the Laplace transform of $[\ ]$

$\mu(t)$	= unit step function
x'	= $x - \bar{x}$, a zero mean variable
$\Phi(t), H(t)$	= Heaviside step function
$S_{xx}(\omega)$	= power spectral density of the function x
$R_{xx}(\tau)$	= auto correlation of the function x
$E[x]$	= the expected value of x
M, a, b	= constants

CHAPTER 4

$\mathbf{x}$	= displacement column vector
$x_{10}, x_{20},$	= initial conditions for the coordinates
$\dot{x}_{10}\ldots$	= $x_1(t)$ and $x_2(t)$, respectively
$\mathbf{x}^T$	= the transpose of $\mathbf{x}$ or a row vector
K	= the stiffness matrix
M	= the mass matrix
C	= the damping matrix
A^T	= the transpose of the matrix A
A^{-1}	= the inverse of the matrix A
$A^{1/2}$	= the matrix square root of the matrix A
$\mathbf{u}$	= a constant column vector also $\mathbf{u}_1, \mathbf{u}_2$, etc.
I	= the identity matrix
$\tilde{K}$	= $M^{-1/2}KM^{-1/2}$
$\tilde{C}$	= $M^{-1/2}CM^{-1/2}$
$M^{-1/2}$	= the inverse of the matrix square root of the matrix M
$\mathbf{q}(t)$	= a vector of generalized coordinates
$q_i(t)$	= the ith generalized coordinate
$\mathbf{v}$	= an eigenvector of $\tilde{K}$
λ, λ_i	= the eigenvalues
α	= a scalar

$\mathbf{w}, \mathbf{w}_i$	= mass-normalized eigenvector of K
P	= modal matrix
δ_{ij}	= Kronecker delta (not to be confused with the Dirac delta or the log decrement)
Λ	= the diagonal matrix of eigenvalues of a matrix
$\mathbf{r}(t)$	= modal coordinate vector
$r_i(t)$	= the ith modal coordinate
S	= $M^{-1/2}P$, a mass weighted modal matrix
$\mathbf{0}$	= the zero vector
$\theta_1, \theta_2, \theta_3$	= rotational displacements
$\det(A)$	= the determinant of the matrix A
d_i	= expansion coefficients (scalars)
ζ_I	= modal damping ratios
β	= a scalar
$\mathbf{F}(t)$	= a vector of external forces
ω_{di}	= $\omega_i\sqrt{1-\zeta_i}$ the ith damped natural frequency
$\mathbf{f}(t)$	= $P^T M^{-1/2}\mathbf{F}(t)$
Q_i	= a generalized force or moment
L	= $T-U$ = Lagrangian, also the Cholesky factor of M

CHAPTER 5

T.R.	= transmissibility ratio
R	= $1 - T.R.$ = reduction in transmissibility
δ_s	= static deflection
k_a	= stiffness of the absorber
m_a	= mass of the absorber
x_a	= displacement coordinate of absorber mass

X_a	= magnitude of the displacement of the absorber mass
μ	= m_a/m ratio of absorber mass to primary mass
ω_p	= natural frequency of primary system without absorber
ω_a	= natural frequency of absorber without primary system

β $= \omega_a/\omega_p$, a frequency ratio

X $= [X \quad X_a]^T$, where X is the magnitude of the response of the spring mass and X_a is that of the absorber mass

r $= \omega_r/\omega_p$, a frequency ratio

x_m $=$ the value of x that causes $f(x_m)$ to be an extreme (critical point)

f_x $=$ partial derivatives of f with respect to x

f_y $=$ partial derivatives of f with respect to y

f_{xx} $=$ 2nd partial derivatives of f with respect to x

f_{yy} $=$ 2nd partial derivatives of f with respect to y

γ $= k_2/k_1$, a stiffness ratio

$\overline{\eta}$ $=$ loss factor

k^* $=$ complex stiffness

E^* $=$ complex modulus

e_2 $=$ ratio of modulus $= E_2/E_1$

h_2 $=$ ratio of height $= H_2/H_1$

g_1, g_2 $=$ feedback control gain

G' $=$ shear modulus (real part)

E' $=$ real part of elastic modulus $2(1 + v)G'$, v Poisson's ratio

σ_s $=$ shear stress

ε_s $=$ strain

η_s $=$ shear loss factor

G^* $=$ complex shear modulus

E'', G'' $=$ imaginary part of elastic and shear modulus respectively (G'' called a loss modulus)

k' $=$ shear stiffness

T_R $=$ transmissibility ratio at resonance

G'_ω $=$ dynamic shear modulus

CHAPTER 6

$w(x, t)$ $=$ displacement

$f(x, t)$ $=$ applied force

τ $=$ string tension

$\dfrac{\partial w}{\partial x}$ $=$ partial derivatives also denoted by w_x

l $=$ length

c $= \sqrt{\tau/\rho}$, wave speed also $\sqrt{E/\rho}$, depending on the structure

$X(x)$ $=$ a spatial function of x only

$T(t)$ $=$ a temporal function of t only

σ $=$ a constant

$a_1, a_2,$ $=$ constants of integration

a, b $=$ constants of integration

$A_n, B_n,$ $=$ constants of integration

c_n, d_n $=$ constants of integration

$w_0(x)$ $=$ the initial displacement

$\dot{w}_0(x)$ $=$ the initial velocity

m, n $=$ indices, integers

β $=$ separation constant

λ $=$ eigenvalues

$X_n(x)$ $=$ eigenfunctions

$A(x)$ $=$ cross-sectional area of a beam

ω_n $=$ the natural frequency of the nth node

$\theta(x, t)$ $=$ angular rotation of a shaft

τ $=$ torque when used with a rod

G $=$ shear modulus

J $=$ polar moment of area

J_0 $=$ polar moment of inertia (often ρ^J)

γ $=$ torsional constant

$\Theta(x)$ $=$ spatial function of angular rotation

$M(x, t)$ $=$ bending moment

$I(x)$ $=$ cross-sectional area moment of inertia

$f(x, t)$ $=$ distributed force per unit area

$V(x, t)$ $=$ shear force

β_n $=$ solution to a transcendental equation

ψ $=$ total angle in bending

κ $=$ shear coefficient

$w(x, y, t)$ $=$ membrane displacement in the z direction

∇^2 $=$ Laplace operator

∇^4 $=$ harmonic operator

D_E $=$ the flexural rigidity of the plate

$\gamma, \ \beta$ $=$ viscous-damping coefficients

CHAPTER 7

ω_c	= cutoff frequency
x_k	= $x(t_k)$ = value of $x(t)$ at the discrete time t_k
a_0, a_i, b_i	= digital spectral coefficients
C	= matrix of digital Fourier coefficients
a	= a vector of spectral coefficients
N	= size of data set
$R_{xf}(\tau)$	= cross correlation function
$S_{xf}(\omega)$	= cross spectral density function
γ^2	= coherence function
$\alpha(\omega)$	= α mobility frequency response function and has a different value for the various transfer functions: mobility, receptance, etc. (also used to denote an angle in Nyquist plots)
$\alpha(\omega)$	= receptance matrix $= (K - \omega_r^2 M + j\omega C)^{-1}$

CHAPTER 8

c_1, c_2, etc.	= constants of integration
$u_i(t)$	= ith nodal displacement (local)
$T(t)$	= kinetic energy
$V(t)$	= potential energy
$U_i(t)$	= global nodal displacement
θ	= angle between global and local coordinates
Γ	= coordinate transformation matrix between local and global coordinates
K_e	= element stiffness matrix (local coordinates)
$K_{()}$	= element stiffness matrix (global coordinates)
$K'_{()}$	= element stiffness matrix (global coordinates) expanded to fit global vector size
M_{ii}, K_{ii}	= partitions of the mass as stiffness matrices, respectively
Q	= reduction transformation matrix

F | Codes and Websites

The best and most current source of information on using the three math codes used in the text can be found by using a search engine and typing in the name of the code. Many academics and users keep up-to-date websites on using these codes, and there are many very good tutorials available on the Internet. In addition, the syntax and commands in codes are updated frequently, so if you have copied a code out of the text and it does not work, try checking the code's homepage for updates. The codes printed here were updates as of August 2012.

Each of the codes mentioned in the text may be used to solve vibration problems. Each is best approached for the first time by trying something simple, like plotting a known function or performing a simple calculation. The purpose of using these codes is to enhance understanding, to visualize, and to replace tedious hand calculations with more accurate machine calculations. Any of these codes are straightforward to use by just copying the solutions given in Sections 1.9, 1.10, 2.8, 2.9, 3.8, 3.9, 4.9, and 4.10. In addition, it is helpful to have access to a manual or tutorial. However, the codes presented in the text are enough to get started.

It is important to note that the output of any of these codes is only as trust-worthy and accurate as the input given to the code. Just because an answer appears, does not mean it is a correct solution to the problem.

A search engine (such as Google) can be used to find other useful information related to the text material on vibration. In particular, some interesting movies of vibration can be downloaded. Searching phrases such as "vibration movies," "resonance," "spring constants," etc., will reveal very useful visuals on vibration phenomenon.

G Engineering Vibration Toolbox and Web Support

Web Support

Web support will be available through www.pearsonhighered.com/inman and will be updated often.

MATLAB Engineering Vibration Toolbox

Dr. Joseph C. Slater of Wright State University has authored a MATLAB Toolbox keyed to this text. At the end of each chapter, problems are listed that may be solved with the Engineering Vibration Toolbox (EVT). In addition, the EVT may be used to help solve the homework problems suggested for computer usage in Sections 1.9, 1.10, 2.8, 2.9, 3.8, 3.9, 4.9, and 4.10. The Toolbox is organized by chapter and may be used to solve the Toolbox problems found at the end of each chapter. MATLAB and the EVT are interactive and intended to assist in learning, analysis, parametric studies, and design, as well as in solving homework problems. The EVT is updated and improved regularly.

The EVT contains sets of M-files and data files for use with current versions of MATLAB and can be downloaded for free. Go to the Engineering Vibration Toolbox home page at http://www.cs.wright.edu/vtoolbox.

This site includes editions that run on earlier versions of MATLAB, as well as the most recent version. An email list of instructors who use EVT is maintained so that users can receive notification of the latest updates. The EVT is designed to run on any platform supported by MATLAB and is regularly updated to maintain compatibility with the current version of MATLAB. A brief introduction to MATLAB and UNIX is available on the home page as well. Please read the file `readme.txt` to get started and type `help vtoolbox` to obtain an overview.

Use the path command in MATLAB to make sure that a path has been set to an installed directory containing the Vibration Toolbox. The path command will list all directories available to MATLAB. One of them should end in "vtoolbox." If not, see your system administrator if working on a multiuser system. If you are installing on a personal computer, please consult the `readme.txt` file and your MATLAB manual on how to set paths. Please note that using the path command from the MATLAB prompt sets the path only for the current session. Once the vibration toolbox is installed and the paths are set properly in MATLAB, typing `help vtoolbox` will provide a table of contents of the toolbox. Likewise, typing `help vtbON` will provide a table of contents for the files related to chapter "N." Typing help codename will provide help on the particular code. Note that the filename is `codename.m`. Don't use the `.m` from within MATLAB. The Engineering Vibration Toolbox commands can be run by typing them with the necessary arguments just as any other MATLAB commands/functions. For instance, vtb1_1 can be run by typing vtb1_1(1,.1,1,1,0,10). On a UNIX system you may need to set the DISPLAY variable properly to view the results (consult your local user's guide or system support). Many functions have multiple forms of input. The help for each function shows this flexibility when it exists. Updates to the Engineering Vibration Toolbox are made as needed, and other enhancements occur from time to time. The vtbud command can be used to view the current revision status on your system. If MATLAB is configured to do so, it will also download a copy of vtbud.m from the site containing the current revision status on the site. Thus you can download incremented upgrades at your convenience.

References

BANDSTRA, J. P., 1983, "Comparison of Equivalent Viscous Damping and Nonlinear Damping in Discrete and Continuous Vibrating Systems," *Journal of Vibration, Acoustics, Stress, Reliability in Design,* Vol. 105, pp. 382–392.

BERT, C. W., 1973, "Material Damping: An Introductory Review of Mathematical Models, Measures and Experimental Techniques," *Journal of Sound and Vibration,* Vol. 29, No. 2, pp. 129–153.

BLEVINS, R. D., 1987, *Formulas for Natural Frequencies and Mode Shapes,* R. E. Krieger, Melbourne, Fla.

BLEVINS, R. D., 1990, *Flow-Induced Vibration,* Van Nostrand Reinhold, New York.

BOYCE, W. E., and DiPRIMA, P. C., 2008, *Elementary Differential Equations and Boundary Value Problems,* 9th ed., Wiley, New York.

CANNON, R. M., 1967, *Dynamics of Physical Systems,* McGraw-Hill, New York.

CAUGHEY, T. K., and O'KELLY, M. E. J., 1965, "Classical Normal Modes in Damped Linear Dynamic Systems," *ASME Journal of Applied Mechanics,* Vol. 49, pp. 867–870.

CHILDS, D. W., 1993, *Turbomachinery Rotordynamics: Phenomena, Modeling, and Analysis,* Wiley, New York.

CHURCHILL, R. V., 1972, *Operational Mathematics,* 3rd ed., McGraw-Hill, New York.

COOLEY, J. W., and TUKEY, J. W., 1965, "An Algorithm for the Machines Calculation of Complex Fourier Series," *Mathematics of Computation,* Vol. 19, No. 90, pp. 297–311.

COOK, P. A., 1986, *Nonlinear Dynamical Systems,* Prentice Hall, Englewood Cliffs, N.J.

COWPER, G. R., 1966, "The Shear Coefficient in Timoshenko's Beam Theory," *ASME Journal of Applied Mechanics,* Vol. 33, pp. 335–340.

CUDNEY, H. H., and INMAN, D. J., 1989, "Determining Damping Mechanisms in a Composite Beam," *International Journal of Analytical and Experimental Modal Analysis,* Vol. 4, No. 4, pp. 138–143.

DATTA, B. N., 1995, *Numerical Review Algebra and Application,* Brooks Cole Publishing, Pacific Grove, Calif.

DOEBELIN, E. O., 1980, *System Modeling and Response,* Wiley, New York.

DONGARRA, J., BUNCH, J. R., MOLER, C. B., and STEWART, G. W., 1978, *LINPACK Users' Guide,* SIAM Publications, Philadelphia.

EHRICH, F. F., ed., 1992, *Handbook of Rotor Dynamics,* McGraw-Hill, New York.

EWINS, D. J., 2000, *Modal Testing: Theory and Practice,* 2nd ed., Research Studies Press, distributed by Wiley, New York.

FIGLIOLA, R. S., and BEASLEY, D. C., 1991, *Theory and Design for Mechanical Measurement,* Wiley, New York.

FORSYTH, G. E., MALCOLM, M. A., and MOLER, C. B., 1977, *Computer Methods for Mathematical Computation,* Prentice Hall, Englewood Cliffs, N.J.

GOLUB, G. H., and VAN LOAN, C. F., 1996, *Matrix Computations,* 3rd ed., Johns Hopkins University Press, Baltimore.

GUYAN, R. I., 1965, "Reduction of Stiffness and Mass Matrices," *AIAA Journal,* Vol. 3, No. 2, p. 380.

INMAN, D. J., 1989, *Vibration with Control Measurement and Stability,* Prentice Hall, Englewood Cliffs, N.J.

INMAN, D. J., 2006, *Vibration and Control,* Wiley, Chichester, U.K.

KENNEDY, C. C., and PANCU, C. D. P., 1947, "Use of Vectors in Vibration Measurement and Analysis," *Journal of Aeronautical Science,* Vol. 14, No. 11, pp. 603–625.

MAGRAB, E. B., 1979, *Vibration of Elastic Structural Members,* Sijthoff & Noordhoff, Winchester, Mass.

MANSFIELD, N. J., 2005, *Human Response to Vibration,* CRC Press, Boca Raton, Fla.

MEIROVITCH, L., 1995, *Principles and Techniques of Vibration,* Prentice Hall, Upper Saddle River, N.J.

NASHIF, A. D., JONES, D. I. G., and HENDERSON, J. P., 1985, *Vibration Damping,* Wiley, New York.

NEWLAND, D. E., 1993, *Random Vibration and Spectral Analysis,* 2nd ed., Longman, New York.

OTNES, R. K., and ENCOCHSON, L., 1972, *Digital Time Series Analysis,* Wiley, New York.

REISMANN, H., 1988, *Elastic Plates: Theory and Application,* Wiley, New York.

REISMANN, H., and PAWLIK, P. S., 1974, *Elastokinetics,* West Publishing, St. Paul, Minn.

SHAMES, I. H., 1980, *Engineering Mechanics: Statics and Dynamics,* 3rd ed., Prentice Hall, Englewood Cliffs, N.J.

SHAMES, I. H., 1989, *Introduction to Solid Mechanics,* 2nd ed., Prentice Hall, Englewood Cliffs, N.J.

SNOWDEN, J. C., 1968, *Vibration and Shock in Damped Mechanical Systems,* Wiley, New York.

WEAVER, W., JR., TIMOSHENKO, S. P., and YOUNG, D. H., 1990, *Vibration Problems in Engineering,* 5th ed., Wiley, New York.

WOWK, V., 1991, *Machinery Vibration: Measurement and Analysis,* McGraw-Hill, New York.

Answers to Selected Problems

(1.1) **1.1.**   9.81 N m

1.5.   1.635 N/m

1.16.   **(a)** 0.05 Hz   **(b)** 0.07 Hz

1.22.   0.91 Hz

(1.2) **1.27.**   0.0396 m

1.29.   4.63 N/m

1.37.   **(a)** 15.5 rad/s   **(b)** 14.6 rad/s

1.40.   20 rad/s

(1.3) **1.44.** $x(t) = e^{-t}\sin t$ mm

1.56. $\omega_n = 5.477$ rad/s, $\zeta = 0.251$ (oscillates), $\omega_d = 4.74$ rad/s

1.57. $\omega_n = \sqrt{\dfrac{kl^2}{J + ml^2}}$ rad/s

(1.4) **1.65.** $\omega_n = \sqrt{\dfrac{3m + 6m_t}{2m + 6m_t}}\sqrt{\dfrac{g}{l}}$ rad/s

1.67. $\left(\dfrac{J}{r^2} + m\right)\ddot{x} + \left(k_2 + \dfrac{k_1}{r^2}\right)x = 0,$ $\omega_n = \sqrt{\dfrac{k_1 + r^2 k_2}{J + mr^2}}$

1.71. $c = (0.02)m\sqrt{g\ell^3}$

1.73. $\omega_d = 0.645$ rad/s

(1.5) **1.82.** $\omega_n = 1.49 \times 10^4$ rad/s

1.87. $k_{eq} = 248$ N/m

1.91. $\zeta = 0.1885$, underdamped

(1.6) **1.97.** $E = 5.64 \times 10^{11}$ N/m^2

1.99. $\zeta = 0.215$

(1.7) **1.102.** $c = 87.2$ kg/s

1.109. $\dfrac{l^3}{bh^3} = \dfrac{\Delta E}{4mg}$

(1.8) **1.113.** $\dfrac{kl}{2} > \left(\dfrac{m_1}{2} + m_2\right)g$

(1.10) **1.124.** $\Delta t = 4.68$ s

1.130. Two: $\begin{bmatrix} x_1 \\ x_2 \end{bmatrix} = \begin{bmatrix} 0 \\ 0 \end{bmatrix}$ and $\begin{bmatrix} x_1 \\ x_2 \end{bmatrix} = \begin{bmatrix} -\dfrac{\omega_n^2}{\beta} \\ 0 \end{bmatrix}$

(2.1) **2.5.** $\omega_n = 55$ rad/s, $\omega = 45$ rad/s

2.6. $T_b = 20$ s

2.9. $x(t) = 0.002795 \sin 4.472t + 0.04 \cos 4.472t - 0.00125 \sin 10t$

2.12. $k = 5.996 \times 10^4$ N/m, $E = 1.859 \times 10^5$ Pa

2.15. $\dfrac{30J}{\pi}\left(\dfrac{2M_0}{J} + \dfrac{\pi\omega^2}{30}\right) < k$

(2.2) **2.21.** $X = 0.133$ m, $\theta = -\pi/2$ rad

2.27. $\omega_n = \sqrt{\dfrac{kl_1^2 + mgl}{ml^2}}$

2.31. $c = 55.7$ kg/s

2.35. $\theta(t) = 0.334 \cos(2\pi t - 6.196)$ rad

(2.3) **2.41.** 0.00388 m

(2.4) **2.47.** $X = 10.002$ cm, $F_T = 4500.9$ N

2.52. $c = 730.297$ kg/s, $F_T = 320$ N

2.55. $c = 1162$ kg/s

2.60. $X = 0.498$ m

(2.5) **2.61.** $X = 8.53$ mm

2.62. $\zeta = 0.05$

2.67. **(a)** $X = 2.71 \times 10^{-5}$ m **(b)** $e = 2.344 \times 10^{-3}$ m

(2.6) **2.69.** $k = 98{,}696$ N/m, $c = 87.956$ Ns/m

(2.7) **2.73.** $X = 1.8$ mm

2.74. $F_0 = 1481.6$ N

2.82. $F_0 = 294$ N

(3.1) **3.1.** $x(t) = 0.071e^{-0.1t} \sin(1.411t)$

3.8. $x(t) = 3.16 \times 10^4 e^{-39.72t} \sin 316.24t$

3.16. $k = \dfrac{1}{m}\left(\dfrac{m_b v}{|X|}\right)^2$

(3.2) **3.20.** $|Z(v)| \approx \left| \dfrac{Y\left(\dfrac{\pi}{\ell}\right)^3 v^3}{\omega_n(\omega_n^2 - \omega_b^2)} \right|$

3.23. $x(t) = 0.5t - 0.05 \sin(10t)$ m

3.25. $t_p = \dfrac{\pi}{\omega_d}$

3.27. $\omega_n = 3.26 \text{ rad/s}, \quad \zeta = 0.268$

(3.3) **3.32.** $F(t) = \displaystyle\sum_{n=1}^{\infty} b_n \sin nt$

where $b_n = \begin{cases} 0 & n \text{ even} \\ \dfrac{4}{\pi n} & n \text{ odd} \end{cases}$

3.36. $x(t) = 0.0965 e^{-5t} \sin(31.22t + 0.104) + 0.0501 \cos(1.414t - 1.57) \text{ m}$

(3.4) **3.39.** $x(t) = \dfrac{F_0}{k} - \dfrac{F_0}{k\sqrt{1 - \zeta^2}} e^{-\zeta\omega_n t} \sin\left(\omega_n \sqrt{1 - \zeta^2}\, t + \cos^{-1}(\zeta)\right)$

3.42. $x(t) = \dfrac{1}{5\sqrt{5}} \sin \sqrt{20}\, t$

(3.5) **3.44.** $\overline{x^2} = S_0 \displaystyle\int \left| \dfrac{10}{5 + 3j\omega} \right|^2 d\omega$

3.47. $E[x^2] = 11.333$

(3.6) **3.49.** $x(t) - y(t) = \dfrac{A}{m\omega_n^2}\left[1 - \dfrac{1}{t_0} + \dfrac{1}{t_0\omega_n} \sin \omega_n t - \cos \omega_n t \right] - \dfrac{A}{2}t^2 - \dfrac{A}{6t_0}t^3 \quad 0 \le t \le 2t_0$

$x(t) - y(t) = \dfrac{A}{m\omega_n^2}\left[\dfrac{1}{t_0\omega_n}(\sin \omega_n t - \sin \omega_n(t - 2t_0)) - \cos \omega_n t - \cos \omega_n(t - 2t_0) \right] \quad t > 2t_0$

(3.7) **3.55.** $\dfrac{X(s)}{F(s)} = \dfrac{1}{as^4 + bs^3 + cs^2 + ds + e}$

3.58. $\omega_n = 3 \text{ rad/s}, \quad \zeta = 0.227, \quad c = 3.03 \text{ kg/s}, \quad m = 2.22 \text{ kg}, \quad k = 20 \text{ N/s}$

(3.8) **3.62.** $x(t) = 1 - e^{-t} < 1$ therefore bounded

(4.1) **4.3.** $\mathbf{u}_1 = \begin{bmatrix} 1 \\ 2.6 \end{bmatrix}$ and $\mathbf{u}_2 = \begin{bmatrix} 1 \\ 5 \end{bmatrix}$

4.7. $\omega_1 = 0, \quad \omega_2 = 5/\sqrt{2} \text{ rad/s}$

4.8. $\omega_1 = 0.255 \text{ rad/s}, \quad \omega_2 = 1.239 \text{ rad/s}$

4.14. $x_1(t) = 0.025 - 0.025 \cos 16.73t$

$x_2(t) = 0.025 + 0.025 \cos 16.73t$

(4.2) **4.20.** $M^{1/2} = \begin{bmatrix} 3 & -4 \\ -4 & 5 \end{bmatrix}$

4.24. $P^T \hat{K} P = \begin{bmatrix} 0.382 & 0 \\ 0 & 2.618 \end{bmatrix}$

4.27. $\Lambda = \text{diag}(\lambda_i) = \begin{bmatrix} 0.454 & 0 \\ 0 & 220.05 \end{bmatrix}$

$P = [\mathbf{v}_1 \quad \mathbf{v}_2] = \begin{bmatrix} 0.9999 & -0.0144 \\ 0.0144 & 0.9999 \end{bmatrix}$

4.33. $a = -2$

(4.3) **4.36.** $\ddot{r}_1(t) + 0.465\, r_1(t) = 0$
$\ddot{r}_2(t) + 2.868\, r_2(t) = 0$

   4.37. $\theta(t) = \begin{bmatrix} 0.2774\cos\omega_1 t - 0.2774\cos\omega_2 t \\ 0.3613\cos\omega_1 t + 0.6387\cos\omega_2 t \end{bmatrix}$

   where $\omega_1 = 0.4821\sqrt{\dfrac{k}{J_2}}$ and $\omega_2 = 1.1976\sqrt{\dfrac{k}{J_2}}$

   4.41. $\mathbf{x}(t) = \begin{bmatrix} -0.630\cos 0.2845t + 0.175\cos 1.2426t \\ 0.111\cos 0.2845t + \cos 1.2426t \end{bmatrix}$ mm

(4.4) **4.47.** $\mathbf{x}(t) = 0.9468t\begin{bmatrix} 1 \\ 1 \\ 1 \end{bmatrix} + \begin{bmatrix} -0.1398 \\ -0.0570 \\ 0.00539 \end{bmatrix}\sin 9.672t + \begin{bmatrix} 0.01388 \\ -0.02346 \\ 0.000485 \end{bmatrix}\sin 20.844t$ m

(4.5) **4.64.** $\mathbf{x}(t) = \begin{bmatrix} 0.01709 \\ -0.01859 \\ 0.01709 \end{bmatrix} e^{-2.0771\times 10^{-6}t}\sin(2.0770\times 10^{-4}t - 1.5808)$

   $\qquad + \begin{bmatrix} 0.01744 \\ 0.03206 \\ 0.01709 \end{bmatrix} e^{-1.8142\times 10^{4}t}\sin(8.8877\times 10^{-4}t + 1.3694)$ m

   4.68. $\alpha = 0.1966$
   $\beta = 0.2778$

(4.6) **4.73.** $\mathbf{x}(t) = \begin{bmatrix} 0.0394e^{-0.1t}\sin 1.4106t + 0.0279e^{-0.2t}\sin 1.9899t \\ 0.118e^{-0.1t}\sin 1.4106t - 0.0834e^{-0.2t}\sin 1.9899t \end{bmatrix}$

   4.78. $x_2(t) = 2.221\times 10^{-4}t - 8.606\times 10^{-5}e^{-0.259t}\sin 2.581t$

(5.1) **5.3.** ≈ 11 Hz, $\bar{v} = \omega_n\bar{v} = \sqrt{2}\times 10^{-3}$ m/s
   5.4. $\omega_n = 96.4$ Hz. For an arm displacement of 10 μm the vibration is unsatisfactory.
(5.2) **5.10.** $c = 30.265$ kg/s, $k = 916$ N/m
   5.16. 2.58×10^{-3} m
   5.26. $\zeta = 0.167$

(5.4) **5.46.** $\dfrac{Xk}{M_0} = 150.6$

   5.48. With $\dfrac{Xk}{F_0} = 0.372$, $\zeta = 0.767$

   5.51. $X = 0.00147$ m
(5.5) **5.56.** $\zeta_{op} = 0.908$, $\omega = r\omega_n = 0.908\omega_n$
   5.62. $k_2 = 8000$ N/m, $c = 10.83$ kg/s
(5.6) **5.68.** $E = 7.136\times 10^{10}$ N/m^2
(5.7) **5.74.** **(a)** 61.67 rpm **(b)** 0.25 m

(6.2) **6.5.** $\omega_n = \dfrac{(2n-1)\pi}{2l}\sqrt{\dfrac{\tau}{\rho}}$ $X_n = \sin\dfrac{(2n-1)\pi x}{2l}$ for $n = 1, 2, 3, \ldots$

6.6. $c_n = 0$,

   $d_n = 0$ n even

   $d_n = \dfrac{8}{n^2\pi^2}\sin\left(\dfrac{n\pi}{2}\right)$, n odd

(6.3) **6.11.**   Steel

6.28. $w(x, t) = 0.04 \displaystyle\sum_{n=1}^{\infty} (-1)^{n+1} \sin[1367(2n-1)\pi t]\sin\left[\dfrac{(2n-1)}{1.2}\right]\pi x$

(6.4) **6.32.** $\omega_1 = 400.55$ Hz

   $\omega_2 = 1201.64$ Hz

   $\omega_3 = 2002.65$ Hz

6.36.   Torsion: $\omega_{tn} = 3162\,\dfrac{(2n-1)\pi}{2l}$ for $n = 1, 2, 3, \ldots$

   Longitudinal: $\omega_{ln} = 5164\,\dfrac{(2n-1)\pi}{2l}$ for $n = 1, 2, 3, \ldots$

6.38. $\phi_n(x) = \cos\dfrac{n\pi x}{l}$ $\omega_n = \sqrt{\dfrac{G}{\rho}}\,\dfrac{n\pi}{l}$ for $n = 1, 2, 3, \ldots$

(6.5) **6.42.** $\omega_n = \sqrt{\dfrac{\beta_n^4 EI}{\rho A}}$ where $\cos(\beta_n l) = -\dfrac{1}{\cosh(\beta_n l)}$

   $\phi_n(x) = -\left(\dfrac{\cos(\beta_n l) + \cosh(\beta_n l)}{\sin(\beta_n l) + \sinh(\beta_n l)}\sin(\beta_n x) + \cos(\beta_n x)\right)$

   $\quad + \left(\dfrac{\cos(\beta_n l) + \cosh(\beta_n l)}{\sin(\beta_n l) + \sinh(\beta_n l)}\sinh(\beta_n x) - \cosh(\beta_n x)\right)$ for $n = 1, 2, 3, \ldots$

6.45. $\omega_n = \left(\dfrac{n\pi}{l}\right)^2\sqrt{\dfrac{EI}{\rho A}}$ $\phi_n(x) = \sin\dfrac{n\pi x}{l}$

   for $n = 1, 2, 3, \ldots$, and $w(x, t) = \cos(\omega_2 t)\sin\dfrac{2\pi x}{l}$

(6.6) **6.50.** $\omega_{mn} = \sqrt{(2m-1)^2 + 4n^2}\sqrt{\dfrac{\tau}{\rho}}\dfrac{\pi}{2}$ for $m, n = 1, 2, 3, \ldots$

(6.7) **6.58.**   **(a)** $\zeta_n = \dfrac{\gamma}{2\sqrt{\rho\tau}}\dfrac{n\pi}{l}$ $n = 1, 2, 3, \ldots$

   (b) $\zeta_{mn} = \dfrac{\gamma l}{2\sqrt{\rho\tau(m^2 + n^2)}}$ $m, n = 1, 2, 3, \ldots$

6.59. $\dfrac{\text{kg}}{\text{m}\cdot\text{s}}$

(6.8) **6.68.** $w(x,t) = \sum\limits_{n=1}^{\infty} \Bigl\{ C_{1n} \sin \omega_n t + C_{2n} \cos \omega_n t$

$$+ \frac{(-1)^{n-1}}{\rho A} \left(\frac{F_0}{\omega_n^2 - \omega^2} \right) \sin \omega t \Bigr\} \sin \frac{(2n-1)\pi x}{2l}$$

where $\omega_n = \sqrt{\dfrac{E}{\rho}} \dfrac{(2n-1)\pi}{2l} \quad n = 1, 2, 3, \ldots$

(7.2) **7.3.** delta input: $f(t) = \delta(t) \quad F(s) = 1$

triangle input: $f(t) = \dfrac{1}{2} - \dfrac{4}{\pi^2} \sum\limits_{n=1,3,5,\ldots} \dfrac{1}{n^2} \cos \dfrac{2\pi n}{T} t$

$$F(s) = \frac{1}{2s} - \frac{4}{\pi^2} \sum\limits_{n=1,3,5,\ldots} \frac{1}{n^2} \frac{s}{s^2 + a^2} \qquad a = \frac{2\pi n}{T}$$

7.4. $\text{Error} = \sqrt{\dfrac{300}{(12 + m_s)}} - \sqrt{\dfrac{300}{12}} \quad 0.5 \le m_s \le 5.0 \; kg$

7.5. $\omega_T = 1 \; \text{rad/s}$
$a_0 = a_2 = 0 \qquad a_1 = -1 \qquad a_n = 0 \qquad n > 2$
$b_1 = -2 \qquad b_2 = 4 \qquad b_n = 0 \qquad n > 2$

(7.3) **7.8.** $x(t) = -\dfrac{5}{\pi} \sum\limits_{n=1}^{\infty} \dfrac{1}{n} (1 - 2 \cos(n\pi) + \cos(2n\pi)\sin(nt))$

(7.4) **7.10.** The system has eight modal peaks with the approximate natural frequencies (in Hz) of:

$$\begin{array}{cccc} \omega_1 \approx 2 & \omega_2 \approx 4 & \omega_3 \approx 10 & \omega_4 \approx 15 \\ \omega_5 \approx 22 & \omega_6 \approx 29 & \omega_7 \approx 36 & \omega_8 \approx 47 \end{array}$$

7.14. $\Delta \zeta = 0.5$

7.17. $\dfrac{X_1(s)}{F(s)} = \dfrac{s^2 + cs + 2}{s^4 + cs^3 + 4s^2 + 2cs + 3}$

$$\frac{X_2(s)}{F(s)} = \frac{1}{s^4 + cs^3 + 4s^2 + 2cs + 3}$$

(7.5) **7.24.** $\zeta = 0.05$

(7.6) **7.25.** $\alpha(\omega) = \dfrac{1}{\det(A)} \begin{bmatrix} 2 - \omega^2 + j\omega & 2 + j\omega \\ 2 + j\omega & 6 - 2\omega^2 + j3\omega \end{bmatrix}$

$\det(A) = 2\omega^4 - 12\omega^2 + 8 + j(-5\omega^3 + 8\omega)$

(8.1) **8.2.** $\omega_1 = 0, \quad \omega_2 = \sqrt{\dfrac{\pi^2 E}{\rho l^2}}$ which is 10.2% off from the FEM value.

(8.2) **8.13.** $\omega_1 = 5459 \; \text{rad/s}, \quad \omega_2 = 19{,}630 \; \text{rad/s}$

8.15. $\omega_1 = 43{,}895 \; \text{rad/s}, \quad \omega_2 = 91{,}528 \; \text{rad/s}$

8.17. $\omega_1 = \dfrac{1.984}{l} \sqrt{\dfrac{E}{\rho}}$

(8.3) **8.26.** $\omega_1 = 3.52 \dfrac{1}{l^2} \sqrt{\dfrac{EI}{\rho A}}$, $\omega_2 = 20.49 \dfrac{1}{l^2} \sqrt{\dfrac{EI}{\rho A}}$, $\omega_3 = 34.98 \dfrac{1}{l^2} \sqrt{\dfrac{EI}{\rho A}}$

8.28. $\omega_1 = 1132\ \text{rad/s}$, $\omega_2 = 3123\ \text{rad/s}$, $\omega_3 = 6122\ \text{rad/s}$

8.30. $\omega = 20.49 \dfrac{1}{l^2} \sqrt{\dfrac{EI}{\rho A}}\ \text{rad/s}$

(8.4) **8.34.** $\omega_1 = 6670\ \text{rad/s}$, $\omega_2 = 13{,}106\ \text{rad/s}$

(8.5) **8.47.** $M = l \begin{bmatrix} 0.9 & 0 \\ 0 & 0.9 \end{bmatrix}$, $K = \dfrac{1}{l} \begin{bmatrix} 1.89 & 0.48 \\ 0.48 & 0.36 \end{bmatrix}$

(8.6) **8.52.** $M_r = \begin{bmatrix} 2 & 0 \\ 0 & 2 \end{bmatrix}$, $K_r = \begin{bmatrix} 19.95 & -0.15 \\ -0.15 & 36.55 \end{bmatrix}$

Index

A

Accelerance transfer function, 272–273
Acceleration, simple harmonic motion, 22, 25w
Accelerometers, 178f, 179f, 590
Aircraft
    base excitation example, 169–171
    control tab, 125f
    foot pedal model, 117f
    jet engine with transverse vibration, 302f
    landing system, 118f
    steering-gear mechanism, 116f
    vibration-induced fatigue, 491
    wing
        distributed-parameter system, 514
        engine mount, 119f
        harmonic excitation example, 181–182
        impulse response function, 300f
        stability example, 83
        torsional vibration, 432f
        vibration examples, 61–64, 385–387
        vibration models, 212f, 435f
Air damping, 191
Airfoil, 216f
Algebraic eigenvalue problem, 412, 414
Aliasing, 593–594
Amplitude, 20
Angular motion, 538f
Angular natural frequency, 20
Arbitrary input, general forced response, 238–247
    examples, 240–247
    problems, 302–304
Arbitrary periodic input, general forced response, 247–254
    examples, 249–254
    problems, 305–306
Argand plane plots, 603
Assumed mode method, 578

B

Banded matrix, 686
Bar
    distributed-parameter systems, 531–537
        examples, 532–536
        problems, 579–581
    finite element method, 631–637
        example, 636–637
        problems, 661–662
    two materials, 580f
    See also Cantilevered bar
Baseball bat, 54
Base excitation, 163–172
    examples, 168–172
    harmonic excitation, 163f
    problems, 217–220
Beam bending vibration, 544–556
    examples, 549–556
    problems, 582
Beam elements, finite element method for, 642–650
    examples, 646–650
    problems, 664–665
Beam–mass model, 138

Asymmetric eigenvalue problem, 344
Asymptotic stability, 81, 277, 279
Autocorrelation function, 261
Automobiles
    base excitation example, 169–171
    brake pedal model, 214f
    drive train vibration analysis, 435f
    frequency response function, 453f
    single-degree-of-freedom model, 452f
    tires and resonance, 129
    vibration isolation example, 458–460, 459f
    vibration response example, 452–453
    See also Suspension systems; Vehicles
Average value, 32

Beams
    Euler–Bernoulli, 545–552, 545f
    shear deformation, 553f
    single-finite-element model, 643f
    Timoshenko, 552–556, 553f
    tip mass, 580f
    transverse vibration, 545f, 551t
Bearing housing displacement, 621f
Beat, 136
Beats, two-degree-of-freedom system, 329
Bell-shaped curve, 265
Bernoulli–Euler beams. See Euler–Bernoulli beams
BIBO (bounded-input, bounded-output stability), 275, 277, 279
Biharmonic operator, 561
Bilinear systems, 95
Borel's theorem, 258
Boundary value problem, 518
Bounded-input, bounded-output (BIBO) stable, 275, 277, 279
Bridge, 663f
Broadband vibration absorption, 481f
Buildings
    ground motion, 219f
    horizontal vibration example, 360–363, 361f, 363f
    machine with rotating unbalance, 572f, 583f

C

Cable vibration, 516–519
    example, 519
Camera mount, 234f
Cantilevered bar
    finite element grids, 632f
    longitudinal vibration, 567f, 632w
    one-element model, 661f
    three-element four-node model, 632f
Cantilevered beam, 545
    applied axial force, 661f
    driving points, 611f

Cantilevered beam (*Continued*)
  measurement points, 614f
  spring-mass system attached, 664f
  two-element, three-node
    mesh, 649f
Center of percussion, 51, 52
Characteristic equation, 35, 323, 521
Cholesky decomposition, 330,
  401–404
Circle fitting, 603–607
  problems, 625–626
Clamped beam, 665f
Clamped–clamped bar, 567f, 652f
Clamped–free bar. *See*
  Cantilevered bar
Clamped–free beam. *See*
  Cantilevered beam
Clamped two-element beam
  system, 666f
Clamped two-step aluminum
  beam, 665f
Coherence function, 599, 600f
Coiled spring, stiffness of, 64–65, 65f
Complex arithmetic, 478w
Complex modes, 416
Complex modulus, 190
Complex stiffness, 190, 491, 492f
Computational eigenvalue problems
  for vibration in MDOF
  systems, 401–419
Computer-controlled vibration
  endurance test, 619f
Computer disk drive motor, 464f
Computer software
  eigenvalues, 404
  numerical simulation, 84, 192
Consistent-mass matrices, 650
Constrained-layer damping, 494
Continuous systems. *See*
  Lumped-parameter systems
Conversation of energy
  equations, 45
Convolution integral, 239, 240w, 245
Cooling fan, 495f
Coulomb damping, 93–94
Coulomb friction
  free response, 96, 97f, 98f
  harmonic excitation, 182–185
  vibration, 93–100
    examples, 97–100
    problems, 126–127
Coupling device, 429f
Critical damping coefficient, 35

Critically damped motion, 39–43
  response, 40f
Critical points, 483
Critical speeds, for vibration
  suppression on rotating
  disks, 497–503
  examples, 500–503
  problems, 512–513
Cross-correlation function, 597
Cross-spectral density, 598
Cutoff frequency, 589

D

Damped natural frequency, 36
Damped systems
  eigenvalue problems, 408–411,
    414–419
  harmonic excitation, 142–156
    examples, 146–156
    problems, 213–216
  single-degree-of-freedom, 35f
  two-degree-of-freedom, 440f
Damping
  air, 191
  Coulomb, 93–94, 182–185
  distributed-parameter systems,
    562–567
    examples, 563–567
    problems, 582–583
  harmonic excitation, 182–192
    examples, 185–192
    problems, 222–223
  hysteretic, 188
  modal, 368–374
  models, 192t
  proportional, 374
  vibration absorber with, 475f
  vibration absorption, 475–482
  viscous, 33–43
Damping coefficient, 34, 72–73
  amplitude of vibration example,
    76–77
Damping ratio, 36, 72–73
Dashpot, 34f
Decibel (dB), 32
Decoupling equations of motion
  using modal analysis, 347f
Degree of freedom, 16
  See also Multiple-degree-of-
    freedom (MDOF) systems;
    Single-degree-of-freedom
    systems

Design, definition of, 448
Design considerations
  harmonic excitation example,
    197–199
  modal approach, 419
  range of, 460
  robustness, 80, 475
  rotor system example, 500–501
  vibration, 75–80
    examples, 76–80
    problems, 123–124
  vibration, acceptable levels of, 454
  vibration absorbers, 470
  vibration suppression, 447–513
Diagnostics, vibration testing for,
  618–621
  example, 619–620
Digital Fourier analyzer, 591
Digital Fourier transform
  (DFT), 591
Digital representations of
  signals, 593f
Digital signal processing, 591–596
  example, 594
  problems, 623
Digital spectral coefficients, 594
Dirac delta function, 231
Discrete systems. *See*
  Lumped-parameter systems
Discretization, 630
Disk drive motor of personal
  computer, 464f
Disk–shaft system, 20w
  critical speeds for vibration
    suppression, 497–503
  example, 49
  harmonic excitation example,
    161–162
  torsional vibration, 59f
Displacement
  simple harmonic motion,
    22, 23w
  vibration, 449t
Displacement transmissibility, 165,
  166f, 168, 169f, 454, 455w
Distributed-parameter systems,
  514–584
  bar vibration, 531–537
  beam bending vibration,
    544–556
  cable vibration, 516–519
  damping models, 562–567
  explanation, 515

forced response modal analysis, 568–578
  membrane vibration, 556–562
  modes, 520–530
  natural frequencies, 520–530
  plate vibration, 556–562
  rod vibration, 531–537
  string vibration, 516–519
  torsional vibration, 537–544
Divergent instability, 81
Divergent response, 81, 81f
Diving board, 124f
Dot product, 318
Double pendulum, with generalized coordinates, 381f
Driving frequency, 130, 136
Driving point, 611
Duhamel integral, 240
Dynamically coupled systems, 387
  eigenvalue problems, 401–419

E

Effective mass, 47
Eigenfunctions, 522
Eigenvalue problems
  computational, 401–419, 442–443
  damped systems, 414–419
    example, 416–419
  dynamically coupled systems, 401–404
    example, 403–404
  two-degree-of-freedom system example, 336–338
  using codes, 404–411
Eigenvalues
  distributed-parameter systems, 522
  MDOF systems, 330–344
    examples, 332–344
    problems, 430–432
Eigenvectors, 332w, 333, 336–338
Elastic damper, 488f
Elastic modulus, 58
  complex data, 493t
  measurement, 70
  stress–strain curve, 71f
  temperatures, 493f
Electric motor mount, 505f
Electronic cabinet with cooling fan, 495f

Endurance, vibration testing for, 618–621
Energy methods for modeling vibration, 43–58
  examples, 46–58
  problems, 116–120
Engineering Vibration Toolbox
  distributed-parameter systems, 584
  eigenvalue problems, 401
  finite element method, 668
  general forced response, 313–314
  harmonic excitation, 226–227
  MDOF systems, 445
  Runge–Kutta method, 89
  vibration, 127–128
  vibration testing, 627–628
Ensemble average, 265
Equilibrium points/positions, 93, 95, 99–101, 101f
Equivalent viscous damping, 297f
Euler–Bernoulli beam model, 555
Euler–Bernoulli beams, 545–552, 545f
Euler method
  linear and nonlinear equations, 102–103
  numerical solution, 86–90
  single-degree-of-freedom system, 84
Euler relations, 30, 36, 37w
Exciters, 587–588
Expansion method, 358–363
  examples, 360–363
Expansion theorem, 359, 578
Expected value, 265
Experimental modal analysis. *See* Vibration testing

F

Fast Fourier transform (FFT), 591, 595
FEM. *See* Finite element method
Finite dimensional systems. *See* Lumped-parameter systems
Finite element analysis (FEA), 631
Finite element mesh/grid, 630
Finite element method (FEM), 629–668
  bar, 631–637
  beam elements, 642–650
  lumped-mass matrices, 650–653

model reduction, 658–661
  three-element bar, 637–642
  trusses, 653–658
Finite element model, 631
Finite elements (FE), 630
First mode shape, 326, 360
Flexural vibrations, 544
Floor-mounted compressor, 506f
Fluid systems
  example, 47
  natural frequency example, 65–67
Flutter instability, 81, 82f, 279
Forced response. *See* General forced response
Forced response modal analysis
  distributed-parameter systems, 568–578
  examples, 569–578
  problems, 583–584
  MDOF systems, 374–381
  examples, 376–381
  problems, 440–441
Force summation method, 44
Force transmissibility, 167, 169f, 454, 455w
Forcing frequency, 130
Formula error, 87–88
Fourier coefficients, 248
Fourier representations of signals, 593f
Fourier series, 248, 591–592, 592w
Fourier transforms, 258, 591
Fragility, 448
Free response, 17, 22
  Coulomb friction, 96, 97f, 98f
  numerical simulation of time response, 84–93
Free vibration, 13
Frequency
  cutoff, 589
  importance of concept, 27
  two-degree-of-freedom system, 329
  vibration, 449
Frequency response approach to harmonic excitation, 158–160
  problems, 217–218
Frequency response curves for mode shapes, 612–613f
Frequency response function, 159, 600, 600f
Friction coefficients, 94f, 94t

G

Gaussian distribution function, 265
General forced response, 228–314
  arbitrary input response, 238–247
  arbitrary periodic input response,
      247–254
  impulse response function,
      229–238
  nonlinear response properties,
      291–299
  numerical simulation, 279–291
  random input response, 259–267
  shock spectrum, 267–271
  stability, 274–279
  transfer functions, 271–274
  transform methods, 254–259
Generalized eigenvalue
    problem, 411
Generalized symmetric eigenvalue
    problem, 343
Geometric approach to harmonic
    excitation, 157–158
  problems, 217–218
Gibbs phenomenon, 250
Global condition, 101
Global coordinate system, 653
Global mass matrix, 639
Global stiffness matrix, 639
Gravity, spring problems and,
    28–29

H

Hammer
  center of percussion, 54
  impact, 588–590, 589f
  impulse, 588
  instrumented, 233, 234
Hanning window, 595, 596f
Harmonic excitation, 129–227
  base excitation, 163–172
  damped systems, 142–156
  damping, forms of, 182–192
  design considerations, 196–199
  explanation, 130
  frequency response approach,
      158–160
  geometric approach, 157–158
  measurement devices, 178–182
  nonlinear response properties,
      200–209
  numerical simulation, 192–199
  rotating unbalance, 172–177

transform approach, 160–163
  undamped systems, 130–142
Harmonic motion
  examples, 28–33
  problems, 111–113
  representations, 31w
  vibration, 25–27
Heaviside step functions, 236–237,
    269, 256, 284–285
Helical spring
  spring–mass system natural
      frequency, 78
  stiffness, 60
Helicopter. *See* Rotorcraft
Hertz (Hz), 27
Hooke's law, 71
Houdaille damper, 481f
Humans
  forearm vibration model, 211f
  longitudinal vibration, 211f
Hysteresis loop, 187, 187f
Hysteretic damping, 188
Hysteretic damping constant, 188

I

Impact, 233, 235f, 238f, 589f
Impact hammer, 588–590
Impulse, definition of, 230
Impulse hammer, 588
Impulse response function
  general forced response, 228–238
  examples, 233–238
  problems, 299–314
Inconsistent-mass matrices, 650
Inertia force, 44
Inertia matrix, 320
Infinite-dimensional systems, 515, 527
  See also Lumped-parameter
      systems
Initial conditions, 22f
Inner product, 319
Input frequency, 130
International Organization of
    Standards (ISO), 448, 449
Inverted pendulum, 124–125,
    125f, 278
Isolation problems, 454

K

Kelvin–Voigt damping, 566
Kinetic energy, 14
Kronecker delta, 339w

L

Lagrange's equations, 44
  energy method, 55–56
  example, 55–57
  MDOF systems, 381–389
  examples, 383–389
  problems, 441–443
Lagrange stability, 277
Laplace operator, 557
Laplace transforms
  common, 256t
  convolution type evaluations, 245
  Fourier transforms versus, 259
  general forced response, 254–259
  harmonic excitation, 160–163
Laptop computers, 465
Lathe
  MDOF system example, 389–393
  moving parts, 390f
Leaf spring, transverse vibration of,
    61, 61f
Leakage, 595, 596f
Legs, vibration example, 40–43
Levers, vibration model of
    coupled, 384f
Linear systems, 19
Local coordinate direction, 653
Local stability, 102
Logarithmic decrement, 72
Longitudinal motion, 58
Longitudinal vibration, 211f, 531f,
    536t, 537t, 567f
Loss coefficient, 186
Loss factor, 186
Lumped-mass matrices, 650–653
  problems, 666–667
Lumped-parameter systems, 515, 536
  example, 651–653

M

Machinery
  acceptable vibration levels, 450f
  health monitoring, 619–620
  rotating unbalance examples,
      172f, 221f, 504f
  rubber mount, 218f
  vibration absorbers, 467
  vibration isolation, 488f
  vibration model, 383f
Marginal stability, 80
Mass, frequency of oscillation for
    measuring, 74–75

Mass condensation, 660
Mass loading, 588
Mass matrix, 320
Mass moment of inertia, 70
Mass normalized stiffness, 331
Mass ratio versus natural
    frequency, 473f
Mathcad
    eigenvalues, 405–406, 417–419
    general forced response,
        281–283, 285, 286, 288, 294,
        297–299
    harmonic excitation, 193–194, 197
    linear and nonlinear equations,
        102–105, 207–209
    MDOF systems, 420–421,
        422–427
    Runge–Kutta method, 91
Mathematica
    eigenvalues, 406–408, 410–411,
        418–419
    general forced response, 283,
        285–286, 288, 290, 295–296,
        298–299
    harmonic excitation, 199
    linear and nonlinear equations,
        104–105, 205, 208–209
    MDOF systems, 422, 424, 426–427
    Runge–Kutta method, 91
MATLAB
    eigenvalues, 405, 409, 417–419
    Engineering Vibration Toolbox,
        89, 127–128, 226, 313, 401,
        445, 513, 584, 627, 668
    general forced response, 282,
        284–286, 287, 289, 294, 298
    harmonic excitation, 195–196, 198
    linear and nonlinear equations,
        103–104, 204, 208
    MDOF systems, 421, 423, 426
    Runge–Kutta method, 89
Matrix inverse, 322
Matrix of mode shapes, 347, 347f
Matrix square root, 330
MDOF systems. See Multiple-
    degree-of-freedom (MDOF)
    systems
Measurement
    hardware, 586f, 587–591, 612f
        problems, 623
    harmonic excitation, 178–182
        example, 181–182
        problems, 222

transfer functions, 272–274
    vibration
        examples, 71–75
        problems, 122–123
Membrane vibration, 556–562, 557f
    example, 558–562
    problems, 582
Method of undetermined
    coefficients, 132
Mindlin–Timoshenko theory, 562
Min-max problem, 486
Mobility frequency response
    function, 604f, 604w
Modal analysis
    forced response distributed
        parameter systems, 568–578
    forced response MDOF systems,
        374–381
    MDOF systems, 344–352
        problems, 432–434
    See also Vibration testing
Modal coordinate system, 346, 348
Modal damping, 368–371, 373w,
    416, 562–567
Modal data extraction, 600–603
    example, 602–603
    problems, 624–625
Modal equations, 346, 348, 563
Modal participation factors, 360
Modal testing. See Vibration testing
Modeling, definition of, 43
Modeling methods, vibration, 43–58
    examples, 46–58
    problems, 116–120
Model reduction, 658–661
    example, 660–661
    problems, 667
Modes, 367
    distributed-parameter systems,
        520–530
        examples, 524–530
        problems, 578–579
Mode shapes
    clamped–pinned beam, 550f
    definition, 316
    eigenvectors, 338
    explanation, 367
    first, 326, 360
    longitudinal vibration, 537f
    measurement for vibration
        testing, 608–618
        examples, 611–618
        problems, 626–627

nodes, 363
    normalizing, 342–343
    resonance, 378–379
    second, 326
    torsional vibration, 544t
    vibrating string, 527f
Mode summation method
    distributed-parameter systems,
        534–536
    forced response, 379–381
    modal analysis, 358–363
        examples, 360–363
    modal damping, 370–373, 373w
Modulus data, 59t
Mounting bracket, 575f
Mounts
    aircraft wing engine, 119f
    base excitation and, 163–172
    electric motor, 505f
Multiple-degree-of-freedom
    (MDOF) systems, 315–446
    computational eigenvalue
        problems, 401–419
    eigenvalues, 330–344
    examples, 389–401
    forced response modal analysis,
        374–381
    Lagrange's equations, 381–389
    modal analysis, 344–352, 358–363
    more than two degrees of
        freedom, 352–358, 353f
        examples, 355–358
        problems, 434–438
    natural frequencies, 330–344
    numerical simulation, 419–427
    two-degree-of-freedom model
        (undamped), 316–330
    viscous damping, 368–374

N

Natural frequency
    aircraft wing, 61–62
    angular, 20
    damped, 36
    distributed-parameter systems,
        520–530
        examples, 524–530
        problems, 579–581
    energy method, 54
    fluid system, 50
    human leg, 41–42
    longitudinal vibration, 537t

Natural frequency (*Continued*)
  mass ratio versus, 472f
  MDOF systems, 316, 330–344
    examples, 332–344
    problems, 430–432
  pendulum, 29, 52–54
  spring–mass system, 21, 28, 49,
    67–68
  torsional system, 60
  torsional vibration, 544t
  wheel, 46–47
n-degree-of-freedom system, 353f
Neutral plane/surface, 561
Newton's laws, 21–22, 44
Nodes, in finite element analysis,
  630, 631
Nodes of a mode, 363, 527
Nonlinear response properties
  general forced response, 291–299
    examples, 292–299
  harmonic excitation, 200–209
    examples, 201–209
    problems, 225–226
Nonlinear systems, 19
  Coulomb friction, 93–100
  general forced response
    problems, 311–313
  nonlinear pendulum equations,
    101–107
Nonperiodic forces, 229–230
Normalization of eigenvectors,
  336–338
Nose cannon, 510f
Numerical simulation
  general forced response, 279–291
    examples, 281–291
    problems, 310–311
  harmonic excitation, 192–199
    examples, 193–199
    problems, 223–225
  MDOF systems, 419–427
    examples, 420–427
    problems, 443–445
  vibration and free response, 84–93
    examples, 86–93
    problems, 126–127
Numerical solutions
  concept of, 84–85
  Euler method examples, 85–89
  sources of error, 87
Nyquist circles, 603
Nyquist frequency, 594
Nyquist plots, 603, 606f

O

Operational deflection shape (ODS)
  measurement, 621–623
Optical table with vibration
  absorber, 468f
Optimization in vibration
  suppression
    examples, 486–491
    problems, 510–512
  vibration suppression, 483–491
Orthogonality, 248, 334, 337, 533
Orthogonal matrices, 338
Orthonormal vectors, 334, 338, 338w
Oscillation
  decay in, 33
  frequency of, for measuring mass
    and stiffness, 73–75
  natural frequency examples,
    46–47, 50–51, 60
Oscillatory motion, 22
Overdamped motion, 38–39, 39f
Overshoot, 242

P

Package, vibration model of
  dropped, 301f
Parts sorting machine, 508f
Peak amplitude method, 602f
Peak frequency, 155
Peak time, 242
Peak value, 31
Pendulum
  compound, 51–54, 52f, 53f
  damped, 214f
  double, with generalized
    coordinates, 381f
  equilibrium positions, 101f
  examples, 14–16, 47–48, 51–54
  inverted, 82–83, 278
  nonlinear systems, 101–106
    problems, 127
  swinging, 20w
Performance robustness, 475
Periodic forces, 229
Period T, 27, 136
Personal computer disk drive
  motor, 464f
Phase, 20
Physical constants for common
  materials, 59t
Piezoelectric accelerometers, 179f,
  181, 590

Pinned beam, 547
Pitch, 353f
Pivot point, 54
Plate vibration, 556–562
Positive definite matrix, 402w
Positive semidefinite matrix, 402w
Potential energy, 14
Power-line pole with
  transformer, 308f
Power spectral density (PSD),
  261–263, 597
Printed circuit board, 507f
Probability density function, 265
Proportional damping, 374
Pulse input function, 293f
Punch press
  base excitation example, 171
  machine schematic, 398f
  MDOF system example, 397–401
  three-element-bar problem, 663f
  vibration model, 398f

Q

Quadratic damping, 191
Quadrature peak picking
  method, 601f

R

Radial saw, 473f
Radius of deflection, 500f
Radius of gyration, 52
Random input, general forced
  response, 259–267
  examples, 264–267
  problems, 307
Random signal analysis, 596–599
Random vibration analysis, 597w
Rattle space, 460
Rayleigh dissipation function, 119
Receptance matrix, 608–609
Receptance transfer function, 604f
Rectilinear system, 58t
Reduced-order modeling. *See*
  Model reduction
Resonance
  damped systems, 150, 155
  distributed systems, 515
  explanation, 129–130
  importance of concept, 133
  MDOF systems, 378–379
  modal testing, 587
  undamped systems, 137, 137f

Response
 divergent, 81, 81f
 free, 17, 22, 84–93, 96, 97f, 98f
 steady-state, 145, 149–150
 transient, 145, 149–150
 See also Forced response modal
  analysis; General forced
  response
Response spectrum. *See* Shock
 spectrum
Rigid-body modes, 364–367, 392
 example, 364–367
Robustness, of designs, 80, 475
Rod vibration, 531–537
 examples, 532–536
 problems, 581–582
Roll, 353f
Rolling disk vibration model, 120f
Root-mean-square value, 32, 261, 449
Rotating disk critical speeds,
 497–503, 497f
Rotating unbalance
 equation, 498w
 harmonic excitation, 172–177
  examples, 175–177
  problems, 220–222
 model of disk–shaft system, 497f
 model of machine, 172f, 221f, 504f
 model of machine in building,
  572f, 583f
 model of motor, 221f
Rotational kinetic energy, 46
Rotational system, 58t
Rotorcraft
 and resonance, 129
 rotating unbalance example,
  176–177
 thrust directions, 176f
Round-off error, 87
Runge–Kutta method, 87–88
 examples, 89–92
 general forced response, 284–290
 linear and nonlinear equations, 104
 Mathematica, 91
 MDOF systems, 422–424
 single-degree-of-freedom
  system, 84

S

Saddle point, 484, 485f
Sample function, 259–267
Sampling theorem, 594

Scalar product, 319
Scanning laser doppler vibrometer
 (SLDV), 591
Second mode shape, 326
Seismic accelerometer, 178f
Self-excited vibrations, 81
Semidefinite systems, 392
Separation of variables, 520
 solutions method, 530w
Settling time, 242
Shaft and disk. *See* Disk–shaft
 system
Shaker, 272
Shakers, 587–588
Shannon's sampling theorem, 594
Shape functions, 633
Shear coefficient, 553
Shear modulus, 553f
Ship, fluid system example, 65–66
Shock, 267, 453–454, 454f
Shock loading, 229
Shock pulse, 458
Shock spectrum, general forced
 response, 267–271
 examples, 268–271
 problems, 307–308
Signal conditioners, 590
Signal processing. *See* Digital signal
 processing
Signals, representations of, 593f
Signum function, 95
Simple harmonic motion, 22, 25w
Simple harmonic oscillator, 22
Simple machine part, vibration
 model of, 383f
Simple sine function, 260f
Simply supported beam, 547
Sine function, 13
Single-degree-of-freedom curve
 fit, 600
Single-degree-of-freedom systems,
 17, 231w
 compliance frequency response
  function, 601f
 damped, 35f
 example, 20w
 external force, 131f
 response, 231w
 undamped, 22
Sinusoidal vibration, acceptable
 limits, 450f
Sliding boundary, 547
Sloshing, 51

Software
 eigenvalues, 404
 numerical simulation, 84, 192
Solid damping, 188
Specific damping capacity, 186
Spectral matrix, 340
Spring–mass–damper system
 deterministic and random
  excitations, 266w, 267
 excitation response, 598w
 general applied force, 292f
 magnitude plot, 273f
 potentially nonlinear
  elements, 201f
 square input, 284f
 total time response, 254f
 truck suspension system
  example, 244–245
Spring–mass system, 20w
 examples, 23, 49–50, 55
 gravitational field, 45f
 harmonic excitation examples,
  134–142, 153–154
 helical spring natural frequency, 78
 kinetic coefficient of friction, 94f
 natural frequency example, 67–69
 nonnegligible mass, 49f
 problems, 107–110
 response of, 21f
 vibration, 17–25
 vibration absorbers, 467
Springs
 coiled, 64–65, 65f
 constants, 65t, 68
 helical
  natural frequency, 78
  stiffness, 63
 leaf, 61–62, 61f
 manufacture of, 68
 static deflection, 18f, 69, 79
 stiffness calculation rules, 67f, 68
Stability
 asymptotic, 81, 277, 278
 BIBO, 275, 277, 279
 general forced response, 274–279
  examples, 278–279
  problems, 310
 local, 102
 marginal, 80
 vibration, 80–83, 275w
  examples, 82–83
  problems, 124–125
  response, 81f

State matrix, 89, 413
State variables, 89
State vector, 89, 413
Static coupling, 387
Static deflection of spring, 18f, 69, 79
Stationary signals, 260f
Steady-state response, 145, 149–150
Steam-pipe system with
 absorber, 507f
Steel, elastic modulus, 71f
Step function, 240, 241f
Stereo turntable
 frequency response function, 453f
 single-degree-of-freedom
  model, 452f
 vibration response example,
  452–453
Stiffness, 58–69
 calculation rules for parallel and
  series springs, 67f, 68
 coiled spring, 61–62, 62f
 definition, 18
 examples, 60–69
 frequency of oscillation for
  measuring, 74–75
 helical spring, 60
 problems, 120–122
 twist, 60
Stiffness matrix, 320
Stinger, 588
Strain gauges, 590
Strain rate damping, 566
Stress–strain curve for elastic
 modulus, 71f
String equation, 518
String vibration, 516–519, 516f
Structural damage
 acceptable vibration levels, 450f
 vibration measurements, 621
Structural damping, 188
Structural Health Monitoring
 (SHM), 586
Subway car coupling device, 429f
Superposition, 107, 130, 229, 648
Support motion. *See* Base excitation
Suspension systems
 arbitrary input response, 244
 base excitation, 171
 chassis dynamometer, 301f
 damped, 120f
 design of, 54
 examples, 23, 60–61, 79–80, 170
 harmonic excitation, 162

mass of occupants, 219f
model of, 301f
multiple-degree-of-freedom
 systems, 352
speed bump, 303f
torsional, 161–162
torsion rod, 112f
trifilar, 70f
two-degree-of-freedom
 model, 429f
vertical, 79–80
Symmetric eigenvalue problem,
 333–334, 332w, 344, 413
Symmetric matrix, 320
Synchronous whirl, 499

T

Tangent line method, 85
Telephone lines, 468
Tennis racket, 54
Tensile test, 71
Thin plate theory, 562
3-dB down point, 601
Three-element bar, finite element
 method for, 637–642
 examples, 639–642
 problems, 664–665
Time history of impulse force, 230f
Time response
 lumped- versus distributed-
  parameter systems, 536
 MDOF systems, 419–427
 vibration and free response, 84–93
  examples, 86–93
Time to peak, 242
Timoshenko beam model, 555
Timoshenko beams, 552–556, 553f
Timoshenko shear coefficient, 553
Tires, and resonance, 129
Torsional constant, 539–540
Torsional motion, 58
Torsional system
 natural frequency, 60–61
 two degrees of freedom, 428f
Torsional vibration
 boundary conditions, 540t
 distributed-parameter systems,
  537–544
  examples, 540–544
  problems, 581–582
 shaft, 59f
 transform approach, 161–162

Transducers, 178, 587, 590
Transfer functions, 161, 272t
 general forced response, 272–274
 problems, 308–309
Transformations, 344
Transform methods
 general forced response, 254–259
  examples, 255–259
  problems, 306–307
 harmonic excitation, 160–163
  problems, 216–217
Transient response, 145, 149
Transmissibility
 base excitation example, 169–170
 displacement, 165, 166f, 167, 169f
 force, 167, 169f
 formulas, 455w
Transmissibility ratio, 454–467, 458f
Transmission lines, 468
Transpose of matrix, 320
Transverse motion, 58
Transverse vibrations, 544
Truck
 hitting object, 583f
 loading dirt, 244f
 pipe stacking, 117f
 spring–mass–damper system
  example, 244–245
Trusses, 653–658
 problems, 666–667
 three-element, 667f
Turntable. *See* Stereo turntable
Twist, 60
Two-degree-of-freedom model
 damped, 440f
 MDOF systems, 316–330,
  317f, 340f
  examples, 319–330
  problems, 427–430
 rigid-body translation, 364f
 vehicle example, 394–397
 viscous damping, 373f
Two-member framed structure, 654f

U

Undamped motion, 80–81
Undamped systems
 harmonic excitation, 130–142
  examples, 134–142
  problems, 209–213
 Lagrange stability, 277
 single-degree-of-freedom, 22

two-degree-of-freedom, 316–330
  examples, 319–330
Underdamped motion, 36–38
Underdamped solution, 37w
Underdamped systems
  forced response, 152w, 376w
  response, 38f, 72f
  vibration example, 42–43
Unrestrained degree of
  freedom, 364
U-tube manometer, 50f

V

Valve and rocker arm
  system, 114f
Variance, 261
Vector equation, 319
Vehicles
  side section, 394f
  two-degree-of-freedom system
    example, 394–397
  See also Automobiles
Velocity, simple harmonic motion,
  25, 26w
Velocity-squared damping, 191,
  205–206, 296–297
Vibration, 13–128
  acceptable levels, 448–454
    examples, 450–454
    problems, 503
  consequences of, 448
  Coulomb friction, 93–107
  description, 13–14
  design considerations, 75–80
  displacement amplitude, 449t
  energy methods, 43–58
  explanation, 14
  frequency ranges, 449t
  harmonic motion, 25–33
  measurement, 70–75
  modeling methods, 43–58
  nonlinear, 93
  nonlinear pendulum equations,
    101–106

numerical simulation of time
  response, 84–93
performance standards, 448
shock versus, 454
spring–mass model, 17–25
stability, 80–83
stiffness, 58–69
viscous damping, 33–43
Vibration absorbers, 467–475
  damping, 475f
  examples, 473–474
  problems, 507–508
  viscous, 481f
Vibration absorption damping,
  475–482
  problems, 508–510
Vibration dampers, 475
Vibration isolation, 454–467
  examples, 458–467
  optimization, 488–490
  problems, 503–506
  transmissibility formulas, 455w
Vibration model
  airplane wing, 62f, 385f
  coupled levers, 384f
  punch press, 398f
  simple machine part, 383f
Vibration suppression, 447–513
  optimization, 483–491
    rotating disk critical speeds,
      497–503
  vibration absorbers, 467–475
  vibration absorption damping,
    475–482
  vibration isolation, 454–467
Vibration testing, 585–628
  circle fitting, 603–607
  digital signal processing, 591–596
  endurance and diagnostics,
    618–621, 619f
  measurement hardware, 587–591,
    587f, 612f
  modal data extraction, 600–603
  mode shape measurement,
    608–618

operational deflection shape
  (ODS) measurement,
  621–623
random signal analysis, 596–599
uses of, 586
Virtual displacements, 382
Virtual work, 382
Viscoelastic, definition of, 492
Viscous damping, 33–43
  critically damped motion,
    39–43
  equivalent, 297f
  examples, 39–43
  MDOF systems, 368–374, 388
    examples, 369–374, 388–389
    problems, 438–440
  overdamped motion, 38–39
  problems, 113–115
  two-degree-of-freedom
    system, 373f
  underdamped motion, 36–38
  vibration, 33–43
  See also Damped systems;
    Undamped systems;
    Underdamped systems
Viscous vibration absorbers,
  481f, 482f

W

Washing machine, 504f
Wave hitting seawall, 304f
Wave speed, 518
Whirling, 497, 499
Window function, 595, 596f
Wing. *See under* Aircraft

Y

Yaw, 353f
Young's modulus. *See* Elastic
  modulus

Z

Zero mode, 392